Lecture Notes in Computer Science 9491

Commenced Publication in 1973
Founding and Former Series Editors:
Gerhard Goos, Juris Hartmanis, and Jan van Leeuwen

More information about this series at http://www.springer.com/series/7407

Sabri Arik · Tingwen Huang
Weng Kin Lai · Qingshan Liu (Eds.)

Neural Information Processing

22nd International Conference, ICONIP 2015
Istanbul, Turkey, November 9–12, 2015
Proceedings, Part III

 Springer

Organization

General Chair

Sabri Arik   Istanbul University, Turkey

Honorary Chair

Shun-ichi Amari   Brain Science Institute, RIKEN, Japan

Program Chairs

Tingwen Huang   Texas A&M University at Qatar, Qatar
Weng Kin Lai   School of Technology, Tunku Abdul Rahman College
                                      (TARC), Malaysia
Qingshan Liu   Huazhong University of Science Technology, China

Advisory Committee

P. Balasubramaniam   Deemed University, India
Jinde Cao   Southeast University, China
Jonathan Chan   King Mongkut's University of Technology, Thailand
Sung-Bae Cho   Yonsei University, Korea
Tom Gedeon   Australian National University, Australia
Akira Hirose   University of Tokyo, Japan
Tingwen Huang   Texas A&M University at Qatar, Qatar
Nik Kasabov   Auckland University of Technology, New Zealand
Rhee Man Kil   Korea Advanced Institute of Science and Technology
                                      (KAIST), Korea
Irwin King   Chinese University of Hong Kong, SAR China
James Kwok   Hong Kong University of Science and Technology,
                                      SAR China
Weng Kin Lai   School of Technology, Tunku Abdul Rahman College
                                      (TARC), Malaysia
James Lam   The University of Hong Kong, Hong Kong,
                                      SAR China
Kittichai Lavangnananda   King Mongkut's University of Technology, Thailand
Minho Lee   Kyungpook National University, Korea
Andrew Chi-Sing Leung   City University of Hong Kong, SAR China
Chee Peng Lim   University Sains Malaysia, Malaysia
Derong Liu   The Institute of Automation of the Chinese Academy
                                      of Sciences (CASIA), China

Chu Kiong Loo	University of Malaya, Malaysia
Bao-Liang Lu	Shanghai Jiao Tong University, China
Aamir Saeed Malik	Petronas University of Technology, Malaysia
Seichi Ozawa	Kobe University, Japan
Hyeyoung Park	Kyungpook National University, Korea
Ju. H. Park	Yeungnam University, Republic of Korea
Ko Sakai	University of Tsukuba, Japan
John Sum	National Chung Hsing University, Taiwan
DeLiang Wang	Ohio State University, USA
Jun Wang	Chinese University of Hong Kong, SAR China
Lipo Wang	Nanyang Technological University, Singapore
Zidong Wang	Brunel University, UK
Kevin Wong	Murdoch University, Australia

Program Committee Members

Syed Ali, India
R. Balasubramaniam, India
Tao Ban, Japan
Asim Bhatti, Australia
Jinde Cao, China
Jonathan Chan, Thailand
Tom Godeon, Australia
Denise Gorse, UK
Akira Hirose, Japan
Lu Hongtao, China
Mir Md Jahangir Kabir, Australia
Yonggui Kao, China
Hamid Reza Karimi, Norway
Nik Kasabov, New Zealand
Weng Kin Lai, Malaysia
S. Lakshmanan, India
Minho Lee, Korea
Chi Sing Leung, Hong Kong, SAR China
Cd Li, China

Ke Liao, China
Derong Liu, USA
Yurong Liu, China
Chu Kiong Loo, Malaysia
Seiichi Ozawa, Japan
Serdar Ozoguz, Turkey
Hyeyoung Park, South Korea
Ju Park, North Korea
Ko Sakai, Japan
Sibel Senan, Turkey
Qianqun Song, China
John Sum, Taiwan
Ying Tan, China
Jun Wang, Hong Kong, SAR China
Zidong Wang, UK
Kevin Wong, Australia
Mustak Yalcin, Turkey
Enes Yilmaz, Turkey

Special Sessions Chairs

Zeynep Orman	Istanbul University, Turkey
Neyir Ozcan	Uludag University, Turkey
Ruya Samli	Istanbul University, Turkey

Publication Chair

Selcuk Sevgen   Istanbul University, Turkey

Organizing Committee

Emel Arslan   Istanbul University, Turkey
Muhammed Ali Aydin   Istanbul University, Turkey
Eylem Yucel Demirel   Istanbul University, Turkey
Tolga Ensari   Istanbul University, Turkey
Ozlem Faydasicok   Istanbul University, Turkey
Safak Durukan Odabasi   Istanbul University, Turkey
Sibel Senan   Istanbul University, Turkey
Ozgur Can Turna   Istanbul University, Turkey

Contents – Part III

Design of an Adaptive Support Vector Regressor Controller for a Spherical Tank System

Kemal Uçak[✉] and Gülay Öke Günel

Faculty of Electrical-Electronics, Department of Control and Automation
Engineering, Istanbul Technical University, Ayazaga Campus, 34469 Istanbul, Turkey
{kemal.ucak,gulay.oke}@itu.edu.tr
http://www.kontrol.itu.edu.tr

Abstract. In this study, an adaptive support vector regressor (SVR) controller which has previously been proposed [1] is applied to control the liquid level in a spherical tank system. The variations in the cross sectional area of the tank depending on the liquid level is the main cause of nonlinearity in system. The parameters of the controller are optimized depending on the future behaviour of the system which is approximated via a seperate online SVR model of the system. In order to adjust controller parameters, the "closed-loop margin" which is calculated using the tracking error has been optimized. The performance of the proposed method has been examined by simulations carried out on a nonlinear spherical tank system, and the results reveal that the SVR controller together with SVR model leads to good tracking performance with small modeling, transient state and steady state errors.

Keywords: Model based adaptive control · Online support vector regression · Spherical tank system · SVR controller · SVR model identification

1 Introduction

Changing of living organisms' characteristics physically or behaviorally to enhance their resistance against alternating environmental aspects is called "adaptation" [2]. Inspired by this feature of living organisms, adaptation capability can be interfused to conventional controllers which are especially essential for nonlinear systems that are hard to control using only fixed parameter controllers. Adaptation of controller parameters to fluxional dynamics of closed-loop system is required to obtain acceptable control performance. For this purpose, intelligent systems such as ANN (Artificial Neural Networks), ANFIS (Adaptive Neuro-Fuzzy Inference Systems) and SVR (Support Vector Regression) can be utilized to design adjustable controllers for nonlinear systems.

Adaptive controller structures based on SVR have proved to be effective controller design methods among other intelligent methods such as ANN, ANFIS because of their superior generalization capabilities, in the last decade. The major

© Springer International Publishing Switzerland 2015
S. Arik et al. (Eds.): ICONIP 2015, Part III, LNCS 9491, pp. 1–8, 2015.
DOI: 10.1007/978-3-319-26555-1_1

strength of SVR is that it ensures global minimum owing to its convex objective function and linear constraints, which avoids getting stuck at local minima.

In technical literature, various controllers based on SVR have been proposed for nonlinear systems such as adaptive PID controller, inverse controller and model predictive control(MPC). Iplikci [3], Shang et al. [4] and Zhao et al. [5] have utilized SVR model of the system to update the parameters of PID controllers. Yuan et al. [6] have proposed a control law based on SVR which is derived via Taylor expansion of system model. Liu et al. [7], Wang et al. [8] and Yuan et al. [9] have deployed SVR as an inverse controller to identify inverse dynamics of the controlled system. In MPC, first and second order derivatives of system output with respect to control input are required. In order to increase accuracy of the required information about system, Iplikci [10,11], Du and Wang [12] and Shin et al. [13] have proposed to utilize SVR in MPC framework.

In this study, an adaptive online SVR controller previously proposed in [1] is used to control the liquid level of a spherical tank system. Two separate SVRs are employed in the control architecture, one for estimating the system model and the other for calculating the control input. The paper is organized as follows: Sect. 2 describes the working prensiples of adaptive SVR controller. In Sect. 3, optimization problem for SVR controller is constructed. In Sect. 4, the effectiveness of the proposed controller has been examined on nonlinear spherical tank system and performance analysis of the controller is given. The paper ends with a brief conclusion in Sect. 5.

2   Adaptive Online SVR Controller

The tuning mechanism of the adaptive SVR controller based on estimated system model is depicted in Fig. 1. The proposed mechanism has two SVR structures; $\mathrm{SVR_{controller}}$ generates the control input to be applied to the system and $\mathrm{SVR_{model}}$ is utilized to approximate system behaviour. The control signal produced by online $\mathrm{SVR_{controller}}$ is computed as:

$$u_n = \sum_{k \in SV_{\mathrm{controller}}} \alpha_k K_{\mathrm{controller}}(\mathbf{\Pi_c}, \mathbf{\Pi_k}) + b_{\mathrm{controller}}. \tag{1}$$

where $\mathbf{\Pi_c}$ is input vector, $K_{\mathrm{controller}}(,.,)$ is the kernel, α_k, $\mathbf{\Pi_k}$ and $b_{\mathrm{controller}}$ are the parameters of the controller to be tuned at time index n. The future behaviour of the controlled system is estimated via $\mathrm{SVR_{model}}$ as

$$\hat{y}_{n+1} = \sum_{j \in SV_{\mathrm{model}}} \lambda_j K_{\mathrm{model}}(\mathbf{M_c}, \mathbf{M_j}) + b_{\mathrm{model}} \tag{2}$$

where K_{model} is the kernel matrix of the system model, $\mathbf{M_c}$ is current input, and λ_j, $\mathbf{M_j}$ and b_{model} are the parameters of the system model to be adjusted.

The estimation of system model and computation of control input are carried out in two consecutive phases at each sampling interval, namely the training and application phases. In training phase of the controller, $\mathrm{SVR_{model}}$ is employed to

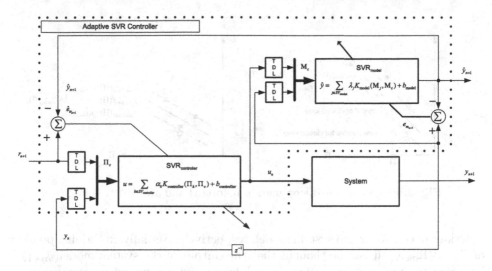

Fig. 1. Adaptive SVR$_{controller}$ mechanism.

observe the impact of the tuned controller parameters on closed-loop system performance, and SVR$_{controller}$ can be optimized depending on approximated tracking error ($\hat{e}_{tr_{n+1}} = r_{n+1} - \hat{y}_{n+1}$). $\hat{y}_{n+1}$ is the system output estimate calculated by SVR$_{model}$. Therefore, while SVR$_{controller}$ is in training phase, SVR$_{model}$ is in application phase. After the training phase of SVR$_{controller}$ is completed, the control signal(u_n) is computed and applied to the system(y_{n+1}) in the application phase of the controller. Thus, the training data pair for SVR$_{model}$ ($\mathbf{M_c}$,y_{n+1}) is obtained for training phase of the system model. The parameters of SVR$_{model}$ are adjusted via modelling error $e_{m_{n+1}} = y_{n+1} - \hat{y}_{n+1}$. The training algorithm for the overall architecture is explained in [1,14].

The regressor margins of SVR$_{controller}$ and SVR$_{model}$ of the closed-loop system are illustrated in Fig. 2 where $f_{controller}$ and f_{model} denote the regression

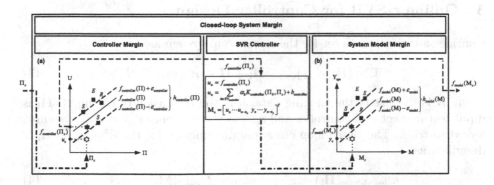

Fig. 2. Margins of SVR$_{controller}$(**a**) and SVR$_{model}$(**b**).

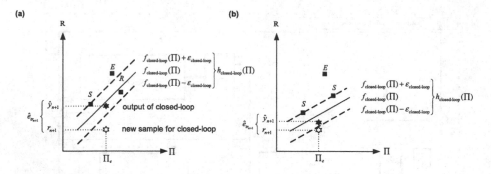

Fig. 3. Projected closed loop margin before(a) and after(b) training.

functions of controller and system model, respectively. Actually, in training phase of the SVR$_{\text{model}}$, it can be thought that the output of the system model($\hat{y}_{n+1}$) is forced to track system output (y_{n+1}) by optimizing model parameter and using ($\mathbf{M_c}$,y_{n+1}) as the input-output training data pair. Therefore, the axes for SVR$_{\text{model}}$ regresion surface are given as $\mathbf{M}$ and Y_{sys} in Fig. 2 (b). Since the control signal that minimize tracking error is unknown, the parameters of SVR$_{\text{controller}}$ can not be obtained directly. For this reason, closed-loop margin notion which is emerged by combining controller and system model margins has been proposed to optimize controller parameters in [1]. If the margins of the controller and system model are fused, the combined closed-loop margin is projected onto R-$\mathbf{\Pi}$ axes as in Fig. 3. Since the aim in controller design is to force closed-loop system output (y_{n+1}) to track reference signal(r_{n+1}), ($\mathbf{\Pi_c}$,r_{n+1}) data pair has been utilized to optimize closed-loop margin. For this reason, the input-output axes for closed-loop system are defined as $\mathbf{\Pi}$ and R with respect to input-output data pair of closed-loop system as in Fig. 3. That is, the axis R which denotes the reference signal is used in place of Y_{sys} for closed-loop system as in Fig. 3. For more detailed information, it can be consulted to [1].

3   Online ε-SVR for Controller Design

Consider a training data set for the closed-loop system as:

$$\mathbf{T} = \{\mathbf{\Pi_i}, r_{i+1}\}_{i=1}^{N} \quad \mathbf{\Pi_i} \in \mathbf{\Pi} \subseteq R^n, r_{i+1} \in R \tag{3}$$

where N is the size of the training data, n is the dimension of the input, $\mathbf{\Pi_i}$ is input feature vector of controller and r_{i+1} is the reference signal that system is forced to track. The closed-loop error margin function for the ith sample $\mathbf{\Pi_i}$ is described as:

$$h_{\text{closed-loop}}(\mathbf{\Pi_i}) = \hat{y}_{i+1} - r_{i+1} = f_{\text{model}}(\mathbf{M_i}) - r_{i+1} \tag{4}$$

where

$$\hat{y}_{i+1} = f_{\text{model}}(\mathbf{M_i}) = \sum_{j \in SV_{\text{model}}} \lambda_j K_{\text{model}}(\mathbf{M_j}, \mathbf{M_i}) + b_{\text{model}}$$

$$\mathbf{M_i} = [u_i \cdots u_{i-n_u}, y_i \cdots y_{i-n_y}]$$

$$u_i = f_{\text{controller}}(\mathbf{\Pi_i}) = \sum_{k \in SV_{\text{controller}}} \alpha_k K_{\text{controller}}(\mathbf{\Pi_k}, \mathbf{\Pi_i}) + b_{\text{controller}}$$

$$\mathbf{\Pi_i} = [r_i \cdots r_{i-n_r}, y_i \cdots y_{i-n_y}, u_{i-1} \cdots u_{i-n_u}]$$

and $\hat{e}_{tr_{i+1}}$ is approximated tracking error. As mentioned before, SVR_{model} is utilized to approximate system behaviour, the system model is fixed and system model parameters are known in training phase of the controller. Therefore, the closed loop margin in (4) can be rewritten as

$$h_{\text{closed-loop}}(\mathbf{\Pi_i}) = \hat{y}_{i+1} - r_{i+1} = f_{\text{closed-loop}}(\mathbf{\Pi_i}) - r_{i+1} = -\hat{e}_{tr_{i+1}} \qquad (5)$$

with respect to an input-output data pair of closed-loop system $(\mathbf{\Pi_i}, r_{i+1})$ where $f_{\text{controller}}$ is the approximated output of the closed-loop system. The main aim is to adjust the unknown parameters of $SVR_{\text{controller}}$ $(\alpha_k, b_{\text{controller}})$ for the given training samples $(\mathbf{\Pi_i}, r_{i+1})$. Using $(\mathbf{\Pi_i}, r_{i+1})$ data pair and closed-loop error margin defined in (4), online learning rules for the parameters of $SVR_{\text{controller}}$ $(\alpha_k, b_{\text{controller}})$ can be acquired. The basic idea is to change the coefficient α_c corresponding to the new sample $\mathbf{\Pi_c}$ in a finite number of discrete steps until it meets the KKT conditions while ensuring that the existing samples in $\mathbf{T}$ continue to satify the KKT conditions at each step [14]. The derivation of update rules for controller design are described in detail in [1].

4   Simulation Results

The performance of the controller has been examined on the spherical tank system which is pictured in Fig. 4. Dynamics of the spherical tank system are defined with the following set of differential equation:

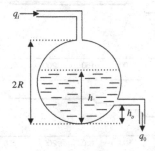

Fig. 4. Spherical tank system.

$$\frac{dh(t)}{dt} = \frac{q_i(t-d) - q_o(t)}{\pi R^2 (1 - \frac{R^2 - h(t)}{R^2})}, \quad q_0(t) = \sqrt{2g(h(t) - h_0)} \tag{6}$$

where R is the radius of spherical tank, $q_i(t)$ is the input flow rate and control signal, $h(t)$ is the level of the liquid system and controlled output of the system, $q_o(t)$ is the outlet flow rate and d indicates the delay in system. In simulations, the dynamics of the system are defined via fourth order Runge-Kutta method with 0.1 s sampling period, system parameters are chosen as $d = 0$ s, $R = 1$ m, $h_0 =$

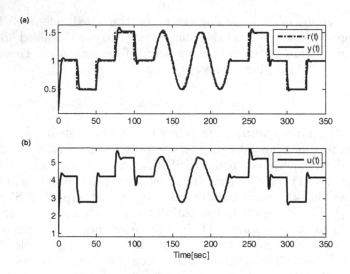

Fig. 5. System (a) and controller output (b) with no measurement noise.

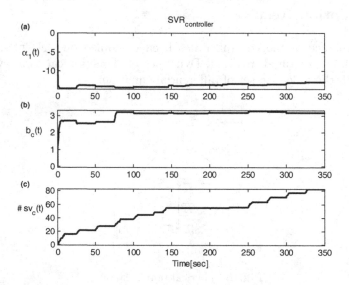

Fig. 6. Adaptation of $SVR_{controller}$ parameters.

0.1 m, magnitude of the control signal is allowed to vary between $u_{min} = 0$ and $u_{max} = 6$; and its duration is kept constant at $\tau_{min} = \tau_{max} = 0.5$ s. The input feature vector for SVR$_{\text{controller}}$ is selected as: $\mathbf{\Pi_c} = [r_n, P_n, I_n, D_n, y_n, u_{n-1}]^T$ where $P_n = e_n - e_{n-1}$, $I_n = e_n$, $D_n = e_n - 2e_{n-1} + e_{n-2}$ and $e_n = r_n - y_n$. In order to identify the dynamics of the controlled system, SVR$_{\text{model}}$ with $\mathbf{M_c} = [u_n \cdots u_{n-n_u}, y_n \cdots y_{n-n_y}]^T$ as the input feature vector where $n_u = n_y = 1$ is utilized. The closed-loop tracking performance of the controller and the control signal are illustrated in Fig. 5 (a) and (b) respectively. It can be deduced that the closed-loop system has very small transient-state and steady state errors. The first Lagrange multiplier and bias of SVR$_{\text{controller}}$ are depicted in Fig. 6 (a) and (b) respectively to exemplify the adaptation of the SVR$_{\text{controller}}$ in order to capture new dynamics [1]. In Fig. 6 (c), the number of the support vectors are illustrated to demonstrate the evolution of SVR$_{\text{controller}}$.

5 Conclusion

In this paper, liquid level of a spherical tank system has been controlled by an adaptive architecture based on SVR. The control mechanism is composed of two seperate SVR structures where SVR$_{\text{controller}}$ and SVR$_{\text{model}}$ are concurrently utilized to compute the control input signal and estimate the system model. The proposed mechanism adjusts SVR$_{\text{controller}}$ parameters without an explicit knowledge of the control signal applied to the system. The results indicate that the closed-loop system can be successfully forced to track reference signal with small transient and steady-state errors. In future works, new SVR type adaptive controllers can be developed for nonlinear liquid level systems.

References

1. Uçak, K., Günel, G.Ö.: An adaptive support vector regressor controller for nonlinear systems. Soft Comput. (2015). (Article in Press)
2. Aström, K.J., Wittenmark, B.: Adaptive Control. Dover Publications, Mineola (2008)
3. Iplikci, S.: A comparative study on a novel model-based PID tuning and control mechanism for nonlinear systems. Int. J. Robust Nonlinear Control **20**(13), 1483–1501 (2010). doi:10.1002/rnc.1524
4. Shang, W.F., Zhao, S.D., Shen, Y.J.: Adaptive PID controller based on online LSSVM identification. In: IEEE/ASME International Conference on Advanced Intelligent Mechatronics (AIM 2008), pp. 694–698. IEEE Press, Xian (2008). doi:10.1109/AIM.2008.4601744
5. Zhao, J., Li, P., Wang, X.S.: Intelligent PID controller design with adaptive criterion adjustment via least squares support vector machine. In: 21st Chinese Control and Decision Conference (CCDC 2009), pp. 7–12. IEEE Press, Guilin (2009)
6. Yuan, X.F., Wang, Y.N., Wu, L.H.: Composite feedforward-feedback controller for generator excitation system. Nonlinear Dyn. **54**(4), 355–364 (2008). doi:10.1007/s11071-008-9334-6

7. Liu, X., Yi, J., Zhao, D.: Adaptive inverse control system based on least squares support vector machines. In: Wang, J., Liao, X.-F., Yi, Z. (eds.) ISNN 2005. LNCS, vol. 3498, pp. 48–53. Springer, Heidelberg (2005)
8. Wang, H., Pi, D.Y., Sun, Y.X.: Online SVM regression algorithm-based adaptive inverse control. Neurocomputing **70**(4–6), 952–959 (2007). doi:10.1016/j.neucom.2006.10.021
9. Yuan, X.F., Wang, Y.N., Wu, L.H.: Adaptive inverse control of excitation system with actuator uncertainty. Neural Process. Lett. **27**(2), 125–136 (2008). doi:10.1007/s11063-007-9064-7
10. Iplikci, S.: Online trained support vector machines-based generalized predictive control of non-linear systems. Int. J. Adapt. Control Signal Process. **20**(10), 599–621 (2006). doi:10.1002/acs.919
11. Iplikci, S.: Support vector machines-based generalized predictive control. Int. J. Robust Nonlinear Control **16**(17), 843–862 (2006). doi:10.1002/rnc.1094
12. Du, Z.Y., Wang, X.F.: Nonlinear generalized predictive control based on online SVR. In: 2nd International Symposium on Intelligent Information Technology Application, pp. 1105–1109. IEEE Press, Shanghai (2008)
13. Shin, J., Kim, H.J., Park, S., Kim, Y.: Model predictive flight control using adaptive support vector regression. Neurocomputing **73**(4–6), 1031–1037 (2010). doi:10.1016/j.neucom.2009.10.002
14. Ma, J., Theiler, J., Perkins, S.: Accurate online support vector regression. Neural Comput. **15**(11), 2683–2703 (2003). doi:10.1162/089976603322385117

Robust Tracking Control of Uncertain Nonlinear Systems Using Adaptive Dynamic Programming

Xiong Yang[1](✉), Derong Liu[2], and Qinglai Wei[1]

[1] The State Key Laboratory of Management and Control for Complex Systems, Institute of Automation, Chinese Academy of Sciences, Beijing 100190, China
{xiong.yang,qinglai.wei}@ia.ac.cn
[2] School of Automation and Electrical Engineering, University of Science and Technology Beijing, Beijing 100083, China
derong@ustb.edu.cn

Abstract. In this paper, we develop an adaptive dynamic programming-based robust tracking control for a class of continuous-time matched uncertain nonlinear systems. By selecting a discounted value function for the nominal augmented error system, we transform the robust tracking control problem into an optimal control problem. The control matrix is not required to be invertible by using the present method. Meanwhile, we employ a single critic neural network (NN) to approximate the solution of the Hamilton-Jacobi-Bellman equation. Based on the developed critic NN, we derive optimal tracking control without using policy iteration. Moreover, we prove that all signals in the closed-loop system are uniformly ultimately bounded via Lyapunov's direct method. Finally, we provide an example to show the effectiveness of the present approach.

Keywords: Adaptive dynamic programming · Robust control · Tracking control · Uncertain nonlinear system

1 Introduction

During the past several decades, robust tracking control of nonlinear systems has attracted considerable attention [1–3]. Many significant approaches have been proposed. Among these methods, the feedback linearization approach is often employed. However, to use the feedback linearization method, the control matrix needs to be invertible. This requirement is usually hard to satisfy in applications.

Recently, adaptive dynamic programming (ADP) [4] is applied to give the optimal tracking control of nonlinear systems. In [5], Heydari and Balakrishnan proposed a single network adaptive critic architecture to obtain the optimal tracking control for continuous-time (CT) nonlinear systems. By employing the architecture, the control matrix was no longer required to be invertible. After that, Modares and Lewis [6] introduced a discounted value function for the CT constrained-input optimal tracking control problem. By proposing an ADP algorithm, the optimal tracking control was obtained without requiring

© Springer International Publishing Switzerland 2015
S. Arik et al. (Eds.): ICONIP 2015, Part III, LNCS 9491, pp. 9–16, 2015.
DOI: 10.1007/978-3-319-26555-1_2

the control matrix to be invertible either. Though the aforementioned literature provides important insights into deriving optimal tracking control for CT nonlinear systems, the ADP-based robust tracking control for CT uncertain nonlinear systems is not considered.

In this paper, an ADP-based robust tracking control is developed for CT matched uncertain nonlinear systems. By choosing a discounted value function for the nominal augmented error dynamical system, the robust tracking control problem is transformed into an optimal control problem. The control matrix is not required to be invertible in the present method. Meanwhile, a single critic neural network (NN) is used to approximate the solution of the Hamilton-Jacobi-Bellman (HJB) equation. Based on the developed critic NN, the optimal tracking control is obtained without using policy iteration. In addition, all signals in the closed-loop system are proved to be uniformly ultimately bounded (UUB) via Lyapunov's direct method.

The rest of the paper is organized as follows. Preliminaries are presented in Sect. 2. The problem transformation is given in Sect. 3. Approximating the HJB solution via ADP is shown in Sect. 4. Simulation results are provided in Sect. 5. Finally, several conclusions are given in Sect. 6.

2   Preliminaries

Consider the CT uncertain nonlinear system given by

$$\dot{x}(t) = f(x(t)) + g(x(t))u(t) + \Delta f(x(t)) \tag{1}$$

where $x(t) \in \mathbb{R}^n$ is the state vector available for measurement, $u(t) \in \mathbb{R}^m$ is the control vector, $f(x(t)) \in \mathbb{R}^n$ and $g(x(t)) \in \mathbb{R}^{n \times m}$ are known functions with $f(0) = 0$, and $\Delta f(x(t)) \in \mathbb{R}^n$ is an unknown perturbation. $f(x) + g(x)u$ is Lipschitz continuous on a compact set $\Omega \subset \mathbb{R}^n$ containing the origin, and system (1) is assumed to be controllable.

Assumption 1. *There exists a constant $g_M > 0$ such that $0 < \|g(x)\| \le g_M$ $\forall x \in \mathbb{R}^n$. $\Delta f(x) = g(x)d(x)$, where $d(x) \in \mathbb{R}^m$ is unknown function bounded by a known function $d_M(x) > 0$. Moreover, $d(0) = 0$ and $d_M(0) = 0$.*

Assumption 2. *$x_d(t)$ is the desired trajectory of system (1). Meanwhile, $x_d(t)$ is bounded and produced by the command generator model $\dot{x}_d(t) = \eta(x_d(t))$, where $\eta \colon \mathbb{R}^n \to \mathbb{R}^n$ is a Lipschitz continuous function with $\eta(0) = 0$.*

Objective of Control: Without the requirement of the control matrix $g(x)$ to be invertible, a robust control scheme based on ADP is developed to keep the state of system (1) following the desired trajectory $x_d(t)$ to a small neighborhood of the origin in the presence of the unknown term $d(x)$.

3 Problem Transformation

Define the tracking error as $e_{\mathrm{err}}(t) = x(t) - x_d(t)$. Then, the tracking error dynamics system is derived as

$$\dot{e}_{\mathrm{err}}(t) = f(x_d(t) + e_{\mathrm{err}}(t)) + g(x_d(t) + e_{\mathrm{err}}(t))u(t)$$
$$- \eta(x_d(t)) + \Delta f(x_d(t) + e_{\mathrm{err}}(t)). \tag{2}$$

In this sense, the robust tracking control can be obtained by giving a control such that, without the requirement of $g(\cdot)$ to be invertible, system (2) is stable in the sense of uniform ultimate boundedness and the ultimate bound is small.

Denote $z(t) = [e_{\mathrm{err}}^{\mathsf{T}}(t), x_d^{\mathsf{T}}(t)]^{\mathsf{T}} \in \mathbb{R}^{2n}$. By Assumption 2 and using (2), we derive an augmented system for the error dynamics as

$$\dot{z}(t) = F(z(t)) + G(z(t))u(t) + \Delta F(z(t)) \tag{3}$$

where $F \colon \mathbb{R}^{2n} \to \mathbb{R}^{2n}$ and $G \colon \mathbb{R}^{2n} \to \mathbb{R}^{2n \times m}$ are, respectively, defined as

$$F(z(t)) = \begin{bmatrix} f(x_d(t) + e_{\mathrm{err}}(t)) - \eta(x_d(t)) \\ \eta(x_d(t)) \end{bmatrix}, \quad G(z(t)) = \begin{bmatrix} g(x_d(t) + e_{\mathrm{err}}(t)) \\ 0 \end{bmatrix}$$

and $\Delta F(z(t)) = G(z(t))d(z(t))$ with $d(z(t)) \in \mathbb{R}^m$ and $\|d(z(t))\| \le d_M(z(t))$.

In what follows we show that the robust tracking control problem can be transformed into the optimal control problem with a discounted value function for the nominal augmented error system (i.e., system (3) without uncertainty). The nominal augmented system is given as

$$\dot{z}(t) = F(z(t)) + G(z(t))u(t). \tag{4}$$

The value function for system (4) is described by

$$V(z(t)) = \int_t^\infty e^{-\alpha(\tau - t)} \big[\rho d_M^2(z(\tau)) + \bar{U}\big(z(\tau), u(\tau)\big) \big] \mathrm{d}\tau \tag{5}$$

where $\alpha > 0$ is a discount factor, $\rho = \lambda_{\max}(R)$, and $\lambda_{\max}(R)$ denotes the maximum eigenvalue of R, $\bar{U}(z, u) = z^{\mathsf{T}}\bar{Q}z + u^{\mathsf{T}}Ru$ with $\bar{Q} = \mathrm{diag}\{Q, 0_{n \times n}\}$, and $Q \in \mathbb{R}^{n \times n}$ and $R \in \mathbb{R}^{m \times m}$ are symmetric positive definite matrices.

According to [7], the optimal control for system (4) with the value function (5) is

$$u^*(z) = -(1/2)R^{-1}G^{\mathsf{T}}(z)V_z^* \tag{6}$$

where $V_z^* = \partial V^*(z)/\partial z$ and $V^*(z)$ denotes the optimal value of $V(z)$ given in (5). Meanwhile, the corresponding HJB equation is derived as

$$\rho d_M^2(z) + V_z^{*\mathsf{T}} \big(F(z) + G(z)u^* \big) - \alpha V^*(z) + z^{\mathsf{T}}\bar{Q}z + u^{*\mathsf{T}}Ru^* = 0. \tag{7}$$

Theorem 1. *Consider the CT nominal system described by (4) with the value function (5). Let Assumptions 1 and 2 hold. Then, the optimal control $u^*(x)$ given in (6) ensures system (2) to be stable in the sense of uniform ultimate boundedness and the ultimate bound can be kept small.*

Proof. Taking the derivative of $V^*(z)$ along the system trajectory $\dot{z} = F(z) + G(z)u^* + \Delta F(z)$, we have $\dot{V}^*(z) = V_z^{*\mathsf{T}}\big(F(z) + G(z)u^*\big) + V_z^{*\mathsf{T}}\Delta F(z)$. Noticing that $V_z^{*\mathsf{T}}\Delta F(z) = -2u^{*\mathsf{T}}Rd(z)$ and by (7), we obtain $\dot{V}^*(z) = -\rho d_M^2(z) - z^{\mathsf{T}}\bar{Q}z - u^{*\mathsf{T}}Ru^* - 2u^{*\mathsf{T}}Rd(z) + \alpha V^*(z)$. Then it can be rewritten as $\dot{V}^*(z) = -\rho d_M^2(z) - e_{\mathrm{err}}^{\mathsf{T}}Qe_{\mathrm{err}} - \big(u^* + d(z)\big)^{\mathsf{T}}R\big(u^* + d(z)\big) + d^{\mathsf{T}}(z)Rd(z) + \alpha V^*(z)$. Observing that $\rho = \lambda_{\max}(R)$ and $\|d(z)\| \leq d_M$, we derive $\dot{V}^*(z) \leq -\lambda_{\min}(Q)\|e_{\mathrm{err}}\|^2 + \alpha V^*(z)$. Because u^* is actually an admissible control, there exists a constant $b_{v^*} > 0$ such that $\|V^*(z)\| \leq b_{v^*}$. Thus, $\dot{V}(z) \leq -\lambda_{\min}(Q)\|e_{\mathrm{err}}\|^2 + \alpha b_{v^*}$. So, $\dot{V}(z) < 0$ as long as e_{err} is out of the set $\tilde{\Omega}_{e_{\mathrm{err}}} = \{e_{\mathrm{err}} \colon \|e_{\mathrm{err}}\| \leq \sqrt{\alpha b_{v^*}/\lambda_{\min}(Q)}\}$. By Lyapunov Extension Theorem [8], we obtain that the optimal control u^* guarantees $e_{\mathrm{err}}(t)$ to be UUB with ultimate bound $\sqrt{\alpha b_{v^*}/\lambda_{\min}(Q)}$. Moreover, if α is selected to be very small, then $\sqrt{\alpha b_{v^*}/\lambda_{\min}(Q)}$ can be kept small enough.

From Theorem 1, we can find that the robust tracking control can be obtained by solving the optimal control problem (4) and (5). In other words, we need get the solution of (7). In what follows, a novel ADP-based control scheme is developed to obtain the approximate solution of (7). Before proceeding further, we present an assumption used in [9,10].

Assumption 3. *Let $L_1(z) \in C^1$ be a Lyapunov function candidate for system (4) and satisfied $\dot{L}_1(z) = L_{1z}^{\mathsf{T}}\big(F(z) + G(z)u^*\big) < 0$ with L_{1z} the partial derivative of $L_1(z)$ with respect to z. In addition, there exists a symmetric positive definite matrix $\Lambda(z) \in \mathbb{R}^{2n \times 2n}$ such that $L_{1z}^{\mathsf{T}}\big(F(z) + G(z)u^*\big) = -L_{1z}^{\mathsf{T}}\Lambda(z)L_{1z}$.*

4   Approximate the HJB Solution via ADP

By using the universal approximation property of NNs, $V^*(z)$ given in (7) can be represented by a single-layer NN on a compact set $\tilde{\Omega}$ as

$$V^*(z) = W_c^{\mathsf{T}}\sigma(z) + \varepsilon(z) \tag{8}$$

where $W_c \in \mathbb{R}^{N_o}$ is the ideal NN weight, $\sigma(z) = [\sigma_1(z), \sigma_2(z), \ldots, \sigma_{N_0}(z)]^{\mathsf{T}} \in \mathbb{R}^{N_o}$ is the activation function with $\sigma_j(z) \in C^1(\tilde{\Omega})$ and $\sigma_j(0) = 0$, the set $\{\sigma_j(z)\}_1^{N_0}$ is often selected to be linearly independent, N_0 is the number of neurons, and $\varepsilon(z)$ is the NN function reconstruction error.

Substituting (8) into (6), we have

$$u^*(z) = -(1/2)R^{-1}G^{\mathsf{T}}(z)\nabla\sigma^{\mathsf{T}}W_c + \varepsilon_{u^*} \tag{9}$$

where $\nabla\sigma = \partial\sigma(z)/\partial z$ and $\varepsilon_{u^*} = -(1/2)R^{-1}G^{\mathsf{T}}(z)\nabla\varepsilon$. Meanwhile, by using (8), (7) becomes

$$W_c^{\mathsf{T}}\nabla\sigma F - \alpha W_c^{\mathsf{T}}\sigma + z^{\mathsf{T}}\bar{Q}z + \rho d_M^2(z) - (1/4)W_c^{\mathsf{T}}\nabla\sigma\mathcal{A}\nabla\sigma^{\mathsf{T}}W_c = \varepsilon_{\mathrm{HJB}} \tag{10}$$

where $\mathcal{A} = G(z)R^{-1}G^{\mathsf{T}}(z)$ and $\varepsilon_{\mathrm{HJB}} = -\nabla\varepsilon^{\mathsf{T}}F + \alpha\varepsilon + (1/2)W_c^{\mathsf{T}}\nabla\sigma\mathcal{A}\nabla\varepsilon + (1/4)\nabla\varepsilon^{\mathsf{T}}\mathcal{A}\nabla\varepsilon$ is the HJB approximation error [11].

Due to the unavailability of W_c, $u^*(z)$ given in (9) cannot be implemented in real control process. Therefore, we use a critic NN to approximate $V^*(z)$ as

$$\hat{V}(z) = \hat{W}_c^{\mathsf{T}} \sigma(z) \tag{11}$$

where $\hat{W}_c$ is the estimated weight of the ideal weight W_c. The weight estimation error for the critic NN is defined as $\tilde{W}_c = W_c - \hat{W}_c$.

Using (11), the estimated value of optimal control $u^*(z)$ is

$$\hat{u}(z) = -(1/2)R^{-1}G^{\mathsf{T}}(z)\nabla\sigma^{\mathsf{T}}\hat{W}_c. \tag{12}$$

Combining (7), (11) and (12), we obtain the residual error as $\delta = \hat{W}_c^{\mathsf{T}}\nabla\sigma F - \alpha\hat{W}_c^{\mathsf{T}}\sigma + z^{\mathsf{T}}\bar{Q}z + \rho d_M^2(z) - (1/4)\hat{W}_c^{\mathsf{T}}\nabla\sigma\mathcal{A}\nabla\sigma^{\mathsf{T}}\hat{W}_c$. By utilizing (10), we have $\delta = -\tilde{W}_c^{\mathsf{T}}\phi + (1/4)\tilde{W}_c^{\mathsf{T}}\nabla\sigma\mathcal{A}\nabla\sigma^{\mathsf{T}}\tilde{W}_c + \varepsilon_{\mathrm{HJB}}$ with $\phi = \nabla\sigma(F(z) + G(z)\hat{u}) - \alpha\sigma(z)$.

To get the minimum value of δ, we develop a novel critic NN tuning law as

$$\dot{\hat{W}}_c = -\gamma\bar{\phi}\Big(Y(z) + \rho d_M^2(z) - (1/4)\hat{W}_c^{\mathsf{T}}\nabla\sigma\mathcal{A}\nabla\sigma^{\mathsf{T}}\hat{W}_c\Big) + \frac{\gamma}{2}\Sigma(z,\hat{u})\nabla\sigma\mathcal{A}L_{1z}$$

$$+ \gamma\Big((K_1\varphi^{\mathsf{T}} - K_2)\hat{W}_c + (1/4)\nabla\sigma\mathcal{A}\nabla\sigma^{\mathsf{T}}\hat{W}_c\frac{\varphi^{\mathsf{T}}}{m_s}\hat{W}_c\Big) \tag{13}$$

where $\bar{\phi} = \phi/m_s^2$, $\varphi = \phi/m_s$, $m_s = 1 + \phi^{\mathsf{T}}\phi$, $Y(z) = \hat{W}_c^{\mathsf{T}}\nabla\sigma F - \alpha\hat{W}_c^{\mathsf{T}}\sigma + z^{\mathsf{T}}\bar{Q}z$, L_{1z} is given as in Assumption 3, K_1 and K_2 are tuning parameter matrices with suitable dimensions, and $\Sigma(z,\hat{u})$ is an indicator function defined as

$$\Sigma(z,\hat{u}) = \begin{cases} 0, & \text{if } L_{1z}^{\mathsf{T}}\big(F(z) + G(z)\hat{u}\big) < 0, \\ 1, & \text{otherwise.} \end{cases} \tag{14}$$

Then, we obtain the weight estimation error dynamics of the critic NN as

$$\dot{\tilde{W}}_c = \gamma\bar{\phi}\Big(-\tilde{W}_c^{\mathsf{T}}\phi + (1/4)\tilde{W}_c^{\mathsf{T}}\nabla\sigma\mathcal{A}\nabla\sigma^{\mathsf{T}}\tilde{W}_c + \varepsilon_{\mathrm{HJB}}\Big) - \frac{\gamma}{2}\Sigma(z,\hat{u})\nabla\sigma\mathcal{A}L_{1z}$$

$$- \gamma\Big((K_1\varphi^{\mathsf{T}} - K_2) - (1/4)\nabla\sigma\mathcal{A}\nabla\sigma^{\mathsf{T}}(W_c - \tilde{W}_c)\frac{\varphi^{\mathsf{T}}}{m_s}\Big)(W_c - \tilde{W}_c). \tag{15}$$

In what follows we develop a theorem to show the stability of all signals in the closed-loop system. Before proceeding further, an assumption is provided as follows.

Assumption 4. W_c is bounded by a known constant $W_M > 0$. There exist constants $b_\varepsilon > 0$ and $b_{\varepsilon z} > 0$ such that $\|\varepsilon(z)\| < b_\varepsilon$ and $\|\nabla\varepsilon(z)\| < b_{\varepsilon z}$ $\forall z \in \tilde{\Omega}$. There exists a constant $b_{\varepsilon_{u^*}} > 0$ such that $\|\varepsilon_{u^*}\| \le b_{\varepsilon_{u^*}}$. In addition, there exist constants $b_\sigma > 0$ and $b_{\sigma z} > 0$ such that $\|\sigma(z)\| \le b_\sigma$ and $\|\nabla\sigma(z)\| \le b_{\sigma z}$ $\forall z \in \tilde{\Omega}$.

Theorem 2. Consider the CT nominal system given by (4) with associated HJB equation (7). Let Assumptions 1–4 hold and take the control input for system (4) as given in (12). Meanwhile, let the critic NN weight tuning law be (13). Then, the function L_{1z} and the weight estimation error $\tilde{W}_c$ are guaranteed to be UUB.

Proof. We provide an outline of the proof due to the space limit. Consider the Lyapunov function candidate $L(t) = L_1(z) + (1/2)\tilde{W}_c^\mathsf{T}\gamma^{-1}\tilde{W}_c$. Taking the time derivative of $L(t)$, we have $\dot{L}(t) = L_{1z}^\mathsf{T}\big(F(z) + G(z)\hat{u}\big) + \dot{\tilde{W}}_c^\mathsf{T}\gamma^{-1}\tilde{W}_c$. By using (15) and simplification, and noticing that $\dot{z} = F(z) + G(z)\hat{u}$, we obtain

$$\dot{L}(t) \leq L_{1z}^\mathsf{T}\dot{z} - \lambda_{\min}(M)\|\mathcal{Z}\|^2 + b_N\|\mathcal{Z}\| - (1/2)\Sigma(z,\hat{u})L_{1z}^\mathsf{T}\mathcal{A}\nabla\sigma^\mathsf{T}\tilde{W}_c \qquad (16)$$

where $\mathcal{Z} = \big[\tilde{W}_c^\mathsf{T}\varphi, \tilde{W}_c^\mathsf{T}\big]^\mathsf{T}$, b_N is the upper bound of $\|N\|$, M and N are, respectively, given as

$$M = \begin{bmatrix} I & -\dfrac{W_c^\mathsf{T}\mathcal{B}}{8m_s} - \dfrac{K_1^\mathsf{T}}{2} \\ -\dfrac{\mathcal{B}W_c}{8m_s} - \dfrac{K_1}{2} & K_2 - \dfrac{\varphi^\mathsf{T}W_c\mathcal{B}}{4m_s} \end{bmatrix}, \quad N = \begin{bmatrix} \dfrac{\varepsilon_{\mathrm{HJB}}}{m_s} \\ \dfrac{\mathcal{B}W_c\varphi^\mathsf{T}W_c}{4m_s} + K_2W_c - K_1\varphi^\mathsf{T}W_c \end{bmatrix}$$

with $\mathcal{B} = \nabla\sigma\mathcal{A}\nabla\sigma^\mathsf{T}$. Due to the definition of $\Sigma(z,\hat{u})$ given in (14), we divide (16) into the following two cases for discussion:

(i) $\Sigma(z,\hat{u}) = 0$. In this circumstance, we have $L_{1z}^\mathsf{T}\dot{z} < 0$. By employing *dense property* of $\mathbb{R}$ [12], we can obtain a positive constant β such that $0 < \beta \leq \|\dot{z}\|$ implies $L_{1z}^\mathsf{T}\dot{z} \leq -\|L_{1z}\|\beta < 0$. Then (16) can be developed as $\dot{L}(t) \leq -\|L_{1z}\|\beta + (1/4)b_N^2/\lambda_{\min}(M) - \lambda_{\min}(M)\big(\|\mathcal{Z}\| - (1/2)b_N/\lambda_{\min}(M)\big)^2$. Notice that $\|\mathcal{Z}\| \leq \sqrt{1 + \|\varphi\|^2}\|\tilde{W}_c\| \leq (\sqrt{5}/2)\|\tilde{W}_c\|$. Therefore, $\dot{L}(t) < 0$ is valid only if $\|L_{1z}\| > b_N^2/(4\beta\lambda_{\min}(M))$ or $\|\tilde{W}_c\| > 2b_N/(\sqrt{5}\lambda_{\min}(M))$.

(ii) $\Sigma(z,\hat{u}) = 1$. In this case, (16) can be developed as $\dot{L}(t) \leq L_{1z}^\mathsf{T}\big(F(z) + G(z)u^*\big) + L_{1z}^\mathsf{T}G(z)(\hat{u} - u^*) - \lambda_{\min}(M)\|\mathcal{Z}\|^2 + b_N\|\mathcal{Z}\| - \frac{1}{2}L_{1z}^\mathsf{T}\mathcal{A}\nabla\sigma^\mathsf{T}\tilde{W}_c$. By using Assumption 3 and similar with (i), we can obtain that $\dot{L}(t) < 0$ is valid only if $\|L_{1z}\| > g_M b_{\varepsilon_{u^*}}/(2\lambda_{\min}(\Lambda(z))) + \sqrt{\ell/\lambda_{\min}(\Lambda(z))}$ or $\|\tilde{W}_c\| > b_N/(\sqrt{5}\lambda_{\min}(M)) + \sqrt{4\ell/(5\lambda_{\min}(M))}$, where $\ell = g_M^2 b_{\varepsilon_{u^*}}^2/(4\lambda_{\min}(\Lambda(z))) + b_N^2/(4\lambda_{\min}(M))$.

Combining (i) and (ii) and using the standard Lyapunov Extension Theorem [8], we derive that the function L_{1z} and the weight estimation error $\tilde{W}_c$ are UUB.

5   Simulation Results

Consider the CT uncertain nonlinear system given by

$$\dot{x}_1 = -x_1 + x_2$$
$$\dot{x}_2 = -(x_1 + 1)x_2 - 49x_1 + u + q\cos^3(x_1)\sin(x_2) \qquad (17)$$

where $x = [x_1, x_2]^\mathsf{T} \in \mathbb{R}^2$, and the uncertain term $d(x) = q\cos^3(x_1)\sin(x_2)$ with unknown parameter $q \in [-1, 1]$. We choose $d_M(x) = \|x\|$. The reference trajectory x_d is generated by $\dot{x}_{1d} = x_{2d}$ and $\dot{x}_{2d} = -49x_{1d}$ with the initial condition $x_d(0) = [0.2, 0.4]^\mathsf{T}$. Then the augmented tracking error system is derived as

$$\dot{z} = C(z) + D(z)(u + d(z)) \qquad (18)$$

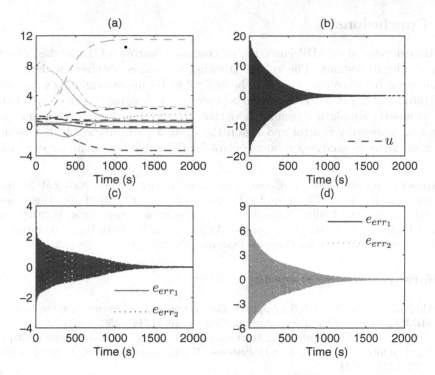

Fig. 1. (a) Convergence of critic NN weights (b) Control input u (c) Evolution of tracking errors $e_{err_i}(t)$ ($i = 1, 2$) during NN learning process (d) Tracking errors e_{err_i} ($i = 1, 2$) between the state of system (17) and the desired trajectory x_d under the approximate optimal control

where $z = [z_1, z_2, z_3, z_4]^{\mathsf{T}} = [e_{err_1}, e_{err_2}, x_{1d}, x_{2d}]^{\mathsf{T}}$ with $e_{err_i} = x_i - x_{id}$, and $D(z) = [0, 1, 0, 0]^{\mathsf{T}}$, and $C(z) = [-z_1 + z_2 - z_3; -(z_1 + z_3)(z_2 + z_4) - 49z_1 - z_2 - z_4; z_4; -49z_3]$. The nominal augmented system is $\dot{z} = C(z) + D(z)u$ with $C(z)$ and $D(z)$ is given in (18). The cost function $V(z)$ for nominal augmented error system is given as (5), where $R = 1$ and $Q = 2I_2$. The activation function of the critic NN is chosen with $N_0 = 10$ as $\sigma(x) = \left[z_1^2, z_2^2, z_3^2, z_4^2, z_1 z_2, z_1 z_3, z_1 z_4, z_2 z_3, z_2 z_4, z_3 z_4\right]^{\mathsf{T}}$, and the weight of the critic NN is written as $\hat{W}_c = [W_{c1}, W_{c2}, \ldots, W_{c10}]^{\mathsf{T}}$.

The initial system state is $x(0) = [0.5, -0.5]^{\mathsf{T}}$, and the initial weight for the critic NN is selected randomly within an interval of $[0, 1]$, which implies that no initial stabilizing control is required. In addition, $\alpha = 0.15$ and $\gamma = 0.5$. The developed control algorithm is implemented via (12) and (13). The computer simulation results are shown by Fig. 1(a)–(d). Figure 1(a) and (b) indicate convergence of critic NN weights and control input u, respectively. Figure 1(c) shows the evolution of tracking errors e_{err_i} ($i = 1, 2$) during NN learning process. Figure 1(d) illustrates tracking errors e_{err_i} ($i = 1, 2$) between the state of system (17) and the desired trajectory x_d under the approximate optimal control. From simulation results, it is observed that the state $x(k)$ tracks the desired trajectory $x_d(k)$ very well, and all signals in the closed-loop system are bounded.

6   Conclusions

We have developed an ADP-based robust tracking control for CT matched uncertain nonlinear systems. The robust tracking control is obtained without the requirement of the control matrix to be invertible. By using Lyapunov's method, the stability of the closed-loop system is proved, and all signals involved are UUB. The computer simulation results show that the developed control scheme can perform successfully control and attain the desired performance. In our future work, we focus on studying robust control for CT nonaffine nonlinear systems.

Acknowledgments. This work was supported in part by the National Natural Science Foundation of China under Grants 61233001, 61273140, 61304086, and 61374105, in part by Beijing Natural Science Foundation under Grant 4132078, and in part by the Early Career Development Award of the he State Key Laboratory of Management and Control for Complex Systems (SKLMCCS).

References

1. Godbole, D.N., Sastry, S.S.: Approximate decoupling and asymptotic tracking for MIMO systems. IEEE Trans. Autom. Control **40**(3), 441–450 (1995)
2. Chang, Y.C.: An adaptive H_∞ tracking control for a class of nonlinear multiple-input multiple-output (MIMO) systems. IEEE Trans. Autom. Control **46**(9), 1432–1437 (2001)
3. Liu, D., Yang, X., Wang, D., Wei, Q.: Reinforcement-learning-based robust controller design for continuous-time uncertain nonlinear systems subject to input constraints. IEEE Trans. Cybern. **45**(7), 1372–1385 (2015)
4. Werbos, P.J.: Beyond Regression: New Tools for Prediction and Analysis in the Behavioral Sciences. Ph.D. Thesis, Harvard University, Cambridge, MA (1974)
5. Heydari, A., Balakrishnan, S.: Fixed-final-time optimal tracking control of input-affine nonlinear systems. Neurocomputing **129**, 528–539 (2014)
6. Modares, H., Lewis, F.L.: Optimal tracking control of nonlinear partially-unknown constrained-input systems using integral reinforcement learning. Automatica **50**(7), 1780–1792 (2014)
7. Liu, D., Yang, X, Li, H.: Adaptive optimal control for a class of nonlinear partially uncertain dynamic systems via policy iteration. In: 3rd International Conference on Intelligent Control and Information Processing, Dalian, China, pp. 92–96 (2012)
8. Lewis, F.L., Jagannathan, S., Yesildirak, A.: Neural Network Control of Robot Manipulators and Nonlinear Systems. Taylor & Francis, London (1999)
9. Yang, X., Liu, D., Wei, Q.: Online approximate optimal control for affine nonlinear systems with unknown internal dynamics using adaptive dynamic programming. IET Control Theor. Appl. **8**(16), 1676–1688 (2014)
10. Dierks, T., Jagannathan, S.: Optimal control of affine nonlinear continuous-time systems. In: American Control Conference, Baltimore, MD, USA, pp. 1568–1573 (2010)
11. Abu-Khalaf, M., Lewis, F.L., Huang, J.: Neurodynamic programming and zero-sum games for constrained control systems. IEEE Trans. Neural Netw. **19**(7), 1243–1252 (2008)
12. Rudin, W.: Principles of Mathematical Analysis. McGraw-Hill Publishing Co., New York (1976)

Moving Target Tracking Based on Pulse Coupled Neural Network and Optical Flow

Qiling Ni, Jianchen Wang, and Xiaodong Gu[✉]

Department of Electronic Engineering, Fudan University,
Shanghai 200433, China
{12210720035,14210720044,xdgu}@fudan.edu.cn

Abstract. Video contains a large number of motion information. The video–particularly video with moving camera – is segmented based on the relative motion occurring between moving targets and background. By using fusion ability of pulse coupled neural network (PCNN), the target regions and the background regions are fused respectively. Firstly using PCNN fuses the direction of the optical flow fusing, and extracts moving targets from video especially with moving camera. Meanwhile, using phase spectrums of topological property and color pairs (red/green, blue/yellow) generates attention information. Secondly, our video attention map is obtained by means of linear fusing the above features (direction fusion, phase spectrums and magnitude of velocity), which adds weight for each information channel. Experimental results shows that proposed method has better target tracking ability compared with three other methods– Frequency-tuned salient region detection (FT) [5], visual background extractor (Vibe) [6] and phase spectrum of quaternion Fourier transform (PQFT) [1].

Keywords: PCNN · Optical flow · Visual attention · Target tracking

1 Introduction

On June 10, 2014, the document [8] of the Cisco Visual Networking Index predicts that it would take an individual over 5 million years to watch the amount of video that will cross global IP networks each month in 2018. Every second, nearly a million minutes' video content will cross the network by 2018. Video data generation rate is considerably larger than video data analysis rate. Video-based target tracking has drawn increasing interest for its highly applications [9], such as video surveillance, traffic control, machine intelligence, biological medical, etc.

In this paper, we are committed at designing a simple simulation system of human vision by combining static information and motion information. Static information means phase information [1] of color pairs and topological property [4]. Motion information means PCNN fusion based on optical flow. The following sections detail how they contribute to mapping targets.

Figure 1 illustrates the structure of proposed model. Firstly, we get three channels (color pair RG, color pair BY and topological property) from video frames. Then, phase information can be obtained from phase spectrum of two color pairs and topological

© Springer International Publishing Switzerland 2015
S. Arik et al. (Eds.): ICONIP 2015, Part III, LNCS 9491, pp. 17–25, 2015.
DOI: 10.1007/978-3-319-26555-1_3

property by inverse Fourier transform. Secondly, PCNN is utilized as fusion tool of motion features by optical flow direction. Pulse generates from outside to inside based on directional difference in the video frame until high enough difference value happens. Thirdly, saliency map is computed by smoothing linear fusion of phase information, magnitude and direction fusion.

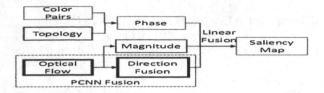

Fig. 1. Structure of proposed model

2 Related Work

This section includes PCNN fusion based on optical flow and topological information extraction. Optical flow and PCNN applied in our model will be briefly introduced before computational process descriptions of PCNN fusion and topological information are given in detail.

2.1 Optical Flow

Optical flow is the pattern of apparent motion of objects, surfaces, and edges in a visual scene caused by the relative motion between an observer (an eye or a camera) and the scene. The concept of optical flow was introduced by the American psychologist James J. Gibson [2] in the 1940 s to describe the visual stimulus provided to animals moving through the world.

The optical flow methods try to calculate the motion between two image frames which are taken at times t and $t + \delta t$ at every pixel. For a 2-dimensional case, a pixel at location (x, y, t) with intensity $I(x, y, t)$ will have moved by $\delta x, \delta y$ and δt between the two image frames, and the following brightness constancy constraint can be given:

$$I(x, y, t) = I(x + \delta x, y + \delta y, t + \delta t) \tag{1}$$

Assuming the movement to be small, the image constraint at $I(x, y, t)$ with Taylor series can be developed to get:

$$I(x + \delta x, y + \delta y, t + \delta t) = I(x, y, t) + \delta x \frac{\partial I}{\partial x} + \delta y \frac{\partial I}{\partial y} + \delta t \frac{\partial I}{\partial t} + e(0) \tag{2}$$

Equations (1) and (2) result in following Eq. (3), in which

$$I_x = \partial I / \partial x, I_y = \partial I / \partial y, I_t = \partial I / \partial t, V_x = \delta x / \delta t, V_y = \delta y / \delta t, \quad I_x V_x + I_y V_y + I_t = 0 \tag{3}$$

The equation with two unknowns is known as the aperture problem of the optical flow algorithms and cannot be solve. To find the optical flow another set of equations is needed, given by some additional constraint. The Horn-Schunck [10] algorithm assumes smoothness in the flow over the whole image. The flow is formulated as a global energy functional which is then sought to be minimized.

$$ E = \iint [(I_x V_x + I_y V_y + I_t)^2 + \alpha^2 (\|\Delta V_x\|^2 + \|\Delta V_y\|^2)] dx dy \tag{4} $$

In Eq. (3), the parameter α is a regularization constant. Larger values of α lead to a smoother flow. This functional can be minimized by solving the associated multi-dimensional Euler-Lagrange equations.

$$ \alpha^2 \Delta V_X = I_x^2 V_X + I_y I_x V_Y + I_t I_x, \alpha^2 \Delta V_Y = I_x I_y V_X + I_y^2 V_Y + I_t I_y \tag{5} $$

where subscripts again denote partial differentiation and $\Delta = \partial^2 / \partial x^2 + \partial^2 / \partial y^2$ denotes the Laplace operator. In practice the Laplacian is approximated numerically using finite differences, and may be written $\Delta V_x = \bar{V}_x - V_x, \Delta V_y = \bar{V}_y - V_y$, where $\bar{V}_x$ and $\bar{V}_y$ is a weighted average of $\bar{V}_x$ and $\bar{V}_y$ calculated in a neighborhood around the pixel at location (x, y). Using this notation the above equation system may be written:

$$ (I_x^2 + \alpha^2) V_x + I_x I_y V_y = \alpha^2 \bar{V}_x - I_t I_x, I_x I_y V_x + (I_y^2 + \alpha^2) V_y = \alpha^2 \bar{V}_y - I_t I_y \tag{6} $$

However, since the solution depends on the neighboring values of the flow field, it must be repeated once the neighbors have been updated. The following iterative scheme is derived:

$$ V_x^{n+1} = \bar{V}_x^n - I_x \frac{I_x \bar{V}_x^n + I_y \bar{V}_y^n + I_t}{\partial^2 + I_x^2 + I_y^2}, V_y^{n+1} = \bar{V}_y^n - I_y \frac{I_x \bar{V}_x^n + I_y \bar{V}_y^n + I_t}{\partial^2 + I_x^2 + I_y^2} \tag{7} $$

where the superscript $n + 1$ denotes the next iteration, which is to be calculated and n is the last calculated result.

2.2 Pulse-Coupled Neural Network

Figure 2 illustrates the structure of unit-linking PCNN, quoted from Literature [3].
The Unit-linking PCNN architecture can be described by the following formula:

$$ F_j = dif\_dir_j, L_j = step \left[\sum_{k \in N(j)} Y_k(t) \right] = \begin{cases} 1 & \sum_{k \in N(j)} Y_k(t) > 0 \\ 0, & else \end{cases} \tag{8} $$

$$ U_j = F_j(1 + \beta_j L_j), Y_j = step(U_j - \theta_j) = \begin{cases} 1 & if\ U_j(t) \geq \theta_j(t) \\ 0 & else \end{cases} $$

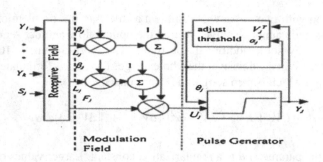

Fig. 2. Unit-linking PCNN architecture [3]

2.3 PCNN Fusion Based on Optical Flow

We extract moving targets from the optical flow, which respectively use the dimension of its amplitude and direction. PCNN has the fusion feature, this section combined the PCNN fusion characteristics with the quantitative optical flow direction information which has been pretreated, besides, it respectively fused the target and the background which has the same moving characteristic, so as to realize the separation of foreground and background, and then segment the optical flow moving targets, the calculation process is following 3 steps:

Step1: Optical flow field pre-process: for each pixel, optical flow cushioned with a background vector against the maximum optical flow direction, and value 1/10 of the magnitude of the maximum optical flow.

Step2: Direction difference quantization: pixel (x, y) and its optical flow value is (u, v), the direction of optical flow is expressed as $Ang = \mathrm{atan2}(u, v)/\pi$, this will use -1 ~ 1 to indicate the value direction, the distribution as shown in Fig. 3.

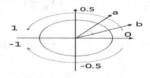

Fig. 3. Direction difference quantization

The above-mentioned direction difference is shown as formula 9, the result are determined by the current pixel direction value and the mean value of its four neighborhood which has been fired, is the absolute value of their difference. Then through the formula 9 processing. We get the 0 ~ 1 monotonically increasing direction difference dif_dir_i.

$$difAng_i = \left| Ang_i - \mathrm{mean}(Ang_j) \right|, j \in \Omega; dif\_dir_i = \min(difAng_i, 2 - difAng_i) \qquad (9)$$

Step3: Unit-linking PCNN fusing direction features: input to F channel is direction difference dif_dir_i between current pixel's direction value Ang_i and the mean of its 4-neighborhood direction values whose fire set is 1 (shown in Eq. (9)). L channel collects fire information of the current pixel' 4-neighborhood.

Figure 4 is the PCNN fusion effect chart using PCNN fusion method.

Fig. 4. PCNN fusion effect

According to [6], the metric most widely used in computer vision to assess the performance of a binary classifier is the percentage of correct classification (PCC), which combines four values - the number of true positives (TP), which counts the number of correctly detected foreground pixels; the number of false positives (FP), which counts the number of background pixels incorrectly classified as foreground; the number of true negatives (TN), which counts the number of correctly classified background pixels; and the number of false negatives (FN), which accounts for the number of foreground pixels incorrectly classified as background.

$$PCC = TP + TN / TP + TN + FP + FN \qquad (10)$$

As shown in Fig. 5, the classification accuracy of PCNN fusion movement information almost all above 95 %.

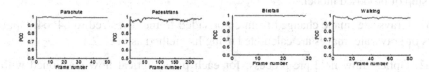

Fig. 5. The effect of SVM classification

2.4 Topological Property

Chen Lin proposed topological perception theory in 1982 [7]. A stimulus was separated into different global wholes (a figure and a background), dependent only on global properties. These global properties can be described mathematically as topological properties, such as connectivity. Literature [4] applied connectivity of topological perception into visual attention. We improve topological algorithm in literature [4] as following 3 steps to avoid the problem of selecting segmentation threshold and benefit the filtering of two or more color tones of background.

Step1: Grayscale image changed from color video frame is resized to 64*64.

Step2: Unit-linking PCNN extracting topological connectivity: input to F channel is intensity difference dif_I_i between current pixel's intensity I_i and the mean of its 4-neighborhood direction values whose fire set is 1. L channel collects fire information of the current pixel' 4-neighborhood.

$$F_j = dif\_I_j, L_j = step\left[\sum_{k \in N(j)} Y_k(t)\right] = \begin{cases} 1 & \sum_{k \in N(j)} Y_k(t) > 0 \\ 0, & else \end{cases} \tag{11}$$

Step3: The binary image computed from step2 PCNN_filter is the input of topological channel which expressed connectivity.

In Fig. 6, $2th$ image and $3th$ image are original topological channel and improved topological channel of $1th$ image separately. Different tones (grass and load) cannot be filtered out all in the original topological channel, which leads to pedestrians' sinking into grass. By inputting intensity difference into F channel of PCNN, improved topological channel filter grass and load successfully. Improvement is done to more accurately represent connectivity of topology.

Fig. 6. Examples of improved topological channel

3 Algorithm Structure

The step of proposed model.

Step1: Grayscale image changed from color video frame is resized to 64*64. Optical flows of grayscale images are calculated using hs method as Sect. 2.1.

Step2: Optical flow field pre-process: for each pixel, optical flow cushioned with a background vector against the maximum optical flow direction, and value 1/10 of the magnitude of the maximum optical flow. PCNN fuse pre-processed optical flow as Sect. 2.3.

Step3: Topological channel T is computed by using PCNN as Sect. 2.4. Two color pairs [1] is $RG = R - G, BY = B - Y$, where $R = r - (g + b)/2$, $G = g - (r + b)/2$, $B = b - (r + g)/2$, $Y = (r + g)/2 - |r - g|/2 - b$, and r, g, b are separately red channel, green channel, blue channel of color image.

Step4: Phase spectrum can be obtained by normalizing Fourier transform of T, RG and BY. Then, phase information is obtained from phase spectrum by inverse Fourier transform.

$$p = f^{-1}\{P[f(RG, BY, 0.4 * T)]\} \tag{12}$$

Step5: Saliency map is computed by smoothing linear fusion of phase information p, magnitude |OF| and direction fusion *fus*. In Eq. (13), $\omega_1 = 1.0, \omega_2 = 1.2$, $\omega_3 = 1.5, \sigma = 8$.

$$S\_Map = G(\sigma) * \left[\omega_1 * p + \omega_2 * |OF| + \omega_3 * fus\right]^2 \tag{13}$$

4   Experimental Results

The proposed algorithm is implemented on our video database and compared with FT [5], Vibe [6], PQFT [1].

4.1   Database

See Table 1.

Table 1. Information of database

Database	Frame number	Size	Characteristics
Parachute	50	414*352	moving camera
Birdfall	28	259*327	static camera, background disturbance
Pedestrians	239	360*240	static camera, multiple targets
Walking	94	1280*720	static camera, single target

4.2   Attention Detection Effects

In our experiments, one widely used saliency detection algorithm (FT [5]) and two common target tracking algorithms (Vibe [6], PQFT [1]) are compared with our algorithm. In Fig. 7, We show the results of these three methods and proposed method using the above figures.

As can be seen in Fig. 7, the results of proposed detection get more salient targets and darker background compared with FT, Vibe, and PQFT. For video Parachute with moving camera, FT focuses more on brighter light through the hole than flying parachute because of the lack of motion information. Vibe concerns with the edge of flying parachute and light, because the light is "moving" in the screen. The proposed model elegantly solves the problem of target tracking with moving camera by utilizing motion direction difference. For those videos with static camera, Vibe' effect is quite good sometimes such as the result of 84*th* frame in Walking. However, inexplicable target just happen frequently such as the middle pedestrian of Pedestrian' 116*th* frame. Although PQFT focuses on moving targets correctly, some of its results are incomplete and its background distracts us.

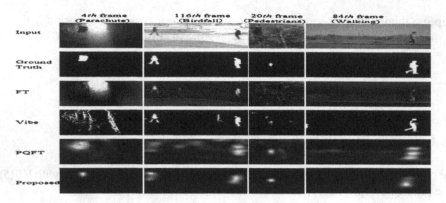

Fig. 7. Video frames and their saliency maps

4.3 Comparison of Attention Detection Models

To further illustrate the effectiveness of the proposed algorithm which combines the visual attention model with PCNN and optical flow, we select commonly used evaluation index F-Measure to compare proposed model to FT [5], Vibe [6], PQFT [1]. Assuming G is ground truth regions, S is saliency regions:

$$\text{Precision} = G \cap S / S, \text{Recall} = G \cap S / G, \text{F-Measure} = 2 * p_{thr} * r_{thr} / \left(p_{thr} + r_{thr} \right) \qquad (13)$$

This paper further compares the proposed algorithm with FT [5], Vibe [6], PQFT [1] on the evaluation index F-Measure with different samples. The result is shown in Table 2. For videos with moving camera or background disturbance (Parachute and Birdfall), F-Measures of proposed model are biggest of four models. Dealing with video Pedestrians and video Walking, proposed model is more effective than FT and PQFT. And our model is effective and comparable with Vibe.

Table 2. F-Measure

Algorithm	Parachute	Bird fall	Pedestrians	Walking
FT [5]	0.0340	0.0026	0.2271	0.1260
Vibe [6]	0.0700	0.0850	**0.5176**	**0.8022**
PQFT [1]	0.2367	0.1426	0.2420	0.2641
Proposed model	**0.5697**	**0.2782**	**0.4395**	**0.4902**

5 Conclusion

This paper proposed a moving target tracking algorithm, which combines the visual attention model with pulse coupled neural network and optical flow, has better tracking performance compared with traditional algorithms. Based on the relative motion occurring between moving targets and background, the target regions and the background are fused respectively by using fusion ability of PCNN. Meanwhile, the improved topological channel benefits the filtering of more color tones of background. Experimental results show that proposed method has higher detection rate and better ability of suppressing background.

Acknowledgements. This work was supported in part by National Natural Science Foundation of China under grant 61371148.

References

1. Guo, C.L., Ma, Q., Zhang, L.M.: Spatio-temporal saliency detection using phase spectrum of quaternion Fourier transform. In: IEEE Conference on Computer Vision and Pattern Recognition, pp, 1–8(2008)
2. Eckhorn, R., Reitboeck, H.J., Arndt, M., et al.: Feather linking via synchronization among distributed assemblies: simulation of results from cat cortex. Neural Comput. 2(3), 293–307 (1990)
3. Gu, X.D., Yu, D.H., Zhang, L.M.: Image shadow removal using pulse coupled neural network. IEEE Trans. Neural Networks 5, 692–698 (2005)
4. Gu, X.D., Fang, Y., Wang, Y.Y.: Attention selection using global topological properties based on pulse coupled neural network. Comput. Vis. Image Underst. 117, 1400–1411 (2013)
5. Achanta, R., Hemami, S., Estrada, F., Susstrunk, S.: Frequency-tuned salient region detection. In: IEEE CVPR, pp. 1597–1604 (2009)
6. Barnich, Olivier, Van Droogenbroeck, Marc: Vibe: a universal background subtraction algorithm for video sequences. IEEE Trans. Image Process. 20(6), 1709–1724 (2011)
7. Chen, L.: Topological structure in visual perception. Science 218, 699–700 (1982)
8. Cisco VNI, Cisco visual networking index: forecast and methodology, 2013–2018 [EB/OL]. http://www.cisco.com/c/en/us/solutions/collateral/service-provider/ip-ngn-ip-next-generation-network/white_paper_c11-481360.html. 10–14 June 2014
9. Kim, W., Kim, C.: Spatiotemporal saliency detection using textural contrast and its applications. IEEE Trans. Circuits Syst. Video Technol. 24, 646–659 (2014)
10. Horn, B., Schunch, B.: Detemining optical flow. Artif. Intell. 17, 185–203 (1981)

Efficient Motor Babbling Using Variance Predictions from a Recurrent Neural Network

Kuniyuki Takahashi[1,2]([✉]), Kanata Suzuki[3], Tetsuya Ogata[3], Hadi Tjandra[1], and Shigeki Sugano[1]

[1] School of Creative Science and Engineering, Waseda University, Tokyo, Japan
{takahashi,tjandra}@sugano.mech.waseda.ac.jp
[2] Research Fellow of Japan Society for the Promotion of Science, Tokyo, Japan
[3] School of Fundamental Science and Engineering, Waseda University, Tokyo, Japan
suzuki@idr.ias.waseda.ac.jp, {ogata,sugano}@waseda.jp

Abstract. We propose an exploratory form of motor babbling that uses variance predictions from a recurrent neural network as a method to acquire the body dynamics of a robot with flexible joints. In conventional research methods, it is difficult to construct real robots because of the large number of motor babbling motions required. In motor babbling, different motions may be easy or difficult to predict. The variance is large in difficult-to-predict motions, whereas the variance is small in easy-to-predict motions. We use a Stochastic Continuous Timescale Recurrent Neural Network to predict the accuracy and variance of motions. Using the proposed method, a robot can explore motions based on variance. To evaluate the proposed method, experiments were conducted in which the robot learns crank turning and door opening/closing tasks after exploring its body dynamics. The results show that the proposed method is capable of efficient motion generation for any given motion tasks.

Keywords: Recurrent Neural Network · Flexible joint robot · Motor babbling

1 Introduction

Recently, various humanoid robots, especially those with flexible joints, have been developed to coexist with humans [1]. The characteristics of flexible joints are adaptation, safety, and dynamics. In terms of adaptation, the presence of noise and environmental change is absorbed by flexible joints. As a result, the robots are able to avoid breaking themselves and the environment. In terms of safety, the flexible joints absorb shocks without any control when the robot collides with a human. The dynamic nature of flexible joints makes it possible to realize dynamic motion that utilizes the inertia of the robot's body.

With the development of robotic technology, robots have become very complex and have an increased number of degrees of freedom (DOFs). Therefore, it is difficult to control a robot considering the complexity of their dynamics. To overcome the problems encountered in past studies, various methods have

© Springer International Publishing Switzerland 2015
S. Arik et al. (Eds.): ICONIP 2015, Part III, LNCS 9491, pp. 26–33, 2015.
DOI: 10.1007/978-3-319-26555-1_4

been suggested: attractors [2,3], oscillators [4], and search trees [5]. Attractors and oscillators force the trajectory to return to a pre-designed trajectory when deviations occur. The trajectory for the task is called the "motion primitive." Search trees find suboptimal motion. These methods have three problems: (1) Attractors and oscillators generate only the designed motion, (2) Search trees require all of the robot model information, and (3) Search trees take a long time to search. To overcome these problems, we use recurrent neural networks (RNNs) and motor babbling to acquire body dynamics [6]. (1) RNNs can predict and generate the next state from the current state. RNNs can learn various motions. (2) This enables robots to acquire body dynamics from motor babbling, which is the robotics concept of the movement process that infants use to acquire their own body model [7]. To perform motor babbling without pre-determined motion, the robot acquires body dynamics by itself. (3) RNNs can also generate associative motion, and do not incur long search times.

Motor babbling is used to identify parameters [8] and acquire a forward inverse model [6]. In the parameter identification stage, the robot has an ideal equation of motion. The ideal and realistic parameters of this equation will differ. Therefore, for the excitation of motion, the robot modifies the parameter values. As a result, accurate motion is calculated from the equation of motion. In the acquisition of a forward inverse model, the robot has no information about the sensory-motor relationship. Through the experience of motor babbling, the robot learns a forward inverse model as its body dynamics. As a result, the robot generates motion directly using body dynamics. In this research, we use a method based on RNNs. This makes it possible to adjust multiple flexible joint models, and does not require other calculations to generate motion.

In research reported by Takahashi et al. [6], random motion was used as motor babbling. Random motion is difficult to apply to an actual robot because of the enormous number of possible motions. Motor babbling is considered to be exploratory movement [9]. Some motions are easy to predict, whereas others are considerably more difficult. In this research, our aim is to propose an exploratory form of motor babbling to explore and learn motions that are difficult to predict, enabling body dynamics to be acquired efficiently.

2   Exploratory Motor Babbling

2.1   Stochastic Continuous Time-Scales Recurrent Neural Networks

In motor babbling, there are easy-to-predict motions and those that are more difficult to predict. If a robot has more experience of a motion, it will be easier to predict. This means that the variance will be small. If a robot has less experience of a motion, it will be more difficult to predict. This means that the variance will be large. The robot can evaluate the prediction accuracy of its motion from the variance. Therefore, to learn efficiently, the robot needs to explore and learn motions that are difficult to predict.

To realize this, we use a Stochastic Continuous Timescale Recurrent Neural Network (S-CTRNN). S-CTRNNs can predict and generate the next state from the current state, and determine the prediction accuracy from the variance [10]. The likelihood function L is expressed by the following equation:

$$\frac{\partial InL}{\partial v_{t,i}} = -\frac{1}{2v_{t,i}} + \frac{(y_{t,i} - \hat{y}_{t,i})^2}{2v_{t,i}^2}, \tag{1}$$

where v is the variance, y is the output signal, and $\hat{y}$ is the teaching signal.

The network generates v as an estimate of the prediction error. The squared error is divided by the variance; therefore, it is possible to avoid unstable learning. Namely, if the variance is large, the influence of the prediction error will decrease. If the variance is small, the influence of the prediction error will increase. Learning is performed according to the maximum likelihood using the gradient descent method. Therefore, the variance is calculated from the input signal without a teaching signal.

2.2 Learning Process of Exploratory Motor Babbling

This section describes the learning process of exploratory motor babbling. This method is composed of two steps. First, a robot learns the random motor babbling and its variance. The number of motor babbling motions is N_r. Then, the basic body dynamics are acquired.

In the next step, the robot performs exploratory motor babbling by applying noise to its joint angles based on the predicted variance acquired in the previous step. The number of exploratory motor babbling motions with this added noise is N_e. The noise added to the joint angle G is expressed by the following equation based on a Gaussian distribution:

$$G = \frac{1}{\sqrt{2\pi\sigma^2}} exp\left(-\frac{(1- \mid \bar{v} - v_i \mid -\mu)^2}{2\sigma^2}\right), \tag{2}$$

where σ^2 and μ are the variance and mean, respectively, of the Gaussian distribution. In this study, we set $\sigma^2 = 15.0$ and $\mu = 0.0$. $\bar{v}$ is the average variance of all motions, and v_i is the variance used to apply noise.

Since the variance is large in the vicinity of unpredictable motion, the noise added to the joint angle is increased. Therefore, the robot will gain more experience of motions that are similar to those that are difficult to predict. By doing so, the robot acquires body dynamics with higher prediction accuracy.

3   Experimental Setup

3.1   Robot Model in Simulation

To evaluate the proposed method for exploratory motor babbling with a flexible-joint robot, a humanoid robot model was constructed in the OpenHRP3 robotics simulator. The model's size and DOFs were based on the humanoid robot ACTROID. The range of motion of the model's joint angles was based on that of a human. Motion generation for the teaching signal was used to train the S-CTRNN, and the torque of this motion was generated by the flexible joints according to the following proportional derivative (PD) control:

$$\tau_s = (\theta_d - \theta)Pgain + (\dot{\theta}_d - \dot{\theta})Dgain$$
$$+d(\dot{\theta} - \dot{\theta}_d) + k(\theta - \theta_d) + \tau_f, \tag{3}$$

where τ_s is the input torque, θ_d is the target angle, θ is the current angle, $\dot{\theta}_d$ is the target angular velocity, d is damping, $\dot{\theta}$ is the angular velocity, k is the spring term, and τ_s is the input torque. We set $\tau_f = 0.2$, $d = 10.0$, and $k = 10.0$.

The simulations were performed on the Ubuntu 12.04 LTS operating system with an Intel Core i7-4790 CPU.

3.2   Design of Motor Babbling

To evaluate the effectiveness of the proposed approach, the robot's movement was confined to two-dimensional movements with only three of the seven DOFs in the robot's arm. During the motion, the robot's joint angles, angular velocity, joint torque, and arm tip positions were measured. The initial position of the arm was set randomly. The robot performed motor babbling for 30 steps over a period of 2.07 [s]. The number of random motor babbling motions N_r was 15. After learning the random motor babbling and variance, the robot performed exploratory motor babbling, which was produced by adding a noise term based on the predicted variance in the joint angle. In this research, noise was added to all random motor babbling motions. Therefore, the number of exploratory motor babbling motions with added noise N_e was 15. Figure 1 illustrates the motor babbling, where the upper figure shows random motor babbling and the lower part shows exploratory motor babbling.

To learn motor babbling and the task described in the next section, we used an S-CTRNN. Table 1 describes the construction of the CTRNN. The probability distribution p of the teaching data was assumed to be 0.01.

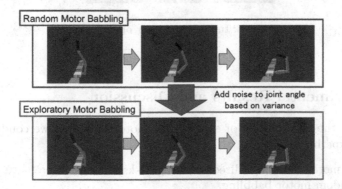

Fig. 1. Exploratory motor babbling

Table 1. Construction of S-CTRNN

Node Name	No. of Nodes	Time Constant
Angle Input Nodes	3	2
Angular Velocity Input Nodes	3	2
Torque Input Nodes	3	2
Arm Tip Input Nodes	2	2
Fast Context Nodes	20	5
Slow Context Nodes	2	70

3.3 Experimental Evaluation

To evaluate the effectiveness of performing motor babbling in advance, crank-turning and door opening/closing tasks were conducted (Fig. 2). We chose these tasks because they require smooth and dynamical motion without precise control. Crank turning is a simple rotational motion. In the motion of door opening/closing, the direction of motion changes.

The crank had a diameter of 0.3 [m]. The robot turned the crank five times. Each rotation consisted of 240 steps over a period of 12.0 [s]. Thus, the total motion consisted of 1200 steps, taking 60.0 [s]. The radius of the arc of the door's trajectory was 0.55 [m]. The robot opened and closed the door 10 times. Each open/close cycle had 120 steps, and took 6.0 [s]. Thus, there were a total of 1200 steps over 60.0 [s].

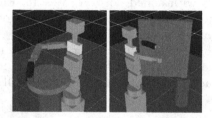

Fig. 2. Crank turning and door opening/closing

4   Experimental Results and Discussion

To evaluate the effectiveness of exploratory motor babbling, we conducted the following experiments:

– Without motor babbling (direct learning tasks).
– With random motor babbling.
– With exploratory motor babbling.

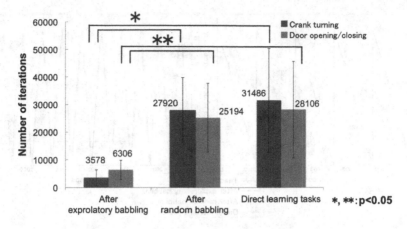

Fig. 3. Learning cycles for completing the task of crank turning and door opening/closing

Figure 3 shows the number of iterations that were required to learn the tasks. This number of iterations was needed by the robot to generate the motion of crank turning and door opening/closing. Each experiment was performed five times with different initial parameters for the neural network, and the figures show the mean and variance over these repetitions. For the crank turning task, the number of learning cycles with exploratory motor babbling was 3578, whereas with random motor babbling this increased to 27920, and without any motor babbling it was even higher at 31486 cycles. Thus, our method produced a reduction of 87.2 % in terms of learning cycles compared with random babbling. We confirmed that this difference was significant using a t-test ($p < 0.05$). The mean computation time of the learning cycle with exploratory motor babbling was 273.5 [s], whereas random motor babbling with the same number of motion patterns required 2453.0 [s]. Without any motor babbling, the mean computation time was 2650.2 [s]. For the door opening/closing task, the number of learning cycles with exploratory motor babbling was 6306, whereas with random motor babbling this rose to 25194, and without motor babbling it increased further to 28106. This represents a reduction of almost 75.0 % when using the proposed method compared with random babbling. We confirmed that this difference was again significant using a t-test ($p < 0.05$). The mean computation time of the learning cycle with exploratory motor babbling was 535.0 [s]; random motor babbling and the same number of motion patterns took 1916.3 [s], and without any motor babbling the learning cycle required 2130.7 [s]. In summary, Fig. 3 indicates that exploratory motor babbling drastically decreased the learning time.

Figure 4 shows the angle of crank turning and door opening/closing after 2350 and 3540 iterations, respectively. These iteration numbers correspond to one of the evaluated times in Fig. 3 at which the target motions had been generated correctly with exploratory motor babbling. The joint angles of the teaching signal

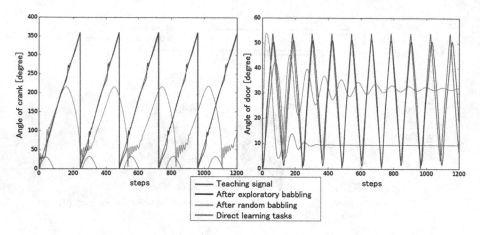

Fig. 4. Generated motions of crank turning and door opening/closing after 2350 and 3540 iterations (Left: crank turning, Right: door opening/closing)

and the motion learned with exploratory motor babbling are similar. In the case of crank turning with random motor babbling and without motor babbling, the robot could not turn the crank, and performed forward and reverse rotation repeatedly. In the case of crank turning with random motor babbling and without motor babbling, the angle of door opening and closing became gradually smaller. It is clear that the robot can perform these motions correctly after a relatively small number of iterations using exploratory motor babbling.

5   Conclusion

The objective of this research was to examine the benefits of exploratory motor babbling with a flexible-joint robot. First, the robot performed random motor babbling and learned the motion and variance. Next, the robot performed exploratory motor babbling by adding noise to the learned motions based on the predicted variance in the joint angles. In the evaluation experiments, we used the OpenHRP3 robotics simulator to perform crank turning and door opening/closing tasks. The results showed that the learning time was reduced compared to that without motor babbling and with random babbling.

In future work, we plan to increase the number of DOFs and verify the effective range by increasing the number of tasks.

Acknowledgments. The work has been supported by the Program for Leading Graduate Schools, "Graduate Program for Embodiment Informatics" of the Ministry of Education, Culture, Sports, Science, and Technology; JSPS Grant-in-Aid for Scientific Research (S)(2522005); "Fundamental Study for Intelligent Machine to Coexist with Nature" Research Institute for Science and Engineering, Waseda University; MEXT Grant-in-Aid for Scientific Research (A) 15H01710; and Scientific Research on Innovative Areas "Constructive Developmental Science" 24119003.

References

1. Asano, Y., Mizoguchi, H., Kozuki, T., Motegi, Y., Urata, J., Nakanishi, Y., Okada, K., Inaba, M.: Achievement of twist squat by musculoskeletal humanoid with screw-home mechanism. In: IEEE/RSJ International Conference on Intelligent Robots and Systems (IROS 2013), pp. 4649–4654 (2013)
2. Okada, M., Osato, K., Nakamura, Y.: Motion emergency of humanoid robots by an attractor design of a nonlinear dynamics. In: IEEE International Conference on Robotics and Automation (ICRA 2005), pp. 18–23 (2005)
3. Ijspeert, A.J., Nakanishi, J., Schaal, S.: Movement imitation with nonlinear dynamical systems in humanoid robots. In: IEEE International Conference on Robotics and Automation (ICRA 2002), pp. 1398–1403 (2002)
4. Miyakoshi, S., Taga, G., Kuniyoshi, Y., Nagakubo, A.: Three dimensional bipedal stepping motion using neural oscillators-towards humanoid motion in the real world. In: IEEE/RSJ International Conference on Intelligent Robots and Systems, vol. 1, pp. 84–89 (1998)
5. Kim, C.H., Tsujino, H., Sugano, S.: Online motion selection for semi-optimal stabilization using reverse-time tree. In: IEEE/RSJ International Conference on Intelligent Robots and Systems (IROS 2011), pp. 3792–3799 (2011)
6. Takahashi, K., Ogata, T., Tjandra, H., Yamaguchi, Y., Sugano, S.: Tool-body assimilation model based on body babbling and neuro-dynamical system. Math. Probl. Eng., August 2014. Article ID 837540
7. Saegusa, R., Metta, G., Sandini, G., Sakka, S.: Active motor babbling for sensorimotor learning. In: IEEE International Conference on ROBIO Robotics and Biomimetics, vol. 2008, pp. 794–799 (2009)
8. Ogawa, Y., Venture, G., Ott, C.: Dynamic parameters identification of a humanoid robot using joint torque sensors and/or contact forces. In: 14th IEEE-RAS International Conference on IEEE Humanoid Robots (Humanoids 2014), pp. 457–462 (2014)
9. von Hofsten, C.: An action perspective on motor development. Trends Cogn. Sci. 8(6), 266–272 (2004)
10. Murata, S., Yamashita, Y., Arie, H., Ogata, T., Sugano, S., Tani, J.: Learning to reproduce fluctuating behavioral sequences using a dynamic neural network model with time-varying variance estimation mechanism. In: Proceedings of the Third Joint IEEE International Conference on Development and Learning and Epigenetic Robotics, pp. 1–6 (2013)

Distributed Control for Nonlinear Time-Delayed Multi-Agent Systems with Connectivity Preservation Using Neural Networks

Hongwen Ma[1], Derong Liu[2]([✉]), and Ding Wang[1]

[1] The State Key Laboratory of Management and Control for Complex Systems, Institute of Automation, Chinese Academy of Sciences, Beijing 100190, China
{mahongwen2012,ding.wang}@ia.ac.cn
[2] School of Automation and Electrical Engineering, University of Science and Technology Beijing, Beijing 100083, China
derong@ustb.edu.cn

Abstract. Nonlinear time-delayed multi-agent systems with connectivity preservation are investigated in this paper. For each agent, the distributed controller is divided into five different parts which are designed to meet the requirements of the nonlinear time-delayed multi-agent systems, such as preserving connectivity, learning the unknown dynamics, eliminating time delays and reaching consensus. In addition, a σ-function technique is utilized to avoid the singularity in the developed distributed controller. Finally, simulation results demonstrate the effectiveness of the developed control protocol.

Keywords: Distributed control · Connectivity preservation · Nonlinear multi-agent systems · Neural networks · Time-delay

1 Introduction

With the growth of scale in large networked multi-agent systems, it is difficult to design an appropriate centralized controller to maintain the performance of the whole systems. Thus, distributed control is a good choice for solving this problem. Distributed control for multi-agent systems has been a hot topic in the last decade [1–3]. In [4], an output-based distributed control was proposed for nonlinear multi-agent systems with small-cyclic theorem. In [5], an online optimal learning approach was added to the decentralized control for a class of continuous-time nonlinear interconnected systems. To the best of authors' knowledge, it is the first time to investigate second-order nonlinear time-delayed multi-agent systems with connectivity preservation. In [6], in order to drive a group of ocean vessels to track a moving target and maintain the connectivity, adaptive neural network region tracking control was proposed. In [7], by virtue of neural networks, a decentralized robust adaptive control was designed to achieve consensus. In [8], a rendezvous protocol was proposed for the double-integrator multi-agent systems with preserved network connectivity. However, none of them

© Springer International Publishing Switzerland 2015
S. Arik et al. (Eds.): ICONIP 2015, Part III, LNCS 9491, pp. 34–42, 2015.
DOI: 10.1007/978-3-319-26555-1_5

takes time-delay into consideration. Thus, in this paper a Lyapunov-Krasovskii functional method is borrowed from [9,10] to eliminate the negative effect of time delays. Moreover, a σ-function is developed to circumvent the singularity in the distributed controller.

The remainder of this paper is given as follows. In Sect. 2, fundamental preliminaries and the problem statement are introduced. In Sect. 3, the distributed control protocol is developed which guarantees the achievement of consensus. Simulation example and conclusion are given in Sects. 4 and 5, respectively.

Notations: $(\cdot)^{\mathsf{T}}$ represents the transpose of a matrix. $\text{tr}(\cdot)$ is the trace of a given matrix and $\|\cdot\|$ is the Frobenius norm or Euclidian norm.

2 Preliminaries

2.1 Graph Theory

A triplet $\mathcal{G} = \{\mathcal{V}, \mathcal{E}, \mathcal{A}\}$ is called a weighted graph if $\mathcal{V} = \{1, 2, \ldots, N\}$ is the set of N nodes, $\mathcal{E} \subseteq \mathcal{V} \times \mathcal{V}$ is the set of edges, and $\mathcal{A} = (\mathcal{A}_{ij}) \in \mathbb{R}^{N \times N}$ is the $N \times N$ matrix of the weights of $\mathcal{G}$. Here we denote $\mathcal{A}_{ij}$ as the element of the ith row and jth column of the matrix $\mathcal{A}$. The ith node in graph $\mathcal{G}$ represents the ith agent, and a directed path from node i to node j is denoted as an ordered pair $(i, j) \in \mathcal{E}$, which means that agent i can directly transfer its information to agent j. $\mathcal{L} = \mathcal{D} - \mathcal{A}$ is the Laplacian matrix, where $\mathcal{D}$ is the $N \times N$ diagonal matrix whose diagonal elements are $d_i = \sum_{j \in \mathcal{N}_i} \mathcal{A}_{ij}$, $i = 1, 2, \ldots, N$ and $\mathcal{N}_i = \{j \in \mathcal{V} | (j, i) \in \mathcal{E}\}$ is the set of neighbour nodes of node i.

2.2 Radial Basis Function Neural Network

In this paper, radial basis function neural networks (RBFNNs) are used for approximating the unknown dynamics of the multi-agent systems. If $h(x)$ is a continuous unknown nonlinear function, then it can be approximated by RBFNNs as follows:

$$h(x) = W^{*\mathsf{T}}\Phi(x) + \theta, \tag{1}$$

where x is the input vector, W^* is the ideal weight matrix with suitable dimensions and θ is the approximating error with $\|\theta\| < \theta_N$. $\Phi(x) = [\gamma_1(x), \gamma_2(x), \ldots, \gamma_p(x)]^{\mathsf{T}}$ is the activation function vector and

$$\gamma_i(x) = \exp\left[\frac{-(x - \mu_i)^{\mathsf{T}}(x - \mu_i)}{\alpha_i^2}\right], \quad i = 1, 2, \ldots, p, \tag{2}$$

where α_i is the width of Gaussian function, p is the number of neurons and μ_i is the center of the receptive field. We denote $\hat{W}$ as the estimation of the ideal weight matrix W^*. Thus, the estimation of $h(x)$ can be written as $\hat{h}(x) = \hat{W}^{\mathsf{T}}\Phi(x)$, where $\hat{W}$ can be updated online. The online updating algorithm is provided in Sect. 3.

2.3 Problem Statement

In this paper, the second-order nonlinear time-delayed multi-agent system is modeled as follows:

$$\dot{p}_i = v_i,$$
$$\dot{v}_i = u_i + f_i(p_i(t), v_i(t)) + g_i(v_i(t - \tau_i)), \quad i = 1, 2, \ldots, N, \tag{3}$$

where $p_i \in \mathbb{R}^2$ is the position of agent i, $v_i \in \mathbb{R}^2$ is the velocity of agent i, τ_i is the unknown time delay of agent i, $f_i(\cdot)\colon \mathbb{R}^2 \to \mathbb{R}^2$ and $g_i(\cdot)\colon \mathbb{R}^2 \to \mathbb{R}^2$ are continuous but unknown nonlinear vector functions, and $u_i(\cdot) \in \mathbb{R}^2$ is the control input. For simplicity, we will ignore time expression t in case there is no confusion.

Assume that all the agents have a common sensing radius R and we adopt the hysteresis function in [8] to avoid measurement noise. When the distance between two agents is greater than R, we say that the two agents lose connectivity. Our control objective is to make all the agents reach consensus with connectivity preservation. That is, $\forall i, j \in \mathcal{V}$,

$$\begin{cases} \lim_{t \to \infty} \|p_i(t) - p_j(t)\| = 0, \\ \lim_{t \to \infty} v_i(t) = v_j(t) = 0, \end{cases} \tag{4}$$

and no agent will lose connection with its neighbors. We adopt the definition of the potential function in [8] which is given as follows:

$$\varphi(\|p_{ij}\|) = \frac{\|p_{ij}\|^2}{R - \|p_{ij}\| + \dfrac{R^2}{\hat{P}}}, \quad \|p_{ij}\| \in [0, R], \tag{5}$$

where R is the radius of communication range, $\|p_{ij}\| = \|p_i(t) - p_j(t)\|$ and $\hat{P} > 0$ is a large constant. It should be noted that we utilize $\mathcal{A}(t)$, $\mathcal{N}(t)$ and $\mathcal{L}(t)$ to represent the switching topology.

3 Distributed Control for Nonlinear Time-Delayed Multi-Agent Systems

Before proceeding, we introduce two important assumptions for demonstrating our main theorem.

Assumption 1. $g_i(v_i(t - \tau_i)), i = 1, 2, \ldots, N$, are unknown smooth nonlinear functions. The inequalities $\|g_i(v_i(t))\| \leq \phi_i(v_i(t)), i = 1, 2, \ldots, N$, hold, where $\phi_i(\cdot), i = 1, 2, \ldots, N$, are known positive smooth scalar functions. Furthermore, $g_i(0) = 0$ and $\phi_i(0) = 0$, $i = 1, 2, \ldots, N$.

Assumption 2. The unknown time delays $\tau_i, i = 1, 2, \ldots, N$, are bounded by a known constant $\tau_{\max}$, i.e., $\tau_i \leq \tau_{\max}, i = 1, 2, \ldots, N$.

In order to avoid the singularity induced by the denominator of the developed distributed controller, we define $\sigma(\cdot)$ as follows:

$$\sigma(v_i) = \begin{cases} 1, & \text{if } v_i = 0, \\ 0, & \text{if } v_i \neq 0. \end{cases} \tag{6}$$

In order to eliminate the effect of time delays, we introduce a Lyapunov-Krasovskii functional as follows:

$$V_U(t) = \frac{1}{2} \sum_{i=1}^{N} \int_{t-\tau_i}^{t} U_i(v_i(\zeta)) \mathrm{d}\zeta, \tag{7}$$

where $U_i(v_i(t)) = \phi_i^2(v_i(t))$. Then, the developed distributed controller is divided into five parts and they are given as follows:

$$u_i(t) = u_{i1}(t) + u_{i2}(t) + u_{i3}(t) + u_{i4}(t) + u_{i5}(t), \tag{8}$$

$$u_{i1}(t) = -\sum_{j \in \mathcal{N}_i(t)} \nabla_{p_i} \varphi(\|p_{ij}\|),$$

$$u_{i2}(t) = -\sum_{j \in \mathcal{N}_i(t)} \mathcal{A}_{ij}(t)(v_i - v_j),$$

$$u_{i3}(t) = -\frac{1}{2} \frac{v_i}{\|v_i\|^2 + \sigma(v_i)} \phi_i^2(v_i(t)),$$

$$u_{i4}(t) = -k_i(t)v_i,$$

$$u_{i5}(t) = -\hat{W}_i^\mathsf{T} \Phi_i(p_i, v_i),$$

$$k_i(t) = k_{i0} + 1 + \frac{1}{\omega_i} \left(\frac{1}{2} + \frac{\int_{t-\tau_{\max}}^{t} \frac{1}{2} U_i(v_i(\zeta)) \mathrm{d}\zeta}{\|v_i\|^2 + \sigma(v_i)} + \frac{\sum_{j \in \mathcal{N}_i(t)} \varphi(\|p_{ij}\|)}{\|v_i\|^2 + \sigma(v_i)} \right). \tag{9}$$

The online updating algorithm for the weight matrix of RBFNN is given as follows:

$$\dot{\hat{W}}_i = \begin{cases} \chi_i \Phi_i(p_i, v_i) v_i^\mathsf{T}, & \text{if } \mathrm{tr}\left(\hat{W}_i^\mathsf{T} \hat{W}_i\right) < W_i^{\max}, \text{ or} \\ & \text{if } \mathrm{tr}\left(\hat{W}_i^\mathsf{T} \hat{W}_i\right) = W_i^{\max} \text{ and } v_i^\mathsf{T} \hat{W}_i^\mathsf{T} \Phi_i(p_i, v_i) < 0, \\ \chi_i \Phi_i(p_i, v_i) v_i^\mathsf{T} - \chi_i \dfrac{v_i^\mathsf{T} \hat{W}_i^\mathsf{T} \Phi_i(p_i, v_i)}{\mathrm{tr}\left(\hat{W}_i^\mathsf{T} \hat{W}_i\right)} \hat{W}_i, & \text{otherwise,} \end{cases} \tag{10}$$

where $\tilde{W}_i = W_i^* - \hat{W}_i$ and χ_i is the updating rate. Moreover, the initial values of $\hat{W}_i$ should satisfy $\mathrm{tr}\left(\hat{W}_i^\mathsf{T}(0)\hat{W}_i(0)\right) \leq W_i^{\max}$. Before proceeding, we define the potential energy function as follows:

$$P_i(t) = \sum_{j \in \mathcal{N}_i(t)} \varphi(\|p_{ij}\|) + \frac{1}{2} v_i^\mathsf{T} v_i + \frac{1}{2} \int_{t-\tau_i}^{t} U_i(v_i(\zeta)) \mathrm{d}\zeta + \frac{1}{2} \mathrm{tr}\left(\frac{1}{\chi_i} \tilde{W}_i^\mathsf{T} \tilde{W}_i\right). \tag{11}$$

Then, the total potential energy function is $P(t) = \sum_{i=1}^{N} P_i(t)$.

Theorem 1. *The multi-agent system (3) consists of N agents and all the agents are driven by the distributed controller (8). With Assumptions 1 and 2, if the initial network topology $\mathcal{G}(0)$ is connected and undirected, and the initial energy $P(0)$ is finite, then the consensus of the multi-agent system (3) can be achieved while preserving connectivity.*

Proof. The derivative of $P(t)$ is

$$
\begin{aligned}
\dot{P}(t) =& \sum_{i=1}^{N} \sum_{j \in \mathcal{N}_i(t)} \dot{\varphi}(\|p_{ij}\|) + \sum_{i=1}^{N} v_i^{\mathsf{T}} \dot{v}_i + \dot{V}_U(t) - \sum_{i=1}^{N} \mathrm{tr}\left(\frac{1}{\chi_i} \tilde{W}_i^{\mathsf{T}} \dot{\hat{W}}_i\right) \\
=& \sum_{i=1}^{N} v_i^{\mathsf{T}} \sum_{j \in \mathcal{N}_i(t)} \nabla_{p_i} \varphi(\|p_{ij}\|) + \frac{1}{2} \sum_{i=1}^{N} \left(\phi_i^2(v_i(t)) - \phi_i^2(v_i(t - \tau_i))\right) \\
&+ \sum_{i=1}^{N} v_i^{\mathsf{T}} \Bigg(- k_i(t) v_i - \sum_{j \in \mathcal{N}_i(t)} \nabla_{p_i} \varphi(\|p_{ij}\|) - \sum_{j \in \mathcal{N}_i(t)} \mathcal{A}_{ij}(t)(v_i - v_j) \\
&- \frac{1}{2} \frac{v_i}{\|v_i\|^2 + \sigma(v_i)} \phi_i^2(v_i(t)) - \hat{W}_i^{\mathsf{T}} \Phi_i(\cdot) + g_i(v_i(t - \tau_i)) + W_i^{*\mathsf{T}} \Phi_i(\cdot) + \theta_i \Bigg) \\
&- \sum_{i=1}^{N} \mathrm{tr}\left(\frac{1}{\chi_i} \tilde{W}_i^{\mathsf{T}} \dot{\hat{W}}_i\right),
\end{aligned}
\tag{12}
$$

where we denote $\Phi_i(\cdot)$ as $\Phi_i(p_i, v_i)$ for short. If $\mathrm{tr}\left(\hat{W}_i^{\mathsf{T}}(0)\hat{W}_i(0)\right) \leq W_i^{\max}$, it is easy to demonstrate that $\mathrm{tr}\left(\hat{W}_i^{\mathsf{T}}(t)\hat{W}_i(t)\right) \leq W_i^{\max}$. Thus, according to the updating algorithm (10), the inequality $\mathrm{tr}\left(\tilde{W}_i^{\mathsf{T}}\left(\frac{1}{\chi_i}\dot{\hat{W}}_i - \Phi_i(\cdot)v_i^{\mathsf{T}}\right)\right) \geq 0$ holds. Merge the polynomial (12) we can obtain

$$
\begin{aligned}
\dot{P}(t) =& \frac{1}{2} \sum_{i=1}^{N} (\phi_i^2(v_i(t)) - \phi_i^2(v_i(t - \tau_i))) + \sum_{i=1}^{N} \mathrm{tr}\left(\frac{1}{\chi_i} \tilde{W}_i^{\mathsf{T}} \dot{\hat{W}}_i\right) + \sum_{i=1}^{N} v_i^{\mathsf{T}} \Bigg(- k_i(t) v_i \\
&- \sum_{j \in \mathcal{N}_i(t)} \mathcal{A}_{ij}(t)(v_i - v_j) - \frac{1}{2} \frac{v_i}{\|v_i\|^2} \phi_i^2(v_i(t)) - \tilde{W}_i^{\mathsf{T}} \Phi_i(\cdot) + g_i(v_i(t - \tau_i)) + \theta_i \Bigg) \\
=& - \sum_{i=1}^{N} k_i(t)\|v_i\|^2 - v^{\mathsf{T}}(\mathcal{L}(t) \otimes I_2)v - \sum_{i=1}^{N} \mathrm{tr}\left(\tilde{W}_i^{\mathsf{T}}\left(\frac{1}{\chi_i}\dot{\hat{W}}_i - \Phi_i(\cdot)v_i^{\mathsf{T}}\right)\right) \\
&- \frac{1}{2} \sum_{i=1}^{N} \phi_i^2(v_i(t - \tau_i)) + \sum_{i=1}^{N} v_i^{\mathsf{T}} (g_i(v_i(t - \tau_i)) + \theta_i)
\end{aligned}
$$

$$\leq -\sum_{i=1}^{N} k_i(t)\|v_i\|^2 - \frac{1}{2}\sum_{i=1}^{N} \phi_i^2(v_i(t-\tau_i)) + \frac{1}{2}\sum_{i=1}^{N} \left(\|v_i\|^2 + \|g_i(v_i(t-\tau_i))\|^2\right)$$

$$+ \frac{1}{2}\sum_{i=1}^{N}\left(\|v_i\|^2 + \|\theta_i\|^2\right) \quad \text{(with Assumption 1)}$$

$$\leq -\sum_{i=1}^{N}(k_i(t)-1)\|v_i\|^2 + \epsilon,$$

where $\epsilon = \frac{1}{2}\sum_{i=1}^{N} \theta_{N_i}^2$, $v = [v_1, v_2, \ldots, v_N]^\mathsf{T}$ and $\|\theta_i\| < \theta_{N_i}$. Thus, with Assumption 2, we have

$$\dot{P}(t) \leq -\sum_{i=1}^{N} k_{i0}\|v_i\|^2 - \sum_{i=1}^{N}\frac{1}{\omega_i}\int_{t-\tau_{\max}}^{t}\frac{1}{2}U_i(v_i(\zeta))\mathrm{d}\zeta - \sum_{i=1}^{N}\frac{1}{2\omega_i}\|v_i\|^2$$

$$-\sum_{i=1}^{N}\frac{1}{\omega_i}\sum_{j\in\mathcal{N}_i(t)}\varphi(\|p_{ij}\|) - \sum_{i=1}^{N}\frac{2W_i^{\max}}{\omega\chi_i} + \sum_{i=1}^{N}\frac{2W_i^{\max}}{\omega\chi_i} + \theta$$

$$\leq -\frac{1}{\omega}P(t) + \sum_{i=1}^{N}\frac{2W_i^{\max}}{\omega\chi_i} + \epsilon,$$

where $\omega = \max\limits_{i\in\mathcal{V}}\omega_i$ and $k_{i0} > 0$. Then, according to Lemma 1 in [7], we have

$$P(t) \leq P(0)e^{-\frac{1}{\omega}t} + \nu\left(1 - e^{-\frac{1}{\omega}t}\right), \tag{13}$$

where $\nu = \sum\limits_{i=1}^{N}\frac{2W_i^{\max}}{\chi_i} + \omega\epsilon$.

The number of agents is finite, thus the switching times of the network topology are finite. We denote the switching times as $t_0, t_1, \ldots$, where t_0 is the initial time. By choosing appropriate parameters in (13) when $t \in [t_0, t_1)$, $P(t) \leq P(0) < P_{\max}$ holds. Therefore, network will not lose connectivity at t_1 and new edges will be added at t_1 because of the decrease of $P(t)$. Following the similar proof steps in the above analysis, when $t \in [t_{k-1}, t_k)$, $k = 2, 3, \ldots$, the connectivity can be guaranteed. In summary, if the initial undirected network topology $\mathcal{G}(0)$ is connected and the initial energy $P(0)$ is finite, the connectivity for $t > 0$ can be preserved. Then, we restrict the following discussion when the network has been fixed. Since every term in $P(t)$ is positive and bounded, all the terms in $P(t)$ will approach to zero, that is, $\lim\limits_{t\to\infty} p_1 = p_2 = \cdots = p_N$ and $\lim\limits_{t\to\infty} v_1 = v_2 = \cdots = v_N = 0$. Furthermore, $\lim\limits_{t\to\infty}\|W_i^* - \hat{W}_i\| = 0$ shows that RBFNNs can learn the unknown dynamics of each agent. $\square$

4   Simulation Example

We choose a multi-agent system with five agents which moves on a two-dimensional plane. We set the sensing radius $R = 2.5\,\mathrm{m}$ and choose the initial positions and velocities randomly from $[0, 4\,\mathrm{m}] \times [0, 4\,\mathrm{m}]$ and $[0, 2\,\mathrm{m/s}] \times [0, 2\,\mathrm{m/s}]$, respectively. Assume that all the existing communication weights are 1 and the time-delay vector is $\tau = [0.1, 0.05, 0.13, 0.08, 0.15]$. $\begin{bmatrix} l_1 \\ l_2 \end{bmatrix} = \begin{bmatrix} 0.4, & -0.65, & 0.5, & -0.75, & 0.1 \\ 0.5, & 0.45, & -0.6, & 0.4, & 1 \end{bmatrix}$ and $\begin{bmatrix} m_1 \\ m_2 \end{bmatrix} = \begin{bmatrix} 0.9, & 1.2, & -1.1, & 0.7, & 0.6 \\ 1.2, & -0.8, & 0.6, & 0.3, & 0.8 \end{bmatrix}$ are the coefficients of $f(\cdot)$ and $g(\cdot)$, respectively. The dynamics of time-delay term is given as follows:

$$g_i(v_i(t)) = \begin{bmatrix} m_{i1}v_{i1}(t)\cos(v_{i2}(t)) \\ m_{i2}v_{i2}(t)\sin(v_{i1}(t)) \end{bmatrix}. \tag{14}$$

Then, $\phi_i(v_i(t)) = \sqrt{(m_{i1}v_{i1}(t))^2 + (m_{i2}v_{i2}(t))^2}$. The unknown dynamics is chosen to be

$$f_i(p_i(t), v_i(t)) = \begin{bmatrix} l_{i1}p_{i1}(t)\sin(p_{i2}(t))v_{i1}v_{i2} \\ l_{i2}p_{i2}(t)\cos(p_{i1}(t))\sin(v_{i1}v_{i2}) \end{bmatrix}. \tag{15}$$

Suppose that all the five agents have the same parameters, $\hat{P} = 1000$, $\tau_{\max} = 0.15$, $k_{i0} = 10$, $\omega_i = 50$, $W_i^{\max} = 100$ and $\chi_i = 100$. The number of neurons for each RBFNN is 16 and $\alpha_i^2 = 2$. μ_i is distributed uniformly among the range $[0, 4] \times [0, 4]$.

In Fig. 1, the red asterisks are the initial positions and the blue arrows are the directions of initial velocities. In Fig. 2, the red pentagram is the final position showing that consensus can be achieved with the developed distributed

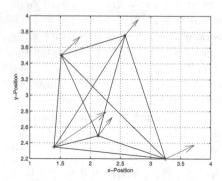

Fig. 1. Initial topology (Color figure online)

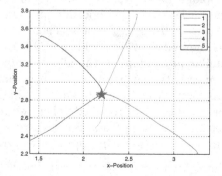

Fig. 2. Consensus of the five agents (Color figure online)

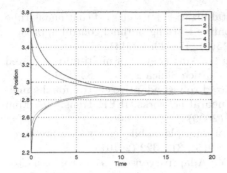
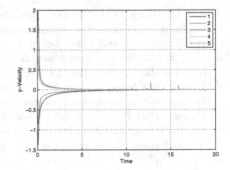

Fig. 3. Trajectories of position in y-axis   **Fig. 4.** Trajectories of velocity in y-axis

controller. We choose to show the trajectories of positions and velocities in y-axis in Figs. 3 and 4, respectively.

5   Conclusion

Nonlinear time-delayed multi-agent systems are investigated in this paper. The distributed controller is divided into five parts. By using RBFNNs, the distributed controller can learn the unknown nonlinear dynamics online. Furthermore, by introducing Lyapunov-Krasovskii functional, the effect of time delays can be eliminated. Finally, connectivity preservation can be guaranteed by designing a high-threshold potential function. Simulation results show the effectiveness of the developed distributed controller.

Acknowledgement. This work was supported in part by the National Natural Science Foundation of China under Grants 61233001, 61273140, 61304086, and 61374105, in part by Beijing Natural Science Foundation under Grant 4132078, and in part by the Early Career Development Award of SKLMCCS.

References

1. Ren, W., Beard, R.W., Atkins, E.M.: A survey of consensus problems in multi-agent coordination. In: Proceedings of the American Control Conference, pp. 1859–1864. Portland, OR, USA (2005)
2. Olfati-Saber, R., Fax, J.A., Murray, R.M.: Consensus and cooperation in networked multi-agent systems. Proc. IEEE **95**(1), 215–233 (2007)
3. Zhang, H., Zhang, J., Yang, G.H., Luo, Y.: Leader-based optimal coordination control for the consensus problem of multiagent differential games via fuzzy adaptive dynamic programming. IEEE Trans. Fuzzy Syst. **23**(1), 152–163 (2015)
4. Liu, T.F., Jiang, Z.P.: Distributed output-feedback control of nonlinear multi-agent systems. IEEE Trans. Autom. Control **58**(11), 2912–2917 (2013)

5. Liu, D., Wang, D., Li, H.: Decentralized stabilization for a class of continuous-time nonlinear interconnected systems using online learning optimal control approach. IEEE Trans. Neural Netw. Learn. Syst. **25**(2), 418–428 (2014)
6. Sun, X., Ge, S.S.: Adaptive neural region tracking control of multi-fully actuated ocean surface vessels. IEEE/CAA J. Automatica Sinica **1**(1), 77–83 (2014)
7. Hou, Z.G., Cheng, L., Tan, M.: Decentralized robust adaptive control for the multiagent system consensus problem using neural networks. IEEE Trans. Sys. Man Cybern. Part B: Cybern. **39**(3), 636–647 (2009)
8. Su, H., Wang, X., Chen, G.: Rendezvous of multiple mobile agents with preserved network connectivity. Syst. Control Lett. **59**(5), 313–322 (2010)
9. Chen, C., Wen, G.X., Liu, Y.J., Wang, F.Y.: Adaptive consensus control for a class of nonlinear multiagent time-delay systems using neural networks. IEEE Trans. Neural Netw. Learn. Syst. **25**(6), 1217–1226 (2014)
10. Ge, S., Hong, F., Lee, T.: Adaptive neural control of nonlinear time-delay systems with unknown virtual control coefficients. IEEE Trans. Syst. Man Cybern. Part B: Cybern. **34**(1), 499–516 (2004)

Coevolutionary Recurrent Neural Networks for Prediction of Rapid Intensification in Wind Intensity of Tropical Cyclones in the South Pacific Region

Rohitash Chandra[1,2](✉) and Kavina S. Dayal[3]

[1] School of Computing Information and Mathematical Sciences,
University of South Pacific, Suva, Fiji
c.rohitash@gmail.com
[2] Artificial Intelligence and Cybernetics Research Group,
Software Foundation, Nausori, Fiji
[3] School of Agricultural, Computational and Environmental Sciences,
University of Southern Queensland, Springfield 4300, Australia
kavinadayal@gmail.com

Abstract. Rapid intensification in tropical cyclones occur where there is dramatic change in wind-intensity over a short period of time. Recurrent neural networks trained using cooperative coevolution have shown very promising performance for time series prediction problems. In this paper, they are used for prediction of rapid intensification in tropical cyclones in the South Pacific region. An analysis of the tropical cyclones and the occurrences of rapid intensification cases is assessed and then data is gathered for recurrent neural network for rapid intensification predication. The results are promising that motivate the implementation of the system in future using cloud computing infrastructure linked with mobile applications to create awareness.

Keywords: Cooperative coevolution · Neuro-evolution · Recurrent neural networks · Tropical cyclones · Rapid intensification

1   Introduction

Rapid intensification occurs when a tropical cyclone intensifies dramatically within a short period of time [1]. Previous studies have shown that operational forecasting models are more skillful in predicting tropical cyclone tracks whereas predicting cyclone intensity remains one of the major challenges in tropical weather forecasting [2]. Forecasting rapid intensification has been another challenge, which is partly due to our limited understanding of the physical mechanisms of tropical cyclone intensity change in general [2,3]. Previous efforts in studying individual tropical cyclone have identified some conditions that are

© Springer International Publishing Switzerland 2015
S. Arik et al. (Eds.): ICONIP 2015, Part III, LNCS 9491, pp. 43–52, 2015.
DOI: 10.1007/978-3-319-26555-1_6

favourable for rapid intensification. For instance, in efforts to understand change in tropical cyclone intensity, it has been shown that warm ocean temperatures [4,5] and warm-ocean eddies [6] influence the rapid intensification of tropical cyclones.

The existing literature defines rapid intensification in various ways. For instance, [1] defined rapid deepening of tropical cyclones when the systems pressure drops by ≥ 42 millibar in 24-h. The rapid intensification for Northern Hemisphere tropical cyclones according to National Hurricane Centre is an increase in the maximum sustained winds of a tropical cyclone by at least 30-knots in a 24-h period. This definition has been employed in [7] who define rapid intensification as an approximate of the 95th percentile of all 24-h over-water intensity change of tropical cyclones in the North Atlantic basins by 30-knots over 24-h period.

Cooperative coevolution (CC) is a evolutionary computation method which divides a large problem into subcomponents and solves them using evolutionary algorithms [8]. CC has been effective for neuro-evolution of feedforward and recurrent neural networks [9–13]. Problem decomposition is an important procedure in cooperation coevolution that determines how the subcomponents are decomposed [9]. Cooperative neuro-evolution of recurrent neural networks have given very promising performance for time series problems [13,14] and also have been successfully applied for cyclone wind-intensity and track prediction problems for the South Pacific Ocean [15,16].

In this paper, cooperative neuro-evolution of recurrent networks is applied for prediction of rapid intensification in tropical cyclones in the South Pacific region. Rapid intensification cases are detected and collected for recurrent neural network for training and testing. We capture the time series during the cyclone for one and two days ahead that led to rapid intensification cases. Although other machine learning methods can be used, we specially chose coevolutionary recurrent neural networks as they showed promising performance in cyclone wind-intensity and track prediction [15,16].

The rest of the paper is organised as follows. Section 2 gives background in cyclone wind-intensity prediction and computational intelligence methods for time series prediction. In Sect. 3, the proposed method is discussed in detail while in Sect. 4, experiments and results are given. Section 5 concludes the paper with discussion of future work.

2 Coevolutionary Recurrent Networks for Rapid Intensification

2.1 Recurrent Network Architecture

Recurrent neural networks are suitable for modelling temporal sequences. Elman recurrent neural networks use context units to store the output of the state neurons from computation of the previous time steps [17]. The context layer is used for computation of present states as they contain information about the previous states as shown in Fig. 1. The dynamics of the change of hidden state neuron activation's in Elman style recurrent networks is given by Eq. (1).

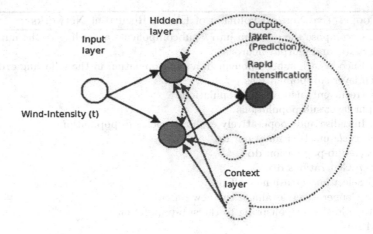

Fig. 1. Elman recurrent neural network used for prediction of rapid intensification in cyclones. We use the wind-intensity to predict the strength of rapid intensification. The recurrent neural network has 1 neuron in the input layer and 1 output neuron assigned for the prediction of rapid intensification.

$$y_i(t) = f\left(\sum_{k=1}^{K} v_{ik}\, y_k(t-1) + \sum_{j=1}^{J} w_{ij}\, x_j(t-1)\right) \tag{1}$$

where $y_k(t)$ and $x_j(t)$ represent the output of the context state neuron and input neurons respectively. v_{ik} and w_{ij} represent their corresponding weights. $f(.)$ is a sigmoid transfer function.

2.2 Cooperative Neuro-Evolutionary Recurrent Networks

Algorithm 1 gives details for the cooperative neuro-evolution method used for training Elman recurrent neural networks shown in Fig. 1.

In Algorithm 1, the recurrent neural network is decomposed in k subcomponents using neural level problem decomposition method [13]. k is equal to the total number of hidden, context and output neurons. Each subcomponents contains all the weight links from the previous layer connecting to a particular neuron. Each hidden neuron also acts as a reference point for the recurrent (state or context) weight links connected to it. Therefore, the subcomponents for a recurrent network with a single hidden layer is composed as follows:

1. Hidden layer subcomponents: weight-links from each neuron in the *hidden(t)* layer connected to all *input(t)* neurons and the bias of *hidden(t)*, where t is time.
2. State (recurrent) neuron subcomponents: weight-links from each neuron in the *hidden(t)* layer connected to all hidden neurons in previous time step *hidden(t − 1)*.
3. Output layer subcomponents: weight-links from each neuron in the *output(t)* layer connected to all *hidden(t)* neurons and the bias of *output(t)*.

Alg. 1. Cooperative Neuro-Evolution of Elman Recurrent Networks

---

Step 1: Decompose the problem into k subcomponents according to the number of Hidden, State, and Output neurons
Step 2: Encode each subcomponent in a sub-population in the following order:
i) Hidden layer sub-populations
ii) State (recurrent) neuron sub-populations
iii) Output layer sub-populations
Step 3: Initialise and cooperatively evaluate each sub-population
for each *cycle* until termination **do**
  for each Sub-population **do**
    for n Generations **do**
      i) Select and create new offspring
      ii) Cooperatively evaluate the new offspring
      iii) Add the new offspring to the sub-population
    end for
  end for
end for

The subcomponents are implemented as sub-populations that employ the generalised generation gap with parent-centric crossover operator genetic algorithm [18]. A *cycle* is completed when all the sub-populations are evolved for a fixed number of generations.

A major concern in this proposed method is the cooperative evaluation of each individual in every sub-population. There are two main phases of evolution in the cooperative coevolution framework. The first is the *initialisation phase* and second is the *evolution phase*.

Cooperative evaluation in the initialisation phase is given in Step 3. In the initialisation stage, the individuals in all the sub-populations do not have a fitness. In order to evaluate the ith individual of the kth sub-population, arbitrary individuals from the rest of the sub-populations are selected and combined with the chosen individual and cooperatively evaluated. The best individual is chosen once fitness has been assigned to all the individuals of a particular sub-population [8]. Cooperative evaluation in the evolution phase is shown in Step 3 (ii). This is done by concatenating the chosen individual from a sub-population k with the single best individual from the rest of the sub-populations. The algorithm halts if the termination condition is satisfied. The termination criteria is a specified fitness is achieved which is given by mean absolute error on the validation data set. Another termination condition is when the maximum number of function evaluations has been reached.

The G3-PCX (generalised generation gap with parent-centric crossover operator) algorithm is used in the sub-populations of cooperative coevolution [18].

G3-PCX has been used in sub-populations of cooperative coevolution methods in our past research that includes cooperative coevolutionary recurrent neural networks for time series prediction [13], memetic cooperative coevolution [11] and competitive cooperative coevolution for time series prediction [14] and also application for cyclone wind-intensity prediction [15]. It gave promising results when compared to related methods from the literature.

2.3   Application Problem: Rapid intensification in Cyclones

The case of rapid intensification involves data pre-processing where the wind-intensity of all the tropical cyclones in a region is examined. The definition in literature for rapid intensification is when there is an increase in wind-intensity by 30 knots in 24 h [7]. In order to make our proposed prediction method more robust, we catered for other cases of rapid intensification by the following rules:

- Case 1: Between 20 – 30 knots
- Case 2: Between 30 – 40 knots
- Case 3: More than 40 knots

The cyclones are from the South Pacific and Indian Ocean region [19] and contains the wind-intensity and cyclone track information in terms of the longitude and latitude.

We pre-processed the cyclone data taking into account two major configurations in order to investigate if the track information of cyclones has major impact in terms of determining rapid intensification. The purpose of this approach is to find out if the track information is important in order to determine cases of rapid intensification.

We used 30 h, 5 data points - i.e., take 5 previous points when the rapid intensification is detected. Therefore, the recurrent neural network would be able to predict rapid intensification when 5 readings (every six hours are given).

3   Experiments and Results

This section presents the results of experiments for cooperative neuro-evolution of recurrent neural networks for prediction of rapid intensification in tropical cyclones in the South Pacific region. Initially, an analysis of the tropical cyclones and the occurrences of rapid intensification cases is assessed and then data is collected for recurrent neural networks training and testing.

3.1   Analysis of the Dataset

We implemented an algorithm that checked the occurrences of the cases of rapid intensification. The definition of rapid intensification from literature is when a tropical cyclones changes its speed by more than 30 knots in 24 h [7].

The Southern Hemisphere tropical cyclone best-track data from Joint Typhoon Warning Centre [19] recorded every 6-h are used. Only the austral summer tropical cyclone season, November to April, from 1980 to 2012 data is analysed in the current study. The South Indian basin domain is taken to be 0–30°S, 30°E-130°E and South Pacific domain is 0–30°S, 130°E-130°W.

We divided the original data of tropical cyclone wind intensity in the South Pacific [19] into training and testing set as follows:

- Training Set: Cyclones from 1985 – 2005 (219 Cyclones)
- Testing Set: Cyclones from 2006 – 2013 (71 Cyclones)

Table 1. Cases of Rapid Intensification in the South Pacific

Dataset	Case-1	Case-2	Case-3	Total
Testing Set	66	6	1	73
Training Set	259	103	52	414

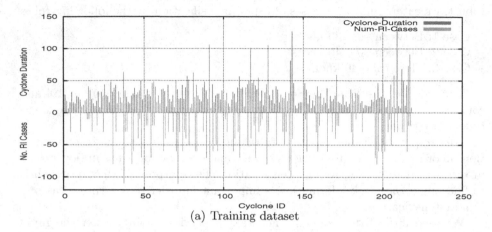

(a) Training dataset

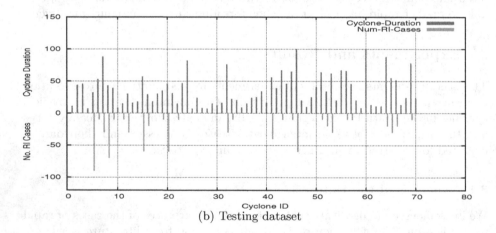

(b) Testing dataset

Fig. 2. Number of Rapid Intensification cases (x 10) and duration of each cyclone over the cyclone identification number (ID). Each point of cyclone duration in y axis represents 6 h. In certain cyclones, there is no case of rapid intensification.

Table 1 gives the details about the occurrences of different cases of rapid intensification in each dataset.

Figure 2 show the details of the duration of each cyclone in the training and testing dataset for different cyclones given by their identification number (ID) in the x axis. Note that each point of duration in the y axis represents 6 h. The negative bars in the histograms shows the number of cases of rapid intensification for the corresponding cyclones. Note that for visualisation purpose, we multiplied each of the case by a factor of 10. For instance, if there is a cyclone ID that shows −50 on the y axis, it represents 5 rapid intensification cases.

3.2 Data Pre-processing

In order to effectively use neural networks for time series prediction, measures need to be taken to pre-process the raw time series data and arranged in a specific way so that it can be used to train the Elman recurrent network. In the cyclone wind-intensity data, a number of missing values were present for cyclones before the year 1985. The set of experiments in this paper used cyclones from the year 1985 and onward.

3.3 Results

The performance and results of the method were evaluated by using different number of hidden neurons (H) and compared with standalone cooperative coevolution.

The maximum training time was given by number of function evaluations by cooperative coevolution (20 000). The G3-PCX evolutionary algorithm [18] was used in sub-populations of cooperative coevolution with fixed parameters such as population size (200), 2 offspring and 2 parents for parent centric crossover operator as used in previous works [13]. The root mean squared error (RMSE) and mean absolute error (MAE) are used to evaluate the performance of the proposed method for cyclone wind-intensity prediction.

These are given in Eq. 2 (RMSE) and Eq. 3 (MAE).

$$RMSE = \sqrt{\frac{1}{N} \sum_{i=1}^{N} (y_i - \hat{y}_i)^2} \qquad (2)$$

$$MAE = \frac{1}{N} \sum_{i=1}^{N} |(y_i - \hat{y}_i)| \qquad (3)$$

where y_i and $\hat{y}_i$ are the observed and predicted data, respectively. N is the length of the observed data. These two performance measures are used in order to compare the results with the literature.

The results are given in Table 2 where the RMSE and MAE have been used as the main performance measures. We observe that there is larger training error than the prediction. The large training error is due to possible inconsistencies and noise in the time series analysis of the rapid intensification dataset snapshot taken for the last 30 h (5 points).

A typical performance on the training and test set is given in Fig. 3. We can observe that the recurrent neural network has good performance for most of the cases except for the extreme cases of rapid intensification where the wind-intensity change is more than 30 knots in 24 h.

Table 2. Results: Wind-Intensity for Rapid Intensification in South Pacific

H	RMSE (Train)	RMSE (Test)	Best	MAE (Train)	MAE (Test)	Best
3	0.1621 ± 0.0005	0.1237 ± 0.0036	0.1112	4.8979 ± 0.0221	4.2122 ± 0.1576	3.6593
5	0.1612 ± 0.0006	0.1205 ± 0.0014	0.1132	4.8613 ± 0.0253	4.0401 ± 0.0659	3.7008
7	0.1615 ± 0.0005	0.1227 ± 0.0019	0.1160	4.8808 ± 0.0219	4.1652 ± 0.0857	3.8001
9	0.1614 ± 0.0005	0.1214 ± 0.0016	0.1089	4.8812 ± 0.0239	4.1032 ± 0.0755	3.6841

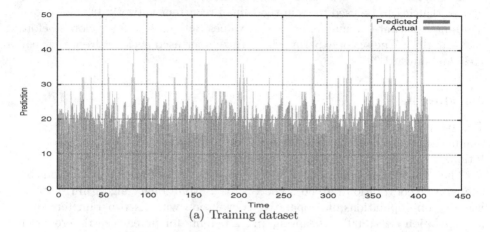

(a) Training dataset

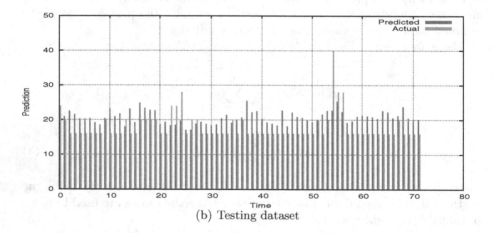

(b) Testing dataset

Fig. 3. Rapid Intensification prediction for all cyclones in the South Pacific

3.4   Discussion

Although the results are promising, they need to be improved further as the training errors are quite high which suggests that the data sets have noise or contractions and hence the recurrent neural network had difficulty to converge to lower errors. There seems to be local convergence in the training as there is not significant changes to the error after 1000 function evaluations in all the experiments. The maximum time was 20 000 function evaluations. One problem is the lack of the past data points. We only considered 5 data points which is taken every 6 h and spans for 30 h. We can have better convergence when more data points are given, i.e., if readings are taken every 3 or 2 h, we will have more information and hence the recurrent neural network can resolve contractions and go towards better convergence and prediction.

Further information along with the wind-intensity can also be incorporated into the system, i.e., if more features of the cyclone is recorded such as humidity, pressure and sea surface temperature, then the system could be more accurate.

The knowledge gained from current analysis can be used to improve our understanding of the process of rapid intensification by identifying useful predictors, hence help improve seasonal and intra-seasonal prediction of rapid intensification activity. Moreover, online web services and mobile applications can be developed for awareness and warning.

We concentrated on predicting the intensity of rapid intensification which is essentially a time series prediction problem. The problem can also be viewed as pattern classification problem, where instead of the intensity, the occurrence of rapid intensification could be predicted. This means that the system would be able to determine if a cyclone will rapidly intensify and the one proposed in this paper will predict the intensity.

4   Conclusions and Future Work

We have been successful in providing an analysis of the number of cases and types of rapid intensification in the South Pacific region over the last three decades. The proposed system based on co-evolutionary recurrent neural networks has been able to give prediction with reasonable errors between actual and predicted wind intensity change. However, more accuracy is desired in order for full implementation.

In future work, we would like to use more data points in terms of readings about the cyclones and features in order to build a more accurate system. We would also like to check other data readings such as the sea surface temperature, humidity and pressure levels and check their relationship with the cases of rapid intensification. Other neural network architectures such as feedforward networks can also be used for prediction of rapid intensification with different training algorithms. The rapid intensification problem can also be approached as a pattern classification problem where the occurrence of rapid intensification is predicted rather than its value of intensification.

References

1. Holliday, C.R., Thompson, A.H.: Climatological characteristics of rapidly intensifying typhoons. Mon. Weather Rev. **107**, 1022–1034 (1979)

2. DeMaria, M., Zehr, R.M., Kossin, J.P., Knaff, J.A.: The use of goes imagery in statistical hurricane intensity prediction. In: 25th Conference on Hurricanes and Tropical Meteorology, San Diego, CA, pp. 120–121 (2002)

3. Gross, J.M.: North atlantic and east pacific track and intensity verification for 2000. In: 55th Interdepartmental Hurricane Conference, Miami, FL. Office of the Federal Coordinator for Meteorological Services and Supporting Research, NOAA, B12B15, pp. 120–121 (2002)

4. Miller, B.I.: On the maximum intensity of hurricanes. J. Meteorol. **15**, 184–195 (1958)

5. Malkus, J.S., Riehl, H.: On the dynamics and energy transformations in steady-state hurricanes. Tellus **12**, 1–20 (1960)

6. Shay, L.K., Goni, G.J., Black, P.G.: Effects of a warm oceanic feature on hurricane opal. Mon. Weather Rev. **128**, 1366–1383 (2000)

7. Kaplan, J., DeMaria, M.: Large-scale characteristics of rapidly intensifying tropical cyclones in the north atlantic basin. Weather Forecast. **18**, 1093–1108 (2003)

8. Potter, M., De Jong, K.: A cooperative coevolutionary approach to function optimization. In: Davidor, Y., Männer, R., Schwefel, H.-P. (eds.) PPSN 1994. LNCS, vol. 866, pp. 249–257. Springer, Heidelberg (1994)

9. Potter, M.A., De Jong, K.A.: Cooperative coevolution: an architecture for evolving coadapted subcomponents. Evol. Comput. **8**(1), 1–29 (2000)

10. García-Pedrajas, N., Ortiz-Boyer, D.: A cooperative constructive method for neural networks for pattern recognition. Pattern Recogn. **40**(1), 80–98 (2007)

11. Chandra, R., Frean, M.R., Zhang, M.: Crossover-based local search in cooperative co-evolutionary feedforward neural networks. Appl. Soft Comput. **12**(9), 2924–2932 (2012)

12. Gomez, F., Mikkulainen, R.: Incremental evolution of complex general behavior. Adapt. Behav. **5**(3–4), 317–342 (1997)

13. Chandra, R., Zhang, M.: Cooperative coevolution of Elman recurrent neural networks for chaotic time series prediction. Neurocomputing **186**, 116–123 (2012)

14. Chandra, R.: Competition and collaboration in cooperative coevolution of Elman recurrent neural networks for time-series prediction. In: IEEE Transactions on Neural Networks and Learning Systems, In Press (2015)

15. Chandra, R., Dayal, K.: Cooperative coevolution of Elman recurrent networks for tropical cyclone wind-intensity prediction in the South Pacific region. In: IEEE Congress on Evolutionary Computtaion, Japan, Sendai, pp. 1784–1791, May 2015

16. Chandra, R., Dayal, K., Rollings, N.: Application of cooperative neuro-evolution of Elman recurrent networks for a two-dimensional cyclone track prediction for the South Pacific region. In: International Joint Conference on Neural Networks (IJCNN), Killarney, Ireland, pp. 721–728, July 2015

17. Elman, J.L.: Finding structure in time. Cogn. Sci. **14**, 179–211 (1990)

18. Deb, K., Anand, A., Joshi, D.: A computationally efficient evolutionary algorithm for real-parameter optimization. Evol. Comput. **10**(4), 371–395 (2002)

19. JTWC, Tropical Cyclone Best Track Data. http://www.usno.navy.mil/NOOC/ nmfc-ph/RSS/jtwc/best_tracks/shindex.php. Accessed 02 February, 2015

Nonlinear Filtering Based on a Network with Gaussian Kernel Functions

Dong-Ho Kang[1,2] and Rhee Man Kil[1(✉)]

[1] College of Information and Communication Engineering,
Sungkyunkwan University, 2066, Seobu-ro, Jangan-gu, Suwon,
Gyeonggi-do 440-746, Korea
rmkil@skku.edu
[2] Multimedia R&D Group, Samsung Electronics, 129, Samsung-ro,
Yeongtong-gu, Suwon, Gyeonggi-do 443-742, Korea
dho.kang@samsung.com

Abstract. This paper presents a new method of nonlinear finetwork with Gaussian kernel functions. In practice, signal enhancement filters are usually adopted as a preprocessor of signal processing system. For this purpose, an approach of nonlinear filtering using a network with Gaussian kernel functions is proposed for the efficient enhancement of noisy signals. In this method, the condition for signal enhancement is obtained by using the phase space analysis of signal time series. Then, from this analysis, the structure of nonlinear filter is determined and a network with Gaussian kernel functions is trained in such a way of obtaining the clean signal. This procedure can be repeated to obtain the multilayer (or deep) structure of nonlinear filters. As a result, the proposed nonlinear filter has demonstrated significant merits in signal enhancement compared with other conventional preprocessing filters.

Keywords: Nonlinear filtering · Noisy signals · Regression model · Gaussian kernel functions · Deep structure

1 Introduction

In practical problems of signal processing system, signals are usually contaminated by noise and this causes the degradation of the system performance. In this respect, signal enhancement filters are usually adopted as a preprocessor of signal processing system. One way of solving this problem is to use a linear filter such as the finite impulse response (FIR) filter in which the clean signals are estimated from the input of noisy signals. However, if we use a fixed form of linear filters, it may not enhance the untrained noisy signals since many signals usually are not generated by the linear systems. To cope with this problem, an optimal filter such as the Kalman filter [1] that can fully exploit the speech and noise statistics can be constructed. However, in this filter, the parameters should be adjusted for every case of noisy signals. From this point of view, the Kalman filter is not so appropriate as a preprocessor of signal processing system.

© Springer International Publishing Switzerland 2015
S. Arik et al. (Eds.): ICONIP 2015, Part III, LNCS 9491, pp. 53–60, 2015.
DOI: 10.1007/978-3-319-26555-1_7

Another way of signal enhancement mainly in speech data is using the power spectrum subtraction (SS) [2] in which the peaks of power spectrum are used to estimate speech signals. This method is quite frequently used as a preprocessor in many speech processing systems. However, this method has the problem with the case of complex interactions between the speech signal and noise. Recently, deep structure networks become popular because of their efficiency of describing complex functions associated with classification and regression problems. The origin of this approach for noise reduction problems is using multilayer perceptrons (MLPs) [3,4] and recently the multilayer structure using restricted Boltzmann machines (RBMs) [5] is also proposed. However, in these methods of signal enhancement, the optimal structure of multilayer networks is not known in advance for the given data. In this respect, this paper propose a model of nonlinear filters with Gaussian kernel Functions (GKFs). In this method, the condition for signal enhancement of the proposed nonlinear filter is obtained by using the phase space analysis of signal time series. Then, from this analysis, the structure of nonlinear filter is determined and a network with Gaussian kernel functions is trained in such a way of obtaining the clean signal. This procedure can be repeated to obtain the multilayer (or deep) structure of nonlinear filters. To show the effectiveness of the proposed approach, the experiments for speech enhancement using the data sets from the TIMIT and NOISEX-92 database were performed and demonstrated the merits in signal enhancement compared with other conventional preprocessing filters.

The rest of this paper is organized as follows: in Sect. 2, the condition for nonlinear filters for signal enhancement is derived, Sect. 3 presents the method of determining the structure of nonlinear filters, Sect. 4 presents the algorithm for constructing nonlinear filters, Sect. 5 shows simulation results for signal enhancement, and finally, Sect. 6 presents the conclusion.

2 Nonlinear Filters for Signal Enhancement

The goal of signal enhancement is to recover the clean signal from a series of noisy signal data at regular time intervals. For this purpose, signal filters in which the input is noisy signal and the output is recovered clean signal, are required. Such mapping of filters may not be linear in many cases and the necessity of nonlinear filtering emerges. In our case, the goal of signal enhancement is to predict the clean signal from a series of current and past data observed at regular time intervals. Moreover, we usually consider the sampled time series data $x(t_k)$; that is,

$$x(t_k) = x(t_0 + k\Delta t) \quad k = 0, 1, 2, \cdots \tag{1}$$

where t_0, k, and Δt represent the initial time, sampling step, and sampling interval, respectively. Here, let us construct E-dimensional vectors

$$\boldsymbol{x}_{\tau,E}(k) = [x(k), x(k-\tau), \ldots, x(k-(E-1)\tau)]^T, \quad k = (E-1)\tau, \cdots \tag{2}$$

where τ and E represent the delay time measured as the unit of Δt and embedding dimension, respectively.

Here, our problem of nonlinear filtering is to recover clean signal $\tilde{x}(t_k)$ from a sequence of noisy signal values $\boldsymbol{x}_{\tau,E}(k)$; that is,

$$\tilde{x}(t_k) = f(\boldsymbol{x}_{\tau,E}(k)) = f(\tilde{\boldsymbol{x}}_{\tau,E}(k) + \epsilon(k)) \tag{3}$$

where f represents a nonlinear filter for the generation of clean signal $\tilde{x}(t_k)$ and $\epsilon(k)$ represents the kth noise vector in which each element is an $i.\ i.\ d.$ random variable with mean 0 and variance σ^2.

Here, let us denote f_k as $f(\boldsymbol{x}_{\tau,E}(k))$ and $\hat{f}_k$ represent an estimated filter model. Then, the mean square error (MSE) of $\hat{f}_k$ is determined by

$$
\begin{aligned}
MSE(\hat{f}, x_k) &= E[(\hat{f}_k - \tilde{x}(t_k))^2] \\
&= E[(\hat{f}_k - E[\hat{f}_k] + E[\hat{f}_k] - \tilde{x}(t_k))^2] \\
&= E[(\hat{f}_k - E[\hat{f}_k])^2] + E[(E[\hat{f}_k] - \tilde{x}(t_k))^2] \\
&\quad + 2E[(\hat{f}_k - E[\hat{f}_k])(E[\hat{f}_k] - \tilde{x}(t_k))] \\
&= E[(\hat{f}_k - E[\hat{f}_k])^2] + (E[\hat{f}_k] - \tilde{x}(t_k))^2.
\end{aligned}
\tag{4}
$$

In the above equation, the first and second terms represent the variance and bias square of $\hat{f}_k$. Here, $\hat{f}_k$ can be approximated by the following form using the Taylor series expansion:

$$\hat{f}_k = \hat{f}(\tilde{\boldsymbol{x}}_{\tau,E}(k) + \epsilon(k)) \approx \hat{f}(\tilde{\boldsymbol{x}}_{\tau,E}(k)) + \epsilon(k) \cdot \nabla \hat{f}(\tilde{\boldsymbol{x}}_{\tau,E}(k)) \tag{5}$$

Then, (4) can be described by

$$MSE(\hat{f}, x_k) \approx \sigma^2 \sum_{i=k}^{k-(E-1)} \left(\frac{\partial \hat{f}}{\partial x_i} \left(\tilde{\boldsymbol{x}}_{\tau,E}(k) \right) \right)^2 + \left(\hat{f} \left(\tilde{\boldsymbol{x}}_{\tau,E}(k) \right) - \tilde{x}(t_k) \right)^2. \tag{6}$$

This equation implies that the filter model $\hat{f}_k$ should recover the clean signal $\tilde{x}(t_k)$ for the given signal vector $\tilde{\boldsymbol{x}}_{\tau,E}(k)$ under the constrain that

$$\sum_{i=k}^{k-(E-1)} \left(\frac{\partial \hat{f}}{\partial x_i} \left(\tilde{\boldsymbol{x}}_{\tau,E}(k) \right) \right)^2 < 1. \tag{7}$$

In other words, the filter model should be smooth enough to reduce the noise component. From this point of view, the structure of filter model is determined and trained for the given signal data.

3   Phase Space Analysis of Noisy Signals

For the modeling of such nonlinear filters, the input and output structure should be determined. To determine the structure of nonlinear filters, the dynamic structure of signal time series should be analyzed. For this purpose, we usually rely on the theory called the delay coordinate embedding proposed by Packard et al. [6]

and Takens [7]. Their purpose is to find the appropriate dimension of attractors generated by the dynamical system. To describe the dynamical system, let us assume that there exists a D dimensional state vector $\boldsymbol{x}(t)$, $t \in \mathbb{R}$; that is,

$$\boldsymbol{x}(t) = [x(t), x^{(1)}(t), \cdots, x^{(D-1)}(t)]^T \tag{8}$$

where $x^{(i)}(t)$ represents the ith derivative of $x(t)$ with respect to t. This vector can be produced by a dynamical system,

$$\dot{\boldsymbol{x}} = \boldsymbol{F}(\boldsymbol{x}, t) \tag{9}$$

where $\boldsymbol{F}$ represents the D dimensional mapping function from the state vector $\boldsymbol{x}$ to the vector $\dot{\boldsymbol{x}}$ at time t. With no information on the state space (or phase space), we can only observe $x(t)$.

In our case, the structure of nonlinear filter should be determined from the feasibility of mapping between the noisy and clean signals. For this purpose, let us define $\tilde{\boldsymbol{x}}_{\tau,E}^r$ as the rth nearest neighbor vector of $\tilde{\boldsymbol{x}}_{\tau,E}$. Then, the gradient between $\tilde{\boldsymbol{x}}_{\tau,E}^r$ and $\tilde{\boldsymbol{x}}_{\tau,E}$ is determined by

$$\Delta\tilde{x}_{\tau,E}(k) = \frac{|f(\tilde{\boldsymbol{x}}_{\tau,E}(k)) - f(\tilde{\boldsymbol{x}}_{\tau,E}^r(k))|}{\|\tilde{\boldsymbol{x}}_{\tau,E}(k) - \tilde{\boldsymbol{x}}_{\tau,E}^r(k)\|} \tag{10}$$

where f represents a model of nonlinear filter.

To find out embedding dimension and delay time properly of nonlinear filters, the condition of noise reduction of (7) is investigated. For the estimation of this condition, a measurement of mean square gradient (MSG) for N samples of signal data is defined by

$$\overline{(\Delta\tilde{x}_{\tau,E})^2} = \frac{1}{N - (E-1)\tau + 1} \sum_{k=(E-1)\tau}^{N} \left(\Delta\tilde{x}_{\tau,E}(k)\right)^2. \tag{11}$$

Then, the region of (τ, E) space in which the MSG values are less than 1, is searched. In this computation of MSG values, the sequences of noisy and clean signals are used as the input and output of filtering network, respectively to ensure that the trained network satisfies the condition of noise reduction of (7). As an example, the MSG values of (11) for a female speech selected from the TIMIT data set mixed with white Gaussian noise are calculated and illustrated in Fig. 1. In this example, the MSG values are less than 1 when the embedding dimension is greater than 7. However, for the safety factor, the structure of nonlinear filters; that is, (τ, E) is determined by $(2, 20)$ where the values of MSG is less than 0.05.

4   Nonlinear Filters with Gaussian Kernel Functions

As a nonlinear filtering model for signal enhancement, a network with Gaussian kernel functions (GKFs) is selected since this network is able to perform nonlinear and nonparametric estimation and good for incremental learning due to the

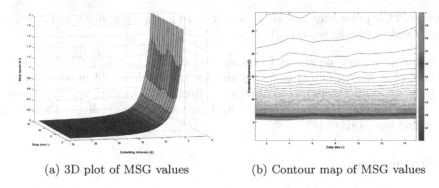

(a) 3D plot of MSG values   (b) Contour map of MSG values

Fig. 1. The MSG values of speech data mixed with white Gaussian noise

locality of kernel functions. The suggested model of a nonlinear filter $\hat{f}$ with m kernel functions is described by

$$\hat{f}(\boldsymbol{x}) = \sum_{i=1}^{m} w_i \psi_i(\boldsymbol{x}), \quad \psi_i(\boldsymbol{x}) = e^{-||\boldsymbol{x}-\boldsymbol{\mu}_i||^2/2\sigma_i^2}, \tag{12}$$

where w_i represents the connection weight between the output and the ith kernel function ψ_i in which $\boldsymbol{\mu}_i$ and σ_i represent the mean and standard deviation, respectively. In (12), we need to determine the parameters of m and also kernel related parameters $w_i, \boldsymbol{\mu}_i, \sigma_i$. For kernel related parameters, [8] suggested an efficient estimation method in such a way of minimizing the mean square error (MSE). In our approach, the structure of nonlinear filter is determined by the condition of (7) and the nonlinear filter model of (3) for the given signal data is implemented. The detailed algorithm of the proposed approach is described as follows:

Construction of a Nonlinear Filter with GKFs

Step 1. From the clean signal time series, determine the values of MSG of (11). Then, identify the region of the embedding dimension E and delay time τ in which the values of MSG are smaller than 1.

Step 2. Determine the input structure of nonlinear filter model of (3) from the identified region of (τ, E). In practice, the input structure of (τ, E) is determined in which the value of MSG is much less than 1 to obtain the enough capability of noise suppression, for example, the MSG values are less than 0.05.

Step 3. For the given signal data, the nonlinear filter model of (12) is trained by using an incremental learning algorithm. In our learning algorithm, the necessary number of GKFs are recruited and the parameters of filter model is further trained in such a way of minimizing the MSE. For the detailed description of this algorithm, refer to [8].

Step 4. For the training of multilayer networks, the output of previous layer becomes the input of the next layer. In this training, the input structure is maintained and the filter network is trained as the same way of the previous layer. This procedure can be repeated until no further noise reduction is achieved by the filter network.

In this algorithm, nonlinear filters are constructed by using GKFs and the structure of nonlinear filters is determined by the condition of noise reduction of (7). As a result, the proposed nonlinear filters are achieving signal enhancement by minimizing the noise variance and recovering the clean signal; that is, minimizing the MSE of (6).

5   Simulation

To demonstrate the effectiveness of the proposed nonlinear filter model, the simulation for speech enhancement was performed for the data sets of TIMIT and NOISEX-92 database. In this simulation, 7 pairs of noisy and clean speech files were used for the training of signal enhancement filters. These files were consisted of 7 different sentences which were spoken by the same (female) speaker. In this case of training signal enhancement filters, the signal data were mixed with white Gaussian noise. For the comparison of the proposed method, the linear filter such as the FIR filter and also the SS filter were trained by using the same training data. In the case of testing signal enhancement filters, 3 pairs of noisy and clean files were used in which the sentences were different with the training data. To test the variation of noisy environments, the Babble, Factory, Pink, Car (Volvo), and white Gaussian noises were mixed with clean speech data. Then, the average performances for these 5 types of noise were determined as the simulation results for signal enhancement. As the performance measure, the noise reduction ratio (NRR) and the normalized mean square error (NMSE) were used. Here, NRR is determined by

$$NRR = SNR_o - SNR_i, \tag{13}$$

where SNR_i and SNR_o represent the input and output signal-to-noise ratios in dB scale; that is,

$$SNR_i = 10 \log \frac{\sum_{k=(E-1)\tau}^{N} \tilde{x}^2(t_k)}{\sum_{k=(E-1)\tau}^{N} (x(t_k) - \tilde{x}(t_k))^2},$$

where $\tilde{x}(t_k)$ and $x(t_k)$ represent the clean and noisy signals, respectively, and

$$SNR_o = 10 \log \frac{\sum_{k=(E-1)\tau}^{N} \tilde{x}^2(t_k)}{\sum_{k=(E-1)\tau}^{N} (\hat{x}(t_k) - \tilde{x}(t_k))^2},$$

where $\hat{x}(t_k)$ represents the filtered signal.

As a measure of signal distortion, the NMSE is determined by

$$NMSE = \frac{\sum_{k=(E-1)\tau}^{N} (\hat{x}(t_k) - \tilde{x}(t_k))^2}{\sum_{k=(E-1)\tau}^{N} (\tilde{x}(t_k) - \bar{\tilde{x}}(t_k))^2}, \tag{14}$$

where $\bar{\tilde{x}}(t_k)$ represents the sample mean of clean signals.

In these performance measures, if the noisy signal is enhanced by the filter, the value of NRR is increasing while the value of NMSE is close to 0. The simulation results for speech enhancement were listed in Tables 1, 2 and 3. In these tables, the GKFN$_1$ and GKFN$_2$ represented the proposed nonlinear with one and two layer Gaussian kernel function networks (GKFNs), respectively. These simulation results have shown that (1) the proposed nonlinear filters using GKFNs outperformed the other methods from the view points of NRRs and also NMSEs on the average, (2) the two layer GKFNs improved the performances of signal enhancement compared with the one layer GKFNs, and (3) the trained nonlinear filters using GKFNs were quite robust to the variations of speakers and also noise environments.

Table 1. Simulation results for speech enhancement

Measurement	FIR	SS	GKFN$_1$	GKFN$_2$
NRR	−4.02	2.78	9.31	**11.10**
NMSE	1.97	0.50	0.34	**0.30**

Table 2. Simulation results for speech enhancement with different speakers

(a) NRRs for various signal enhancement filters				
Speaker	FIR	SS	GKFN$_1$	GKFN$_2$
Male	−6.27	2.07	6.36	**6.91**
Female	−4.02	2.78	9.31	**11.10**
(b) NMSEs for various signal enhancement filters				
Speaker	FIR	SS	GKFN$_1$	GKFN$_2$
Male	1.84	**0.33**	0.45	0.43
Female	1.97	0.50	0.34	**0.30**

Table 3. Simulation results for speech enhancement with different SNRs

(a) NRRs for various signal enhancement filters				
Noise Power	FIR	SS	GKFN$_1$	GKFN$_2$
SNR 0 dB	−4.02	2.78	9.31	**11.10**
SNR 5 dB	−8.18	0.98	9.02	**10.83**
SNR 10 dB	−11.69	0.28	5.90	**7.92**
(b) NMSEs for various signal enhancement filters				
Noise Power	FIR	SS	GKFN$_1$	GKFN$_2$
SNR 0 dB	1.97	0.50	0.34	**0.30**
SNR 5 dB	1.62	0.35	0.22	**0.18**
SNR 10 dB	1.50	**0.17**	0.20	**0.17**

6 Conclusion

A new method of nonlinear filtering for signal enhancement was proposed based on a network with Gaussian kernel functions. In the proposed method, the conditions for signal enhancement were derived and the construction method of nonlinear filters with Gaussian kernel functions was described. Through the simulation for filtering noisy signals of speech data, the effectiveness of the proposed method has been demonstrated from the view points of noise reduction ratio and also signal distortion. In practice, the proposed nonlinear filter can be applied to any preprocessor of signal processing system which requires signal enhancement in the first place.

References

1. Haykin, S.: Kalman Filtering and Neural Networks. John Wiley & Sons, New York (2001)
2. Boll, S.: Suppression of acoustic noise in speech using spectral subtraction. IEEE Trans. Acoust. Speech Signal Process. **27**, 113–120 (1979)
3. Tamura, S., Waibel, A.: Noise reduction using connectionist models. In: International Conference on Acoustics, Speech, and Signal Processing pp. 553–556 (1988)
4. Tamura, S., Nakamura, M.: Improvements to the noise reduction neural network. In: International Conference on Acoustics, Speech, and Signal Processing, pp. 825–828 (1990)
5. Xu, Y., Du, J., Dai, L., Lee, C.: An experimental study on speech enhancement based on deep neural networks. IEEE Signal Process. Lett. **21**(1), 65–68 (2014)
6. Packard, N.H., Crutchfield, J.P., Farmer, J.D., Shaw, R.S.: Geometry from a time series. Phys. Rev. Lett. **45**, 712–716 (1980)
7. Takens, F.: Detecting strange attractors in turbulence. In: Rand, D.A., Young, L.S. (eds.) Dynamical Systems and Turbulence, Warwick 1980. Lecture Notes in Mathematics, vol. 898, pp. 366–381. Springer-Verlag, Berlin (1981)
8. Kil, R.: Function approximation based on a network with kernel functions of bounds and locality. ETRI J. **15**, 35–51 (1993)

Computing Skyline Probabilities
on Uncertain Time Series

Guoliang He[1(✉)], Lu Chen[1], Zhijie Li[1],
Qiaoxian Zheng[2], and Yuanxiang Li[1]

[1] State Key Lab of Software Engineering, College of Computer Science,
Wuhan University, Wuhan, China
glhe@whu.edu.cn
[2] School of Computer Science and Information Engineering,
Hubei University, Wuhan, China

Abstract. In this paper, we model the skyline queries on uncertain time series, and develop a two-step procedure to answer the probabilistic skyline queries on uncertain time series. First, two effective pruning techniques are proposed to obtain the skyline in the interval. Next, two simple methods are proposed to compute the probability of each uncertain time series in the skyline. Experiments verify the effectiveness of probabilistic skylines and the efficiency and scalability of our algorithms.

Keywords: Skyline query · Time series · Uncertainty

1 Introduction

Time series widely exist in many areas, such as traffic flow management, remote sensing, etc. Meanwhile, the observed value at each timestamp exhibits various degree of uncertainty due to the limitations of measuring equipment, imcompleteness of data, environmental influence, etc. Due to the importance of these applications, analyzing uncertain time series has become an important task. Particularly, some uncertain time series dominating others in a time interval are interesting and challenging.

To the best of our knowledge, no prior work studied probabilistic skyline queries over uncertain time series. This is the first study about skyline analysis on uncertain time series. In this paper, we make several contributions. First, we model the skyline queries on uncertain time series and define some notations. Second, based on two novel pruning techniques, we propose two methods to answer the probabilistic skyline queries. Experiments show that our proposed probabilistic skyline answering algorithm is effective.

The rest of the paper is organized as follows. We review the related work in Sect. 2. In Sect. 3, we propose notions of probabilistic skylines on uncertain time series. In Sect. 4, we develop effective pruning techniques and algorithms for probabilistic skyline computation. A systematic performance study is reported in Sect. 5. We conclude the paper in Sect. 6.

© Springer International Publishing Switzerland 2015
S. Arik et al. (Eds.): ICONIP 2015, Part III, LNCS 9491, pp. 61–71, 2015.
DOI: 10.1007/978-3-319-26555-1_8

2 Related Work

Since Borzsonyi *et al.* introduced the concept of skylines into the database field in 2001 [1], skyline queries [2, 3] have been attracted by lots of researchers and many algorithms have been advanced to answer skyline queries on traditional certain data.

Pei *et al.* [4] first tackled the problem of skyline queries on uncertain data, and advanced bottom-up and top-down algorithms based on possible worlds and discrete density distribution to compute probabilistic skylines on uncertain data sets. After that, probability theory has been widely introduced to answer the issue of skyline queries on the uncertain data. For instance, to deal with the problem of supporting skyline queries for uncertain data with maybe confidence, Yong *et al.* defined the skyline probability, based on which they identified top-k skyline tuples with highest probability [5].

Recently, skyline queries on data stream has been attracted many researchers, and lots of efficient methods and techniques have been proposed to continuously monitor the changes in the skyline according to the arrival of new tuples and expiration of old ones. For example, Xuemin *et al.* presented efficient incremental techniques for on-line skyline computation over the most recent N elements in a rapid data stream [6]. Moreover, some effective methods were advanced to handle uncertain data stream [7–9]. However, uncertain time series is different from uncertain data stream, existing methods for skyline queries on uncertain stream is impractical for our problem.

To handle the issue of skyline queries on time series, Jiang et al. advocated a novel type of time series analysis queries, interval skyline queries [10]. Moreover, they introduced an on-the-fly method and a view-materialization method to online answer interval skyline queries on time series. Wang et al. proposed a analytical method called α/β-Dominant-Skyline on multivariate time series [11]. However, above methods assume the time series data is certain, and uncertainty in time series data is inherent in many applications. Therefore, the existing skyline methods are unable to answer the skyline queries on uncertain time series.

3 Background of Skyline Queries

3.1 Skylines on Certain Time Series

A time series s is a sequence of real values $s_1, s_2, \ldots$. These values are ordered with respect to the timestamps. We denote the value of s at timestamp i by s_i. The time series s is said to dominate time series x in the time interval $\omega = [i:j]$, denoted by $s \prec_\omega x$, if $\forall k \in \omega$, $s_k \geq x_k$; and $\exists l \in \omega$, $s_l > x_l$ (For simplicity, here we assume a larger value is better, which could not affect our research).

Given a dataset D of time series and a time interval $\omega = [i:j]$, the skyline is the subset of time series that are not dominated by other time series in D in the interval ω. It is denoted by

$$Sky_\omega(D) = \{s \in D \mid \nexists s' \in D, s' \prec_\omega s\}$$

In other words, the skyline queries are to seek some time series those are not dominated by other time series in the dataset D in the interval ω.

3.2 Skylines on Uncertain Time Series

In this subsection we define some notations used in this paper, and we focus on uncertain time series where uncertainty is localized and limited to the elements.

Definition 1. Uncertain time series. An uncertain time series object X of length n can be denoted as an ordered sequence $(X_1, X_2, ..., X_n)$. At the timestamp t_i, X_i is the random variable modeling the real valued number, which could be conceptually described by a probability density function f in the domain space R. In this paper, we model an uncertain element X_i as a set of multiple points in the domain space, denoted by $X_i = \{x_{i1}, ..., x_{im}\}$. And the number of possible points for X_i is represented as $|X_i| = m$. Each point x_{ij} appears with a probability p_{ij}, and $\sum_{j=1}^{m} p_{ij} = 1$. It can be regarded as discrete probability distribution.

For simplicity, we assume that uncertain elements in an uncertain time series are independent, and for an uncertain element, each value carries the same probability to happen.

Table 1. Three uncertain time series U, V and S

Timestamp	1	2	3
U	{3,5}	{2,6}	{5,6}
V	{4}	{3,7}	{2,5}
S	{4}	{1,5}	{2,5}

Example 1. Three uncertain time series U, V and S of length 3 are shown in Table 1.

Definition 2. Instances of an uncertain time series. Given an uncertain time series $X = \{X_1, X_2,..., X_n\}$, any value of element X_i can be selected with the same probability $1/|X_i|$. Therefore, an instance of the uncertain time series S is the combination of a value of different elements. The number of instance is $\prod_{i=1}^{n} |X_i|$.

Example 2. Consider three uncertain time series in the time interval [1:3] listed in Table 1, we enumerate all instances of U, V and S. U has 8 instances: $u^1 = (3, 2, 5)$, $u^2 = (3, 2, 6)$, $u^3 = (3, 6, 5)$, $u^4 = (3, 6, 6)$, $u^5 = (5, 2, 5)$, $u^6 = (5, 2, 6)$, $u^7 = (5, 6, 5)$, $u^8 = (5, 6, 6)$; V has 4 instances: $v^1 = (4, 3, 2)$, $v^2 = (4, 3, 5)$, $v^3 = (4, 7, 2)$, $v^4 = (4, 7, 5)$; S has 4 instances: $s^1 = (4, 1, 2)$, $s^2 = (4, 1, 5)$, $s^3 = (4, 5, 2)$, $s^4 = (4, 5, 5)$.

Definition 3. Probabilistic Dominance. Consider two uncertain time series U and V. $\{u^1, ..., u^{|U|}\}$ and $\{v^1, ..., v^{|V|}\}$ are the sets of instances of U and V in the time interval $\omega = [i:j]$, respectively. V dominates U at a certain probability if not all instances of U dominate any instance of V in the interval ω. Formally, the probability that V dominates U is

$$Pr(V \prec_\omega U) = \sum_{i=1}^{|U|} \frac{1}{|U|} * \frac{|\{v^j \in V | v^j \prec_\omega u^i\}|}{|V|} = \frac{1}{|U| * |V|} \sum_{i=1}^{|U|} |\{v^j \in V | v^j \prec_\omega u^i\} \quad (1)$$

Lemma 1. Transitivity. Consider three time series u, v and s in the time interval ω. If u $\prec_{[i:j]}$ v and v $\prec_\omega$ s, then u $\prec_\omega$ s.

Now we define the probability that an uncertain time series U is in the skyline in a certain interval.

Definition 4. Probabilistic skyline. Given the uncertain time series data set D and an sample U ∈ D, U is in the probabilistic skyline on the query interval ω = [i:j] if there does not exist another uncertain time series S ∈ D such that U does not probabilistically dominate S in the interval ω. Formally, the probability that U is in the skyline in the time interval ω is

$$\begin{aligned}
Pr_\omega(U) &= \frac{1}{|U|} \sum_{i=1}^{|U|} \prod_{V \neq U, V \in D} (1 - Pr(V \prec_\omega u^i)) \\
&= \frac{1}{|U|} \sum_{i=1}^{|U|} \prod_{V \neq U, V \in D} (1 - \frac{|\{v^j \in V | v^j \prec_\omega u^i\}|}{|V|})
\end{aligned} \quad (2)$$

Moreover, we could a more general expression as following if the data set D consists of multiple uncertain time series.

$$Pr_\omega(U) = \frac{1}{|U|} \sum_{i=1}^{|U|} \prod_{V \neq U, V \in D} (1 - Pr_\omega(V \prec_\omega u^i))$$

According to Definition 4, the probability that an instance u^i is the skyline is as following

$$Pr_\omega(u^i) = \prod_{V \neq U, V \in D} (1 - Pr_\omega(V \prec_\omega u^i)) = \prod_{V \neq U, V \in D} (1 - \frac{|\{v \in V | v \prec_\omega u^i\}|}{|V|}) \quad (3)$$

As a consequence,

$$Pr_\omega(U) = \frac{1}{|U|} \sum_{u^i \in U} Pr_\omega(u^i) \quad (4)$$

4 Probabilistic Skyline Answering Algorithm

Due to many instances for an uncertain time series, it is time-consuming to answer the probabilistic skyline queries on the dataset based on the Definition 4. To improve its efficiency, we use two stages to handle this issue. First, we delete some uncertain time series which are not in the probabilistic skyline in the interval ω. Next, the skyline probability of each uncertain time series in the interval ω is computed.

4.1 Obtaining the Skyline

We first define the maximum and minimum boundary of an uncertain time series.

Boundary Instance: Given an uncertain time series $U = (U_1, U_2, \ldots, U_n)$, where $U_i = \{u_{i1}, u_{i2}, \ldots, u_{i|U_i|}\}$. Let $U_{max} = (\max_{i=1}^{|U_1|}\{u_{1i}\}, \ldots, \max_{i=1}^{|U_n|}\{u_{ni}\})$ and $U_{min} = (\min_{i=1}^{|U_1|}\{u_{1i}\}, \ldots, \min_{i=1}^{|U_n|}\{u_{ni}\})$ be the maximum and the minimum boundary of U, respectively. U_{max} and U_{min} are two instances of U.

For an uncertain time series U or its instance u, if $Pr(U) = 0$ or $Pr(u) = 0$, it means U (u) is impossible in the skyline and could be pruned in advance. Following with Lemma 1, we immediately have the following rules to delete some uncertain time series or instances that they are not in the skyline.

Pruning Rule 1. For two uncertain time series U and V and the interval ω, let $Min(V_{min}) = Min\{\min_{i=1}^{|V_1|}\{v_{1i}\}, \ldots, \min_{i=1}^{|V_n|}\{v_{ni}\}\}$ be the minimal element of the instance V_{min} and $Max(U_{max}) = Max\{\max_{i=1}^{|U_1|}\{u_{1i}\}, \ldots, \max_{i=1}^{|U_n|}\{u_{ni}\}\}$ be the maximal element of the instance U_{max}. If $Min(V_{min}) > Max(U_{max})$, then $Pr_\omega(U) = 0$.

Pruning Rule 2. Let U and V be two different uncertain time series. If $V_{min} \prec_\omega U_{max}$, then $Pr_\omega(U) = 0$.

The pseudocode of obtaining the skyline of the data set D in the interval ω using rule 1 and rule 2 is as following.

Algorithm 1: Obtaining_skyline
Input: the set of uncertain time series D
Output: the skyline $Sky_\omega(D)$
1.   for each uncertain time series $U \in D$ do
2.   generate two instances U_{max} and U_{min}
3.   compute $Max\{U_{max}\}$ and $Min\{U_{min}\}$
4.   end for
5.   MAX_MIN = $Max\{Min\{U_{min}\}, U \in D\}$
6.   for each uncertain time series $U \in D$ do   // pruning rule 1
7.   if $(Max(U_{max}) < MAX\_MIN)$
8.   delete U;
9.   end for
10.   for each uncertain time series $U \in D$ do   // pruning rule 2
11.   for each $V \in D$ && $V \neq U$
12.   if $(V_{min} \prec_\omega U_{max})$
13.   delete U;
14.   end for
15.   end for

4.2 Computing the Skyline Probability

The second step is to compute the probability of each uncertain time series in the skyline. We will introduce two simple methods to answer this issue.

(1) Naïve Method

The most straightforward way is to produce instances of all uncertain time series in the skyline. Then, we calculate probabilistic skyline of each uncertain time series by comparing its instances with instances of other uncertain time series. The Naïve algorithm is described in the following.

> **Algorithm 2.** Naïve Method for probabilistic skyline query (NA)
> Input: the skyline $Sky_\omega(D)$
> Output: the skyline probability of each uncertain time series
> 1.   for each uncertain time series $U \in Sky_\omega(D)$ do
> 2.   generate all instances of U
> 3.   for $V \in D$ && $V \neq U$
> 4.   generate all instances of V
> 5.   compute the probability that V dominates U
> 6.   end for
> 7.   compute the skyline probability of U
> 8.   end for

In Algorithm 2, we use Definition 4 to compute the probabilistic skyline of each instance of an uncertain time series U. Finally, the probabilistic skyline of U is obtained by averaging the probabilities of all instances.

(2) Bounding Method

In the Naïve Method, an instance of an uncertain time series is compared with all instances of other uncertain time series in the process of computing its skyline probability. It is usually time-consuming because an uncertain time series has many instances. In this process, it is actually unnecessary to compare it with all instances of other uncertain time series. Now we give an example to illustrate our idea.

Example 3. Consider the set of uncertain time series in Table 1. We compute the skyline probability of U in the interval $\omega = [1:3]$.

Firstly, we compute the probability that U dominates V. To reduce the computation time, we rank the values of each uncertain element in descending order, as shown in Figs. 1 and 2. We note that the maximum (5) of U_1 is greater than the maximum (4) of V_1 at the timestamp 1. At the time stamp 3, the maximum (6) of U_3 is greater than the maximum (5) of V_3. Therefore, the probability that the instance u^2, u^4, u^5, u^6, u^7 and u^8 being dominated by V is obviously 0 and is not necessary to compare because they include the value 5 at timestamp 1 or the value 6 at timestamp 3. We only need to compute the probability of u^1 and u^3 as following. For an element u_k^1 of $u^1 = (3, 2, 5)$ at timestamp k, we count the number of observations in V_k which are greater than u_k^1. In detail, only the value $v_1^1 = 4$ of V_1 is greater than $u_1^1 = 3$; both v_1^2 and v_2^2 (7 and 2) of V_2 are greater than u_2^1; and only one value v_1^3 (5) of V_3 is greater than u_3^1. That is, in Fig. 2, there is a bounding (bold) line above which values are greater than the value of u^1 at each timestamp. Therefore, the number of instances of V that dominate u^1 is 1*2*1.

timestamp	1	2	3
U	5	6	6
	3	2	5

Fig. 1. The ordered values of U at each timestamp

timestamp	1	2	3
V	4	7	5
		3	2

Fig. 2. The ordered values of V at each timestamp (the values above the bold line dominate the value of u^1 at each timestamp)

As a consequence, instead of producing 8 instances of U, we only need to produce 2 instances in the process of computing the probability that U dominates V. That is, the values above the bold line in Fig. 1 are unnecessary to be used to produce instances of U during this process. It could largely reduce the computation time of the probabilistic dominance between U and V.

In the same way, we only compute the probability that u^1 dominates S. And the number of instances of S that dominate u^1 is $1*1*1$.

From Example 3 we could see that the bounding techniques could effectively improve the efficiency of computing the probabilistic dominance between two uncertain time series. Now we present our idea in formal.

Lemma 2. Let $U = (U_1, U_2, ..., U_n)$ and $V = (V_1, V_2, ..., V_n)$ be two uncertain time series in the time interval ω. For $U_i = \{u_{i1}, u_{i2}, .., u_{i|Ui|}\}$ and $Max(V_i) = Max\{v_{i1}, v_{i2}, .., v_{i|Vi|}\}$, if $u_{ij} > Max(V_i)$, an instance of U including u_{ij} cannot be dominated by any instance of V in the interval ω.

Lemma 3. Consider two uncertain time series $U = (U_1, U_2, ..., U_n)$ and $V = (V_1, V_2, ..., V_n)$ in the time interval ω, Let u be an instance of U and $V_i = \{v_{i1}, v_{i2}, .., v_{i|Vi|}\}$. If $v_{ij} < u_i$, then an instance of V including v_{ij} at timestamp t_i does not dominate u in the time interval ω.

According to Lemmas 2 and 3, in the process of computing the probability that U dominates V, it is not necessary to compare each instance of U with all instances of V in advance. We design simple ordering and bounding techniques to avoid lot of unnecessary comparisons between instances of U and V by counting the number of values (N_i) of V_i which are greater than the value of u_i at each timestamp t_i. As a consequence, $Pr(u \prec_\omega V) = 1 - \prod_{i=1}^{|V|}(N_i/|V_i|)$. The refined algorithm is described in Algorithm 3.

Algorithm 3. Boundary Method for probabilistic skyline query (BM)
Input: the skyline $Sky_\omega(D)$
Output: the skyline probability of each uncertain time series
1.   for each uncertain time series $U \in Sky_\omega(D)$ do
2.   rank the values of U_i in descending order at each timestamp t_i
3.   end for
4.   for each uncertain time series $U \in Sky_\omega(D)$ do
5.   for each uncertain time series $V \in D$ && $V \neq U$
6.   remove values of U_i which dominate $Max\{V_i\}$ at each timestamp t_i using
          Lemma 2
7.   generate instances of U based on the remaining values at each timestamp
8.   compute the probability that each of the remaining instances of U is
          dominated by V using Lemma 3
9.   end for
10.   compute the skyline probability of U
11.   end for
12.   return the probabilistic skyline

5   Experiment Study

In this section, we firstly describe the constitution of the experimental data. Then, we use a synthetic dataset to examine the effectiveness and efficiency of probabilistic skyline analysis on uncertain time series. All of the experimental results are obtained by using a PC computer with Intel Core i5 2.80 GHz CPU and 4 GB of main memory running Professional Windows 7 operating system. Each experiment is implemented 50 times and the average result is reported in this paper.

A synthetic data set consists of n time series. We randomly generate the means $\mu_i(1 \leq i \leq n)$ of all n certain time series using a standard normal distribution $N(0,1)$. Then each certain time series s_i follows a normal distribution $N(\mu_i,\sigma)$. To generate uncertain time series S_i with s_i, we first generate a region u_{ij} for each vaule s_{ij}, and its boundary follows a normal distribution $N(s_{ij}, \sigma')$, where s_{ij} is the value of the certain time series s_i at the timestamp t_j. Values of uncertain time series S_i at the timestamp t_j distributed uniformly in the region u_{ij}, and the number of values at each timestamp follows uniform distribution in range [1, 10]. In our experiments, we vary the number of time series n, the standard deviation σ and the base interval ω as shown in the Table 2. By default, n = 3000, $\sigma = 0.3$ and $\omega = 10$. At the same time, we vary another standard deviation σ' from 0.1 to 0.5 to obtain the diversity of the uncertainty, and 0.3 is the default value. Table 2 summaries the above experimental parameters and the default values are shown in bold font.

Figure 3 show skyline percentage of the dataset according to various parameters. In Fig. 3(a), the skyline percentage decreases when the cardinality of the dataset increases. It shows that as the size of the dataset increasing, a time series is easy to be dominated by other time series. Figure 3(b) shows that the skyline percentage increases linearly with respect to the length of the query interval. When the length of the interval is larger, a time series is becoming difficult to be dominated by other time series in the interval.

From Fig. 3(c) we see that the skyline percentage increases quickly when the standard deviation increases. The larger the standard deviation σ, the weaker the correlation among uncertain time series examples. Therefore, there are more chances that a time series is not dominated by other time series in the interval when the standard

Table 2. A summary of experiment parameters on synthetic data

Parameter	Values
n	1000, 2000, **3000**, 4000, 5000
σ	0.1,0.2,**0.3**,0.4,0.5
σ'	0.1,0.2,**0.3**,0.4,0.5
ω	8,9,**10**,11,12

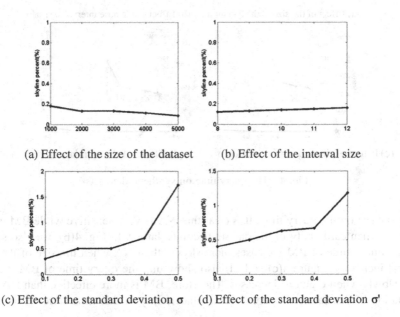

(a) Effect of the size of the dataset   (b) Effect of the interval size

(c) Effect of the standard deviation σ   (d) Effect of the standard deviation σ'

Fig. 3. The skyline percentage on synthetic data sets

deviation σ becomes larger. On the other hand, the size of the skyline for the correlated data is smaller while that for the anti-correlated data is larger. This is similar to the situations of skylines on certain time series.

Figure 3(d) represents that the skyline percentage increases quickly when the standard deviation σ' increases. The larger the standard deviation σ', the higher the degree of uncertainty. To some extent, an uncertain time series is difficult to be dominated by others when the differences among instances of an uncertain time series are larger. It is caused by the higher uncertainty of this time series. As a consequence, more uncertain time series could not be dominated by others and they are in the skyline.

Query Efficiency: We compare the query efficiency of Naïve Method (NA) and Boundary Method (BM) proposed in Sect. 4. Figure 4 plots the running time of two algorithms in different parameter settings. To better describe the tendency of BM, it is magnified in the middle of Fig. 4(b), (c) and (d). Figure 4(a) shows the effect of the size

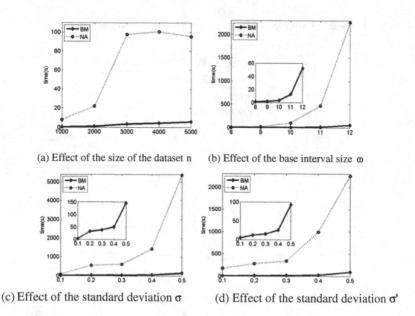

(a) Effect of the size of the dataset n   (b) Effect of the base interval size ω

(c) Effect of the standard deviation σ   (d) Effect of the standard deviation σ'

Fig. 4. The query time on synthetic data sets

of the dataset on the query time. It is clear that NA is very sensitive while BM is more effective when cardinality of the dataset becomes larger. In Fig. 4(b), we also see that that the query time of BM increases more slowly than NA when the size of the query interval increases. Figure 4(c) and (d) also show that the query time of BM increases more slowly when σ and σ' increases. Therefore, BM is more effective than NA as the correlation of the dataset is weaker or the uncertainty of this time series is higher.

6 Conclusions

In this paper, we extended the well-known skyline analysis to uncertain time series, and developed effective techniques to tackle the problem of computing probabilistic skylines on uncertain time series. Using synthetic and real data sets, we illustrated the effectiveness of our proposed probabilistic skyline answering algorithm. Moreover, we also discussed the efficiency and scalability of our algorithms.

References

1. Borzsonyi, S., Kossmann, D., Stocker, K.: The skyline operator. In: The 17th International Conference on Data Engineering (ICDE), Heidelberg, Germany, pp. 421–430 (2001)
2. Chung, Y.-C., Su, I.-F., Lee, C.: Efficient computation of combinatorial skyline queries. Inf. Syst. **38**(3), 369–387 (2013)

3. Lee, J., You, G.-w., Hwang, S.-w., Selke, J., Balke, W.-T.: Interactive skyline queries. Inf. Sci. **211**, 18–35 (2012)
4. Pei, J., Jiang, B., Lin, X., Yuan, Y.: Probabilistic skylines on uncertain data. In: VLDB, pp. 15–26 (2007)
5. Yong, H., Lee, J., Kim, J., Hwang, S.-w.: Skyline ranking for uncertain databases. Inf. Sci. **273**, 247–262 (2014)
6. Lin, X., Yuan, Y., Wang, W., Lu, H.: Stabbing the sky: efficient skyline computation over sliding windows. In ICDE, pp. 502–513 (2005)
7. Bai, M., Xin, J., Wang, G.: Probabilistic reverse skyline query processing over uncertain data stream. In: Lee, S.-g., Peng, Z., Zhou, X., Moon, Y.-S., Unland, R., Yoo, J. (eds.) DASFAA 2012, Part II. LNCS, vol. 7239, pp. 17–32. Springer, Heidelberg (2012)
8. Zhang, W., Li, A., Cheema, M.A., Zhang, Y., Chang, L.: Probabilistic n-of-N skyline computation over uncertain data streams. In: Lin, X., Manolopoulos, Y., Srivastava, D., Huang, G. (eds.) WISE 2013, Part II. LNCS, vol. 8181, pp. 439–457. Springer, Heidelberg (2013)
9. Nagendra, M., Candan, K.S.: SkySuite: a framework of skyline-join operators for static and stream environments. In: VLDB (2013)
10. Bin, J., Jian, P.: Online interval skyline queries on time series. In: International Conference on Data Engineering, Los Alamitos, pp. 1036–1047 (2009)
11. Wang, H., Wang, C.-K., Ya-Jun, X., Ning, Y.-C.: Dominant skyline query processing over multiple time series. J. Comput. Sci. Technol. **28**(4), 625–635 (2013)

Probabilistic Prediction of Chaotic Time Series Using Similarity of Attractors and LOOCV Predictable Horizons for Obtaining Plausible Predictions

Shuichi Kurogi[✉], Mitsuki Toidani, Ryosuke Shigematsu, and Kazuya Matsuo

Kyushu Institute of Technology, Tobata, Kitakyushu, Fukuoka 804-8550, Japan
{kuro,matsuo}@cntl.kyutech.ac.jp,
{toidani,shigematsu}@kurolab.cntl.kyutech.ac.jp
http://kurolab.cntl.kyutech.ac.jp/

Abstract. This paper presents a method for probabilistic prediction of chaotic time series. So far, we have developed several model selection methods for chaotic time series prediction, but the methods cannot estimate the predictable horizon of predicted time series. Instead of using model selection methods employing the estimation of mean square prediction error (MSE), we present a method to obtain a probabilistic prediction which provides a prediction of time series and the estimation of predictable horizon. The method obtains a set of plausible predictions by means of using the similarity of attractors of training time series and the time series predicted by a number of learning machines with different parameter values, and then obtains a smaller set of more plausible predictions with longer predictable horizons estimated by LOOCV (leave-one-out cross-validation) method. The effectiveness and the properties of the present method are shown by means of analyzing the result of numerical experiments.

Keywords: Probabilistic prediction · Attractors of chaotic time series · Leave-one-out cross-validation · Prediction of time series · Estimation of predictable horizon

1 Introduction

This paper presents a method of probabilistic prediction of chaotic time series. So far, we have developed several model selection methods for chaotic time series prediction [1,2]. The method in [1] uses moments of predictive deviation as ensemble diversity measures for model selection in time series prediction, and achieves better performance from the point of view of mean square prediction error (MSE) than the conventional holdout method. The method in [2] uses direct multi-step ahead (DMS) prediction to apply the out-of-bag (OOB) estimate of the MSE. However, both methods cannot be used for estimating the predictable horizons of predicted time series but for estimating the MSE, which is owing

© Springer International Publishing Switzerland 2015
S. Arik et al. (Eds.): ICONIP 2015, Part III, LNCS 9491, pp. 72–81, 2015.
DOI: 10.1007/978-3-319-26555-1_9

mainly to the fact that the MSE is affected by short term predictability and long term unpredictability of chaotic time series (see [2] for the analysis and [3] for properties of chaotic time series).

Instead of using model selection methods employing the estimation of the MSE, we present a method to obtain probabilistic prediction. Here, from [4], we can see that the probabilistic prediction has come to dominate the science of weather and climate forecasting. This is mainly because the theory of chaos at the heart of meteorology shows that for a simple set of nonlinear equations (or Lorenz's equations shown below) with initial conditions changed by minute perturbations, there is no longer a single deterministic solution and hence all forecasts must be treated as probabilistic. From another perspective, the forecast probabilities allow the user to take appropriate action within a proper understanding of the uncertainties. For probabilistic weather forecast, a number of ensemble methods have been employed and progressed, where the forecast uncertainties arised from perturbations to the initial conditions and model parameters are examined.

In this article, we try to utilize learning machines for probabilistic prediction, which indicates that forecast uncertainty arises from model uncertainty. We also try to obtain a set of plausible predictions by means of using the similarity of training time series and the time series predicted by a number of learning machines with different parameter values. Furthermore, we introduce LOOCV (leave-one-out cross-validation) method for estimating the predictable horizon to obtain a smaller set of more plausible predictions. We show the method of probabilistic prediction in Sect. 2, experimental results and analysis in Sect. 3, and the conclusion in Sect. 4.

2 Probabilistic Prediction of Chaotic Time Series

2.1 Point Prediction of Chaotic Time Series

Let $y_t (\in \mathbb{R})$ denote a chaotic time series for a discrete time $t = 0, 1, 2, \cdots$ satisfying

$$y_t = r(\boldsymbol{x}_t) + e(\boldsymbol{x}_t), \tag{1}$$

where $r(\boldsymbol{x}_t)$ is a nonlinear target function of a vector $\boldsymbol{x}_t = (y_{t-1}, y_{t-2}, \cdots, y_{t-k})^T$ for the embedding dimension k generated by the delay embedding from a chaotic differential dynamical system (see [3] for the theory of chaotic time series). Here, y_t is obtained not analytically but numerically, and then y_t involves an error $e(\boldsymbol{x}_t)$ owing to an executable finite calculation precision. This indicates that there are a number of plausible target functions $r(\boldsymbol{x}_t)$ with allowable error $e(\boldsymbol{x}_t)$. Furthermore, the time series y_t for numerical experiments shown below is one of the time series generated from the original chaotic dynamical system with a high precision, which we denote ground truth time series $y_t^{[\text{gt}]}$ depending on the necessity in the context.

Let $y_{t:h} = y_t y_{t+1} \cdots y_{t+h-1}$ denote a time series with the initial time t and the horizon h. For a given and training time series $y_{t_g:h_g}(= y_{t_g:h_g}^{[\text{train}]})$, we are supposed to predict succeeding time series $y_{t_p:h_p}$ for $t_p \geq t_g + h_g$. Then, we make the training dataset $D^{[\text{train}]} = \{(\boldsymbol{x}_t, y_t) \mid t \in I^{[\text{train}]}\}$ for $I^{[\text{train}]} = \{t \mid t_g \leq t < t_g + h_g\}$ to train a learning machine. After the learning, the machine executes iterated prediction by

$$\hat{y}_t = f(\hat{\boldsymbol{x}}_t) \tag{2}$$

for $t = t_p, t_{p+1}, \cdots$, recursively, where $f(\hat{\boldsymbol{x}}_t)$ denotes the prediction function of $\hat{\boldsymbol{x}}_t = (x_{t1}, x_{t2}, \cdots, x_{tk})$ whose elements are given by

$$x_{tj} = \begin{cases} y_{t-j} & (t - j < t_p) \\ \hat{y}_{t-j} & (t - j \geq t_p). \end{cases} \tag{3}$$

Here, we suppose that y_t for $t < t_p$ is known for making the prediction $\hat{y}_{t_p:h_p}$ as the initial state.

2.2 Probabilistic Prediction

For probabilistic prediction, we firstly make a number of predictions $\hat{y}_{t_p:h_p}$ or $y_{t_p:h_p}^{[\theta_m]}$ generated from (2) and (3) by means of learning machines with parameter values $\theta_m \in \Theta$, where Θ indicates the set of parameter values of learning machines. Here, we suppose that there are a number of plausible prediction functions $f(\cdot) = f^{[\theta_m]}(\cdot)$ for the chaotic time series, and we have to remove implausible ones. To have this done, we select the following set of plausible predictions $y_{t_p:h_p}^{[\theta_m]}$,

$$Y_{t_p:h_p}^{[S_{\text{th}}]} = \left\{ y_{t_p,h_p}^{[\theta_m]} \;\middle|\; S\left(y_{t_p,h_p}^{[\theta_m]}, y_{t_g:h_g}\right) \geq S_{\text{th}} \right\} \tag{4}$$

where

$$S\left(y_{t_p,h_p}^{[\theta_m]}, y_{t_g:h_g}^{[\text{train}]}\right) \triangleq \frac{\sum_i \sum_j a_{ij}^{[\theta_m]} a_{ij}^{[\text{train}]}}{\sqrt{\sum_i \sum_j \left(a_{ij}^{[\theta_m]}\right)^2} \sqrt{\sum_i \sum_j \left(a_{ij}^{[\text{train}]}\right)^2}} \tag{5}$$

denotes the similarity of two-dimensional attractor (trajectory) distributions $a_{ij}^{[\theta_m]}$ and $a_{ij}^{[\text{train}]}$ of time series $y_{t_p,h_p}^{[\theta_m]}$ and $y_{t_g:h_g}^{[\text{train}]}$, respectively, and S_{th} is a threshold. Here, the two-dimensional attractor distribution, a_{ij}, of a time-series $y_{t:h}$ is given by

$$a_{ij} = \sum_{s=t}^{t+h-1} \mathbf{1}\left\{ \left\lfloor \frac{y_s - v_0}{\Delta_a} \right\rfloor = i \wedge \left\lfloor \frac{y_{s+1} - v_0}{\Delta_a} \right\rfloor = j \right\}, \tag{6}$$

where v_0 is a constant less than the minimum value of y_t for all time series and Δ_a indicates a resolution of the distribution. Furthermore, $\mathbf{1}\{z\}$ is an indicator function equal to 1 if z is true, and to 0 if z is false, and $\lfloor \cdot \rfloor$ indicates the floor function.

Next, we try to select predictions with longer predictable horizons from the plausible predictions $Y_{t_p:h_p}^{[S_{\mathrm{th}}]}$. To have this done, let us calculate the predictable horizon between two predictions $y_{t_p:h_p}^{[\theta_m]}$ and $y_{t_p:h_p}^{[\theta_n]}$ in $Y_{t_p:h_p}^{[S_{\mathrm{th}}]}$ given by

$$h\left(y_{t_p:h_p}^{[\theta_m]}, y_{t_p:h_p}^{[\theta]}\right) = \max\left\{h \mid \forall s \le h \le h_p; |y_{t_p+s}^{[\theta_m]} - y_{t_p+s}^{[\theta_n]}| \le e_y\right\} \qquad (7)$$

where e_y indicates the threshold of prediction error to determine the horizon. Then, we have the mean of $h\left(y_{t_p:h_p}^{[\theta_m]}, y_{t_p:h_p}^{[\theta_n]}\right)$ between an prediction $y_{t_p:h_p}^{[\theta_m]}$ and the other predictions $y_{t_p:h_p}^{[\theta_n]}$ in the set of plausible predictions, $Y_{t_p:h_p}^{[S_{\mathrm{th}}]}$, given by

$$\tilde{h}_{t_p}^{[\theta_m]} = \frac{1}{\left|Y_{t_p:h_p}^{[S_{\mathrm{th}}]}\right| - 1} \sum_{y_{t_p:h_p}^{[\theta_n]} \in Y_{t_p:h_p}^{[S_{\mathrm{th}}]}\backslash\{y_{t_p:h_p}^{[\theta_m]}\}} h\left(y_{t_p:h_p}^{[\theta_m]}, y_{t_p:h_p}^{[\theta_n]}\right). \qquad (8)$$

where $\left|Y_{t_p:h_p}^{[S_{\mathrm{th}}]}\right|$ denotes the number of elements in $Y_{t_p:h_p}^{[S_{\mathrm{th}}]}$. Note that the above method to obtain the horizon $\tilde{h}_{t_p}^{[\theta_m]}$ is based on the leave-one-out cross-validation (LOOCV), and the prediction with a longer horizon $\tilde{h}_{t_p}^{[\theta_m]}$ is considered to provide a "better" prediction in predicting all plausible predictions $y_{t_p:h_p}^{[\theta]} \in Y_{t_p:h_p}^{[S_{\mathrm{th}}]}$ on average. Thus, we sort the predictable horizons by their lengths as $\tilde{h}_{t_p}^{[\theta_{\sigma(i)}]} \ge \tilde{h}_{t_p}^{[\theta_{\sigma(i+1)}]}$, where $\sigma(i)$ denotes the order for $i = 1, 2, \cdots, |Y_{t_p:h_p}^{[S_{\mathrm{th}}]}|$. Let a subset of plausible predictions with longer predictable horizons be

$$Y_{t_p:h_p}^{[H_{\mathrm{th}},S_{\mathrm{th}}]} = \left\{y_{t_p:h_p}^{[\theta_{\sigma(i)}]} \,\middle|\, \frac{i}{|Y_{t_p:h_p}^{[S_{\mathrm{th}}]}|} \le H_{\mathrm{th}}\right\}, \qquad (9)$$

where the threshold H_{th} ($0 < H_{\mathrm{th}} \le 1$) indicates the ratio of the numbers of elements in $Y_{t_p:h_p}^{[H_{\mathrm{th}},S_{\mathrm{th}}]}$ and $Y_{t_p:h_p}^{[S_{\mathrm{th}}]}$, or $H_{\mathrm{th}} = \left|Y_{t_p:h_p}^{[H_{\mathrm{th}},S_{\mathrm{th}}]}\right| / \left|Y_{t_p:h_p}^{[S_{\mathrm{th}}]}\right|$. Now, we derive the probability $p(y_t)$ of the prediction y_t for $t_p \le t < t_p + h_p$ as

$$p\left(v_i \le y_t < v_{i+1}\right) = \frac{1}{\left|Y_{t_p:h_p}^{[H_{\mathrm{th}},,S_{\mathrm{th}}]}\right|} \sum_{\theta \in \Theta^{[H_{\mathrm{th}},,S_{\mathrm{th}}]}} \mathbf{1}\left\{\left\lfloor \frac{y_t^{[\theta]} - v_0}{\Delta_v} \right\rfloor = i\right\}, \qquad (10)$$

where $\Theta^{[H_{\mathrm{th}},,S_{\mathrm{th}}]}$ is the set of parameters θ of learning machines which have generated the time series $y_{t_p:h_p}^{[\theta]} \in Y_{t_p:h_p}^{[H_{\mathrm{th}},S_{\mathrm{th}}]}$, and Δ_v is a constant representing the resolution of y_t, and $v_i = i\Delta_v + v_0$ for $i = 0, 1, 2, \cdots$. Note that the probability $p(y_t)$ depends on the threshold H_{th}. Namely, the probability $p(v_i \le y_t \le v_{i+1})$ indicates how much the plausible predictions in $Y_{t_p:h_p}^{[H_{\mathrm{th}},S_{\mathrm{th}}]}$ take the values in between

v_i and v_{i+1}, where $Y_{t_p:h_p}^{[H_{\text{th}},S_{\text{th}}]}$ consists of $H_{\text{th}} \times |Y_{t_p:h_p}^{[S_{\text{th}}]}|$ predictions with longer predictable horizons among all plausible predictions in $Y_{t_p:h_p}^{[S_{\text{th}}]}$. As a special case for $H_{\text{th}} = \left|Y_{t_p:h_p}^{[S_{\text{th}}]}\right|^{-1}$, we have $Y_{t_p:h_p}^{[H_{\text{th}},S_{\text{th}}]}$ consisting of only one prediction $y_{t_p:h_p}^{[\theta_{\sigma(1)}]}$ with the longest predictable horizon $\tilde{h}_{t_p}^{[\theta]}$ which we assume the most plausible prediction and we call it representative prediction among plausible predictions. By means of the above probabilistic prediction, we hope that the representative prediction $y_{t_p:h_p}^{[\theta_{\sigma(1)}]}$ has a longer predictable horizon in predicting the ground truth time series $y_t^{[\text{gt}]}$. Furthermore, we would like to tune H_{th} $\left(> \left|Y_{t_p:h_p}^{[S_{\text{th}}]}\right|^{-1}\right)$ to provide more conservative and safe probabilistic prediction so that $p(y_t)$ may have a positive value for the ground truth $y_t = y_t^{[\text{gt}]}$. Then, after the tuning of H_{th}, we can provide an expected predictable horizon $\hat{h}_{t_p}^{[\theta_{\sigma(1)}]}$ of the representative prediction $y_{t_p:h_p}^{[\theta_{\sigma(1)}]}$ in predicting the ground truth time series $y_{t_p:h_p}^{[\text{gt}]}$ by

$$\hat{h}_{t_p}^{[\theta_{\sigma(1)}]} = \max\left\{h \mid \forall s \leq h, \forall y_{t_p:h_p}^{[\theta]} \in Y_{t_p:h_p}^{[H_{\text{th}},S_{\text{th}}]}; |y_{t_p+s}^{[\theta_{\sigma(1)}]} - y_{t_p+s}^{[\theta]}| \leq e_y\right\}. \quad (11)$$

3   Numerical Experiments and Analysis

3.1   Experimental Settings

We use the Lorenz time series, as shown in Fig. 1 and [2], obtained from the original differential dynamical system given by

$$\frac{dx_c}{dt_c} = -\sigma x_c + \sigma y_c, \quad \frac{dy_c}{dt_c} = -x_c z_c + r x_c - y_c, \quad \frac{dz_c}{dt_c} = x_c y_c - b z_c, \quad (12)$$

for $\sigma = 10$, $b = 8/3$, $r = 28$. Here, we use t_c for continuous time and t ($= 0, 1, 2, \cdots$) for discrete time related by $t_c = tT$ with sampling time T. We have generated the time series $y(t) = x_c(tT)$ for $t = 1, 2, \cdots, 5,000$ from the initial state $(x_c(0), y_c(0), z_c(0)) = (-8, 8, 27)$ with $T = 25\,\text{ms}$ via Runge-Kutta method with 128 bit precision of GMP (GNU multi-precision library). As a result of preliminary experiments as shown in [2], $y(t)$ for each duration of time less than 1,200 steps ($= 30\,\text{s}/25\,\text{ms}$) in Fig. 1, or $y_{t_0:1200}$ for each initial time $t_0 = 0, 1, 2, \cdots$ with initial state $(x(t_0), y(t_0), z(t_0))$, is supposed to be correct, while cumulative computational error may increase exponentially after the duration.

We use $y_{0:2000}$ for training a learning machine, and execute multistep prediction of $y_{t_p:h_p}$ with the initial input vector $x_{t_p} = (y(t_p - 1), \cdots, y(t_p - k))$ for prediction start time $t_p = 2000 + 100i$ ($i = 0, 1, 2, \cdots, 19$). As a learning machine, we use CAN2 (see A and [5] for details), where the model complexity is the number of units, N, or the number of piecewise linear regions for approximating the target function $r(\cdot)$. We show the results with the embedding dimension $k = 8$ and we use the parameter of the learning machines as $\theta = N$. We set the thresholds as $S_{\text{th}} = H_{\text{th}} = 0.8$.

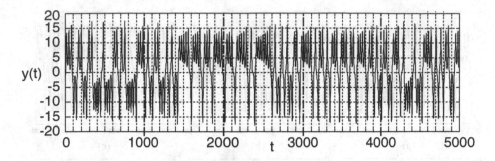

Fig. 1. Lorenz time series $y(t)$ for $t = 0, 1, 2, \cdots, 4999$, or the ground truth time series $y_{0:5000}^{[gt]}$.

For probabilistic prediction, we firstly make a number of predictions $y_{t_p:h_p}^{[\theta]}$ by means of using CAN2s with different number of units as $\theta = N = 10, 12, \cdots, N_{max}$, where we use $N_{max} = 250 = 2000/k$, because CAN2 employs N piecewise linear regions and each piecewise linear region requires more than k independent data from $2000 - k$ data in the training dataset $D^{[train]} = \{(\boldsymbol{x}_t, y_t)|t = k = 8, k+1, \cdots, 1999\}$.

3.2   Results and Analysis

In order to intuitively see how the present method works, we show examples of obtained representative prediction $y_{t_p:h_p}^{[\theta_{\sigma(1)}]}$ and the probability $p(y_t)$ in Fig. 2(a), (b) and (c). In order to see how the probability $p(y_t)$ is obtained, the superimposed plausible predictions $y_{t_p:h_p}^{[\theta]} \in Y_{t_p:h_p}^{[H_{th}, S_{th}]}$ are shown in Fig. 2(d), (e) and (f).

Now, let us examine the present method step by step. Firstly, in Fig. 3, we show the attractor distributions of (a) training and (b) predicted time series and (c) the similarity vs. prediction steps. From (c), we can see that the similarity of the attractors changes with the increase of prediction steps, and $h_p = 1000$ seems necessary for the convergence of the change. This indicates that the similarity have to be calculated after this period much bigger than predictable horizons whose mean values are less than 300 as shown in Fig. 4(a) explained below. This finding should be noted because usual methods for time series prediction (e.g. [1–4]) do not examine the property of the prediction after the predictable horizon so much.

Next, we have examined predictable horizons h, $\hat{h}$, $\tilde{h}$, h_S and h^* as shown in Fig. 4(a). Here, h indicates the predictable horizon achieved by the representative prediction, $\hat{h}$ the expected predictable horizon (see (11)), $\tilde{h}$ the LOOCV predictable horizon (see (8)), h_S the predictable horizon achieved by the learning machine which has generated the largest similarity of attractors, and h^* the longest predictable horizon among the horozons achieved by the learning machines for all N. We can see that h^* has achieved longer predictable horizons than others on average while h^* as well as h and h_S cannot be obtained until the

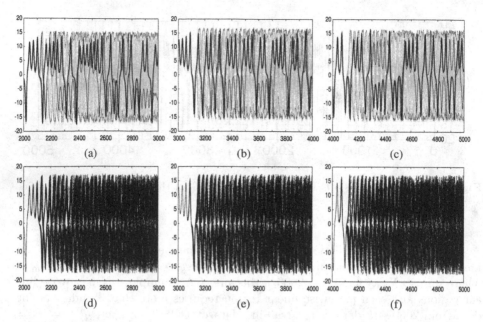

Fig. 2. (a), (b) and (c) show the grand truth time series $y^{[\mathrm{gt}]}_{t_p:h_p}$ (thickest line), representative prediction $y^{[\theta_{\sigma(1)}]}_{t_p:h_p}$ (2nd thickest line) and the probability $p(y_t)$ with positive values (gray area enveloped by thin lines) for $t_p = 2000$, 3000 and 4000, respectively, and $h_p = 1000$. By means of a close look at the difference between the thickest, the second thickest and thin lines growing greater than $e_y = 5$ in (a), (b) and (c), we can see that the pairs of expected and actual predictable horizons of the representative prediction $y^{[\theta_{\sigma(1)}]}_{t_p:h_p}$ are $(\hat{h}^{[\theta_{\sigma(1)}]}_{t_p}, h^{[\theta_{\sigma(1)}]}_{t_p}) = (102, 121)$, $(92, 234)$, $(118, 278)$ for $t_p = 2000$, 3000 and 4000, respectively. (d), (e) and (f) show superimposed plausible predictions $y^{[\theta]}_{t_p:h_p} \in Y^{[H_{\mathrm{th}}, S_{\mathrm{th}}]}_{t_p:h_p}$ for $t_p = 2000$, 3000 and 4000, respectively.

ground truth is obtained. We can see that the estimated predictable horizon $\hat{h}$ is smaller than or equal to the actual predictable horizon h achieved by the representative prediction, and the mean is $\langle \hat{h} \rangle = 82$ [steps] while $\langle h \rangle = 156$ [steps]. This result of estimation can be said conservative and safe, and is obtained by using $H_{\mathrm{th}} = 0.8$. By means of using $H_{\mathrm{th}} = 0.2$, we have $\langle \hat{h} \rangle = 152$ [steps] nearer to $\langle h \rangle = 156$ [steps], but there are a number of $\hat{h}$ bigger than h. This indicates that we can tune H_{th} for the necessary degree of conservativeness in the estimation of predictable horizon.

Next, we can see that h has achieved better result than h_S. This indicates that the effectiveness of the LOOCV predictable horizon $\tilde{h}$ embedded in the present method. The relationship between $\tilde{h}$ and h for all plausible predictions $Y^{[S_{\mathrm{th}}]}_{t_p:h_p}$ at $t_p = 2000$ are shown in Fig. 4(b), where we can see that there are a number of data on or near the line $\tilde{h} = h$ while there are data far from the line. As a result, we have the correlation between $\tilde{h}$ and h being 0.197, which does not indicate

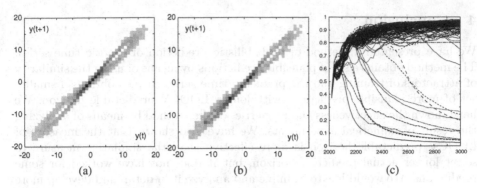

(a)   (b)   (c)

Fig. 3. (a) shows two dimensional training attractor distribution $a_{ij}^{[\mathrm{train}]}$ and (b) shows the predicted distribution $a_{ij}^{[\theta]}$ for $\theta = N = 172$ at $t = 2999$ with $S\left(y_{t_p,h_p}^{[\theta]}, y_{t_g:h_g}^{[\mathrm{train}]}\right) = S\left(y_{2000,1000}^{[\theta]}, y_{0:2000}^{[\mathrm{train}]}\right) = 0.85$. Here, note that $N = 172$ has generated the representative prediction shown in Fig. 2(a). The resolution of the distributions is $\Delta_a = (v_{\max} - v_0)/40 = (18.5 - (-18.5))/40 = 0.925$. (c) shows the similarity $S\left(y_{2000,1000}^{[\theta]}, y_{0:2000}^{[\mathrm{train}]}\right)$ vs. prediction steps for each $\theta = N = 10, 12, \cdots, 250$ for the increase of prediction steps. The predictions $y_{t_p:h_p}^{[\theta]} = y_{2000:1000}^{[\theta]}$ for $\theta = N = 10, 12, 14, 16, 20, 56, 58, 64, 132, 160, 162, 164$ are removed because $S\left(y_{t_p,h_p}^{[\theta]}, y_{t_g:h_g}^{[\mathrm{train}]}\right) < S_{\mathrm{th}} = 0.8$.

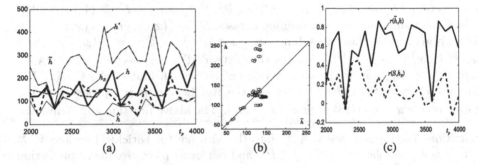

(a)   (b)   (c)

Fig. 4. (a) Predictable horizons h, $\hat{h}$, $\tilde{h}$, h_S and h^* (see the text for details) vs. t_p. The averages are $\langle h \rangle = 156$, $\langle \hat{h} \rangle = 82$, $\langle \tilde{h} \rangle = 139$, $\langle h_S \rangle = 118$ and $\langle h^* \rangle = 273$ [steps]. (b) shows the relationship between $\tilde{h}$ and h for plausible predictions $Y_{t_p:h_p}^{[S_{\mathrm{th}}]}$ at $t_p = 2000$. (c) shows the correlations $r(\tilde{h}, h)$ and $r(S, h_S)$ vs. t_p.

high relationship. From Fig. 4(c), we can see that the correlation between $\tilde{h}$ and h, which we denote $r(\tilde{h}, h)$, is big for some t_p and not so big for other t_p. On the other hand, the correlation between the similarity $S = S\left(y_{t_p,h_p}^{[\theta]}, y_{t_g:h_g}^{[\mathrm{train}]}\right)$ and the predictable horizon h_S derived from S, which we denote $r(S, h_S)$, is much smaller than $r(\tilde{h}, h)$. These results indicate that $\tilde{h}$ may be more effective than S for selecting the predictions achieving longer predictable horizons but it does not work for some predictions.

4 Conclusion

We have presented a method of probabilistic prediction of chaotic time series. The method obtains a set of plausible predictions by means of using the similarity of attractors of training and the predicted time series. It also obtains a smaller set of more plausible predictions with longer LOOCV predictable horizon. We have shown the effectiveness and properties of the method by means of analyzing the result of numerical experiments. We have also shown that the measure of LOOCV predictable horizon is more effective than the similarity measure to select longer actual predictive horizon, but it does not have worked for some predictions. We would like to examine and analyze it in detail and develop more reliable method in our future research studies.

Appendix

A CAN2

The CAN2 (competitive associative net 2) is an artificial neural net for learning efficient piecewise linear approximation of nonlinear function by means of the following schemes (See [5] for details): A single CAN2 has N units. The jth unit has a weight vector $\boldsymbol{w}_j \triangleq (w_{j1}, \cdots, w_{jk})^T \in \mathbb{R}^{k \times 1}$ and an associative matrix (or a row vector) $\boldsymbol{M}_j \triangleq (M_{j0}, M_{j1}, \cdots, M_{jk}) \in \mathbb{R}^{1 \times (k+1)}$ for $j \in I^N \triangleq \{1, 2, \cdots, N\}$. The CAN2 after learning the training dataset $D^n = \{(\boldsymbol{x}_i, y_i) | y_i = r(\boldsymbol{x}_i) + e_i, i \in I^n\}$ approximates the target function $r(\boldsymbol{x}_i)$ by $\widehat{y}_i = \widetilde{y}_{c(i)} = \boldsymbol{M}_{c(i)} \widetilde{\boldsymbol{x}}_i$, where $\widetilde{\boldsymbol{x}}_i \triangleq (1, \boldsymbol{x}_i^T)^T \in \mathbb{R}^{(k+1) \times 1}$ denotes the (extended) input vector to the CAN2, and $\widetilde{y}_{c(i)} = \boldsymbol{M}_{c(i)} \widetilde{\boldsymbol{x}}_i$ is the output value of the $c(i)$th unit of the CAN2. The index $c(i)$ indicates the unit who has the weight vector $\boldsymbol{w}_{c(i)}$ closest to the input vector $\boldsymbol{x}_i$, or $c(i) \triangleq \underset{j \in I^N}{\operatorname{argmin}} \|\boldsymbol{x}_i - \boldsymbol{w}_j\|$. The above function approximation partitions the input space $V \in \mathbb{R}^k$ into the Voronoi (or Dirichlet) regions $V_j \triangleq \{\boldsymbol{x} \mid j = \underset{i \in I^N}{\operatorname{argmin}} \|\boldsymbol{x} - \boldsymbol{w}_i\|\}$ for $j \in I^N$, and performs piecewise linear prediction for the function $r(\boldsymbol{x})$.

References

1. Kurogi, S., Ono, K., Nishida, T.: Experimental analysis of moments of predictive deviations as ensemble diversity measures for model selection in time series prediction. In: Lee, M., Hirose, A., Hou, Z.-G., Kil, R.M. (eds.) ICONIP 2013, Part III. LNCS, vol. 8228, pp. 557–565. Springer, Heidelberg (2013)
2. Kurogi, S., Shigematsu, R., Ono, K.: Properties of direct multi-step ahead prediction of chaotic time series and out-of-bag estimate for model selection. In: Loo, C.K., Yap, K.S., Wong, K.W., Teoh, A., Huang, K. (eds.) ICONIP 2014, Part II. LNCS, vol. 8835, pp. 421–428. Springer, Heidelberg (2014)
3. Aihara, K.: Theories and Applications of Chaotic Time Series Analysis. Sangyo Tosho, Tokyo (2000)

4. Slingo, J., Palmer, T.: Uncertainty in weather and climate prediction. Phil. Trans. R. Soc. A **369**, 4751–4767 (2011)
5. Kurogi, S., Sawa, M., Tanaka, S.: Competitive associative nets and cross-validation for estimating predictive uncertainty on regression problems. In: Quiñonero-Candela, J., Dagan, I., Magnini, B., d'Alché-Buc, F. (eds.) MLCW 2005. LNCS (LNAI), vol. 3944, pp. 78–94. Springer, Heidelberg (2006)

Adaptive Threshold for Anomaly Detection Using Time Series Segmentation

Mohamed-Cherif Dani[1,2]($\boxtimes$), François-Xavier Jollois[1], Mohamed Nadif[1], and Cassiano Freixo[2]

[1] LIPADE – Université Paris Descartes,
45 Rue des Saints-Pères, 75006 Paris, France
{mohamed-cherif.dani,franccois-xavier.jollois,
mohamed.nadif}@parisdescartes.fr
[2] Airbus, Toulouse Airbus RT, Route de Bayonne, 31000 Toulouse, France
{cassiano.freixo,mohamed-cherif.dani}@airbus.com

Abstract. Time series data are generated from almost every domain and anomaly detection becomes extremely important in the last decade. It consists in detecting anomalous patterns through identifying some new and unknown behaviors that are abnormal or inconsistent relative to most of the data. An efficient anomaly detection algorithm has to adapt the detection process for each system condition and each time series behavior. In this paper, we propose an adaptive threshold able to detect anomalies in univariate time series. Our algorithm is based on segmentation and local means and standard deviations. It allows us to simplify time series visualization and to detect new abnormal data as time series jumps within different time series behavior. On synthetic and real datasets the proposed approach shows good ability in detecting abnormalities.

Keywords: Anomaly detection · Time series · Unsupervised detection

1 Introduction

The increasing use of time series data has initiated several research and challenges [1]. The most of the algorithms inspect the irregularities across the time. The nature of data (Univariate or Multivariate) and the type of anomalies (collective, point anomaly) define the approach to apply. Many studies and researches interested in anomaly detection appeared and several techniques have been specifically developed for certain application domains. A large number of surveys, reviews were performed on anomaly detection. Recently Gupta et al. [2], Chandola et al. [3] structured and overviewed different algorithms by research areas, application domains and data nature.

Mining abnormal patterns is complex and challenging problem since the boundary between normal and abnormal in many cases are not precise. The existing solutions such as exceedance detection techniques are confronted to

© Springer International Publishing Switzerland 2015
S. Arik et al. (Eds.): ICONIP 2015, Part III, LNCS 9491, pp. 82–89, 2015.
DOI: 10.1007/978-3-319-26555-1_10

many constraints, like unlabeled data, noises, new behaviors, false alarms and sensitivity problem when for example, the abnormal patterns take a normal attitude.

In this paper, we propose an efficient algorithm for anomaly detection in univariate time series. The proposed algorithm that relies on the *Adaptive Piecewise Constant Approximation* (APCA) representation is based on segmentation and local standard deviations.

The paper is organized as follow. In Sect. 2, we review the commonly methods used in anomaly detection context. Section 3 is devoted to the presentation of our approach. In Sect. 4, numerical experiments on synthetic and real datasets are performed and assessed. We end this work by demonstrating the relevance of the presented work with some concluding remarks and directions for future work.

2   Segmentation and Anomaly Detection

Anomaly detection depends on various domains and objectives and generally it requires appropriate algorithms. For instance, in [4] we adapted a segmentation algorithm and used a density clustering algorithm, in order to detect anomalies in multivariate time series recorded by the aircraft. The same approach was applied by Mart for the Turbomachinery system by combining a clustering and segmentation algorithm [5]. For both techniques, the segmentation allows to reduce the complexity of time series for an efficient anomaly detection. Some other algorithms such as Gecko [6] transforms the time series into segments (states) and then find their characteristics and the logic transition between the states to build a finite state automaton. Gecko splits the time series into clusters using a method based on the knee of the curve. The *Symbolic Aggregate approXimation* (SAX) [7], is based on a symbolic representation of the time series. The approach is used by many algorithms for anomaly detection and unusual subsequence. The algorithm maps all segments produced by the *Piecewise Aggregate Approximation* (PAA) representation, replaces segments by symbols (alphabet) and then measures the distance between symbols to find the most distant subsequence. The *Local Outlier Factor* (LOF) [8] computes the distances between all the points by using the Kth nearest neighbor and density reachability. LOF Detects anomalies by sorting the scores and considering the extreme values as abnormal. An outlier is defined when the density is lower than their neighbors. [9] described another algorithm based on ARIMA model. To find anomalies, t-statistic for each time point is applied on a regression model and the points with critical values are selected as anomalies.

Furthermore, the segmentation can be obtained by using the K means algorithm is also commonly used for anomaly detection. Depending on the domain area (fraud detection, intrusion, aerospace, etc.), several case studies and reviews have been proposed. For instance, Gerhard [10] built normal and abnormal clusters and then computed the minimum distance between new arrival points and cluster centers. Gupta et al. [11] used also Kmeans PCA-based similarity measures for multivariate time series anomaly detection. Minimum information about data are required to be able to distinguish between abnormal and normal data. In such approaches all data points should belong to a cluster.

Finally, there are other simple and efficient techniques able to propose interesting solutions. One of such methods, we the *Extreme Studentized Deviation* method (ESD) also known as *Extreme Studentized Deviation* [12]. ESD considers a point as an outliers, if this point is more than α standard deviation from the mean, where α usually equal to 3. This approach is defined by $\mu \pm \alpha\sigma$. In general α varies between $[1,3]$. Miller [13] suggest to vary the 3σ and use $2,5$ or even 2. The $\mu \pm 3\sigma$ region contains about 96 % of data which makes three-sigma edit rule as a good and simple statistic method to remove anomalies. The problem with this approach is the sensitivity to the extreme values and noises.

Drawing inspiration from this kind, in the sequel and relying on APCA, we propose an effective way to detect anomalies.

3   Adaptive Threshold for Anomaly Detection (ATAD)

We introduced some algorithms for anomaly detection in the previous section. Some of them are complex and difficult to use for other datasets. Most are for specific data. The advantage of our approach (ATAD) is the simplicity and adaptability for each dataset, anomaly, and each context.

3.1   Adaptive Piecewise Constant Approximation

The APCA [14] representation of a time series is obtained by segmenting the time series into segments of unequal length based on data. Long segments are used to represent data regions of low activity, and short segments are used to represent regions of high activity. This representation derives from the Haar wavelet [15]. It takes the problem to a wavelet compression problem for which a minimal solution exists. The algorithm sorts the coefficients and truncates the smaller in order to get an optimal representation. It replaces the approximate segment mean values with the exact ones for a valid APCA representation. Each segment is then represented by its mean value and the index of the right end point. An univariate time series $t_1, \ldots, t_n$ is then represented by $(\mu_1, r_1), \ldots, (\mu_m, r_m)$, where $m \leq n$ is the number of segments, μ_i and r_i are respectively the mean value and the right endpoint of the ith segment.

3.2   Description of the Method

To overcome the ESD sensitivity, we compute ESD during the segments building, rather than computing only the mean. The advantage of having low and high segments increase the detection efficiency. The approach is illustrated in the following Fig. 1. Our method ATAD produces a dynamic threshold called for the entire of time series depending on the sensor activity.

The local standard deviation is computed thanks to the r_i which defines the limits of the segment and the concerned points in the time series. Hence, the

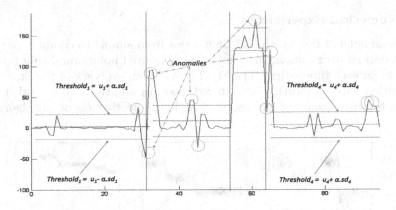

Fig. 1. Threshold calculations (μ_i, σ_i, r_i), O's denote anomalous data

new presentation ATAD is $(\mu_1, \sigma_1, r_1) \ldots (\mu_i, \sigma_i, r_i) \ldots (\mu_m, \sigma_m, r_m)$, where $1 \leq i \leq m$, σ_i denotes the standard deviation of the i_{th} segment.

By grouping data of same behaviors in high and low activity regions. The detection approach that we propose will not be influenced by the noise or anomalies, since the same behaviors are grouped in high and low activity regions. The following Algorithm 1 summarizes the major steps for anomaly detection using adaptive threshold and segmentation.

Algorithm 1. Adaptive threshold for anomaly detection algorithm.

Input:
T : Time series $t_1, \ldots, t_n$, α : SD coefficient, m : Number of segments
Output: Anomalies
Begin

1. Discrete Wavelet Transform on T until zero resolution.
2. Compute the detail coefficients using differencing process
3. Normalize coefficients by deviding each coefficient by $2^{(2/\ell)}$ where ℓ is the haar resolutions level and retain the m highest coefficients
4. Reconstruct the segments approximaton (APCA Representation) from the retained coefficients
5. Merge the segments if the number of segments is greater than m
6. For each segment, compute the threshold $mean \pm \alpha SD$

  if $(t_i < mean \pm \alpha SD)$ then t_i is anomaly

7. labeled t_i as anomalous

End

3.3 Numerical Experiments

We used simulated (i.e. synthetic) time series from simple to complex patterns using a normal distribution, then we added abnormal points randomly (around 10 points for each time series). The Fig. 2 provides an overview of the time series composition with abnormal points in red lines. In addition, we use other data sources, to prove the detection flexibility and the effectiveness of our algorithm.

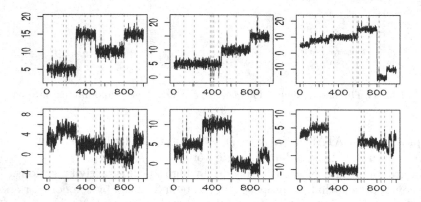

Fig. 2. Example of simulated data with anomalies in red lines (Color figure online).

The second dataset, is recently shared to the public by Yahoo [16]. These data were previously used to develop and to test many anomaly detection algorithms. A part of these data are real traffic from Yahoo services. The anomalies are labeled manually. Another part of data are synthetic and generated automatically.

We will test some anomaly detection algorithms to compare the detection results. The results of our experiments are reported in Table 1. As mentioned, anomalous points generated from a different distribution are known and labeled. They are neglected during the detection, but used for evaluation.To evaluate the detection, we have opted to use a classic evaluation criteria. During the detection process, we used four outcomes for evaluation: True Positive (TP), False Negative (FN), True Negative (TN), and False Positive (FP). To measure the performance of our approach, we use F-score that expresses the performance of an anomaly detection method using the precision P and sensitivity R measures where $P = TP/(TP+FP)$ and $R = TP/(TP+FN)$. The F-measure is denoted by $F = 2(P \times R)/(P+R)$ where $F \simeq 1$ indicates a good performance and $F \simeq 0$ means that the method does not detect any of anomaly in the data.

When data change behavior, new anomalies are considered normal regarding to entire time series. We met this problem with LOF and Chen algorithm (tsoutliers). Both algorithms detect well global anomalies as jumps, but the local anomalies that occurred after a behavior change are rarely detected.

Our method has shown a good performance, less false alarms between 6% and 17% and about 83% of the generated anomalies was detected. We used different

Table 1. Comparison among ATAD, LOF and tsoutliers in terms of false alarm, correct detection, precision and recall, and F-measure.

		Algorithms		
		ATAD	LOF K5,7,3	tsoutliers
Simulated Data	Original number of outliers	450		
45 time series	False Alarm	56	285	719
	Correct Detection	420	165	387
	Precision and Recall	(0,87;0,93)	(0,36;0,36)	(0,34;0,86)
	F-Mesures	0,90	0,36	0,49
Simulated Data II	Original number of outliers	300		
10 time series	False Alarm	17	123	51
	Correct Detection	273	27	176
	Precision and Recall	(0,63;0,91)	(0,19;0,13)	(0,36;0,58)
	F-Mesures	0,74	0,15	0,66
Yahoo Real Data	Original number of outliers	1653		
(67 time series)	False Alarm	57	1030	2918
	Correct Detection	1610	369	744
	Precision and Recall	(0,96;0,96)	(0,26;0,37)	(0,20;0,45)
	F-Measure	0,96	0,26	0,27
Yahoo Synthetic	Original number of outliers	465		
data (A2Benchmark)	False Alarm	58	554	477
	Correct Detection	404	135	224
	Precision and Recall	(0,97;0,82)	(0,19;0,28)	(0,31;0,94)
	F-Measure	0,95	0,23	0,47
Yahoo Synthetic	Original number of outliers	933		
data (A3Benchmark)	False Alarm	60	400	2299
	Correct Detection	571	163	922
	Precision and Recall	(0,90;0,60)	(0,28;0,17)	(0,28;0,97)
	F-Measure	0,73	0,22	0,44
Yahoo Synthetic	Original number of outliers	825		
data (A4Benchmark)	False Alarm	148	125	5822
	Correct Detection	478	408	744
	Precision and Recall	(0,76;0,57)	(0,23;0,15)	(0,11;0,90)
	F-Measure	0,65	0,18	0,20

inputs for each dataset depending on anomalies type. For example, with Yahoo real dataset, the number of segments m is 750 approximately $n/2$ where n is the length of time series. The coefficient α is roughly equal to 1.5. For A2Benchmark dataset is equal to 2 and α to 3 in the most of the time series with $m = 750$ and $\alpha = 1,9$ for some time series with local anomalies. For the sake of brevity, m is small and the coefficient $3 \leq \alpha \leq 2$ when global anomalies are numerous. Otherwise, However, when local anomalies are numerous the coefficient α is ≤ 1.9 and $m = n/2$. Where n is the length of time series, the coefficient α is approximately equal to 1.5. For A2Benchmark dataset, m is equal to 2 and α to 3 in the most of the time series with $m = 750$ and $\alpha \simeq 1,9$ for some time series with local anomalies. For short, m is small and the coefficient $3 \leq \alpha \leq 2$, when global anomalies are numerous. Otherwise, when local anomalies are numerous the coefficient α is ≤ 1.9 and m is approximately $n/2$. On two time series, the Fig. 3 illustrates the detection procedure and how the number of segment is selected.

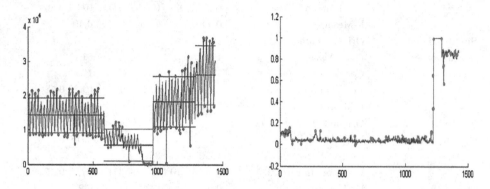

Fig. 3. Yahoo Time series for anomaly detection using ATAD.

Both Recall and Precision are important. It allows us to get the percentage of a good detection. We note the same performance for both LOF and tsoutlier $\simeq 35\%$ comparing to $\simeq 88\%$ using ATAD. With complexity of the time series and the generated noise, our approach shows good detection with less false alarm 56 data points comparing to 285 for LOF and 719 for tsoutliers. By applying ATAD in Yahoo real data we obtain 96% for recall and precision with only 57 false alarms and 59 missed detection. For the rest of the data (*A3Benchmark, A4Benchmark*), F-measures vary between $\simeq 45\%$ for tsoutliers and $\simeq 22\%$ for LOF as illustrated in the Table 1. The precision and recall vary respectively between 22%, 18% and 17%, 15% for LOF while for tsouliers 28%, 11% for precision and 97%, 90% for recall.

4   Conclusion and Future Works

Finding anomalies with a high precision degree is an important task in monitoring services. Many industrial and social applications generate a huge number of sensors and time series. Many algorithms adapted for each case exist. In this paper, we proposed a generic algorithm able to find anomalies in univariate time series, by adapting statistic threshold for each time series behavior. It gives good performances using synthetic and real data. We have demonstrated that for local anomalies the number of segments increases and for global anomalies the number of segments is smaller. For an autonomous anomaly detection it will be interesting to investigate the problem of the number of segments.

References

1. Esling, P., Agon, C.: Time-series data mining. ACM Comput. Surv. **45**(1), 34 (2012)
2. Gupta, M., Gao, J., Aggarwa, C.C., Han, J.: Outlier detection for temporal data: a survey. IEEE Trans. Knowl. Data Eng. **26**, 2250–2267 (2014)
3. Chandola, V., Banerjee, A., Kumar, V.: Anomaly detection: a survey. ACM Comput. Surv. **41**(3), 58 (2009). Article 15
4. Dani, M.-C., Freixo, C., Jollois, F.-X., Nadif, M.: Unsupervised anomaly detection for aircraft condition monitoring system. In: IEEE Aeroconf 2015 (2015)
5. Martí, L., Sanchez-Pi, N., Molina, J.M., Bicharra Garcia, A.C.: YASA: yet another time series segmentation algorithm for anomaly detection in big data problems. In: Polycarpou, M., de Carvalho, A.C.P.L.F., Pan, J.-S., Woźniak, M., Quintian, H., Corchado, E. (eds.) HAIS 2014. LNCS, vol. 8480, pp. 697–708. Springer, Heidelberg (2014)
6. Salvador, S., Chan, P.: Learning states and rules for detecting anomalies in time series. Appl. Intell. **23**(3), 241–255 (2005)
7. Keogh, E., Lin, J., Ada, F.: HOT SAX: efficiently finding the most unusual time series subsequence. In: Proceedings of the Fifth IEEE International Conference on Data Mining (ICDM 2005), pp. 226–233. IEEE Computer Society, Washington, DC (2005)
8. Breunig, M.M., Kriegel, H.-P., Ng, R.T., Sander, J.: LOF: identifying density-based local outliers. SIGMOD Rec. **29**(2), 93–104 (2000)
9. Chen, C., Liu, L.-M.: Joint estimation of model parameters and outlier effects in time series. J. Am. Stat. Assoc. **88**, 284–297 (1993)
10. Mnz, G., Li, S., Carle, G.: Traffic anomaly detection using k-means clustering. In: Proceedings of Leistungs-, Zuverlssigkeits-und Verlsslichkeitsbewertung von Kommunikationsnetzen und Verteilten Systemen, GI/ITG-Workshop MMBnet, September 2007
11. Gupta, M., Sharma, A.B., Chen, H., Jiang, G.: Context-aware time series anomaly detection for complex systems. In: Proceedings of the SDM Workshop (2013)
12. Rosner, B.: Percentage points for a generalized ESD many-outlier procedure. Technometrics **25**(2), 165–172 (1983)
13. Miller, J.: Reaction time analysis with outlier exclusion: bias varies with sample size. Q. J. Exper. Psychol. **43**(4), 907–912 (1991)
14. Chakrabarti, K., Keogh, E., Mehrotra, S., Pazzani, M.: Locally adaptive dimensionality reduction for indexing large time series databases. ACM Trans. Database Syst. **27**(2), 188–228 (2002)
15. Haar, A.: Zur Theorie der orthogonalen Funktionensysteme. Math. Ann. 38–53 (1911). German
16. Yahoo! Research Webscope datasets, A Labeled Anomaly Detection Dataset, version 1.0'. http://webscope.sandbox.yahoo.com/catalog.php?datatype=s

Neuron-Synapse Level Problem Decomposition Method for Cooperative Neuro-Evolution of Feedforward Networks for Time Series Prediction

Ravneil Nand[1,2]([⊠]) and Rohitash Chandra[1,2]

[1] School of Computing Information and Mathematical Sciences,
University of South Pacific, Suva, Fiji
[2] Artificial Intelligence and Cybernetics Research Group,
Software Foundation, Nausori, Fiji
ravneiln@yahoo.com, c.rohitash@gmail.com

Abstract. A major concern in cooperative coevolution for neuro-evolution is the appropriate problem decomposition method that takes into account the architectural properties of the neural network. Decomposition to the synapse and neuron level has been proposed in the past that have their own strengths and limitations depending on the application problem. In this paper, a new problem decomposition method that combines neuron and synapse level is proposed for feedfoward networks and applied to time series prediction. The results show that the proposed approach has improved the results in selected benchmark data sets when compared to related methods. It also has promising performance when compared to other computational intelligence methods from the literature.

1 Introduction

Cooperative coevolution evolutionary algorithm which divides a larger problem into smaller counterparts called sub-components are evolved in isolation and cooperatively evaluated [1]. Cooperative neuro-evolution is a form of machine learning that employs cooperative coevolution (CC) for training [2–5]. Cooperative neuro-evolution has been seen in solving real world problems such as pattern classification [3,6], time series prediction [6–9] and control [10].

It can be generalized that the two established problem decomposition methods for cooperative neuro-evolution are synapse level (SL) [9,11] and neuron level (NL) [6,10,12]. In SL, the network is decomposed to its lowest level of granularity where the number of sub-components depends on the number of weight-links in neural network. In NL problem decomposition, the number of subcomponents consists of the total number of hidden and output neurons. In the case of time series prediction, both NL and SL give very competitive performance [9] while in the case of pattern classification, SL is unable to perform [6]. SL views the network training as a fully separable problem and performs well

© Springer International Publishing Switzerland 2015
S. Arik et al. (Eds.): ICONIP 2015, Part III, LNCS 9491, pp. 90–100, 2015.
DOI: 10.1007/978-3-319-26555-1_11

in problems involving neural network applications where the problem contain less inter-dependencies. It failed in the case of pattern classification as it has difficulty to decompose the problem and therefore NL performed better as it appeals to partially-separable problems [6].

In recent developments, the combination of NL and SL through a competitive island cooperative coevolution method (CICC) gave better results [7,13]. In CICC, a problem decomposition method is seen as an island that competes with other islands or problem decomposition methods in different phases of evolution. The winner island at each phase injects its solution to the losing island. This essentially means that the evolutionary processes takes advantage of features of both problem decomposition methods.

There is potential for incorporation of different problem decomposition methods that can share its strengths to solve the problem. CICC used the different problem decomposition features to guide its search. Another way to incorporate them is to use architectural properties of the neural network, i.e. use synapse level in the region where more diversity in search is needed and neuron level where decision making is required.

In this paper, we combine neuron and synapse problem decomposition to form a hybrid problem decomposition called Neuron-Synapse Level (NSL) problem decomposition. NSL is intended for training feedforward networks that are chaotic time series problems. NSL problem decomposition enables more subpopulations than NL method and lower number than SL method. The performance of the proposed approach is compared with standalone neuron and synapse level [7].

The rest of the paper is organized as follows. In Sect. 2, the proposed method is discussed in detail while in Sect. 3, experiments and results are given with discussion. Section 4 ends the paper with a discussion of future extensions of the paper.

2   Neuron-Synapse Level Problem Decomposition

In cooperative coevolution, problem decomposition is based on the architectural properties of the neural network. Synapse level problem decomposition decomposes the neural network having highest number of subcomponents where each interconnected weight becomes a subcomponent [11]. Whereas, neuron level problem decomposition employs neurons as a reference point. The weights attached to neurons become the subcomponent [14].

In the proposed NSL decomposition method, each subcomponent consists of incoming and outgoing connections associated with neurons in the hidden layer. It is similar to method to cooperative coevolutionary model for evolving artificial neural networks (COVNET) [15] where all weight connections are treated as incoming weights. The difference lies in the breaking down the network further in the weights connected by hidden-output layer where the decomposition is at Synapse Level.

The calculation of the actual output is the sum of all the outputs generated as in all the methods mentioned earlier. NSL employs a single subcomponent for each neuron that groups interacting variables (synapses) that are connected to the hidden neuron. Therefore, each subcomponent for a layer is shown in Fig. 1 and composed as follows:

1. Firstly, hidden layer subcomponents: all neurons given in the hidden layer j. Hidden layer sub-populations comprises of all the weight connection from input neuron i to the hidden layer j. The bias of j is also included.
2. The hidden-output layer subcomponents: all weight-links from each neuron in the hidden j layer connected to all output k neurons and the bias of output k. Hidden-Output layer sub-populations is restricted to 1 as done in CoSyNE [11].

   The total number of sub-components is equal to total number of hidden neurons plus the number of weights and biases within hidden and output neuron in the neural network. The sub-components are implemented as sub-populations.
   The proposed method is used in training feedforward network and is shown in Algorithm 1. In Step 1, the problem is broken down in the number of subcomponents based on the decomposition technique used. In Step 2, the encoding of the problem takes place based on hidden layer. Here the subcomponents in the sub-population is evolved.
   Once the network has been encoded it moves to Step 3 where evolution take place using genetic operators for each sub-population. Here the sub-populations are appended with new offspring based on genetic operators. Evaluation of the fitness of each individual for a particular sub-population is done cooperatively with the fittest individuals from the other sub-populations [1].
   Cooperative evaluation for an individual in a particular sub-population is done by concatenating the fittest individuals from the rest of the sub-populations. All the sub-populations are evolved for a fixed number of generations. Once the network has been evolved for the method trains the network and tests out the accuracy of the proposed method with test data set.

Algorithm 1. NSL for Training Feedforward Networks

Step 1: Decompose the problem into subcomponents according to NSL.
Step 2: Encode each subcomponent in a sub-population according to hidden layer.
Step 3: Initialise and cooperatively evaluate each sub-population.
foreach *Cycle until termination* **do**
    foreach *Sub-population* **do**
        foreach *Depth of n Generations* **do**
            Select and create new offspring using genetic operators
            Cooperative Evaluation the new offspring
            Add new offspring's to the sub-population
        end
    end
end

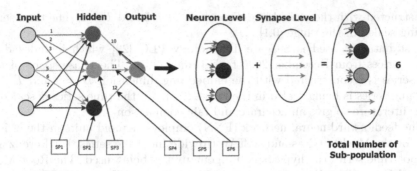

Fig. 1. Neuron-synapse level problem decomposition. Shows how the neuron-synapse methods sub population is decomposed into sub-population.

As mentioned in [16], it was observed for Enforced Sub-population (ESP) that during training of the network, too many or too few hidden units can seriously affect learning and generalization. Therefore, in the proposed method the aim is to see whether the right increase in number of sub-population can improve the prediction or not. Even it was seen in Neuron Level decomposition method, the number of sub-components are higher than what was in ESP and it had good prediction in [6].

3   Experiments, Results and Discussion

This section reports on the results and discussions based on the experiments conducted on chaotic time series problems using cooperative coevolution for training feed forward networks. The proposed method is called Neuron-Synapse level (NSL) problem decomposition method. The two well established problem decomposition methods are used for comparison.

3.1   Experimental Setup

We used five different time series data sets to train and test the proposed method. Taken's embedding theorem [17] is used to reconstruct the data set before data sets can be used as seen in [7–9,18].

Firstly, *Mackey Glass time series* [19] is a benchmark data set which is simulated data set taken out from [19]. The phase space of this original time series is reconstructed with the embedding dimension $D = 3$ and $T = 2$. The time series is scaled in the range [0,1]. Secondly, *Lorenz time series* [20] is being used. The Lorenz time series has been scaled in the range of [-1,1]. The phase space of this time series is reconstructed with the embedding dimension $D = 3$ and $T = 2$.

Thirdly, *Sunspot time series* [21] is a real world data set problem. This time series is being scaled in the range [-1,1]. The phase space of this time series is reconstructed with the embedding dimension $D = 5$ and $T = 2$. Fourth data set used is the ACI Worldwide Inc. [22]. The phase space of this time series is reconstructed with the embedding dimension $D = 5$ and $T = 2$. The time series is being scaled in the range [0,1].

Last data set used is Seagate Technology PLC [22] which includes daily closing prices from December 2006 to February 2010. The phase space of this time series is reconstructed with the embedding dimension $D = 5$ and $T = 2$. The time series is being scaled in the range [0,1]. All the time series is scaled as in the literature to give an accurate and fair comparison.

The feedforward neural network (FNN) employs sigmoid units in the hidden layer for the Mackey Glass and ACI Worldwide Inc. time series. For Lorenz and Sunspot time series the hyperbolic tangent unit is being used. The Root Mean Squared Error (RMSE) and Normalised Mean Squared Error (NMSE) are used to measure the prediction performance of the proposed method as done in [7,9].

The maximum number of function evaluations used was 50,000 with 50 individual runs. The G3-PCX algorithm uses the *generation gap model* [23] for selection in which a pool size of 2 parents and 2 offspring is placed as seen in literature [7,9]. As for the population size, 300 was used as used in literature in order to provide a fair comparison. The algorithm terminates as the maximum number of function evaluations has been reached.

3.2 Results

In Tables 1, 2, 3, 4 and 5, the results are shown for different number of hidden neurons using the NSL, NL and SL method. The results of NSL is compared with the results of standalone cooperative coevolution methods.

The results in the Tables 1, 2, 3, 4 and 5 are based on 95 % confidence interval on RMSE and shows the best run from different numbers of hidden neurons based on different methods. The *Training* shows the train average with train error sum while *Generalisation* is based on test average with test error sum and lastly *Best* shows the best test rmse. The best results for each method are highlighted in bold.

In Table 1, the Mackey-Glass time series problem is being evaluated. It was seen that NSL has performed much better than synapse method. The method recorded better generalization performance and best training value with seven hidden neurons. The NSL method was unable to outperform NL method.

In Table 2, the Lorenz time series problem is being evaluated. It also shows that the NSL has performed much better than the SL and gave competitive result with NL. It has been also observed that the generalisation performance of the NSL and the other two methods deteriorates as the number of the hidden neuron increases due to over fitting. The best result was seen for three hidden neuron for NSL.

In Table 3, the sunspot time series problem is being evaluated. The time series is real time series where noise is present. In this time series, the NSL method outperforms one of the standalone methods. The 5 hidden neurons have given best result for NSL whereas 3 hidden neurons for the other two methods.

In Table 4, the ACI time series problem is being evaluated. The time series is real time series where noise is present as in sunspot time series. Even for this time

Table 1. The prediction training and generalisation performance (RMSE) of NL, SL and NSL for the Mackey-Glass time series

Methods	H	Training	Generalisation	Best
FNN-NL	3	0.0107 ± 0.00131	0.0107 ± 0.00131	0.005
	5	0.0089 ± 0.00097	0.0088 ± 0.00097	**0.0038**
	7	**0.0078 ± 0.00079**	**0.0078 ± 0.00079**	0.0040
FNN-SL	3	0.0237 ± 0.0023	0.0237 ± 0.0023	0.0125
	5	0.0195 ± 0.0012	0.0195 ± 0.0012	0.0124
	7	**0.0177 ± 0.0009**	**0.0178 ± 0.0009**	**0.0121**
FNN-NSL	3	0.0119 ± 0.00089	0.0119 ± 0.00090	**0.0049**
	5	0.0107 ± 0.00081	0.0107 ± 0.00081	0.0056
	7	**0.0100 ± 0.00055**	**0.0100 ± 0.00055**	0.0066

Table 2. The prediction training and generalisation performance (RMSE) of NL, SL and NSL for the Lorenz time series

Methods	H	Training	Generalisation	Best
FNN-NL	3	**0.0170 ± 0.0031**	**0.0176 ± 0.0031**	0.0043
	5	0.0249 ± 0.0062	0.0271 ± 0.0067	**0.0021**
	7	0.0379 ± 0.0093	0.0416 ± 0.0092	0.0024
FNN-SL	3	0.0680 ± 0.0325	**0.0452 ± 0.0229**	0.0153
	5	**0.0526 ± 0.0084**	0.0546 ± 0.0084	0.0082
	7	0.0574 ± **0.0075**	0.0605 ± **0.0074**	**0.0079**
FNN-NSL	3	**0.0350 ± 0.00914**	**0.0357 ± 0.0093**	**0.0023**
	5	0.0523 ± 0.00978	0.0547 ± 0.00999	0.0099
	7	0.0459 ± 0.0090	0.0494 ± 0.00923	0.0032

Table 3. The prediction training and generalisation performance (RMSE) of NL, SL and NSL for the Sunspot time series

Methods	H	Training	Generalisation	Best
FNN-NL	3	**0.0207 ± 0.0035**	**0.0538 ± 0.0091**	**0.015**
	5	0.0289 ± 0.0039	0.0645 ± 0.0093	0.017
	7	0.0353 ± 0.0048	0.0676 ± 0.0086	0.021
FNN-SL	3	**0.5391 ± 0.0261**	**0.4998 ± 0.0238**	**0.210**
	5	0.5601 ± 0.0208	0.5210 ± 0.0177	0.302
	7	0.5682 ± **0.0178**	0.5250 ± **0.0132**	0.344
FNN-NSL	3	0.0403 ± 0.0088	0.0953 ± 0.01443	**0.013**
	5	**0.0356 ± 0.0074**	**0.0842 ± 0.0098**	0.022
	7	0.0396 ± 0.0080	0.0940 ± 0.00987	0.019

Table 4. The prediction training and generalisation performance (RMSE) of NL, SL and NSL for the ACI Worldwide Inc. time series

Methods	H	Training	Generalisation	Best
FNN-NL	3	0.0214 ± 0.00039	0.0215 ± 0.00039	0.020
	5	0.0203 ± 0.00047	0.0212 ± 0.00041	0.019
	7	**0.0201 ± 0.00038**	**0.0208 ± 0.00033**	**0.019**
FNN-SL	3	0.4666 ± 0.0399	0.4112 ± 0.0362	0.080
	5	**0.4135 ± 0.0388**	**0.3902 ± 0.0378**	**0.042**
	7	0.4491 ± **0.0279**	0.4244 ± **0.0270**	0.134
FNN-NSL	3	0.0224 ± 0.00113	0.0197 ± 0.00119	0.015
	5	0.0215 ± 0.000364	**0.0185 ± 0.00092**	**0.015**
	7	**0.0209 ± 0.000364**	0.0192 ± 0.0010	0.015

Table 5. The prediction training and generalisation performance (RMSE) of NL, SL and NSL for the Seagate time series

Methods	H	Training	Generalisation	Best
FNN-NL	3	**0.02530 ± 0.05582**	**0.1809 ± 0.03548**	0.032
	5	0.02313 ± 0.07403	0.2261 ± 0.04811	**0.031**
	7	0.02189 ± **0.08061**	0.2408 ± 0.05306	0.053
FNN-SL	3	0.4129 ± 0.03401	**0.3492 ± 0.02977**	0.089
	5	0.3820 ± **0.03381**	0.3727 ± 0.03371	**0.088**
	7	**0.3816 ± 0.04425**	0.4183 ± 0.04306	0.094
FNN-NSL	3	0.02187 ± 0.00078	**0.1988 ± 0.04394**	**0.025**
	5	**0.01862 ± 0.00029**	0.2562 ± **0.04221**	0.028
	7	0.01896 ± **0.00027**	0.3059 ± 0.04504	0.035

series, the NSL has performed much better than the SL and gave competitive result with NL. Five hidden neurons have given the best result for NSL similar to Lorenz time series problem.

In Table 5, the Seagate time series problem is being evaluated. The time series is real time series where noise is present as in sunspot and ACI time series. For this time series, the NSL method outperforms the other two methods. The 3 hidden neurons have given best result for NSL.

Tables 6, 7, 8, 9 and 10, compares the best results from Tables 1, 2, 3, 4 and 5 with some of the related methods in literature. The RMSE best run together with NMSE are used for the comparison. The proposed NSL method has given better performance when compared to some of the methods in the literature.

In Table 6, the best result of Mackey-Glass time series problem is being compared. The proposed method outperformed all the methods except CICC-RNN

Table 6. A comparison with the results from literature on the Mackey time series

Prediction method	RMSE	NMSE
AMCC-RNN [8]	7.53E-03	3.90E-04
Locally linear neuro-fuzzy model - Locally linear model tree (LLNF-LoLiMot) (2006) [24]	9.61E-04	
SL-CCRNN [9]	6.33E-03	2.79E-04
NL-CCRNN [9]	8.28E-03	4.77E-04
CICC-RNN [7]	3.99E-03	1.11E-04
Proposed FNN-NSL	4.86E-03	4.48E-05

Table 7. A comparison with the results from literature on the Lorenz time series

Prediction method	RMSE	NMSE
Radial basis network with orthogonal least squares (RBF-OLS)(2006) [24]		1.41E-09
Locally linear neuro-fuzzy model - Locally linear model tree (LLNF-LoLiMot) (2006) [24]		9.80E-10
SL-CCRNN [9]	6.36E-03	7.72E-04
NL-CCRNN [9]	8.20E-03	1.28E-03
CICC-RNN [7]	3.55E-03	2.41E-04
Proposed FNN-NSL	2.34E-03	2.87E-05

Table 8. A comparison with the results from literature on the Sunspot time series

Prediction method	RMSE	NMSE
Radial basis network with orthogonal least squares (RBF-OLS)(2006) [24]		4.60E-02
Locally linear neuro-fuzzy model - Locally linear model tree (LLNF-LoLiMot) (2006) [24]		3.20E-02
SL-CCRNN [9]	1.66E-02	1.47E-03
NL-CCRNN [9]	2.60E-02	3.62E-03
CICC-RNN [7]	1.57E-02	1.31E-03
Proposed FNN-NSL	1.33E-02	5.38E-04

result. Due to competition and collaboration, the method used in CICC-RNN has performed better than the proposed method.

The Table 7 below shows the best result on Lorenz time series problem that is being compared to other computational intelligence methods in the literature. The proposed method outperformed all the methods in terms of the RMSE but was unable to outperform NMSE of two methods, RBF-OLS and LLNF-LoLiMot. These methods have additional enhancements such as the optimization

Table 9. A comparison with the results from literature on the ACI time series

Prediction method	RMSE	NMSE
CICC-RNN [7]	1.92E-02	
FNN-SL [18]	1.92E-02	
FNN-NL [18]	1.91E-02	
MO-CCFNN-T=2 [25]	1.94E-02	
MO-CCFNN-T=3 [25]	1.470E-02	
Proposed FNN-NSL	1.51E-02	1.24E-03

Table 10. A comparison with the results from literature on the Seagate time series

Prediction method	RMSE	NMSE
FNN-SL [18]	3.74E-02	
FNN-NL [18]	2.24E-02	
Proposed FNN-NSL	2.45E-02	3.56E-03

of the embedding dimensions and strength of architectural properties of hybrid neural networks with residual analysis [24].

In Table 8, the best result of the Sunspot time series problem is compared with results in the literature where the proposed method has shown to outperform the rest of the methods.

In Table 9, the best result of the ACI time series problem is being compared with results in the literature. The proposed method could not outperform multi-objective method having T=3, however, the results are better when compared to other methods from the literature.

In Table 10, the best result of the Seagate time series problem is compared with results in the literature. The proposed method has been not able to outperform NL method.

3.3   Discussion

The results obtained were very promising when compared to other methods from literature involving five different data sets. The proposed method (NSL) has given better performances in nearly all benchmark data sets and financial data set used. It is being compared to similar evolutionary methods namely training neural fuzzy networks [24] and competitive cooperative coevolution methods [7]. It creates lower number of subcomponents for the problem when compared to synapse level (SL) and higher number of subcomponents when compared to neuron level (NL). The increase in the number of sub-populations when compared to NL seems to provide more diversity in cooperative coevolution and improve the prediction performance.

NSL incorporates two problem decomposition (NL and SL) to use architectural properties of the neural network, where NL appeals to partially-separable problems where decision making is required and SL views the network training as a fully separable problem and is applied to region where more diversity in search is needed. Therefore, NSL performs better than other methods.

4 Conclusions

In this paper, neuron-synapse level problem decomposition method has been proposed for feedforward networks with application to time series prediction. The proposed method uses the strength of both the problem decomposition methods. It fulfills the limitations faced by a single problem decomposition method. It can be further enhanced by involving competition in it as done in competitive island cooperative coevolution method where different problem decomposition methods compete and collaborate during evolution.

The method has given promising performance on the different benchmark problems and has outperformed several methods from the literature. The method performs better for real work time series problems when compared to simulated ones. This is an advantage as real work time series problems contains noise that makes prediction models difficult to train and generalise.

In future work, the proposed method can be extended to recurrent neural networks and pattern classification problems.

References

1. Potter, M., De Jong, K.: A cooperative coevolutionary approach to function optimization. In: Davidor, Y., Schwefel, H.-P., Mnner, R. (eds.) PPSN III. LNCS, vol. 866, pp. 249–257. Springer, Heidelberg (1994)
2. Chandra, R., Frean, M.R., Zhang, M.: Crossover-based local search in cooperative co-evolutionary feedforward neural networks. Appl. Soft Comput. **12**(9), 2924–2932 (2012)
3. García-Pedrajas, N., Ortiz-Boyer, D.: A cooperative constructive method for neural networks for pattern recognition. Pattern Recogn. **40**(1), 80–98 (2007)
4. Lehman, J., Miikkulainen, R.: Neuroevolution. Scholarpedia **8**(6), 30977 (2013)
5. Potter, M.A., De Jong, K.A.: Cooperative coevolution: an architecture for evolving coadapted subcomponents. Evol. Comput. **8**(1), 1–29 (2000)
6. Chandra, R., Frean, M., Zhang, M.: On the issue of separability for problem decomposition in cooperative neuro-evolution. Neurocomputing **87**, 33–40 (2012)
7. Chandra, R.: Competitive two-island cooperative coevolution for training Elman recurrent networks for time series prediction. In: International Joint Conference on Neural Networks (IJCNN), Beijing, China, pp. 565–572, July 2014
8. Chandra, R.: Adaptive problem decomposition in cooperative coevolution of recurrent networks for time series prediction. In: International Joint Conference on Neural Networks (IJCNN), Dallas, TX, USA, pp. 1–8, August 2013
9. Chandra, R., Zhang, M.: Cooperative coevolution of Elman recurrent neural networks for chaotic time series prediction. Neurocomputing **186**, 116–123 (2012)

10. Gomez, F., Mikkulainen, R.: Incremental evolution of complex general behavior. Adapt. Behav. **5**(3–4), 317–342 (1997)
11. Gomez, F., Schmidhuber, J., Miikkulainen, R.: Accelerated neural evolution through cooperatively coevolved synapses. J. Mach. Learn. Res. **9**, 937–965 (2008)
12. Chandra, R., Frean, M., Zhang, M.: An encoding scheme for cooperative coevolutionary feedforward neural networks. In: Li, J. (ed.) AI 2010. LNCS, vol. 6464, pp. 253–262. Springer, Heidelberg (2010)
13. Chandra, R.: Competition and collaboration in cooperative coevolution of Elman recurrent neural networks for time-series prediction. IEEE Trans. Neural Netw. Learn. Syst. (2015). (in press)
14. Chandra, R., Frean, M., Zhang, M., Omlin, C.W.: Encoding subcomponents in cooperative co-evolutionary recurrent neural networks. Neurocomputing **74**(17), 3223–3234 (2011)
15. Garcia-Pedrajas, N., Hervas-Martinez, C., Munoz-Perez, J.: COVNET: a cooperative coevolutionary model for evolving artificial neural networks. IEEE Trans. Neural Netw. **14**(3), 575–596 (2003)
16. Gomez, F.J.: Robust non-linear control through neuroevolution. Ph.D. Thesis, Department of Computer Science, The University of Texas at Austin, Technical Report AI-TR-03-303 (2003)
17. Takens, F.: Detecting strange attractors in turbulence. In: Rand, D., Young, L.-S. (eds.) Dynamical Systems and Turbulence, Warwick 1980. LNM, vol. 898, pp. 366–381. Springer, Heidelberg (1981)
18. Chand, S., Chandra, R.: Cooperative coevolution of feed forward neural networks for financial time series problem. In: International Joint Conference on Neural Networks (IJCNN), Beijing, China, pp. 202–209, July 2014
19. Mackey, M., Glass, L.: Oscillation and chaos in physiological control systems. Science **197**(4300), 287–289 (1977)
20. Lorenz, E.: Deterministic non-periodic flows. J. Atmos. Sci. **20**, 267–285 (1963)
21. SILSO World Data Center, The International Sunspot Number (1834–2001), International Sunspot Number Monthly Bulletin and Online Catalogue, Royal Observatory of Belgium, Avenue Circulaire 3, 1180 Brussels, Belgium. http://www.sidc.be/silso/. Accessed 02 February 2015
22. NASDAQ Exchange Daily: 1970–2010 Open, Close, High, Low and Volume. http://www.nasdaq.com/symbol/aciw/stock-chart. Accessed 02 February 2015
23. Deb, K., Anand, A., Joshi, D.: A computationally efficient evolutionary algorithm for real-parameter optimization. Evol. Comput. **10**(4), 371–395 (2002)
24. Gholipour, A., Araabi, B.N., Lucas, C.: Predicting chaotic time series using neural and neurofuzzy models: a comparative study. Neural Process. Lett. **24**, 217–239 (2006)
25. Chand, S., Chandra, R.: Multi-objective cooperative coevolution of neural networks for time series prediction. In: International Joint Conference on Neural Networks (IJCNN), Beijing, China, pp. 190–197, July 2014

Prediction Interval-Based Control of Nonlinear Systems Using Neural Networks

Mohammad Anwar Hosen[(✉)], Abbas Khosravi, Saeid Nahavandi,
and Douglas Creighton

Centre for Intelligent System Research (CISR), Deakin University,
Waurn Ponds Campus, Geelong, Australia
{anwar.hosen,abbas.khosravi,saeid.nahavandi,
douglas.creighton}@deakin.edu.au

Abstract. Prediction interval (PI) is a promising tool for quantifying uncertainties associated with point predictions. Despite its informativeness, the design and deployment of PI-based controller for complex systems is very rare. As a pioneering work, this paper proposes a framework for design and implementation of PI-based controller (PIC) for nonlinear systems. Neural network (NN)-based inverse model within internal model control structure is used to develop the PIC. Firstly, a PI-based model is developed to construct PIs for the system output. This model is then used as an online estimator for PIs. The PIs from this model are fed to the NN inverse model along with other traditional inputs to generate the control signal. The performance of the proposed PIC is examined for two case studies. This includes a nonlinear batch polymerization reactor and a numerical nonlinear plant. Simulation results demonstrated that the proposed PIC tracking performance is better than the traditional NN-based controller.

Keywords: Prediction interval · PI-based model · PI-based controller · Neural network · Polymerization reactor

1 Introduction

It is already established that PI-based modelling technique is superior over point forecast-based models to quantify the uncertainties and disturbances [11]. In this technique, the model predicts an interval, called lower and upper bound. The condition is that the target values should lie between these two intervals. According to [13], PIs covers both uncertainties and disturbances associated with model mis-match and noises in data.

The most considerable advantage of this technique over traditional point forecast-based model is this model can be developed with a predefined confidence level (CL), and this CL corresponds to the model accuracy. Let's say, for a 90 % CL, the PIs should cover at least 90 % of targets. Due to enormous advantages over traditional modelling techniques, the application of PI-based modelling technique can be found in different field of application. These include,

© Springer International Publishing Switzerland 2015
S. Arik et al. (Eds.): ICONIP 2015, Part III, LNCS 9491, pp. 101–110, 2015.
DOI: 10.1007/978-3-319-26555-1_12

chemical processes [6, 7], manufacturing processes [2], food industry [15], transportation [12] and power generation industries [9, 16, 17, 19].

Though there are vast applications of PIs in modelling field to model nonlinear systems, the use of PIs in control application is still limited. It is notable that, to date, all control techniques are dealing with the point forecast values. The use of informative PIs might improve the controller's tracking performance. However, the use of PIs in control application is not a trivial task. The most challenging part to integrate the PIs in control application is how to extract the information from PIs and feed to the control system. As described in [11], the following outputs can be found from the PI-based model:

1. Upper bound $(\overline{y}(t))$;
2. Lower bound $(\underline{y}(t))$; and
3. Interval $(\overline{y}(t) - \underline{y}(t))$

These model outputs mentioned above can be directly used as effective inputs for the model-based controllers, and PIs can also be used to optimize the controller's signal. Development of fuzzy rules for fuzzy logic controller or development of NN-based controller using PIs may be a viable solution to integrate the informative PIs in control systems. In the present work, upper bound and lower bound of PIs are used as additional inputs along with other traditional inputs to develop NN-based controller.

NN inverse model has been widely used to control the nonlinear systems due to their excellent fitting ability. Moreover, the mathematical complexities are considerably low to utilize the NN controller in real systems. In this work, NN inverse model is used as PIC controller within IMC structure as NN-based IMC performance is more better than the direct NN-based controllers. Three NN models are developed for PIC control system. These include, feed-forward NN model (FFNN), PI-based NN model and PI-based NN inverse model. Finally, the proposed PIC control system is examined for two case studies. These include, a nonlinear polystyrene (PS) polymerization batch reactor and a nonlinear numerical plant.

2   PI-based Controller

As described in Sect. 1, NN-based controller is used to integrate the PIs in control system. Usually, NN inverse model is used as NN controller. A NN inverse model can easily be developed if the system's inputs and outputs are available. In this technique, NN inverse model predict the plant input (control signal) based on plant outputs and previous plant inputs including other effective variables to get the desired plant output (setpoint). Therefore, the performance of the NN-based controller is directly depends on the accuracy of the inverse model. As PIs bears extra information, the performance of the NN inverse model can be improved by adding PIs as the additional inputs to train the inverse model. This means that the PIs may help to improve the controller's performance. The basic inverse model for NN (NN_I) can be defined as:

$$u(k) = NN_I\left(y(k), y(k-1), \cdots y(k-i), u(k-1), u(k-2), \cdots, u(k-j)\right). \quad (1)$$

where y, u, and k are the model output, input and sample index, respectively. i and j are the number of past or lagged samples. In Eq. 1, the PIs can be added as additional inputs for NN_I if the PIs $[\overline{y}(t), \underline{y}(t)]$ are available for every sample instance. It is assumed that the PIs are available from a dynamic PI-based model for every sample time. Then, the Eq. 1 can be re-written as follows:

$$u(k) = NN_{PI}\{y(k), y(k-1), \cdots y(k-i), u(k-1), u(k-2), \cdots, u(k-i)),$$
$$\overline{y}(k), \overline{y}(k-1), \cdots, \overline{y}(k-j), \underline{y}(k), \underline{y}(k-1), \cdots, \underline{y}(k-l)\}. \quad (2)$$

In this work, the PI-based inverse model as Eq. 2 is used as PIC controller in the IMC structure.

It is well known that NN controller's performance within IMC structure is better than the direct inverse NN controller [1]. In IMC structure, a forward model is added to the parallel of the plant as seen if Fig. 1, and compared the plant output and forward model output. The difference between plant and forward model output is considered as system disturbances, and subtracted these disturbances from the setpoint to get the latest or corrected setpoint for the controller. In this way, the controller is able to capture the system disturbances, and produce offset-free stable setpoint tracking.

For PI-based NN inverse controller, a PI-based model is needed to get the PIs for every sample instance. In this connection a PI-based model is added in parallel to the PI-based inverse model (PIC controller) as seen in Fig. 2. Here PI-based model acts as an online estimator for PIs. The PIs from this model are fed to the PIC along with other effective inputs to get the control signal. As seen in Fig. 2, three models are used in this control system. These include, a forward model, a PI-based model and a PI-based inverse model. All models are developed by employing NN modelling strategies.

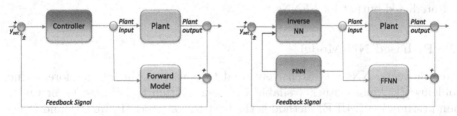

Fig. 1. Basic structure of IMC.   **Fig. 2.** Basic structure of PIC system.

3 Proposed Methodology

As described in Sect. 2, three NN models are developed for PIC control system. Initially, N_{total} NN models are developed by changing the structure of the NN for each class of NNs. 10 different sizes of hidden neurons ($N_u = 2:2:20$) are used to diversify the structure of NN. Then, selected N_{best} NNs to aggregate the NN forecasts as ensemble NNs produce better prediction than individual NNs [8]. The development procedure of these NN models are described as follows:

3.1 Feed-Forward NN Model

The FFNN training process is initiated with the pre-processing of the available training data set. Firstly, the raw training data are rearranged according to the structure of FFNN. The FFNN model can be written as follows:

$$y(k) = NN_{FF} \{(y(k-1), y(k-2), \cdots y(k-i),$$
$$u(k), u(k-1) \cdots u(k-j)\}. \tag{3}$$

where the data structure for FFNN is $[y(k-1), y(k-2), \cdots y(k-i), u(k), u(k-1) \cdots u(k-j); y(k)]$.

The total number of lagged values, i (for plant output or controlled variable) and j (for plant input or manipulated variable) in Eq. 3 are determined by partial correlation analysis. The data are then normalized in the range of -1 to 1, and split the data for training (S_{train}), testing (S_{test}) and validation (S_{vald}) data sets. Single hidden layer and 10 different hidden neuron sizes are used to train the FFNN model. Mean square error (MSE) is used as the cost function for FFNN. Levenberg-Marquardt optimization algorithm is used to optimize the NN parameters by minimizing the MSE.

Total N_{total} FFNN models are developed using S_{train} data set. Then, followed "mean mechanism" to ensemble the NNs as described by Hosen et al. (2014) [6]. In this mechanism, the developed NN models are ranked based on MSE. S_{test} data set are used to calculate the MSE for NNs ranking process, and selected top ranked N_{best} FFNN models for ensemble process. The other models are discarded. The mean predicted value from these N_{best} NNs is considered as the final predicted output for FFNNs.

3.2 PI-based NN Model

In recent years, NNs are extensively used to construct the PIs for forecasting problems [7]. Among other available PIs construction techniques, lower upper bound estimation (LUBE) method is the best to construct the quality and informative PIs [11]. In this work, LUBE method is used to develop the PI-based NN (PINN) models for a 90 % CL.

The same data that used to develop FFNNs are employed in the LUBE method to generate the PINN models. Despite the model output, the training data structure for both PINN and FFNN is same. The model for PINN can be defined mathematically as follows:

$$y_{PI}(k)[\underline{y}(k), \overline{y}(k)] = NN_{LUBE}\{(y(k), y(k-1), \cdots y(k-i),$$
$$u(k), u(k-1) \cdots u(k-j)\}. \tag{4}$$

where the data structure for FFNN is $[y(k-1), y(k-2), \cdots y(k-i), u(k), u(k-1) \cdots u(k-j); y(k)]$.

In contrast to traditional error-based cost function, LUBE method used a PI-based cost function to optimize the NN parameters as the prior values of PI are unavailable. The PI-based cost function, namely coverage width criterion (CWC) can be defined as [10]:

$$CWC = PINAW + (PICP)e^{-\eta(PICP-\varphi)}. \tag{5}$$

where $PICP$ is the PIs coverage probability, $PINAW$ is the normalise average width of the PIs, φ is correspond to the nominal CL, and η is the hyperparameter that magnify the penalty if $PICP <$ nominal CL. Here $PICP$ and $PINAW$ indicate the quality of the PIs, and these two terms can be defined mathematically as follows:

$$PICP = \frac{1}{n}\sum_{j=1}^{n} c_j \tag{6}$$

where

$$c_j = \begin{cases} 1, tr_j \in [\underline{y}_j, \overline{y}_j] \\ 0, tr_j \notin [\underline{y}_j, \overline{y}_j], \end{cases} \quad \text{and}$$

$$PINAW = \frac{1}{R}\left(\frac{1}{n}\sum_{j=1}^{n}(\overline{y}_j - \underline{y}_j)\right). \tag{7}$$

where R is the range of underlying target ($max(t_r) - min(t_r)$), and j is the sample index of PIs.

In a recent study, Hosen et al. (2014) elaborately described the LUBE method and its cost function to develop the PINN model [6]. This work followed the same procedure to develop N_{total} PINN models. Simulated annealing (SA) is used to minimize the CWC. After developing N_{total} PINN models using S_{train} data set, the N_{best} PINN models are selected based on the cost function, CWC, and ensemble the N_{best} models as mentioned in Subsection 3.1.

3.3 PI-based NN Inverse Model (PIC)

As seen in Eq. 2, PI-based inverse model is a function of plant inputs, output and PIs. The structure of this NN model is $[y(k), y(k-1), \cdots y(k-i), u(k-1), u(k-2), \cdots, u(k-i), \overline{y}(k), \overline{y}(k-1), \cdots, \overline{y}(k-j), \underline{y}(k), \underline{y}(k-1), \cdots, \underline{y}(k-l); u(k)]$. The PIs for PIC are generated from the developed PINN models in Sect. 3.2. The same cost function, training algorithm and NN ensemble procedure that used in the development of FFNN are used to develop the ensemble PIC models. The ensemble PIC models are then used as a PIC.

4   Case Studies

The proposed PIC system is examined for two case studies. The first case study is a PS polymerization batch reactor. In a recent study, Hosen et al. (2011) developed a hybrid model for this reactor, and validated the model experimentally [4]. This hybrid model is used to generate the training data. According to [3,5], reactor temperature, T_r is the controlled variable (reactor output) and heater duty, Q is the manipulated variable (reactor input) for this reactor. Therefore, reactor temperature profile is generated by varying the heater power. In this work, same training data that generated in [7] to develop PINN models for PS reactor are used to train the FFNN and PINN models.

The second case study is a nonlinear numerical plant model (NNPM). The mathematical model of this plant is as follows:

$$y(k+1) = \frac{y(k)}{1 + y(k)^2} + u(k)^3. \tag{8}$$

This plant model previously used to examine the performance of the different type of controller due to its nonlinearity behaviour [14,18]. In this work, the model in Eq. 8 is used to generate the training data. The data are generated by using the following setpoint:

$$y_{set} = 2.5 \sin\left(\frac{10\pi k}{N}\right) + 2.5 \sin\left(\frac{4\pi k}{N}\right). \tag{9}$$

where N is the total number of samples and k is the sampling time. White noise (variance=1) is added in the plant output as unknown disturbances. Total 5000 data are collected (where sampling time= 1) by simulating Eq. 8. The raw data are then rearranged according to the data structure of FFNN and PINN.

For both case studies, after developing the PINN model PIs are constructed from this model to train the PIC.

5   Results and Discussion

In this work, three NN models are developed for PIC system. A traditional NN controller within IMC structure is also developed to compare the results with the PIC. The data structure for traditional NN inverse model (NNC) is $[y(k), y(k-1), \cdots y(k-i), u(k-1), u(k-2), \cdots, u(k-j); u(k)]$. The training procedure for NNC is same as the PIC as described in Sect. 3.3.

Firstly, set the total number of NNs, $N_{total} = 10$, also, set the number of ranked NNs to be used in ensemble process, N_{best} =5 for each class of NNs. As mentioned in Sect. 3, N_u are set to 2:2:20. This means that 10 different hidden neuron sizes are used to develop 10 NNs (one NN for each size of hidden neurons). After pre-processing and normalising the training data, randomly split the structured data into S_{train} (60 % of total data), S_{vald} (20 % of total data) and S_{test} (20 % of total data). The parameters used for LUBE and SA optimization methods are taken from [8].

After preprocessing the training data and initializing the training parameters, 10 NNs are developed for each class of NNs (FFNN, PINN, PIC and NNC). Table 1 depicts the prediction performance of NNs for all classes. As described in Sect. 3, S_{test} data set are used to evaluate the performance of NNs. In Table 1, CWC indicates the performance of PINN models and MSE indicates the performance of FFNN, PIC and NNC models. As seen in Table 1, PINN with 8,14,16,18 and 20 hidden neurons produce low CWC among others. Therefore, these five PINNs are selected for ensemble process and used in PIC system. According to MSE value, the prediction performance of all FFNN, PIC and NNC models are excellent as MSE values are less than $6\times^{-10}$. Based on MSE values, NNs with [2 14 16 18 20], [2 12 14 18 20] and [4 12 14 18 20] hidden neurons produce better results among others for FFNN, PIC and NNC, respectively. Therefore, these NNs are selected for ensemble process. Finally, ensemble NNs are used for PIC system.

Table 1. Prediction performance of NN models

Neuron size	Performance criterion			
	CWC (PINN)	**MSE** (FFNN)	**MSE** (PIC)	**MSE** (NNC)
2	11.09	**3.93E-07**	**7.89E-07**	8.24E-07
4	9.74	4.11E-07	7.96E-07	**7.76E-07**
6	9.96	4.07E-07	8.05E-07	8.15E-07
8	**9.72**	4.00E-07	7.95E-07	8.11E-07
10	10.14	4.10E-07	7.98E-07	8.05E-07
12	12.56	3.98E-07	**7.18E-07**	**7.98E-07**
14	**4.25**	**3.97E-07**	**7.84E-07**	**7.91E-07**
16	**5.47**	**3.89E-07**	7.96E-07	8.03E-07
18	**5.08**	**3.97E-07**	**7.79E-07**	**7.93E-07**
20	**5.62**	**3.80E-07**	**7.94E-07**	**7.97E-07**

Now, the performance of the PI-based controller is examined for PS batch reactor and NNPM systems. Integral absolute error (IAE) is used as the performance criterion of the controllers. For case 1, PS reactor, optimum temperature profile is used as setpoint since PS reactor usually operate with pre-specified optimum temperature profile to get the desired polymer products [5]. Figure 3 depicts the tracking performance of PIC and NNC for PS reactor. The figure clearly demonstrates that both controllers performance are quite good in terms of offset and overshoot, however, little fluctuations are observed for NNC. A considerable improvement of 80.80 % is observed for PIC (IAE= 43.34) compared to the NNC (IAE= 226.00) in terms of IAE as seen in Table 2.

For case 2, NNPM system, constant setpoint is used to check the tracking performance of PIC. The setpoint tracking profiles of PIC and NNC for NNPM are

Table 2. Tracking performances of NNC and PIC in terms of IAE

Cases	Type of setpoint	Controllers	IAE	Improvement (%)
Case 1	Optimum	NNC	226.00	80.80
		PIC	43.34	
	Optimum with disturbances	NNC	349.70	68.86
		PIC	108.90	
Case 2	Constant	NNC	351.50	22.40
		PIC	296.00	
	Step Changes	NNC	1204.00	17.60
		PIC	992.70	

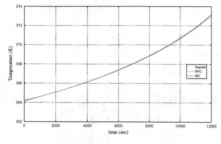

Fig. 3. Optimum setpoint tracking for PS polymerization reactor.

Fig. 4. Constant setpoint tracking for NNPM.

shown in Fig. 4. As seen in Fig. 4, smooth tracking is observed for PIC in terms of fluctuation, offset and overshoot where significant overshoots and fluctuations are noticed for NNC at the initial stage of experiment. The PIC performance for NNPM is also better than the NNC in terms of IAE, where IAE for PIC is 296.00 and IAE for NNC is 381.50. 22.40 % improvement can observed for constant setpoint tracking using the proposed PIC than that NNC.

The stability and robustness of PIC also test in the presence of disturbances. For PSPR, the disturbances are made by varying the effective parameters (coolant flow rate and coolant temperature) during the optimum setpoint tracking as previous work done in [5]. Figure 5 shows the tracking performance of PIC and NNC in the presence of disturbances. The performance of the PIC and NNC is also good in this case as well. However, the PIC tracking is better than NNC in terms of IAE, where IAEs are 108.90 and 349.70 for PIC and NNC, respectively.

For case 2, the stability and robustness of the proposed PIC are examined by step changes of setpoint. Upward and backward step changes are made in the experiment. Figure 6 demonstrates the tracking performance of PIC and NNC for step changes setpoint. As seen in Fig. 6, PIC quickly merged the new setpoints when step changes are made. Smooth tracking is obesrved for PIC, however, considerable fluctuation and overshoot are observed for NNC for every

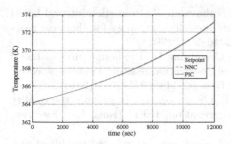
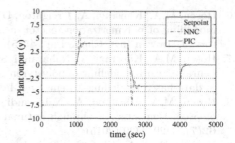
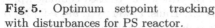

Fig. 5. Optimum setpoint tracking with disturbances for PS reactor.

Fig. 6. Setpoint tracking with step changes for NNPM.

step changes. According to Table 1, PIC improved the tracking performance in this particular case by 17.60 % over NNC.

By evaluation above results, it can be concluded that the PIC performance outforms over NNC. Even overshoot, offset and fluctuation free setpoint tracking is observed for PIC in the presence of disturbances.

6 Conclusion

It is well established that PIs bear more information than traditional forecasts. With the consideration of informative PIs, a new controller, called PIC is proposed and developed for nonlinear systems. NN-based controller within IMC structure is used to develop PI-based controller. Three NN models are developed for this control system, these include, FFNN, PINN and PI-based inverse NN. Finally, the performance of the proposed controller is examined for two case studies. Simulation results revealed that the proposed PIC performance is outform than traditional NN-based controllers.

References

1. Deepa, N., Arulselvi, S.: Design and implementation of neuro controllers for a two-tank interacting level process. Int. J. ChemTech Res. **6**(12), 4948–4959 (2014)
2. Ho, S., Xie, M., Tang, L., Xu, K., Goh, T.: Neural network modeling with confidence bounds: a case study on the solder paste deposition process. IEEE Trans. Electron. Packag. Manuf. **24**(4), 323–332 (2001)
3. Hosen, M.A., Hussain, M.A., Mjalli, F.S.: Control of polystyrene batch reactors using neural network based model predictive control (nnmpc): an experimental investigation. Control Eng. Pract. **19**(5), 454–467 (2011)
4. Hosen, M.A., Hussain, M.A., Mjalli, F.S.: Hybrid modelling and kinetic estimation for polystyrene batch reactor using artificial neutral network (ann) approach. Asia-Pacific J. Chem. Eng. **6**(2), 274–287 (2011)
5. Hosen, M.A., Hussain, M.A., Mjalli, F.S., Khosravi, A., Creighton, D., Nahavandi, S.: Performance analysis of three advanced controllers for polymerization batch reactor: an experimental investigation. Chem. Eng. Res. Des. **92**(5), 903–916 (2014)

6. Hosen, M.A., Khosravi, A., Creighton, D., Nahavandi, S.: Prediction interval-based modelling of polymerization reactor: a new modelling strategy for chemical reactors. J. Taiwan Inst. Chem. Eng. **45**(5), 2246–2257 (2014)
7. Hosen, M.A., Khosravi, A., Nahavandi, S., Creighton, D.: Prediction interval-based neural network modelling of polystyrene polymerization reactor: a new perspective of data-based modelling. Chem. Eng. Re. Des. **92**(11), 2041–2051 (2014)
8. Hosen, M.A., Khosravi, A., Nahavandi, S., Creighton, D.: Improving the quality of prediction intervals through optimal aggregation. IEEE Trans. Ind. Electron. **62**(7), 4420–4429 (2015)
9. Kavousi-Fard, A., Khosravi, A., Nahavandi, S.: A new fuzzy-based combined prediction interval for wind power forecasting. IEEE Trans. Power Syst. **99**, 1–9 (2015)
10. Khosravi, A., Nahavandi, S.: Combined nonparametric prediction intervals for wind power generation. IEEE Trans. Sustainable Energy **4**(4), 849–856 (2013)
11. Khosravi, A., Nahavandi, S., Creighton, D., Atiya, A.: Lower upper bound estimation method for construction of neural network-based prediction intervals. IEEE Trans. Neural Netw. **22**(3), 337–346 (2011)
12. Khosravi, A., Mazloumi, E., Nahavandi, S., Creighton, D., Van Lint, J.: A genetic algorithm-based method for improving quality of travel time prediction intervals. Transp. Res. Part C: Emerg. Technol. **19**(6), 1364–1376 (2011)
13. Khosravi, A., Nahavandi, S., Creighton, D., Srinivasan, D.: Optimizing the quality of bootstrap-based prediction intervals. In: The 2011 International Joint Conference on Neural Networks (IJCNN), pp. 3072–3078. IEEE (2011)
14. Narendra, K.S., Parthasarathy, K.: Identification and control of dynamical systems using neural networks. IEEE Trans. Neural Netw. **1**(1), 4–27 (1990)
15. Pierce, S.G., Worden, K., Bezazi, A.: Uncertainty analysis of a neural network used for fatigue lifetime prediction. Mech. Syst. Signal Process. **22**(6), 1395–1411 (2008)
16. Quan, H., Srinivasan, D., Khosravi, A.: Incorporating wind power forecast uncertainties into stochastic unit commitment using neural network-based prediction intervals. IEEE Trans. Neural Netw. Learn. Syst. **99**, 1–1 (2014)
17. Quan, H., Srinivasan, D., Khosravi, A.: Uncertainty handling using neural network-based prediction intervals for electrical load forecasting. Energy **73**, 916–925 (2014)
18. Salman, R.: Neural networks of adaptive inverse control systems. Appl. Math. Comput. **163**(2), 931–939 (2005)
19. Shrivastava, N., Khosravi, A., Panigrahi, B.: Prediction interval estimation of electricity prices using pso-tuned support vector machines. IEEE Trans. Ind. Inform. **11**(2), 322–331 (2015)

Correcting a Class of Complete Selection Bias with External Data Based on Importance Weight Estimation

Van-Tinh Tran[✉] and Alex Aussem

LIRIS, UMR 5205 University of Lyon 1, 69622 Lyon, France
{van-tinh.tran,aussem}@univ-lyon1.fr

Abstract. We present a practical bias correction method for classifier and regression models learning under a general class of selection bias. The method hinges on two assumptions: (1) a feature vector, X_s, exists such that S, the variable that controls the inclusion of the samples in the training set, is conditionally independent of (X, Y) given X_s; (2) one has access to some external samples drawn from the population as a whole in order to approximate the unbiased distribution of X_s. This general framework includes covariate shift and prior probability shift as special cases. We first show how importance weighting can remove this bias. We also discuss the case where our key assumption about X_s is not valid and where X_S is only partially observed in the test set. Experimental results on synthetic and real-world data demonstrate that our method works well in practice.

Keywords: Selection bias · Importance weighting · Graphical models

1 Introduction

Selection bias, which occurs when training and test joint distributions are different, i.e. $P_{tr}(x, y) \neq P_{te}(x, y)$, is pervasive in almost all empirical studies, including Machine Learning, Statistics, Social Sciences, Economics, Bioinformatics, Biostatistics, Epidemiology, Medicine, etc. It is therefore highly desirable to devise algorithms that remain effective under such distribution shifts. In general, the estimation problem with two different distributions $P_{tr}(x, y)$ and $P_{te}(x, y)$ is unsolvable, as the two terms could be arbitrarily far apart. However, when $P_{tr}(x, y)$ and $P_{te}(x, y)$ differ only in $P_{tr}(x)$ and $P_{te}(x)$ (known as *covariate shift*) or only in $P_{tr}(y)$ and $P_{te}(y)$ (known as *prior probability shift*), effective adaptation is possible. In this paper, we present a practical correction method for classifier and regression model learning under a more general class of selection bias. We assume implicitly that there exists a joint probability distribution $P(x, y, s)$ that satisfies: $P_{te}(x, y) = P(x, y) = \sum_{s} P(x, y, s)$ and $P_{tr}(x, y) = P(x, y|s = 1)$, where the variable S controls the selection of examples in the training set (1 means the example is selected, 0 means the example is not selected). While $P_{te}(x, y)$

© Springer International Publishing Switzerland 2015
S. Arik et al. (Eds.): ICONIP 2015, Part III, LNCS 9491, pp. 111–118, 2015.
DOI: 10.1007/978-3-319-26555-1_13

and $P_{tr}(x,y)$ are derived from the same distribution $P(x,y,s)$, we assume no independence assumption holds between X, Y, and S. This is termed *complete selection bias* in the literature. In this case, we have to resort to some additional information on the mechanism by which the samples were preferentially selected to the data set to correct the bias.

The recent paper by Bareinboin et al. [1] has been very influential in our thinking. Mirroring their work, we show that, if we have a combination of biased data and unbiased data and qualitative probabilistic assumptions that are deemed plausible about our sampling mechanism, our problem becomes solvable. More specifically, we assume we have access to a S-control feature vector, X_s, and some additional sample of the form (x_s) that is drawn from the population as a whole, such that S is conditionally independent of (X,Y) given X_s. Despite being limited to specific or idealized situations, this framework includes covariate shift and prior probability shift as special cases. We also consider the case where X_s is not fully measured in the target population. This situation typically arises in various clinical studies or epidemiological scenarios, where some variables are too difficult or costly to measure in the target population.

We show that one may account for the difference between $P_{tr}(x,y)$ and $P_{te}(x,y)$ by reweighting the training points using the so-called importance weight, denoted as $\beta(x_s)$. If the selection process is explicitly known, then $\beta(x_s)$ is simply given by $\frac{P(s=1)}{P(s=1|x_s)}$, otherwise, we resort to an external (bias-free) data set of X_s in order to estimate $\beta(x_s)$ directly [3,5]. Furthermore, as one usually has a partial understanding of the sampling mechanism, we investigate whether covariate shift and prior probability shift corrections may help reduce complete selection bias despite not being valid.

2   Bias Correction

In this section, we investigate the interplay between two types of variables, V_B and V_P, where V_B are variables collected under selection bias, $P(V_B|S=1)$, and V_P are variables collected in the population-level, $P(V_P)$[1]. We assume that $Y \in V_B$. In [1], Bareinbom et al. provide a sufficient condition for $P(x,y)$ to be recoverable when no data is gathered over X and Y in the population level. We extend slightly their result by considering also the case where either Y or some input variables in X are not only collected under selection bias, but also in the population-level, (i.e. $\{X,Y\} \cap V_P \cap V_B \neq \emptyset$),

Theorem 1. *The bias-free distribution $P(x,y)$ is recoverable from a S-bias training samples if there exists a set of variables $X_s \subseteq V_B \cap V_P$, such that*

---

[1] Upper-case letters in italics denote random variables (e.g., X, Y) and lower-case letters in italics denote their values (e.g., x, y). X denotes the input variables and Y the target.

$S \perp\!\!\!\perp (X,Y)|X_s$ *and the support of* $P(x_s|s=1)$ *contains the support of* $P(x_s)$.
Let $\beta(x_s) = \frac{P(s=1)}{P(s=1|x_s)}$, $P(x,y)$ *is then given by the formula,*

$$P(x,y) = \sum_{x_s \setminus \{x,y\}} P(x,y,x_s|s=1)\beta(x_s) \tag{1}$$

$\beta(x_s)$ can be reformulated as $\beta(x_s) = \frac{P(x_s)}{P(x_s|s=1)}$. So $\beta(x_s)$ may be estimated from a combination of biased and unbiased data. Theorem 1 relies on qualitative assumptions (X_s controls S over (X,Y)) that may appear difficult to satisfy in practice. However, in certain domains like epidemiology, information about the selection process can sometimes be expressed and modeled in a communicable scientific language (e.g., graphs or structural equations) by the domain experts. The selection bias mechanisms depicted in Fig. 1 are common examples[2]. in epidemiology [2]. The directed acyclic graphs should be regarded as graphical structures encoding conditional independence between X, Y, and S which may involve other variables as as well variables, like M, that is not observed in the target domain, and thus that is not included as input variable to the model.

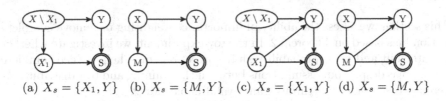

(a) $X_s = \{X_1, Y\}$ (b) $X_s = \{M, Y\}$ (c) $X_s = \{X_1, Y\}$ (d) $X_s = \{M, Y\}$

Fig. 1. Examples of complete selection bias mechanisms depicted graphically. The S-control vector is shown along each plot. X_1 is some input variable to the model, M denotes a variables that is not observed in the target domain, and thus not used a input variable.

Theorem 2. *Given that condition of Theorem 1 is satisfied, if $\hat{P}$ is a distribution such that:* $\hat{P}(x,y,x_s,s) = P(x,y,x_s,s)\beta(x_s)$ *then* $\hat{P}(x,y|s=1) \equiv P(x,y)$.

Proof.

$$\hat{P}(x,y,x_s,s=1) = P(x,y,x_s,s=1)\beta(x_s) = P(x,y,x_s,s=1)\frac{P(x_s)}{P(x_s|s=1)}$$

$$= P(s=1)P(x_s|s=1)P(x,y|x_s,s=1)\frac{P(x_s)}{P(x_s|s=1)} = P(s=1)P(x,y,x_s)$$

Thus, $\hat{P}(x,y,x_s,s=1) = P(x,y,x_s)P(s=1)$. If we sum this expression over x,y,x_s we obtain $\hat{P}(s=1) = P(s=1)$. Therefore,

$$\hat{P}(x,y,x_s|s=1) = \frac{\hat{P}(x,y,x_s,s=1)}{\hat{P}(s=1)} = \frac{P(x,y,x_s)P(s=1)}{P(s=1)} = P(x,y,x_s)$$

---

[2] We assume the reader is familiar with the concepts of d-separation [6].

Finally, $\hat{P}(x, y | s = 1) = \sum_{x_s \setminus \{x, y\}} \hat{P}(x, y, x_s | s = 1) = P(x, y)$. □

Theorem 2 states that an unbiased training sample can be obtained by weighting each training example by $\beta(x_s) = \frac{P(x_s)}{P(x_s | s = 1)}$. Note however that the support of $P(x_s)$ should be contained in the support of $P(x_s | s = 1)$ for the $\beta(x_s)$ to be always defined. A similar technique applied to covariate shift was discussed in [9]. The unbiased expected loss of the model follows:

Corollary 1. *Given that condition of Theorem 1 is satisfied, and $\hat{P}$ in Theorem 2, for all classifier h, all loss function $l = l(h(x), y)$,*

$$E_{x, y \sim P}(l) = E_{x, y \sim \hat{P}}(l | s = 1)$$

$E_{x, y \sim P}(l)$ is the loss that we would like to minimize and $E_{x, y \sim \hat{P}}(l | s = 1))$ is the loss that may be estimated from the new biased sample drawn from weighted distribution $\hat{P}$.

3   Experiments

In this section, we assess the ability of importance weighting to remove complete selection bias based on Theorem 2. In the toy experiment, we investigate whether covariate shift and prior probability shift corrections may help reduce complete selection bias despite our assumptions between the training and test distributions difference being violated (through an invalid choice for X_s).

When the selection process is explicitly known, $\beta(x_s)$ is simply given by $\frac{P(s=1)}{P(s=1|x_s)}$. Otherwise, we resort to an external (bias-free) data set of X_s in order to estimate $\beta(x_s)$ as $\frac{P(x_s)}{P(x_s | s = 1)}$. In this study, we use the Kernel Mean Matching (KMM) [3, 8] estimator for $\beta(x_s)$ denoted as KMM(X_s). As one usually has a partial understanding of the sampling mechanism, we investigate whether covariate shift (i.e., $\beta(x) = \frac{P(x)}{P(x | s = 1)}$) and prior probability shift (i.e., $\beta(y) = \frac{P(y)}{P(y | s = 1)}$), corrections may help reduce complete selection bias despite not being valid. These strategies are denoted as KMM(X), KMM(Y). We first apply our method to a simple toy problem and then compare KMM to another estimator called the Unconstrained Least-Square Importance Fitting (uLSIF) [5] on a variety of regression and classification benchmarks from the UCI Archive.

3.1   Toy Problem

Consider the S-bias mechanism displayed in Fig. 1b, where the feature X has a uniform distribution in $[0, 1]$: $P(X) \sim \mathcal{U}(0, 1)$. Note that the influence of M on Y is mediated by $\{X, S\}$. The observations are generated according to $y = 1 - 0.5x$ and are observed in Gaussian noise with standard deviation 0.5 (see Fig. 2c; the black solid line is the noise-free signal). The intermediate variable M, between X and S, is generated according to $M = X + \mathcal{N}(0, 0.3^2)$. As M is only measured in the training set, it is not used as an input variable in our regression model.

Therefore, we investigate a case where X_s is partially missing in the test set. The probability of a given example being included in the training set depends on Y and M and is given by

$$P(S = 1|m, y) \sim \begin{cases} y - m, & \text{if } 0.1 \leq (y - m) \leq 1 \\ 0.1, & \text{if } (y - m) \leq 0.1 \\ 1, & \text{otherwise} \end{cases}$$

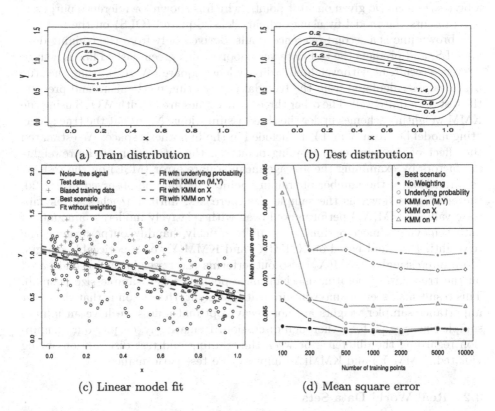

(a) Train distribution   (b) Test distribution

(c) Linear model fit   (d) Mean square error

Fig. 2. Toy regression problem 1. (a) and (b) Contour plots X-Y on training and test sets; (c) Polynomial models of degree 1 fit with OLS and WOLS; (d) Average performances of four WOLS methods and OLS on the test data as a function of the number of training points.

Note that the minimum value of $P(S = 1|m, y)$ needs to be greater than 0 so that the support of $P(m, y)$ is contained in the support of $P(m, y|s = 1)$, as required by Theorem 1. The choice of $P(m, y)$ is intended to induce a noticeable discrepancy between $P(y|x, s = 1)$ and $P(y|x)$. We sampled 200 training (red crosses in Fig. 2c) and testing (grey circles) points from P_{tr} and P_{te} respectively. The bias is clearly noticeable from the X-Y contour plots in Fig. 2a and b. The bias-free distribution $P(x, y)$ is recoverable from the S-bias training samples since

$\{M, Y\}$ satisfies Theorem 1. Thus we use Theorem 2, to remove selection bias by weighting each example by the importance ratio:

$$\beta(x_s) = \beta(m, y) = \frac{P(m, y)}{P(m, y|s = 1)} = \frac{P(s = 1|m, y)}{P(s = 1)}$$

where $P(s = 1|m, y)$ and $P(s = 1)$ may be obtained from the known selection mechanism shown above or directly estimated by KMM. We attempted to model the observations with a degree 1 polynomial. The black dashed line in Fig. 2c is a best-case scenario given our test points, which is shown for reference purposes: it represents the model fit using ordinary least squared (OLS) on the test set. The brown line is a second reference result, derived only from the training data via OLS, and predicts the test data very poorly. The green dashed line is a third reference result, fit with weighted ordinary least square (WOLS), using the true $\beta(x_s)$ values calculated from the true data generating mechanism, and predicts the test data quite well. The other three dashed lines are fit with WOLS using the KMM weighting schemes under the three assumptions. Note that the true generating model between X and Y is included in the hypothesis space. We estimated the effect of the number of training points on the estimation of the reweighting factors by examining the average mean square error (MSE) on the test set as a function of the number of training points. As may be observed in Fig. 2d, the error goes down as the sample size increases, until it reaches an asymptotic value. KMM(X_s) performs well even with relatively moderate amounts of data achieving almost optimal error quite quickly, handily outperforming the reweighting method based on KMM(X) and KMM(Y) by a noticeable margin. More interestingly, KMM(X_s) also outperforms the reweighting method based on the true data generating mechanism, especially when sample size is small. This result may seem counter-intuitive at first sight: the reason is that the exact importance-sampler weights are not always optimal unless we have an infinite sample size. See [7] for a thorough discussion. Remarkably, despite our assumption regarding the difference between the training and test distributions being violated, KMM(Y) and KMM(X) improve the test performance.

3.2   Real-World Data Sets

We now examine whether using importance weighting can reduce selection bias in 10 UCI data sets with 5 classification tasks and 5 regression tasks. We employ three methods to estimate importance weighting: ratio of underlying probability, KMM and uLSIF and compare their performance against the baseline unweighted method. For each data set, X_s is chosen to be the label Y and the most correlated input variable to Y (denoted as X_1 for simplicity). The selection bias mechanism is illustrated in Fig. 1c. The selection variable S for each training example is determined according to two scenarios depending on whether it is regression or classification problem. For regression problem, we use $P(s = 1|x_1, y) = exp(ax_1 + by + c)/[1 + exp(ax_1 + by + c)]$, where a, b, c, are parameters that determine the bias. For binary classification problem, we use:

$$P(s = 1|x_1, y) = \begin{cases} 0.5 & \text{if } x_1 > mean(x_1) \text{ and } y = 1 \\ 1 & \text{if } otherwise. \end{cases}$$

For each data set, we then train 4 predictive models learned under the four weighting schemes discussed above and a model learned from the unbiased data (baseline) using SVM-light [4] which allows importance weighting to be fed directly to SVM. All classifiers are trained with the common Radial Basis Function (RBF), with a kernel size σ chosen through a 5-fold cross validation. This procedure is repeated 100 times for each data set.

Table 1. Mean test error averaged over 100 trials of different weighting schemes on UCI data set. Data sets marked with * are for regression problems

Data set	No weighting	KMM	uLSIF	Underlying P	Unbiased model
India diabetes	0.338 ± 0.049	0.266 ± 0.040	0.332 ± 0.053	0.287 ± 0.055	0.258 ± 0.035
Ionosphere	0.069 ± 0.039	0.066 ± 0.039	0.067 ± 0.040	0.067 ± 0.039	0.065 ± 0.036
BreastCancer	0.044 ± 0.016	0.039 ± 0.015	0.043 ± 0.017	0.040 ± 0.016	0.038 ± 0.015
Haberman	0.264 ± 0.069	0.262 ± 0.071	0.263 ± 0.070	0.262 ± 0.071	0.262 ± 0.071
GermanCredit	0.300 ± 0.044	0.298 ± 0.046	0.298 ± 0.045	0.298 ± 0.046	0.295 ± 0.046
Airfoil self noise*	0.534 ± 0.104	0.470 ± 0.122	0.475 ± 0.082	0.445 ± 0.081	0.403 ± 0.059
Abanlone*	0.526 ± 0.048	0.484 ± 0.054	0.521 ± 0.057	0.466 ± 0.041	0.456 ± 0.036
Computer Hardware*	0.326 ± 0.308	0.321 ± 0.304	0.321 ± 0.299	0.319 ± 0.307	0.305 ± 0.201
Auto MGP*	0.268 ± 0.148	0.298 ± 0.192	0.212 ± 0.129	0.203 ± 0.128	0.129 ± 0.063
Boston Housing*	0.323 ± 0.110	0.327 ± 0.127	0.349 ± 0.133	0.332 ± 0.127	0.298 ± 0.112

The results are reported in Table 1. As may be seen, all importance weighting schemes achieve lower prediction error with respect to the baseline unweighted scheme. The underlying probability weighting scheme performs pretty good. Curiously, on the Boston Housing data set, all three weighting schemes perform worse than the baseline unweighted method. It seems therefore that the effectiveness of bias correction based on importance weighting is data dependent. In order to better assess the overall results obtained for each of the 4 weighting schemes, a non-parametric Friedman test was firstly used to evaluate the rejection of the hypothesis that all the models perform equally well (except the unbiased model of course) at significant level 5 %. Statistically significant differences were observed. So we proceeded with the Nemenyi post hoc test. The results along with the average rank diagrams are shown in Fig. 3. The ranks are depicted on the axis, in such a manner that the best ranking algorithms are at the rightmost side of the diagram. The algorithms that do not differ significantly (at $p = 0.05$) are connected with a line. As may be observed in Fig. 3, contrary to uLSIF, KMM is significantly better than no weighting.

4 Discussion and Conclusion

The aim of this paper was to elaborate on the idea of exploiting the assumptions that are deemed plausible about the sampling mechanism to correct or reduce

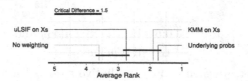

Fig. 3. Post hoc analysis

selection bias in machine learning tasks. The method hinges on the existence of a S-control feature vector, X_s, and an additional (biased-free) sample that allows us to estimate the distribution of X_s. We showed experimentally that direct weighting estimation is able to achieve significant improvements in accuracy over the unweighted method, even in situations where our key assumption is not valid (assuming covariate shift and prior probability shift instead of complete selection bias). However the gain in accuracy is data dependent. In fact, all conclusions are extremely sensitive to which variables we choose for X_s. As the choice of X_s usually reflects the investigator's subjective and qualitative knowledge of statistical influences in the domain, the data analyst must weight the benefit of reducing selection bias against the risk of introducing new bias carried by unmeasured covariates even where none existed before. Nevertheless, we hope this study will convince others about the importance of selection bias correction methods in practical studies and suggest relevant tools which can be used to achieve that goal.

References

1. Bareinboim, E., Tian, J., Pearl, J.: Recovering from selection bias in causal and statistical inference. In: Brodley, C.E., Stone, P. (eds.) AAAI (2014)
2. Hernán, M.A., Hernández-Díaz, S., Robins, J.M.: A structural approach to selection bias. Epidemiol. **15**(5), 615–625 (2004)
3. Huang, J., Smola, A.J., Gretton, A., Borgwardt, K.M., Schölkopf, B.: Correcting sample selection bias by unlabeled data. In: Schölkopf, B., Platt, J.C., Hoffman, T. (eds.) NIPS, pp. 601–608. MIT Press (2006)
4. Joachims, T.: Making large-scale svm learning practical. LS8-Report 24, Universität Dortmund, LS VIII-Report (1998)
5. Kanamori, T., Hido, S., Sugiyama, M.: A least-squares approach to direct importance estimation. J. Mach. Learn. Res. **10**, 1391–1445 (2009)
6. Pearl, J.: Probabilistic Reasoning in Intelligent Systems - Networks of Plausible Inference. Morgan Kaufmann, Morgan Kaufmann series in representation and reasoning (1989)
7. Shimodaira, H.: Improving predictive inference under covariate shift by weighting the log-likelihood function. J. Stat. Plann. Infer. **90**(2), 227–244 (2000)
8. Yu, Y., Szepesvári, C.: Analysis of kernel mean matching under covariate shift. In: Langford, J., Pineau, J. (eds.) ICML (2012)
9. Zadrozny, B.: Learning and evaluating classifiers under sample selection bias. In: Brodley, C.E. (ed.) ICML, pp. 114–121 (2004)

Lagrange Programming Neural Network for the l_1-norm Constrained Quadratic Minimization

Ching Man Lee, Ruibin Feng, and Chi-Sing Leung$^{(\boxtimes)}$

Department of Electronic Engineering, City University of Hong Kong,
Kowloon Tong, Hong Kong
rfeng4-c@my.cityu.edu.hk, eeleungc@cityu.edu.hk

Abstract. The Lagrange programming neural network (LPNN) is a framework for solving constrained nonlinear programm problems. But it can solve differentiable objective/contraint functions only. As the l_1-norm constrained quadratic minimization (L1CQM), one of the sparse approximation problems, contains the nondifferentiable constraint, the LPNN cannot be used for solving L1CQM. This paper formulates a new LPNN model, based on introducing hidden states, for solving the L1CQM problem. Besides, we discuss the stability properties of the new LPNN model. Simulation shows that the performance of the LPNN is similar to that of the conventional numerical method.

Keywords: Stability · Sparse approximation · LPNN

1 Introduction

Throughout the years, there are many studies on solving constrained nonlinear programming problems [1–3] based on neural circuit approach. When the real-time solutions are needed, the analog neural circuit is preferred. Although many neural models [4,5] were introduced, they are designed for a particular kind of problems only. The Lagrange programming neural network (LPNN) [3,6,7] is a framework for solving constrained nonlinear programming problems. But it can solve differentiable objective/contraint functions only.

In signal processing, one of the important topics is sparse approximation [8,9]. Sparse approximation aims at recovering an unknown sparse signal from the measurements. Sparse approximation has many potential applications, such as channel estimation in MIMO wireless communication channels [10] and image denoising [11]. However, in sparse approximation, many problems involve nondifferentiable objective or constraints. By introducing the concept of soft threshold function, the local competition algorithm (LCA) [12], being an analog method, is able to solve the basis pursit denoising problem [9]. which is a unconstrained nonlinear programming problem. However, the LCA is not designed to handle constrained optimization problems.

© Springer International Publishing Switzerland 2015
S. Arik et al. (Eds.): ICONIP 2015, Part III, LNCS 9491, pp. 119–126, 2015.
DOI: 10.1007/978-3-319-26555-1_14

This paper focuses on l_1-norm constrained quadratic minimization (L1CQM) for sparse approximation. Section 2 reviews the basic concept of LPNN, sparse approximation, and LCA. Section 3 presents the proposed LPNN model for solving the L1CQM. Section 4 presents some properties of the LPNN. Simulation results are then presented in Sect. 5.

2   Background

LPNN: The LPNN approach considers a general constrained nonlinear programming problem:

$$\text{EP: } \min f(\boldsymbol{x}) \text{ s.t. } \boldsymbol{h}(\boldsymbol{x}) = \boldsymbol{0}, \tag{1}$$

where $\boldsymbol{x} \in \Re^n$ is the state vector, $f : \Re^n \to \Re$ is the objective function, and $\boldsymbol{h} : \Re^n \to \Re^m$ $(m < n)$ describes the m equality constraints. The objective function f and constraints $\boldsymbol{h}$ are assumed to be twice differentiable. It should be noticed that the LPNN approach can solve inequality constraints by introducing dummy variables.

In LPNN, a Lagrangian function is set up, given by

$$\mathcal{L}_{ep} = f(\boldsymbol{x}) + \boldsymbol{\lambda}^{\mathrm{T}} \boldsymbol{h}(\boldsymbol{x}), \tag{2}$$

where $\boldsymbol{\lambda} = [\lambda_1, \cdots, \lambda_m]^{\mathrm{T}}$ is the Lagrange multiplier vector. There are two kinds of neurons: variable neurons and Lagrange neurons. The variable neurons hold the variable vector $\boldsymbol{x}$, while the Lagrange neurons hold the multiplier vector $\boldsymbol{\lambda}$. The LPNN dynamics are given by

$$\frac{1}{\epsilon}\frac{d\boldsymbol{x}}{dt} = -\frac{\partial \mathcal{L}_{ep}}{\partial \boldsymbol{x}}, \text{ and } \frac{1}{\epsilon}\frac{d\boldsymbol{\lambda}}{dt} = \frac{\partial \mathcal{L}_{ep}}{\partial \boldsymbol{\lambda}}, \tag{3}$$

where ϵ is the time constant of the circuit. In this paper, ϵ is considered as equal to 1 in regard to generality.

Sparse Approximation: In sparse approximation, we would like to estimate a sparse solution $\boldsymbol{x} \in \Re^n$ from the measurement $\boldsymbol{b} = \boldsymbol{\Phi}\boldsymbol{x} + \boldsymbol{\xi}$, where $\boldsymbol{b} \in \Re^m$ is the observation vector, $\boldsymbol{\Phi} \in \Re^{m \times n}$ is the measurement matrix with a rank of m, $\boldsymbol{x} \in \Re^n$ is the unknown sparse vector $(m < n)$, and ξ_i 's are the measurement noise. The recovery can be formulated as the following programming problem:

$$\min |\boldsymbol{b} - \boldsymbol{\Phi}\boldsymbol{x}|_2^2, \text{ s.t. } |\boldsymbol{x}|_1 \leq \psi, \tag{4}$$

where $\psi > 0$. This problem is called as l_1-norm Constrained Quadratic Minimization (L1CQM). In this problem, we would like to minimize the residue subject to the sum of absolute of signal elements is less than a value. Since the constraint function $|\boldsymbol{x}|_1 - \psi$ is nondifferentiable, the conventional LPNN is unable to solve L1CQM directly.

Subdifferential: Subdifferential was developed to handle nondifferentiable functions. Their definitions are stated in the following.

Definition 1. *Given a convex function f, the subgradient ρ of f at $\boldsymbol{x}$ is*

$$f(\boldsymbol{y}) \geq f(\boldsymbol{x}) + \rho^{\mathrm{T}}(\boldsymbol{y} - \boldsymbol{x}), \forall \boldsymbol{y}. \tag{5}$$

Definition 2. *The subdifferential $\partial f(\boldsymbol{x})$ at $\boldsymbol{x}$ is the set of all subgradients:*

$$\partial f(\boldsymbol{x}) = \left\{ \rho \,|\, \rho^{\mathrm{T}}(\boldsymbol{y} - \boldsymbol{x}) \leq f(\boldsymbol{y}) - f(\boldsymbol{x}), \forall \boldsymbol{y} \right\}. \tag{6}$$

Note that when $f(\cdot)$ is differentiable at $\boldsymbol{x}_o$, its subdifferential at $\boldsymbol{x}_o$ is equal to the conventional partial derivative. Let us use the absolute function $f(x) = |x|$ as an example to example the subdifferential:

$$\partial |x| = \begin{cases} [-1, 1] & x = 0, \\ \mathrm{sign}(x) & x \neq 0. \end{cases} \tag{7}$$

Concept of LCA: The LCA is designed to handle the following unconstrained optimization problems:

$$\mathcal{L} = \frac{1}{2} |\boldsymbol{\Phi}\boldsymbol{x} - \boldsymbol{b}|_2^2 + \lambda |\boldsymbol{x}|_1, \tag{8}$$

where λ is a trade-off parameter. In the LCA, there are n neurons. The neuron outputs are denoted as $\boldsymbol{x}$ and their internal states are denoted as $\boldsymbol{u}$. By a threshold function, the mapping from $\boldsymbol{u}$ to $\boldsymbol{x}$ is stated as below

$$x_i = T_\lambda(u_i) = \begin{cases} 0, & \text{for } |u_i| \leq \lambda, \\ u_i - \lambda \mathrm{sign}(u_i), & \text{for } |u_i| > \lambda. \end{cases} \tag{9}$$

The forward mapping T_λ from u_i to x_i is one-to-one for $|u_i| > \lambda$ and is many-to-one for $|u_i| \leq \kappa$. The inverse mapping T_λ^{-1} from x_i to u_i is one-to-one for $x_i \neq 0$, while it is one-to-many for $x_i = 0$. It implies that $T_\lambda^{-1}(0)$ is equal to a set: $[-\lambda, \lambda]$. The LCA defines the dynamics on $\boldsymbol{u}$ rather than $\boldsymbol{x}$, given by

$$\frac{d\boldsymbol{u}}{dt} = -\partial_{\boldsymbol{x}} \mathcal{L}_{lca} = -\lambda \partial |\boldsymbol{x}|_1 + \boldsymbol{\Phi}^{\mathrm{T}}(\boldsymbol{b} - \boldsymbol{\Phi}\boldsymbol{x}). \tag{10}$$

With the property of $T_\lambda(\cdot)$ and the definition subdifferential [12], the LCA replaces "$\kappa \partial |\boldsymbol{x}|_1$" with "$\boldsymbol{u} - \boldsymbol{x}$". The dynamics become

$$\frac{d\boldsymbol{u}}{dt} = -\boldsymbol{u} + \boldsymbol{x} + \boldsymbol{\Phi}^{\mathrm{T}}(\boldsymbol{b} - \boldsymbol{\Phi}\boldsymbol{x}). \tag{11}$$

If we do not introduce the internal state vector $\boldsymbol{u}$, $\partial |\boldsymbol{x}|_1$ (may be equal to a set) is not implementable.

3 LPNN for L1CQM

Our aim is to solve the programming problem:

$$\min |\boldsymbol{b} - \boldsymbol{\Phi}\boldsymbol{x}|_2^2, \text{ s.t. } |\boldsymbol{x}|_1 \leq \psi. \tag{12}$$

From the convex optimization theory, one can obtain the following theorem.

Theorem 1. *For the programming problem (12), $\boldsymbol{x}^\star$ is an optimal solution, iff, there exists a $\lambda^\star$ (Lagrange multiplier) and*

$$0 \in -2\boldsymbol{\Phi}^{\mathrm{T}}(\boldsymbol{b} - \boldsymbol{\Phi}\boldsymbol{x}^\star) + \lambda^\star(\partial|\boldsymbol{x}^\star|_1), \tag{13a}$$
$$|\boldsymbol{x}^\star|_1 - \psi \leq 0, \tag{13b}$$
$$\lambda^\star \geq 0, \tag{13c}$$
$$\lambda^\star(|\boldsymbol{x}^\star|_1 - \psi) = 0. \tag{13d}$$

where $\lambda^\star$ is the optimal dual variable (Lagrange multiplier).

Since the problem (12) has an inequality constraint, we cannot directly use the LPNN. However, if ψ^2 is less than a certain value, given by $\psi^2 < \frac{|\boldsymbol{\Phi}^{\mathrm{T}}\boldsymbol{b}|_2^2}{|\boldsymbol{\Phi}^{\mathrm{T}}\boldsymbol{\Phi}|_2^2}$, then the inequality constraint becomes an equality ones. Therefore, Theorem 1 becomes the following theorem.

Theorem 2. *If $\psi^2 < \frac{|\boldsymbol{\Phi}^{\mathrm{T}}\boldsymbol{b}|_2^2}{|\boldsymbol{\Phi}^{\mathrm{T}}\boldsymbol{\Phi}|_2^2}$, then the optimization problem (13) becomes*

$$\min |\boldsymbol{b} - \boldsymbol{\Phi}\boldsymbol{x}|_2^2, \ s.t. \ |\boldsymbol{x}|_1 = \psi. \tag{14}$$

And $\boldsymbol{x}^\star$ is the optimal solution, iff, there exists a $\lambda^\star$ (Lagrange multiplier) and

$$0 \in -2\boldsymbol{\Phi}^{\mathrm{T}}(\boldsymbol{b} - \boldsymbol{\Phi}\boldsymbol{x}^\star) + \lambda^\star(\partial|\boldsymbol{x}^\star|_1), \tag{15a}$$
$$|\boldsymbol{x}^\star|_1 - \psi = 0, \tag{15b}$$
$$\lambda^\star > 0. \tag{15c}$$

Note that (15) summarizes the KKT conditions (necessary and sufficient).

Proof: Suppose $(\boldsymbol{x}^\star, \lambda^\star)$ is an optimal solution of Theorem 1. That means, $\lambda^\star \geq 0$. Firstly, we will use contradiction to prove that $\lambda^\star$ cannot be equal to zero when $\psi^2 < \frac{|\boldsymbol{\Phi}^{\mathrm{T}}\boldsymbol{b}|_2^2}{|\boldsymbol{\Phi}^{\mathrm{T}}\boldsymbol{\Phi}|_2^2}$.

Since $(\boldsymbol{x}^\star, \lambda^\star)$ are optimal, they satisfy the KKT conditions (14) of Theorem 1. If $\lambda^\star = 0$, then from (13a) we have

$$\boldsymbol{\Phi}^{\mathrm{T}}\boldsymbol{\Phi}\boldsymbol{x}^\star = \boldsymbol{\Phi}^{\mathrm{T}}\boldsymbol{b} \Rightarrow |\boldsymbol{\Phi}^{\mathrm{T}}\boldsymbol{\Phi}|_2|\boldsymbol{x}^\star|_2 \geq |\boldsymbol{\Phi}^{\mathrm{T}}\boldsymbol{b}|_2 \Rightarrow |\boldsymbol{x}^\star|_2^2 \geq \frac{|\boldsymbol{\Phi}^{\mathrm{T}}\boldsymbol{b}|_2^2}{|\boldsymbol{\Phi}^{\mathrm{T}}\boldsymbol{\Phi}|_2^2}. \tag{16}$$

From (13b),

$$\psi^2 \geq |\boldsymbol{x}^\star|_1^2 = (\sum_{i=1}^{n} |x_i^\star|)^2 = \sum_{i=1}^{n} x_i^{\star 2} + \sum_{i=1}^{n}\sum_{i \neq j}^{n} |x_i^\star||x_j^\star| \geq \sum_{i=1}^{n} x_i^{\star 2} \geq |\boldsymbol{x}^\star|_2^2 \tag{17}$$

From (16) and (17),

$$\psi^2 \geq |\boldsymbol{x}^\star|_2^2 \Rightarrow \psi^2 \geq \frac{|\boldsymbol{\Phi}^{\mathrm{T}}\boldsymbol{b}|_2^2}{|\boldsymbol{\Phi}^{\mathrm{T}}\boldsymbol{\Phi}|_2^2}. \tag{18}$$

The above contradicts the assumption of $\psi^2 < \frac{|\boldsymbol{\Phi}^{\mathrm{T}}\boldsymbol{b}|_2^2}{|\boldsymbol{\Phi}^{\mathrm{T}}\boldsymbol{\Phi}|_2^2}$. As a result, it proves that $\lambda^* > 0$. This means, the KKT conditions of Theorem 1 can be rewritten as (15). Besides, the inequality in (12) can be removed. Thus, the optimization problem can be written as (14). The proof is complete. ∎

With Theorem 2, we can define the Lagrangian function:

$$\mathcal{L}_n = |\boldsymbol{b} - \boldsymbol{\Phi}\boldsymbol{x}|_2^2 + \lambda(|\boldsymbol{x}|_1 - \psi). \tag{19}$$

Since the traditional LPNN model cannot handle nondifferentiable constraints, direct implementing the neuron dynamics is impossible. Using the concept of the LCA, we introduce the hidden state vector $\boldsymbol{u}$ as the internal state vector and $\boldsymbol{x}$ as the corresponding neuron outputs. Besides, from the property of $T_\kappa(\cdot)$ and the definition subdifferential [12], the dynamics of $\boldsymbol{u}$ and λ is given by

$$\frac{d\boldsymbol{u}}{dt} = 2\boldsymbol{\Phi}^{\mathrm{T}}(\boldsymbol{b} - \boldsymbol{\Phi}\boldsymbol{x}) - \lambda(\boldsymbol{u} - \boldsymbol{x}) \tag{20a}$$

$$\frac{d\lambda}{dt} = |\boldsymbol{x}|_1 - \psi. \tag{20b}$$

The role of (20a) is used to minimize the objective value, while the role of (20b) is used to constraint $\boldsymbol{x}$ in the feasible region.

We can use Fig. 1 to illustrate the idea of LPNN for L1CQM. In Fig. 1(a), there is a 1D sparse signal. The length of this signal is 128. There are five nonzero elements. The number of measurement is 30. The measurement matrix is an ± 1 random matrix. In Fig. 1(b), we show the recovery from the pseudoinverse. Clearly it is not our expected signal. In Fig. 1(c), we show the recovery signal

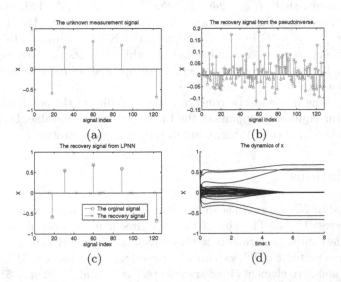

(a)   (b)

(c)   (d)

Fig. 1. Illustration example for the LPNN. (a) The original signal. (b) Recovery from pseudoinverse. (c) Recovery from LPNN. (d) The dynamics of LPNN.

from the LPNN, which is pretty close to the original signal. Figure 1(d) shows the dynamics of x. It can be seen that the LPNN can settle down in five characteristic time.

4   Properties of LPNN

This section discusses the properties of the proposed LPNN for L1CQM. Firstly, we will show that the equilibrium point of the LPNN approach is the global minimum point of the L1CQM.

Theorem 3. *Let $(u^\star, \lambda^\star)$, with $\lambda^\star > 0$ and $u^\star \neq 0$, be an equilibrium point of the LPNN dynamics (Eq. (20a)) and $x^\star$ be the corresponding output vector. The KKT conditions in Theorem 2 are satisfied at this equilibrium point. And the corresponding output vector $x^\star$ is the optimal solution of the problem.*

Proof. In the proof, we will prove that $(u^\star, \lambda^\star)$, with $\lambda^\star > 0$ and $u^\star \neq 0$ satisfies the KKT conditions in Theorem 2. At the equilibrium point, from (20b), we have

$$2\Phi^{\mathrm{T}}(b - \Phi x^\star) - \lambda^\star(u^\star - x^\star) = 0, \tag{21}$$

$$|x^\star|_1 - \psi = 0. \tag{22}$$

For (21) and by $\partial|x|_1 = u - x$,

$$2\Phi^{\mathrm{T}}(b - \Phi x^\star) - \lambda^\star(u^\star - x^\star) = 0 \Rightarrow -2\Phi^{\mathrm{T}}(b - \Phi x^\star) + \lambda^\star(\partial|x^\star|_1) = 0.$$

The above means that "$0 \in -2\Phi^{\mathrm{T}}(b - \Phi x^\star) + \lambda^\star(\partial|x^\star|_1)$". This satisfies the KKT condition (15a). Based on the assumption that $\lambda^\star > 0$, (15c) is satisfied. Also for "$|x^\star|_1 - \psi = 0$" in (22), it satisfies the KKT condition (15b). As the KKT conditions are necessary and sufficient, which implies that any equilibrium point $(u^\star, \lambda^\star)$, with $u^\star \neq 0$ and $\lambda^\star > 0$, is an optimal solution to the problem. The proof is complete.   ∎

Another thing needed to be concern is the stability of the equilibrium point. Otherwise, the equilibrium points of the LPNN are not achievable. Based on the approach in [3,6], the equilibrium point of (20) is an asymptotically stable point.

5   Simulations

The proposed LPNN approach is undergoing some experiments by using the standard configures [13,14]. The aim of the experiment is to verify if the proposed LPNN has the similar performance to the conventional numerical method LASSO. For the sparse vector $x \in \Re^n$, we consider two signal lengths: $n = 512, 4096$. The numbers of non-zero elements in x are selected as 15 and 25 for $n = 512$, and 75 and 125 for $n = 4096$, respectively. For the non-zero elements, they are uniformly distributed random numbers, either in between -5 to -1 or in between 1 to 5.

In the tests, the measurement matrix is an ± 1 random matrix, which is then normalized with the signal length. In the experiments, we vary the number m of measurement signals. For each setting, we repeat the experiments with 100 times using different measurement matrices and sparse signals. The variances of Gaussian noise introduced in the measured signal are $\sigma^2 = \{0.05^2, 0.025^2, 0.005^2\}$.

In order to compare the proposed LPNN approach with digital method, the LASSO algorithm from SPGL1 [13,14] is applied for recovering the sparse signals. During comparison, the mean square error (MSE) values of recovery signals are recorded. The results are shown in Figs. 2 and 3. From the figures, the performance of the LPNN is similar to that of the LASSO algorithm. In addition, as shown in Fig. 2, for $n = 512$ with 15 non-zero data points, around 80 measurements are required to recover the sparse signal. For 25 non-zero data points, around 120 measurements are required. As shown in Fig. 3, for $n = 4096$ with 75 non-zero data points, around 500 measurements are required to recover the sparse signal. For 125 non-zero data points, around 700 measurements are required.

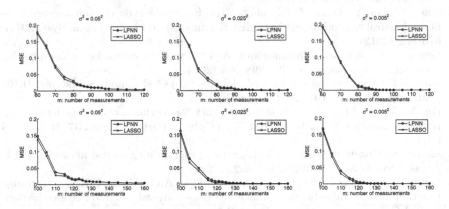

Fig. 2. The MSE of the recovery signals from noisy measurement for signal length $n = 512$. The signals contain 15 and 15 non-zero data points in first and second row.

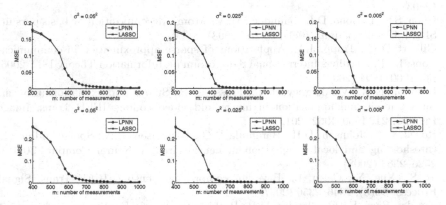

Fig. 3. The MSE of the recovery signals from noisy measurement for signal length $n = 4096$. The signals contain 75 and 125 non-zero data points in first and second row.

6   Conclusion

This paper proposed a new LPNN model for solving the L1CQM problem. In the theoretical side, we proved that the equilibrium points of the LPNN is the optimal solution. Besides, stimulations are carried out to verify the effectiveness of the LPNN. There are some possible extensions of our works. From the simulation, the network always converges. However, we do not theoretically show that the LPNN is global stable. Hence it is interesting to theoretically study the global stability.

Acknowledgement. The work was supported by a research grant from City University of Hong Kong (Project No.: 7004233).

References

1. Chua, L.O., Lin, G.N.: Nonlinear programming without computation. IEEE Trans. Circuits Syst. **31**, 182–188 (1984)
2. Xiao, Y., Liu, Y., Leung, C.S., Sum, J., Ho, K.: Analysis on the convergence time of dual neural network-based kwta. IEEE Trans. Neural Netw. Learn. Syst. **23**(4), 676–682 (2012)
3. Leung, C.S., Sum, J., Constantinides, A.G.: Recurrent networks for compressive sampling. Neurocomputing **129**, 298–305 (2014)
4. Gao, X.B.: Exponential stability of globally projected dynamics systems. IEEE Trans. Neural Netw. **14**, 426–431 (2003)
5. Hu, X., Wang, J.: A recurrent neural network for solving a class of general variational inequalities. IEEE Trans. Syst. Man Cybern. Part B Cybern. **37**(3), 528–539 (2007)
6. Zhang, S., Constantinidies, A.G.: Lagrange programming neural networks. IEEE Trans. Circuits Syst. II **39**(7), 441–452 (1992)
7. Leung, C.S., Sum, J., So, H.C., Constantinides, A.G., Chan, F.K.W.: Lagrange programming neural network approach for time-of-arrival based source localization. Neural Comput. Appl. **12**, 109–116 (2014)
8. Donoho, D.L., Elad, M.: Optimally sparse representation in general (nonorthogonal) dictionaries via l_1 minimization. Proc. Nat. Acad. Sci. **100**(5), 2197–2202 (2003)
9. Chen, S.S., Donoho, D.L., Saunders, M.A.: Atomic decomposition by basis pursuit. SIAM J. Sci. Comput. **20**(1), 33–61 (1998)
10. Gilbert, A.C., Tropp, J.A.: Applications of sparse approximation in communications. In: Proceedings International Symposium on Information Theory ISIT 2005, pp. 1000–1004 (2005)
11. Rahmoune, A., Vandergheynst, P., Frossard, P.: Sparse approximation using m-term pursuit and application in image and video coding. IEEE Trans. Image Process. **21**, 1950–1962 (2012)
12. Rozell, C.J., Johnson, D.H., Baraniuk, R.G., Olshausen, B.A.: Sparse coding via thresholding and local competition in neural circuits. Neural Comput. **20**(10), 2526–2563 (2008)
13. Ji, S., Xue, Y., Carin, L.: Bayesian compressive sensing. IEEE Trans. Signal Process. **56**(6), 2346–2356 (2007)
14. van den Berg, E., Friedlander, M.P.: Probing the Pareto frontier for basis pursuit solutions. SIAM J. Sci. Comput. **31**(2), 890–912 (2008)

Multi-Island Competitive Cooperative Coevolution for Real Parameter Global Optimization

Kavitesh K. Bali[1,2](✉) and Rohitash Chandra[1,2]

[1] School of Computing Information and Mathematical Sciences,
University of South Pacific, Suva, Fiji
[2] Artificial Intelligence and Cybernetics Research Group,
Software Foundation, Nausori, Fiji
{bali.kavitesh,c.rohitash}@gmail.com

Abstract. Problem decomposition is an important attribute of cooperative coevolution that depends on the nature of the problems in terms of separability which is defined by the level of interaction amongst decision variables. Recent work in cooperative coevolution featured competition and collaboration of problem decomposition methods that was implemented as islands in a method known as competitive island cooperative coevolution (CICC). In this paper, a multi-island competitive cooperative coevolution algorithm (MICCC) is proposed in which several different problem decomposition strategies are given a chance to compete, collaborate and motivate other islands while converging to a common solution. The performance of MICCC is evaluated on eight different benchmark functions and are compared with CICC where only two islands were utilized. The results from the experimental analysis show that competition and collaboration of several different island can yield solutions with a quality better than the two-island competition algorithm (CICC) on most complex multi-modal problems.

1 Introduction

Coevolutionary algorithms have gained popularity as a vital extension to the traditional evolutionary algorithms [1]. Cooperative coevolution (CC) is one such evolutionary computation method which solves a problem by dividing it into subcomponents [2]. Essentially, cooperative coevolution has the ability to simplify the complexities of a problem through decomposition [3]. However, the performance of CC algorithms are highly sensitive to problem decomposition strategy [4]. Variable interaction [5] is a major constraint that governs the decomposition of a problem [6]. It is generally believed that placement of interacting variables into separate subcomponents degrades the optimization performance significantly [7,8]. Inter-dependencies exists amongst decision variables specifically in non-separable and partially separable functions [2,9,10]. Grouping of interacting variables accurately into separate subcomponents hence becomes a challenge to CC. An efficient problem decomposition technique identifies and

© Springer International Publishing Switzerland 2015
S. Arik et al. (Eds.): ICONIP 2015, Part III, LNCS 9491, pp. 127–136, 2015.
DOI: 10.1007/978-3-319-26555-1_15

groups variables with inter-dependencies together [9]. For this reason, several decomposition mechanisms have been proposed that automatically capture and group interacting variables together [6,11–13].

There is no unique decomposition strategy for some classes of problems such as fully-separable functions, fully-non separable or overlapping functions [14]. For instance, in a fully-separable function, all of the decision variables can be optimized independently, hence any decomposition is viable. Some partially separable functions may also contain a relatively high dimensional fully-separable subcomponent. Poor decomposition of such subcomponents may affect the optimization process and the solution quality [4]. Unfortunately, attempting to determine an effective decomposition strategy for these different classes of functions is a laborious task, which requires extensive experimentation [4].

It was shown that in spite of such automated decomposition strategies, it is still possible to identify a set of near optimal *static* decomposition [4]. However, it comes with the expense of elaborate empirical studies. To remedy the need for identifying the optimal decomposition, a very simple reinforcement learning approach to dynamically adapt the decomposition strategy was utilized [4].

An alternative method called competitive island-based cooperative coevolution (CICC) was proposed recently to eliminate the need for finding an optimal decomposition [15–17]. CICC has been originally designed for training recurrent neural networks on chaotic time series problems [15,16]. In such a scenario, neural level and synapse level problem decomposition methods are implemented as islands that compete and collaborate with each other. The competition algorithm ensures that different problem decomposition methods are given an opportunity to compete in different phases during the course of evolution [17]. It was shown that CICC is a promising approach for different classes of global optimization problems [17]. By enforcing competition and collaboration of different problem decomposition strategies, the CICC framework could yield solutions with a quality better than individual decomposition strategies used in isolation [17]. Initially, the CICC method incorporated competition amongst only two different problem decomposition methods [17].

In this paper, a multi-island competitive cooperative coevolution (MICCC) algorithm is proposed in which several different problem decomposition strategies are given a chance to compete, collaborate and motivate other islands while converging to a common solution. The performance of MICCC using three and five islands are evaluated on eight different benchmark functions and compared with CICC where only two islands were used.

The organization of the rest of this paper is as follows. Section 2 describes the proposed method and its application to different classes of problems. Experimental results and their analysis are provided in Sect. 3. Finally, Sect. 4 concludes the paper and outlines possible future extensions.

2   Multi-Island Competitive Cooperative Coevolution

In this section, we extend the two-island competition algorithm (CICC) applied in [15–17] to a multi-island cooperative coevolution (MICCC) algorithm which enforces competition and collaboration between various different problem decomposition strategies that are implemented as islands. In competitive coevolution, individuals of a species show its competitive ability through its fitness scores. The individuals of higher fitness win in the competition and continuously improve their performance through evolution of populations [18].

Algorithm 1. Multi-Island Competitive Cooperative Coevolution

---

Stage 1: Initialization:
while *Island-n* ≤ *MaxNumIslands* **do**
  | Cooperatively evaluate Island-n
end

Stage 2: Evolution:
while *FE* ≤ *Global-Evolution-Time* **do**
  | **while** *Island-n* ≤ *MaxNumIslands* **do**
  | | **while** *FE* ≤ *Island-Evolution-Time* **do**
  | | | **foreach** *Sub-population at Island-n* **do**
  | | | | **foreach** *Cycle in Max-Cycles* **do**
  | | | | | **foreach** *Generation in Max-Generations* **do**
  | | | | | | Create new individuals using genetic operators
  | | | | | | Cooperative Evaluation of Island-n
  | | | | | **end**
  | | | | **end**
  | | | **end**
  | | **end**
  | **end**

  | **Stage 3:** Competition: Compare and mark the island with the best fitness.

  | **Stage 4:** Collaboration: Inject the best individual from the *winner* island into all the other islands.
end

In MICCC [17], several different uniform problem decomposition decompositions are constructed as islands that compete and collaborate to optimize a function. These islands are evolved in isolation by independent G3-PCX [19] algorithm. The islands enforce competition by comparing their solutions after a fixed time (implemented as fitness evaluations), and exchange the best solution between the islands. The solution migration occurs between the different islands when evolutionary processes carry on for defined fitness evaluations or generations; by migrating feasible solutions of the winner island into those who lose the competition. The key aspects of the proposed MICCC algorithm are initialization, evolution, competition and collaboration.

2.1 Initialization

Each different problem decomposition strategy (island) is constructed with uniform problem decomposition strategies. To enforce an unbiased competition, all the different islands begin search with the same genetic materials in their sub-population.

Initially, all the sub-populations of Island One are initialized with random-real number values from a domain specified in Table 1. Next, these real values (from Island One) are copied into the sub-populations of the rest of the islands that are constructed with unique problem decompositions. In MICCC algorithm, the number of fitness evaluations depends on the number of sub-populations implemented in an island. An island with higher number of subcomponents will basically acquire more fitness evaluations for each cycle. Since each island is simultaneously evolved in isolation for complete cycles, the number of fitness evaluations cannot be exactly the same for each island due to unique problem decomposition strategies.

2.2   Cooperative Coevolution

After initialization, each of the islands are evolved in isolation simultaneously for a predefined time in the round robin fashion of the cooperative co-evolution framework. This predefined time is the *island-evolution time* that is established by the number of cycles that makes the required number of fitness evaluations for each of the respective islands. A single cycle completes when all the sub-populations of the respective island have been cooperatively evolved for a depth of n generations. The individuals from each of the sub-populations are cooperatively evaluated by concatenating the chosen individual from a given sub-population with the best individuals from the rest of the sub-populations [8].

2.3   Competition

In the competition phase of the MICCC algorithm, fitness ranking and comparison of all the cooperative evolved islands are conducted. A ranking process is adapted in order to identify the better and poor performing islands. This is done by quantifying the fitness of each of the islands at certain time intervals. The islands with higher fitness are ranked high while the poor performing islands with lower fitness are ranked lower. In MICCC, the island producing the highest fitness is the winner and ranked as the best island at that point of check. For the case where the fitness is the same, the winner island is randomly selected to encourage a fair competition.

2.4   Collaboration - Solution Migration

Collaboration is the core feature of the MICCC framework whereby the actual interaction and migration between different islands occur. Here, the best solution of the winner island is copied and injected into to the runner-up islands. This migration of the best feasible solution is able assist and motivate the other islands to compete fairly in the next round.

The transfer of best solutions from one island to the rest is done via the context vector [20]. As an island wins, the best individuals from each of the subcomponents need to be carefully concatenated into a context vector. The best solutions are then split from the context vector and are then injected into the respective subcomponents of each of the runner-up islands. The runner-up island which receives the best (injected) solution is cooperatively evaluated to ensure that the newly injected solution has a fitness. The best fitness of the winning island is also transferred alongside the best solution to the rest of the islands. Moreover, since the fitness of the best solution from the last sub-population carries a stronger solution, this fitness value is transferred and is used to override the fitness of the best solutions of all the sub-populations of the runner-up islands.

3   Simulation and Analysis

In this section, we evaluate the performances the multi-island instances; three and five island algorithms of MICCC and compare them with the standalone CC implementations. Next, we compare these multi-island instances with the original two island algorithm [17].

3.1   Benchmark Problems and Configuration

The experimental results in this paper are based on eight benchmark functions used in [17]. These functions are selected considering the level of difficulty, the scope of separability and the nature of problem,i.e. unimodal or multi-modal listed by Table 1. Furthermore, we use different problem decomposition strategies of 100 dimensions as inputs for competition in the multi-island algorithms. According to [21], these values that represent low, medium and high dimensional subcomponent sizes allow us to approximately determine the optimal subcomponent size.

The generalized generation gap with parent-centric crossover evolutionary algorithm (G3-PCX) [19] is used as the subcomponent optimizer. We use a pool size of 2 parents and 2 offspring as presented in [19]. The mean and standard deviation of function errors $(f(x)-f(*x))$ of 25 runs for each of the experiments are reported in the next subsection. The maximum number of fitness evaluations was set to 1500000 as suggested in [17]. The number of individuals in each of the respective sub-populations are fixed at 100.

3.2   Results and Analysis

In this section, we evaluate the performance of the three island algorithm on the different classes of functions f_1 to f_8. We observe that the proposed method has produced better quality solutions than the respective standalone CC counterparts according to Table 2. It has improved the solution quality of the unimodal and fully-separable functions f_1-f_3. It has performed fairly well for the multi-modal non-separable Rosenbrock instances of f_4 and f_5. The three island algorithm performed better than the standalone CC implementations on f_5. For the multi-modal, separable Rastrigin functions f_6 and f_7, the three island algorithm

Table 1. Problem Definitions based on [10, 22, 23]

Problem	Name	Optimum	Range	Multi-modal	Fully Separable
f_1	Ellipsoid	0	[-5,5]	No	Yes
f_2	Shifted Sphere	−450	[-100,100]	No	Yes
f_3	Schwefel's Problem 1.2	0	[-5,5]	No	Yes
f_4	Rosenbrock	0	[-5,5]	Yes	No
f_5	Shifted Rosenbrock	390	[-100,100]	Yes	No
f_6	Rastrigin	0	[-5,5]	Yes	Yes
f_7	Shifted Rastrigin	−330	[-5,5]	Yes	Yes
f_8	Shifted Griewank	−180	[-600,600]	Yes	No

Table 2. Comparison of MICCC-3 Island results against individual decomposition strategies used in isolation (CC)

Functions	Stats.	Standard CC			MICCC - 3 Island
		20×5	10×10	4×25	
f_1	Mean	6.83e+00	4.80e−98	3.76e−98	2.51e−101
	StDev	1.83e+00	1.70e-98	1.82e-98	1.41e−101
f_2	Mean	3.43e+05	9.03e−03	1.08e−12	4.84e−13
	StDev	1.40e+04	2.03e−03	1.06e−12	1.31e−13
f_3	Mean	5.00e−03	7.98e−51	1.13e−50	0.00e+00
	StDev	1.00e-04	2.01e-50	1.12e-50	0.00e+00
f_4	Mean	2.59e+02	5.16e+01	1.11e+02	5.90e+01
	StDev	2.49e+02	2.01e+00	1.01e+01	1.02e+00
f_5	Mean	7.95e+10	3.80e+01	9.01e+01	0.00e+00
	StDev	1.01e+09	1.02e+01	1.60e+01	0.00e+00
f_6	Mean	1.87e+01	2.70e+02	4.86e+02	0.60e+01
	StDev	1.08e+00	1.02e+01	1.89e+01	0.66e+00
f_7	Mean	8.83e+02	5.02e+02	7.56e+02	3.53e+02
	StDev	5.04e+01	2.02e+01	4.02e+01	1.04e+01
f_8	Mean	2.81e+3	5.11e-13	3.63e−03	1.98e−13
	StDev	2.08e+02	1.14e−13	1.07e−03	2.03e−14

achieved better results than the standalone counterparts and recorded similar performance for f_8.

We observe a similar trend while analyzing the performance of the five island algorithm. The results in Table 3 suggest that we attain better quality solutions while competing the five islands than each of those evolved in isolation as a standalone CC. This can be observed for all the test functions f_1-f_8. The five island competition has shown improved and near optimal solutions for the uni-modal f_1-f_3, and multi-modal functions f_5 and f_8. Moreover, it generated better

Table 3. Comparison of MICCC-5 Island results against individual decomposition strategies used in isolation (CC)

Functions	Stats.	Standard CC					MICCC - 5 Island
		20 × 5	10 × 10	4 × 25	5 × 20	50 × 2	
f_1	Mean	6.83e+00	4.80e−98	3.76e−98	5.62e−98	2.01e−98	5.59e−99
	StDev	2.03e+00	2.36e−98	1.45e−98	1.63e−98	1.70e−98	1.02e−99
f_2	Mean	3.43e+05	9.03e−03	1.08e−12	8.98e−13	0.27e+01	1.71e−13
	StDev	1.45e+04	1.03e−03	2.03e−13	1.21e−13	2.34e−01	1.03e−14
f_3	Mean	5.00e−03	7.98e−51	1.13e−50	1.18e−50	1.54e−50	0.00e+00
	StDev	1.00e−03	6.98e−51	1.03e−50	1.07e−50	2.01e−50	0.00e+00
f_4	Mean	2.59e+02	5.16e+01	1.11e+02	6.31e−01	1.57e+02	7.43e+01
	StDev	2.30e+01	1.01e+01	1.30e+01	2.03e−01	1.34e+01	0.78e+01
f_5	Mean	7.95e+10	3.80e+01	9.01e+01	3.71e+01	5.67e+04	0.00e+00
	StDev	1.96e+10	1.02e+01	1.20e+01	0.34e+01	3.49e+03	0.00e+00
f_6	Mean	1.87e+01	2.70e+02	4.86e+02	4.73e+02	1.10e+02	9.95e−01
	StDev	0.43e+01	1.20e+02	2.21e+02	1.03e+02	0.70e+02	1.45e-01
f_7	Mean	8.83e+02	5.02e+02	7.56e+02	6.91e+02	1.22e+02	9.75e+01
	StDev	2.23e+02	1.34e+02	2.78e+02	0.45e+02	0.94e+02	1.45e+01
f_8	Mean	2.81e+03	5.11e−13	3.63e−03	5.12e−13	0.43e+01	8.53e−14
	StDev	1.02e+03	2.09e−13	1.67e−03	2.87e−13	0.23e+01	1.04e−14

solutions than the standalone CCs' for the Rastrigin instances f_6 and f_7. For, f_4, the five island algorithm provided better solution quality (error) than 4 out of 5 standalone decompositions tested.

3.3 Discussion

This paper proposed multi-island competitive cooperative coevolution that involves increasing the number of islands in the original competitive island cooperative coevolution, CICC [17] which was limited to two islands. The experimental results show that we get better quality solutions than the CCs' with standalone decomposition strategies. Table 4 provides a set of comparative data for two, three and five island setups of CICC tested on eight benchmark functions f_1-f_8. According to Table 4, it is evident that all the three different setups performed equally well on unimodal and fully separable functions f_1-f_3 as they recorded similar solution errors. However, we observe an interesting trend with the complex multi-modal problems f_4, f_6, f_7 and f_8. We notice that the quality and precision of the solutions improve while utilizing more islands in the competition. The five island algorithm outperformed the three and two island algorithms on the multi-modal problems. Moreover, the three island algorithm also performed better than two island algorithm for the same set of problems. The performance improvement with the multi-island algorithm is mainly because many different problem decomposition strategies are given a chance to compete and at the same time collaborate to help and motivate the poorly performing problem decompositions through solution migrations of the winning island. For instance, few standalone problem decomposition strategies such as (25×5) and (50×2) in Table 3 have been inefficient and possibly been victims of premature convergence for most multi-modal problems $(f_5\text{-}f_8)$ because it recorded poor quality solutions overall when evolved in isolation as a standalone CC. However, if the same low-performing problem decomposition strategies are incorporated as competition inputs to the multi-island algorithm, it performs better and recorded substantial improvement in the quality of fitness solutions. As we increase the number of islands, different problem decomposition strategies compete and collaborate with the exchange of the best genetic materials from the winning islands during the evolution. The performance of the island with sub-populations with a lower diversity is also revived during the migration of individuals. To maintain high quality solutions, it is essential to have distinct problem decomposition strategies implemented as islands.

Table 4. Comparison of CICC against MICCC

Functions	Stats.	CICC Versions		
		CICC [17]	MICCC - 3 Island	MICCC - 5 Island
f_1	Mean	3.76e−99	2.51e−101	5.59e−99
	StDev	2.06e−99	1.41e−101	1.02e−99
f_2	Mean	7.78e−13	4.84e−13	1.73e−13
	StDev	1.34e−13	1.31e−13	1.03e−14
f_3	Mean	0.00e+00	0.00e+00	0.00e+00
	StDev	0.00e+00	0.00e+00	0.00e+00
f_4	Mean	7.93e+01	5.90e+01	7.43e+01
	StDev	1.78e+01	1.02e+01	0.78e+01
f_5	Mean	0.00e+00	0.00e+00	0.00e+00
	StDev	0.00e+00	0.00e+00	0.00e+00
f_6	Mean	1.40e+01	0.60e+01	9.95e−01
	StDev	1.34e+00	0.66e+00	1.45e−01
f_7	Mean	3.92e+02	3.53e+02	9.75e+01
	StDev	2.09e+01	1.04e+01	1.45e+01
f_8	Mean	1.99e−13	1.98e−13	8.53e−14
	StDev	1.98e−13	2.03e−14	1.04e−14

4   Conclusions and Future Work

In this paper, we extended the two-island competitive cooperative coevolution algorithm (CICC) to multiple island approaches where we evaluated the performances of the competitive methods for three and five islands with substantial analyses. The experimental results show that enforcing competition with a wider pool of problem decomposition strategies can considerably improve the performance during the course of the optimization phase, and also shows substantial enhancements in the quality of the overall fitness solutions. As we increase the number of islands for competition, more diversity is introduced. Different islands compete and cooperate through the transfer of the best genetic materials from the winner island. We observe an appealing trend with MICCC for most multimodal problems, whereby we attain higher quality solutions and basically escape the vulnerable fitness stagnation trap. In future, we are interested in further improving the solutions by allowing more individuals to share their solutions which would improve diversity and intensify selection pressure. This proposed multi-island algorithm can also be applied to large scale global optimization and extended to combinatorial optimization problems.

References

1. Bäck, T., Fogel, D.B., Michalewicz, Z. (eds.): Handbook of Evolutionary Computation. Institute of Physics Publishing, Bristol, and Oxford University Press, New York (1997)
2. Potter, M.A., De Jong, K.: A cooperative coevolutionary approach to function optimization. In: Davidor, Y., Männer, R., Schwefel, H.-P. (eds.) PPSN 1994. LNCS, vol. 866, pp. 249–257. Springer, Heidelberg (1994)
3. Chandra, R., Frean, M., Zhang, M.: On the issue of separability for problem decomposition in cooperative neuro-evolution. Neurocomputing **87**, 33–40 (2012)
4. Omidvar, M.N., Mei, Y., Li, X.: Effective decomposition of large-scale separable continuous functions for cooperative co-evolutionary algorithms. In: Proceeding of IEEE Congress on Evolutionary Computation, pp. 1305–1312 (2014)
5. Salomon, R.: Reevaluating genetic algorithm performance under coordinate rotation of benchmark functions - a survey of some theoretical and practical aspects of genetic algorithms. Biosyst. **39**, 263–278 (1995)
6. Omidvar, M., Li, X., Mei, Y., Yao, X.: Cooperative co-evolution with differential grouping for large scale optimization. IEEE Trans. Evol. Comput. **18**(3), 378–393 (2014)
7. Liu, Y., Yao, X., Zhao, Q., Higuchi, T.: Scaling up fast evolutionary programming with cooperative coevolution. In: Proceedings of the 2001 Congress on Evolutionary Computation, vol. 2. IEEE, pp. 1101–1108 (2001)
8. Potter, M.A., De Jong, K.: A cooperative coevolutionary approach to function optimization. In: Davidor, Y., Männer, R., Schwefel, H.-P. (eds.) PPSN 1994. LNCS, vol. 866, pp. 249–257. Springer, Heidelberg (1994)
9. Salomon, R.: Re-evaluating genetic algorithm performance under coordinate rotation of benchmark functions. a survey of some theoretical and practical aspects of genetic algorithms. BioSystems **39**(3), 263–278 (1996)
10. Li, X., Tang, K., Omidvar, M.N., Yang, Z., Qin, K., China, H.: Benchmark functions for the CEC 2013 special session and competition on large-scale global optimization. gene **7**(33), 8 (2013)
11. Mahdavi, S., Shiri, M.E., Rahnamayan, S.: Cooperativeco-evolution with a new decomposition method for large-scale optimization. In: Proceedings of the IEEE Congress on Evolutionary Computation, CEC 2014, pp. 1285–1292 (2014)

12. Chen, W., Weise, T., Yang, Z., Tang, K.: Large-scale global optimization using cooperative coevolution with variable interaction learning. In: Schaefer, R., Cotta, C., Kołodziej, J., Rudolph, G. (eds.) PPSN XI. LNCS, vol. 6239, pp. 300–309. Springer, Heidelberg (2010)

13. Omidvar, M.N., Li, X., Yao, X.: Cooperative co-evolution with delta grouping for large scale non-separable function optimization. In: Proceeding of IEEE Congress on Evolutionary Computation, pp. 1762–1769 (2010)

14. Omidvar, M.N., Li, X., Tang, K.: Designing benchmark problems for large-scale continuous optimization. Inf. Sci. **316**, 419–436 (2015)

15. Chandra, R.: Competition and collaboration in cooperative coevolution of Elman recurrent neural networks for time-series prediction. IEEE Trans. Neural Netw. Learn. Syst. (2015). doi:10.1109/TNNLS.2015.2404823. http://ieeexplore.ieee.org/stamp/stamp.jsp?tp=&arnumber=7055352&isnumber=6104215

16. Chandra, R.: Competitive two-island cooperative coevolution for training Elman recurrent networks for time series prediction. In: International Joint Conference on Neural Networks (IJCNN), Beijing, China, pp. 565–572, July 2014

17. Chandra, R., Bali, K.: Competitive two island cooperative coevolution for real parameter global optimization. In: IEEE Congress on Evolutionary Computation, Sendai, Japan, pp. 93–100, May 2015

18. Li, W., Wang, L.: A competitive-cooperative co-evolutionary optimizationalgorithm based on cloud model. In: Fourth International Workshop on Advanced Computational Intelligence (IWACI 2011), pp. 662–669. IEEE (2011)

19. Deb, K., Anand, A., Joshi, D.: A computationally efficient evolutionary algorithm for real-parameter optimization. Evol. Comput. **10**(4), 371–395 (2002)

20. Van den Bergh, F., Engelbrecht, A.P.: A cooperative approach to particle swarm optimization. IEEE Trans. Evol. Comput. **8**(3), 225–239 (2004)

21. Omidvar, M.N., Mei, Y., Li, X.: Effective decomposition of large-scale separable continuous functions for cooperative co-evolutionary algorithms. In: IEEE Congress on Evolutionary Computation (CEC 2014), pp. 1305–1312. IEEE (2014)

22. Tang, K., Yao, X., Suganthan, P.N., MacNish, C., Chen, Y.P., Chen, C.M., Yang, Z.: Benchmark functions for the CEC'2008 special session and competition on large scale global optimization. Nature Inspired Computation and Applications Laboratory, USTC, China, Technical report (2007). http://nical.ustc.edu.cn/cec08ss.php

23. Herrera, F., Lozano, M., Molina, D.: Test suite for the special issue of soft computing on scalability of evolutionary algorithms and other metaheuristics for large scale continuous optimization problems (2010)

Competitive Island-Based Cooperative Coevolution for Efficient Optimization of Large-Scale Fully-Separable Continuous Functions

Kavitesh K. Bali[1,2]([✉]), Rohitash Chandra[1,2], and Mohammad N. Omidvar[3]

[1] School of Computing Information and Mathematical Sciences,
University of South Pacific, Suva, Fiji
[2] Artificial Intelligence and Cybernetics Research Group,
Software Foundation, Nausori, Fiji
{bali.kavitesh,c.rohitash}@gmail.com
[3] School of Computer Science and IT, RMIT University Melbourne,
Melbourne, Australia
mohammad.omidvar@rmit.edu.au

Abstract. In this paper, we investigate the performance of introducing competition in cooperative coevolutionary algorithms to solve large-scale fully-separable continuous optimization problems. It may seem that solving large-scale fully-separable functions is trivial by means of problem decomposition. In principle, due to lack of variable interaction in fully-separable problems, any decomposition is viable. However, the decomposition strategy has shown to have a significant impact on the performance of cooperative coevolution on such functions. Finding an optimal decomposition strategy for solving fully-separable functions is laborious and requires extensive empirical studies. In this paper, we use a competitive two-island cooperative coevolution in which two decomposition strategies compete and collaborate to solve a fully-separable problem. Each problem decomposition has features that may be beneficial at different stages of optimization. Therefore, competition and collaboration of such decomposition strategies may eliminate the need for finding an optimal decomposition. The experimental results in this paper suggest that competition and collaboration of suboptimal decomposition strategies of a fully-separable problem can generate better solutions than the standard cooperative coevolution with standalone decomposition strategies. We also show that a decomposition strategy that implements competition against itself can also improve the overall optimization performance.

1 Introduction

Various meta-heuristic algorithms have been developed for continuous global function optimization. However, one of the core issues associated with these techniques is their scalability to higher dimensions [1]. *Divide-and-conquer* is an

© Springer International Publishing Switzerland 2015
S. Arik et al. (Eds.): ICONIP 2015, Part III, LNCS 9491, pp. 137–147, 2015.
DOI: 10.1007/978-3-319-26555-1_16

effective technique for solving large-scale complex problems. Cooperative Coevolution (CC) [2] is an explicit means of problem decomposition in the context of evolutionary algorithms (EAs) [3].

A major challenge in using CC for large-scale optimization is decomposition of a given problem into smaller sub-problems. Variable interaction [4] is a major constraint that governs the decomposition of a problem [5]. It is generally believed that placement of interacting variables into separate subcomponents degrades the optimization performance significantly [2,6]. For this reason, many decomposition methods have been proposed for automatic variable interaction detection [5,7–9].

For some classes of problems such as fully-separable functions or overlapping functions [10], there is no unique decomposition. In principle, for a fully-separable function, all of the decision variables can be optimized independently, hence any decomposition is viable. This may suggest that a complete decomposition in which each variables is placed in a separate subcomponent is the most efficient decomposition. However, a recent study [11] showed that the performance of CC is very sensitive to decomposition, even on fully-separable problems. Some partially separable functions may also contain a relatively high dimensional fully-separable subcomponent. Poor decomposition strategies of such subcomponents may hinder the convergence of CC to a high quality solutions [11]. Unfortunately, finding an effective decomposition strategy for fully-separable functions is a laborious task, which requires extensive experimentation [11]. To alleviate the need for finding the optimal decomposition, Omidvar et al. used a very simple reinforcement learning approach to dynamically adapt the decomposition strategy [11].

Recently, competitive island-based cooperative coevolution (CICC) algorithm was proposed for global optimization problems that gave promising results [12]. CICC has been originally designed for training recurrent neural networks on chaotic time series problems [13,14]. Neuron and synapse level problem decomposition strategies were implemented as islands that competed and collaborated with each other. CICC has shown to be a promising approach for solving large scale fully-separable functions for which there is no unique decomposition strategy [12].

In this paper, we apply CICC algorithm to eliminate the need for finding an optimal decomposition in the context of fully-separable problems. We speculate that competition and collaboration of decomposition strategies exhibiting various features can yield solutions with a quality better than individual decompositions used in isolation. In particular, the aim of this paper is to answer the following questions:

- How effective is the CICC algorithm when applied to large-scale fully-separable function optimization?
- Can CICC with two *same* effective decomposition strategies adapted from [11], competing against itself improve the overall optimization performance?

– Can competition and collaboration of two *different* suboptimal decomposition strategies yield solutions with a quality better than the near-optimal stand-alone decomposition strategies used in isolation?

The organization of the rest of this paper is as follows. Section 2 describes the proposed method and its application to large-scale fully-separable continuous functions. Experimental results and their analyses are provided in Sects. 3 and 4. Section 5 concludes the paper and outlines possible future extensions.

2   Competitive Island Cooperative Coevolution for Fully-Separable Continuous Functions

In this section, we provide details of cooperative coevolution method that features competition and collaboration with species, motivated by evolution in nature. In nature, competition in an environment of limited resources is mandatory for survival. Collaboration enforces interaction and sharing of resources between the different species having distinct characteristics for adaptation with respect to challenges such as environmental changes [13,14]. Interaction and migration of genetic material or information between the sub-populations can be advantageous in the evolutionary process. Hence, competition and collaboration are vital aspects of the evolutionary process where different groups of species compete for resources in the same environment.

Algorithm 1. Competitive Two-Island Cooperative Coevolution algorithm CICC [12].

Stage 1: Initialization:
i. Cooperatively evaluate Island One
ii. Cooperatively evaluate Island Two
Stage 2: Evolution:
while *FE ≤ Global-Evolution-Time* **do**
    while *FE ≤ Island-Evolution-Time* **do**
        foreach *Sub-population at Island-One* **do**
            foreach *Depth of n Generations* **do**
                Create new individuals using genetic operators
                Cooperative Evaluation of Island One
            end
        end
    end
    while *FE ≤ Island-Evolution-Time* **do**
        foreach *Sub-population at Island-Two* **do**
            foreach *Depth of n Generations* **do**
                Create new individuals using genetic operators
                Cooperative Evaluation of Island Two
            end
        end
    end
    Stage 3: Competition: Compare and mark the island with best fitness.
    Stage 4: Collaboration: Inject the best individual from the island with better fitness into the other island.
    if *ErrorIslandOne ≤ ErrorIslandTwo* **then**
        Inject Island One's best individual into Island Two.
    end
    else
        Inject Island Two's best individual into Island One.
    end
end

In the proposed competitive algorithm, two problem decomposition strategies are implemented as separate islands and evolved by an independent evolutionary algorithm. These islands enforce competition by comparing their solutions after a fixed time (fitness evaluations), and exchange the best solution between the islands [12–14]. Interaction between the two islands occur after separate evolutionary processes are executed in phases that are defined by fitness evaluations or generations. After a phase of evolution is completed, the algorithm migrates feasible solutions from the winner island into the others. The proposed CICC method for fully-separable problems is presented in Algorithm 1 where the key aspects are initialization, evolution, competition and collaboration.

For this paper, we focus on a two-island competition algorithm [12, 14]. This algorithm can be extended to more islands in further studies.

2.1   Initialization

In CICC, a problem decomposition strategy is implemented as an island. To enforce an unbiased competition, we ensure that both islands (Island One and Island Two) begin search with the same genetic materials in the sub-populations and cooperatively evolve them in isolation.

Initially, all the sub-populations of Island One are initialized with random-real number values from a domain specified in Table 1. These real values (from Island One) are copied into the sub-populations of Island Two. A problem decomposition (configuration) for an island can either have same sized *(uniform)* or varied sized *(non-uniform)* subcomponents. Since we are utilizing the problem decomposition strategies from [11], we employ uniform subcomponent sizes for this study. The highest level of decomposition for an island would have one subcomponent for each variable. A study has concluded that such extreme decompositions do not quite perform well as the rest of the effective decomposition configurations and they should be avoided [11]. In CICC, the number of fitness evaluations depend on the number of sub-populations used in the island. Therefore, an island with higher number of subcomponents will acquire more fitness evaluations for each cycle. We would like to assign each island with the similar time for evolution and encourage a fair competition. Since each island is simultaneously evolved for complete cycles, the number of fitness evaluations cannot be exactly the same for each island if they are defined by different problem decomposition strategies. Therefore, different islands adapt to different times (fitness evaluations) because the search difficulties along different dimensions of each island are different [12]. The islands compete and collaborate with each other to optimize a problem until the termination criteria is reached.

2.2   Coevolution in CICC

Once both islands have been initialized with the same search space, they are evolved simultaneously for a predefined time in the usual round robin fashion of the cooperative coevolution algorithm. According to Algorithm 1, this predefined time is termed as *island-evolution time*. The island evolution time is established by the number of cycles that makes the required number of fitness evaluations for each of the two islands. Basically, a cycle is complete when all the sub-populations of an island have been cooperatively evolved for n number of generations in the conventional round-robin fashion of the CC algorithm. Cooperative evaluation of individuals in the respective sub-populations is done by concatenating the chosen individual from a given sub-population with the best individuals from the rest of the sub-populations [2].

2.3 Competition and Collaboration

In Stage 3 of the CICC algorithm, a simple yet efficient competition strategy is implemented. After evolution of each of the islands, through a ranking process, the algorithm marks the island with the best fitness. The island producing the minimum fitness error is the winner island and the individual with the best fitness is copied to the other islands. The migration of the best feasible solution is able to assist and motivate the other islands to compete fairly in the next phase of competition.

As an island wins, the best individuals from each of the subcomponents need to be carefully concatenated into a context vector [15]. The best solutions are then split from the context vector and are then injected into one of the runner-up island(s). The algorithm must be implemented in such a way that it ensures that the solutions are transferred without losing any genotype to phenotype mapping [13,14].

In the conventional CC algorithm, each sub-population contains individuals that each have a unique fitness, which are cooperatively evaluated with the best solutions from the rest of the subcomponents. Taking that into consideration, there can be many distinct fitness values for the best solutions in each of the different sub-populations. Since the fitness of the best solution from the last sub-population carries a stronger solution, this fitness value is transferred (migrated) and is used to override the fitness of the best solutions of all the sub-populations of the runner-up islands.

3 Simulation and Analysis

In this section, we compare the performance of the competition enforced algorithm, CICC against the standalone CC algorithm for fully-separable problems.

3.1 Problem Decomposition Strategies

Two sets of experiments are conducted in this paper. Firstly, the best effective static decomposition strategies (near optimal) are selected from [11] and implemented as a potential island. The best effective decomposition strategies that have been identified empirically through previous studies are shown in Table 1. In this scenario, the two islands of CICC algorithm are constructed with the same problem decomposition strategies (best) which compete and collaborate to optimize a fully-separable function. It should be noted that the problem decomposition strategy for each island does not need to be the same as highlighted in [12,13]. In the next set of experiments, we extend the study by competing two different problem decomposition strategies.

3.2   Benchmark Problems and Parameter Settings

The experimental results in this paper are based on eight fully-separable functions taken from previous work [11] and listed in Table 1. Functions f_1 and f_2 were selected from De Jong suite [16], and the remaining are commonly used functions for benchmarking continuous optimization algorithms defined in [17–19]. Functions f_1, f_3 and f_7 are uni-modal and the remaining five functions are multi-modal [11]. The total number of fitness evolutions is set to 3×10^6. The number of individuals in each of the respective sub-populations are fixed at 100.

The generalized generation gap with parent-centric crossover evolutionary algorithm (G3-PCX) [20] is used as the sub-population optimization algorithm. We use a pool size of 2 parents and 2 offspring as presented in [20]. In this generalized generation gap model selection criteria, several individuals are replaced at every generation and only those that are replaced are evaluated.

Table 1. A list of fully-separable and scalable benchmark problems.

Function	Equation	Domain	Optimum	Best Decomp. [11]
Sphere function	$f_1(\mathbf{x}) = \sum_{i=1}^{n} x_i^2$	$[-100, 100]^n$	$\mathbf{x}^* = \mathbf{0}, f_1(\mathbf{x}^*) = 0$	10×100
Quadratic funciton	$f_2(\mathbf{x}) = \sum_{i=1}^{n} i x_i^4 + \mathcal{N}(0, 1)$	$[-100, 100]^n$	$\mathbf{x}^* = \mathbf{0}, f_2(\mathbf{x}^*) = 0$	5×200
Elliptic function	$f_3(\mathbf{x}) = \sum_{i=1}^{n} 10^{6 \frac{i-1}{n-1}} x_i^2$	$[-100, 100]^n$	$\mathbf{x}^* = \mathbf{0}, f_3(\mathbf{x}^*) = 0$	10×100
Rastrigin's function	$f_4(\mathbf{x}) = \sum_{i=1}^{n} \left[x_i^2 - 10\cos(2\pi x_i) + 10 \right]$	$[-5, 5]^n$	$\mathbf{x}^* = \mathbf{0}, f_4(\mathbf{x}^*) = 0$	500×2
Ackley's function	$f_5(\mathbf{x}) = -20\exp\left(-0.2\sqrt{\frac{1}{n}\sum_{i=1}^{n} x_i^2}\right) - \exp\left(\frac{1}{n}\sum_{i=1}^{n}\cos(2\pi x_i)\right) + 20 + e$	$[-32, 32]^n$	$\mathbf{x}^* = \mathbf{0}, f_5(\mathbf{x}^*) = 0$	10×100
Schwefel's function	$f_6(\mathbf{x}) = 418.9829n - \sum_{i=1}^{n} x_i \sin(\sqrt{|x_i|})$	$[-512, 512]^n$	$\mathbf{x}^* = \mathbf{1}, f_6(\mathbf{x}^*) = 0$	500×2
Different powers	$f_7(\mathbf{x}) = \sum_{i=1}^{n} |x_i|^{i+1}$	$[-1, 1]^n$	$\mathbf{x}^* = \mathbf{0}, f_7(\mathbf{x}^*) = 0$	10×100
Styblinski-Tang	$f_8(\mathbf{x}) = \frac{1}{2}\sum_{i=1}^{n}(x_i^4 - 16x_i^2 + 5x_i) + 38.16599n$	$[-5, 5]^n$	$\mathbf{x}^* = -2.903534 \times \mathbf{1}$ $f_8(\mathbf{x}^*) = 0$	500×2

4   Results and Analyses

4.1   Competition Between Same Problem Decomposition Strategies

In this section, we evaluate if CICC with the best decomposition strategy competing against itself can improve the performance given by the best problem

Table 2. Competition of same problem decomposition strategies against itself (CICC) compared with standalone CC with standalone decomposition strategies.

Functions	Decomposition	Stats	CC	CICC
f_1	10×100	Median	4.21e−23	3.76e−26
		Mean	4.23e−23	8.29e−26
		StDev	9.17e−24	2.69e−26
f_2	5×200	Median	7.77e+03	1.39e−09
		Mean	4.43e+08	1.60e−09
		StDev	3.53e+03	2.73e−03
f_3	10×100	Median	2.53e−18	5.50e−20
		Mean	2.46e−18	5.45e−20
		StDev	3.36e−19	4.06e−20
f_4	500×2	Median	8.74e+02	1.40e+02
		Mean	8.66e+02	1.45e+02
		StDev	3.00e+01	2.02e+01
f_5	10×100	Median	2.86e+00	1.26e+00
		Mean	2.78e+00	1.30e+00
		StDev	1.11e−01	2.40e−01
f_6	500×2	Median	2.26e+04	1.30e+04
		Mean	3.48e+04	1.35e+04
		StDev	8.69e+02	6.17e+02
f_7	10×100	Median	6.61e−01	1.00e−09
		Mean	7.79e−01	4.00e−04
		StDev	1.06e−01	1.01e−04
f_8	500×2	Median	1.20e+01	0.00+e00
		Mean	1.65e+02	0.00+e00
		StDev	1.55e+02	0.00+e00

decomposition strategy used in isolation. The results are given in Table 2 that shows the median, mean and the standard deviation of the final results obtained by 25 independent runs. In this scenario, CICC implements the best problem decomposition strategies for each of the functions determined in [11]. This essentially means that the two islands are constructed having the same problem decomposition that compete with each other. The results of the standalone CC with the same problem decomposition strategy is also presented for comparison. Generally, these results in Table 2 show that the proposed CICC algorithm outperforms the standalone CC in each of the eight fully separable functions f_1–f_8. It can be noted that CICC performed fairly well on the three unimodal functions (f_1, f_3, f_7) recording optimal solutions within the max 3×10^6 fitness evaluations. Additionally, the CICC algorithm performed considerably well on

Table 3. Competition results of two different problem decomposition strategies (CICC) compared to individual decompositions used in isolation (standalone CC)

Functions	Stats	Standard CC		CICC
		100×10	50×20	
f_1	Median	2.02e−47	1.56e−51	1.79e−79
	Mean	1.02e−47	3.21e−51	1.08e−79
	StDev	1.38e−47	3.94e−51	6.04e−80
f_2	Median	5.40e−02	5.42e−03	1.96e−09
	Mean	3.68e−01	1.81e−03	3.11e−10
	StDev	5.50e−01	2.89e−03	9.83e−10
f_3	Median	1.57e−45	3.74e−49	1.26e−78
	Mean	9.25e−46	1.37e−49	9.78e−78
	StDev	4.05e−46	1.19e−49	1.42e−77
f_4	Median	3.53e+03	3.70e+03	3.47e+03
	Mean	3.50e+03	3.85e+03	3.86e+03
	StDev	1.39e+02	2.23e+02	5.39e+01
f_5	Median	9.64e−02	9.78e−01	4.44e−16
	Mean	1.60e+00	1.37e+00	4.44e−16
	StDev	1.20e+00	7.52e−01	1.00e−16
f_6	Median	1.19e+05	1.17e+05	1.06e+05
	Mean	1.26e+05	1.14e+05	1.06e+05
	StDev	3.27e+03	3.97e+03	2.37e+03
f_7	Median	5.36e−03	1.60e+00	3.28e−04
	Mean	1.38e−03	1.48e+00	4.00e−04
	StDev	7.75e−05	2.93e−01	1.00e−04
f_8	Median	4.82e+03	5.80e+03	4.11e+03
	Mean	4.12e+03	5.78e+03	3.75e+03
	StDev	1.89e+02	2.06e+02	1.08e+02

the multi-modal Quadratic function- f_2 than the standalone CC counterpart. CICC recorded better solutions for f_4, f_5, f_6 and outperformed its respective CC configurations. For the Styblinski-Tang function - f_8, CICC performed substantially better.

The competition of the two islands with the same decomposition scheme does not follow the motivation of the original CICC method [12,13], where only different decomposition strategies were competing in order to exchange their unique features during evolution. In the case of competition using the same decomposition strategies, it can be noted that each island has features or solutions at different landscape of the problem that may be beneficial to the other island which may be struggling in a local optimum. These features are acquired through diversity and execution of genetic operators in the different islands that evolve in isolation for short span of time until there is comparison and then collaboration.

4.2 Competition Between Different Problem Decomposition Strategies

To further evaluate the efficiency of the CICC framework, we run an experiment with a set of two different problem decomposition strategies (100×10) and (50×20) that competes with each other to optimize a function. The results are presented in Table 3. Once again, it is clear that using CICC to compete two different sets of problem decomposition strategies generates better solutions than the standard cooperative coevolution with standalone decompositions strategies. CICC performed better on the fully-separable functions f_1-f_8. It generated near-optimum solutions for the uni-modal functions f_1 and f_3 and performed significantly better than the standalone CC for f_7. Multi-modal functions such as f_5 and f_2 were well optimized by CICC. The CICC algorithm managed to outperform the standalone CC implementations of f_4, f_6, f_8 and generated better quality solutions for these multi-modal functions.

In summary, it can be observed that CICC has performed well without having the need to find an optimal decomposition strategy to optimize large-scale fully-separable functions. It has generated solutions of equal quality and at times better than those found through optimal decomposition schemes. If a competition algorithm, competing problem decomposition strategies can perform better, then we do not have to empirically find the best decomposition strategy in the first instance. This can help save time and computational resources.

5 Conclusions and Future Work

In this paper, we have applied an island-based competitive cooperative coevolution algorithm to large-scale fully-separable continuous optimization problems. The results show that CICC can significantly improve the performance when compared to optimal problem decomposition strategies of standalone cooperative coevolution method. We found that competition and collaboration of two different suboptimal problem decomposition strategies of a fully-separable problem can also generate better solutions than the standard cooperative coevolution with standalone decomposition strategies.

Furthermore, we have shown that if competition of problem decomposition strategies can perform equally better than that of the best performing decomposition, then there is not a need to empirically find the best decomposition strategy. In other words, CICC helps us to eliminate the need for finding the optimal decomposition strategy and yet we can have solutions with similar quality.

In future work, CICC can be extended to a wide range of problems such as partially-separable functions as well as the recently introduced overlapping functions [10]. Further improvement of the results can also be achieved by implementing a multi-island CICC algorithm where more than two islands are considered.

References

1. Weise, T., Chiong, R., Tang, K.: Evolutionary optimization: pitfalls and booby traps. J. Comput. Sci. Technol. (JCST) **27**(5), 907–936 (2012). Special Issue on Evolutionary Computation
2. Potter, M.A., De Jong, K.A.: A cooperative coevolutionary approach to function optimization. In: Davidor, Y., Männer, R., Schwefel, H.-P. (eds.) PPSN 1994. LNCS, vol. 866, pp. 249–257. Springer, Heidelberg (1994)
3. Bäck, T., Fogel, D.B., Michalewicz, Z. (eds.): Handbook of Evolutionary Computation. Institute of Physics Publishing, Bristol, Oxford University Press, New York (1997)
4. Salomon, R.: Reevaluating genetic algorithm performance under coordinate rotation of benchmark functions - a survey of some theoretical and practical aspects of genetic algorithms. BioSystems **39**, 263–278 (1995)
5. Omidvar, M., Li, X., Mei, Y., Yao, X.: Cooperative co-evolution with differential grouping for large scale optimization. IEEE Trans. Evol. Comput. **18**(3), 378–393 (2014)
6. Liu, Y., Yao, X., Zhao, Q., Higuchi, T.: Scaling up fast evolutionary programming with cooperative coevolution. In: Proceedings of the 2001 Congress on Evolutionary Computation, vol. 2, pp. 1101–1108. IEEE (2001)
7. Mahdavi, S., Shiri, M.E., Rahnamayan, S.: Cooperative co-evolution with a new decomposition method for large-scale optimization. In: Proceedings of the IEEE Congress on Evolutionary Computation, CEC 2014, pp. 1285–1292 (2014)
8. Chen, W., Weise, T., Yang, Z., Tang, K.: Large-scale global optimization using cooperative coevolution with variable interaction learning. In: Schaefer, R., Cotta, C., Kołodziej, J., Rudolph, G. (eds.) PPSN XI. LNCS, vol. 6239, pp. 300–309. Springer, Heidelberg (2010)
9. Omidvar, M.N., Li, X., Yao, X.: Cooperative co-evolution with delta grouping for large scale non-separable function optimization. In: Proceedings of IEEE Congress on Evolutionary Computation, pp. 1762–1769 (2010)
10. Omidvar, M.N., Li, X., Tang, K.: Designing benchmark problems for large-scale continuous optimization. Inf. Sci. **316**, 419–436 (2015)
11. Omidvar, M.N., Mei, Y., Li, X.: Effective decomposition of large-scale separable continuous functions for cooperative co-evolutionary algorithms. In: Proceedings of IEEE Congress on Evolutionary Computation, pp. 1305–1312 (2014)
12. Chandra, R., Bali, K.: Competitive two island cooperative coevolution for real parameter global optimization. In: IEEE Congress on Evolutionary Computation, Sendai, Japan, pp. 93–100, May 2015
13. Chandra, R.: Competition and collaboration in cooperative coevolution of Elman recurrent neural networks for time-series prediction. IEEE Trans. Neural Netw. Learn. Syst. (2015). doi:10.1109/TNNLS.2015.2404823. http://ieeexplore.ieee.org/stamp/stamp.jsp?tp=&arnumber=7055352&isnumber=6104215
14. Chandra, R.: Competitive two-island cooperative coevolution for training Elman recurrent networks for time series prediction. In: International Joint Conference on Neural Networks (IJCNN), Beijing, China, pp. 565–572, July 2014

15. Van den Bergh, F., Engelbrecht, A.P.: A cooperative approach to particle swarm optimization. IEEE Trans. Evol. Comput. 8(3), 225–239 (2004)
16. De Jong, K.A.: Analysis of the behavior of a class of genetic adaptive systems (1975)
17. Li, X., Tang, K., Omidvar, M.N., Yang, Z., Qin, K.: Benchmark functions for the CEC 2013 special session and competition on large-scale global optimization. RMIT University, Melbourne, Australia, Technical Report (2013). http://goanna. cs.rmit.edu.au/xiaodong/cec13-lsgo
18. Hansen, N., Finck, S., Ros, R., Auger, A., et al.: Real-parameter black-box optimization benchmarking 2009: Noiseless functions definitions (2009)
19. Molga, M., Smutnicki, C.: Test functions for optimization needs (2005). http:// www.zsd.ict.pwr.wroc.pl/files/docs/functions.pdf
20. Deb, K., Anand, A., Joshi, D.: A computationally efficient evolutionary algorithm for real-parameter optimization. Evol. Comput. 10(4), 371–395 (2002)

Topic Optimization Method Based on Pointwise Mutual Information

Yuxin Ding[(⊠)] and Shengli Yan[(⊠)]

Department of Computer Sciences and Technology,
Key Laboratory of Network Oriented Intelligent Computation,
Harbin Institute of Technology Shenzhen Graduate School, Shenzhen, China
yxding@hitsz.edu.cn, yans_hitsz@hotmail.com

Abstract. Latent Dirichlet Allocation (LDA) model is biased to draw high-frequency words to describe topics. This affects the accuracy of the representation of topics. To solve this issue, we use point-wise mutual information (PMI) to estimate the internal correlation between words and documents and propose the LDA model based on PMI. The proposed model draws words in a topic according to the mutual information. We also propose three measures to evaluate the quality of topics, which are readability, consistency of topics, and similarity of topics. The experimental results show that the quality of the topics generated by the proposed topic model is better than that of the LDA model.

Keywords: Latent Dirichlet allocation · Mutual information · Topic model

1 Introduction

The bag-of-words representation of text documents has been widely used in natural language processing. In the text analysis, the bag-of-words representation is unsatisfactory because words are supposed independent. In recent years, a new class of generative models called Topic Model has quickly become more popular in some text-related tasks. Topic Model supposes documents and corpus composed of mixture topics and then documents can be thought of "bag of topics". Thus, these models can handle the problem effectively about terms dependency. Topics can be view as a probability distribution over words, where the distribution implies semantic coherence. The classic topic models include LSA [1] model, PLSA [2] topic model, and LDA [3] model.

In present topic models have been widely used in practice. For example, the paper [4] applied topic model to analyze instant messages. Griffiths et al. [5] applied topic models to find the "popular" research fields, and the "unpopular" research fields. Topic models have also been used in social network. Usually the text in Twitter is short, has a big noise, and is updated very quickly. To better analyze the short text, the paper [6, 7] proposed Author Topic Model (ATM) [6] and Twitter-LDA [7].

One of the most successful generative topic models is latent Dirichlet allocation (LDA), and many improved versions of LDA have been proposed, which include: correlated topic model [8], online learning topic model [9], interactive topic model and supervised topic model [11].

© Springer International Publishing Switzerland 2015
S. Arik et al. (Eds.): ICONIP 2015, Part III, LNCS 9491, pp. 148–155, 2015.
DOI: 10.1007/978-3-319-26555-1_17

When constructing the topics of a corpus, LDA considers all words equally important. So, LDA is likely to draw high-frequency words to describe topics, which results many low-frequency words, but related with the documents, are ignored. This affects the readability of the topics [10] and results the words in topics cannot better represent the content of a document. This also causes some topics extracted by LDA are similar to each other, which is named the topic overlapping.

To solve this issue, in this paper we propose a LDA model based on the point-wise mutual information, namely PMI-LDA topic model. The point-wise mutual information (PMI) represents the relation between words and documents, and the proposed model uses it to evaluate the important of words when assigning topics to words. The experiments show that topics extracted by PMI-LDA can better describe the content of documents.

2 LDA Based on Point-Wise Mutual Information (PMI-LDA)

2.1 Introduction of the LDA Topic Model

LDA model is a three-probability generation model, which is shown in Fig. 1. Suppose a corpus contains M documents, each document as a mixing generated by K topics, we assume that K is known and topics are independent of each other. V is the size of a word dictionary. Each topic contains a group of words which obey the polynomial distribution. In Fig. 1 word w is the only observed variable. w_{dn} represents the n-th word of document d, $w_{dn} \in V$; z_{dn} represents the topic which generates w_{dn}; α, β is the parameters of the priori probability distribution of the topics in a corpus; θ_d denotes a K-dimensional probability vector that is the parameters of the multinomial distribution, and represents the topic distribution of document d, with $\theta_d \sim Dir(\theta_d|\alpha)$; $\phi_{1:k}$ represents the distribution of the K topics on the words in a document; N is the total number of words of document d (contain duplicated words).

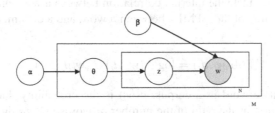

Fig. 1. LDA topic model

The words of the documents are drawn as follows:

1. A K dimensional vector θ_d is randomly selected from the Dirichlet distribution $p(\theta|\alpha)$. 2. A word w_{dn} in document d is drawn from the distribution $p(w_{dn}|\theta_d, \phi_{1:k})$.

One method for finding the LDA model by inference is via Gibbs sampling [5]. In LDA the goal is to estimate the distribution $p(z|w)$. The topic assignment of word w_i is sampled according to formula (1). Knowing z, the variables θ and ϕ are estimated as the formulas (2) and (3).

$$p(z_j|z_{N\backslash j}, w_j) \propto \frac{n_{z_j,N\backslash j}^{(w_j)} + \beta}{n_{z_j,N\backslash j}^{(.)} + V\beta} \frac{n_{z_j,N\backslash j}^{d_i} + \alpha}{n_{(.),N\backslash j}^{d_i} + K\alpha} \tag{1}$$

$$\phi_{z,w} = (n_{z,w} + \beta)/(n_z + V\beta) \tag{2}$$

$$\theta_{d,z} = (n_{d,z} + \alpha)/(n_d + K\alpha) \tag{3}$$

Where $z_{N\backslash j} = (z_1, \ldots, z_{j-1}, z_{j+1}, \ldots, z_N)$; $n_{z_j,N\backslash j}^{(w_i)}$ is the number of times word w_i is assigned to topic z_j; $n_{z_j,N\backslash j}^{(.)}$ is the total number of words assigned to topic z_j; $n_{z_j,N\backslash j}^{d_i}$ is the number of times words in document d_i is assigned to topic z_j; $n_{(.),N\backslash j}^{d_i}$ is the total number of words in document d_i. All the counts are taken over words 1 through N, excluding the word at position i itself (hence the $N\backslash j$ subscripts). ϕ_{zw} is the probability of word w in topic z; θ_{dz} is the probability that the document d contains topic z. $n_{d,z}$ is the number of words in document d with topic assignment z, $n_{z,w}$ is the number of words w in the whole corpus with topic assignment z, n_d is the length of document d and n_z is the number of all words in the corpus with topic assignment z.

2.2 PMI-LDA Topic Model

In general, the point-wise mutual information (PMI) measures the relation between the two specific events. It is defined as formula (4). If $I(x, y) > 0$, it means events x and y are highly correlated. If $I(x, y) = 0$, it means events x and y are highly independent. If $I(x, y) < 0$, events x and y are complementary distribution.

$$I(x, y) = \log(p(x, y)/p(x)p(y)) \tag{4}$$

We use PMI to estimate the internal correlation between a word and a document in our study, the definition of the   PMI   between a word and a document   is defined as follows.

$$\text{PMI}(w_i, d_j) = \log(P(w_i, d_j)/P(w_i)P(d_j)) \tag{5}$$

Where w_i represents a word in a corpus, $p(w_i)$ is the probability that w_i occurs in a corpus. $p(d_i)$ is defined as the ratio of the number of words in d_j to the total number of words in the corpus; $p(w_i, d_i)$ is defined as the ratio of the number of word w_i in d_j to the total number of the words in the corpus.

In general the distribution of the words in a corpus obeys the power-law distribution, so different from LDA, in the PMI-LDA topic model words in a corpus are not regarded as equally important. When generating a document, the model considers the correlation between the word and the document, and it generates a word according to the mutual information between the word and the document. We assign the point-wise mutual information between a word and a document as the weight of words. So Eq. (1) can be translated into Eq. (6).

$$p(z_j|z_{N\backslash j}, w_j) \propto \frac{M(w_j)n_{z_j,N\backslash j}^{(w_j)} + \beta}{\sum_{w \in z_j} M(w)n_{z_j,N\backslash j}^{(w)} + V\beta} \frac{\sum_{w \in d_i} m(w, d_i)n_{z_j,N\backslash j}^{d_i} + \alpha}{\sum_{w \in d_i} m(w, d_i)n_{(.),N\backslash j}^{d_i} + K\alpha} \qquad (6)$$

$m(w,d)$ represents the point-wise mutual information between word w and document d which contains w. $M(w)$ is define as the average point-wise mutual information of word w, which is defined as Eq (7).

$$M(w) = (\sum_{d_i} m(w, d_i))/n \qquad (7)$$

In (7), n is number of documents in a corpus. If $m(w,d) = 1$, then formula (6) equals to formula (1). From Eq. (6) we can see when generating words in a topic, PMI-LDA not only considers the frequency of words in a corpus, but also considers the relations between words and documents. Therefore, PMI-LDA can effectively reduce the affect of high-frequency words on topics.

3 Topic Evaluation

We evaluate the quality of a topic from three aspects, readability, consistency, and similarity. Readability refers that we can get the basic meaning of a topic by observing the top K words with the highest probabilities in the topic.

The topic consistency refers to the correlation degree of the words in a topic. The higher the correlation degree of the words, the better the topic is. We use the following formula to calculate the consistency:

$$C(t; v^{(t)}) = \sum_{m=2}^{M} \sum_{l=1}^{m-1} \log((D(v_m^{(t)}, v_l^{(t)}) + 1)/D(v_l^{(t)})) \qquad (8)$$

In Eq. (8) $D(v)$ represents the number of documents containing the word v. $D(v,v')$ represents the number of documents which include words v and v', M represents the number of words in a topic, t represent a topic, $C(t, V^{(t)})$ means the consistency value of the topic t, $V^{(t)}$ represents the words in topic t.

The similarity of topics is the similarity between each two topics. The high similarity of topics represents the discrimination power of a topic is low. The most optimal topic structure should be the one with the lowest similarity. The similarity of two topics can be calculated as formula (9).

$$\text{corre}(Z_i, Z_j) = \text{corre}(\beta_i, \beta_j) = (\sum_{v=0}^{V} \beta_{iv} \times \beta_{jv})/\sqrt{\sum_{v=0}^{V} \beta_{iv}^2 \sum_{v=0}^{V} \beta_{jv}^2} \qquad (9)$$

4 Experiments

We use the benchmark dataset 20Newsgroup to evaluate the performance of topic models. The corpus includes about 20000 news. We select the political news as the experimental data. The data set includes 2622 pieces of news. We set the topic number as 20. We use LDA and PMI-LDA to extract the topics of the corpus, respectively. We select the top seven words with the highest probability from each topic and use them to describe each topic. Tables 1 and 2 show the words used to describe each topic.

Table 1. Topic words extracted by LDA model

Topic	Word List						
1	univers	professor	istanbul	histori	ankara	turkey	ottoman
2	muslim	nazi	bosnian	serb	war	write	religion
3	write	articl	don	apr	love	david	clinton
4	jew	israel	arab	jewish	isra	palestin	countri
5	cramer	write	articl	optilink	gai	homosexual	sex
6	govern	peopl	right	law	constitut	power	human
7	adl	american	polic	report	media	san	anti
8	write	article	stratu	cdt	fbi	jim	new
9	armenia	turkish	armenia	turk	peopl	soviet	russian
10	peopl	didn	armenian	don	time	start	kill
11	tax	insur	health	govern	privat	care	system
12	fire	compound	batf	people	koresh	tank	fbi
13	israel	isra	write	arab	kill	article	attack
14	gun	file	firearm	weapon	bill	law	control
15	war	kuwait	iran	military	nuclear	island	south
16	presid	don	ms	job	myer	people	talk
17	greek	turkish	greec	govern	turkey	write	br
18	write	article	apr	book	cs	org	post
19	gun	crime	don	kill	peopl	death	weapon
20	drug	legal	peopl	article	write	don	illeg

We can observe that some topics in Table 1 have good readability, for example, the main content of topic three and topic thirteen is associated with "Israel"; the main content of topic five is associated with "gay"; the main content of topic fourteen is associated with "law"; the main content of topic nineteen is associated with "crime". On the other hand, the readability of some topics is medium, for example, some content of topic one is associated with "Turkey"; some content of topic two is associated with "Islam" and "religion"; some content of topic eleven is associated with "government welfare"; some content of topic fifteen is associated with "Middle East" and "nuclear weapons". However, there are still some topics have poor readability, and we cannot judge the meaning of the topics from the words they contain, for example, the topics three, seven, nine, sixteen, eighteen. From the table, we can find that some words appear in many topics, for example, the word "write" appears in the topic two, three, five, eight, thirteen, seventeen, eighteen; the word "people" appears in the topic six, nine, ten, twelve, sixteen, nineteen, and twenty; the word "article" appears in the topic three, five, eight, thirteen, and eighteen. These words are high frequency words in the corpus, and most of them cannot describe the concrete content of a document. This results the poor readability of the topics. From this sample, we can see the high frequency words affect the readability of the topics.

Table 2. Topic words extracted by PMI-LDA model

Topic	Word List						
1	insur	tax	health	parti	privat	care	system
2	safeti	auto	revolv	cop	semi	kratz	glock
3	israel	isra	arab	palestinian	jew	jewish	kill
4	fire	batf	fbi	koresh	compound	atf	tank
5	drug	law	legal	firearm	court	bill	gun
6	presid	myer	ms	decis	senat	package	hous
7	war	garrett	nuclear	secret	island	georgia	naval
8	didn	azerbaijani	apart	door	neighbor	shout	armenian
9	file	adl	militia	amend	bullock	gerard	francisco
10	govern	kuwait	power	program	peopl	human	iran
11	muslim	bosnian	serb	bosnia	islam	croat	yugoslavia
12	azerbaijan	phd	hojali	dead	propos	baku	hai
13	god	uiuc	religion	jewish	Indiana	cultur	jew
14	gun	rate	crime	death	crimin	handgun	firearm
15	sex	homosexual	male	sexual	partner	cramer	optilink
16	stratu	cdt	write	sw	cs	stove	articl
17	greek	turkish	turkey	greec	istanbul	turk	henrik
18	weapon	gun	licens	bullet	rifl	dan	militari
19	armenian	Turkish	Armenia	soviet	turk	russian	azeri
20	professor	univers	nazi	book	histori	jew	holocaust

Table 2 shows the topics extracted by the point-mutual information LDA model. We found that almost all topics have different words, and we do not observe that high frequency words appear in several topics. The average frequency of the words in each topic is shown in Fig. 3. In Fig. 3 the vertical axis represents the frequency, and the horizontal axis represents the topics. We can see that the frequency of words generated by PMI-LDA is obviously lower than that of the words generated by LDA. This shows that compared with LDA, high frequency words do not make an important affect on PMI-LDA. Just from the view of the readers' understanding, the number of topics with higher readability in Table 2 is greater than that of the topics in Table 1. This shows that the quality of the topics generated by PMI-LDA is better than that of the topics generated by LDA (Fig. 2).

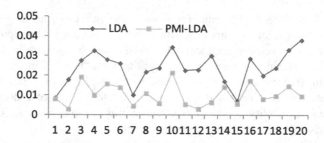

Fig. 2. Average frequency of the words generated by the two models

To evaluate the quality of topics, we calculate the consistency of topics according to Eq. (8). We sort the topics in the descending order of their consistency values. Figure 3 shows the consistency of topics of the two models. We only show the top 30 percent topics with the highest consistent value. In Fig. 3 the vertical axis represents topics, and the horizontal axis represents the consistency value. We can see that the consistency of topics extracted by PMI-LDA model is superior to that of the topics extracted by the LDA model.

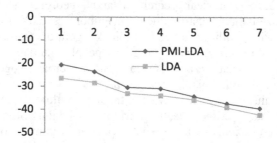

Fig. 3. Comparison of consistency of topics

The similarity of topics measures the overall performance of all topics. We calculate the average similarity between each two topics. The average similarity between each two topics generated from LDA is 0.16, while the average similarity between each two topics generated from PMI-LDA is 0.08. So compared with LDA, there is less topic overlapping in the topics generated by PMI-LDA.

5   Conclusion

In LDA model, words are regarded as equally important, so high frequency words are likely to be draw to describe topics. This results LDA generates many similar and hard readable topics. To solve this issue, we use the mutual information between a word and a document to evaluate the importance of words, and make the Gibbs algorithm draw words with high mutual information to describe topics. In addition, we propose three measures, topic readability, consistency of topics, and topic similarity, to evaluate the quality of topics generated by topic models. The experiments on the benchmark dataset show that compared with LDA, the quality of the topics extracted by the PMI-LDA is improved.

Acknowledgments. This work was partially supported by Scientific Research Foundation in Shenzhen (Grant No. JCYJ20140627163809422), Scientific Research Innovation Foundation in Harbin Institute of Technology (Project No. HIT.NSRIF2010123), State Key Laboratory of Computer Architecture, Chinese Academy of Sciences and Key Laboratory of Network Oriented Intelligent Computation (Shenzhen).

References

1. Thomas, K.L., Peter, W.F., Darrell, L.: An introduction to latent semantic analysis. Discourse Process **25**, 259–284 (1998)
2. Hofmann, T.: Probabilistic latent semantic indexing. In: Special Interest Group on Information Retrieval, pp. 50–57, Berkeley, CA, USA (1999)
3. Blei, D.M., Ng, A.Y., Jordan, M.I.: Latent Dirichlet allocation. J. Mach. Learn. Res. **3**(1), 993–1022 (2003)
4. Ding, Y., Meng, X., Chai, G., Tang, Y.: User identification for instant messages. In: Lu, B.-L., Zhang, L., Kwok, J. (eds.) ICONIP 2011, Part III. LNCS, vol. 7064, pp. 113–120. Springer, Heidelberg (2011)
5. Griffiths, T.L., Steyvers, M.: Finding scientific topics. Proc. Natl. Acad. Sci. **101**, 5228–5235 (2004)
6. Michal, R.Z., Griffiths, T., Steyvers, M., et al.: The author-topic model for authors and documents. In: Proceedings of the 20th Conference on Uncertainty in Artificial Intelligence, pp. 487–494 (2004)
7. Zhao, W.X., Jiang, J., Weng, J., He, J., Lim, E.-P., Yan, H., Li, X.: Comparing twitter and traditional media using topic models. In: Clough, P., Foley, C., Gurrin, C., Jones, G.J., Kraaij, W., Lee, H., Mudoch, V. (eds.) ECIR 2011. LNCS, vol. 6611, pp. 338–349. Springer, Heidelberg (2011)
8. Blei, D.M., Lafferty, J.D.: Correlated topic models.. In: International Conference on Machine Learning, pp. 113–120 (2006)
9. Canini, K.R., Shi, L., Griffiths, T.L.: Online inference of topics with latent Dirichlet allocation. In: International Conference on Artificial Intelligence and Statistics, pp. 41–48, Clearwater Beach, Florida, USA (2009)
10. David, M., Wallach, H.M., Talley, E., et al.: Optimizing semantic coherence in topic models. In: Empirical Methods in Natural Language Processing, pp. 262–272 (2011)
11. Blei, D.M., Jon, D.: McAuliffe. supervised topic models. In: NIPS (2007)

Optimization and Analysis of Parallel Back Propagation Neural Network on GPU Using CUDA

Yaobin Wang[1,3(✉)], Pingping Tang[2,3], Hong An[3], Zhiqin Liu[1],
Kun Wang[1], and Yong Zhou[1]

[1] Department of Computer Science and Technology,
Southwest University of Science and Technology, Mianyang 621010, China
wyb1982@mail.ustc.edu.cn,
{lzq,wangkun,zhouyong}@swust.edu.cn
[2] Department of Material Science and Engineering,
Southwest University of Science and Technology, Mianyang 621010, China
pingping@mail.ustc.edu.cn
[3] Department of Computer Science and Technology,
University of Science and Technology of China, Hefei 230027, China
han@ustc.edu.cn

Abstract. Graphic Processing Unit (GPU) can achieve remarkable performance for dataset-oriented application such as Back Propagation Network (BPN) under reasonable task decomposition and memory optimization. However, advantages of GPU's memory architecture are still not fully exploited to parallel BPN. In this paper, we develop and analyze a parallel implementation of a back propagation neural network using CUDA. It focuses on kernels optimization through the use of shared memory and suitable blocks dimensions. The implementation was tested with seven well-known benchmark data sets and the results show promising 33.8x to 64.3x speedups can be realized compared to a sequential implementation on a CPU.

Keywords: Back propagation · GPU · Network density · Producer-consumer locality · Bank conflict

1 Introduction

The field of artificial neutral network (ANN) has evolved over the past years. ANNs are widely used for data mining and pattern recognition. Training and testing networks for such large scale dataset-oriented applications is computationally expensive and time consuming. Therefore, both fast convergent algorithm and parallel algorithm [1] are researched. Back Propagation Network (BPN) uses supervised learning for training. It has one input layer, one or more hidden layers and one output layer. Neurons in the same layer are independent. To obtain the appropriate weights between neurons, the training process is composed of thousands of iterations. In each iteration, the input vectors of layers are calculated in forward pass while local gradient calculation and weight matrixes updating are processed in backward pass. Researchers have proposed many methods to build parallel BPN on different architecture such as cluster [1] and stream processor [6].

© Springer International Publishing Switzerland 2015
S. Arik et al. (Eds.): ICONIP 2015, Part III, LNCS 9491, pp. 156–163, 2015.
DOI: 10.1007/978-3-319-26555-1_18

Graphic Processing Unit (GPU) has become increasingly competitive because of considerable parallel computational power and programmability by Compute Unified Device Architecture (CUDA). Researches show that the general purpose GPU is suitable for applications with compute-intensive, Data level parallel (DLP) and producer-consumer locality characteristics [2]. GPUs have been already employed in the training of neural networks using different algorithms and programming languages [3–5, 11]. However, advantages of GPU's memory architecture are still not fully exploited to parallel BPN.

In this paper, we describe a parallel implementation of a back-propagation neural network (BPNN) on GPU using CUDA. Aiming at making full use of GPU's memory architecture, it focuses on kernels optimization through the use of shared memory and suitable blocks dimensions. In our method, parallel data structures are designed, which can be naturally mapped onto on-chip shared memory and accessed without bank conflict. They're stored partially in shared-memory to minimize the cost of data transfers. Moreover, we optimized block matrix-vector multiplication algorithm by using GPU-adapting suitable block size. Intermediate vectors are stored in shared memory in order to avoid swap vector in and out frequently. Finally, the implementation was tested with seven standard benchmark data sets. The parallel speedup and performance influence factor analysis are demonstrated by experiments. It shows the parallel BPNN can get ideal 33.8x to 64.3x speedup under GPU's memory optimizations toward specific applications.

The rest of the paper is organized as follows. Section 2 analyzes Back Propagation algorithm parallelization on GPU. An optimized program framework based on shared memory and constant memory is described in Sect. 3. Section 4 demonstrates the experimental results. Finally we conclude in Sect. 5.

2   Parallel Back Propagation Algorithm on GPU

GPU operates as a highly multi-threaded co-processor (device) to CPU (host). While CPU is responsible for sequential computing and logical transaction, GPU is specialized for compute-intensive, highly parallel computation. Parallel sections of an application are executed on GPU as CUDA kernels. Many lightweight threads execute the same kernel code but on different data based on its threadID. These concurrent threads are grouped into parallel and independent blocks. On the other hand, shared memory is a kind of read-write per-block on-chip memory, which is divided into 16 equally-sized banks that can be accessed simultaneously by each thread. When n addresses memory reference fall in n distinct memory banks, it can achieve an effective bandwidth that is n times as high as the bandwidth of a single module.

The original BP algorithm is describes in detail [8]. BPN is constructed of one input layer, one output layer and one or more hidden layers with connection between adjacent layers. A vector of neurons and a matrix of weights together with an activation function are called a layer. All layers except the input layer produce outputs in two steps [3].

- Get inner product of weights and input vectors:

$$x_j = \sum_i w_{ji} y_i + b_j \tag{1}$$

- Activate function:

$$y_j = (1 + e^{-x_j})^{-1} \tag{2}$$

The subscript j indexes neurons in the current layer to be calculated and i indexes neurons of the front layer connected with the j neurons. W_{ji} denotes the weight of the connection from i to j neurons. y denotes output of a neuron and bj denotes bias [8].

These steps can be replaced with matrix-vector multiplication to suit GPU. Particularly, the output of one layer is consumed as input vector in following layer to produce output vector. In order to capture this kind of producer-consumer locality, intermediate vectors are stored in shared memory. Moreover, output vectors are also consumed to get local gradient in backward pass, so we copy them to global memory because shared memory is limited.

After getting output vector of the output layer, the total error (E) is calculated by comparing the actual and desired output vector [8]. E is defined as

$$E = \frac{1}{2} \sum_e \sum_j (y_{j,e} - d_{j,e})^2$$

To minimize E by gradient descent it is necessary to compute the partial derivative of E with respect to each weight. Now, the backward pass which propagates derivative starts.

For a W_{ji}, from i to j the derivative is

$$\frac{\partial E}{\partial w_{ji}} = \frac{\partial E}{\partial x_j} * \frac{\partial x_j}{\partial w_{ji}} = \frac{\partial E}{\partial x_j} * y_i \tag{3}$$

$$\frac{\partial E}{\partial x_j} = \frac{\partial E}{\partial y_j} * \frac{dy_j}{dx_j} \tag{4}$$

In output layer, $\partial E/\partial y$ is directly. In other layers,

$$\frac{\partial E}{\partial y_j} = \sum_i \frac{\partial E}{\partial x_j} * w_{ij} \tag{5}$$

This equation takes into account all connections emanating from neutron j. When every neuron in a layer is ready, all $\partial E/\partial y$ of the front layer can be calculated simultaneously. Actually, Eq. 5 is equivalent to vector-matrix multiplication. The result vector of vector-matrix multiplication is consumed in next layer from Eqs. 4 and 5. It is stored in shared memory to exploit the producer consumer locality.

Weight matrixes are updated after calculating $\partial E/\partial x$ by

$$\Delta w(t) = -\varepsilon * \frac{\partial E}{\partial w(t)} + \alpha * \Delta w(t-1) \tag{6}$$

This procedure is replaced with many 2-dimensional matrixes assignment naturally. At last, we summarize following solutions or principles:

(1) Data structures. Parallel vector and parallel matrix are designed to make parallel computing convenient.
(2) Producer-consumer locality. (1) and (2) are recursive as the same as (4) and (5). There are producer-consumer localities in these forward and backward propagations. Intermediate vectors are stored circularly in shared memory to capture producer-consumer locality and save memory bandwidth. They are saved back to global memory when necessary. Shared memory is 16 KB [7] and can store 4096 floats. At any time, it is adequate to store only one intermediate vector of a layer.
(3) Constant memory. ε and α in (6) and network topology are stored in constant memory.
(4) Weight matrix access. The access pattern of weight matrix is different in (1) and (5). It reads elements by row in (1), but by column in (5). The algorithm of matrix-vector multiplication and vector-matrix multiplication must avoid non-coalesced global memory accesses.
(5) In our application, the number of neurons in output layer is few, about 1 to 3. So E is calculated by CPU.

3 Optimized Program Framework on GPU

This optimized parallel BPNN absorbs advantages in kernel design of [3, 4, 9]. In [3], the input vector includes the bias in Eq. 1 to embed additional summation into uniform matrix-vector multiplication. This method can simplify programming and utilize processing units more fully. In [4], local gradient calculation and weight updating are separated to two kernels, which alleviate pressure of shared memory by utilizing global memory. In [9], kernels are subdivided by layer. It can reorganize grid and block dimension according to the number of neurons in a layer. The size of weight matrix from ith layer to jth layer is n*m, m and n are the number of neurons in ith and jth layer respectively. Using cudaMemcpy2DToArray API, weight matrixes are transferred to global memory in turn.

Figure 1 shows an overall flow chart of the program based on Sect. 2. Three kernels are created to decompose BP algorithm: Fire, LocalGradient and WeightUpdate.

In the Fire kernel, output vectors of each layer are calculated layer by layer. In fact, small matrix-vector multiplication kernels [5] make up Fire kernel. Similarly, LocalGradient kernel is composed of small vector-matrix multiplication kernels. To utilize producer-consumer locality directly [2], we organize the forward and backward pass into two adequate big kernels, not organize each layer into a kernel.

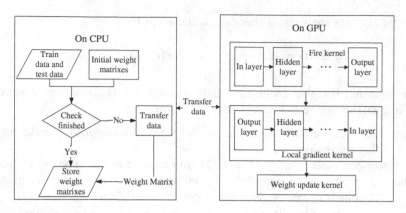

Fig. 1. Overall flow of BPN on GPU

Host transfers weights to GPU all at once, but train data and test data packet by packet to feed device memory. After network is stable, weight matrixes are copy back to CPU and stored.

4   Experimental Results and Discussion

4.1   Data Sets for Experiments

We use 7 real-world data sets from http://archive.ics.uci.edu/ml/datasets.html for both classification and regression tasks. Table 1 summarizes data sets properties such as area, task type, the number of patterns and data size. In addition, we figure out relatively suitable network topology for each data set.

Table 1.   Data sets properties and neural network topology

Database	Area	Patterns	Data size (KB)	Network topology	Task type
Abalone	Life	4177	192	8-25-1-1	Classification
Breast cancer	Medicine	569	119	30-25-2	Regression
Glass identify	Physics	214	12	10-16-3	Classification
Letter recognition	Image	20000	729	16-25-1	Classification
Wine quality	Business	4898	263	11-25-1	Classification Regression
Satellite statlog	Physics	6435	518	36-20-1	Classification
Musk	Chemistry	476	312	166-26-1	Classification

4.2   Results of Experiments

Our experiments were conducted on a PC with Intel Core 2 Duo E6700 2.67 GHz processor and 4 GB RAM running Windows XP, the GPU is NVIDIA GeForce GTX 275 with 30 SM(240 cores). We train each neural network 100000 iterations/pattern in different block size: 2×16, 4×16 and 16×16.

The results are presented on Table 2: Root Mean Square Error (RMS), time for each pattern (100000 iterations) in ms, and speedup of different block design. The RMS-1 and Time-1(ms) are used for CPU version vs. RMS-2 and time-2(ms) are used for CUDA version. Three kinds of blocks: 2×16, 4×16 and 16×16 have similar RMS but different time cost. The RMS is sufficient for each application. Links are calculated from network topology stand for the density. Best speedup is given at the last column. It shows that most of the chosen well-known benchmarks can achieve 33.84x to 64.32x speedup compared with a CPU version on our platform.

Table 2.   Experimental results

Database	Links	RMS-1	Time-1 (ms)	RMS-2	Time-2 (ms)			Best	
					2×16	4×16	16×16	Block	Speedup
Abalone	453	0.00364	1963	0.00355	58	62	174	2×16	33.84
Breast cancer	827	0.00363	4193	0.00372	101	86	92	4×16	48.76
Glass identify	227	0.00066	1795	0.00068	90	100	325	2×16	19.94
Letter recognition	451	0.00674	2524	0.00601	87	60	70	4×16	42.07
Wine Quality	326	0.00327	2070	0.00323	67	55	71	4×16	37.64
Satellite Statlog	761	0.00553	3029	0.00595	90	65	69	4×16	46.6
Musk	2689	0.00370	39299	0.00403	1077	653	611	16×16	64.32

4.3   Discussion

Q: which block dimension is suitable for GPU accelerating toward specific application?

Figure 2 shows time cost values fluctuate when use different block size to implement the block matrix-vector multiplication algorithm. Actually, this situation reflects task decomposition and parallelism granularity balancing on GPU. The dimension of weight matrix is decided by neural network topology, which affects the design of block and grid dimension. On the one hand, block with more threads can take full advantage of execution units. On the other hand, more blocks can hide pipelined delays effectively. We must balance them according to NN topology. For maximum speedup, we select those block resulting in minimum time cost in our experiments. Our choices are given in Table 2.

Both widths of these kinds of blocks are 16, the block algorithm accesses intermediate vectors sequential and threads in a half-warp request respective bank without bank conflict. This strategy improves overall performance of our experiments.

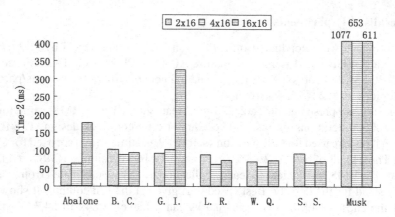

Fig. 2. Time cost fluctuate with block size

5   Conclusions

This paper presents parallel BP algorithm design toward specific application using GPU-adaptive operations. Block matrix-vector and vector-matrix multiplication algorithm is used to exploit various granularity parallelism and avoid non-coalesced global memory accesses. On-chip memory is used to capture producer-consumer locality and save bandwidth. The results show max 64.3x speedup to CPU stand-alone implementation. The GPU implementation by CUDA make train phase of BPN faster. These methods are useful to other algorithm and applications. Understanding specific hardware architecture includes memory hierarchy and computational units' organization, programmers can map it efficiently on the basis of characteristics of application.

From [13], we get that shared memory lives with kernel. Intermediate results of producer kernel and consumer kernel have to be transferred back and force. No persistent state in shared memory data cache results in less efficient communication between producer and consumer kernels [2]. In our experiments, we sacrifice reorganization of block in each layer for direct producer-consumer locality. It's worthwhile because block size in output layer has little or no impact on performance due to limited output neurons (1 to 3) and kernel for layer is low computation. We adopt forward and backward pass as basic kernel unit to reduce data transfers by 1/5 and improve computing resource utilization.

Acknowledgement.   This work is supported financially by the National Natural Science Foundation of China grants 61202044, the National Basic Research Program of China under contract 2011CB302501, the National Hi-tech Research and Development Program of China under contracts 2012AA010902 and 2012AA010303, the Research Fund of Southwest University of Science and Technology 10zx7119.

References

1. Suresh, S., Omkar, N., Mani, V.: Parallel implementation of back-propagation algorithm in networks of workstations. IEEE Trans. Parallel Distrib. Syst. (TPDS) **16**(1), 24–34 (2005)
2. Che, S., et al.: A performance study of general-purpose applications on graphics processors using CUDA. J. Parallel Distrib. Comput. (JPDC) **68**(10), 1370–1380 (2008)
3. Jang, H., Park, A., Jung, K.: Neural network implementation using CUDA and OpenMP. In: Digital Image Computing: Techniques and Applications (DICTA 2008), pp. 155–161 (2008)
4. Lopes, N., Ribeiro, B.: GPU implementation of the multiple back-propagation algorithm. In: Corchado, E., Yin, H. (eds.) IDEAL 2009. LNCS, vol. 5788, pp. 449–456. Springer, Heidelberg (2009)
5. Fujimoto, N.: Faster matrix-vector multiplication on GeForce 8800GTX. In: IEEE International Symposium on Parallel and Distributed Processing (IPDPS 2008), pp. 1–8 (2008)
6. Furedi, L., Szolgay, P.: CNN model on stream processing platform. In: European Conference on Circuit Theory and Design (ECCTD 2009), pp. 843–846 (2009)
7. NVIDIA CUDA Programming Guide 2.3. NVIDIA Corporation (2009)
8. Rumelhart, D.E., Hinton, G.E., Williams, R.J.: Learning internal representations by error propagation. Readings Cogn. Sci. **1**, 399–421 (1985)
9. Kavinguy, B.: A Neural Network on GPU. http://www.codeproject.com/KB/graphics/GPUNN.aspx
10. Lopes, N., Ribeiro, B.: An evaluation of multiple feed-forward networks on GPUs. Int. J. Neural Syst. **21**(1), 31–47 (2011)
11. Xavier, S.-C., Madera-Ramirez, F., Uc-Cetina, V.: Parallel training of a back-propagation neural network using CUDA. In: Ninth IEEE International Conference on Machine Learning and Applications (ICMLA), pp. 307–312 (2010)
12. Scanzio, S., Cumani, S., Gemello, R., et al.: Parallel implementation of artificial neural network training for speech recognition. Pattern Recogn. Lett. **31**(11), 1302–1309 (2010)
13. Ryoo, S., et al.: Optimization principles and application performance evaluation of a multithreaded GPU using CUDA. In: IEEE Symposium on Principles and Practice of Parallel Programming (PPoPP 2008), pp. 73–82. ACM (2008)
14. Li, X., Han, W., Liu, G., Hong A., et al.: A speculative HMMER search implementation on GPU. In: 2012 IEEE 26th International Parallel and Distributed Processing Symposium Workshops (IPDPSW). IEEE Computer Society (2012)
15. Sun, T., Hong, A., et al.: CRQ-based fair scheduling on composable multicore architectures. In: 26th ACM SIGARCH International Conference on Supercomputing (ICS), Venice, Italy, pp. 173–184 (2012)
16. Liu, G., Han, W., Hong, A., et al.: FlexBFS: A parallelism-aware implementation of breadth-first search on GPU. In: 17th ACM SIGPLAN Symposium on Principles and Practice of Parallel Programming (PPoPP), New Orleans, USA (2012)
17. Liu, G., Hong, A., et al.: A program behavior study of block cryptography algorithms on GPGPU. In: 2009 International Conference on Frontier of Computer Science and Technology (FCST), pp. 33–39 (2009)
18. Yao, P., Hong, A., Wang, Y., et al.: CuHMMer: a load-balanced cpu-gpu cooperative bioinformatics application. In: The 2010 International Conference on High Performance Computing and Simulation (HPCS 2010). IEEE Computer Society Press, Caen, France (2010)

Objective Function of ICA with Smooth Estimation of Kurtosis

Yoshitatsu Matsuda$^{(\boxtimes)}$ and Kazunori Yamaguchi

Department of General Systems Studies, Graduate School of Arts and Sciences,
The University of Tokyo, 3-8-1, Komaba, Meguro-ku, Tokyo 153-8902, Japan
{matsuda,yamaguch}@graco.c.u-tokyo.ac.jp

Abstract. In this paper, a new objective function of ICA is proposed by a probabilistic approach to the quadratic terms. Many previous ICA methods are sensitive to the sign of kurtosis of source (sub- or super-Gaussian), where the change of the sign often causes a large discontinuity in the objective function. On the other hand, some other previous methods use continuous objective functions by using the squares of the 4th-order statistics. However, such squared statistics often lack the robustness because they magnify the outliers. In this paper, we solve this problem by introducing a new objective function which is given as a summation of weighted 4th-order statistics, where the kurtoses of sources are incorporated "smoothly" into the weights. Consequently, the function is always continuously differentiable with respect to both the kurtoses and the separating matrix to be estimated. In addition, we propose a new ICA method optimizing the objective function by the Givens rotations under the orthonormality constraint. Experimental results show that the proposed method is comparable to the other ICA methods and it outperforms them especially when sub-Gaussian sources are dominant.

1 Introduction

Independent component analysis (ICA) is a useful method in signal processing [4] and feature extraction [8]. It can solve blind source separation under the assumptions that source signals are statistically independent of each other according to any non-Gaussian distributions. In the linear model (given as $x^m = As^m$), ICA estimates the $N \times N$ mixing matrix $A = (a_{ij})$ and the N-dimensional source signals $s^m = (s_i^m)$ from only the N-dimensional observed signals $x^m = (x_i^m)$, where each m corresponds to a sample ($m = 1, \cdots, M$). N and M are the number of signals and the sample size, respectively. The estimation of A (or the separating matrix $W = A^{-1}$) is carried out by optimizing an objective function of ICA, which is generally defined by the function of some higher-order statistics.

In this paper, we propose a new objective function of ICA. Many widely-used ICA methods directly utilize higher-order statistics such as kurtosis or $\log\left(\cosh\left(u\right)\right)$ of signals (e.g. fast ICA [7] and the extended InfoMax algorithm [9]). Though such type of functions are simple, the estimation of the sign of kurtosis (corresponding to the super-Gaussian (positive) or the sub-Gaussian

© Springer International Publishing Switzerland 2015
S. Arik et al. (Eds.): ICONIP 2015, Part III, LNCS 9491, pp. 164–171, 2015.
DOI: 10.1007/978-3-319-26555-1_19

(negative) source) is crucial because the change of the sign causes a large discontinuity. Some other widely-used ICA methods avoid the discontinuity by employing the square of higher-order statistics (e.g. Comon's method [5] and JADE [2,3]). However, such methods often lack the robustness because the square operation magnifies the outliers [2]. On the other hand, the proposed objective function is a weighted summation of the 4th-order statistics without any square operation, where the weights depend on the kurtosis estimators. In addition, the function is smooth with respect to the kurtosis estimators even at the points where their signs change. This paper is organized as follows. In Sect. 2, a new objective function of ICA is derived. The optimization method of the proposed function is described in Sect. 3. Section 4 shows the experimental results on artificial datasets and an image separation problem. Lastly, Sect. 5 concluded this paper.

2   Derivation of Objective Function of ICA

Here, a new objective function of ICA is derived by a probabilistic approach to the quadratic terms of estimated sources. We have developed "a probabilistic approach to cumulants in ICA" for improving the robustness and the efficiency of JADE [10–12]. By estimating theoretically the "true" probability distribution of some cumulants or polynomials under the conditions that the sources are accurately estimated, the approach could give an objective function of ICA as its log-likelihood. In this paper, the probabilistic approach is applied to the quadratic terms of observed signals and those of "true" sources. In what follows, (1) the preliminary conditions and notations are described; (2) the "true" probability distribution is theoretically estimated by the Gaussian approximation; (3) the objective function is derived as its log-likelihood.

2.1   Preliminaries

A noiseless linear ICA model $x^m = As^m (x_i^m = \sum_j a_{ij} s_j^m)$ in the real domain is assumed, where A is an invertible square matrix. In addition, the mean and the variance of each independent source s_i^m are 0 and 1, respectively. In other words, $E_s(s_i^m) = 0$ and $E_s(s_i^m s_j^m) = \delta_{ij}$ where $E_s()$ is the expectation operator over s^m and δ_{ij} is the Kronecker delta. Note that the above conditions are generally assumed in many ICA methods. The quadratic terms of x^m are defined as $\kappa_{ij}^m = x_i^m x_j^m$. A matrix notation is introduced by $\Gamma^m = (\kappa_{ij}^m)$. Furthermore, $\tilde{\Gamma}^m = (\tilde{\kappa}_{ij}^m)$ is defined as the "true" matrix of the quadratic terms of the sources, where $\tilde{\kappa}_{ij}^m = s_i^m s_j^m$. It is easily shown that $\tilde{\Gamma}^m = W \Gamma^m W^T$ holds where $W = A^{-1}$.

2.2   Estimation of True Distribution

Here, the "true" probability distribution on all the elements of $\Gamma^m = (\kappa_{ij}^m)$ for each sample m is approximately estimated.

First, using the Gaussian approximation, we estimate the distribution of the "true" matrix on the sources $\tilde{\boldsymbol{\Gamma}}^m = \left(\tilde{\kappa}_{ij}^m = s_i^m s_j^m\right)$. The first and second moments of $\tilde{\kappa}_{ij}^m \; (i \leq j)$ are estimated as follows. It is easily shown that the first moment $E_s\left(\tilde{\kappa}_{ij}^m = s_i^m s_j^m\right)$ is δ_{ij}. The second-order moment of $\tilde{\kappa}_{ij}^m (i \leq j)$ and $\tilde{\kappa}_{kl}^m (k \leq l)$ is given as

$$E_s\left(\tilde{\kappa}_{ij}^m \tilde{\kappa}_{kl}^m\right) = E_s\left(s_i^m s_j^m s_k^m s_l^m\right) = \kappa_{ijkl}^s + \delta_{ij}\delta_{kl} + \delta_{ik}\delta_{jl} + \delta_{il}\delta_{jk}$$

where κ_{ijkl}^s is the 4th-order cumulant of s^m for each quadruplet (i, j, k, l). As each s_i^m is an independent source, every κ_{ijkl}^s is 0 except κ_{iiii}^s. Thus,

$$E_s\left(\tilde{\kappa}_{ij}^m \tilde{\kappa}_{kl}^m\right) = \begin{cases} \alpha_i + 1 & (i = j = k = l), \\ 1 & (i = k, j = l, i < j) \text{ or } (i = j, k = l, i \neq k), \\ 0 & (\text{otherwise.}) \end{cases} \quad (1)$$

where $\alpha_i = \kappa_{iiii}^s + 2$. Note that $(i \leq j)$ and $(k \leq l)$ are assumed here. The set of the unknown parameters is given as a vector $\boldsymbol{\alpha} = (\alpha_i)$. Then, the Gaussian approximation is applied to the estimated first and second moments of $\tilde{\kappa}_{ij}^m (i \leq j)$. Here, $\tilde{\kappa}_{ji}^m (i < j)$ is equal to $\tilde{\kappa}_{ij}^m$ by algebraic symmetry. As the derived log-likelihood is not essentially changed even if any fixed prior distributions are employed for $\tilde{\kappa}_{ji}^m (i < j)$, an independently and identically uniform distribution $u(x) = K$ is assumed for simplicity in this paper. Thus, the Gaussian approximated distribution of $\tilde{\boldsymbol{\Gamma}}^m$ is given as follows:

$$P^{\tilde{\Gamma}}\left(\tilde{\boldsymbol{\Gamma}}^m\right) = K^{N(N-1)/2} \prod_{i,j>i} g_{off}^{ij}\left(\tilde{\kappa}_{ij}^m\right) g_{on}\left(\text{diag}\left(\tilde{\boldsymbol{\Gamma}}^m\right)\right) \quad (2)$$

where $g_{off}^{ij}(\tilde{\kappa})$ is a Gaussian distribution of each off-diagonal elements $(i < j)$ given by

$$g_{off}^{ij}(\tilde{\kappa}) = \exp\left(-\tilde{\kappa}^2/2\right)/\sqrt{2\pi} \quad (3)$$

whose mean and variance are 0 and 1, respectively (see the second row in Eq. (1)). $g_{on}(\tilde{\kappa})$ is a multidimensional Gaussian distribution of the N-dimensional vector of the on-diagonal elements $\text{diag}\left(\tilde{\boldsymbol{\Gamma}}^m\right) = (\tilde{\kappa}_{ii}^m)$ by

$$g_{on}\left(\text{diag}\left(\tilde{\boldsymbol{\Gamma}}^m\right)\right) = \frac{\exp\left(-\frac{1}{2}\left(\text{diag}\left(\tilde{\boldsymbol{\Gamma}}^m\right) - \mathbf{1}\right)' \boldsymbol{\Sigma}^{-1} \left(\text{diag}\left(\tilde{\boldsymbol{\Gamma}}^m\right) - \mathbf{1}\right)\right)}{\sqrt{(2\pi)^N |\boldsymbol{\Sigma}|}} \quad (4)$$

where $\mathbf{1}$ is the N-dimensional ones vector, which means that the mean of each $\tilde{\kappa}_{ii}^m$ is 1. $|\boldsymbol{\Sigma}|$ is the determinant of $\boldsymbol{\Sigma}$. By the first and second rows in Eq. (1), $\boldsymbol{\Sigma} = (\sigma_{ij})$ is given as

$$\sigma_{ij} = E_s\left(\tilde{\kappa}_{ii}^m \tilde{\kappa}_{jj}^m\right) - 1 = \begin{cases} \alpha_i & (i = j), \\ 0 & (\text{otherwise.}) \end{cases} \quad (5)$$

Therefore, Σ^{-1} is a diagonal matrix, each diagonal element of which is given as $1/\alpha_i$. $K^{N(N-1)/2}$ corresponds to the off-diagonal symmetric elements. Since Σ is determined by α, the derived distribution can be denoted as the conditional distribution $P^{\tilde{\Gamma}}\left(\tilde{\Gamma}^m|\alpha\right)$.

Second, the distribution of Γ^m is estimated from the above $P^{\tilde{\Gamma}}\left(\tilde{\Gamma}^m|\alpha\right)$. By utilizing the transformation $\tilde{\Gamma}^m = W\Gamma^m W^T$, the linear transformation matrix from the vectorized elements of Γ^m to those of $\tilde{\Gamma}^m$ is given as the Kronecker product $W \otimes W$. Therefore, the distribution of Γ^m is given as

$$P^{\Gamma}\left(\Gamma^m|W,\alpha\right) = |W \otimes W|P^{\tilde{\Gamma}}\left(\tilde{\Gamma}^m|\alpha\right) = |W|^{2N}P^{\tilde{\Gamma}}\left(\tilde{\Gamma}^m|\alpha\right). \qquad (6)$$

This is the estimated "true" distribution of $\Gamma^m = (\kappa_{ij})$ whose conditional parameters are W and α.

2.3 Objective Function

In order to estimate the optimal parameters for W and α, the objective function is defined as the log-likelihood of $P^{\Gamma}\left(\Gamma^m|W,\alpha\right)$ in Eq. (6), which is given as

$$\ell(W,\alpha) = \sum_m \log P^{\Gamma}\left(\Gamma^m|W,\alpha\right)$$

$$= -\sum_m \sum_{i,j>i} \left(\tilde{\kappa}_{ij}^m\right)^2 - M\sum_i \log\alpha_i - \sum_m \sum_i \frac{\left(\tilde{\kappa}_{ii}^m - 1\right)^2}{\alpha_i} + 4NM\log|W| \qquad (7)$$

where some constants are omitted and the factor $\frac{1}{2}$ is removed. Note that $\tilde{\kappa}_{ij}^m$ is no longer the "true" value on the sources but an estimated value for the current W. Therefore, the estimated $\tilde{\kappa}_{ij}^m$ can be rewritten as $y_i^m y_j^m$ where $y^m = (y_i^m) = Wx^m$ is the current estimator of s^m. Consequently, Eq. (7) is rewritten as

$$\ell(W,\alpha) = -\sum_i \log\alpha_i - \sum_{i,j}\left(\frac{1-\delta_{ij}}{2} + \frac{\delta_{ij}}{\alpha_i}\right)\frac{\sum_m \left(y_i^m y_j^m - \delta_{ij}\right)^2}{M} + 4N\log|W| \qquad (8)$$

where the factor M is removed. This is the proposed objective function of ICA.

3   Optimization Method

Here, we describe the optimization method for maximizing the objective function ℓ in Eq. (8). An alternating algorithm is employed where α and W are optimized alternately.

Regarding the optimization of ℓ with respect to α for a fixed W, its derivative gives the optimal $\hat{\alpha}_i$ analytically as follows:

$$\hat{\alpha}_i = \frac{\sum_m \left(\left(y_i^m\right)^2 - 1\right)^2}{M}. \qquad (9)$$

In order to optimize ℓ with respect to $\boldsymbol{W}$ for a fixed $\boldsymbol{\alpha}$, $\boldsymbol{x}^m$ is pre-whitened and $\boldsymbol{W}$ is limited to be orthonormal in this paper. Though this orthonormality constraint is not necessary in the original form of ℓ, it is employed here for the simple optimization. Then, the terms $|\boldsymbol{W}|$, $\sum_i (y_i^m)^2$, and $\sum_{i,j} (y_i^m y_j^m)^2$ in Eq. (8) are constant irrespective of $\boldsymbol{W}$. Therefore, Eq. (8) is simplified into

$$\ell(\boldsymbol{W}) = \sum_i v_i (y_i^m)^4 \tag{10}$$

where each weight v_i is given by

$$v_i = \frac{1}{2} - \frac{1}{\alpha_i}. \tag{11}$$

It is a weighted summation of the kurtoses. As $\alpha_i = \kappa_{iiii}^s + 2$ is equal to 2 when the source distribution is Gaussian, the weight v_i is positive for super-Gaussian source, or negative otherwise. It is consistent with the standard ICA framework. Note that this weight is smooth and causes no discontinuity. In order to optimize $\ell(\boldsymbol{W})$, the Jacobi method is employed in the similar way as in JADE [3] and Comon's method [5], which repeats a simple Givens rotation of each pair of signals in order to optimize the total objective function. For a pair (y_i^m, y_j^m), the rotated pair $(\tilde{y}_i, \tilde{y}_j^m)$ with an angle θ is given as $\begin{pmatrix} \tilde{y}_i^m \\ \tilde{y}_j^m \end{pmatrix} = \begin{pmatrix} \cos\theta & \sin\theta \\ -\sin\theta & \cos\theta \end{pmatrix} \begin{pmatrix} y_i^m \\ y_j^m \end{pmatrix}$.
Let ℓ_{ij} be the sum of the i-th and j-th terms in ℓ. After some transformations, $\ell_{ij}(\theta)$ is given as

$$\ell_{ij}(\theta) = \beta_1 \sin 4\theta + \beta_2 \cos 4\theta + \beta_3 \sin 2\theta + \beta_4 \cos 2\theta + \beta_5 \tag{12}$$

where

$$\beta_1 = \frac{(v_i + v_j) \sum_m \mu_{ij}^m \nu_{ij}^m}{2}, \tag{13}$$

$$\beta_2 = \frac{(v_i + v_j) \sum_m \left((\nu_{ij}^m)^2 - 4 \sum_m (\mu_{ij}^m)^2 \right)}{8}, \tag{14}$$

$$\beta_3 = (v_i - v_j) \sum_m \mu_{ij}^m \xi_{ij}^m, \tag{15}$$

$$\beta_4 = \frac{(v_i - v_j) \sum_m \nu_{ij}^m \xi_{ij}^m}{2}, \tag{16}$$

and β_5 is a constant. Here, μ_{ij}^m, ν_{ij}^m, and ξ_{ij}^m are defined as

$$\mu_{ij}^m = y_i^m y_j^m, \tag{17}$$

$$\nu_{ij}^m = (y_i^m)^2 - (y_j^m)^2, \tag{18}$$

$$\xi_{ij}^m = (y_i^m)^2 + (y_j^m)^2. \tag{19}$$

Then, the optimal $\hat{\theta}$ is given as $\hat{\theta} = \operatorname{argmax}_\theta \ell_{ij}(\theta)$ which increases the total ℓ monotonically. Though the optimum of $\ell_{ij}(\theta)$ can be given analytically by solving a quartic equation, it is optimized numerically in this paper because the numerical approach is easily used for such a periodic function with a single parameter. A simple MATLAB function "fminbnd" was employed. Regarding the termination for each pair optimization, a small threshold for $\hat{\theta}$ was used.

The complete process of the optimization method is given as follows:

1. *Initialization.* Whiten given observed signals x^m, let y^m be x^m, and set W to the identity matrix.
2. *Optimization.* For every pair (i, j) of signals,
   (a) Estimate $\hat{\alpha}_i$ and $\hat{\alpha}_j$ by Eq. (9).
   (b) Update W and y^m by the optimal Givens rotation minimizing $\ell_{ij}(\theta)$ of Eq. (12)
3. *Convergence decision.* If no pair has been rotated in Step 2, end. Otherwise, go back to Step 2.

This process is guaranteed to converge to a minimum because it monotonically increases the objective function $\ell(W, \alpha)$.

4   Results

The proposed method was compared with the following three widely-used ICA algorithms: fast ICA [7] (using the two types of objective function: the kurtosis-based one and the log(cosh)-based one), the extended InfoMax [3] (using the log(cosh)-based objective function with adaptive estimation of kurtosis) and JADE (using the square of the 4th-order cumulants). We solved five blind source separation problems using artificially generated sources and one natural image separation problem. The number of sources N was set to 12. Regarding artificial data, each source is generated according to the Laplace distribution (super-Gaussian) or the uniform one (sub-Gaussian). The ratio of sub-Gaussian sources was set to 0%(=0/12), 25%, 50%, 75%, and 100% (=12/12). Regarding the image separation, 12 images were selected from the USC-SIPI database. They consisted of 7 super-Gaussian images (with positive kurtosis) and 5 sub-Gaussian ones (with negative kurtosis). The square mixing matrix A was randomly generated. The size of samples M was set from 100 to 10000. The separating errors were measured by the averages of Amari's separating errors [1] over 10 runs. In fast ICA, (A) $g(u) = u^3$ (corresponding to the kurtosis-based objective function) or (B) $g(u) = \tanh(u)$ (corresponding to the log(cosh)-based one) was employed as the non-linear function. The implementation of the extended Info-Max algorithm was given in EEGLAB [6].

Figure 1 shows the log-log curves of the separating errors along the sample size for various sources. First of all, they show that the proposed method gave comparable results to the three widely-used methods: fast ICA (A) using kurtosis, fast ICA (B) using log cosh (u), and JADE. In the cases that the super-Gaussian sources are dominant (Fig. 1(a) and (b)), the proposed method is slightly inferior to fast ICA (B). It is probably because log(cosh)-based objective function is well suited for the Laplace distribution. Nevertheless, the proposed method gave the second best results. In Fig. 1(c) (the neutral mixtures of super- and sub-Gaussian sources), the results of the proposed method and fast ICA (B) are almost the same. In the cases that the sub-Gaussian sources are dominant (Fig. 1(d) and (e)), the proposed method gave the best results for almost all sample sizes. Especially if all the sources are sub-Gaussian for small sample sizes (the left part of Fig. 1(e)), the proposed method clearly outperformed the other methods. Though JADE gave similar good results for large sample sizes in Fig. 1(e), it is not robust for small sample sizes. In Fig. 1(f) (an actual dataset in the image separation problem), the proposed method is slightly inferior to fast ICA (A) and (B) for small sample sizes. However, when the sample size is large, the proposed method gave the best results. In summary, the proposed method gave at least the second best results in all the cases and it showed no significant deterioration in performance. This verifies the robustness of the proposed method which depends on neither the source distributions nor the sample size.

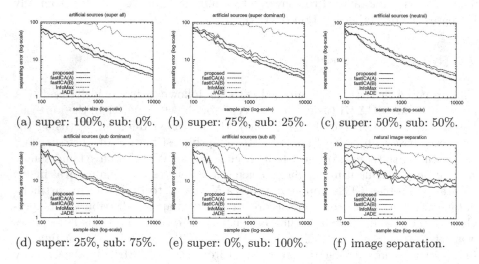

(a) super: 100%, sub: 0%.   (b) super: 75%, sub: 25%.   (c) super: 50%, sub: 50%.

(d) super: 25%, sub: 75%.   (e) super: 0%, sub: 100%.   (f) image separation.

Fig. 1. Comparison of the separating errors along the sample size (in log-log scale): In (a)–(e), artificial sources (according to the Laplace distribution (super-Gaussian) or the uniform one (sub-Gaussian)) were used where the ratio of sub-Gaussian sources was changed from 0 % to 100 %. In (f), natural images were used as the sources in an image separation problem. Each log-log graph shows the curves of the separating errors along the sample size by the five ICA methods: the proposed method (solid curve), fast ICA (A) using the kurtosis-based objective function (thin dashed), fast ICA (B) using the log (cosh)-based one (thick dashed), the extended InfoMax (dotted), and JADE (dot-dashed).

5   Conclusion

In this paper, we propose a new objective function of ICA and its optimization method. The objective function is smooth with respect to the kurtosis estimators. The experimental results show that the proposed method gave comparable results to other ICA methods and outperformed them when the sub-Gaussian sources are dominant. Moreover, the results shows the robustness of the proposed method in various artificial source separations and an image separation. We are now planning to develop more efficient optimization methods by the other techniques such as the stochastic gradient. Furthermore, we are planning to extend the proposed method to non-orthogonal cases. We are also planning to apply the proposed method to other practical datasets. This work is partially supported by Grant-in-Aid for Young Scientists (KAKENHI) 26730013.

References

1. Amari, S., Cichocki, A.: A new learning algorithm for blind signal separation. In: Touretzky, D., Mozer, M., Hasselmo, M. (eds.) Advances in Neural Information Processing Systems, vol. 8, pp. 757–763. MIT Press, Cambridge (1996)
2. Cardoso, J.F.: High-order contrasts for independent component analysis. Neural Comput. **11**(1), 157–192 (1999)
3. Cardoso, J.F., Souloumiac, A.: Blind beamforming for non Gaussian signals. IEE Proceedings-F **140**(6), 362–370 (1993)
4. Cichocki, A., Amari, S.: Adaptive Blind Signal and Image Processing: Learning Algorithms and Applications. Wiley, New York (2002)
5. Comon, P.: Independent component analysis - a new concept? Signal Process. **36**, 287–314 (1994)
6. Delorme, A., Makeig, S.: EEGLAB: an open source toolbox for analysis of single-trial EEG dynamics including independent component analysis. J. Neurosci. Methods **134**(1), 9–21 (2004)
7. Hyvärinen, A.: Blind source separation by nonstationarity of variance: a cumulant-based approach. IEEE Trans. Neural Netw. **12**(6), 1471–1474 (2001)
8. Hyvärinen, A., Karhunen, J., Oja, E.: Independent Component Analysis. Wiley, New York (2001)
9. Lee, T.W., Girolami, M., Sejnowski, T.J.: Independent component analysis using an extended infomax algorithm for mixed subgaussian and supergaussian sources. Neural Comput. **11**(2), 417–441 (1999)
10. Matsuda, Y., Yamaguchi, K.: An adaptive threshold in joint approximate diagonalization by assuming exponentially distributed errors. Neurocomputing **74**, 1994–2001 (2011)
11. Matsuda, Y., Yamaguchi, K.: A robust objective function of joint approximate diagonalization. In: Villa, A.E.P., Duch, W., Érdi, P., Masulli, F., Palm, G. (eds.) ICANN 2012, Part II. LNCS, vol. 7553, pp. 205–212. Springer, Heidelberg (2012)
12. Matsuda, Y., Yamaguchi, K.: Ensemble joint approximate diagonalization by an information theoretic approach. In: Lee, M., Hirose, A., Hou, Z.-G., Kil, R.M. (eds.) ICONIP 2013, Part III. LNCS, vol. 8228, pp. 309–316. Springer, Heidelberg (2013)

FANet: Factor Analysis Neural Network

Jiawen Huang[✉] and Chun Yuan

Graduate School at Shenzhen, Tsinghua University Shenzhen,
Guangdong 518055, China
huangjw13@mails.tsinghua.edu.cn, yuanc@sz.tsinghua.edu.cn

Abstract. A cascaded factor analysis network is proposed in this paper, which is suitable for extracting distributed semantic representations to various problems ranging from digit recognition and image classification to face recognition. There are two key points in this novel model: 1. simplify and accelerate the deep convolution networks with competitive accuracy even state-of-the-art for many general image tasks; 2. combine a statistical methodfactor analysis with neural networks for excellent automatically learning ability and abundant semantic information. Experiments on many benchmark visual datasets demonstrate that this simple network performs efficiently and effectively while attaining competitive accuracy to the current state-of-the-art methods.

Keywords: Convolutional neural network · Factor analysis · Image classification

1 Introduction

Over the last decades, numerous efforts have been devoted to designing appropriate feature descriptors for tasks from image classification to object recognition. Some designed manual low-level features to solve this problem, such as SIFT and HOG features. However, these features are designed for specific task and data. When applied for other tasks or data, the features become ineffective. Some models utilize two or more successive layers of such feature extractors, fol-lowed by a supervised classifier. One of the representatives of multi-layers models is deep neural networks (DNNs), which attempts to learn higher level features by hierarchical features extracting. It can learn higher levels of latent semanteme automatically, containing some complex functions mapping the input data to the output, without using manual features. In DNNs, Convolution Neural Networks (CNNs) [10] is one of the most popular structures for image classification. CNNs has been successfully applied in considerable complex image datasets, which include digit recognition (MNIST dataset [10]), object recognition (NORB dataset [11]), and image classification (CIFAR-10 dataset [8]). It is based on two key concepts: local receptive fields, and tied weights. The utility of local receptive fields is imitated from human visual system, which only looks at a localized region of the image. It makes CNNs more efficient in training and more stable in

ⓒ Springer International Publishing Switzerland 2015
S. Arik et al. (Eds.): ICONIP 2015, Part III, LNCS 9491, pp. 172–181, 2015.
DOI: 10.1007/978-3-319-26555-1_20

transformations. Moreover, weight-tying forces the weights in one filter sharing the same weight, which observably reduces the number of learnable parameters.

However, despite the super success of these deep network architectures, experience of weight-tuning and extra training tricks are necessities for these deep networks. Whats more, the lack of fundamental theory makes it hard to train and understand.

Therefore, some researchers start pondering their properties by designing novel optimal configurations of them. Reference [5] used a single-layer network in unsupervised feature learning with huge amounts of units, which achieve performance beyond all previous results on the CIFAR-10 and NORB. Reference [14] demonstrated the importance of convolution pooling architectures, which can be inherently frequency selective and translation invariant, even with random weights. The first instance that has led to clear mathematical justification was the wavelet scattering networks [2]. With the prefixed convolution filters, ScatNet has demonstrated superior performance over CNNs in several challenging vision tasks. Then another simple deep network appeared, [7] applied principal component analysis (PCA) into deep networks, earning new records for many classification tasks. However, PCA put more emphasis on dimensionality reduction and lack of the capacity of explaining the relationship between features and data. Moreover, PCANet may be unpractical to some complex deformation even though it may be invariant to some linear translation. Non-linear deformations also induce important variability within image classes. For instance, image deformations due to noise must be taken into account.

In order to address these problems, a new neural network model, named as FANet, is introduced by using the factor analysis algorithm. Factor analysis is a statistical method used to describe variability among observed, correlated variables in terms of a potentially lower number of unobserved variables called factors, which are regarded as features in machine learning. The FANet, a tractable multiple layer model, has fast approximate inference and still retains simplicity of the operation towards training. Considerable experiments indicate that this simple and fast network is enough to obtain excellent performance comparing with some time-consuming complicated models even though without common training tricks.

The main contributions of the paper are (1) To the best of our knowledge, FANet is the first to use factor analysis for CNNs, which combines a statistic method and neural network. This combination reserves the excellent learning ability of CNNs and represents more statistic semanteme of the input data, which makes the theoretical exploration of CNNs more available; (2) In comparison to CNNs, the overall training time is much less since FANet has no use for timeconsuming iteration; and (3) a practical method for considerable domains ranging from digit recognition to face recognition, which earns competitive accuracy.

2   Unsupervised Feature Learning via FANet

In FANet, factor analysis is employed as an unsupervised learning algorithm that extracts features from image patches. A FANet can be described as a

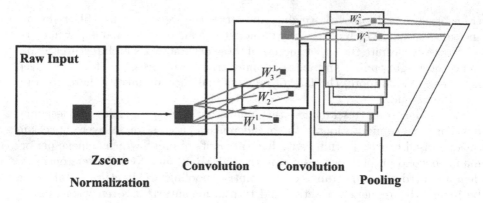

Fig. 1. FANet: Factor Analysis Neural Network (2 convolution layers and 1 pooling layer). W_j^l is denoted as the weight of j_{th} filter in l_{th} layer.

multilayered neural network, with multi-convolution layers followed by a pooling layer Fig. 1. With new-designed operations in convolution layers, the advantage of varied-semantic factor analysis and automatic-learning neural network will be taken full use to extract features. Since the optimization problem in training stage has already been solved by factor analysis, the time-consuming iterative refinement is needless in FANet. This made the model effective while earning superior performance. Moreover, this design of construction will lead to new DNNs models in the future, such as combining proved optimization methods with deep hierarchical networks, or using this simple extractor as an interlayer in training stage. Last but not the least, mathematical analysis and justification of DNNs will be immensely simplified by this theory-existed operation. In the following description, the symbols k is uniformly denoted to be the size of local receptive fields, and L as the amount of filters.

2.1   Normalizing the Input Data

Propose the dataset has N image samples for training. In order to accord with the condition of factor analysis, the given input image Img is preprocessed as follows.

1. every image Img, is taken into $b = kk$ overlapping patches;
2. z-score normalization is used for every patch X, by subtracting its mean, and dividing its standard deviation.

The normalized patch is denoted as $\bar{X}$.

2.2   Convolution Layers

Factor analysis is an algorithm, which focuses on obtaining some latent factors F, the amount of which is much smaller than the raw input size of data. Every variable in the input data can be drawn by the combination of the factors F. And the specificity of the variable will be denoted as the distributed error terms ε.

$$\bar{X} = AF + \varepsilon, \bar{X} \in R^{b \times 1}, A \in R^{b \times L}, F \in R^{L \times 1}, \varepsilon \in R^{b \times 1}, \tag{1}$$

where A is the factor weight matrix, quantifiably representing the relationship between $\bar{X}$ and F. The target of the algorithm is to find a suitable A for working out this optimization problem,

$$min \|\bar{X} - AF\|^2, s.t.\ D(F) = I_L \tag{2}$$

For the reason of information lossless, with the distribution unchanged as assumed, it is obvious that $D(\bar{X}) = D(AF)$, and the solution of Eq. 2 is clear,

$$Corrcoef(\bar{X}) = AA' \tag{3}$$

where $Corrcoed(\bar{X})$ is the correlation coefficient matrix of all normalized patches $\bar{X}$. In hence, this optimization problems of Eq. 3 is directly worked out without any iterative refinement. So when L factors are selected to describe the input variables, the factor weight matrix A can be estimated as

$$\widetilde{A} = (\sqrt{\lambda_1}e_1, \sqrt{\lambda_2}e_2, \ldots, \sqrt{\lambda_L}e_L), s.t.\ L \leq b \tag{4}$$

e_j is the eigenvectors of $Corrcoef(\bar{X})$, λ_j is its corresponding eigenvalue. For the purpose of getting weight matrix to convolution operation, the relationship between F and $\bar{X}$ must be quantified.

$$F_j = \beta_{j1}\bar{X}_1 + \beta_{j2}\bar{X}_2 + \cdots + \beta_{jL}\bar{X}_L, j = 1, 2, \ldots, L \tag{5}$$

where β are described as factor scores in factor analysis. Thomas regression method can be employed to estimate these scores,

$$W = \widetilde{A}'Corrcoef(\bar{X})^-1 \tag{6}$$

So the output of this layer is,

$$Output_l = W * Output_{l-1} \tag{7}$$

in which l is standing for the l_{th} layers, $*$ is the inner-product operation, and $Output_1 = Input$.

2.3   Pooling Layers

The pooling layer is set in much the same way as PCANet [4]. The value of $Output$ is changed into binary data (by changing the positive value into 1 and the negative value into 0). And then weight of every filter is assigned incrementally in the order of the eigenvalue got from the convolution layers, and sum these value in block-size. The operation can be described as,

$$Total = \sum_{j=1}^{L} 2^j Bin(Output_{l-1}) \tag{8}$$

Where the function *Bin* is the binary-like function change the positive value into 1 and the negative ones into 0. Finally, all *Total* of the given image will be concentrated as a vector. This setting is much like the full connection layers in CNNs but without any learnable parameters.

2.4 Analytical Characterization of FANet

In convolution layers, the main factors of all variables are extracted, then the relationship between variables and learned features is estimated by switching the weight matrix to factor scores. So it is easy to visualize the importance of every variable in these learned features. The main function of this layer is to quantify the relationship between variables and features while extracting the main direction of the data (the filters learned by FANet on FERET are shown in Fig. 2 RIGHT). Multiple convolution layers are set in FANet before the pooling layer in order to take the advantage of cascading to reinforce the feature filters.

The design of FANet taked over the advantage of CNNs [10], such as the layer catalogue and the concept of local receptive fields. However the training algorithm is totally changed. The back propagation learning algorithm (BP), which is common used in training Neural Networks including CNNs, Contractive AutoEncoder (CAEs) [10], Deep Belief Networks (DBNs) [6], can be divided into two phase propagation and weight update. Time-consuming iteration and fine-tuning experience of the learning rate and momentum is necessary for performance in classification or recognition tasks. FANet used a worked-out algorithm to solving these problems. Without repeating iteration of weight update, the efficiency is enhanced remarkably because FANet has no weight update phase and only onetime propagation. It is a useful way to simplify CNNs while keeping powerful performance. Comparing with the filters learned from traditional CNNs (Fig. 2 LEFT), it is apparent that the filters learned by FANet are more orderly. Main directions of the change in frequency are extracted by these filters. Hence the architecture is frequency selective. Additionally, FANet can be used as a previous layer or inter-layer in complex CNNs models for speed-up convergence.

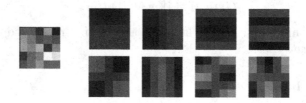

Fig. 2. Filters learning from FERET dataset, LEFT: by CNNs, RIGHT: by FANet

3   Experiments

In this section, FANet model is evaluated as a feature extraction method for recognition and classification. To demonstrate the adaption for different domains, three datasets are obtained - MNIST handwritten digit database and its variations [9], and FERET facial recognition database [12]. Without special illustrations, the filter amount is fixed as 8, for the reason of the eight direction in image filtering, and the classification is studied with linear SVM classifiers. Since the size of the input image is small in following experiments, two convolution layers are enough for evaluation. One should note that, for larger image input, the more layers will be more capable to hold the extraction of features.

3.1 Facial Recognition on FERET

The FERET database comprises of 14051 gray-scale images including 1199 individuals. The images contain variations in lighting, facial expressions, pose angle etc. These facial images can be divided into five sets followings, Fa set (as gallery images), containing frontal images of 1196 people; fb set (1195 images) is different in illusion from fa set; fc set (194 images) was variant lighting-condition image; dup-I set (722 images) was taken later in time; dup II set (234 images) is a subset of the dup-I set containing those images that were taken at least a year after the corresponding gallery image. These gray-scale images are cropped into the size of 150 × 90 pixels. The results are displayed in Table 1. The FANet achieves the state-of-the-art accuracies in almost all tasks and reaches at the accuracy of 97.40 % on average.

Table 1. Error rate on FERET dataset (%)

Algorithms	fb	fc	dup-I	dup-II	Aver
LBP [1]	7	49	39	50	36.25
P-LBP [16]	2	2	10	15	7.25
POEM [18]	0.4	0.5	11.2	15	6.77
SPOEM [17]	1	0	6.8	6	3.45
G-LQP [7]	**0.1**	1	6	9	4.02
LGBP-LGXP [19]	0.3	0	5.1	7	3.1
GOM [3]	0.1	0	4.3	6.9	2.82
PCANet-2 [4]	0.33	0.52	**4.16**	**5.98**	2.75
FANet	0.25	0	**4.16**	**5.98**	**2.6**

Not only to challenge ourselves but also to prove the superiority in recognition for images sustaining more complex deformations, the subset of FERET is utilized, which obtains 1400 images from 200 individuals (each individual has 7 images, including a front image and its variations in facial expression, illumination, 15 and 30 pose). To start with, the amount of training samples is changed from 1 per individual to 6 per individual, and the rest of samples are used for testing. The training samples are selected at random from the FERET subset

Table 2. Accuracy on FERET subset with different proportion between numbers of training samples and testing samples

Proportion	1:6	2:5	3:4	4:3	5:2	6:1
PCANet-2 [4]	28.7	50.0	63.5	74.0	78.5	84.0
FANet	31.9	50.8	66.4	75.0	79.8	86.0

Table 3. Accuracy on FERET subset with generalized noise

Algorithms	GN	OO	MGN	RN
PCANet-2 [4]	43.75	44.75	50.12	48.75
FANet	44.25	45.75	51.12	51.12

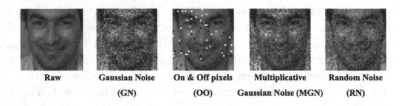

Raw   Gaussian Noise   On & Off pixels   Multiplicative   Random Noise
(GN)   (OO)   Gaussian Noise (MGN)   (RN)

Fig. 3. Visualization of varying kind of an added-noise test image in FERET subset

and then these experiments are repeated for 10 times and accuracies are averaged. The results are given in Table 2. FANet outperforms the PCANet [4] in all different proportions. Whats more, the max gap is 3.17 %, and the average gap is 1.85 %. Then, in order to testify the consistence of model for more complicated image deformation, some kinds of noise are added into the test samples (3 samples from every individual are randomly picked for training and the rest for testing), such as Gaussian noise and random noise. The deformation samples are shown in Fig. 3 and the results are shown in Table 3. No matter which kind of noise was added to the images, the performance of FANet is better than the PCANet [4].

3.2 Handwritten Digit Recognition on MNIST and MNIST Variations

In MNIST database of handwritten digits, there is a training set of 60,000 examples, and a testing set of 10,000 examples. In order to make a thorough inquiry with the capacity of FANet while confronting transformation which contains many factors of variation, the MNIST variation datasets are used, which have generated datasets from MNIST by adding diverse perturbations and including 9 classification tasks. The visualization of samples are given in Fig. 4, and the description of the validation datasets is as follows. Rot, the digits were rotated by an angle generated uniformly; back-rand, a random background was inserted in the digit image; back-image, a patch from a black and white image was used

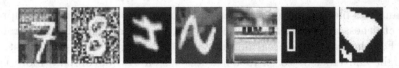

Fig. 4. Samples of MNIST validations, left to right: back-img, back-rand, rot, bg-img-rot, rect-img, rectangle, convex

Table 4. Error rate on MNIST and MNIST validations (%)

Data-set	CAE [13]	PGBM+DN-1 [15]	ScatNet-2 [2]	PCANet-2 [4]	FANet
MNIST	1.14	-	**0.43**	0.66	0.60
basic	2.83	-	1.27	1.06	**0.05**
back-img	16.68	12.25	12.3	6.19	**5.98**
back-rand	13.57	**6.08**	18.4	10.95	10.54
rot	11.59	-	7.48	7.37	**6.99**
bg-img-rot	48.10	36.76	50.48	35.48	**34.54**
retangles	1.48	-	0.01	0.24	0.19
rect-img	21.86	-	**8.02**	14.08	12.85
convex	-	-	6.5	4.36	**3.81**

as the background for the digit image; rot-back-image (bg-img-rot), the perturbations used in rot and back-image were combined; rectangles, the pixels corresponding to the border of the rectangle has a value of 255, the rest are 0, the height and width of the rectangles were sampled uniformly; rectangles-image (rect-img), the border and inside of the rectangles corresponds to an image patch and a background patch is also sampled; convex, consisting of a single convex region with pixels of value 255. All of these images from MNIST and MNIST variations are 28 × 28 grayscale image. The size of local receptive fields is set to be 7 pixels. The pooling size is firstly set as 8, and then tuned by a cross-validation for MNIST and its validation sets (one-fifth size of the training sets). To be fair, the results obtained by only using original training samples are displayed. The results of this series of experiments are given in Table 4. While the error rates in MNIST are very close, it is apparent that FANet gains state-of-the-art accuracies in half of the validation datasets. Additionally, in all of the MNIST validation datasets, FANet outperforms the PCANet at most 1.23 %. These results proves the description mention above, FANet can cover more complex transformation of the input data since the statistics ability which comes from the factor analysis algorithm.

4   Conclusion

In this paper, a novel feature learning model, called factor analysis neural network, is proposed for general image classification. By making the best of factor analysis method into deep neural networks suitably, this model improved the effectiveness and efficiency of the training stages in convolution neural networks. For the filter maps learning does not involve iterative refinement, it is so straight and efficient to extract features which also encompass more useful statistic semanteme for further theoretical analysis. In future work, we expect to explore the theoretical reasons for the excellent performance in deep network construction base on this simple but powerful model.

References

1. Ahonen, T., Hadid, A., Pietikainen, M.: Face description with local binary patterns: application to face recognition. IEEE Trans. Pattern Anal. Mach. Intell. **28**(12), 2037–2041 (2006)
2. Bruna, J., Mallat, S.: Invariant scattering convolution networks. IEEE Trans. Pattern Anal. Mach. Intell. **35**(8), 1872–1886 (2013)
3. Chai, Z., Sun, Z., Mendez-Vazquez, H., He, R., Tan, T.: Gabor ordinal measures for face recognition. IEEE Trans. Inf. Forensics Secur. **9**(1), 14–26 (2014)
4. Chan, T.-H., Jia, K., Gao, S., Lu, J., Zeng, Z., Ma, Y.: Pcanet: A simple deep learning baseline for image classification? (2014). arXiv preprint arXiv:1404.3606
5. Coates, A., Ng, A.Y., Lee, H.: An analysis of single-layer networks in unsupervised feature learning. In: International Conference on Artificial Intelligence and Statistics, pp. 215–223 (2011)
6. Hinton, G.E., Osindero, S., Teh, Y.-W.: A fast learning algorithm for deep belief nets. Neural Comput. **18**(7), 1527–1554 (2006)
7. Hussain, S.U., Napoléon, T., Jurie, F.: Face recognition using local quantized patterns. In: British Machive Vision Conference, 11 p. (2012)
8. Krizhevsky, A., Hinton, G.: Learning multiple layers of features from tiny images (2009)
9. Larochelle, H., Erhan, D., Courville, A., Bergstra, J., Bengio, Y.: An empirical evaluation of deep architectures on problems with many factors of variation. In: Proceedings of the 24th International Conference on Machine Learning, pp. 473–480. ACM (2007)
10. LeCun, Y., Bottou, L., Bengio, Y., Haffner, P.: Gradient-based learning applied to document recognition. Proc. IEEE **86**(11), 2278–2324 (1998)
11. LeCun, Y., Huang, F.J., Bottou, L.: Learning methods for generic object recognition with invariance to pose and lighting. In: Proceedings of the 2004 IEEE Computer Society Conference on Computer Vision and Pattern Recognition, 2004. CVPR 2004, vol. 2, p. II-97. IEEE (2004)
12. Phillips, P.J., Wechsler, H., Huang, J., Rauss, P.J.: The feret database and evaluation procedure for face-recognition algorithms. Image Vis. Comput. **16**(5), 295–306 (1998)
13. Rifai, S., Vincent, P., Muller, X., Glorot, X., Bengio, Y.: Contractive auto-encoders: explicit invariance during feature extraction. In: Proceedings of the 28th International Conference on Machine Learning (ICML 2011), pp. 833–840 (2011)
14. Saxe, A., Koh, P.W., Chen, Z., Bhand, M., Suresh, B., Ng, A.Y.: On random weights and unsupervised feature learning. In: Proceedings of the 28th International Conference on Machine Learning (ICML 2011), pp. 1089–1096 (2011)

15. Sohn, K., Zhou, G., Lee, C., Lee, H.: Learning and selecting features jointly with point-wise gated {B} oltzmann machines. In: Proceedings of The 30th International Conference on Machine Learning, pp. 217–225 (2013)
16. Tan, X., Triggs, B.: Enhanced local texture feature sets for face recognition under difficult lighting conditions. IEEE Trans. Image Process. **19**(6), 1635–1650 (2010)
17. Ngoc-Son, V.: Exploring patterns of gradient orientations and magnitudes for face recognition. IEEE Trans. Inf. Forensics Secur. **8**(2), 295–304 (2013)
18. Ngoc-Son, V., Caplier, A.: Enhanced patterns of oriented edge magnitudes for face recognition and image matching. IEEE Trans. Image Process. **21**(3), 1352–1365 (2012)
19. Xie, S., Shan, S., Chen, X., Chen, J.: Fusing local patterns of gabor magnitude and phase for face recognition. IEEE Trans. Image Process. **19**(5), 1349–1361 (2010)

Oscillated Variable Neighborhood Search for Open Vehicle Routing Problem

Bekir Güler and Aişe Zülal Şevkli[✉]

Department of Computer Engineering, Fatih University,
34500 Büyükçekmece, İstanbul, Turkey
{bguler, zsevkli}@fatih.edu.tr

Abstract. Open Vehicle routing problems is a variant of Vehicle Routing Problem, in which vehicles don't return the depot after serving the customers. In this study, we proposed a cluster first-routed second based algorithm. We combined Kmeans and Variable Neighborhood Search in this algorithm. Our proposed algorithm achieves the best know solutions within a reasonable time for all well-known small and medium scale benchmarks.

Keywords: Metaheuristic · Open Vehicle Routing Problem · Variable Neighborhood Search · Deterministic Oscillated Perturbation

1 Introduction

Open Vehicle Routing Problem (OVRP) is a Vehicle Routing Problem (VRP). It differs from VRP by its routing plan. Whereas a vehicle starts and ends at a depot in VRP, the vehicle starts from a depot, but ends at any customer in OVRP. We can explain formally as follows: OVRP consists of n customers, K vehicles, and a depot. Each vehicle has a predefined capacity Q and each customer has a predefined demand (d_i). Each customer must be visited once by one vehicle and the total demand of customers cannot exceed the certain capacity of the vehicle. All customers' demands must be satisfied. Each vehicle starts from the depot and ends at one of the customers after visiting all customers on the vehicle's route. The main purposes of OVRP are to minimize the total distance of routes and reduce the number of vehicles. OVRP is a NP-Hard problem [1], so that the solution of the OVRP is mostly obtained by meta-heuristic algorithms.

OVRP is can be modelled for the companies which don't have the adequate amount or any vehicle(s) for transportation of the good, can outsource (vehicles) the transportation job via third party logistic firms. Due to the fact that in OVRP, vehicles don't return to the depot after delivery, so this will enable companies to save more money.

One of the earliest study on OVRP is proposed by in [2]. In the study an algorithm is proposed that uses two phases minimum spanning tree. In the first phase, customers are assigned to clusters. Then in the second phase, clustered customers are revisited by reordering the least cost ones. Most of the proposed solutions are based on meta-heuristic algorithms, such as simulated annealing, Tabu search etc. A Simulated Annealing algorithm based Backtracking Adaptive Threshold Accepting Algorithm is

S. Arik et al. (Eds.): ICONIP 2015, Part III, LNCS 9491, pp. 182–189, 2015.
DOI: 10.1007/978-3-319-26555-1_21

proposed in [3], in which the threshold values are adjusted according to previous iteration result, if the value is worse than the best, threshold value is increased and vice versa. In [4], a Tabu Search based List Based Threshold Accepting Algorithm is proposed, which uses a list of threshold values to guide local search. In the study, only distance minimization is considered. A deterministic variant of simulated annealing based algorithm, named as Record to Record algorithm is proposed in [5]. A new set of large scale datasets (O series) is also generated in this study. The size of the test problems ranges from 200 to 480. In [6] an Adaptive Large Neighborhood Search algorithm is proposed, which is another variant of simulated annealing. The proposed algorithm performs random node removals, the removed nodes reinserted to the possible cheapest route. In [7], insertion and swapping based neighborhood structures are used for the proposed algorithm. The infeasibilities are handled by using penalization. Static Move Descriptor (SMD) entities are applied to reduce computational time in [8]. Tabu Search is applied to real-world service bus problem in [9].

A multi-start Variable Search Algorithm (VNS) is proposed in [10], which uses path exchanges and reversals. A Unified VNS algorithm to solve VRP and OVRP with time windows in [11]. In [12] a variant of swarm intelligence algorithm is used, named as Bumble Bees Mating Optimization (BBMO). The equation for the drones moving outside hive is replaced with an Iterative Local Search method to enlarge the amount of search space. A Capacitated Arc Routing Problem (OCARP) which is very similar to OVRP is proposed in [13]. OCARP problem is modelled as VRP, then converted in to OVRP. The last study to mention is a Hybrid Iterative Local Search (ILS) algorithm, which is proposed in [14]. The proposed hybrid algorithm contains ILS- RVND (a Variable Neighborhood Descent with Random neighborhood ordering) - SP (set partitioning). The list of studies and proposed algorithm is given in Table 1

In this paper, we proposed a multi-phase oscillated VNS algorithm for OVRP. The proposed algorithm uses Kmeans and Variable Neighborhood Search algorithms. The proposed algorithm uses VNS implementations for single and route pair optimizations. Besides, stacking to a local optima conditions are handled by a deterministic perturbation method that have an oscillating behavior. The remaining part of this study contains proposed algorithm and the experimental results achieved by this algorithm.

2 Proposed Algorithm

Our proposed algorithm (OscillatedVNSforOVRP) follows cluster-first, route-second solution approach. Kmeans and Cheapest Insertion methods are used for the clustering and Variable Neighborhood Search [15] algorithm is used for route optimization. Pseudo code of proposed algorithm is given in Fig. 1. Route optimization contains three parts: VNS implementations for single route and route-pair optimizations and perturbation. The result of the clustering is used as an initial solution (s) for route optimization. After clustering, each route of the s is improved by the **VNSforSingleRouteOptimization** method. If any improvement occurs, the best solution is updated. The next step is improving the route pairs by **VNSforRoutePairOptimization** method. In both methods, once non-improved solution is obtained, the next neighborhood structure is tackled

Table 1. Comparison of VNSforOVRP results with previous studies

Ins	n	m$_{Min}$	A m/Dist	CPU	B m/Dist	CPU	C m/Dist	CPU	D m/Dist	CPU	E m/Dist	CPU	VNSforOVRP m/Dist	CPU	Avg Dist	Sdev
C1	50	5	488.2	0.22	416.1	88.8	416.06	120	416.06	230	416.06	17.6	416.06	0.461	418.79	3.75
C2	75	10	795.3	0.16	574.5	167.5	567.14	360	567.14	530	567.14	29.0	567.14	46.44	574.28	4.20
C3	100	8	815	0.94	641.6	325.3	641.76	850	641.76	1280	639.74	239.6	639.74	12.09	642.01	1.67
C4	150	12	1034.1	0.88	740.8	870.2	733.13	1790	733.13	2790	733.13	585.0	733.13	63.71	734.62	1.53
C5	199	16	1349.7	2.2	953.4	1415	897.93	1240	896.08	2370	905.96	292.1	(17)871.27	596.95	875.81	2.82
C11	120	7	828.3	1.54	683.4	696	682.12	730	682.12	1410	682.12	231.6	682.46	48.86	684.98	2.54
C12	100	10	882.3	0.76	535.1	233.6	534.24	800	534.24	1180	534.24	163.7	534.24	2.94	535.78	1.77
F11	71	4	-	-	177.4	398.1	177	690	177	1040	178.09	140.2	176.99	74.58	179.23	1.7
F12	134	7	-	-	781.2	1000.2	770.17	2370	770.17	3590	769.66	1237.5	770.17	749.38	772.13	1.88
			Pentium 133 MHz		Pentium III 500 MHz		Pentium IV 3 GHz		Athlon 1 GHz		Pentium M 2 Hz		Core i5 2.8 GHz			

Inst	n	m^{Min}	F m/Dist	CPU	G m/Dist	CPU	H m/Dist	CPU	I m/Dist	CPU	VNSforOVRP m/Dist	CPU	Avg	Sdev
C1	50	5	416.06	5.8	416.06	4.8	416.06	1.78	416.06	-	416.06	0.461	418.79	3.75
C2	75	10	567.14	15.1	567.14	19.2	567.14	6.44	567.14	-	567.14	46.44	574.28	4.20
C3	100	8	644.22	46.9	639.74	33	639.74	15.67	639.74	-	639.74	12.09	642.01	1.67
C4	150	12	733.13	117.6	733.13	123	733.13	27.77	733.13	-	733.13	63.71	734.62	1.53
C5	199	16	896.31	431.3	893.39	184.2	883.5	1579.45	908.94	-	(17)871.27	596.95	875.81	2.82
C11	120	7	685.95	92.7	682.12	60.6	682.12	23.54	682.12	-	682.81	48.86	685.92	2.54
C12	100	10	534.24	48.3	534.24	70.2	534.24	5.74	534.24	-	534.24	2.94	535.78	1.77
F11	71	4	177.00	14.0	-	-	177	4.41	-	-	176.99	74.58	179.23	1.7
F12	134	7	773.15	98.4	-	-	769.55	40.51	-	-	770.17	749.38	772.13	1.88
			Core 2 2.53 Ghz		Core 2 DUO 2.66		Core i7 2.93 Ghz		Intel Xeon 2.67		Core i5 2.8 GHz			

A: Minimum Spanning Tree With Penalties Algorithm [2]
B: Tabu Search [7]
C: Adaptive Large Neighborhood Search Algorithm (ALNS) 25 K [6]
D: Tabu Search [5]
E: Multi-Start VNS Algorithm [10]
F: Memetic Algorithm [13]
G: Bumble Bee Mating Algorithm [12]
H: Iterated Local Search- VND with Random Neighborhood Ordering - Set Partition [14]
I: Variable Neighborhood Search [11]

```
1  Procedure  OscillatedVNSforOVRP(ß,µ,α,loopMax,kmax)
2    S=solution_set : nodes(1….n))
3    // Clustering
4    s= Cheapest_Insertion(Kmeans(S))//s:initial
solution is generated
5    maxPerturbationCount=ß  //ß is calculated by dividing
     # of node by 25 and adding 2
6    perturbCounter= maxPerturbationCount
7    k_intra = 1, k_inter=1, sBest=s
8    for  loop_counter=0 to loopMax
9      // Route Optimization
10       s'=VNSforSingleRouteOptimization(s,k_intra)
11       If(f(s')<f(sBest))
12         sBest=s'
13         s=s'
14       else
15           k_intra =(k_intra+1)%2;
16       end if
17       s''=VNSforRoutePairOptimization (s,k_inter)
18       If(f(s'') <f(sBest))
19         sBest=s''
20           s=s''
21       else
22           k_inter=(k_inter+1)%2
23       end if
24       // Perturbation
25       if  (f(s')==f(s''))  //stacked in local optima
26         s=RemoveInsert(s,perturbCounter)
27         perturbCounter—
28         if(perturbCounter<1)
29           perturbCounter= maxPerturbationCount
30         end if
31       end if
32       // Ossilasion
33       if (loop_counter%µ==0) s<-sBest
34       if(loop_counter% foldofMaxPerturb ==0)
35         maxPerturbationCount ---
36       if(maxPerturbationCount<1)
37           maxPerturbationCount= ß
38    end for
39 End Procedure
```

Fig. 1. Pseudo code of proposed algorithm: VNSforOVRP

(Line 15, 22). Switching of neighborhood structures one to another provides a systematical exploration of the search space.

The initial solution is generated by using Kmeans [16] and Cheapest insertion algorithms (Line 4). Clustering with Kmeans doesn't generate a completed solution for OVRP. It just groups the nodes into number of vehicles clusters. Therefore, the cheapest insertion method is used to order nodes in a cluster to generate k routes for an initial solution. Depot node is assigned as the first node for each route of k vehicles. Then, the cheapest insertion method [17] is performed for each node in the cluster.

We performed two VNS algorithms for the route optimization and a deterministic perturbation. Single route optimization which is performed by VNSforSingleRouteOptimization method (Line 10) is applied to each route separately by using multi-threading. To start exploitation with a promising point in the search space. All neighborhoods are applied sequentially to the each route of solution (s). In the following step, VNS starts with the shaking method that uses two neighborhoods which are Double Bridge [18] and k-Shake that applies random route segment relocations. After the shaking phase, the local area is exploited by using three neighborhoods (2-Opt [19], or-Opt [20] and 3-Opt [21]) in this VNS algorithm. Starting index of neighborhood structures is specified by the main algorithm. The index refers to either 2-Opt or or-Opt structure. In case of improvement, the algorithm keeps using the same neighborhoods of shaking and local search to make intensification. Otherwise, current neighborhood in local search is switched to next neighborhood after testing with all shaking neighborhoods. The algorithm is terminated, after getting predefined number of back to back non-improvement results or at a certain number of iteration.

Route-Pair optimization is handled by VNSforRoutePairOptimization method (Line 17) that follows almost the same steps with the method for single route optimization. It differs by the neighborhoods it used in the local search and shaking phases and by using route-pairs as an arguments instead of a single route. We used four neighborhood structures for local search which are 2-Opt* [22], k-ExchangeWith Reverse [10], Relocate [23] and CrossExchange [24] respectively. SubpathExchange neighborhood which chooses two random subpaths from the current route-pair and exchanges are used for the shaking phase with ten times. If an improvement occurs in the local search, the algorithm continues exploitation with the same neighborhood; otherwise, it performs shaking method then local search with the same neighborhood structure. If still there is no improvement, the algorithm switches to the next neighborhood in the local search. This process repeats until a certain number of iteration or a predefined number of back to back non-improvement results.

Local optima is the main drawback of the metaheuristic search. To prevent it, RemoveInsert (Line 26) method is used. This method applies a deterministic perturbation by removing certain number nodes with maximum cost, and reinserting to the place in the solution with minimum cost. If the results of VNSforSingleRouteOptimization and VNSforRoutePairOptimization are the same, this is a kind of sign for stacking in a local optima which is an undesirable situation for searching. This method removes certain amount of nodes with maximum costs. Then, the removed nodes are sorted in descending order. This is necessary to prevent infeasible changes, otherwise after some insertions, remaining node with big demand may not be able to be inserted because of the capacity constraint. As a final step, nodes in the sorted removed nodes are inserted in to the solution by using Cheapest Insertion [25] method. If any un-inserted node is left, non-modified solution is returned, otherwise, new solution is returned.

The maximum number of removed nodes (ß) is determined by according to the size of the problem (Line 5). To decrease the effects of diversification on search performance, the maximum number of removed node *(maxPerturbCounter)* is reduced after every *(foldofMaxPerturb)* (Line 34–35) iterations until a predefined number is found. For the same purpose, the number of removed node *(perturbCounter)* starts from the maximum value of *maxPerturbCounter* and in each stacking condition, it decreases by one until 1 (Line 27–29). Then, it is reassigned to the maximum value (ß). This change on both *maxPerturbationCount* and *perturbCounter* gives oscillation effects on the diversification mechanism of the algorithm. The similar effect is applied on the intensification mechanism of the algorithm by assigning the best solution *(sBest)* to the current solution at every μ iterations in the main loop. The test results of efficiency of the oscillation can be reached from this link.

3 Experimental Results

The proposed algorithm is implemented in C# programming language and tested for several benchmark problems and a real-world problem dataset. Experimentation is performed on Dell computer with i5 2.80 GHz CPU with 4 GB memory. There are 2 different series for OVRP in the literature: C series with 7 problem sets provided by Christofides [26] and F series with 2 problem sets provided by Fisher [27]. C series are symmetric problem instances where the depot located at the center, F and the real-world dataset are asymmetric.

The best results obtained by VNSforOVRP are compared with the best results obtained by the previous studies based on solution quality (minimum total distance) and computational time (in second) in Table 1. As it can be seen from table, although results of A [2] and B [7] results are worse than best results for all series, their CPU times are very low compared with their low CPU configuration. F [13] has the best result for C1, C2, C4, C12 and F11 instances, but results for C3, C11 and F12 instances are worse than literature best results. Other studies generally have the best known results for C and F series. The experimental results for VNSforOVRP in Table 1 shows that our proposed algorithm achieves the best know solutions within a reasonable time for all small and medium scale benchmarks in C and F series. In addition to this, we present an improvement solution (871.27) by increasing the number of vehicles by 1 for C5. The running-time performance of the VNSforOVRP is competitive with other state-of-art algorithms.

4 Conclusion

In this paper, a VNS based algorithm is presented for solving OVRP. The algorithm is designed by using cluster first, route second approach. Clustering is achieved by K-Means clustering algorithm and the route optimization is accomplished by two VNS algorithms dedicated to single and multi-route optimizations respectively. To escape from local optima, oscillated perturbation is applied.

We have tested our algorithm with well-known 9 benchmark instances. The algorithm finds the best results for all benchmark instances. As a future work, this algorithm can be adapted to large scale and real-world OVRP problems.

Acknowledgements. This work is supported by the Scientific Research Fund of Fatih University under the project number P50071503_B.

References

1. Leeuwen, J.V.: Algorithms and complexity. Elsevier [u.a.] (1998)
2. Sariklis, D., Powell, S.: A heuristic method for the open vehicle routing problem. J. Oper. Res. Soc. **51**(5), 564–573 (2000)
3. Tarantilis, C.D., Ioannou, G., Kiranoudis, C.T., Prastacos, G.P.: A threshold accepting aproach to the open vehicle routing problem, RAIRO. Oper. Res. **38**, 345–360 (2004)
4. Tarantilis, C.D., Ioannou, G., Kiranoudis, C.T., Prastacos, G.P.: Solving the open vehicle routeing problem via a single parameter metaheuristic algorithm. J. Oper. Res. Soc. **56**(5), 588–596 (2005)
5. Li, F., Golden, B., Wasil, E.: The open vehicle routing problem: Algorithms, large-scale test problems, and computational results. Comput. Oper. Res. **34**(10), 2918–2930 (2007)
6. Pisinger, D., Ropke, S.: A general heuristic for vehicle routing problems. Comput. Oper. Res. **34**(8), 2403–2435 (2007)
7. Brandão, J.: A tabu search algorithm for the open vehicle routing problem. Eur. J. Oper. Res. **157**(3), 552–564 (2004)
8. Zachariadis, E.E., Kiranoudis, C.T.: An open vehicle routing problem metaheuristic for examining wide solution neighborhoods. Comput. Oper. Res. **37**, 712–723 (2010)
9. Aksen, D., Özyurt, Z., Aras, N.: Open vehicle routing problem with driver nodes and time deadlines. J. Oper. Res. Soc. **58**(9), 1223–1234 (2007)
10. Fleszar, K., Osman, I.H., Hindi, K.S.: A variable neighbourhood search algorithm for the open vehicle routing problem. Eur. J. Oper. Res. **195**(3), 803–809 (2009)
11. Kritzinger, S., Doerner, K.F., Tricoire, F., Hartl, R.F.: Adaptive search techniques for problems in vehicle routing, Part II: A numerical comparison. Yugoslav J. Oper. Res. (2014)
12. Marinakis, Y., Marinaki, M.: A bumble bees mating optimization algorithm for the open vehicle routing problem. Swarm Evol. Comput. **15**, 80–94 (2014)
13. Fung, R.Y.K., Liu, R., Jiang, Z.B.: A memetic algorithm for the open capacitated arc routing problem. Transp. Res. Part E Log. Transp. Rev. **50**, 53–67 (2013). ISSN 1366-5545
14. Subramanian, A., Uchoa, E., Ochi, L.S.: A hybrid algorithm for a class of vehicle routing problems. Comput. Oper. Res. **40**(10), 2519–2531 (2013). ISSN 0305-0548
15. Hansen, P., Mladenovic, N.: Variable neighborhood search: principals and applications. Eur. J. Oper. Res. **130**(3), 449–467 (2001). ISSN 0377-2217
16. MacQueen, J.: Some methods for classification and analysis of multivariate observations. In: Proceedings of the Fifth Berkeley Symposium on Mathematical Statistics and Probability, Volume 1: Statistics, pp. 281–297. University of California Press, Berkeley (1967). http://projecteuclid.org/euclid.bsmsp/1200512992
17. http://en.wikipedia.org/w/index.php?title=K-means_clustering&oldid=655354053. (Accessed 10 April 2015 08:02 UTC 2015)

18. Derbel, H., Jarboui, B., Chabchoub, H., Hanafi, S., Mladenovic, N.: A variable neighborhood search for the capacitated location-routing problem. In: 2011 4th International Conference on Logistics (LOGISTIQUA), pp. 514–519, 31 May 2011–3 June 2011

19. Flood, M.M.: The traveling-salesman problem. Oper. Res. **4**(1), 61–75 (1956)

20. Or, I.: Traveling salesman-type combinatorial problems and their relation to the logistics of blood banking. Northwestern University (1993)

21. Lin, S.: Computer solutions of the traveling salesman problem. Bell Syst. Tech. J. **44**, 2245–2269 (1965)

22. Potvin, J.-Y., Rousseau, J.-M.: An exchange heuristic for routing problems with time windows. J. Oper. Res. Soc. **46**, 1433–1446 (1995)

23. Savelsbergh, M.: The vehicle routing problem with time windows: minimizing route duration. Informs J Comput **4**, 146–154 (1992)

24. Taillard, E., Badeau, P., Gendreau, M., Guertin, F., J-Y, P.: A tabu search heuristic for the vehicle routing problem with soft time windows. Transport Sci. **31**, 170–186 (1997)

25. Rosenkrantz, D.J., Stearns, R.E., Lewis, P.M.: An analysis of several heuristics for the traveling salesman problem. SIAM J. Comput. **6**(3), 563–581 (1977)

26. Christofides, N., Mingozzi, A., Toth, P.: The vehicle routing problem. In: Combinatorial optimization. Wiley (1979)

27. Fisher, M.: Optimal solution of vehicle routing problems using minimum k-trees. Oper. Res. **42**, 626–642 (1994)

Non-Line-of-Sight Mitigation via Lagrange Programming Neural Networks in TOA-Based Localization

Zi-Fa Han[1], Chi-Sing Leung[1(✉)], Hing Cheung So[1], John Sum[2], and A.G. Constantinides[3]

[1] Department of Electronic Engineering, City University of Hong Kong, Kowloon Tong, Hong Kong
eeleungc@cityu.edu.hk
[2] National Chung Hsing University, Taichung, Taiwan
[3] Imperial College, London, UK

Abstract. A common measurement model for locating a mobile source is time-of-arrival (TOA). However, when non-line-of-sight (NLOS) bias error exists, the error can seriously degrade the estimation accuracy. This paper formulates the problem of estimating a mobile source position under the NLOS situation as a nonlinear constrained optimization problem. Afterwards, we apply the concept of Lagrange programming neural networks (LPNNs) to solve the problem. In order to improve the stability at the equilibrium point, we add an augmented term into the LPNN objective function. Simulation results show that the proposed method provides much robust estimation performance.

Keywords: TOA localization · Non-line-of-sight (NLOS) · LPNN

1 Introduction

In the last decade, estimating the location of a mobile source with a number of separate sensors has received a lot of attention in wireless sensor networks and cellular networks [1,2]. It has a lot of applications in navigation, surveillance, and geophysics. Time of arrival (TOA) [3,4] is often used to obtain range measurements between the mobile source and sensors. The source location can be derived based on a set of range measurements. For example, in the line-of-sight (LOS) transmission model (no obstacles), when there are no noise in the TOA range measurements, the coordinates of the mobile source can be obtained by calculating the intersect point of three circles centered at three sensors. In real situations, the range measurements may be contaminated by different kinds of noise. However, in typical environments, the transmission paths between the source and receivers may not follow the transmission model due to the existence of obstacles. In this case, the radio signal may be reflected by obstacles. This kind of error, referred as non-line-of-sight (NLOS) bias error [5], may lead to erroneous location estimation.

© Springer International Publishing Switzerland 2015
S. Arik et al. (Eds.): ICONIP 2015, Part III, LNCS 9491, pp. 190–197, 2015.
DOI: 10.1007/978-3-319-26555-1_22

In the past three decades, analog neural networks have been found as effective and efficient alternatives for optimization. For instance, Hopfield [6] and Chua [7] investigated analog neural circuits for solving nonlinear programming with inequality constraints. Apart from optimization, neural circuits can also be used for searching the maximum of a set of numbers [8,9]. An important model is the Lagrange programming neural network (LPNN) model [2,10,11]. The LPNN approach provides a general framework for solving nonlinear constrained optimization problems.

This paper first describes the NLOS situation in the source localization problem. The localization problem was formulated as a constrained optimization problem. Furthermore, an analog neural network model based on LPNN is proposed to solve this constraint optimization problem.

The rest of this paper is organized as follows. Section 2 provides the concept of LPNNs and formulates the estimation problem of the NLOS situation as a constraint optimization problem. Section 3 discusses how to solve this problem based on LPNN. Besides, we briefly discuss the stability of this LPNN model. The simulation experiments are described in Sect. 4. Lastly, we conclude the paper in Sect. 5.

2 Background

LPNN: The LPNN is used to solve constrained nonlinear optimization problem:

$$\min_{x} f(x), \quad \text{s.t.} \quad h(x) = 0, \tag{1}$$

where $x \in \mathbb{R}^N$ is the variable vector, and $f : \mathbb{R}^N \to \mathbb{R}$ is the nonlinear objective function. The function $h : \mathbb{R}^N \to \mathbb{R}^M$ describes the M equality constraints, and 0 denotes an $M \times 1$ zero vector. We further assume that f and h are twice differentiable. The LPNN approach exploits the following Lagrangian:

$$\mathcal{L}(x, \lambda) = f(x) + \lambda^T h(x), \tag{2}$$

where $\lambda = [\lambda_1, \cdots, \lambda_M]^T$ is the Lagrange multiplier vector.

A LPNN consists of two types of neurons, namely, variable and Lagrangian neurons. The variable neurons hold the state variables x, while the Lagrangian neurons holds the Lagrange multipliers. The dynamics of these neurons are given by

$$\frac{1}{\tau}\frac{dx}{dt} = -\frac{\partial \mathcal{L}}{\partial x}, \text{ and } \frac{1}{\tau}\frac{d\lambda}{dt} = \frac{\partial \mathcal{L}}{\partial \lambda}, \tag{3}$$

where τ is the time constant of the circuit. Without loss of generality, we set $\tau = 1$. As mentioned in [2,10,11], the dynamics of variable neurons are used for minimizing the objective value, while the dynamics of the Lagrange neurons are used for restricting x in the feasible region.

2.1   Problem Formulation

We consider the TOA-based range measurements between the mobile source and N receivers in a two-dimensional space. Let r_i be the distance between the mobile source and the ith sensor. The NLOS prorogation scenario [12] can be modeled as

$$r_i = \|\boldsymbol{c} - \boldsymbol{u}_i\|_2 + n_i + \xi_i, i = 1, \cdots, N, \tag{4}$$

where $\boldsymbol{c} = [c_1, c_2]^T$ is the position of the mobile source, $\boldsymbol{u}_i = [u_{i,1}, u_{i,2}]^T$ is the known position of the ith sensor, n_i is a zero-mean Gaussian measurement error, and ξ_i is a positive bias error caused by the NLOS propagation. In NLOS scenario, the common model for describing the distribution of ξ_i is exponential distribution [12].

Assume that the NLOS bias error is much greater than the Gaussian measurement error [13]. The Gaussian measurement error can be ignored. Based on this assumption, the following propagation model can be formulated:

$$r_i = \|\boldsymbol{c} - \boldsymbol{u}_i\|_2 + \xi_i, i = 1, \cdots, N. \tag{5}$$

With this range measurement model, given $\boldsymbol{u}_i$'s and r_i, our task is to estimate the position $\boldsymbol{c}$ of the mobile source. If we assume that the NLOS bias error ξ_i follows an exponential distribution, then a maximum likelihood (ML) estimation problem can be formulated as

$$
\begin{aligned}
\min_{\boldsymbol{c},\boldsymbol{g},\boldsymbol{\xi}} \ & \sum_{i=1}^{N}(r_i - g_i) \\
s.t. : \ & g_i^2 = \|\boldsymbol{c} - \boldsymbol{u}_i\|_2^2, \ i = 1, \cdots, N, \\
& \xi_i = r_i - g_i, \ i = 1, \cdots, N. \\
& \xi_i \geq 0, \ i = 1, \cdots, N. \\
& g_i \geq 0, \ i = 1, \cdots, N.
\end{aligned}
\tag{6}
$$

where $\boldsymbol{g} = [g_1, \cdots, g_N]^T$ and $\boldsymbol{\xi} = [\xi_1, \cdots, \xi_N]^T$. In the above, some constraints contain inequality constraints, and we cannot directly the LPNN concept to solve the problem. Hence, we need some procedures to the inequality constraints to equality ones.

To guarantee "$\xi_i \geq 0$", we introduce dummy variables b_i's. Also, we set b_i^2 as a surrogate of ξ_i. In this case, "$\xi_i = r_i - g_i$" becomes "$b_i^2 = r_i - g_i$" and "$\xi_i \geq 0$" can be removed. Similarly, to guarantee "$g_i \geq 0$", we introduce dummy variables y_i's and modify "$g_i \geq 0$" to "$g_i = y_i^2$".

Now, the optimization problem becomes

$$
\begin{aligned}
\min_{\boldsymbol{c},\boldsymbol{g},\boldsymbol{b},\boldsymbol{y}} \ & \sum_{i=1}^{N}(r_i - g_i) \\
s.t. : \ & g_i^2 = \|\boldsymbol{c} - \boldsymbol{u}_i\|_2^2, \ i = 1, \cdots, N, \\
& b_i^2 = r_i - g_i, \ i = 1, \cdots, N, \\
& g_i = y_i^2, \ i = 1, \cdots, N.
\end{aligned}
\tag{7}
$$

where $\boldsymbol{y} = [y_1, \cdots, y_N]^T$. With this formulation, we can apply the LPNN concept to solve the optimization problem.

3   Algorithm Development

The Lagrangian function of (7) has the following form:

$$\mathcal{L}(\boldsymbol{x}, \boldsymbol{\lambda}) = \sum_{i=1}^{N}(r_i - g_i) + \sum_{i=1}^{N}\alpha_i(g_i^2 - \|\boldsymbol{c} - \boldsymbol{u}_i\|_2^2)$$

$$+ \sum_{i=1}^{N}\beta_i(r_i - g_i - b_i^2) + \sum_{i=1}^{N}\gamma_i(g_i - y_i^2) \tag{8}$$

where $\boldsymbol{\lambda} = [\alpha_1, \cdots, \alpha_N, \beta_1, \cdots, \beta_N, \gamma_1, \cdots, \gamma_N]^T$ are the Lagrangian multiplier vectors and $\boldsymbol{x} = [\boldsymbol{c}^T, \boldsymbol{g}^T, \boldsymbol{b}^T, \boldsymbol{y}^T]^T$. Based on (8), we may derive the neural dynamics with the use of the concept of LPNN. In this case, $(3N + 2)$ variable neurons and $3N$ Lagrangian neurons hold $\boldsymbol{x}$ and $\boldsymbol{\lambda}$, respectively. However, our preliminary simulation study foud out that the second-order gradient of (7) at an equilibrium point may not be positive, indicating that the neural dynamics can be unstable. It is because the objective function "$\sum_{i=1}^{N}(r_i - g_i)$" is a linear function.

To improve the convexity and stability, we add an augmented term [2,10,11] into the Lagrangian function, given by

$$\frac{C_0}{2}\left[\sum_{i=1}^{N}(g_i^2 - \|\boldsymbol{c} - \boldsymbol{u}_i\|_2^2)^2 + \sum_{i=1}^{N}(r_i - g_i - b_i^2)^2 + \sum_{i=1}^{N}(g_i - y_i^2)^2\right], \tag{9}$$

where C_0 is a positive constant.

The augmented Lagrangian function is given by

$$\mathcal{L}(\boldsymbol{x}, \boldsymbol{\lambda}) = \sum_{i=1}^{N}(r_i - g_i) + \sum_{i=1}^{N}\alpha_i(g_i^2 - \|\boldsymbol{c} - \boldsymbol{u}_i\|_2^2)$$

$$+ \sum_{i=1}^{N}\beta_i(r_i - g_i - b_i^2) + \sum_{i=1}^{N}\gamma_i(g_i - y_i^2)$$

$$+ \frac{C_0}{2}\left[\sum_{i=1}^{N}(g_i^2 - \|\boldsymbol{c} - \boldsymbol{u}_i\|_2^2)^2 + \sum_{i=1}^{N}(r_i - g_i - b_i^2)^2 + \sum_{i=1}^{N}(g_i - y_i^2)^2\right]. \tag{10}$$

Note that this augmented term does not affect the objective value at an equilibrium point [2,10,11]. It is because at an equilibrium point, the constraints are satisfied and the augmented term is equal to zero.

With (10), from (3), the dynamics of LPNN are given by

$$\frac{d\boldsymbol{c}}{dt} = 2C_0\sum_{i=1}^{N}(g_i^2 - \|\boldsymbol{c} - \boldsymbol{u}_i\|^2)(\boldsymbol{c} - \boldsymbol{u}_i) + 2\sum_{i=1}^{N}\alpha_i(\boldsymbol{c} - \boldsymbol{u}_i), \tag{11}$$

$$\frac{dg_i}{dt} = 1 - 2C_0(g_i^2 - \|\boldsymbol{c} - \boldsymbol{u}_i\|^2)g_i + C_0(r_i - g_i - b_i^2) - C_0(g_i - y_i^2)$$
$$-2\alpha_i g_i + \beta_i - \gamma_i, \tag{12}$$

$$\frac{db_i}{dt} = 2C_0 b_i(r_i - g_i - b_i^2) + 2\beta_i b_i, \tag{13}$$

$$\frac{dy_i}{dt} = 2C_0 y_i(g_i - y_i^2) + 2\gamma_i y_i, \tag{14}$$

$$\frac{d\alpha_i}{dt} = g_i^2 - \|\boldsymbol{c} - \boldsymbol{u}_i\|^2, \tag{15}$$

$$\frac{d\beta_i}{dt} = r_i - g_i - b_i^2, \tag{16}$$

$$\frac{d\gamma_i}{dt} = g_i - y_i^2. \tag{17}$$

The dynamics of the LPNN from (11) to (17) contain $3N + 2$ state variable neurons and $3N$ Lagrangian neurons. Now, we begin to analyze the local stability of LPNN approach [10]. Two conditions are required to ensure the local stability of LPNN approach. The first one is convexity, which is achieved by adding augmented term (9) into the Lagrangian function [10]. The second one is that the gradient vectors $\{\nabla_{\boldsymbol{x}} h_i(\boldsymbol{x}^*) : i = 1, \cdots, 3N\}$ of the constraints should be linearly independent at an equilibrium point $\boldsymbol{x}^* = [\boldsymbol{c}^*, \boldsymbol{g}^*, \boldsymbol{b}^*, \boldsymbol{y}^*]^T$. The $3N$ constraints have the form

$$\begin{aligned} h_i(\boldsymbol{x}) &= g_i^2 - \|\boldsymbol{c} - \boldsymbol{u}_i\|_2^2, \ i = 1, \cdots, N, \\ h_{N+i}(\boldsymbol{x}) &= r_i - g_i - b_i^2, \ i = 1, \cdots, N, \\ h_{2N+i}(\boldsymbol{x}) &= g_i - y_i^2, \ i = 1, \cdots, N. \end{aligned} \tag{18}$$

Then, we can get the gradient vectors at $\boldsymbol{x}^*$, which are denoted as $\{\nabla_{\boldsymbol{x}} h_i(\boldsymbol{x}^*) : i = 1, \cdots, N\}$, $\{\nabla_{\boldsymbol{x}} h_{N+i}(\boldsymbol{x}^*) : i = 1, \cdots, N\}$, and $\{\nabla_{\boldsymbol{x}} h_{2N+i}(\boldsymbol{x}^*) : i = 1, \cdots, N\}$. One can verify that if $g_i^* \neq 0, b_i^* \neq 0$ and $b_i^* \neq 0$ for all i, then those gradient vectors are linearly independent. Therefore, the dynamics around the equilibrium point $\boldsymbol{x}^*$ is stable. That means at an equilibrium point, if $g_i^* \neq 0$, $b_i^* \neq 0$ and $b_i^* \neq 0$ for all i, then the equilibrium point $\boldsymbol{x}^*$ is stable.

4   Simulation Results

In this section, computer simulations are carried out to evaluate the performance of the proposed LPNN model for the NLOS situation. We consider two sensor configurations, shown in Fig. 1. In the first configuration, there are $N = 4$ sensors with positions at $[10, 0]^T$, $[0, 10]^T$, $[-10, 0]^T$ and $[0, -10]^T$. In the second configuration, there are $N = 6$ sensors with with positions at $[10, 0]^T$, $[5, 8.66]^T$, $[-5, 8.66]^T$, $[-10, 0]^T$, $[-5, -8.66]^T$ and $[5, -8.66]^T$.

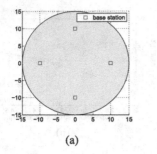
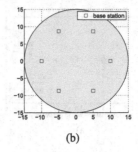

(a)   (b)

Fig. 1. The configure of base stations and mobile source. (a) $N = 4$. (b) $N = 6$.

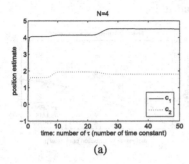
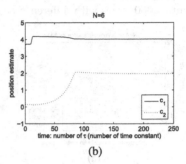

(a)   (b)

Fig. 2. Dynamics of estimated source position. (a) $N = 4$. (b) $N = 6$. The true source position is at $[4, 2]^T$.

The mobile source is at $[4, 2]^T$ and the NLOS bias error follows the exponential distribution [12] with mean equal to 2. In addition, we add Gaussian zero-mean noise with variance 0.1 into the measurement. Figure 2 plots the dynamics of the estimated source position for $N = 4$ and $N = 6$. It can be observed that the network settles down around 30 to 150 time constants.

To demonstrate the effectiveness of our LPNN approach model, we compare our approach with two localization estimators, namely LMS [14] and Huber [15]. Gaussian zero-mean noise with variance 0.1 are also added into the measurements. We vary the mean of NLOS bias error (exponential distribution). We repeated the experiments 1,000 times for each NLOS bias error level. Figure 3 shows the MSE performance for $N = 4$ sensors and $N = 6$ sensors respectively.

From Fig. 3, we can see that the proposed LPNN approach is better than LMS and Huber, especially, for large μ (large NLOS bias error). For example, for $N = 6$, when the NLOS bias error level is 2, the errors of the LMS and Huber approaches are greater than 8.86 dB. When our method is used, the error in our approach is 7.252 dB only. For $N = 6$, when the NLOS bias error level is 3, the errors of the LMS and Huber approaches are greater than 13.8 dB. When our method is used, the error in our approach is 10.6 dB only.

The simulation setting is similar to the previous one. We repeat the experiment 1,000 times with different mobile source positions, which are uniformly

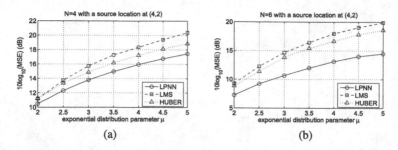

Fig. 3. Performance of various approaches with a fixed source position at $[4, 2]^T$. (a) The number of sensors is $N = 4$. (b) The number of sensors is $N = 6$. We repeated the experiments 1,000 times with difference random noise.

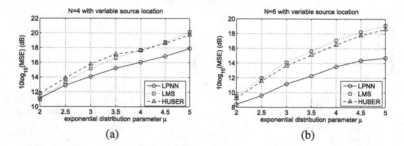

Fig. 4. Performance of various approaches with variable positions. (a) The number of sensors is $N = 4$. (b) The number of sensors is $N = 6$. We repeated the experiments 1,000 times with different source locations and different random noise.

chosen inside a circle centered at the origin with radius 15 at each trial, shown in Fig. 1. The results are plotted in Fig. 4.

For $N = 4$, when the NLOS bias error level is small ($\mu = 2$), our method is a bit poorer than the LMS method. This phenomena meets our expectation as we ignore the Gaussian measurement error in our model which assumes that the NLOS bias error is much larger than Gaussian measurement error. Obviously this assumption does not hold for small NLOS bias error levels.

However, when the NLOS bias error level is greater than or equal to 2.5, the proposed LPNN approach is better than LMS and Huber. For example, for $N = 4$, when the NLOS bias error level is 3, the errors of the LMS and Huber approaches are greater than 15.2 dB. When our method is used, the error in our approach is 14.1 dB only. For $N = 6$, when the the NLOS bias error level is 3, the errors of the LMS and Huber approaches are greater than 13.6 dB. When our method is used, the error in our approach is 11.2 dB only.

5   Conclusion

In this paper, we propose an analog neural network model to mitigate NLOS bias errors in the TOA-based localization. We first relax the NLOS transmission model by ignoring the Gaussian noise, as the NLOS bias error are much larger than Gaussian measurement noise. A ML position estimator was formulated. Then, we apply the LPNN on this optimization problem. To make sure the convexity and the stability at the equilibrium points, an augmented term was added in the LPNN objective function. Through experiments we can see that the proposed method provides much robust estimation performance when large NLOS bias errors happen.

Acknowledgement. The work was supported by GRF from Hong Kong (Project No.: CityU 115612).

References

1. Mao, G., Fidan, B., Anderson, B.D.: Wireless sensor network localization techniques. Comput. Netw. **51**, 2529–2553 (2007)
2. Leung, C.S., Sum, J., So, H.C., Constantinides, A.G., Chan, F.K.: Lagrange programming neural networks for time-of-arrival-based source localization. Neural Comput. Appl. **24**, 109–116 (2014)
3. So, H.C.: Source localization: algorithms and analysis. In: Zekavat, S.A., Buehrer, R.M. (eds.) Handbook of Position Location: Theory, Practice, and Advances. John Wiley & Sons, Inc., Hoboken (2011)
4. Caffery, J.J.: A new approach to the geometry of toa location. In: Proceedings of the IEEE Vehicular Technology Conference 2000, vol. 4, pp. 1943–1949 (2000)
5. Torrieri, D.J.: Statistical theory of passive location systems. J. IEEE Trans. Aerosp. Electron. Syst. **20**(2), 183–197 (1984)
6. Hopfield, J.J.: Neural networks and physical systems with emergent collective computational abilities. Proc. Nat. Acad. Sci. **79**(8), 2554–2558 (1982)
7. Chua, L.O., Lin, G.N.: Nonlinear programming without computation. IEEE Trans. Circuits Syst. **31**, 182–188 (1984)
8. Sum, J., Leung, C.S., Tam, P., Young, G., Kan, W., Chan, L.W.: Analysis for a class of winner-take-all model. IEEE Trans. Neural Networks **10**(1), 64–71 (1999)
9. Xiao, Y., Liu, Y., Leung, C.S., Sum, J., Ho, K.: Analysis on the convergence time of dual neural network-based kwta. IEEE Trans. Neural Netw. Learn. Syst. **23**(4), 676–682 (2012)
10. Zhang, S., Constantinidies, A.G.: Lagrange programming neural networks. IEEE Trans. Circuits Syst. **39**(7), 441–452 (1992)
11. Leung, C.S., Sum, J., Constantinides, A.G.: Recurrent networks for compressive sampling. Neurocomputing **129**, 298–305 (2014)
12. Li, J., Wu, S.: Non-parametric non-line-of-sight identification and estimation for wireless location. In: Proceedings IEEE International Conference on Computer Science and Service System (CSSS 2012), pp. 81–84 (2012)
13. Gezici, S., Sahinoglu, Z.: UWB geolocation techniques for IEEE 802.15.4a personal area networks, Mitsubishi Electric Research Laboratory, Technical report TR-2004-110 (2004)
14. Casas, R., Marco, A., Guerrero, J., Falco, J.: Robust estimator for non-line-of-sight error mitigation in indoor localization. EURASIP J. Appl. Sig. Process. **2006**, 156–156 (2006)
15. Sun, G.L., Guo, W.: Bootstrapping m-estimators for reducing errors due to non-line-of-sight (NLOS) propagation. IEEE Commun. Lett. **8**, 509–510 (2004)

Wave-Based Reservoir Computing
by Synchronization of Coupled Oscillators

Toshiyuki Yamane[1]([⊠]), Yasunao Katayama[1],
Ryosho Nakane[2], Gouhei Tanaka[2], and Daiju Nakano[1]

[1] IBM Research - Tokyo, Kawasaki, Kanagawa 212-0032, Japan
{tyamane,yasunaok,dnakano}@jp.ibm.com
[2] Graduate School of Engineering, The University of Tokyo, Tokyo 113-8656, Japan
nakane@cryst.t.u-tokyo.ac.jp, gouhei@sat.t.u-tokyo.ac.jp

Abstract. We propose wave-based computing based on coupled oscillators to avoid the inter-connection bottleneck in large scale and densely integrated cognitive systems. In addition, we introduce the concept of reservoir computing to coupled oscillator systems for non-conventional physical implementation and reduction of the training cost of large and dense cognitive systems. We show that functional approximation and regression can be efficiently performed by synchronization of coupled oscillators and subsequent simple readouts.

Keywords: Wave-based computing · Inter-connection bottleneck · Learning bottleneck · Reservoir computing · Synchronization · Coupled oscillators

1 Introduction

It never ceases to amaze us how biological systems perform complex cognitive tasks such as memorization, recognition and categorization. Cognitive computing is an approach for realizing such cognitive abilities artificially. However, software implementations at biologically realistic level have proved to be highly power-hungry and computer-intensive so far because of their serial operation and von-Neumann bottleneck of the current architecture. In this respect, hardware implementations are of great interest because of their parallel operation and distributed information representation.

On the other hand, current VLSI technologies also are facing several serious challenges. Complexity of placement and wiring in VLSI design increasingly becomes a major bottleneck of chip performance [5]. Since the flexible and diverse capabilities of brains come from massive synaptic inter-connections rather than the function of individual neurons, current hardware implementation is not necessarily efficient in terms of interconnections.

We need to consider other possibilities for implementation than conventional CMOS-based VLSI technology, such as nano-structures since it is quite unclear whether the integration by CMOS scaling continues in the future [3]. Our main

S. Arik et al. (Eds.): ICONIP 2015, Part III, LNCS 9491, pp. 198–205, 2015.
DOI: 10.1007/978-3-319-26555-1_23

question is that how we can realize interconnection-centric cognitive computing systems on such non-conventional physical systems beyond the bottleneck of hard wiring. To this end, we make use of physical waves as an alternative to hard wiring. The use of wave dynamics for computing is attracting much interest recently in the research pursuing post-CMOS technology. For example, wave-based multi-valued logic by superposition of spin waves was proposed as more power- and area-efficient computing framework than CMOS based digital computing [4]. From our perspective, waves have an attractive property that they can propagate in any direction and at any distance as long as there exists medium transmitting waves, which makes them a promising alternative for hard wiring. This lead us to explore computing systems where computing elements interacting through wave propagation. In this work, we use oscillators as computing elements and realize wave-based computing systems as coupled oscillators. We show the phase locking in a synchronized state of coupled oscillators can approximate a given function.

The cost of modifying inter-connections by learning will become another serious bottleneck if we are going to build cognitive systems on non-conventional physical systems. In this respect, it is useful to apply the framework of reservoir computing, where training is performed only in readout part outside computational part called reservoir [6]. The reservoirs do not need to be traditional neural networks but can be built on a variety of physical systems such as coupled oscillators. We realize the reservoirs as coupled oscillators and show that the phase dynamics of coupled oscillators can be viewed as a special case of reservoir computing in time domain. We also show function approximation problems can be robustly solved by wave-based reservoir computing using synchronization of coupled oscillators.

2   Wave-Based Reservoir Computing

The reservoir computing is an emerging computation framework for design of recurrent neural networks [6]. Generally, reservoir computing systems have two functional components. One is a fixed recurrent neural network, called reservoir, which is a (non-linear) mapping of input data to a high dimensional space and should be complex enough to generate rich dynamical behaviour. The other is adaptive read-out functions which extract desired results from the reservoir output.

In this work, we realize the interconnection in the reservoir as wave propagation to avoid interconnection bottleneck. In addition, we use oscillators as the computational elements because they can naturally interact through waves. Thus, we formulate the phase domain reservoir computing as phase dynamics of coupled oscillators. Even if the dynamics is restricted to phase domain, it can still exhibit very rich behaviour such as phase transition, clustering and synchronization, enough to potentially perform complex computational tasks [8]. For example, coupled oscillators has been applied to associative memory systems [2] and convolutional neural networks [7]. Since the wave-related phenomena and

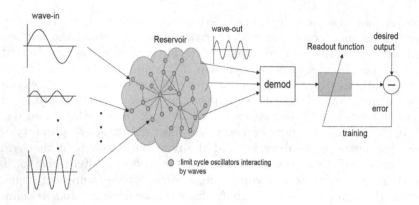

Fig. 1. Proposed architecture of wave-based reservoir computing

synchronization are abundant in the natural world, the wave-based computing is expected to be advantageous for physical implementation. Therefore, we use synchronization dynamics for stable computation in this work.

Figure 1 shows the overall architecture of the proposed wave-based reservoir computing. Multiple waves modulated by analog information are fed into the reservoir. The reservoir consists of coupled oscillators designed for specific computational tasks. Then, read-out functions are applied to obtain desired results after the reservoir dynamics reaches a final state. We can apply adaptive filters such as least mean square filters or simple perceptrons to read-out functions.

3   Phase Dynamics of Coupled Oscillators

3.1   Phase Response Curves

Suppose that a stable limit cycle $X_0(\theta), \theta \in [0, 2\pi) \to \mathbb{R}^n$ is subject to a small impulse stimulus I. Since the limit cycle is neutrally stable along its orbit and stable in the directions orthogonal to its orbit, the small impulse stimulus I results in a small phase shift on the limit cycle. The phase shift caused by I can be described by the following phase response curve $g(\theta; I) = \Delta\theta = \theta(X_0(\theta) + I) - \theta$, where $\theta(X) : X \to [0, 2\pi)$ is mapping from the point on the limit cycle X to its phase θ. Since the strength of stimulus $|I|$ is small, the phase response curve is well approximated by linear response as $g(\theta; I) \cong Z(\theta) \cdot I$. Here, we introduce the linear response coefficient of phase response for small I as $Z(\theta) = \nabla_X \theta(X)|_{X=X_0(\theta)}$. The linear response coefficient $Z(\theta)$ is the 2π-periodic impulse response function for the phase shift and called phase sensitivity function. Though both I and $Z(\theta)$ are n dimensional vectors, we assume hereafter these two functions are scalar functions for simplicity.

Under some assumptions such as weakness of stimulus and linearity of response, the phase response can be understood from the viewpoint of linear system theory [1]. That is, the total phase response caused by input stimulus $I(\theta)$ during one cycle of oscillation is given by the circular linear convolution as

$$H(\theta; I) = \int_0^{2\pi} Z(\theta_0) \cdot I(\theta - \theta_0) d\theta_0. \tag{1}$$

The interactions through the phase response (1) is the counterpart of those in the reservoir of time domain. Since the interactions in the reservoirs of time domain are generally nonlinear, the linearity of response (1) means that the wave-based reservoir computing in phase domain is a special class of reservoir computing in time domain. However, we will show in Sect. 4 that wave-based reservoir can still perform complex tasks such as function approximations.

3.2 Phase Synchronization

Consider a N coupled stable limit cycle oscillators described by the following equation.

$$\frac{dX_i(t)}{dt} = F_i(X_i(t)) + \sum_{j \neq i}^N G_{ij}(X_j(t)), \quad i = 1, \ldots, N. \tag{2}$$

Here, $X_i = (x_i, y_i \ldots)^T \in \mathbb{R}^n$ is the state variables of the oscillator i, F_i is a nonlinear function describing limit cycle oscillation. The function G_{ij} means the interaction waveform from the oscillator j to the oscillator i determined by the physical interactions among the oscillators, for example, chemical materials, electric currents, surface acoustic waves on elastic bodies, spin waves on magnetic materials, etc. If the interaction is small enough, the Eq. (2) can be reduced to the following dynamical phase equations [9]

$$\frac{d\theta_i(t)}{dt} = \omega_i + \sum_{j \neq i}^N H_{ij}(\theta_i - \theta_j), \quad i = 1, \ldots, N. \tag{3}$$

Here, θ_i and ω_i are the phase and the natural frequency of oscillator i, respectively. The function H_{ij} is called phase coupling function and computed as the phase response (1) with the input stimulus replaced with the interaction waveform between two oscillators:

$$H_{ij}(\theta_i - \theta_j) = \int_0^{2\pi} Z(\theta) \cdot G_{ij}(\theta_j - \theta_i + \theta) d\theta, G_{ij}(\theta) = G_{ij}(X_j(\theta)). \tag{4}$$

In a synchronized state, all the oscillators have a common frequency $\dot{\theta}_i = \bar{\omega}$ and their phase locked state are given by $\theta_i(t) = \bar{\omega}t + \phi_i$. Thus, the existence of a phase-locked synchronized state reduces to solving the following set of equations

$$\bar{\omega} = \omega_i + \sum_{j \neq i}^N H_{ij}(\phi_{ij}), \quad \phi_{ij} = \phi_i - \phi_j, \quad i = 1, \ldots, N. \tag{5}$$

4   Function Approximation and Regression by Wave-Based Reservoir Computing

4.1   Function Approximation by Two Coupled Oscillators

Consider a system of two symmetrically coupled oscillators given by

$$\frac{d\theta_1(t)}{dt} = \omega_1 + H(\theta_1 - \theta_2), \quad \frac{d\theta_2(t)}{dt} = \omega_2 + H(\theta_2 - \theta_1), \tag{6}$$

where we assume the phase interaction function is an odd function $H(\theta_2 - \theta_1) = -H(\theta_1 - \theta_2)$. Then, in a synchronized state, it can be easily seen that the common frequency is given by $\bar{\omega} = (\omega_1 + \omega_2)/2$ and the phase difference $\phi = \phi_1 - \phi_2$ satisfies

$$\Delta = -2H(\phi), \quad \Delta = \omega_1 - \omega_2. \tag{7}$$

The stability of the phase difference can be judged by two conditions. The first condition is that $-\Delta/2$ is within the range $[\min H(\phi), \max H(\phi)]$ so that Eq. (7) has at least one solution ϕ^*. The second condition is that the solution ϕ^* satisfies $H'(\phi^*) < 0$ so that the phase difference ϕ^* is stable.

The basic idea for function approximation is that we design the phase coupling function H so that the phase difference after synchronization gives the inverse of the desired function of the difference of two frequencies ω_1 and ω_2. This is a kind of inverse problems where we need to find an interaction waveform which realize the given target function. Since the Eq. (4) is a linear convolution of two functions, we can find the appropriate interaction waveforms for the target phase interaction function by Fourier transforms [9]. Figure 2(a) shows the examples of interaction waveforms for the target function $H(\phi) = a\sqrt{\phi}$ and logit function (inverse of the sigmoid function) defined as $H(\phi) = a + \varepsilon \log(\phi/(2\pi - \phi))$. Here, we extended the square root function as $H(\phi) = -\varepsilon\sqrt{-\phi}, -\pi \leq \phi \leq 0$ and $H(\phi) = \varepsilon\sqrt{\phi}, 0 \leq \phi \leq \pi$ so that $H(\phi)$ becomes a 2π-periodic odd function. In addition, we assumed that the phase sensitivity function is $Z(\theta) = \cos(\theta) - \sin(\theta)$ and applied numerical FFT and IFFT by digitizing H and Z with 100 sampling points equally spaced on $[0, 2\pi)$.

We can realize arithmetic operations such as adder and multiplier based on this function approximation by two coupled oscillators. We can implement the adder $\omega_1 \pm \omega_2$ by reading out $2\bar{\omega}$ of the synchronized oscillators in (6). Similarly, we can implement the square operation $\omega \mapsto \omega^2$ by $\Delta = -2H(\phi) = -\sqrt{2\phi}$. Note that there is only one stable fixed point for $\Delta < 0$ since $-\sqrt{2\phi}$ is a monotonously decreasing function. Using these two operations and the identity $xy = ((x+y)^2 - (x-y)^2)/4$, we can construct a multiplier $\omega_1\omega_2$ as is depicted in Fig. 2(b) under the restriction that input variables ω_1 and ω_2 satisfy $\omega_1 \pm \omega_2 \in [\min H(\phi), \max H(\phi)]$.

4.2   Functional Regression by an Oscillator Reservoir

Though we can realize any target function as the phase interaction function by the method in Sect. 4.1, the problem is that we need to re-design the interactions

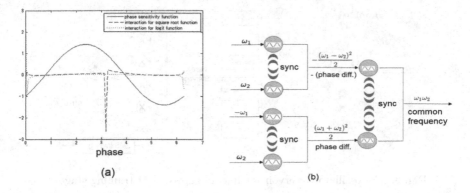

Fig. 2. (a) Phase sensitivity function $Z(\theta) = \sin\theta - \cos\theta$, interaction waveforms for square root function with $a = 0.05$ and logit function with $a = 0, \varepsilon = 0.01$. (b) Coupled oscillators computing the product $\omega_1\omega_2$.

depending on the target function. This is a disadvantage especially for non-conventional implementation such as nano structures. In this subsection, we show that one fixed oscillator reservoir can approximate different target functions. Let us consider the coupled oscillators connected in star-like topology as illustrated in Fig. 3. The phase dynamics is written as

$$\frac{d\theta_0(t)}{dt} = \omega + \sum_{j=1}^{N} H_{0n}(\theta_0 - \theta_n), \quad \frac{d\theta_n(t)}{dt} = H_{n0}(\theta_n - \theta_0), n = 1,\ldots,N. \quad (8)$$

We assume that the phase interaction functions are all odd $H_{0n}(-\phi) = -H_{0n}(\phi)$, the interaction between oscillator 0 and n are symmetric $H_{0n}(\phi) = H_{n0}(\phi)$ and all the oscillators have the same phase sensitivity function $Z(\theta)$. We initialize the coupled oscillators so that only the central one oscillates with frequency ω and the rest are all quiescent. After synchronization state is reached, the oscillators have common frequency $\bar{\omega}$ and relative phase difference $\phi_n - \phi_0$ which satisfies

$$\bar{\omega} = \omega + \sum_{n=1}^{N} H_{0n}(\phi_0 - \phi_n), \quad \bar{\omega} = H_{n0}(\phi_n - \phi_0), n = 1,\ldots,N. \quad (9)$$

It can be easily seen that the common frequency is $\bar{\omega} = \omega/(N+1)$. Assuming $\phi_0 = 0$ and replacing $\bar{\omega}$ with ω, we calculate the n-th basis functions as $f_n(\omega) = H_{n0}^{-1}(\omega)$ by the method in Sect. 4.1. Thus, we can view the reservoir of $N+1$ coupled oscillators as calculating N different basis functions in parallel. We employ sigmoid functions $f_n(\omega) = \frac{1}{1+\exp(-(\omega-p_n)/\varepsilon)}$ as basis functions and set the parameters to $p_n = 2\pi n/N$ and $\varepsilon = 0.1$.

Given a target function $f(\omega)$, we approximate $f(\omega)$ by the linear combination of basis functions $f_1,\ldots,f_N$ as follows

$$\hat{f}(\omega) = \sum_{n=1}^{N} a_n f_n(\omega). \quad (10)$$

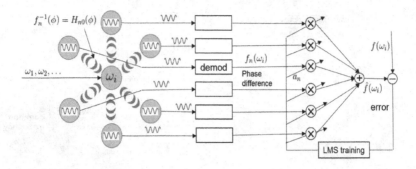

Fig. 3. The oscillator reservoir for linear regression at training stage

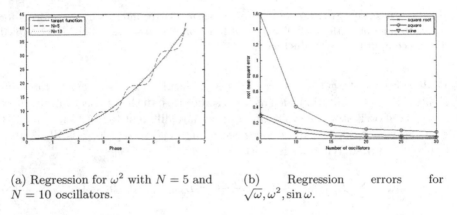

(a) Regression for ω^2 with $N = 5$ and $N = 10$ oscillators.

(b)   Regression   errors   for $\sqrt{\omega}, \omega^2, \sin \omega$.

Fig. 4. Functional regression by oscillator reservoir.

The role of the readout function is to determine the coefficients $\boldsymbol{a} = (a_1, \ldots, a_N)$ by minimizing the mean square error

$$\min_{a_1,\ldots,a_N} \int_0^{2\pi} (f(\omega) - \hat{f}(\omega))^2 d\omega. \tag{11}$$

at training stage and to output the estimated value $\hat{f}(\omega)$ by (10) at operation stage. The minimization of (11) can be achieved by solving the normal equation $C\boldsymbol{a} = \boldsymbol{r}$, where $C = (c_{ij}), c_{ij} = \int_0^{2\pi} f_i(\omega)f_j(\omega)d\omega$ and $\boldsymbol{r} = (r_1, \ldots, r_N)$, $r_i = \int_0^{2\pi} f(\omega)f_i(\omega)d\omega$. Figure 4(a) shows the regression for target function ϕ^2 by solving the normal equation. To perform the training online, we choose randomly ω_i from 100 sampling points equally spaced on $[0, 2\pi)$ and inject wave with frequency ω_i to the central oscillator. After the oscillator reservoir reaches synchronization, the readout updates the coefficients by stochastic gradient descent method as $\boldsymbol{a} \leftarrow \boldsymbol{a} + \alpha e_i \boldsymbol{\phi}_i, e_i = f(\omega_i) - \hat{f}(\omega_i), \boldsymbol{\phi}_i = (f_1(\omega_i), \ldots, f_N(\omega_i))$. We choose the step size parameter $\alpha = 0.01$ and iterate this update 10^4 times. Figure 4(b) shows that how root mean square error is improved by increasing the number of oscillators for $f(\omega) = \sqrt{\omega}, \omega^2, \sin \omega$.

5   Conclusion

We have proposed a framework for wave-based reservoir computing based on synchronization of coupled oscillators. We applied the proposed framework to function approximation and regression and showed that such problems can be solved efficiently due to the parallelism and stability of synchronized state. In summary, wave-based reservoir computing is promising for large scale and densely integrated cognitive computing systems.

References

1. Achuthan, S., Butera, R.J., Canavier, C.C.: Synaptic and intrinsic determinants of the phase resetting curve for weak coupling. J. Comput. Neurosci. **30**(2), 373–390 (2011)
2. Hoppensteadt, F.C., Izhikevich, E.M.: Weakly Connected Neural Networks (Applied Mathematical Sciences 128). Springer, New York (1997)
3. Katayama, Y., Yamane, T., Nakano, D., Nakane, R., Tanaka, G.: Wave-based neuromorphic computing framework for brain-like energy efficiency and integration. In: IEEE NANO 2015 (2015)
4. Khasanvis, S., Rahman, M., Rajapandian, S., Moritz, C.: Wave-based multi-valued computation framework. In: IEEE/ACM International Symposium on Nanoscale Architectures (NANOARCH), pp. 171–176 (2014)
5. Legenstein, R.A., Maass, W.: Wire length as a circuit complexity measure. J. Comput. Syst. Sci. **70**, 53–72 (2005)
6. Lukoševičius, M., Jaeger, H.: Reservoir computing approaches to recurrent neural network training. Comput. Sci. Rev. **3**(3), 127–149 (2009)
7. Nikonov, D.E., Young, I.A., Bourianoff, G.I.: Convolutional networks for image processing by coupled oscillator arrays. http://arxiv.org/abs/1409.4469 (2014)
8. Pikovsky, A., Rosenblum, M., Kurths, J.: Synchronization - A Universal Concept in Nonlinear Science, vol. 1. Cambridge University Press, Cambridge (2001)
9. Rusin, C.G., Kori, H., Kiss, I.Z., Hudson, J.L.: Synchronization engineering: tuning the phase relationship between dissimilar oscillators using nonlinear feedback. Philos. Trans. R. Soc. A **368**, 2189–2204 (2010)

Hybrid Controller with the Combination of FLC and Neural Network-Based IMC for Nonlinear Processes

Mohammad Anwar Hosen[(✉)], Syed Moshfeq Salaken,
Abbas Khosravi, Saeid Nahavandi, and Douglas Creighton

Centre for Intelligent System Research (CISR), Deakin University,
Waurn Ponds Campus, Geelong, Australia
{anwar.hosen,ssalaken,abbas.khosravi,saeid.nahavandi,
douglas.creighton}@deakin.edu.au

Abstract. This work presents a hybrid controller based on the combination of fuzzy logic control (FLC) mechanism and internal model-based control (IMC). Neural network-based inverse and forward models are developed for IMC. After designing the FLC and IMC independently, they are combined in parallel to produce a single control signal. Mean averaging mechanism is used to combine the prediction of both controllers. Finally, performance of the proposed hybrid controller is studied for a nonlinear numerical plant model (NNPM). Simulation result shows the proposed hybrid controller outperforms both FLC and IMC.

Keywords: Fuzzy logic controller · Hybrid controller · Internal model-based controller · Neural network · Forward model · Inverse model

1 Introduction

Nonlinear process control can be difficult due to lack of process knowledge and the existence of nonlinearity in the process nature [21]. Even though first principles models should be developed for reliability in process control, it is often difficult for poorly understood process and hence, a black box model is developed instead by looking at inputs and outputs of the process of interest [21]. As shown by many researchers, any approximation modelling technique with sufficient accuracy can be considered for this purpose without sacrificing reliability [2,3,15]. Neural network and fuzzy logic-based controllers are shown to be good universal approximator for these applications [5,7,10,20]. FLC-based system are easily interpretable due to the rule-base structure and firing methods while neural network (NN)-based controllers are not. However, neural network-based controllers can approximate extremely difficult function mapping without changing its structure while FLCs lost their interpretability, but not accuracy, due to firing of large number of rules in the database. Both FLC and NN-based controller are successfully applied in a number of fields including robotic manipulator [4], biomedical applications [14], machine learning [13,16], process controlling [5], power systems [11], forecasting [6,9,11,17], chemical and manufacturing

© Springer International Publishing Switzerland 2015
S. Arik et al. (Eds.): ICONIP 2015, Part III, LNCS 9491, pp. 206–213, 2015.
DOI: 10.1007/978-3-319-26555-1_24

plants [1,4,8] etc. An advantage of FLC is their capability to accommodate uncertainty and expert knowledge in the process, which is not inherently possible with NN-based controller. Therefore, if combined, they can produce a robust and accurate controller with the ability to manage uncertainty produced from measurements, human interpretation and noise.

In this work, a hybrid controller is proposed with the combination of individual FLC and NN-based IMC. Two inputs against one output is used to developed FLC. For IMC, two NN-based controllers are developed. These include, a forward model and an inverse model. Finally, the performance of the proposed controller is tested for a nonlinear process and compared the results with the individual FLC and IMC in terms of controller's performance criterions, integral absolute error (IAE) and integral square error (ISE).

2   Proposed Hybrid Controller

2.1   Fuzzy Logic Controller

A typical fuzzy logic controller is essentially a fuzzy logic system (FLS). It takes a crisp input, assign a membership degree to that input by the means of fuzzification, utilize a rule-base to complete inference mechanism, defuzzify the resulting fuzzy set and finally produce a crisp output. A block diagram of FLC is shown in Fig. 1. FLC can utilize both type-1 (T1) and type-2 (T2) fuzzy sets in the process. If T1 fuzzy sets are not feasible due to a requirement of more degrees of freedom, usually an interval type-2 (IT2) fuzzy set-based system is used. For a T1 FLC, output processing block contains only defuzzifier. For all other systems, a type reduction mechanism is needed.

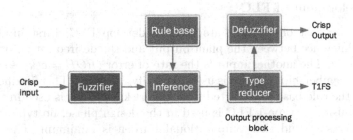

Fig. 1. FLC block diagram

2.2   NN-Based Internal Model Controller

NN-based IMC is popular for its simple structure and offset-free setpoint tracking. In this IMC structure, two NN-based models are used as seen in Fig. 2. One is forward and another one is inverse model. The forward model is added parallel to the plant, and the model output is compare with the plant output. The difference between the plant and forward model outputs is finally compare with the setpoint, and the updated setpoint is feed to the inverse model. In this way, the controller is able to cater the plant mismatch and disturbances.

2.3 Hybrid Controller

The proposed hybrid controller is a combination of FLC and IMC. In this proposed scheme, the set point is passed to the FLC and IMC. Their output is then combined by using mean mechanism, and finally this modified output is passed to plant. The plant output is fed back to FLC and IMC. A schematic diagram for this controller is shown in Fig. 3.

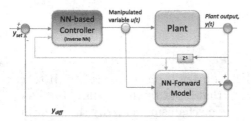

Fig. 2. Basic structure of internal model-based controller (IMC)

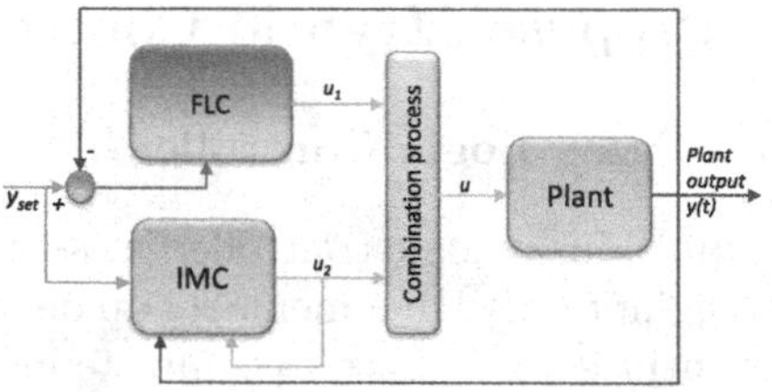

Fig. 3. The structure of the proposed hybrid controller

3   Methodology

3.1 Development of FLC

Two inputs against one output are used to develop FLC. First input is the error, the difference between the plant output and the desired set point ($e(k) = y(k) - y_{set}(k)$). The another input is the rate of error ($ROE = e(k) - e(k-1)$). Triangular membership function is used to fuzzify the input. This fuzzified input, along with the rule-base constructed from expert knowledge, is used in the inference mechanism. As type-1 FLC is used in the design phase, no type reduction block is necessary and the computational burden is minimum. The parameters of the FLC are tuned using IAE as cost function. Output of the FLC is passed to combination block to form a combined control signal. Parameters of FLC are tuned with trial and error method which is also used by several other researchers [5,19].

3.2 Development of NN-Based IMC

IMC usually contains two separate models for the plant, one is forward and the other is inverse model. Firstly, these models are developed independently, and used together to form IMC.

Feed-Forward NN Model. In order to develop the feed-forward model, the raw dataset is first arranged according to the structure of FFNN. The structure for FFNN is $[y(t-1), y(t-2), u(t-1), u(t-2)]; [y(t)]$, where y is the plant output and u is the plant input. The structured data are then rescaled in the range $[-1, 1]$. Afterwards they are partitioned into training, validation and testing sets. Different number of hidden neurons are used in the three layer network with one hidden layer. This way a number of NNs are developed using different hidden layer structure and tested. Initial parameters of NN are randomly initialized. Mean squared error (MSE) is used as cost function throughout the experiment for NNs. Levenberg-Marquardt optimization algorithm is used for parameter tuning. Training is assumed complete when either the maximum number of iteration is reached or the MSE value reach a lower value than predefined threshold. After experimenting with different NNs, the NN with lowest MSE is picked and considered as the final trained model.

Inverse NN Model. The inverse NN model is developed in a similar fashion described above. The cost function, optimization method and training initiation are same as the FFNN. The data structure for this model is $\{[y(t), y(t-1), y(t-2), u(t-1), u(t-2)]; [u(t)]\}$. This model is actually acts as NN-based controller, where model output is the plant input (controlled signal).

4   Case Study and Experimental Data

In this work, proposed controller is used to control a nonlinear numerical plant, NNPM, which can be described by the following equation:

$$y(k+1) = \frac{y(k)}{1 + y(k)^2} + u(k)^3 \qquad (1)$$

Here, $y(k)$ and $u(k)$ are the controlled and manipulated variable respectively. This nonlinear plant is widely used as benchmark for controller performance experiments [12,18].

Equation 1 is used to generate training data for proposed controller using the following setpoint:

$$y_{set} = 2.5 \sin(\frac{10\pi k}{N}) + 2.5 \sin(\frac{4\pi k}{N}) \qquad (2)$$

where N denotes the total number of sample and k denotes the sampling number. Equation 1 allows the setpoint to be limited in the interval $[-5, 5]$, which is infact the nonlinear region of the plant [18].

Finally, this plant is run in the simulation for 5000 s with a sampling interval of 1 s. This provides 5000 instances of raw data which are arranged as follows for forward and inverse model, respectively, and used to train both NNs.

$$FFNN = \{[y(t-1), y(t-2), u(t-1), u(t-2)]; [y(t)]\} \qquad (3)$$

Table 1. Rule-base of type-1 FLC for 25 fuzzy rules

		ROE				
		VN	N	Z	P	VP
error (e)	VN	P	P	VP	VP	VP
	N	Z	Z	P	VP	VP
	Z	P	Z	Z	N	N
	P	Z	Z	N	N	N
	VP	N	Z	VN	VN	VN

Table 2. Optimum range for inputs and output for FLC with 25 fuzzy rules

Inputs and output	Min	Max
error (e)	−0.10	0.10
ROE	−0.05	0.05
Plant output (y)	−10.00	10.00

$$\text{Inverse NN} = \{[y(t), y(t-1), y(t-2), u(t-1), u(t-2)]; [u(t)]\} \qquad (4)$$

$60\,\%, 20\,\%$ and $20\,\%$ of these data are used training, validating and testing purpose.

To develop FLC for this particular system, several FLS structure is tested and checked for performance. After few trial and error, best result is achieved through a FLS with 5 membership functions on each input which produced 25 rules in the rule-base. The rule-base is created with expert knowledge. This approach of designing FLC/FLS is in line with several other researchers [19]. Developed rule-base structure for FLC, along with the range of inputs and outputs, is shown in Tables 1 and 2, respectively. The control surface for FLC is shown in Fig. 4.

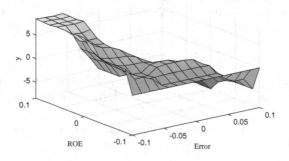

Fig. 4. FLS output surface

5   Results and Discussion

As the hybrid controller is combining the IMC and FLC outputs using a mean averaging mechanism, the settling time for this controller is improved. This can be seen from Figs. 5 and 6. Observe from Fig. 5 that, for constant setpoint, hybrid controller takes the shortest time to reach the setpoint from steady state than other controllers. NN-based IMC controller, denoted as NNC in the figure, is the slowest to reach the setpoint and demonstrates large overshoot and oscillation. FLC is settling much sooner than NNC, but slower than proposed hybrid controller.

The stability and robustness of the proposed controller also tested for a wavy setpoint profile. Observe from Fig. 6, which shows the wavy setpoint tracking performance for 5000sec, that NNC cannot fully track the plant. However, both the FLC and hybrid controller is able to successfully follow the plant setpoint.

Finally, the performance of the proposed controller is evaluated in terms of controller's performance criterions, such as IAE and ISE. Table 3 shows the IAE and ISE values for hybrid controller, FLC and NNC. Observe from this table that both IAE and ISE error indicators are much smaller for hybrid controller than those of NNC and FLC. For the wavy setpoint, NNC performs very poorly. This table clearly shows that hybrid controller is much superior than FLC, and specially NN-based IMC. The combination of IMC and FLC produces much better result.

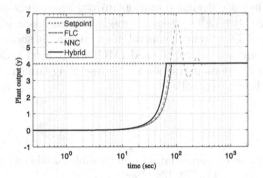

Fig. 5. Constant tracking performance

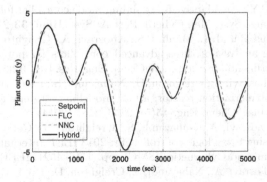

Fig. 6. Wavy tracking performance

Table 3. Tracking performances of FLC, NNC and hybrid controller in terms of IAE and ISE

Type of setpoint	Controllers	IAE	ISE
Constant	FLC	259.80	805.10
	NNC	378.30	945.70
	Hybrid	233.80	664.90
Wavy setpoint	FLC	42.22	0.45
	NNC	1681.00	977.90
	Hybrid	20.68	0.11

6   Conclusion

This research work explores the combined potential of fuzzy logic-based controller and neural network-based internal model controller for the control of nonlinear plant. They are designed and optimized separately and their effect is combined in parallel through a mean averaging mechanism in order to design a hybrid controller. This hybrid control mechanism is shown to outperform both the FLC and neural network-based IMC, when used independently. Proposed hybrid controller demonstrates a smaller settling time and zero overshoot, proving itself as a better controller.

References

1. Bristol, E.: On a new measure of interaction for multivariable process control. IEEE Trans. Autom. Control **11**, 133–134 (1966)
2. Cybenko, G.: Approximation by superpositions of a sigmoidal function. Math. Control, Signals Syst. **2**(4), 303–314 (1989)
3. Girosi, F., Poggio, T.: Networks and the best approximation property. Biol. Cybern. **63**(3), 169–176 (1990)
4. Guo, Y., Woo, P.Y.: An adaptive fuzzy sliding mode controller for robotic manipulators. IEEE Trans. Syst. Man Cybern. Part A: Syst. Hum. **33**(2), 149–159 (2003)
5. Hosen, M.A., Hussain, M.A., Mjalli, F.S., Khosravi, A., Creighton, D., Nahavandi, S.: Performance analysis of three advanced controllers for polymerization batch reactor: an experimental investigation. Chem. Eng. Res. Des. **92**(5), 903–916 (2014)
6. Hosen, M.A., Khosravi, A., Creighton, D., Nahavandi, S.: Prediction interval-based modelling of polymerization reactor: a new modelling strategy for chemical reactors. J. Taiwan Inst. Chem. Eng. **45**(5), 2246–2257 (2014)
7. Hosen, M.A., Khosravi, A., Nahavandi, S., Creighton, D.: Control of polystyrene batch reactor using fuzzy logic controller. In: 2013 IEEE International Conference on Systems, Man, and Cybernetics (SMC), pp. 4516–4521. IEEE (2013)
8. Hosen, M.A., Khosravi, A., Nahavandi, S., Creighton, D.: Prediction interval-based neural network modelling of polystyrene polymerization reactor: a new perspective of data-based modelling. Chem. Eng. Res. Des. **92**(11), 2041–2051 (2014)

9. Hosen, M.A., Khosravi, A., Nahavandi, S., Creighton, D.: Improving the quality of prediction intervals through optimal aggregation. IEEE Trans. Ind. Electron. **62**(7), 4420–4429 (2015)
10. Khosravi, A., Talebi, H., Karrari, M.: A neuro-fuzzy based sensor and actuator fault estimation scheme for unknown nonlinear systems. In: Proceedings. 2005 IEEE International Joint Conference on Neural Networks, IJCNN 2005, vol. 4, pp. 2335–2340, July 2005
11. Khosravi, A., Nahavandi, S.: An optimized mean variance estimation method for uncertainty quantification of wind power forecasts. Int. J. Electr. Power Energy Syst. **61**, 446–454 (2014)
12. Narendra, K.S., Parthasarathy, K.: Identification and control of dynamical systems using neural networks. IEEE Trans. Neural Netw. **1**(1), 4–27 (1990)
13. Nguyen, T., Khosravi, A., Creighton, D., Nahavandi, S.: Eeg signal classification for bci applications by wavelets and interval type-2 fuzzy logic systems. Expert Syst. Appl. **42**(9), 4370–4380 (2015)
14. Nguyen, T., Khosravi, A., Creighton, D., Nahavandi, S.: Fuzzy system with tabu search learning for classification of motor imagery data. Biomed. Signal Process. Control **20**, 61–70 (2015)
15. Park, J., Sandberg, I.W.: Universal approximation using radial-basis-function networks. Neural Comput. **3**(2), 246–257 (1991)
16. Quan, H., Srinivasan, D., Khambadkone, A.M., Khosravi, A.: A computational framework for uncertainty integration in stochastic unit commitment with intermittent renewable energy sources. Appl. Energy **152**, 71–82 (2015)
17. Quan, H., Srinivasan, D., Khosravi, A.: Uncertainty handling using neural network-based prediction intervals for electrical load forecasting. Energy **73**, 916–925 (2014)
18. Salman, R.: Neural networks of adaptive inverse control systems. Appl. Math. Comput. **163**(2), 931–939 (2005)
19. Sanchez, M.A., Castillo, O., Castro, J.R.: Generalized type-2 fuzzy systems for controlling a mobile robot and a performance comparison with interval type-2 and type-1 fuzzy systems. Expert Syst. Appl. **42**(14), 5904–5914 (2015)
20. Wang, L.X., Mendel, J.M.: Fuzzy basis functions, universal approximation, and orthogonal least-squares learning. IEEE Trans. Neural Netw. **3**(5), 807–814 (1992)
21. Zhang, J., Morris, A.J.: Recurrent neuro-fuzzy networks for nonlinear process modeling. IEEE Trans. Neural Netw. **10**(2), 313–326 (1999)

Comparative Study of Web-Based Gene Expression Analysis Tools for Biomarkers Identification

Worrawat Engchuan[1], Preecha Patumcharoenpol[2], and Jonathan H. Chan[1(✉)]

[1] Data and Knowledge Engineering Laboratory, School of Information Technology,
King Mongkut's University of Technology Thonburi, Bangkok, Thailand
{worrawat.eng,jonathan}@sit.kmutt.ac.th
[2] Systems Biology and Bioinformatics Laboratory,
King Mongkut's University of Technology Thonburi, Bangkok, Thailand
preecha.pat@mail.kmutt.ac.th

Abstract. With the flood of publicly available data, it allows scientists to explore and discover new findings. Gene expression is one type of biological data which captures the activity inside the cell. Studying gene expression data may expose the mechanisms of disease development. However, with the limitation of computing resources or knowledge in computer programming, many research groups are unable to effectively utilize the data. For about a decade now, various web-based data analysis tools have been developed to analyze gene expression data. Different tools were implemented by different analytical approaches, often resulting in different outcomes. This study conducts a comparative study of three existing web-based gene expression analysis tools, namely Gene-set Activity Toolbox (GAT), NetworkAnalyst and GEO2R using six publicly available cancer data sets. Results of our case study show that NetworkAnalyst has the best performance followed by GAT and GEO2R, respectively.

Keywords: Gene set activity · Gene expression · Disease classification · Cross-dataset validation · Web-based microarray analysis

1 Introduction

In the era which data is flooding from advancement in data generating technology, it is an opportunity for scientists to discover new findings. In study of genetic disease, many research groups have published their genetic datasets including genome sequence, methylation profile, gene expression profile, etc. [1]. Gene expression profile is the data which captures the activity of thousands of genes in the cell. Studying gene expression data may expose the mechanisms of disease development. Currently, there are many public gene expression databases, i.e. Gene Expression Omnibus (GEO), Array Express, Expression Atlas, and so on [1–3]. Many data analysis approaches have been proposed to analyze the available data. Starting from the simplest one, differentially expressed genes (DEG) analysis utilizes statistical testing to compare expression profiles between two groups of

---

W. Engchuan and P. Patumcharoenpol–These authors contributed equally to this work.

© Springer International Publishing Switzerland 2015
S. Arik et al. (Eds.): ICONIP 2015, Part III, LNCS 9491, pp. 214–222, 2015.
DOI: 10.1007/978-3-319-26555-1_25

population (case/control) [4]. By doing so, the gene with potential of being disease marker can be identified. To improve the performance of DEG, integrative approaches which utilize other biological data or knowledge to help to analyze gene expression data have been proposed [5]. Gene-set-based analysis is an integrative study of gene-set data and gene expression data. Instead of testing each individual gene, this approach tests the impact of each gene-set on disease susceptibility. Thus, the gene-sets are used as disease markers other than single genes [6, 7]. Like Gene-set-based analysis, network-based analysis also tests for the significance of gene subnetwork on disease development. However, this approach also takes the relationship between genes into account [8].

Those proposed approaches are commonly provided as standalone application or program library. For example, MultiExperiment Viewer (MEV) is a common standalone tool for microarray analysis [9], or Bioconductor, which was published as a library for R package [10]. However, with the limitation of computing resources or knowledge in computer programming, this may limit the use of those approaches. To maximize their use, several web-based applications for gene expression analysis have been developed. Back in 2003, Herrero et al. developed a web-based pipeline for microarray gene expression analysis, namely Gene Expression Profile Analysis Suite (GEPAS), but currently, it is no longer available [11]. Later in 2005, Hsiao et al. developed a simple tool for DEG analysis and annotated the significant genes based on Gene Ontology [12]. Instead of being only a database, in 2013, GEO also developed their online analysis tool namely, GEO2R, which generates the R source code for the user along with analysis of the results [1]. In 2014, a comprehensive tool for network-based analysis of gene expression data namely NetworkAnalyst, was developed by Xia et al. [13]. More recently, our research group developed the Gene-set Activity Toolbox (GAT), which implemented the baseline and in-house gene-set-based gene expression data analysis approaches [14].

With a variety of choices of tools and approaches, this work conducts a comparative study of three existing web-based gene expression analysis tools, which are GEO2R, NetworkAnalyst and GAT. The performance of these tools is compared in the aspect of their features, the significance of selected markers and classification performance of their selected markers. Six gene expression datasets from GEO which covered three common cancers; colorectal cancer, lung cancer and breast cancer, are used as a case study.

2 Methodology

2.1 Gene Expression Datasets

This comparative study is conducted on a case study of six microarray datasets of three different cancers. Two independent datasets for each cancer are obtained from Gene Expression Omnibus (GEO) database. These datasets are as following; (1) Two color-ectal datasets of 22 and 64 patients were generated by Hong et al., and Sabates-Bellver et al. [15, 16]. (2) Lung cancer datasets of 187 and 107 patients studied by Spira et al. and Landi et al. [17, 18]. (3) Breast cancer datasets containing expression profiles of 30 and 62 patients were generated by Turashvili et al. and Richardson et al. [19], [20]. Table 1 summarizes the detail of each datasets and notation use to refer to each dataset.

Table 1. Gene expression datasets

Accession no.	Notation	Cancer type	Publication
GSE4107	C1	Colorectal	Hong, et al. 2006 [15]
GSE8671	C2	Colorectal	Sabates-B. et al. 2007 [16]
GSE4115	L1	Lung	Spira, et al. 2007 [17]
GSE10072	L2	Lung	Landi et al. 2008 [18]
GSE5764	B1	Breast	Turashvili et al. 2007 [19]
GSE7904	B2	Breast	Richardson et al. 2006 [20]

2.2 Web-Based Gene Expression Analysis Tools

GEO2R. GEO2R is a web-based tool for performing a DEG analysis on microarray in GEO data repository [1]. It is fully integrated with the GEO website, making it able to be used with all data within GEO. This tool provides simple pre-processing techniques (log transform, data normalization), statistical testing and adjustment. The result can be obtained as a list of significant genes and box plot of expression distribution. For the benefit of the user, this tool also generates R-script for the re-analysis purpose. GEO2R is available at http://ncbi.nim.nih.gov/geo/geo2r/. In this work, GEO2R with a default option (auto apply of log transformation and p-value adjustment by Benjamini and Hochberg (False discovery rate)) was used to produce a list of significant genes. The result was sorted by adjusted p-value and genes with adjusted p-value $< 1E-4$ were selected as potential gene markers. In case too many genes passed the p-value threshold, the top 250 genes were selected.

NetworkAnalyst. NetworkAnalyst is a web-based tool for performing a DEG, subnetwork identification and functional analysis by mapping a set of significant genes into Protein-Protein Interactions (PPI) network [13]. It provides a network viewer and simple functional analysis makes it great for an exploratory analysis. NetworkAnalyst can be accessed via http://networkanalyst.ca. In this study, NetworkAnalyst was used with InnateDB Interactome data to produce an initial list of subnetworks. The big subnetwork ($500 <$ nodes) was resized with "Reduce" function, which will shrink the subnetwork by removing high-order interactions from the subnetwork. Finally, the nodes of all subnetworks (i.e. gene) were exported and those nodes having betweenness > 10 were used as potential gene markers.

Gene-set Activity Toolbox (GAT). GAT is a gene-set-based gene expression analysis toolbox, which provides several gene-set-based analysis methods including Conditional Responsive Genes-based (CORG-based), Negatively Correlated Feature Set (NCFS-i), Analysis of Variance Feature Set (AFS) and Gene Network-based Feature Set (GNFS) [14]. Besides those gene-set-based analysis approaches, GAT is also equipped with data mining tools using WEKA as a library [21]. Four simple feature selections (ReliefF,

Principal Component Analysis, Gain Ratio and Information Gain) and four standard classifiers (Support Vector Machine, Multilayer Perceptron, Random Forest and Linear Regression) are available for selecting gene-sets markers and evaluating those selected gene-sets.

This study used trial and error technique for adjusting choices of gene-set collection, number of gene-sets (1–10), feature selection method, classifier used to build a model. The cross dataset validation scheme was applied here to evaluate the adjustment. The best adjustment having highest AUC will be used to retrieve list of gene markers.

2.3 Evaluation

In order to compare these web-based tools, we focus on two aspects, which are the significant of selected gene markers and the performance on disease classification of those selected markers. We propose the use of Gene prospector as an evaluation tool for measuring the significance of selected gene markers and we apply 5-fold cross-validation as a classification evaluation.

Gene prospector is a web-based tool which helps to search for the disease related genes [22]. This tool is a text-mining-based implementation with a variety of highly curated and updated literature datasets of genetic association studies. In this work, we used search terms such as "Colorectal cancer", "Lung cancer" and "Breast cancer" to retrieve a gene list of each cancer type. Then, the score of each selected gene list is calculated using following Eq. (1).

$$F(X) = \left(\sum_{i=1}^{N} G_i \right) \Big/ N \tag{1}$$

where $F(X)$ is the scoring function, $X = \{X1, X2, \dots XN\}$ is a selected gene list and G_i is a disease-related score of gene X_i.

For classification evaluation, we applied a gene list from one dataset to another independent dataset of the same disease study. Then, 5-fold cross-validation is performed and repeated 10 times to obtain the estimated classification performance. As Support Vector Machine (SVM) has been widely and successfully applied in many fields of study [23], we used SVM as a classifier in this evaluation. Area under Receiver Operating Characteristic (AUROC) was used as a measurement of classification performance because it has been found to be unbiased against imbalanced datasets [24].

3 Results and Discussion

This study compares three web-based microarray analysis tools on three aspects which are feature comparison, significance of selected gene markers and performance on classification using selected gene markers. Table 2 presents the number of selected markers for each dataset and each tool. With parameters and threshold used as mentioned in the previous section, GEO2R could not find any significant genes in B1 datasets, while NetworkAnalyst could not detect any subnetwork in both Lung cancer datasets. The above highlights one limitation of these two web-based tools.

Table 2. Number of selected genes

| | Tool (number of selected genes) | | |
Dataset	GEO2R	NetworkAnalyst	GAT
C1	17	54	10
C2	250	140	22
L1	250	–	10
L2	250	–	14
B1	–	51	45
B2	250	387	17

3.1 Features Comparison

Before going through other comparison results, Table 3 compares the features of three tools. In data gathering, GAT is the most convenient tool as it allows user to upload their datasets and to retrieve datasets from GEO as well. GEO2R has limited use as only datasets available in GEO can be used. For data pre-processing, GEO2R is fully integrated with GEO datasets, thus most of pre-processing steps are automatically done by

Table 3. Features comparison of three tools

| Feature | Tools | | |
	GEO2R	NetworkAnalyst	GAT
Data gathering	- Via GEO	- Upload by user	- Via GEO - Upload by user
Data pre-processing	- Automatic Log transform	- Log transform - Quantile Normalization	- z-transform
Identification of markers	- DEG analysis - P-value adjustment	- DEG analysis - Network-based analysis	- Gene-set-based analysis - Network-based analysis - Feature selection - Classification
Functional analysis	- None	- Gene Ontology	- Pathway analysis
Visualization	- Box plot of distribution	- Network viewer	- KEGG pathway mapper - Network viewer

the system while for NetworkAnalyst and GAT, the user needs to prepare the sample information for each sample as specified in their disease status. Then different data normalization techniques are applied for each tool. Hence, with different pre-processing processes, the data may be altered differently before analysis and can result in selecting different gene markers at the end of analysis. GEO2R and NetworkAnalyst identify gene markers by performing DEG analysis. However, NetworkAnalyst takes the significant genes further to do network-based analysis to identify disease-related subnetworks. Unlike those tools, GAT has both network-based and gene-set-based analysis and uses machine learning approaches to select and evaluate the disease-related genes. The selected gene markers are then annotated by obtaining the function of each marker from a gene database. Finally, network visualization of both NetworkAnalyst and GAT as well as KEGG pathway mapper available with GAT are used to visualize the results to gain more understanding from the interaction between gene markers.

3.2   Gene Markers Comparison

In the first comparison result of our case study, we compared the significance of selected gene markers from each tool. Figure 1 presents the comparison result of disease-related score of gene list selected from each tool. The disease-related score is calculated by retrieving a disease-related gene list from GeneProspector and summarizing the score of all selected gene markers as in (1). The result shows that gene list identified by NetworkAnalyst have quite high disease-related scores compared to the other tools. This may be because of the benefit of selecting high betweenness nodes. Those high betweenness nodes represent the important genes, which have been well studied already. As a result, there will be more evidence to support those genes already. However, there are no significant differences found between each pair of tools. In L1 dataset, 10 genes selected by GAT are not found in GeneProspector result of "Lung cancer" search term.

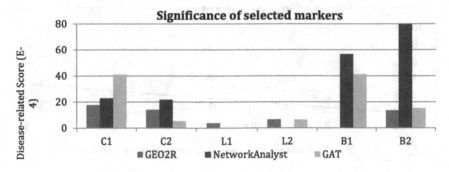

Fig. 1. Comparison of gene markers

3.3   Classification of Selected Gene Markers

GAT achieved a good performance in C1, C2 and B2 datasets (see Fig. 2). GAT performed poorly in datasets L1, L2, and B1.

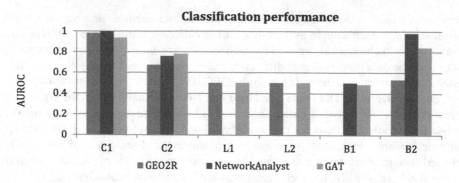

Fig. 2. Comparison of classification performance

In comparison between GAT and other tools, NetworkAnalyst achieved a better AUROC score in a comparison to GAT in C1, B1, and B2 datasets, but got a lower score in C2 dataset. GEO2R achieve a slightly better score in C1, but overall worse or equal to GAT. Finally, t-test between these results shows no significant difference between these results.

It should be noted that all tools performed poorly or were unable to find any significant gene markers at all on L1 and L2 datasets. We suspected that this might be a problem from the dataset itself. Therefore, it is worth investigating further and the result could be used for further improvement.

Furthermore, we also did a correlation analysis between classification performance and disease-related score by applying Pearson's Correlation analysis. The result (Fig. 3) shows that these two measures are decently correlated (Correlation Coefficient = 0.51). In other words, they can be used as a representative of each other.

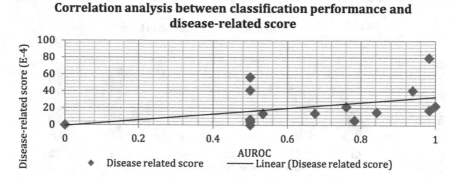

Fig. 3. Correlation analysis between classification performance and disease-related score

In summary, from the comparison results, NetworkAnalyst may be the first choice to be considered followed by GAT and GEO2R, respectively. As GEO2R is a simple analysis tool, it has several limitations such as the dataset should be from GEO only, lack of visualization and functional analysis. However, those disadvantages of GEO2R are overlooked by its easy-to-use user interface. NetworkAnalyst and GAT are comprehensive web-based tools, which are implemented with network-based and gene-set-based analysis approaches. So, these tools provide more systematic view of the results than GEO2R. However, different pre-processing techniques applied may be the cause of difference in selected gene list.

4 Conclusion

This paper conducts a comparative study of current available web-based gene expression analysis tools: GEO2R, NetworkAnalyst and GAT. Six microarray datasets were used in a case study to compare these tools in terms of significance of selected markers and their performance on disease classification. NetworkAnalyst with a comprehensive pipeline achieved the best performance in most cases followed by GAT and GEO2R. Additionally, our GAT is still a beta version. For the next release, we will improve on it to allow user to develop their algorithm as a plugin like R package. By doing so, this will help them to publish their new proposed algorithm in the form of user-friendly web-based application.

Acknowledgment.   The authors would like to thank Ms. Katlin Kreamer-Tonin for proofreading this paper.

References

1. Barrett, B.T., et al.: NCBI GEO: archive for functional genomics data sets—update. Nucleic Acids Res. **41**(Database issue), D991–D995 (2013)
2. Kolesnikov, N., et al.: ArrayExpress update–simplifying data submissions. Nucleic Acids Res. **43**(Database issue), D1113–D1116 (2015)
3. Petryszak, R., et al.: Expression Atlas update—a database of gene and transcript expression from microarray- and sequencing-based functional genomics experiments. Nucleic Acids Res. **42**(Database issue), D926–D932 (2014)
4. Dudoit, S., et al.: Statistical methods for identifying differentially expressed genes in replicated cDNA microarray experiments. Statistica Sinica**12**(1), 111–140 (2002)
5. Mootha, V.K., et al.: PGC-1alpha-responsive genes involved in oxidative phosphorylation are coordinately downregulated in human diabetes. Nat. Genet. **34**, 267–273 (2003)
6. Sootanan, P., Prom-on, S., Meechai, A., Chan, J.H.: Pathway-based microarray analysis for robust disease classification. Neural Comput. Appl. **21**, 649–660 (2012)
7. Engchuan, W., Chan, J.H.: Pathway activity transformation for multi-class classification of lung cancer dataset. Neurocomputing**165**, 81–89 (2015)
8. Doungpan, N., Engchuan, W., Meechai, A., Chan, J.H.: Clustering-based multi-class classification of complex disease. IJCNN 2015 (in press)
9. Saeed, A.I., et al.: TM4: a free, open-source system for microarray data management and analysis. Biotechniques **34**, 374–378 (2003)
10. Huber, W., et al.: Orchestrating high-throughput genomic analysis with bioconductor. Nat. Methods **12**, 115–121 (2015)
11. Herrero, J., Al-Shahrour, F., Diaz-Uriarte, R., Mateos, A., Vaquerizas, J.M., Santoyo, J., Dopazo, J.: GEPAS: a web-based resource for microarray gene expression data analysis. Nucleic Acid Res. **31**, 3461–3467 (2003)

12. Hsiao, A., Ideker, T., Olefsky, J.M., Subramaniam, S.: VAMPIRE microarray suite: a web-based platform for the interpretation of gene expression data. Nucleic Acid Res. **33**, W627–W632 (2005)
13. Xia, A.J., Gill, E.E., Hancock, R.E.W.: NetworkAnalyst for statistical, visual and network-based meta-analysis of gene expression data. Nat. Protoc. **10**, 823–844 (2015)
14. Engchuan, W., Meechai, A., Tongsima, S., Chan, J.H.: Gene-set activity toolbox (GAT): a platform for microarray-based cancer diagnosis using an integrative gene-set analysis approach. http://www.gat.sit.kmutt.ac.th
15. Hong, Y., Ho, K.S., Eu, K.W., Cheah, P.Y.: A susceptibility gene set for early onset colorectal cancer that integrates diverse signaling pathways: implication for tumorigenesis. Clin. Cancer Res. **13**, 1107–1114 (2007)
16. Sabates-Bellver, J., et al.: Transcriptome profile of human colorectal adenomas. Mol. Cancer Res. **5**, 1263–1275 (2007)
17. Spira, A., et al.: Airway epithelial gene expression in the diagnostic evaluation of smokers with suspect lung cancer. Nat. Med. **13**, 361–366 (2007)
18. Landi, M.T., et al.: Gene expression signature of cigarette smoking and its role in lung adenocarcinoma development and survival. PLoS ONE **3**, 1651 (2008)
19. Turashvili, G., et al.: Novel markers for differentiation of lobular and ductal invasive breast carcinomas by laser microdissection and microarray analysis. BMC Cancer **7**, 55 (2007)
20. Richardson, A.L., Wang, Z.C., De Nicolo, A., Lu, X., Brown, M., Miron, A., Liao, X., Iglehart, J.D., Livingston, D.M., Ganesan, S.: X chromosomal abnormalities in basal-like human breast cancer. Cancer Cell **9**, 121–132 (2006)
21. Hall, M., Frank, E., Holmes, G., Pfahringer, B., Reutemann, P., Witten, I.H.: The WEKA data mining software: an update. SIGKDD Explor. **11**, 10–18 (2009)
22. Yu, W., Wulf, A., Liu, T., Khoury, M.J., Gwinn, M.: Gene prospector: an evidence gateway for evaluating potential susceptibility genes and interacting risk factors for human diseases. BMC Bioinform. **9**, 528 (2008)
23. Cortes, C., Vapnik, V.: Support-vector networks. Mach. Learn. **20**, 272–297 (1995)
24. Kotsiantis, S., Kanellopoulos, D., Pintelas, P.: Handling imbalanced dataset: a review. GESTS Int. Trans. ComSci. **30**, 25–36 (2006)

Eye Can Tell: On the Correlation Between Eye Movement and Phishing Identification

Daisuke Miyamoto[1]([✉]), Gregory Blanc[2], and Youki Kadobayashi[3]

[1] Information Technology Center, The University of Tokyo, 2-11-16 Yayoi,
Bunkyo-ku, Tokyo 113-8658, Japan
daisu-mi@nc.u-tokyo.ac.jp
[2] Institut Mines-Télécom/Télécom SudParis CNRS UMR 5157 SAMOVAR,
9 Rue Charles Fourier, 91011 Évry, France
gregory.blanc@telecom-sudparis.eu
[3] Graduate School of Information Science, Nara Institute of Science and Technology,
8916-5 Takayama, Ikoma, Nara 630-0192, Japan
youki-k@is.aist-nara.ac.jp

Abstract. It is often said that the eyes are the windows to the soul. If that is true, then it may also be inferred that looking at web users' eye movements could potentially reflect what they are actually thinking when they view websites. In this paper, we conduct a set of experiments to analyze whether user intention in relation to assessing the credibility of a website can be extracted from eye movements. In our within-subject experiments, the participants determined whether twenty websites seemed to be phishing websites or not. We captured their eye movements and tried to extract intention from the number and duration of eye fixations. Our results demonstrated the possibility to estimate a web user's intention when making a trust decision, solely based on the user's eye movement analysis.

Keywords: Phishing · Cognitive psychology · Eye-tracking

1 Introduction

Phishing is a fraudulent activity defined as the acquisition of personal information by tricking an individual into believing the attacker is a trustworthy entity [1]. Phishing attackers usually lure people through the use of a phishing email, which appears to be sent by a legitimate corporation. The attackers then attract the email recipients to a phishing site, which is a replica of an existing web page, to fool them into submitting personal, financial, and/or password data.

In this paper, we propose a method to estimate which user is going to be victim of phishing on the basis of their eye movement analysis. According to the theory of mind [8], eye movements are different between users with a motivation to find any particular objects and users without. In the context of phishing identification, expert users may gain information from intentionally looking at the

© Springer International Publishing Switzerland 2015
S. Arik et al. (Eds.): ICONIP 2015, Part III, LNCS 9491, pp. 223–232, 2015.
DOI: 10.1007/978-3-319-26555-1_26

browser's address bar, and evaluating this information based on their knowledge. By contrast, novice users would look at the address bar with no particular motivation, simply because they are unable to evaluate this piece of information due to their lack of knowledge. Instead, novice users may intend to gain information from the website's contents even if the contents may not give any meaningful indications with regards to phishing identification.

One major obstacle to that proposal is that the eye movements are affected by many factors - age, eyesight, stress, knowledge, level of vigilance, awareness, and familiarity with the website as well as user's intention. We conjecture that eye movement certainly informs on the user's intention. Our challenge is to assess the user's intention based on the extracted information even if other factors are involved. The results may lead to estimating whether a user is likely to fall victim to phishing.

This paper therefore assesses this hypothesis with a participant based experiment in which 23 participants have their eye movements monitored while taking a test where they need to determine which websites are phish sites among twenty samples and provide their decision's criteria. Based on our experiment, it might be reasonable to consider that the analysis of eye movement is feasible for estimating users' both intention and decision.

The rest of the paper is organized as follows. Section 2 provides theoretical background on extracting human implicit intention, and Sect. 3 proposes our method which aims at estimating user intention while assessing the credibility of the websites. Section 4 explains the conditions for our experiments, and Sect. 5 evaluates the performance of our proposed method. Section 6 shows our follow-up study, and finally, Sect. 7 summarizes our contributions.

2   Related Work

Cognitive psychology is the study of the relationship between internal mental processes and observable behavior. In this paper, observation is carried out with respects to the criteria formulated by Groojten [2] for the evaluation of cognitive methods for supporting operators. These criteria are as follows:

- **Sensitivity to Workload Changes.** We need to employ the behavioral observation methods that can estimate the internal mental model. The methods might also leverage the collected information regardless of the Fear of Negative Evaluation (FNE) [11]; observations are often affected by FNE, in which some people will attempt to conceal their errors. In fact, disclosing mistakes often damage their own self-image and professional standing.
- **Obtrusiveness for the Operator.** The observation should not take much effort to start collecting data or disturb the handling of people during the tasks performance. Furthermore, people will not carry implants, needles or other devices which may hurt them in any way.

– **Availability of Equipment.** The observation should employ the method which is easily applicable to people. Within the context of phishing prevention, the method should be available while users are browsing. Non-contact devices might be preferred.

In this paper, we decided to employ eye movement-based observations since it meets the above requirements. Brain activity, heart measure, and blood pressure are feasible due to the sensitivity to workload changes, but they tend to require much more obtrusive monitoring devices. By contrast, Facial expression [5] and Gesture recognition [3] were often affected by FNE.

According to Leigh et al., eye movement is generally classified into four categories, namely Saccades, Smooth pursuit movements, Fixations, and Vestibulo-ocular reflexes [6]. Saccades are rapid, ballistic movements of the eyes that abruptly change the point of fixation. Tokuda showed that mental workload, the indicator of how mentally busy a person is, can be estimated from Saccadic intrusions [10]. By contrast, Smooth pursuit movements are slow and continuous eye movements that are used to track an object in motion with central vision in order to maintain a clear and continuous perception of it, and they are deficient in schizophrenia patients [9]. Fixation is the eye movement that maintains the visual gaze on a single location, and vestibulo-ocular reflexes stabilize gaze during movement.

Regarding eye fixation, prior studies [4, 7] contributed to show that there may be correlation between eye fixation and intention. The intention refers to an idea or plan of what a person is going to do. The theory of mind states that a person has a natural way to predict, represent and interpret intention expressed explicitly or implicitly [8]. A person expresses explicit intentions using different sequences of actions. For example, during an interaction, a person tends to express intention explicitly through speech, gesture, and facial expression. By contrast, implicit human intentions are subtle, vague and otherwise often difficult to interpret. Since the explicit expression alone may not be enough to understand the intention of a person, it is critical to understand the implicit intention.

According to [4, 7], the implicit intentions can be identified through the following biomedical signals during a visual stimulus.

– *Navigational intention* refers to an idea or plan of a person to find any object in a visual input without a particular motivation.
– *Informational intention* refers to an idea or plan of a person to find a particular object of interest or to behave with a motivation.

The authors have built classifiers for identifying intentions based on Support Vector Machine (SVM), and observed that there were positive correlations between the intentions and eye movement patterns.

3 Proposal

In this paper, we evaluate the feasibility of estimating the user's decision to trust or not the websites, when assessing the credibility of the websites, based on the user's eye movement patterns. The key idea is to apply the analysis of eye fixation, an established technique in the research domain of cognitive psychology, in order to improve security.

We consider that eyes are suitable to monitor personal mental processes. As we mentioned in Sect. 2, eyes can give information on whether a person has the intention to look for something. In the context of phishing prevention, a web user is presented with security indicators and web contents displayed in the browser. Our assumption is that experts have the intention to check security indicators rather than web contents, while novices would do the opposite.

Our primary motivation is identifying users who are likely to become victims of phishing attacks. If a phishing prevention system could find that the user would disclose their personal information, there might be a chance to protect them from phishing. In order to develop such systems, it is necessary to understand what the users are really thinking. In the context of cognitive psychology, such internal mental processes can be estimated by observable and measurable behavior. Thus, we employ the analysis of the eye movement patterns for phishing prevention.

To achieve this goal, we explore a suitable method for recognizing how users do to assess the credibility of websites. We consider such case in which a user assesses a website by its contents, and ignore meaningful signals displayed in the browser's address bar. Figures 1 and 2 respectively show the heat maps of the eye fixation locations and durations on both phish and legitimate website for novice users (a) and expert users (b). The color red denotes the areas that attracted the user's gaze the most and green denotes moderate gaze activity. In the phishing case, the novice looked at the web content but ignored the browser's address bar while assessing credibility, as shown in Fig. 1a. Since the text and visuals in phishing sites are quite similar to the ones in legitimate sites, the novice failed to label the phishing site correctly. In the legitimate case, the novice also only paid attention to the content of a web page as shown in Fig. 2a. By contrast, an expert tends to evaluate the site's URL and/or the browser's SSL indicator rather than the contents of the web page to judge the credibility of the sites, as shown in Figs. 1b and 2b.

We therefore hypothesize that the analysis of eye movement on the particular areas of interest (AoIs) would allow to extract what are the criteria that helped the user in making a trust decision. To assess our hypothesis, we conducted

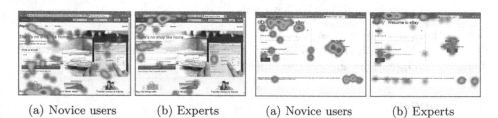

(a) Novice users   (b) Experts   (a) Novice users   (b) Experts

Fig. 1. Eye-tracking in a phishing site   **Fig. 2.** Eye-tracking in a legitimate site

Table 1. Conditions of each site used for recording eye movement

#	Website	Phish	Lang	Description
1	Google	no	JP	SSL
2	Amazon	yes	JP	`tigratami.com.br`, once reported as a compromised host
3	Sumishin Net Bank	no	JP	EV-SSL
4	Yahoo	yes	JP	`kazuki-j.com`, once reported as a compromised host
5	Square Enix	yes	JP	`secure.square-enlix.com`, similar to legitimate URL `secure.square-enix.com`
6	Ameba	no	JP	non-SSL
7	Tokyo Mitsubishi UFJ Bank	yes	JP	`bk.mufg.jp.iki.cn.com`, similar to legitimate URL `bk.mufg.jp`
8	All Nippon Airways	yes	JP	IP address
9	Gree	no	JP	non-SSL
10	eBay	no	EN	EV-SSL
11	Japan Post Holdings	yes	JP	`direct.yucho.org`, SSL
12	Apple	yes	JP	`apple.com.uk.sign.in...`
13	DMM	no	JP	SSL
14	Twitter	yes	JP	`twittelr.com`
15	Facebook	yes	JP	IP address
16	Rakuten Bank	yes	JP	`vrsimulations.com`, once reported as a compromised host
17	Sumitomo Mitsui Card	yes	JP	`www.smcb-card.com`, SSL
18	Jetstar Airways	no	JP	SSL, non pad-lock icon by accessing non-SSL content
19	PayPal	yes	EN	`paypal.com.0.security-c...`
20	Tokyo-Tomin Bank	no	JP	3rd party URL `www2.answer.or.jp`, EV-SSL

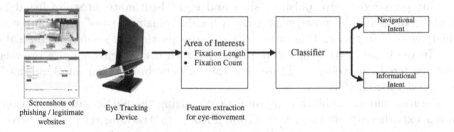

Fig. 3. Block diagram of the experiment

two types of participant-based experiments. In the first experiment, we analyze the correlation between eye movements and decision criteria to confirm whether eye fixations can be used as decision criteria indicators. The second experiment investigates whether the eye movement allow to estimate the likeliness of a user to fall victim to phishing.

4 Experiment Setup

This section introduces the procedures of our experiments. Individuals were recruited through a poster advertisement at a college campus during the period of November 2013 - February 2014. Of the 23 participants, a majority of the participants were males in their twenties.

As ethical issue is a concern in our organization, the participants were told they were participating in a security research study. One of the most important ethical rules was that all participants must give their informed consent before taking part in our experiments. The ethical rules also stipulate the need to explain "Why we observe?", "What we observe?", "How we observe?" and "Who uses the observed data?".

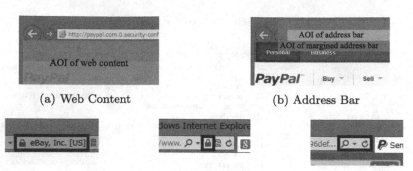

(a) Web Content   (b) Address Bar

(c) Security Icon for EV- (d) Security Icon for SSL (e) Surrounding Area of Se-
SSL websites   websites   curity Icon

Fig. 4. Definition of AOI

Our prepared twelve phishing sites and eight legitimate sites are listed in Table 1. It should be noted that what we call "phishing sites" were not real phishing sites in the wild, in order to avoid any participant's information leakage. Instead, our participants were presented with the screenshots of a browser that rendered the websites. These screenshots have been taken on Windows 7 equipped with IE 10.0.

Figure 3 shows the block diagram for recognizing the participants' intention. In our experiments, we employed a Tobii TX300 eye tracking system to analyze the eye movement data. With their consent, we measure their eye movements after we calibrated the eye tracking device for each participant. The participants were also shown several options to indicate their decision's criteria: "Content of Web page," "URL of the site," "Security Information of Browser," and "Other Reason." The participants were requested to mark all options that applied (multiple answers allowed), and described in details their reason when selecting the "Other Reason" option.

5   Eye Movement Analysis

5.1   Extraction of Implicit Intention

We hereafter examine the feasibility of extracting implicit intention from observing the user's eye movements. Based on the number and duration of fixations in each Area of Interest (AOI) of a given input stimulus image, we construct classifiers with SVM to differentiate the participant's implicit intention into navigational and informational intentions.

At first, we evaluate such hypothesis that *the analysis of the eye movement can extract a user's intentions while watching web pages.* Since novices tend to assess credibility by the "Content of Web page," their eye movements would be different from the eye movements of experts. The feature vectors include the

Table 2. Participants' recognition performance by eye movement analysis

Tyep of Intent		AER	AUC
Content of Web page			
	entire time period	32.4%	0.741
	initial ten seconds period	32.2%	0.759
URL of the site			
	entire time perod	28.0%	0.741
	initial ten seconds period	27.8%	0.759
(removed noise)			
	entire time perod	21.3%	0.890
	initial ten seconds period	19.7%	0.917
Security Information from browser (AOI of the address bar)			
	entire time epriod	14.5%	0.855
	initial ten seconds period	14.3%	0.855
(AOI of the padlock icon)			
	entire time epriod	13.5%	0.841
	initial ten seconds period	13.7%	0.809

number and duration of fixations towards the web content AoI (as shown in Fig. 4a). The objective variable is a binomial value that denotes whether the participant checked the "Content of Web page" option or not in our questionnaire. The average error rate (AER) was 32.4% and the area under the curve (AUC) was 0.741 as shown in Table 2. Additionally, we assumed that some participants would try to find some trustworthiness information as soon as they have begun browsing the websites. From this perspective, we also extracted the fixation count and duration within the first ten seconds. In this case, the AER was 32.2% and the AUC was 0.759. Hereafter, "initial ten seconds period" means the analysis of eye movement within the first ten seconds, and "entire time period" means the analysis of the entire time while making decision.

We wished for the participants to check the browser's address bar intentionally since the browser's attention on address bar gives trustworthy information such as the URL and security related information. The feature vectors are the number and duration of fixations towards the address bar AoI (as shown in Fig. 4b), and the objective variable is a binomial value that denotes whether the participant checked the "URL of the site" option or not in our questionnaire. The AER was 28.0% and the AUC was 0.741 in the case of the entire time period. In the experiment, we found that several participants labeled "URL of the site" as their decision making criteria without actually gazing at the address bar. Even when we redefined the AoI in order to add the surrounding margins, as shown by the green rectangle in Fig. 4b, their eye fixations towards the AoI still could not be accounted for. If we remove such falsely motivated decisions, the AER would be 21.3% and the AUC would be 0.890. Additionally, the margined AoI did not improve the performance: in the case of the entire time period, the AER was 22.1% and the AUC was 0.842.

We also assumed that some participants would choose to look at the address bar to find a security indicator. The feature vectors are the number and duration of fixations for that particular AoI, and the objective variable is a binomial value that denotes whether the participant checked the "Security information of

Table 3. Estimation of participants who were going to be victims of phishing

Area of Interest		AER	AUC
Web Content			
	entire time period	24.8%	0.799
	initial ten seconds period	25.1%	0.818
Address Bar			
	entire time perod	24.1%	0.820
	initial ten seconds period	25.7%	0.815
Security Icons			
	entire time perod	24.4%	0.782
	initial ten seconds period	24.8%	0.761
All types of AOIs			
	entire time perod	20.7%	0.873
	initial ten seconds period	21.1%	0.853

browser" option or not in our questionnaire. The possible AOIs are the address bar and the padlock icon. The AER was 14.5 % and the AUC was 0.855.

We defined the AOIs of the padlock icon, as shown in Fig. 4c, d, and e, for an EV-SSL certificate where the AoI is around the name of the entity as well as the padlock icon, for an SSL certification, it is a rectangle around the padlock icon, and in the case of non-SSL websites, the AOI was a surrounding area for icons displayed in the address bar, respectively. For this last case, we assumed that some participants would check the nonexistence of the SSL certificates. In total, the AER was 13.5 % and the AUC was 0.841.

We found that some participants tend not to check the "Security Information of Browser" option, even when the website displayed an SSL padlock icon. The predictor therefore indicates that all participants did not intentionally look at this AoI. We concluded that the AoIs were not as useful to construct a good predictor, however, the AER was 7.6 % and the AUC was 0.785. In the case of the websites that displayed an EV-SSL padlock icon, the AER was 33.3 % and the AUC was 0.711. When the websites had no certificate, the AER was 10.5 % and the AUC was 0.775.

5.2   Estimation of Participant's Likelihood to be Victim

In this experiment, we hypothesized that *the analysis of the eye movement can estimate whether or not a user is going to fall victim to phishing.* The feature vectors in this scenario are the number and duration of fixations towards the three types of AOIs (web content, address bar, and security icons). The objective variable is a binomial value that denotes whether the participant judged correctly or not.

The results are shown in Table 3. By using the combination of all types of AoIs, we observed that the AER was 20.7 % and the AUC was 0.873, in the case of the entire time period. The lowest error rate was observed at 8.7 % in Websites 10 and 15, and followed by Websites 1, 4, 8, and 9 with 13.0 %. Since Website 10 is displayed in English, and since a significant number of the participants were non-native English speakers, we therefore assume that the participants had attempted to assess the website based on the address bar rather than the content.

Additionally, we performed a 10-fold cross validation with tuning parameters by grid search. The results showed that the AER was 29.3 % in the case of the entire time period, and 30.8 % in the case of the initial ten seconds period.

6   Follow-Up Study

In order to thwart bias, we conducted a follow-up study to our experiments for another set of users The study was conducted in September 2014 with 33 new participants. Of the 33, three were female and the rest were male. Twenty of the participants were in their twenties, eight in their thirties, four in their forties, and one in their teens. All of them were attendees of a domestic workshop held in Japan, and were mainly network researchers. The participants were volunteers, and the experiments were done in a conference hall. The rest of the experimental conditions were the same as in the previous experiments.

The analysis of the follow-up study also found that there is a correlation between their eye movement patterns with regards to the AoIs and the criteria they indicated while assessing the credibility of the websites. In the case of web content, the AER of the predictor was 26.7 % and the AUC was 0.787.

We then analyzed the eye movements with respects to the AoI of the address bar and their assessment based on "URL of the site". The performance for the intention of checking the URL was observed with the AER of 18.7 % while the AUC was 0.837. We also measured the performance for the intention of gaining security information, and found that the pair of the AER and AUC was (18.0 % and 0.806) in the case of the address bar, (17.7 % and 0.779) in the case of the padlock icon.

We finally observed the feasibility of predicting the participant's likelihood to fall victim of phishing. By using the combination of all types of AoI, the AER was 15.2 %, and our ten-fold cross validation raised it to 21.5 %.

7   Conclusion

In this paper, we presented a user study in which we evaluated the correlation between eye movements and phishing identification. We used both the duration and the number of eye fixations with respects to particular AoIs, including the area of the rendered web content, the address bar, the padlock icons and their surrounding area. We categorized eye movement patterns along with the types of human implicit intentions.

We conducted a set of experiments which focused on verifying if the eye movement analysis was able to extract users' intentions for assessing the websites' credibility, and to estimate users who were likely to fall victim to phishing. Our result showed that the average error was 32.4 % if users assessed the credibility of the website by paying attention to web content, 21.3 % if users looked at the URL of the site, and 13.5 % if users checked the security information of the browser. We also verified our ability to predict the likelihood that users may fall victim to phishing attacks on the basis of the analysis of their eye movement patterns, and found that it can be estimated with a probability of 79.3 %.

Although there still remain other factors in decision-making which must be investigated, we believe that this paper proposed a novel prediction methodology for phishing identification and demonstrates its feasibility. We hope that this work can help utilize insights from cognitive psychology in order to help protect people from cyber threats.

Acknowledgment. This research has been supported by the Strategic International Collaborative R&D Promotion Project of the Ministry of Internal Affairs and Communication, Japan, and by the European Union Seventh Framework Programme (FP7/2007-2013) under grant agreement No. 608533 (NECOMA). The opinions expressed in this paper are those of the authors and do not necessarily reflect the views of the Ministry of Internal Affairs and Communications, Japan, or of the European Commission.

References

1. Abad, C.: The economy of phishing: a survey of the operations of the phishing market. First Monday 10(9) (2005)
2. Grootjen, M., Neerincx, M.A., van Weert, J.C.: Task based interpretation of operator state information for adaptive support. Technical report ACI/HFES-2006 (2006)
3. Haag, A., Goronzy, S., Schaich, P., Williams, J.: Emotion recognition using biosensors: first steps towards an automatic system. In: André, E., Dybkjær, L., Minker, W., Heisterkamp, P. (eds.) ADS 2004. LNCS (LNAI), vol. 3068, pp. 36–48. Springer, Heidelberg (2004)
4. Jang, Y.M., Mallipeddi, R., Lee, S., Kwak, H.W., Lee, M.: Human intention recognition based on eyeball movement pattern and pupil size variation. Neurocomputing **128**, 421–432 (2014)
5. van Kuilenburg, H., Wiering, M., den Uyl, M.: A model based method for automatic facial expression recognition. In: Gama, J., Camacho, R., Brazdil, P.B., Jorge, A.M., Torgo, L. (eds.) ECML 2005. LNCS, vol. 3720, pp. 194–205. Springer, Heidelberg (2005)
6. Leigh, R.J., Zee, D.S.: The Neurology of Eye Movements, 4th edn. Oxford University Press, Oxford (1991)
7. Park, U., Mallipeddi, R., Lee, M.: Human implicit intent discrimination using EEG and eye movement. In: Loo, C.K., Yap, K.S., Wong, K.W., Teoh, A., Huang, K. (eds.) ICONIP 2014. LNCS, vol. 8834, pp. 11–18. Springer, Heidelberg (2014)
8. Premacka, D., Woodruffa, G.: Does the chimpanzee have a theory of mind? Behav. Brain Sci. **1**, 515–526 (1978)
9. Slaghuis, W.L., Holthouse, T., Hawkes, A., Bruno, R.: Eye movement and visual motion perception in schizophrenia I: apparent motion evoked smooth pursuit eye movement reveals a hidden dysfunction in smooth pursuit eye movement in schizophrenia. Exp. Brain Res. **182**, 399–413 (2007)
10. Tokuda, S., Obinata, G., Palmer, E., Chaparro, A.: Estimation of mental workload using saccadic eye movements in a free-viewing task. In: Proceedings of the 33rd Annual International Conference of the IEEE Engineering in Medicine and Biology Society, pp. 4523–4529, August 2011
11. Watson, D., Friend, R.: Measurement of social-evaluative anxiety. Consult. Clin. Psychol. **33**, 448–457 (1969)

Gaussian Hamming Distance

De-Identified Features of Facial Expressions

Insu Song[✉]

School of Business/IT, James Cook University Australia,
Singapore Campus, Singapore
insu.song@jcu.edu.au

Abstract. We present new image features for diagnosing general wellbeing states and medical conditions. The new method, called Gaussian Hamming Distance (GHD), generates de-identified features that are highly correlated with general wellbeing states, such as happiness, smoking, and facial palsy. This method allows aid organizations and governments in developing countries to provide affordable medical services. We evaluate the new approach using real face-image data and four classifiers: Naive Bayesian classier, Artificial Neural Network, Decision Tree, and Support Vector Machines (SVM) for predicting general wellbeing states. Its predictive power (over 93 % accuracy) is suitable for providing a variety of online services including recommending useful health information for improving general wellbeing states.

Keywords: Gaussian hamming distance · General Well-being · Facial feature · Feature selection · Mobile diagnostics

1 Introduction

Mobile devices have been used for rapid diagnosis of various health conditions [1], recommending communities for health social networks [2], and improving livelihood of rural areas of developing countries [3]. A US$14 cell phone equipped with a digital camera can capture a facial expression a child. The typical resolution of such devices is about 640 by 480 pixels. A face image of a person can tell a lot about the person's health conditions. For instance, Bell's palsy is well known to be associated with respiratory infections, such as pneumonia, cold and flu [4–6], tuberculosis [7], and herpes simplex [8]. However, photographic images of individuals contain identifiable private and sensitive personal information in its raw form and thus not suitable for online services. Due to the sensitive privacy issues of sharing face images, there have been little previous studies in the use of facial images for assessing health conditions. New de-identified feature extraction approaches will open up new research opportunities.

The House-Brackmann Grading System [4] is used by clinicians to manually assess medical conditions of patients suffering from Bell's Palsy. The assessment is based on measuring misalignments between the left and right hand sides of a patient's face. In telecommunications industry, Hamming distance has been used for measuring the

© Springer International Publishing Switzerland 2015
S. Arik et al. (Eds.): ICONIP 2015, Part III, LNCS 9491, pp. 233–240, 2015.
DOI: 10.1007/978-3-319-26555-1_27

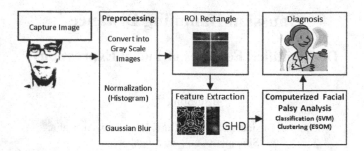

Fig. 1. The overall process of predicting wellbeing from facial images comprises of capturing a face image, normalization, Gaussian blur, ROI detection, GHD feature generation, and prediction.

amount of errors in a code (e.g., a string) transmitted over a telecommunication channel. In this paper, we develop a new method called Gaussian Hamming Distance (GHD) which calculates Hamming distance for face images: the degree of misalignment between the left and right hand sides of a face. GHD allows patients to safely generate de-identified data and send to remote servers for rapid diagnosis. The anonymous data can be safely shared with other medical researchers for studying various global health issues, such as pneumonia, flu, tuberculosis, and flu, and for predicting epidemics. Various data analysis methods can be applied on GHD features. In this paper, we evaluate the predictive power of GHD using regression analysis, weak classifier (Naive Bayesian classifier), and strong classifier (Decision Tree, Artificial Neural Network, and Support Vector Machines). GHD features of 61 facial expressions are used to predict general well-being states, such as feel fit, feel slow, and face numb. Except the weak classifier, all strong classifiers performed well achieving over 90 % of accuracy, precision, specificity, and sensitivity.

The details of the feature extraction algorithms are described in Sect. 3. GHD features are evaluated using statistical data analysis and machine learning approaches in Sect. 4.

2 Background

2.1 Image Analysis for Medical Diagnosis

During the last few years, techniques for automated medical image analysis have been proposed increasingly. Some of the essential methods for medical image analysis are enhancement, segmentation, and visualization [9]. More advanced image analysis include extracting useful patterns from radiology images, such as mammograms, X-ray, MRI, and CT scan images, for classification and prediction of various types of cancer. However, these methods are highly intrusive and require expensive equipment to acquire the necessary image data. Furthermore, patients must travel to hospitals. The ideas of overcoming these methods have been proposed only recently by researchers and aid organizations working to improve livelihood of people living in resource poor environment [2–4, 10, 11]. House-Brackmann Grading System for Bell's Palsy [4] is based on Hamming distance [12].

2.2 Facial Palsy

Facial palsy, clinically diagnosed as facial nerve paralysis, is a facial nerve inflammation that causes facial muscle distortion. Patients who suffer from facial palsy have symptomatic, functional, communicable, and psychological problems [13]. Though not fatal, chronic and severe facial palsy symptoms largely affect patients' social interaction and regular life. In order to provide accurate diagnostic information for facial palsy, studies of facial paralysis patients' database on facial expression have been undertaken. By analyzing and interpreting facial expression changes, image analysis techniques can assist in both prognosis and rehabilitation of facial nerve palsy.

One particular type of acute facial palsy, called Bell's palsy, is well known to be associated with acute infectious diseases, such as pneumonia, cold and flu [4–6], herpes simplex [8], as well as tuberculosis [7]. Conventionally, Bell's palsy is diagnosed using a guideline to assess its degree of progress. It has been shown that computerized image analysis methods utilizing techniques developed for face recognition are able to assess the degree of Bell's palsy using low resolution facial images of patients [4].

3 Gaussian Hamming Distance

Well-being in this paper refers to psychological or medical states, such as feeling happy, energetic, headache, coughing, and sluggish. Our aim is to develop an efficient algorithm that can be used on low-cost mobile devices for extracting de-identified features from facial expressions for predicting general well-being. This is motivated by the findings of previous researchers that facial palsy is linked to various health problems [4]. Bell's palsy or idiopathic facial paralysis (IFP) is a frequent disease worldwide, affecting all ages, without gender distinction and with high rate of remission [14]. It has been frequently observed in various chronic [15] and infectious diseases [6]. It has also been reported as an adverse event following immunization. Therefore, we hypothesize that imbalance in facial expressions will be correlated to general well-being states including cues for various other chronic and infectious diseases. The aim of our paper is to show the extent to which we can ascertain the general well-being from some de-identified features extracted from otherwise sensitive face images in their raw form.

Figure 1 illustrates the overall process of extracting the de-identified features from facial expressions and predicting well-being of a person. Firstly, we need to capture a facial expression of a person, such as smiling. A parent of a child can use his/her mobile phone to take a photo of a child. An application on the phone then detects a face in the photo and extracts the de-identified features, which is then sent to a remote server. Alternatively, the parent can send the photo to a remote trusted-server via MMS, which then extracts the de-identified features and discard any identifiable features. The de-identified features can then be shared with other servers running automated diagnosis methods or medical experts for manual inspection. The de-identified features stored in a central server will allow medical researchers to further study the relation between the features and various medical conditions.

The details of extracting the de-identified features from a facial expression image now follow. The captured image may contain various noises dues to different ambient lighting conditions, image sensors, and image compression algorithms used by mobile phones. These noises will affect the quality of the collected images. Therefore, we first remove color information from the images by converting them into grey scale images. We also normalize the pixel values of each image using a Histogram equalization process. This removes the effects of different lighting conditions, exposure, shutter speed, optical filters, and image processing/compression algorithms of mobile devices.

The captured image could also be rotated or off-centered. These slight rotation and transposition errors need to be removed as they will show as misalignment errors when comparing the left and right hands sides of a face image. Gaussian blur operation is known to smooth images while preserving important feature information, such as edges. For digital image processing, we use a convolution kernel, a square matrix, to approximate the Gaussian function. The Gaussian kernel, $G(h, \sigma)$, has two parameters: kernel size h and the standard deviation σ. The kernel is used to calculate weighted average value for a pixel p at (x, y).

Once the images are cleaned and normalized, we detect if there is a face in the image and detect the region of interest. Once we have detected a face and ROI in the image, we extract the face image using the ROI rectangle. The extracted face image is then rescaled, and down sampled to a 50×50 pixel, gray-scale image. This effectively removes any background images.

The extracted face image can be further processed to extract various features for classification tasks. For example, a Sobel edge detector extracts edges (areas where brightness changes suddenly), such as winkles, nose, mouth, and eye shapes in the face image. Principal Component Analysis (PCA) has been used by face recognition researches for extracting principal components (also called significant eigenvalues of eigenfaces). However, these conventional features contain some degree of identifiable information.

As we are interested in capturing imbalances in facial expressions, we have been looking for some existing methods of measuring imbalances. Hamming distance [12] is a simple, yet effective method for measuring misalignments in signals. In telecommunication industry, Hamming distance has been used for error detection and error correction of codes transmitted over a communication channel. The same analogy can be applied to measure the degree of imbalance in facial expressions using Hamming distance. Unlike edge features and principal components, it is extremely fast to calculate the Hamming distance, and Hamming distance itself does not contain any identifiable information. This is because the Hamming distance is a measure of noise or errors of a transmission line. Hamming distance of a face image is defined as follows:

$$H = |L - \text{flip}(R)| \tag{1}$$

where H is Hamming distance of a face image, L and R are the left and right hand side images of a face, respectively, and $\text{flip}(R)$ is a function that flip an image horizontally. $\text{flip}(R)$ is used to compare the same facial features on both sides of a face. If L and R are black-and-white images (i.e., 1 bit per pixel), $\text{sum}(H)$, the sum of all elements of H, is the number of bits that do not match between the two images. If L and R are grey scale images, sum(H) is the Manhattan distance between the two images.

However, unlike a code or string, a face image captured using a digital camera may contain rotation, distortion, and transposition errors, which are not related to personal well-being states. Therefore, direct application of Hamming distance will result in incorrect measures. The errors captured by H can be decomposed into four error components:

$$H = E_m + E_r + E_t + E_d \qquad (2)$$

where E_m is the misalignment error that we are interested in, and E_r, E_t, and E_d are rotation, transposition, and distortion errors, respectively. To remove E_r, E_t, and E_d, we apply the Gaussian filter on the face image to calculate only the misalignments of blobs and predominant facial features.

$$GHD = |G(L) - G(\text{flip}(R))| \qquad (3)$$

where GHD is the Gaussian Hamming distance image of a face image, G is the Gaussian convolution function that calculates $g_{x,y}$ for each pixel $p_{x,y}$ in the input image, and L and R are the left and right hand sides of a face, respectively. GHD effectively removes effects of minor rotation, distortion, and transposition errors.

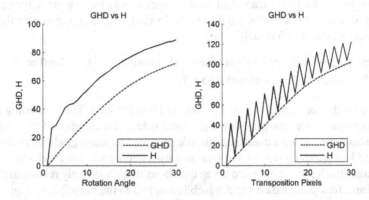

Fig. 2. Comparison of sum(GHD) and sum(H) using an image containing bundles of evenly distributed horizontal lines and vertical lines. *GHD* is less affected by rotation and transposition errors of a face image. The figure also shows that *GHD* is proportional to the true misalignment errors (i.e., the rotation angle and transposition distance).

Figure 2 demonstrates how Gaussian removes such errors (the jagged lines). However, it must be able to capture true alignment errors. Figure 2 demonstrates that *GHD* is proportional to the true misalignment error. Ideally, *GHD* should be close to E_m with some small fixed residual error $r \leq \varepsilon$:

$$H = GHD + E_r + E_t + E_d + r \qquad (4)$$

We can adjust the parameters of the Gaussian blur function in order to achieve the desired tolerance level ε.

4 Properties of GHD Features of Facial Images

In addition to its ability to remove rotation, transposition, and distortion errors, GHD has three other important properties: de-identifiable, proportional to misalignment, and scale invariant. GHD features do not contain identifiable information. It is not possible to reconstruct the original face image from the GHD of the image because errors and misalignments do not depend on the source of signals used for computing the GHD. It depends on the underlying health conditions and diseases.

Other distance measures developed for comparing strings, such as Damerau–Levensh-tein distance and Word ladder, are not suitable for measuring distances of regions of facial images directly as the methods for comparing strings depend on the underlying alphabets of strings. To use such string distance measures, we would need to further process the face images to detect facial features, such as eye, nose, mouth, eyelid, and eyelash, and use the facial features as alphabets. Facial feature detection is both computationally intensive and unreliable, introducing additional noise to the process.

5 Evaluation of GHD

Let us now see if the GHD feature of face images would be any good for predicting well-being states. For our study, we use the following equation to estimate the degree of well-being states of individuals:

$$W = [\text{FeelFit} + \text{PositiveEnergy} + \text{FeelPleased} + (3 - \text{FeelSlow}) + (3 - \text{SoreThroat})$$
$$+ (3 - \text{Cough}) + (3 - \text{FaceNumb})]/(3 \times 7) \tag{5}$$

where W is the degree of well-being: $1 =$ well and $0 =$ not well. The well-being measure is based on seven factors shown on the right hand side of the equation. The values of the factors are estimated using a questionnaire. We recruited 61 undergraduate students, who participated in a photo taking session followed by a survey session conducted using the questionnaire. Each participant rated seven questions that are directly related to the seven factors using a four point Likert scale, which is mapped to the range [0, 3]. Figure 3 (left) shows a scatter plot of GHD of the face images of the 61 students over their well-being values obtained using Eqs. (3) and (5). This regression analysis shows that GHD and well-being states are significantly correlated.

The generated GHD features are used to train machine learning algorithms to eval-uate their predictive power. We selected one weak classifier (Naive Bayesian (NB) classier) and three strong classifiers (Artificial Neural Network (ANN), Decision Tree (DT), and Support Vector Machines (SVM)). Figure 3 (right) shows the performance of predicting well-being using ANN with 425 GHD values of the eye area. Using all 1250 GHD feature values (i.e., the entire face), ANN achieved accuracy of 93.5 %. Using 153 GHD values that correlated with general welling, ANN achieved accuracy of 82.6 %. Similarly, DT and SVM performed well on the GHD features: Area under the curve of ROC of DT and SVM are 0.983 and 0.967, respectively. Except for NB classifier, all other classifiers show curves on the upper left corner, well above the diagonal lines, achieving over 80 % accuracies.

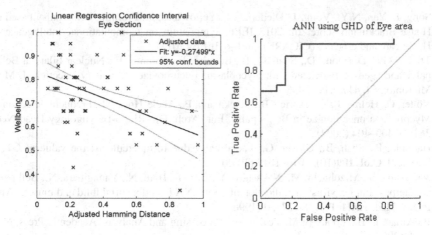

Fig. 3. (Left) Scatter plot of wellbeing versus GHD of the eye area. Linear regression coefficients were significant for six regions ($p \leq 5$ %). (Right) ANN prediction performance predicting wellbeing using 425 GHD values of the eye area. Accuracy = 82.6 %.

6   Conclusion

The results indicate that the automated machine learning algorithms learn to predict personal well-being as well as medical conditions from de-identified features of the facial expressions. The use of automated grading system greatly reduces the duration of diagnosis while increasing consistency, because GHD is proportional to the degree of misalignment of facial features and references of GHD features stored to provide comparisons. The GHD feature extraction is computationally efficient: linear to the size of the input image. As GHD is proportional to the misalignments in facial features, users will get immediate feedback before submitting to a remote server for more detailed analysis. This can provide timely diagnosis to patients situated in not easily accessible areas, emergency situations and areas with high medical cost, and lastly for the under-served populations.

Unlike existing medical diagnostics, such as contact thermal imaging, our method utilizes existing cell phones and infrastructure to provide a robust, non-invasive, and rapid assessment method based on the simple analysis of images. The developed system can be rapidly deployed to many needed communities over the Internet.

References

1. Lei, B., Song, I., Rahman, S.A.: Optimal watermarking scheme for breath sound. In: The 2012 International Joint Conference on Neural Networks (IJCNN), pp 1–6. IEEE (2012)
2. Song, I., Marsh, N.V.: Anonymous indexing of health conditions for a similarity measure. IEEE Trans. Inf. Technol. Biomed. **16**(4), 737–744 (2012)
3. Vong, J., Fang, J., Song, I.: Delivering financial services through mobile phone technology: a pilot study on impact of mobile money service on micro–entrepreneurs in rural Cambodia. Int. J. Inf. Syst. Change Manage. **6**(2), 177–186 (2012)

4. Song, I., Yen, N.Y., Vong, J., Diederich, J., Yellowlees, P.: Profiling bell's palsy based on House-Brackmann score. In: 2013 IEEE Symposium on Computational Intelligence in Healthcare and e-health (CICARE), pp 1–6. IEEE (2013)
5. Trad, S.G.J., Dormont, D., Stankoff, B., Bricaire, F., Caumes, E.: Nuclear bilateral Bell's palsy and ageusia associated with mycoplasma pneumoniae pulmonary infection. J. Med. Microbiol. **54**, 417–419 (2005)
6. Völter, C., Helms, J., Weissbrich, B., Rieckmann, P., Abele-Horn, M.: Frequent detection of Mycoplasma pneumoniae in Bell's palsy. Eur. Arch. Oto-Rhino-Laryngology Head Neck **261**(7), 400–404 (2004)
7. Hadfield, P., Shah, B., Glover, G.: Facial palsy due to tuberculosis: the value of CT. J. Laryngol. Otol. **109**(10), 1010–1012 (1995)
8. Murakami, S., Mizobuchi, M., Nakashiro, Y., Doi, T., Hato, N., Yanagihara, N.: Bell palsy and herpes simplex virus: identification of viral DNA in endoneurial fluid and muscle. Ann. Intern. Med. **124**(1_Part_1), 27–30 (1996)
9. Bankman, I.: Handbook of Medical Image Processing and Analysis. Academic Press, New York (2008)
10. Lei, B., Rahman, S.A., Song, I.: Content-based classification of breath sound with enhanced features. Neurocomputing **141**, 139–147 (2014)
11. Song, I., Vong, J.: Affective core-banking services for microfinance. In: Lee, R. (ed.) Computer and Information Science. SCI, vol. 493, pp. 91–102. Springer, Heidelberg (2013)
12. Hamming, W.R.: Error detecting and error correcting codes. Bell Syst. Tech. J. **29**(2), 147–160 (1950)
13. Terzis, J.K., Konofaos, P.: Nerve transfers in facial palsy. Facial Plast. Surg. **24**(02), 177–193 (2008)
14. Kasse, C.A., Ferri, R.G., Vietler, E.Y.C., Leonhardt, F.D., Testa, J.R.G., Cruz, O.L.M.: Clinical data and prognosis in 1521 cases of Bell's palsy. Int. Congr. Ser. **1240**, 641–647 (2003). doi:10.1016/s0531-5131(03)00757-x
15. House, J.W., Brackmann, D.E.: Facial nerve grading system. Otolaryngol. Head Neck Surg. **93**, 146–147 (1985)

Local Sparse Representation Based Interest Point Matching for Person Re-identification

Mohamed Ibn Khedher[(✉)] and Mounim A. El Yacoubi

Institut Mines-Telecom/Telecom SudParis: CEA Saclay Nano-Innov,
91191 Gif sur Yvette Cedex, France
{mohamed.ibn_khedher,mounim.el_yacoubi}@telecom-sudparis.eu

Abstract. This paper presents a multi-shot person re-identification system from video sequences based on Interest Points (SURFs) matching. Our objective is to improve the Interest Points (IPs) matching using low resolution images in terms of re-identification accuracy and running time. First, we propose a new method of SURF matching via Local Sparse Representation (LSR). Each SURF in the test video sequence is expressed as a sparse representation of a subset of SURFs in the reference dataset. Our approach consists of searching the latter subset from the reference IPs that are located on a similar spatial neighborhood to the query IP. Second, it investigates whether IPs filtering can decrease the re-identification running time. An ensemble of binary classifiers are evaluated. Our approach is assessed on the large dataset PRID-2011 and shown to outperform favorably with current state of the art.

Keywords: Person re-identification · Local sparse representation · Interest point · Filtering · Binary classifier · SURF

1 Introduction

Person re-identification is the task of determining if a person leaving the field of camera A reappears in the field of camera B. Re-identification may be viewed as a soft biometric task since it consists of matching a query input to a reference one by using low resolution information related to the human silhouette.

This work is an extension of the standard IPs matching via Sparse Representation (SR) [12] (It called later Standard SR). The idea behind is to express each query IP as a linear combination vector from a subset of reference IPs (called dictionary). The SR corresponds to a sparse vector whose nonzero entries correspond to the weights of reference IPs. To classify the query IP, a reconstruction error is calculated for each reference identity using only the SR coefficients corresponding to this identity. The query IP is then identified as the reference identity minimizing the reconstruction error. Finally, the reference person obtaining the majority of votes is claimed as the re-identified person.

With respect to standard SR, our contribution is twofold. First, the Standard SR uses all the reference dataset to construct the dictionary regardless of the

© Springer International Publishing Switzerland 2015
S. Arik et al. (Eds.): ICONIP 2015, Part III, LNCS 9491, pp. 241–250, 2015.
DOI: 10.1007/978-3-319-26555-1_28

spatial position of the query IP in the image. Therefore, due to IPs noisiness and ambiguity in uncontrolled conditions, IPs from different image parts may be included in the dictionary, thus making SR unreliable. In this work, we add a spatial constraint related to the position of IPs in the images. Concretely, we propose a Local Sparse Representation (LSR), where for each query IP, a dictionary is selected from its spatial neighborhood reference IPs. Second, rather than considering a binary 1/0 vote after SR matching, we propose to use a continuous vote for each identity, that is related to the weight of its nonzero coefficients in the SR.

The second part of this work is about IPs filtering. Having a large number of IPs per person, re-identification becomes a much time-consuming task. In this context, we propose a new filtering scheme to reject unreliable matched IPs pairs that are probably resulting from matching IPs from different persons or associated with different parts of the silhouette. To do this, we design a binary classifier that learns on a training dataset, made up of pairs of positive IP pairs (each pair {query IP, closest reference IP} is associated with the same person) and negative IP pairs (each pair {query IP, closest reference IP} is associated with different persons). Our motivation is that each IP for which the closest IP belongs to a different person is unreliable and is better to be dropped from SR matching and subsequent voting for re-identification. In this paper, we study the power of filtering of two popular classifiers, Support Vector Machine (SVM) and Random Forest (RF) and investigate the tradeoff between reducing running time and keeping a better re-identification performance.

The rest of the paper is organized as follows. In Sect. 2, a state of the art is presented. The principle of our approach is discussed in Sect. 3. Sections 4, 5 and 6 present respectively the major steps of the re-identification system: feature extraction, matching and filtering. Section 7 is dedicated to the experimental part and finally a conclusion and perspectives are presented.

2   State of the Art

From a learning perspective, the re-identification approaches can be grouped into two categories: supervised approach and unsupervised approach.

Unsupervised Approaches: This category mainly focuses on the way to represent the image. Usually, the latter is represented by a set of either IPs or regions corresponding generally to body parts. Hamdoun et al. [7] collect during a short video a set of SURFs to represent the person. The authors of [10] add a shape information to the standard SIFT to improve the matching step. In the category of region based approaches, Farenzena et al. [4] propose a Symmetry-Driven Accumulation of Local Features (SDALF) by exploiting the symmetry property of the human body and decomposing it into three parts. Authors of [9] combine color and texture features extracted from each rectangular region to form one vector descriptor per image.

Supervised Methods: The learning phase can be related to parameters of the metric used to compare images, or related to the discriminant descriptors selected among all extracted features. Regarding learning metrics, in [6], an "Ensemble of Localized Features" (ELF) is presented to model person signature. The weights of features are learned using the Adaboost algorithm. In [8], the authors propose to learn a metric from pairs of samples from different cameras to take into account the transition between cameras. Authors of [1] propose to measure similarity between two images in a pre-learned space where correlation between images associated with the same person is maximized. As far as discriminative methods are concerned, a handful of works are found in the literature. Authors of [13] introduce a graph-based approach for a non-linear dimensionality reduction. It is applied to extract the most informative color representation to describe the person.

Our local sparse representation (LSR) method lies in the unsupervised category, while our filtering approach lies in the supervised one. LSR takes into account the spatial position of IPs in images contrary to [12]. Our filtering approach is automatic and does not depend on empirically parameters like in [11].

3   Proposed Approach

Our approach basically consists of four stages: (1) Feature extraction (SURFs), (2) SURFs Filtering based on binary classifier, (3) SURF identification via Local

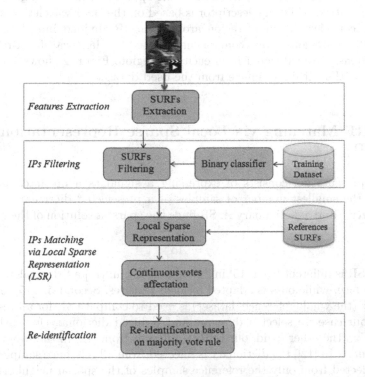

Fig. 1. Flowchart of our approach.

Sparse Representation (LSR) and (4) Person re-identification based on majority vote rule with continuous votes. Figure 1 shows the flowchart of our approach. Its principle is the following: first, each test person is described by a set of SURFs collected from a video sequence. A filtering step is then applied: after matching each test SURF to the closest reference one, we generate a difference vector, obtained by component-wise difference of the matched IP pair descriptors. The pre-learned classifier accepts or rejects the test IP based on the input difference. All retained SURFs are subsequently matched via LSR. To match one SURF, SR coefficients are used to infer a continuous vote for each reference identity based on its associated nonzero coefficients. In this way, a vote vector of dimension equal to the number of reference identities is generated. Finally, the reference person obtaining the majority of votes is claimed as the re-identified person.

4   Features Extraction

SURF is a popular IP descriptor proposed by [2], and used for several computer vision applications including person re-identification. We motivate our use of SURF by its robustness to geometric transformations (angle of view and scale) and to lighting variation and to its fast detection/description compared to others IPs. To compute SURF, two stages are required: SURF detection and SURF description. The detection step is based on the approximation of the determinant of Hessian matrix, while the descriptor is based on the Haar wavelet. The SURF descriptor considers a square region around the IP, divided into 4×4 grids to form 16 sub-regions. Four components related to Haar-wavelet x-responses and y-responses are extracted from each sub-region. Figure 2 shows samples of detected SURFs within an image from the used dataset.

5   SURF Matching via Local Sparse Representation (LSR)

Sparse representation consists of expressing a signal as a linear combination involving the smallest number of samples of a preselected dictionary. Given a query SURF q and a dictionnary A, SR finds the sparsest solution of the equation Eq. 1.

$$y = A\alpha \tag{1}$$

Our LSR is different from [15] in 2 points. First, in [15] a SR is calculated for the whole face, while ours is adapted to local features. Second, in [15], only one dictionary (the whole reference dataset) is used to compute SR for all test faces, while in our case we select a dynamic and reduced dictionary for each query SURF. From the other hand, our LSR is different from [12] in the way to select the dictionary. In [12], the dictionary is selected from all reference samples, while ours is selected from only the reference samples of the spatial neighborhood of the query IP.

The matching of one query SURF via LSR requires three steps: (1) Local dictionary selection, (2) Sparse representation and (3) Identity assignment.

- **Local Dictionary Selection:** the dictionary A is composed of the N closest reference SURFs. A is of dimension (DxN) where $D = 64$, dimension of SURF descriptor and N is empirically set to 200. The N closest reference SURFs are selected from the reference IPs in the spatial neighborhood, in a rectangular region around the query SURF, as shown in Fig. 2. The width of the region is learned on the training dataset; using this optimization, it is set in our experiments to 60 pixels. We use the same region dimensions when evaluating the unsupervised protocol.
- **Sparse Representation:** the Coordinate Descent Algorithm [5] is used to find the sparsest solution of Eq. 1 as shown in Eq. 2. Its advantage is the use of a tuning parameter λ, to adjust the tradeoff between sparsity term $\|\alpha\|_1$ and error reconstruction term $\|\Phi\alpha - y\|_2^2$.

$$\alpha_s = \min_{\alpha}(\|\Phi\alpha - y\|_2^2 + \lambda\|\alpha\|_1) \tag{2}$$

- **Identity Assignment:** the nonzero coefficients of α_s are used to identify the query IP. We propose to use a continuous vote contrary to [12] where a binary vote is generated. In fact, a x_i vector is calculated for each reference identity i having at least one non-zero coefficient:

$$x_i = [0, \ldots 0, \alpha_{i,1}, \alpha_{i,2}, \ldots, \alpha_{i,k_i}, 0, \ldots, 0] \tag{3}$$

x_i is a coefficient vector obtained from α_s with all elements set to zero except those associated with the identity i. For each reference identity i, the associated vote V_i is incremented (Eq. 4) by a value reflecting the weight in the sparse representation of reference identity i.

$$V_i = V_i + \frac{\|x_i\|}{\|\alpha_s\|} \tag{4}$$

Then, the vote vector V is normalized to unit length. Finally, the query person is claimed as the person that gathers the majority of votes.

6   Binary Classifier for SURFs Filtering

The proposed filtering method is based on a supervised binary classifier. Its goal is to classify IPs into two classes: reliable and unreliable IPs. Ideally, the classifier discards unreliable IPs and retains reliable ones. Two classifiers are evaluated: Support Vector Machine (SVM) [14] and Random Forest (RF) [3]. To run a classifier, two stages are required: first, the classifier learns a filtering model which is used in the second step to discard or retain test IPs.

Training Stage: The classifier takes as input two vector sets: S_{Same} (positive vectors associated with class +1) and S_{Diff} (negative vectors associated with

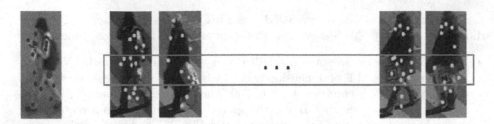

Fig. 2. Local Dictionary Selection: Left (samples of test SURFs), Right (reference dataset). To match a query SURF (green point), a dictionary is selected from all reference SURFs belongs to a rectangular regions around the test SURF (Points in the violet region) (Colour figure online).

class -1). S_{Same} and S_{Diff} model respectively reliable and unreliable IPs. To construct S_{Same} and S_{Diff}, each query SURF is matched to its closet reference one; if the matched pair is associated with the same person, the difference pair descriptor is added to S_{Same}, else it is added to S_{Diff}. In the case of SVM, the training stage consists of finding the hyperplane that separates S_{Same} and S_{Diff}; while for RF, these two sets are used to construct trees by maximizing the variance between the two classes.

Test Stage: To classify a query IP, SVM uses the pre-learned model to assign a probability to each class. The IP is retained if $P(+1) > P(-1)$, where $P(.)$ is a function returning the probability of input class. On the other hand, RF classifies a query IP by running down all of the tree. Then tree decisions (predicted classes) are aggregated to provide a final decision (majority vote rule).

7   Experimental Results

We evaluated our approach on the multi-shot dataset PRID-2011 obtained from two cameras (A and B). The camera-A filmed 749 people and Camera-B filmed 385 people (200 people are common). Two protocols are used in evaluation:

- Unsupervised Protocol consists of identifying the 200 common people filmed by Camera-A in the gallery set (Camera-B) of 749 people.
- Supervised Protocol: PRID-2011 is divided into two parts: training and test. The training set contains two sequences of the first 100 common people. The test set contains the remainder 649 people from Camera-B in reference and the remaining common 100 people from Camera-A in test.

Results are shown in terms of the Cumulative Matching Characteristic (CMC) curve associated with the identification rate. Throughout the rest of this paper, Standard Approach (SA) means that (1) the dictionary is selected from all reference SURFs, (2) binary votes are used and (3) non filtering is applied.

7.1 Contribution of LSR with Continuous Votes

We evaluated our LSR on the PRID-2011 dataset using the two protocols. Starting by the supervised one, results are shown in Fig. 3a and Table 1. Figure 3a shows the obtained CMC (From rank 1 to rank 20) of our approach compared to SA and (SA + continuous votes). Table 1 shows different methods performances (identification rate at rank 1).

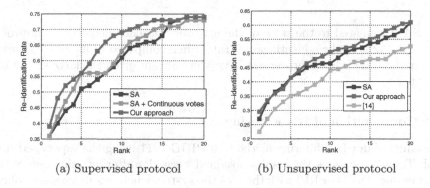

(a) Supervised protocol   (b) Unsupervised protocol

Fig. 3. CMC performance on PRID-2011

Table 1. Results on PRID-2011 (Supervised protocol)

Approach	Re-identification rate (%)
Standard Approach (SA)	36
SA + continuous votes	36
Our approach	39

The PRID-2011 is evaluated in the state of the art only using the unsupervised protocol where all the dataset is used in test. The obtained results with the unsupervised protocol are shown in Fig. 3b and Table 2. Figure 3b shows the obtained CMC (From rank 1 to rank 20) compared to SA and the state of the art. Table 2 shows our performance compared to the state of the art.

For both protocols, the results show that our approach outperforms the standard one (SA). Using the supervised protocol, our approach achieves an improvement of 3 % in the re-identification rate at rank 1. This proves that adding a spatial constraint to construct the dictionary makes sparse representation more effective. Moreover, it proves the efficiency of using continuous votes (soft decisions) rather than binary votes (hard decisions). Using the unsupervised protocol, the results show the benefits of (LSR + continuous votes). Our approach achieves an improvement of 2.5 % in the re-identification rate at rank 1

Table 2. Results on PRID-2011 (Unsupervised protocol)

Approach	Re-identification rate (%)
[9]	19.18
[11]	22.5
[12]	27
Our approach	29.5

w.r.t SA. Compared to the state of the art, our approach achieves an improvement of 12.32 % in the e-identification rate when compared to [9] and 7 % when compared to [11]. This improvement is very significant given the large size of the dataset.

7.2   Contribution of IPs Filtering

We evaluated our IPs' filtering method on PRID-2011 using the supervised protocol. Table 3 compares the results obtained after IPs filtering using one of the two classifiers (SVM or RF) with those of the system where no filtering is applied, according to two performance indicators: re-identification rate and average running time per image.

Table 3. Results of our approach on PRID-2011 (Supervised Protocol)

Classifier	Filtering rate	Re-identification rate	Running time/Image
RF	56.81 %	38 %	1.31(s)
SVM	78.93 %	39 %	0.92(s)
—	No filtering	39 %	2.36(s)

Table 3 shows that by filtering 56.81 % of IPs using RF or 78.93 % using SVM, the accuracy of our approach does not decrease while the processing time becomes much lower. For example, the IPs filtering of SVM achieves an improvement of 61.01 % in average running time per image. These results prove the importance of filtering to reduce running time.

8   Conclusion

This paper has studied IPs matching in uncontrolled conditions for a human re-identification task. It proposed a novel IP matching via Local Spare Representation (LSR). The idea behind is to take into account the spatial distribution of IPs in reference and test images. Our contribution consists of selecting the dictionary from only the reference IPs lying on a learned spatial neighborhood of the query IP. Moreover, we used a soft IPs identification based on continuous votes. On the other hand, we proved the importance of IP filtering to reduce the re-identification running time. The experiment results on the large PRID-2011 database showed that our LSR method performed better in terms of re-identification rate when compared to the state of the art. Moreover it proved the utility of IPs filtering to reduce running time. Using SVM for IPs filtering allows to automatically discard about 80 % of the IPs in the test dataset while keeping the same re-identification accuracy when processing all IPs without filtering. In the future, we will focus on optimizing the Local Dictionary since the latter's size affects significantly the total running time. Moreover, we will study better SR representation schemes for the re-identification task.

References

1. An, L., Kafai, M., Yang, S., Bhanu, B.: Reference-based person re-identification. In: Proceedings of the 10th IEEE International Conference on Advanced Video and Signal Based Surveillance, pp. 244–249 (2013)
2. Bay, H., Tuytelaars, T., Van Gool, L.: SURF: speeded up robust features. In: Leonardis, A., Bischof, H., Pinz, A. (eds.) ECCV 2006. LNCS, vol. 3951, pp. 404–417. Springer, Heidelberg (2006)
3. Breiman, L.: Random forests. Mach. Learn. **45**(1), 5–32 (2001)
4. Farenzena, M., Bazzani, L., Perina, A., Murino, V., Cristani, M.: Person re-identification by symmetry-driven accumulation of local features. In: Conference on Computer Vision and Pattern Recognition, pp. 2360–2367 (2010)
5. Friedman, J.H., Hastie, T., Tibshirani, R.: Regularization paths for generalized linear models via coordinate descent. J. Stat. Softw. **33**, 1–22 (2010)
6. Gray, D., Tao, H.: Viewpoint invariant pedestrian recognition with an ensemble of localized features. In: Forsyth, D., Torr, P., Zisserman, A. (eds.) ECCV 2008. LNCS, vol. 5302, pp. 262–275. Springer, Heidelberg (2008)
7. Hamdoun, O.: Pedestrian detection and re-identification using interest points between non overlapping cameras. Ph.D. thesis, École Nationale Supérieure des Mines de Paris (2010)
8. Hirzer, M., Roth, P., Bischof, H.: Person re-identification by efficient impostor-based metric learning. In: Proceedings of the 9th IEEE International Conference on Advanced Video and Signal-Based Surveillance, pp. 203–208 (2012)
9. Hirzer, M., Beleznai, C., Roth, P.M., Bischof, H.: Person re-identification by descriptive and discriminative classification. In: Heyden, A., Kahl, F. (eds.) SCIA 2011. LNCS, vol. 6688, pp. 91–102. Springer, Heidelberg (2011)
10. Jungling, K., Arens, M.: View-invariant person re-identification with an implicit shape model. In: International Conference on Advanced Video and Signal-Based Surveillance, pp. 197–202 (2011)
11. Khedher, M.I., El-Yacoubi, M.A., Dorizzi, B.: Probabilistic matching pair selection for surf-based person re-identification. In: International Conference of Biometrics Special Interest Group, pp. 1–6 (2012)
12. Khedher, M.I., El-Yacoubi, M.A., Dorizzi, B.: Multi-shot surf-based person re-identification via sparse representation. In: International Conference on Advanced Video and Signal-Based Surveillance (2013)

13. Cong, D.N.T., Achard, C., Khoudour, L., Douadi, L.: Video sequences association for people re-identification across multiple non-overlapping cameras. In: Foggia, P., Sansone, C., Vento, M. (eds.) ICIAP 2009. LNCS, vol. 5716. Springer, Heidelberg (2009)
14. Vapnik, V.N.: Statistical Learning Theory. Wiley-Interscience, New York (1998)
15. Wright, J., Yang, A.Y., Ganesh, A., Sastry, S.S., Ma, Y.: Robust face recognition via sparse representation. IEEE Trans. Pattern Anal. Mach. Intell. **31**, 210–227 (2009)

Behavior Based Darknet Traffic Decomposition for Malicious Events Identification

Ruibin Zhang[1], Lei Zhu[1], Xiaosong Li[1], Shaoning Pang[1(✉)],
Abdolhossein Sarrafzadeh[1], and Dan Komosny[2]

[1] Unitec Institute of Technology, Auckland 1025, New Zealand
{pzhang,lzhu,xli,ppang,hsarrafzadeh}@unitec.ac.nz
[2] Brno University of Technology, 616 00 Brno, Czech Republic
komosny@feec.vutbr.cz

Abstract. This paper proposes a host (corresponding to a source IP) behavior based traffic decomposition approach to identify groups of malicious events from massive historical darknet traffic. In our approach, we segmented and extracted traffic flows from captured darknet data, and categorized flows according to a set of rules that summarized from host behavior observations. Finally, significant events are appraised by three criteria: (a) the activities within each group should be highly alike; (b) the activities should have enough significance in terms of scan scale; and (c) the group should be large enough. We applied the approach on a selection of twelve months darknet traffic data for malicious events detection, and the performance of the proposed method has been evaluated.

1 Introduction

Darknet monitoring, which observes traffic on non-active IP addresses, has become an increasingly important analyzing technique for detecting malicious activities on Internet. Since there are no legitimate hosts exist in darknet space, any observed traffic could be the backscatter packets of Distributed Denial of Service (DDoS) attacks that are using spoofed source addresses; scanning from worms and other network probing; or misconfiguration [1]. Regarding to this fact, the traffic observed on darknet is invaluable for monitoring and analyzing cyber threats. Darknet traffic has been extensively used in observing a wide range of activities which includes denial of service [1–3], worms and scanning [1,4–6], misconfigured network traffic [7], and malware behavior categorizing [3]. In practice, over 50 % of cyber attacks are preceded by some form of network scanning activity according to Panjwani et al. [8], and in most of cases, it is the first stage of an intrusion attempt that enables an adversary to remotely locate, target, or exploit vulnerable systems.

In literature, a number of time series based malicious events identification technique has been reported. In [9], a wavelet decomposition based anomaly detection is proposed. In [10], a dynamic sliding window cumulative sum (CUSUM) algorithm is proposed to identify nested change for detecting anomalous activities.

© Springer International Publishing Switzerland 2015
S. Arik et al. (Eds.): ICONIP 2015, Part III, LNCS 9491, pp. 251–260, 2015.
DOI: 10.1007/978-3-319-26555-1_29

In [11], a method of scan detection through sequential hypothesis testing is presented. In [12], a statistical profile of normal behaviors is modeled to distinguish the anomaly. The traffic time series analysis can only provide temporal knowledge of malicious events. Traffic flow show good potential in passive traffic analysis, but only a few initial works found in this category. Kanda et al. [13] introduce the traffic flow to investigate communication pattern per host, and Kim et al. [14] propose a flow based abnormal traffic detection method.

This paper proposes a constructive flow analysis method for detecting malicious events from darknet observation data. For events identification, we find that most suspect events captured from the time series observation are caused by scan behavior, which follows that, a source IP sends packets to a set of destination IPs and ports in relatively short period of time. Thus we segment the traffic from each source IP into flows, and identify the scans as flows that have their 'width' larger than 'length'. Then we try to identify meaningful event groups, each group follows criterion: (a) the activities in each group should be as similar as possible; (b) the activities should have enough significance, in terms of scan scale; and (c) the group should be large enough. To obtain such groups, ground knowledge and observation on the data are utilized. We argue that the average packet delay ($AvgDly$) is an important feature for event grouping, as it is a reflection of the malicious software routine, thus events generated by the same malicious software should have similar $AvgDly$. As the result, scan is divided into three categories and a set of important event groups are identified in each category.

2   Methodology

2.1   Suspicious Event Observations and Flow Segmenting

The objective of our work is to identify the most significant malicious events and categories them into groups for further studies. For events identification, various studies utilize time series data extracted from captured traffic data for abnormal detection, as the presence of abnormal peaks in the time series is a clear trail of some suspicious activities. Such time series based anomaly detection can only provide temporal feature of events, however additional knowledge is required for better understanding of events' behavior. Therefore we pay special attention to captured traffic data that was transmitted during the period of events.

By analyzing hundreds of similar suspicious events and their corresponding packets data, we summarize that, the source hosts could be easily divided into two types: Type I Hosts have limited population but contribute huge proportion of traffic data, and be active within the period of event; and Type II hosts are great in population, and show minimum recognizable behavior within the period of event. Thus, from the event analysis perspective view, we distinguish activities of type I hosts against type II.

In order to investigate the activities of type I hosts, we conduct flow segmentation for each host using Maximum Packets Interval (MPI). The packet is been counted as the same flow, as long as it falls within the MPI. The procedure of flow segmentation is illustrated in Fig. 1. We adopt 15 s as the size of MPI, as the delay between packets is mostly less than 10 s for type I traffic. In some cases, the source hosts might be spoofed. In our practice, we find that the spoofed hosts are either eliminated due to minor volume (i.e. spoofed once), or taken as candidate for behavior analysis (i.e. spoofed many times).

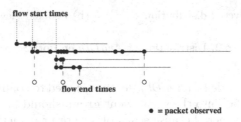

Fig. 1. Procedure of flow segmentation

As the above is completed, we extract the following features for later event grouping: number of packets ($nPkt$), flow duration, number of destination IP addresses ($nDstIP$), number of source port ($nSport$), number of destination port ($nDport$), number of destination IP & destination port combination ($nDipDport$), average delay between packets ($AvgDly$), and the maximum number of packets sent to single destination IP & port combination ($MAXnPktDD$).

2.2 Scan Flow Grouping

As observed from darknet analysis, all suspicious events are caused by scan activity (i.e. a source host send packets to a number of destination IPs and ports). Thus we need to distinguish scan against non-scan flows. In doing that, we measure a flow by its 'width' (i.e. nDipDport, the spread of the victim targets) and 'length' (i.e. MAXnPktDD, the continuity on a single target), and define scan as flows which are active and have their 'width' larger than 'length'.

We divide scan flows into three categories: (a) Port Scan, the flows have only one destination IP, thus the source host scans multiple destination ports on the same destination IP; (b) IP Scan, the flows have only one destination port, the host scans multiple destination IPs; and (c) Hybrid Scan, the flows have more than one destination ports and IPs.

In each scan category, we group the flows using a combination of following 'hard' features: the scanned destination port (ports set), the number of source port used and the number of packet sent. The reason for using these 'hard' feature is that flows with the same value on these features are more likely implies the same activity. Due to the complexity and scale of the data, there is s huge number of groups after the 'hard' feature grouping. We put our focus on those largest ones as we do not expect to capture all potential events but trying to discover event groups of interest.

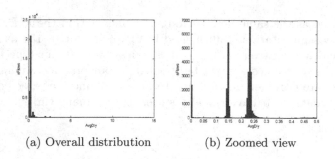

(a) Overall distribution   (b) Zoomed view

Fig. 2. Distribution of *AvgDly* on Port Scan

The average packet delay (*AvgDly*) feature is used to confirm an event group, since the flows that fall into the same event group should have similar *AvgDly*. We adopt this feature here because scans or whatever malicious traffic received in the darknet is mostly initiated by malware, and the packet delay is a reflection of the malware routine. Figure 2a gives an example of *AvgDly* distribution of all port scan flows, and Fig. 2b gives it's zoomed view, while three pikes are identified clearly. One can easily agree that two flows from the same pike (say the one around .15 s) are more likely doing the same or similar things as compared to those from different pikes respectively.

Note that, If event group cannot be formed by 'hard' conditions (i.e. the flows with the same hard feature values are not similar in *AvgDly*), we can apply threshold ('soft') on numerical features such as number of destination IP. The detailed procedure is shown in Sect. 4.

3   Data Description

The datasets we used is a selection of 12 months darknet traffic from January to December 2013. The dataset was captured from the telescope system on a /20 network block, which incorporates 4,096 IP addresses. Table 1 gives statistical information of the datasets in terms of network protocols applied, As seen from the table, TCP traffic is the biggest portion of the datasets, and SYN packets contributes almost half of the TCP data. In our experiment, we conduct data analysis only on TCP/SYN packets, we believe that once a method tested on TCPSYN, it can be also applied to other packets data analysis.

Table 1. Statistical information of data in terms of network protocols

Protocol	Number of packets	Weight	TCP Flag	Number of packets	Weight
TCP(6)	**149,429,974**	**74.53 %**	**SYN**	96,268,005	48.01 %
UDP(17)	45,773,243	22.83 %	SYN/ACK	45,808,500	22.85 %
ICMP(1)	5,304,160	2.64 %	ACK/RST	4,462,157	2.23 %

4   Experiment Results and Discussion

By applying flow segmentation, there are $13,179,281$ flows have been extracted from $96,268,005$ TCP SYN packets. Then according to our scan definition, $1,262,464$ scan flows are identified as they all have larger 'width' than 'length'. Scan flows are further divided into three categories conforming to how the scan width spreads. Table 2 gives a brief statistic of these three scan categories.

Table 2. Statistics of 3 types of scan

Measurement	Port scan	IP scan	Hybrid scan	Total
Number of packets	263018 - 0.35 %	68928739 - 91.86 %	5849029 - 7.79 %	75040786
Number of flows	30856 - 2.44 %	984369 - 77.97 %	247239 - 19.58 %	1262464

4.1   Port Scan

The top 10 popular port scan patterns are listed in Table 3 (the last column nPC is the number of destination port combination). As shown, flows in this category are highly concentrated in the two most popular patterns, which accounts for around three quarters in total number of flows.

Table 3. Top 10 port scan patterns

P#	#Flows	nDport	nSport	nPC
P1	16272	3	2	7
P2	8537	5	2	5
P3	1947	11	2	2
P4	1556	8	1	1
P5	720	2	2	64
P6	611	9	1	2
P7	255	2	1	12
P8	183	7	1	5
P9	158	38	38	1
P10	143	13	13	2

The $AvgDly$ distribution is then studied on frequent pattern level and further on corresponding major destination port combination level, Fig. 3 gives the examples respectively. We found that the group of scan on each destination port combination forms a nearly perfect normal distribution, different $DstPort$ combinations within the same scan pattern may have the similar distribution, and some may not.

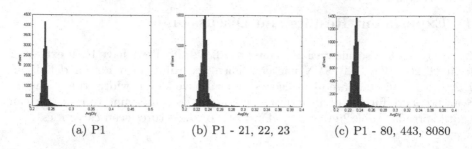

(a) P1   (b) P1 - 21, 22, 23   (c) P1 - 80, 443, 8080

Fig. 3. AvgDly distribution on frequent pattern and port combinations

For the Port Scan category, we conclude that event can be firmly grouped by using a combination of three features: *nDport*, *nSport* and the combination of destination port it scans. We list a set of large event groups found in Table 4.

Table 4. Top 8 event groups in port scan

#*Flows*	*nDport*	*nSport*	*DstPort*	*AvgDly* around
8533	5	2	25 110 143 993 995	.147 s
8385	3	2	21 22 23	.23 s
7865	3	2	80 443 8080	.23 s
1942	11	2	25 80 110 143 443 465 475 587 993 995 6697	.54 s
1556	8	1	80 3128 6666 6669 6675 8080 8909 9415	.0015 s
571	9	1	6160 6666 8012 8048 8080 8081 10000 12882 12883	.002 s
265	2	2	80 443	.8 s
258	2	2	80 8080	1.98 s

4.2 IP Scan

The situation in IP scan becomes much more complicated, we list the most frequent IP scan patterns in Table 5. Here we can see that most of these frequent patterns are very small scaled scans, which we are not much interested. From the perspective of source port usage, we can see two patterns: single source and one source for one *dstIP* (however, when the number of *dstIP* becomes large, source port usage is far from this simple). When we study the *AvgDly* distribution of these patterns as well as heavily scanned destination ports under them, Fig. 4 gives two example. We found that in these very small scaled scans, there are full of randomness, all the known features that we have are unable to further divide these scans into reasonable event group.

Table 5. Top 20 IP scan patterns

P#	#Flows	nPkt	nDip	nSport	P#	#Flows	nPkt	nDip	nSport
P1	356609	2	2	2	P11	10636	5	3	3
P2	124445	3	3	3	P12	10441	5	5	1
P3	123811	2	2	1	P13	7625	8	8	8
P4	59025	4	4	4	P14	6186	4	3	3
P5	42223	3	3	1	P15	6139	6	6	1
P6	32206	5	5	5	P16	6092	4096	4096	1
P7	23097	6	3	3	P17	5423	9	9	9
P8	18938	4	4	1	P18	4771	8	4	4
P9	18503	6	6	6	P19	4701	10	10	10
P10	11453	7	7	7	P20	3935	7	7	1

Then, we turn our focus on large IP scans in terms of the number of IP scanned, as they are more attracted due to the scale significance. Firstly, we look at the 'Full Scans' in which the total of 4,096 IPs in our sensor are all touched. There are 8,937 flows in this set, we divide the full scan set into two subsets according to the number of source port used (1 or more than 1). We found that, the average delay turns to concentrate in very small number within in large scale of scan, mostly less than one second, and for those scans using one source IP, mostly less than .01 s. Through the observation of $AvgDly$ distribution on frequently scanned ports, we found the distribution of each port is concentrated. However, if we consider only the full scan as events in IP scan category, it will results in relatively small number of samples. Thus we consider releasing the full scan condition to a low bounder on number of IP scanned to increase the sample size, meanwhile ensure the samples selected are doing the same thing. To achieve such objective, we scatter flows on the port of interest using the feature $AvgDly$ and $nDstIP$ to determine the low bounder. Figure 5 gives an example of such scatter, for port 5900 case, the $AvgDly$ pattern remains stable until the number of destination IP lower than about 1,000, then we can have those flows with $nDip$ over 1,000 as an event group. Another option is to apply threshold on $AvgDly$, then we can have two groups from 5900: (a) $nDip > 100$, $AvgDly < .01$ and (b) $nDip > 100$, $AvgDly > .1$. Similarly, we can also have two flow groups from port 80 case by applying an $AvgDly$ threshold at around 1s. By the end of this section, we conclude that in IP Scan category, flows with relatively larger scale in terms of $nDip$ show higher regularity and are more useful for further study in terms of significance. To achieve meaningful event groups, a combination of four conditions can be applied, which are: (a) destination port, (b) number of source port used (binary, 1 or more than 1), (c) thresholding on number of destination IP, and (d) thresholding on $AvgDly$. The last two thresholds can be obtained by observing the scatter graph shown in Fig. 5. Some large event groups that we have obtained in our data are listed in Table 6.

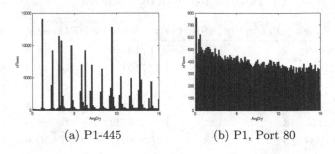

(a) P1-445   (b) P1, Port 80

Fig. 4. Distribution of $AvgDly$ on representative ports

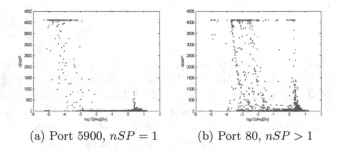

(a) Port 5900, $nSP = 1$ (b) Port 80, $nSP > 1$

Fig. 5. Flow scatter examples

Table 6. Significant event groups in IP scan

#Flow	Dport	SinglenSport?	$Log_{10}AvgDly$	nDip
2526	22	True	<-3.2	>500
513	22	False	<-1.4	>500
1286	23	False	<-1.4	>1000
2160	80	True	<-3	>100
117	110	False	<-1	>1500
3787	1433	True	<-4.3	>500
1987	1998	True	<0.9	>100
2065	3128	True	<-4.2	>100
1221	3389	True	<-2.5	>500
382	3389	False	<-3.5	>500
2126	8080	True	<-4	>100
683	8080	False	<-3.6	>3600
1980	21320	True	<-4.2	>100

4.3 Hybrid Scan

We have found that flows are in small scale are fully filled with randomness, which is similar to what we found in IP Scan. When we filter out flows which touch less than ten IPs, only 5, 628 flows left (less than .4% of the total *HybridScan* flows) belonging to more than 300 destination port combinations. We also found that the binary feature on the number of source port used is no longer discriminated in terms of *AvgDly* distribution. The significant events groups have been found here are listed in Table 7.

Table 7. Large scale event groups within hybrid scan

#Flow	Dport	$Log_{10}AvgDly$	nDip
1154	139; 445	-	>10
435	80; 8080	>1	>10
430	19; 53	-	>10
285	23; 80	-	>10

5   Conclusions

In this paper, we propose a host behavior based traffic decomposition approach for identifying malicious events and categorizing similar events into groups for further studies. In our method, we firstly segment traffic from each source IP into flows, and filter out the scan flows which have their 'width' larger than 'length'. Additionally, the scan flows can be divided into three categories: Port Scan, IP Scan and Hybrid Scan. In most of cases, the 'hard' features will be sufficient for scan flows grouping under each category. However, due to the complexity and scale of the data, occasionally, the threshold('soft') needs to be deployed on numerical features for regulating the degree of similarity in *AvgDly*. According to our experiment results, various significant suspicious events have been grouped and identified by the method proposed in this paper.

References

1. Moore, D., Shannon, C., Brown, D.J., Voelker, G.M., Savage, S.: Inferring Internet denial-of-service activity. ACM Trans. Comput. Syst. **24**, 115–139 (2006)
2. Cooke, E., Jahanian, F., McPherson, D.: The zombie roundup: understanding, detecting, and disrupting botnets. Networks **7**, 39–44 (2005)
3. Kumar, A., Paxson, V., Weaver, N.: Exploiting underlying structure for detailed reconstruction of an internet-scale event. In: Proceedings of the 5th ACM SIG-COMM Conference on Internet Measurement - IMC 2005, p. 1 (2005)
4. Harder, U., Johnson, M.W., Bradley, J.T., Knottenbelt, W.J.: Observing internet worm and virus attacks with a small network telescope. Electron. Notes Theoret. Comput. Sci. **151**(3), 47–59 (2006)

5. Staniford, S., Moore, D., Paxson, V., Weaver, N.: The top speed of flash worms. In: WORM 2004 - Proceedings of the 2004 ACM Workshop on Rapid Malcode, pp. 33–42 (2004)
6. Li, Z., Shi, W., Shi, X., Zhong, Z.: A supervised manifold learning method. Comput. Sci. Inf. Syst. **6**(2), 205–215 (2009)
7. Francois, J., Festor, O., et al.: Tracking global wide configuration errors. In: IEEE/IST Workshop on Monitoring, Attack Detection and Mitigation (2006)
8. Panjwani, S., Tan, S., Jarrin, K.M., Cukier, M.: An experimental evaluation to determine if port scans are precursors to an attack. In: Proceedings of the International Conference on Dependable Systems and Networks, pp. 602–611 (2005)
9. Limthong, K., Kensuke, F., Watanapongse, P.: Wavelet-based unwanted traffic time series analysis. In: Proceedings of the 2008 International Conference on Computer and Electrical Engineering, ICCEE 2008, pp. 445–449 (2008)
10. Ahmed, E., Clark, A., Mohay, G.: Effective change detection in large repositories of unsolicited traffic. In: Fourth International Conference on Internet Monitoring and Protection, ICIMP 2009, pp. 1–6. IEEE (2009)
11. Jung, J., Paxson, V., Berger, A.W., Balakrishnan, H.: Fast portscan detection using sequential hypothesis testing. In: Proceedings of the 2004 IEEE Symposium on Security and Privacy, pp. 211–225. IEEE (2004)
12. Giorgi, G., Narduzzi, C.: Detection of anomalous behaviors in networks from traffic measurements. IEEE Trans. Instrum. Measur. **12**(57), 2782–2791 (2008)
13. Kanda, Y., Fukuda, K., Sugawara, T.: A flow analysis for mining traffic anomalies. In: 2010 IEEE International Conference on Communications (ICC), pp. 1–5. IEEE (2010)
14. Kim, M.-S., Kong, H.-J., Hong, S.-C., Chung, S.-H., Hong, J.: A flow-based method for abnormal network traffic detection. In: 2004 IEEE/IFIP Network Operations and Management Symposium (IEEE Cat. No.04CH37507), vol. 1 (2004)

Statistical Modelling of Artificial Neural Network for Sorting Temporally Synchronous Spikes

Rakesh Veerabhadrappa[1], Asim Bhatti[1(✉)], Chee Peng Lim[1],
Thanh Thi Nguyen[1], S.J. Tye[2], Paul Monaghan[1],
and Saeid Nahavandi[1]

[1] Centre for Intelligent Systems Research,
Deakin University,
Geelong, VIC 3216, Australia
asim.bhatti@deakin.edu.au
[2] Department of Psychiatry & Psychology,
Mayo Clinic, Rochester, MN 55905, USA

Abstract. Artificial neural network (ANN) models are able to predict future events based on current data. The usefulness of an ANN lies in the capacity of the model to learn and adjust the weights following previous errors during training. In this study, we carefully analyse the existing methods in neuronal spike sorting algorithms. The current methods use clustering as a basis to establish the ground truths, which requires tedious procedures pertaining to feature selection and evaluation of the selected features. Even so, the accuracy of clusters is still questionable. Here, we develop an ANN model to specially address the present drawbacks and major challenges in neuronal spike sorting. New enhancements are introduced into the conventional backpropagation ANN for determining the network weights, input nodes, target node, and error calculation. Coiflet modelling of noise is employed to enhance the spike shape features and overshadow noise. The ANN is used in conjunction with a special spiking event detection technique to prioritize the targets. The proposed enhancements are able to bolster the training concept, and on the whole, contributing to sorting neuronal spikes with close approximations.

1 Introduction

Extracellular recordings are a conventional procedure to study and understand the behaviour of living neuronal cells [1, 2]. A biological neural network comprises more than one neurons, and attempts to process its stimuli with dense interconnectivity among neurons. The stimuli could be generally acquired through an in-vivo technique i.e., surgically operating electrodes into a brain region [3–5] or an in-vitro techniques; where the Micro-electrode array (MEA) device is used to record the activity of cells or tissue specimen [6–10]. An example of the recorded voltage in any single channel is demonstrated in Fig. 1.

The complexity of extracellular recordings can be perceived by observing the signal presented in Fig. 1; an example of tetrode recordings from anesthetized rats extracted

© Springer International Publishing Switzerland 2015
S. Arik et al. (Eds.): ICONIP 2015, Part III, LNCS 9491, pp. 261–272, 2015.
DOI: 10.1007/978-3-319-26555-1_30

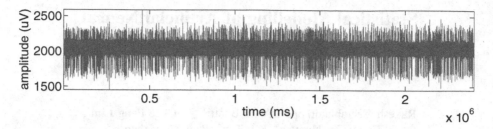

Fig. 1. Single channel voltage information from any extracellular recording unit

from the CA1 hippocampus region [11, 12]. With such obscured information, effective processing techniques are required to identify the number of active neurons that project signals to this channel and the intervals at which they are projecting. These projections are generally referred to as spikes, and are used by analysts to infer conclusions on the behaviour of neurons or a nervous system under certain evoked conditions [13]. A common spike sorting approach associate a distinct pattern for each neuron spike and characterize the interval based on spike shape [14].

An important attribute of these spikes is that the minimum interval between two spikes from the same neuron has to be at least 2.5 ms (ms) comprising of an action potential and refractory interval [15]. But, any two distinct neurons under observation are independent to project their spikes at any interval. One of the aggravating criteria is, if two or more distinct neurons locating at a close distance to the channel electrode spike within the refractory interval, then their respective patterns overlap [16]. This produces a spike shape not is related to neither of their parent neurons; therefore making spike sorting a challenging task.

Neuronal spike sorting procedures described in [5, 13, 14, 17] initially identify spikes through hard thresholding techniques and then use a windowing technique to extract the spike shapes. Clustering procedures are often a choice for partitioning a large number of data into smaller groups. Based on these resulting clusters and understanding their approximate intervals, the prior ground truths can be established for each neuron. Fitting algorithms, a statistical approach formulated pertaining to the prior ground truths, are a choice employed in sorting overlapped spikes. The initial thresholding methods used to identify the spike events do not comprehensively target the spikes. Instead, the method is limited to visual inspection of the data and adjustment of the threshold. This flimsy identification method of spike events fails under low Signal-to-Noise Ratio (SNR) conditions. The clustering results are not able to consider the missed spikes, establishing an improper spike rate and, therefore, incorrect prior estimations.

The algorithm proposed in [18] focuses on addressing the overlap spike sorting issue through a convolution representation of channel voltage and Gaussian distributed noise. However, it does not take into account the distribution of spikes. Additionally, the greedy binary fitting procedure randomly targets the peaks as spike events, thereby identifying too many false positives. As such, the algorithm does not claim the authenticity of sorted spikes. It should also be noted that as the spikes are getting sorted over the time, the established probability of spiking changes accordingly, which impacts the fitting procedure. This effect is ignored in the existing algorithms [13, 17, 18].

In contrast to the aforementioned spike sorting methods, we propose a method that performs initial spike identification using Cepstrum of Bispectrum (CoB) and a denoising procedure, thereby eliminating the drawback of false identification of spike events [19]. Irrespective to the clustering procedure, we employ the maximum likelihood principle to sort the spikes based on the probability established using their current spike counts. To overcome the influence of fluctuating probability, we model our fitting procedure through a novel feedforward ANN with the backpropagation learning algorithm [20]. The initial sequence estimated using clustering is used as the weights for the ANN, while the probability computed with the log likelihood principle is used as the decision criteria. To enhance the effectiveness of the fitting procedure, we employ an optimization technique using background noise, as described in [21]. The process is repeated until the fluctuating probability stabilizes, or with minimal changes, indicating that the neurons are sorted efficiently.

1.1   The Proposed Method

The proposed method assumes the generative model described in [22], in which every neuron is associated with a distinctive spike shape $h(\tau)$ of length T secs. The neuron yields a spike at interval $x(t)$, which assume either a Poisson or Gaussian distribution. Furthermore, the spiking intervals take up binary values such that any time t, a '1' indicates a spike event and '0' indicates no spike. The relationship between $h(\tau)$ and spike interval $x(t)$ form a spike train $y(t)$, which can be derived using a causal LTI system, as shown in Fig. 2(a).

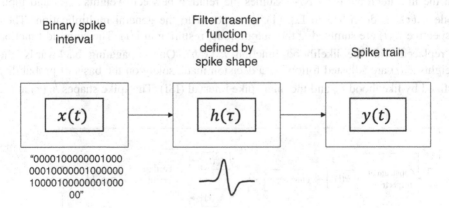

Fig. 2a. LTI system for synthesizing and analysing spike trains

Note that

$$y(t) = \begin{cases} h(\tau) & when\ x(t) = '1' \\ '0' & otherwise \end{cases} \qquad (1)$$

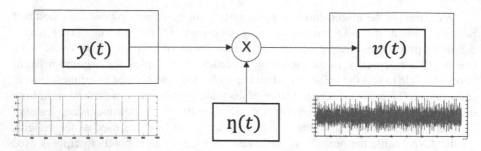

Fig. 2b. Transformation of spike trains into noisy channel voltage [11, 12]

Equation (1) represents spike train $y(t)$ for a single neuron. The final voltage $v(t)$, is the noise modulated version of spike train $y(t)$, as shown in Fig. 2(b).

By applying CoB based spike detection to voltage $v(t)$, it is possible to recover spike interval $x(t)$. Clustering algorithms are capable of recovering the spike waveforms from the detected intervals. After clustering, $x(t)$ is split into $\hat{x}_n(t)$, depending on the number of clusters formed. It should be noted that clustering algorithms do not handle overlapped spikes, and only clearly identical spike waveforms are grouped together.

Figure 3 shows the proposed ANN model, which is capable of efficiently sorting spikes even under poor clustering scenarios. The model presents some improvisations to the regular backpropagation algorithm [20, 23, 24]. The split intervals $\hat{x}_n(t)$ are treated as weights, and probability p is the decision parameter. The ANN defined here has no hidden layer. The network has one output to represent a single channel voltage. On the first iteration, the ANN assumes the relation between weights $\hat{x}_n(t)$ and input node $\hat{h}_n(\tau)$, as described in Eq. (1), as opposed to the general multiplication. Their respective $y_n(t)$ are summed at the output node, resulting in $\hat{v}(t)$. The sigmoid function is replaced with the likelihood function $L_n(\hat{x}, \hat{h})$. On propagating backwards, the weights $\hat{x}_n(t)$ are adjusted following a decision made solely on the basis of probability defined by likelihood L_n and the inter spike interval (ISI). The spike shapes $\hat{h}_n(\tau)$ at the

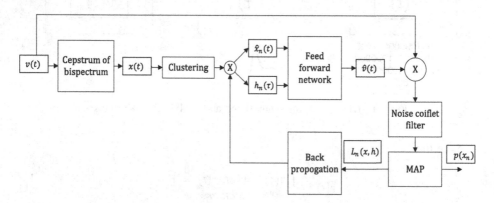

Fig. 3. Spike detection and sorting using artificial neural network

output layer is re-estimated for the new $\hat{x}_n(t)$, and the process is repeated. After several iterations, when the change in probability falls to a minimum level, or the change in probability stabilizes, the ANN is said to be trained. Ideally, the final weights $\hat{x}_n(t)$ should correspond to the exact spiking intervals with all the overlapped spikes completely sorted.

2 Methodology

2.1 Estimating Spike Intervals $x(t)$

The proposed method is able to emphasize the intervals of spiking events $x(t)$. We have adopted the CoB technique considering its higher order spectral analysis (HoS) capability [19, 25]. For our model presented in Fig. 1 and Eq. (1), $h(\tau)$ represents a filter which impacts on the incoming sequence $x(t)$ and yields spike train $y(t)$. For the general notion of LTI systems,

$$Y(z) = H(z).X(z) \tag{2}$$

$$X(z) = \frac{Y(z)}{H(z)} = Y(z).H^{-1}(z) \tag{3a}$$

$$H(z).H^{-1}(z) = 1 \tag{3b}$$

It can be inferred from the relations in Eqs. (2) and (3) that by filtering the spike train $y(t)$ through $h^{-1}(\tau)$ (an inverse filter), it is possible to recover $x(t)$. Since our final term is $v(t)$, the noise obscured version of $y(t)$, as shown in Fig. 2(b), filtering $v(t)$ through the inverse filter $h^{-1}(\tau)$ results in $x(t)$ and its noise obscured term $\tilde{x}(t)$. One possible method of synthesizing the inverse filter is through CoB, as follows [19].

   Given $v(t)$, if $V(l)$ represents the fourier transform at some frequency component l, then its bispectrum $B_v(l, m)$ describes the third moment for a 2-D fourier transform over frequency indices l and m. CoB, through bispectrum estimation, takes the advantage of HoS, a method generally applicable to signal processing, to decay the effect of the Gaussian noise components. $Cep(B_v(l, t))$ is computed by taking the logarithm of bispectrum $B_v(l, m)$, followed by the inverse fourier transform of the log term over time t and frequency index l. Solving Eq. (1), it is possible to compute $H(l)$, i.e., the filter transfer function in the frequency domain. Using Eq. (3b), $H^{-1}(l)$ describes the inverse filter in the frequency domain, and the inverse Fourier transform of which gives $h^{-1}(\tau)$.

2.2 Artificial Neural Network

Simple clustering algorithms are capable of extracting spike waveforms and partitioning a large set of spike waveforms into different groups. As an example; Wave_clus and Klustakwik in [5, 26] perform partitioning based on the feature set, but they do not claim the accuracy of clustering. Additionally, the overlapped spikes are treated as a deformity and are set aside from the rest of the partitions. On the other hand, the ANN is driven by an initial coarse ground truth, and does not depend on accurate results.

Feedforward ANN. Following the clustering results, assuming that the number of clustered partitions reflects the number of neurons N. The feedforward ANN is used to surmise the voltage information, $\hat{v}(t)$, based on some known criteria (clustered results) (Fig. 4).

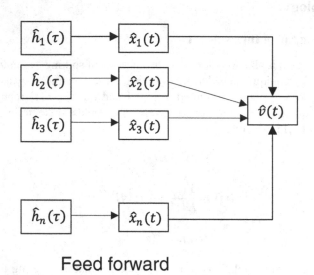

Feed forward

Fig. 4. Feed forward network for deriving voltage $\hat{v}(t)$

The intervals of spike events $x(t)$ detected (as described in Sect. 2.1) are used to identify the individual spike intervals $\hat{x}_n(t)$ for all N neurons, where n is an integer between 1 and N. Assuming that the relationship described in Eq. (1) is commutative i.e., if,

$$y(t) = x(t) * h(\tau) = h(\tau) * x(t) \tag{4}$$

then, as shown in Fig. 2(a), it is possible to synthesize $\hat{v}(t)$ as follows

$$\hat{v}(t) = \sum_{n=1}^{N} \hat{y}_n(t) \tag{5}$$

It should be noted that $\hat{v}(t)$ in Eq. (5) is derived entirely from the clustering results. Therefore, the difference between voltages $\hat{v}(t)$ and $v(t)$ indicates overlapped spike waveforms and noise $\eta(t)$.

Noise Coiflet Wavelet. In image processing, the coiflet wavelet decomposition is employed to decay noise and to enhance the resolution [27]. A similar concept is also described as part of the denoising procedure introduced in the Sect. 2.1. Specifically, a stationary wavelet decomposition method using the first coiflet wavelet is applied to enhance the recovered spike event $x(t)$. Considering that the unresolved spike waveforms and noise $\eta(t)$ contribute to the overall noise from any channel, a coiflet type wavelet which has the characteristic of noise can be modelled. By decomposing $v(t)$ through the noise characterized coiflet wavelet, it is possible to maximise the characterstics of the resolved spike waveforms [21].

The procedure to develop noise coiflet wavelet is to first compute the noise covariance of a finite lag. A toeplitz matrix of covariance lag is computed [28], and the middle row which resembles the characteristics of the coiflet wavelet is chosen.

Backpropagation. With the filtered voltage information $\bar{v}(t)$ and spike event $x(t)$, the overlapped spike follows a log likelihood principle, as described in Eq. (6).

$$L\big(x(t) = \text{'1'}, \hat{h}_n(\tau)\big) = \log[p\{x_n(t), \hat{h}_n(\tau)|\bar{v}(t)\}] \tag{6}$$

If $\hat{h}_n(\tau)$ represents the surmised principle spikes shape for each cluster, the likelihood is estimated for all n events at any time t for an estimated $x(t) = \text{'1'}$.

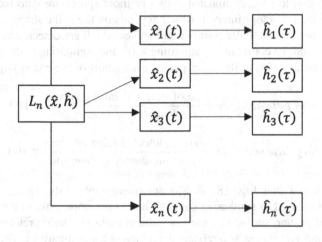

Back propagation

Fig. 5. Back propagation algorithm, re-adjust weights $\hat{x}_n(t)$, compute new spike waveforms $\hat{h}_n(\tau)$

To follow with the backpropagation procedure, the computed L_n component is used as a contention parameter, wherein the nth neuron spike shape with maximum L wins the completion, and $x(t) = \text{'1'}$ reflects the related $\hat{x}_n(t)$. The procedure is repeated for all $x(t) = \text{'1'}$, which have not been previously resolved due to overlapped or jittered spike shapes.

With the new $\hat{x}_n(t)$, the new principle spike shapes $\hat{h}_n(\tau)$ are re-computed, and the procedure repeats. The fluctuation of the probability term is monitored until it stabilizes, or changes minimally at which stage it can be decided that the spikes are resolved. The principle spike shapes $\hat{h}_n(\tau)$ in Fig. 5 can be computed using the least square linear regression.

3 Performance Evaluation

The developed method aims to identify the spike that belongs to a genuine neuron cluster. Some of the evaluation techniques described in [13, 18, 19] comprises the quantification of false positives, false negatives, and sorted overlaps. The quantification can be defined as

$$\% \, of \, refractory \, violations = \frac{no.of \, identified \, overlap \, spike \, events}{total \, number \, of \, spike \, events} \times 100 \quad (7)$$

A spike is said to be contaminated if two or more spikes are detected in a region within a specified refractory interval, which depends on the spike shape, sampling rate and area of recorded data. False positives are spikes which are detected in a no spiking region, which could be computed by summing $\hat{x}_n(t)$ and comparing with $x(t)$. Since not all spikes can be detected at the exact peaks, a variation of 0.5 ms is neglected.

$$\% \, of \, false \, positives = \frac{no.of \, detected \, false \, positves}{total \, number \, of \, spike \, events} \times 100 \quad (8a)$$

$$\% \, of \, false \, negatives = \frac{no.of \, detected \, false \, positves}{total \, number \, of \, spike \, events} \times 100 \quad (8b)$$

It should be noted that Eqs. (8a & 8b) are quantifiable only on the basis of the generative model used to synthesize the data, and is not applicable to the raw data. In the absence of ground truths, i.e., the raw data conditions, the correlation coefficient between $v(t)$ and $\hat{v}(t)$ serves as a reference coefficient for comparison. The later can be derived using Eq. (4) for the sequence $\hat{x}_{n,i}(t)$, its respective surmised spike shapes $\hat{h}_{n,i}(\tau)$, and the ith iteration for which the training is concluded.

4 Results and Discussion

The raw data described in [11, 12] have no information on the number of neurons spiking at any of the tetrode electrodes; therefore it is difficult to analyse through the developed method. On the other hand, synthesizing data sets with known number of neurons and the noise level can provide useful information to compare the outcomes at different phases of the proposed method described in Fig. 3. As such, we use the model described in [22] to synthesize a single channel data set; two distinct neuron spike shapes, 7 overall jittered spike shapes, 15 uncorrelated spike shapes with Gaussian noise of 100 dB used in the synthesis of a 2 s single channel voltage information sampled at 24 kHz. Additionally, both Poisson and Gaussian distributions are used to describe each spike train.

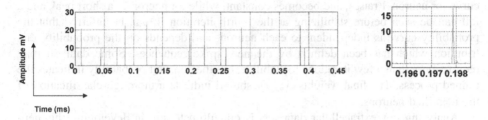

Fig. 6. Segment of CoB output showing the detected spike events

As an example, Fig. 6 shows the performance of CoB, where the peaks indicate the detected spike events. To highlight the capability of CoB under overlapped conditions, a section of the graph around 0.2 s in Fig. 6 is magnified. A second spike event can be detected within the duration of 0.2 ms, which is well within the refractory period of the first spike event. In any case, the second spike event would have gone unnoticed when threshold-based spike detection techniques are used.

The performance of CoB also varies under different noise levels, and it does not guarantee full identification. Since the CoB method was applied to excerpts of data, the evaluated inverse filter $h^{-1}(t)$ varied. To maximise the performance of the inverse filter, it is advisable to use mean of the computed filters for each excerpt.

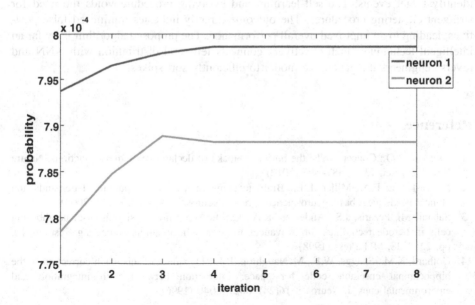

Fig. 7. Probability variation over spread over several iterations

To emulate the results from the clustering procedure, $\hat{x}_n(t)$ were fabricated by ignoring all the overlapped indices. The performance of the ANN could be visualised in Fig. 7. The probability curve for both neurons starts with $\hat{x}_{n,1}(t)$, where "1" indicates the iteration number. As the training procedure repeats through several iterations, the curve for neuron-1 raises, and becomes constant, while for neuron-2, a short peak and a fall can be seen before stabilizing at the fourth iteration. It can be inferred that the probability curve is independent to each neuron, and depends on the probability distribution which has been defined by the new spike sequences. Some data set may require more than a few iterations to stabilize, but the constant probability indicates the trained process. The final weights, $\hat{x}_{n,i}(t)$, should indicate a thorough classification of the identified neurons.

Analysing raw extracellular data sets is our ultimate aim in developing this neuronal spike sorting algorithm. A spatio-temporal realization of the same concept can provide a better platform to analyse and sort the spikes from multi-channel systems, especially MEA's. Some tetrode recordings also accompany intracellular data extracted directly from within the cells, which could be used as references to make better approximations of the developed procedure.

5 Conclusions

The proposed ANN model serves as a useful method for neuronal spike sorting. The result demonstrates the effectiveness of the method for sorting spikes from any single channel data. The challenges ignored in the existing approaches have been carefully scrutinized and addressed in this study. The advantage offered by the CoB over greedy pursuit methods ensures that the ANN concedes reduced irrelevant computations to identify target events. The self-learning and evolving procedure voids the need for stringent clustering procedures. The outcome directly indicates minimised false positives, leading to an improved overall performance. The proposed algorithm is by far an intelligent technique which effectively connects temporal distribution with ANN and reverse engineers the generative model to efficiently sort spikes.

References

1. Quiroga, R.Q.: Concept cells: the building blocks of declarative memory functions. Nature Rev. Neurosci. **13**(8), 587–597 (2012)
2. Mussa-Ivaldi, F.A., Miller, L.E.: Brain machine interfaces: computational demands and clinical needs meet basic neuroscience. Trends Neurosci. **26**(6), 329–334 (2003)
3. Sahani, M., Pezaris, J.S., Andersen, R.A.: On the separation of signals from neighboring cells in tetrode recordings. In: Advances in Neural Information Processing Systems 10, pp. 222–228. MIT Press (1998)
4. Gothard, K.M., Skaggs, W.E., McNaughton, B.L.: Dynamics of mismatch correction in the hippocampal ensemble code for space: interaction between path integration and environmental cues. J. Neurosci. **16**(24), 8027–8040 (1996)

5. Harris, K.D., Henze, D.A., Csicsvari, J., Hirase, H., Buzsáki, G.: Accuracy of tetrode spike separation as determined by simultaneous intracellular and extracellular measurements. J. Neurophysiol. **84**(1), 401–414 (2000)
6. Reinhard, K., Tikidji-Hamburyan, A., Seitter, H., Idrees, S., Mutter, M., Benkner, B., Munch, T.A.: Step-by-step instructions for retina recordings with perforated multi electrode arrays. PLoS ONE **9**(8), e106148 (2014)
7. Wang, Y., Yeung, C.-K., Ingebrandt, S., Offenhaeusser, A., Chan, M.: Multi-electrode arrays (meas) with guided network for cell-to-cell communication transduction. In: IEEE International Electron Devices Meeting, IEDM Technical Digest, p. 3, p. 484, December 2005
8. Pfeffer, L., Ide, D., Stewart, C., Plenz, D.:. A life support system for stimulation of and recording from rodent neuron networks grown on multi-electrode arrays. In: Proceedings of the 17th IEEE Symposium on Computer-Based Medical Systems, CBMS 2004, pp. 473–478, June 2004
9. Hottowy, P., Beggs, J.M., Chichilnisky, E.J., Dabrowski, W., Fiutowski, T., Gunning, D.E., Hobbs, J., Jepson, L., Kachiguine, S., Mathieson, K., et al.: 512-electrode mea system for spatio-temporal distributed stimulation and recording of neural activity. In: Stett, A. (ed.) Proceedings of the 7th International Meeting on Substrate-Integrated Microelectrode Arrays, Reutlingen, Germany, pp. 327–330, June 2010
10. Gaburro, J., Duchemin, J.-B., Bhatti, A., Walker, P., Nahavandi, S.: Neurophysiology of insects using microelectrode arrays: current trends and future prospects. In: Loo, C.K., Yap, K.S., Wong, K.W., Beng Jin, A.T., Huang, K. (eds.) ICONIP 2014, Part III. LNCS, vol. 8836, pp. 493–500. Springer, Heidelberg (2014)
11. Henze, D.A., Harris, K.D., Borhegyi, Z., Csicsvari, J., Mamiya, A., Hirase, H., Sirota, A., Buzsáki, G.: Simultaneous intracellular and extracellular recordings from hippocampus region ca1 of anesthetized rats (2009)
12. Henze, D.A., Borhegyi, Z., Csicsvari, J., Mamiya, A., Harris, K.D., Buzsáki, G.: Intracellular features predicted by extracellular recordings in the hippocampus in vivo. J. Neurophysiol. **84**(1), 390–400 (2000)
13. Prentice, J.S., Homann, J., Simmons, K.D., Tkacik, G., Balasubramanian, V., Nelson, P.C.: Fast, scalable, bayesian spike identification for multi-electrode arrays. PLoS ONE **6**(7), e19884 (2011)
14. Hulata, E., Segev, R., Ben-Jacob, E.: A method for spike sorting and detection based on wavelet packets and shannon's mutual information. J. Neurosci. **117**, 1–12 (2002)
15. Platkiewicz, J., Brette, R.: A threshold equation for action potential initiation. PLoS Comput. Biol. **6**(7), e1000850 (2010)
16. Marre, O., Amodei, D., Deshmukh, N., Sadeghi, K., Soo, F., Holy, T.E., Berry, M.J.: Mapping a complete neural population in the retina. J. Neurosci. **32**(43), 14859–14873 (2012)
17. Ekanadham, C., Tranchina, D., Simoncelli, E.P.: A unified framework and method for automatic neural spike identification. J. Neurosci. Meth. **222**, 47–55 (2014)
18. Pillow, J.W., Shlens, J., Chichilnisky, E.J., Simoncelli, E.P.: A model-based spike sorting algorithm for removing correlation artifacts in multi-neuron recordings. PLoS ONE **8**(5), e62123 (2013)
19. Shahid, S., Walker, J., Smith, L.S.: A new spike detection algorithm for extracellular neural recordings. IEEE Trans. Biomed. Eng. **57**(4), 853–866 (2010)
20. Robert Gordon University. The back propogation algorithm. http://www4.rgu.ac.uk/files/chapter3%20-%20bp.pdf
21. Pouzat, C., Mazor, O., Laurent, G.: Using noise signature to optimize spike-sorting and to assess neuronal classification quality. J. Neurosci. Meth. **122**(1), 43–57 (2002)

22. Smith, L.S., Mtetwa, N.: A tool for synthesizing spike trains with realistic interference. J. Neurosci. Meth. **159**(1), 170–180 (2007)
23. Robert Gordon University. Aritificial neural networks. http://neuron.csie.ntust.edu.tw/homework/98/NN/homework3/M9809103,M9809111,M9809113_3/Methodology.html
24. Robert Gordon University. Aritificial neural networks. http://www4.rgu.ac.uk/files/chapter2%20-%20intro%20to%20ANNs.pdf
25. Feng, M., Kammeyer, K.-D.: Suppression of gaussian noise using cumulants: a quantitative analysis. In: 1997 IEEE International Conference on Acoustics, Speech, and Signal Processing, ICASSP 1997, vol. 5, pp. 3813–3816, April 1997
26. Quiroga, R.Q., Nadasdy, Z., Ben-Shaul, Y.: Unsupervised spike detection and sorting with wavelets and superparamagnetic clustering. Neural Comput. **16**(8), 1661–1687 (2004)
27. Ma, Y., Li, J.: A novel method based on adaptive median filtering and wavelet transform in noise images. In: 2011 IEEE 3rd International Conference on Communication Software and Networks (ICCSN), pp. 626–629, May 2011
28. Yagle, A.E.: A fast algorithm for toeplitz-block-toeplitz linear systems. In: Proceedings of the 2001 IEEE International Conference on Acoustics, Speech, and Signal Processing, ICASSP 2001, vol. 3, pp. 1929–1932 (2001)

A Novel Condition for Robust Stability of Delayed Neural Networks

Neyir Ozcan[1], Eylem Yucel[2]($\boxtimes$), and Sabri Arik[2]

[1] Department of Electrical and Electronics Engineering,
Uludag University, Bursa, Turkey
neyir@uludag.edu.tr
[2] Department of Computer Engineering, Istanbul University,
Avcilar, Istanbul, Turkey
{eylem,ariks}@istanbul.edu.tr

Abstract. This paper presents a novel sufficient condition for the existence, uniqueness and global robust asymptotic stability of the equilibrium point for the class of delayed neural networks by using the Homomorphic mapping and the Lyapunov stability theorems. An important feature of the obtained result is its low computational complexity as the reported result can be verified by checking some well-known properties of some certain classes of matrices, which simplify the verification of the derived result.

Keywords: Neural networks · Lyapunov functionals · Stability analysis

1 Introduction

In recent years, dynamical neural networks have been widely used in solving various classes of engineering problems such as image and signal processing, associative memory, pattern recognition, parallel computation, control and optimization. In such applications, the equilibrium and stability properties of neural networks are of great importance in the design of dynamical neural networks. It is known that in the VLSI implementation of neural networks, time delays are unavoidably encountered during the processing and transmission of signals, which may affect the dynamics of neural networks. On the other hand, some deviations in the parameters of the neural network may also affect the stability properties. Therefore, we must consider the time delays and parameter uncertainties in studying stability of neural networks, which requires to deal with the robust stability of delayed neural networks. Recently, many conditions for global robust stability of delayed neural networks have been reported [1–19]. In this paper, we present a new sufficient condition for the global robust asymptotic stability of neural networks with multiple time delays.

Consider the following neural network model:

$$\frac{dx_i(t)}{dt} = -c_i x_i(t) + \sum_{j=1}^{n} a_{ij} f_j(x_j(t)) + \sum_{j=1}^{n} b_{ij} f_j(x_j(t - \tau_{ij})) + u_i \tag{1}$$

© Springer International Publishing Switzerland 2015
S. Arik et al. (Eds.): ICONIP 2015, Part III, LNCS 9491, pp. 273–280, 2015.
DOI: 10.1007/978-3-319-26555-1_31

where n is the number of the neurons, $x_i(t)$ denotes the state of the neuron i at time t, $f_i(\cdot)$ denote activation functions, a_{ij} and b_{ij} denote the strengths of connectivity between neurons j and i at time t and $t - \tau_{ij}$, respectively; τ_{ij} represents the time delays, u_i is the constant input to the neuron i, c_i is the charging rate for the neuron i.

The parameters a_{ij} and b_{ij} and c_i are assumed to satisfy the conditions

$$
\begin{aligned}
C_I &= [\underline{C}, \overline{C}] = \{C = diag(c_i) : 0 < \underline{c_i} \leq c_i \leq \overline{c_i}, i = 1, 2, ..., n\} \\
A_I &= [\underline{A}, \overline{A}] = \{A = (a_{ij})_{n \times n} : \underline{a_{ij}} \leq a_{ij} \leq \overline{a}_{ij}, i, j = 1, 2, ..., n\} \\
B_I &= [\underline{B}, \overline{B}] = \{B = (a_{ij})_{n \times n} : \underline{b_{ij}} \leq b_{ij} \leq \overline{b}_{ij}, i, j = 1, 2, ..., n\}
\end{aligned}
\tag{2}
$$

The activation functions f_i are assumed to satisfy the condition:

$$
|f_i(x) - f_i(y)| \leq \ell_i |x - y|, \ i = 1, 2, ..., n, \quad \forall x, y \in R, x \neq y
$$

where $\ell_i > 0$ denotes a constant. This class of functions is denoted by $f \in \mathcal{L}$.

The following lemma will play an important role in the proofs:

Lemma 1. [3]: Let A be any real matrix defined by

$$
A \in A_I = [\underline{A}, \overline{A}] = \{A = (a_{ij})_{n \times n} : \underline{a_{ij}} \leq a_{ij} \leq \overline{a}_{ij}, i, j = 1, 2, ..., n\}
$$

Let $x = (x_1, x_2, ..., x_n)^T$ and $y = (y_1, y_2, ..., y_n)^T$. Then, we have

$$
2x^T A y \leq \beta \sum_{i=1}^{n} x_i^2 + \frac{1}{\beta} \sum_{i=1}^{n} p_i y_i^2
$$

where β is any positive constant, and

$$
p_i = \sum_{k=1}^{n} (\hat{a}_{ki} \sum_{j=1}^{n} \hat{a}_{kj}), \ i = 1, 2, ..., n
$$

with $\hat{a}_{ij} = max\{|\underline{a_{ij}}|, |\overline{a}_{ij}|\}, i, j = 1, 2, ..., n$.

2 Global Robust Stability Analysis

In this section, we present the following result:

Theorem 1. For the neural system (1), let the network parameters satisfy (2) and $f \in \mathcal{L}$. Then, the neural network model (1) is globally asymptotically robust stable, if there exist positive constants α and β such that

$$
\varepsilon_i = 2\underline{c_i} - \beta - \frac{1}{\beta} p_i \ell_i^2 - \sum_{j=1}^{n} (\alpha \ell_j + \frac{1}{\alpha} \hat{b}_{ji}^2 \ell_i) > 0, \quad i = 1, 2, ..., n
$$

where $p_i = \sum_{j=1}^{n} (\hat{a}_{ji} \sum_{k=1}^{n} \hat{a}_{jk})$, $i = 1, 2, ..., n$ and $\hat{a}_{ij} = max\{|\underline{a_{ij}}|, |\overline{a}_{ij}|\}$ and $\hat{b}_{ij} = max\{|\underline{b_{ij}}|, |\overline{b}_{ij}|\}$, $i, j = 1, 2, ..., n$.

Proof. In order to prove the existence and uniqueness of the equilibrium point of system (1), we consider the following mapping associated with system (1):

$$H(x) = -Cx + Af(x) + Bf(x) + u \tag{3}$$

Clearly, if x^* is an equilibrium point of (1), then, x^* satisfies the equilibrium equation of (1):

$$-Cx^* + Af(x^*) + Bf(x^*) + u = 0$$

Hence, we can easily see that every solution of $H(x) = 0$ is an equilibrium point of (1). Therefore, for the system defined by (1), there exists a unique equilibrium point for every input vector u if $H(x)$ is homeomorphism of R^n. Now, let x, $y \in R^n$ be two different vectors such that $x \neq y$. For $H(x)$ defined by (3), we can write

$$H(x) - H(y) = -C(x - y) + A(f(x) - f(y)) + B(f(x) - f(y)) \tag{4}$$

For $f \in \mathcal{L}$, first consider the case where $x \neq y$ and $f(x) - f(y) = 0$. In this case, we have

$$H(x) - H(y) = -C(x - y)$$

from which $x - y \neq 0$ implies that $H(x) \neq H(y)$ since C is a positive diagonal matrix. For $f \in \mathcal{L}$, now, consider the case where $x - y \neq 0$ and $f(x) - f(y) \neq 0$. In this case, multiplying both sides of (4) by $2(x - y)^T$ results in

$$
\begin{aligned}
2(x - y)^T(H(x) - H(y)) = & -2(x - y)^T C(x - y) + 2(x - y)^T A(f(x) - f(y)) \\
& + 2(x - y)^T B(f(x) - f(y)) \\
= & -2\sum_{i=1}^{n} c_i(x_i - y_i)^2 + 2(x - y)^T A(f(x) - f(y)) \\
& + 2\sum_{i=1}^{n}\sum_{j=1}^{n} b_{ij}(x_i - y_i)(f_j(x_j) - f_j(y_j))
\end{aligned}
\tag{5}
$$

We first note the following inequality:

$$
\begin{aligned}
& 2\sum_{i=1}^{n}\sum_{j=1}^{n} b_{ij}(x_i - y_i)(f_j(x_j) - f_j(y_j)) \\
& \leq \sum_{i=1}^{n}\sum_{j=1}^{n} 2|b_{ij}||x_i - y_i||f_j(x_j) - f_j(y_j)| \\
& \leq \sum_{i=1}^{n}\sum_{j=1}^{n} 2|b_{ij}|\ell_j|x_i - y_i||x_j - y_j| \\
& \leq \sum_{i=1}^{n}\sum_{j=1}^{n} \ell_j(\alpha(x_i - y_i)^2 + \frac{1}{\alpha}b_{ij}^2(x_j - y_j)^2)
\end{aligned}
$$

$$= \alpha \sum_{i=1}^{n} \sum_{j=1}^{n} \ell_j(x_i - y_i)^2 + \frac{1}{\alpha} \sum_{i=1}^{n} \sum_{j=1}^{n} b_{ij}^2 \ell_j(x_j - y_j)^2$$

$$= \alpha \sum_{i=1}^{n} \sum_{j=1}^{n} \ell_j(x_i - y_i)^2 + \frac{1}{\alpha} \sum_{i=1}^{n} \sum_{j=1}^{n} b_{ji}^2 \ell_i(x_i - y_i)^2$$

$$\leq \sum_{i=1}^{n} \sum_{j=1}^{n} (\alpha \ell_j + \frac{1}{\alpha} \hat{b}_{ji}^2 \ell_i)(x_i - y_i)^2 \tag{6}$$

For any positive constant β, we can also write

$$2(x-y)^T A(f(x) - f(y)) \leq \beta(x-y)^T(x-y) + \frac{1}{\beta}(f(x) - f(y))^T A^T A(f(x) - f(y)) \tag{7}$$

For $f \in \mathcal{L}$, from Lemma 1, we can write

$$(f(x) - f(y))^T A^T A(f(x) - f(y)) \leq \sum_{i=1}^{n} p_i(f_i(x_i) - f_i(y_i))^2$$

$$\leq \sum_{i=1}^{n} p_i \ell_i^2(x_i - y_i)^2 \tag{8}$$

Hence, in the light of (6)–(8), (5) takes the form:

$$2(x-y)^T(H(x) - H(y)) \leq -2 \sum_{i=1}^{n} \underline{c}_i(x_i - y_i)^2 + \beta \sum_{i=1}^{n}(x_i - y_i)^2)$$

$$+ \frac{1}{\beta} \sum_{i=1}^{n} p_i \ell_i^2(x_i - y_i)^2 + \sum_{i=1}^{n} \sum_{j=1}^{n}(\alpha \ell_j + \frac{1}{\alpha} \hat{b}_{ji}^2 \ell_i)(x_i - y_i)^2$$

which is equivalent to

$$2(x-y)^T(H(x) - H(y)) \leq - \sum_{i=1}^{n}(2\underline{c}_i - \beta - \frac{1}{\beta} p_i \ell_i^2 - \sum_{j=1}^{n}(\alpha \ell_j + \frac{1}{\alpha} \hat{b}_{ji}^2 \ell_i))(x_i - y_i)^2$$

$$= - \sum_{i=1}^{n} \varepsilon_i(x_i - y_i)^2 \leq -\varepsilon_m \sum_{i=1}^{n}(x_i - y_i)^2$$

$$= -\varepsilon_m ||x - y||_2^2 \tag{9}$$

where $\varepsilon_m = min\{\varepsilon_i\}, i = 1, 2, ..., n$. Let $x - y \neq 0$ and $\varepsilon_m > 0$. Then,

$$(x-y)^T(H(x) - H(y)) < 0$$

from which we can conclude that $H(x) \neq H(y)$ for all $x \neq y$. In order to show that $||H(x)|| \to \infty$ as $||x|| \to \infty$, we let $y = 0$ in (9), which yields

$$x^T(H(x) - H(0)) \leq -\varepsilon_m ||x||_2^2$$

from which it follows that $||H(x) - H(0)||_1 \geq \varepsilon_m ||x||_2$. Using the property $||H(x) - H(0)||_1 \leq ||H(x)||_1 + ||H(0)||_1$, we obtain $||H(x)||_1 \geq \varepsilon_m ||x||_2 - ||H(0)||_1$. Since $||H(0)||_1$ is finite, it follows that $||H(x)|| \to \infty$ as $||x|| \to \infty$. This completes the proof of the existence and uniqueness of the equilibrium point of (1).

We will now prove the global asymptotic stability of the equilibrium point of system (1). We first shift the equilibrium point x^* of system (1) to the origin. Using $z_i(\cdot) = x_i(\cdot) - x_i^*$, $i = 1, 2, ..., n$, puts the (1) in the form:

$$\dot{z}_i(t) = -c_i z_i(t) + \sum_{j=1}^{n} a_{ij} g_j(z_j(t)) + \sum_{j=1}^{n} b_{ij} g_j(z_j(t - \tau_{ij})) \tag{10}$$

where $g_i(z_i(\cdot)) = f_i(z_i(\cdot) + x_i^*) - f_i(x_i^*)$. Note that $f \in \mathcal{L}$ implies that $g \in \mathcal{L}$ with

$$|g_i(z)| \leq \ell_i |z|, \quad \text{and} \quad g_i(0) = 0, \quad i = 1, 2, ..., n$$

Since $z(t) \to 0$ implies that $x(t) \to x^*$, the asymptotic stability of $z(t) = 0$ is equivalent to that of x^*. In order to prove the global asymptotic stability of $z(t) = 0$, we will employ the following positive definite Lyapunov functional:

$$V(z(t)) = \sum_{i=1}^{n} z_i^2(t) + \sum_{i=1}^{n} \sum_{j=1}^{n} (\gamma + \frac{1}{\alpha} \ell_j \hat{b}_{ij}^2) \int_{t-\tau_{ij}}^{t} z_j^2(\xi) d\xi$$

where α and γ are some positive constants. The time derivative of the functional along the trajectories of system (10) is obtained as follows

$$
\dot{V}(z(t)) = -2 \sum_{i=1}^{n} c_i z_i^2(t) + 2 \sum_{i=1}^{n} \sum_{j=1}^{n} a_{ij} z_i(t) g_j(z_j(t))
$$
$$
+ 2 \sum_{i=1}^{n} \sum_{j=1}^{n} b_{ij} z_i(t) g_j(z_j(t - \tau_{ij}))
$$
$$
+ \sum_{i=1}^{n} \sum_{j=1}^{n} \frac{1}{\alpha} \ell_j \hat{b}_{ij}^2 z_j^2(t) - \sum_{i=1}^{n} \sum_{j=1}^{n} \frac{1}{\alpha} \ell_j \hat{b}_{ij}^2 z_j^2(t - \tau_{ij})
$$
$$
+ \gamma \sum_{i=1}^{n} \sum_{j=1}^{n} z_j^2(t) - \gamma \sum_{i=1}^{n} \sum_{j=1}^{n} z_j^2(t - \tau_{ij}) \tag{11}
$$

We have

$$-\sum_{i=1}^{n} c_i z_i^2(t) \leq -\sum_{i=1}^{n} \underline{c}_i z_i^2(t) \tag{12}$$

For any positive constant β, we can write

$$2 \sum_{i=1}^{n} \sum_{j=1}^{n} a_{ij} z_i(t) g_j(z_j(t)) \leq \beta z^T(t) z(t) + \frac{1}{\beta} g^T(z(t)) A^T A g(z(t)) \tag{13}$$

From Lemma 1, we obtain:

$$g^T(z(t)) A^T A g(z(t)) \leq \sum_{i=1}^{n} p_i g_i^2(z_i(t))$$

Since $|g_i(z_i(t))| \leq \ell_i |z_i(t)|$, $(i = 1, 2, ..., n)$, (14) can be written as

$$g^T(z(t))A^T Ag(z(t)) \leq \sum_{i=1}^{n} p_i \ell_i^2 z_i^2(t) \tag{14}$$

Using (14) in (13) results in

$$2z^T(t)Ag(z(t)) \leq \beta \sum_{i=1}^{n} z_i^2(t) + \frac{1}{\beta} \sum_{i=1}^{n} p_i \ell_i^2 z_i^2(t) \tag{15}$$

We also note that

$$2\sum_{i=1}^{n}\sum_{j=1}^{n} b_{ij} z_i(t) g_j(z_j(t - \tau_{ij})) \leq \sum_{i=1}^{n}\sum_{j=1}^{n} 2|b_{ij}||z_i(t)||g_j(z_j(t - \tau_{ij}))|$$

$$\leq \sum_{i=1}^{n}\sum_{j=1}^{n} 2\ell_j |b_{ij}||z_i(t)||z_j(t - \tau_{ij})|$$

$$\leq \sum_{i=1}^{n}\sum_{j=1}^{n} \ell_j(\alpha z_i^2(t) + \frac{1}{\alpha}\hat{b}_{ij}^2 z_j^2(t - \tau_{ij})) \tag{16}$$

where α is a positive constant. Using (12), (15) and (16) in (11), we obtain

$$\dot{V}(z(t)) \leq -2\sum_{i=1}^{n} \underline{c}_i z_i^2(t) + \sum_{i=1}^{n} \beta z_i^2(t) + \frac{1}{\beta}\sum_{i=1}^{n} p_i \ell_i^2 z_i^2(t)$$

$$+ \sum_{i=1}^{n}\sum_{j=1}^{n} \ell_j \alpha z_i^2(t) + \sum_{i=1}^{n}\sum_{j=1}^{n} \frac{1}{\alpha}\ell_i \hat{b}_{ji}^2 z_i^2(t)$$

$$+ \gamma \sum_{i=1}^{n}\sum_{j=1}^{n} z_j^2(t) - \gamma \sum_{i=1}^{n}\sum_{j=1}^{n} z_j^2(t - \tau_{ij})$$

which can be written as

$$\dot{V}(z(t)) \leq -\sum_{i=1}^{n}(2\underline{c}_i - \beta - \frac{1}{\beta}p_i\ell_i^2 - \sum_{j=1}^{n}(\alpha\ell_j + \frac{1}{\alpha}\hat{b}_{ji}^2\ell_i))z_i^2(t)$$

$$+ \gamma \sum_{i=1}^{n}\sum_{j=1}^{n} z_j^2(t) - \gamma \sum_{i=1}^{n}\sum_{j=1}^{n} z_j^2(t - \tau_{ij})$$

$$= -\sum_{i=1}^{n} \varepsilon_i z_i^2(t) + \gamma \sum_{i=1}^{n}\sum_{j=1}^{n} z_j^2(t) - \gamma \sum_{i=1}^{n}\sum_{j=1}^{n} z_j^2(t - \tau_{ij})$$

$$\leq -\sum_{i=1}^{n} \varepsilon_m z_i^2(t) + \gamma \sum_{i=1}^{n}\sum_{j=1}^{n} z_j^2(t)$$

$$= -\varepsilon_m ||z(t)||_2^2 + n\gamma ||z(t)||_2^2 = -(\varepsilon_m - n\gamma)||z(t)||_2^2 \tag{17}$$

In (17), $\gamma < \frac{\varepsilon_m}{n}$ implies that $\dot{V}(z(t))$ is negative definite for all $z(t) \neq 0$. Now let $z(t) = 0$. Then, $\dot{V}(z(t))$ is of the form:

$$\dot{V}(z(t)) = -\frac{1}{\alpha} \sum_{i=1}^{n} \sum_{j=1}^{n} \ell_j \hat{b}_{ij}^2 z_j^2(t - \tau_{ij}) - \sum_{i=1}^{n} \sum_{j=1}^{n} \gamma z_j^2(t - \tau_{ij})$$

$$\leq -\sum_{i=1}^{n} \sum_{j=1}^{n} \gamma z_j^2(t - \tau_{ij})$$

in which $\dot{V}(z(t)) < 0$ if there exists at least one nonzero $z_j(t - \tau_{ij})$, implying that $\dot{V}(z(t)) = 0$ if and only if $z(t) = 0$ and $z_j(t - \tau_{ij}) = 0$ for all i, j, and $\dot{V}(z(t)) < 0$ otherwise. Also note that, $V(z(t))$ is radially unbounded since $V(z(t)) \to \infty$ as $\|z(t)\| \to \infty$. Hence, the origin system (10), or equivalently the equilibrium point of system (1) is globally asymptotically stable.

3   Conclusions

By employing Homomorphic mapping theorem and Lyapunov stability theorem, we have derived a new result for the existence, uniqueness and global robust stability of equilibrium point for neural networks with constant multiple time delays with respect to the Lipschitz activation functions. The key contribution of this paper is to establish some new relationships between the upper bound absolute values of the elements of the interconnection matrix, which is given in Lemma 1. The obtained condition is independently of the delay parameters and establishes a new a relationship between the network parameters of the system.

References

1. Bao, G., Wen, S., Zeng, Z.: Robust stability analysis of interval fuzzy Cohen-Grossberg neural networks with piecewise constant argument of generalized type. Neural Netw. **33**, 32–41 (2012)
2. Deng, F., Hua, M., Liu, X., Peng, Y., Fei, J.: Robust delay-dependent exponential stability for uncertain stochastic neural networks with mixed delays. Neurocomputing **74**(10), 1503–1509 (2011)
3. Arik, S.: An improved robust stability result for uncertain neural networks with multiple time delays. Neural Netw. **54**, 1–10 (2014)
4. Faydasicok, O., Arik, S.: A new upper bound for the norm of interval matrices with application to robust stability analysis of delayed neural networks. Neural Netw. **44**, 64–71 (2013)
5. Guo, Z., Huang, L.: LMI conditions for global robust stability of delayed neural networks with discontinuous neuron activations. Appl. Math. Comput. **215**(3), 889–900 (2009)
6. Huang, T.: Robust stability of delayed fuzzy Cohen-Grossberg neural networks. Comput. Math. Appl. **61**(8), 2247–2250 (2011)
7. Kao, Y.K., Guo, J.F., Wang, C.H., Sun, X.Q.: Delay-dependent robust exponential stability of Markovian jumping reaction-diffusion Cohen-Grossberg neural networks with mixed delays. J. Franklin Inst. **349**(6), 1972–1988 (2012)

8. Kwon, O.M., Park, J.H.: New delay-dependent robust stability criterion for uncertain neural networks with time-varying delays. Appl. Math. Comput. **205**(1), 417–427 (2008)
9. Liao, X.F., Wong, K.W., Wu, Z., Chen, G.: Novel robust stability for interval-delayed Hopfield neural networks. IEEE Trans. Circuits Syst. I **48**(11), 1355–1359 (2001)
10. Luo, M., Zhong, S., Wang, R., Kang, W.: Robust stability analysis for discrete-time stochastic neural networks systems with time-varying delays. Appl. Math. Comput. **209**(2), 305–313 (2009)
11. Pan, L., Cao, J.: Robust stability for uncertain stochastic neural network with delay and impulses. Neurocomputing **94**(1), 102–110 (2012)
12. Shen, T., Zhang, Y.: Improved global robust stability criteria for delayed neural networks. IEEE Trans. Circuits Syst. II: Express Briefs **54**(8), 715–719 (2007)
13. Wang, Z., Liu, Y., Liu, X., Shi, Y.: Robust state estimation for discrete-time stochastic neural networks with probabilistic measurement delays. Neurocomputing **74**(1–3), 256–264 (2010)
14. Wang, Z., Zhang, H., Yu, W.: Robust exponential stability analysis of neural networks with multiple time delays. Neurocomputing **70**(1315), 2534–2543 (2007)
15. Yang, R., Gao, H., Shi, P.: Novel robust stability criteria for stochastic Hopfield neural networks with time delays. IEEE Trans. Syst. Man Cybern. Part B: Cybern. **39**(2), 467–474 (2009)
16. Zhang, H., Wang, Z., Liu, D.: Robust stability analysis for interval Cohen-Grossberg neural networks with unknown time-varying delays. IEEE Trans. Neural Netw. **19**(11), 1942–1955 (2008)
17. Zhang, Z., Zhou, D.: Global robust exponential stability for second-order Cohen-Grossberg neural networks with multiple delays. Neurocomputing **73**(13), 213–218 (2009)
18. Zheng, M., Fei, M., Li, Y.: Improved stability criteria for uncertain delayed neural networks. Neurocomputing **98**(3), 34–39 (2012)
19. Zhu, S., Shen, Y.: Robustness analysis for connection weight matrices of global exponential stability of stochastic recurrent neural networks. Neural Netw. **38**, 17–22 (2013)

Robust L_2E Parameter Estimation of Gaussian Mixture Models: Comparison with Expectation Maximization

Umashanger Thayasivam[1], Chinthaka Kuruwita[2],
and Ravi P. Ramachandran[1(✉)]

[1] Rowan University, Glassboro, NJ, USA
{thayasivam,ravi}@rowan.edu
[2] Hamilton College, Clinton, NY, USA
ckuruwit@hamilton.edu

Abstract. The purpose of this paper is to discuss the use of L_2E estimation that minimizes integrated square distance as a practical robust estimation tool for unsupervised clustering. Comparisons to the expectation maximization (EM) algorithm are made. The L_2E approach for mixture models is particularly useful in the study of big data sets and especially those with a consistent numbers of outliers. The focus is on the comparison of L_2E and EM for parameter estimation of Gaussian Mixture Models. Simulation examples show that the L_2E approach is more robust than EM when there is noise in the data (particularly outliers) and for the case when the underlying probability density function of the data does not match a mixture of Gaussians.

Keywords: Robust L_2E estimation · Gaussian mixture model · Expectation maximization · Unsupervised learning · Big data

1 Introduction and Motivation

Mixture models and in particular Gaussian Mixture Models (GMM), are commonly used for density estimation and classification. In this era of Big Data and everyday, the data is highly complex and enormous in size. Mixture models offer a powerful and flexible way to represent the data. A comprehensive discussion on mixture models can be found in [1,2].

When the number of mixture components is known and the component densities are assumed to belong to a specified parametric family, the popular Expectation Maximization (EM) algorithm [3] based on Maximum Likelihood Estimation (MLE) is often used to estimate the GMM parameters. However, when there is a small perturbation in one of the component densities, MLE becomes significantly biased and very sensitive to outliers [4]. Furthermore, when the data is not Gaussian, the EM method may not cluster a set of data points to a Gaussian with a meaningful mean vector and covariance matrix. The EM based approach

© Springer International Publishing Switzerland 2015
S. Arik et al. (Eds.): ICONIP 2015, Part III, LNCS 9491, pp. 281–288, 2015.
DOI: 10.1007/978-3-319-26555-1_32

is not robust when the underlying probability density function of the data does not match a mixture of Gaussians (known as a data/model mismatch).

To overcome this limitation, Scott [5–8] introduced an alternative minimum distance estimation method based on the integrated squared error criterion (termed L_2E) which avoids the use of nonparametric kernel density estimators. The L_2E approach is a special case of a general method introduced in [9] that is based on a whole continuum of divergence estimators that begin with MLE and interpolate to the L_2E estimator. Markatou [10] used the weighted likelihood estimation approach to address the effects of data/model mismatch on parameter estimates.

In this paper, the focus is on the L_2E as an alternative to the EM for parameter estimation of models with a known finite number of mixtures. A discussion of the EM and L_2E approaches are given. Simulation results specific to GMM are shown to depict the robustness property of the L_2E method with respect to noise in the data (particularly outliers) and data/model mismatch [11–13].

The basic notation in this paper is as follows. Let $f_{\boldsymbol{\theta}_m}(x)$ denote a general mixture probability density function with m components as given by

$$f_{\boldsymbol{\theta}_m}(\boldsymbol{x}) = \sum_{i=1}^{m} \pi_i f(\boldsymbol{x}|\boldsymbol{\phi}_i) \tag{1}$$

where $\boldsymbol{\theta}_m = (\pi_1, \ldots, \pi_{m-1}, \pi_m, \boldsymbol{\phi}_1{}^T, \ldots, \boldsymbol{\phi}_m{}^T)^T$, the weights $\pi_i > 0$, $\sum_{i=1}^{m} \pi_i = 1$ and $f(\boldsymbol{x}|\boldsymbol{\phi}_i)$ is a probability density function with parameter vector $\boldsymbol{\phi}_i$. In theory, the $f(\boldsymbol{x}|\boldsymbol{\phi}_i)$ could be any parametric density, although in practice they are often from the same parametric family (usually Gaussian).

2   EM Algorithm

The Expectation-Maximization (EM) algorithm [3] is broadly based on the iterative computation of MLE. The EM method alternates between two steps:

1. Expectation (E) step: Computes an expectation of the likelihood by including the latent variables as if they were observed and a
2. Maximization (M) step: computes the maximum likelihood estimates of the parameters by maximizing the expected likelihood found in the E step.

The parameters found in the M step are then used to begin another E step and the process is repeated.

For finite mixture models, the observed data samples $\boldsymbol{X} = \{\boldsymbol{x}_1, \cdots, \boldsymbol{x}_n\}$ are viewed as incomplete. The complete data is obtained as $\boldsymbol{Z} = \{\boldsymbol{x}_i, \boldsymbol{y}_i\}$ for $i = 1$ to n where $\boldsymbol{y}_i = (\boldsymbol{y}_{1i}, \cdots, \boldsymbol{y}_{mi})^T$ is a latent (unobserved or missing) indicator vector with $\boldsymbol{y}_{ij} = 1$ if $\boldsymbol{x}_i$ is from the mixture component j and zero otherwise. The log-likelihood of $\boldsymbol{Z}$ is defined by

$$L(\boldsymbol{\theta}_m|\boldsymbol{Z}) = \sum_{i=1}^{n} \sum_{j=1}^{m} y_{ij} \log y_{ij} \log[\pi_j f(\boldsymbol{x}_i|\boldsymbol{\phi}_j)] \tag{2}$$

The EM algorithm obtains a sequence of estimates $\boldsymbol{\theta}^{(t)}, t = 0, 1, \cdots$ by alternating the E-Step and the M-Step until some convergence criterion is met.

1. **E-Step:** Calculate the Q function, the conditional expectation of the complete log-likelihood, given $\boldsymbol{X}$ and the current estimate $\boldsymbol{\theta}^{(t)}$.
2. **M-Step:** Update the estimate of the parameters by maximizing the Q function.

In the case of GMM, maximizing Q provides an explicit solution. In most instances, EM has the advantages of reliable global convergence, low cost per iteration, economy of storage, ease of programming and heuristic appeal. However, its convergence can be very slow in simple problems which are often encountered in practice. Also, when there is a small perturbation in one of the component densities due to noise in the data, the MLE estimates become highly unstable due to the lack of robustness to outliers. For the case of GMM [14], this can be seen easily as maximization of the likelihood function under an assumed Gaussian distribution is equivalent to finding the least-squares solution, whose lack of robustness is well known. As a robust alternative we discuss an approach based on the minimization of the integrated square distance, namely L_2E.

3   Robust L_2E Estimator

The integrated squared distance has been used as the goodness-of-fit criterion in nonparametric density estimation for a long time. In the classic papers of Scott [6,7], an alternative minimum distance estimation method based on the integrated squared error criterion, termed L_2E, was introduced and has the following attributes.

1. The use of nonparametric kernel density estimators is avoided.
2. The L_2E is especially suited for parameter-rich models such as mixture models.
3. The genesis of Scott the L_2E approach, which can be traced to the pioneering work of Rudemo [15] and Bowman [16], is computationally feasible and leads to robust estimators.
4. The L_2E is a special class of robust estimators like the median-based estimators, which sacrifice some asymptotic efficiency for substantial computational benefits in difficult estimation problems.
5. The L_2E estimator performs much better than other robust estimators such as minimum Hellinger estimates (MHD) under severe data contamination.

The L_2E estimator belongs to the family of minimum density power divergence ($MDPD$) estimators introduced in [9] with the tuning parameter $\alpha = 1$. The tuning parameter α in an $MDPD$ estimator controls the trade-off between robustness and efficiency. It is also shown that the robustness of the L_2E estimator is achieved at a fairly stiff price in asymptotic efficiency [9]. For the normal, exponential and Poisson distributions with small values of $\alpha \leq 0.10$, the $MDPD$ has strong robustness properties and retains high asymptotic relative efficiency

(ARE) with respect to MLE. However, within the family of density-based power divergence measures, the L_2E approach has the distinct advantage that a key integral can be computed in closed form, especially for Gaussian mixtures.

3.1 L_2E Algorithm

Given the true probability density $g(\boldsymbol{x})$ and the finite mixture with m components, $f_{\boldsymbol{\theta}_m}(\boldsymbol{x})$, consider the L_2 distance between $f_{\boldsymbol{\theta}_m}$ and $g(\boldsymbol{x})$ as given by

$$L_2(f_{\boldsymbol{\theta}_m}, g(\boldsymbol{x})) = \int_{-\infty}^{\infty} [f_{\boldsymbol{\theta}_m}(\boldsymbol{x}) - g(\boldsymbol{x})]^2 d\boldsymbol{x}. \tag{3}$$

The aim is to derive an estimate of $\boldsymbol{\theta}_m$ that minimizes the L_2 distance [5–7, 11–13]. Expanding Eq. (3) gives

$$L_2(f_{\boldsymbol{\theta}_m}, g(\boldsymbol{x})) = \int_{-\infty}^{\infty} f_{\boldsymbol{\theta}_m}^2(\boldsymbol{x}) d\boldsymbol{x} - 2 \int_{-\infty}^{\infty} f_{\boldsymbol{\theta}_m}(\boldsymbol{x}) g(\boldsymbol{x}) d\boldsymbol{x}$$
$$+ \int_{-\infty}^{\infty} g(\boldsymbol{x})^2 d\boldsymbol{x} \tag{4}$$

where the last integral is a constant with respect to $\boldsymbol{\theta}_m$ and therefore, may be ignored for the minimization. The first integral in Eq. (4) is often available as a closed form expression that, for Gaussian mixtures, may be evaluated for any specified value of $\boldsymbol{\theta}_m$ as shown later in Eq. (7). The second integral in Eq. (4) is simply the average height of the density estimate, which may be estimated as $-2n^{-1} \sum_{i=1}^{n} f_{\boldsymbol{\theta}_m}(\boldsymbol{X}_i)$ where $\boldsymbol{X}_i$ is a sample observation. Based on the above analysis, the L_2E estimator of $\boldsymbol{\theta}_m$ is given by

$$\hat{\boldsymbol{\theta}}_m^{L_2E} = \arg\min_{\boldsymbol{\theta}_m} \left[\int_{-\infty}^{\infty} f_{\boldsymbol{\theta}_m}^2(\boldsymbol{x}) d\boldsymbol{x} - 2n^{-1} \sum_{i=1}^{n} f_{\boldsymbol{\theta}_m}(\boldsymbol{X}_i) \right], \tag{5}$$

3.2 GMM Models

For multivariate Gaussian mixtures,

$$f(\boldsymbol{x}|\boldsymbol{\phi}_i) = \phi(\boldsymbol{x}|\ \boldsymbol{\mu}_i, \Sigma_i) \tag{6}$$

where $\boldsymbol{\mu}_i$ is the mean vector and Σ_i is the covariance matrix for component i. In this case, the problem reduces to finding the L_2E estimator for a Gaussian Mixture Model (GMM). Now, the first integral in Eq. (4) reduces to

$$\int_{-\infty}^{\infty} f_{\boldsymbol{\theta}_m}^2(\boldsymbol{x}) d\boldsymbol{x} = \sum_{k=1}^{m} \sum_{l=1}^{m} \pi_k \pi_l \ \phi(\boldsymbol{\mu}_k - \boldsymbol{\mu}_l|\ 0, \Sigma_k + \Sigma_l), \tag{7}$$

thereby making Eq. (4) tractable for minimization and significantly reducing the computations involved in getting the L_2E estimator. Since this is a computationally feasible closed-form expression, estimation of the GMM parameters by the L_2E procedure may be performed by any standard nonlinear optimization algorithm [5,6,11–13]. In this work, we used the 'nlminb' nonlinear minimization routine in [17].

4   Experimental Results

4.1   Performance Due to Data Contamination (Outliers)

In this section, simulations using EM and L_2E parameter estimates are compared when there is no data contamination and when there is (with and without the presence of outliers/noise).

Gaussian Mixture Model with No Outliers: A GMM model $f(x)$ with two components, each being a univariate Gaussian density $\phi(x)$ is simulated as given by

$$f(x) = 0.75\phi(x|\ \mu_1 = 0, \sigma_1^2 = 1) + 0.25\phi(x|\ \mu_2 = 1, \sigma_2^2 = 1). \tag{8}$$

The variable μ denotes the mean and the variable σ^2 denotes the variance. A total of 10000 sample points from the above Gaussian mixture (see Eq. (8)) are generated and parameter estimation is performed. A total of 100 Monte Carlo simulations are performed to evaluate consistency and efficiency.

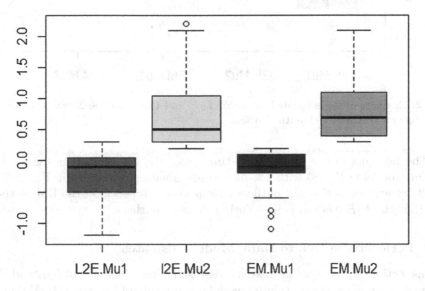

Fig. 1. Boxplots of the estimated mean for L_2E and EM from 100 Monte Carlo Simulations of a GMM Model With No Outliers

The boxplots of the parameter estimates of the component means for the mixture model in Eq. (8) with no data contamination are shown in Fig. 1. The results clearly show that both solutions are comparable and close to the true estimates. Note that the average of the 100 Monte Carlo estimates of the L_2E and EM means are close to the true value.

Gaussian Mixture Model with Outliers: The second simulation extends our study by adding outliers to illustrate the robustness property of L_2E against outliers. In this case, 9900 sample points from the above Gaussian mixture in Eq. (8) are contaminated by adding 100 sample points (outliers) simulated from $\phi(x|\ \mu = 5, \sigma^2 = 1)$. Once again, 100 Monte Carlo simulations are performed to evaluate the performance of L_2E and EM for consistency and efficiency.

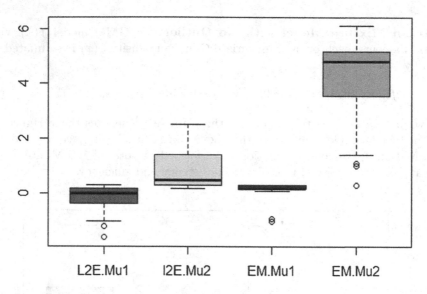

Fig. 2. Boxplots of the estimated mean for L_2E and EM from 100 Monte Carlo Simulations of a GMM Model With Outliers

The boxplots of the parameter estimates of the component means for the mixture model in Eq. (8) with 1% data contamination are shown in Fig. 2. The results clearly show that the outliers have a great influence on the EM method and that the L_2E method is inherently robust to outliers.

4.2 Performance Due to Data/Model Mismatch

In this section, data/model mismatch is assessed. The robustness of L_2E and EM is investigated when the postulated model is a mixture of Gaussians (GMM) but the data are generated from a mixture with symmetric departure from component normality. The setup as described in [12,18] is considered for the parameter estimation. More specifically, for the simulation study, a mixture with two components given by

$$f_{\theta_2}(x) = \pi f_1(x) + (1 - \pi)f_2(x), \tag{9}$$

is considered. Note that f_1 is the density associated with a random variable $X_1 = aY_1$ ($a = 1$ chosen for the simulation) and Y_1 is a Student's $t(df)$-random variable with a degree of freedom $df = 1$. Also, f_2 is the density associated with a random variable $X_2 = Y_2 + b$ ($b = 2$ chosen for the simulation) and Y_2 is a Student's $t(df)$-random variable with degrees of freedom $df = 4$. A total of 100 data points were generated and 50 Monte Carlo simulations were conducted to evaluate the performance of L_2E and EM for consistency and efficiency by calculating the Bias and Mean Square Error (MSE).

Suppose $T(X)$ is an estimate of θ. The Bias and MSE of T are defined as

$$Bias(\theta) = E_\theta T - \theta \tag{10}$$
$$MSE(\theta) = E_\theta (T - \theta)^2 = Var_\theta(T) + Bias^2(\theta) \tag{11}$$

Note that the general shapes of such a two-component postulated (Gaussian mixture) model and a two-component t-mixture model from which the data are generated are different and further, the component densities in the sampling model have a much heavier tail than those in the postulated (Gaussian) mixture model. Table 1 depicts the bias and the mean square error for the mean estimates provided by the $L_2(E)$ and EM algorithms. The results show that the L_2E is more robust than the EM approach with respect to data/model mismatch.

Table 1. Simulation results for data/model mismatch

Estimation method	Component 1		Component 2	
	Bias	MSE	Bias	MSE
L_2E	0.4	1.57	0.11	0.84
EM	−0.43	9.66	1.15	16.55

5 Summary and Conclusions

The L_2E estimation technique can be easily constructed and applied to GMM and is a viable alternative to EM. Simulation studies revealed that the L_2E mean estimates are robust to both outliers and data/model mismatch. The competitive performance of L_2E make it stand out as an attractive alternative to EM for practical applications.

Acknowledgment. This work was supported by the National Science Foundation through Grant DUE-1122296.

References

1. Titterington, D.M., Smith, A.F.M., Markov, U.E.: Statistical Analysis of Finite Mixture Distributions. Wiley, New York (1985)
2. McLachlan, G.J., Peel, D.: Finite Mixture Models. Wiley, New York (2000)

3. Dempster, A.P., Laird, N.M., Rubin, D.B.: Maximum-likelihood from incomplete data via the EM algorithm. J. Roy. Stat. Soc. Ser. B **39**, 1–38 (1977)
4. Aitkin, M., Wilson, G.T.: Mixture models, outliers, and the EM algorithm. Technometrics **22**, 325–331 (1980)
5. Scott, D.W.: On fitting and adapting of density estimates. Comput. Sci. Stat. **30**, 124–133 (1998). (Weisberg, S., ed.)
6. Scott, D.W.: Remarks on fitting and interpreting mixture models. Comput. Sci. Stat. **31**, 104–109 (1999). (Berk, K., Pourahmadi, M., eds.)
7. Scott, D.W.: Parametric statistical modeling by minimum integrated square error. Technometrics **43**, 274–285 (2001)
8. Scott, D.W.: Outlier detection and clustering by partial mixture modeling. In: COMPSTAT Symposium. Physica-Verlag/Springer (2004)
9. Basu, A., Harris, I.R., Hjort, H.L., Jones, M.C.: Robust and efficient estimation by minimizing a density power divergence. Biometrika **85**, 549–560 (1998)
10. Markatou, M., Basu, A., Lindsay, B.G.: Weighted likelihood estimating equations with a bootstrap root search. J. Am. Stat. Assoc. **93**, 740–750 (1998)
11. Thayasivam, U., Sriram, T.N.: L_2E estimation for mixture complexity for count data. Comput. Stat. Data Anal. **53**, 4243–4254 (2009)
12. Thayasivam, U., Sriram, T.N., Lee, J.: Simultaneous robust estimation in finite mixtures: the continuous case. J. Indian Stat. Assoc. **50**, 277–295 (2012)
13. Thayasivam, U., Shetty, S., Kuruwita, C., Ramachandran, R.P.: Detection of anomalies in network traffic using L2E for accurate speaker recognition. In: 55th International Midwest Symposium on Circuits & Systems, Boise, pp. 884–887 (2012)
14. Kai, Y., Dang, X., Bart, H., Chen, Y.: Robust model-based learning via Spatial-EM algorithm. IEEE Trans. Knowl. Data Eng. **27**, 1670–1682 (2015)
15. Rudemo, M.: Empirical choice of histograms and kernel density estimators. Scand. J. Statist. **9**, 65–78 (1982)
16. Bowman, A.W.: An alternative method of cross-validation for the smoothing of density estimates. Biometrika **71**, 353–360 (1984)
17. R: A Language and Environment for Statistical Computing, R Development Core Team, R Foundation for Statistical Computing, Vienna, Austria, (2011). http://www.R-project.org/
18. Woodward, W.A., Parr, W.C., Schucany, W.R., Lindsay, H.: A comparison of minimum distance and maximum likelihood estimation of a mixture proportion. J. Am. Stat. Assoc. **79**, 590–598 (1984)

Real-Time Robust Model Predictive Control of Mobile Robots Based on Recurrent Neural Networks

Shuzhan Bi, Guangfei Zhang, Xijun Xue, and Zheng Yan[⊠]

Shannon Lab, Huawei Technologies Co., Ltd., Shenzhen, Guangdong, China
yanzheng@huawei.com

Abstract. This paper presents a novel model predictive control (MPC) approach to tracking control of mobile robots based on recurrent neural networks (RNNs). The tracking control problem is firstly formulated as a sequential dynamic optimization problem in framework of MPC. Then a novel neurodynamic approach is developed for computing the optimal control signals in real time, where multiple RNNs are applied in a collective fashion. The proposed approach enables MPC of mobile robots to be synthesized in real time. Simulation results are provided to substantiate the effectiveness of the proposed approach.

1 Introduction

Tracking control of mobile robots has attracted much attention in the past two decades due to its widespread applications [1]. Tracking control aims to make the position and orientation of a mobile robot converge to references. Challenges of tracking control arise from several factors such as nonlinearity, nonholonomic constraints, and random disturbances. Moreover, most existing results are based on kinematic models where the trajectory is controlled by linear and angular velocities. However, kinematic control implies a strong assumption on perfect velocity tracking [2]. To make the control scheme more practical, it is expedient to directly design torques as control inputs.

Model predictive control (MPC) is an optimization based control strategy where control signals are generated by iteratively minimizing a performance index subject to a prediction model over a moving time window [3]. Specifically, MPC has several distinctive features for motion control of mobile robots. For example, MPC with implicit discontinuous feedback enables it to deal with nonholonomic constraints [4], and MPC can explicitly handle input and state constraints. MPC also shows advantages in achieving robustness against disturbances [7].

Principles of MPC have been intensively investigated and many theoretic results are available in the literature. However, the computational issues were somewhat overlooked. Many MPC schemes simply assume that an efficient optimization solver is available off the shelf. In practice, nevertheless, this assumption can hardly be valid due to two challenges. The first is the nonconvexity of

© Springer International Publishing Switzerland 2015
S. Arik et al. (Eds.): ICONIP 2015, Part III, LNCS 9491, pp. 289–296, 2015.
DOI: 10.1007/978-3-319-26555-1_33

the resulting optimization problem, and the second one is the requirement to computing optimal control signals in real time.

Since the pioneering work of Hopfield network [8], neurodynamic optimization based on recurrent neural networks (RNNs) emerges as a promising and competent approach to real time optimization, where RNNs are designed as goal seeking computational models. Recent advances in neuromorphic engineering offer an availability of hardware/firmware implementation of RNNs [9], which can make the computation truly parallel and distributed. Several studies on RNNs-based MPC have been carried out [10]. One limitation of previoous results is that they require a substantial problem reformulation so as to cast the MPC problems as generalized convex optimization. The reformulation naturally results in suboptimal control signals.

Inspired by the principles of swarm intelligence in humans, Yan and Wang proposed a collective neurodynamic approach to nonconvex optimization problems with box constraints, where multiple RNNs are exploited collaboratively to search for the global optima [11]. In this paper, we extend the collective neurodynamic approach for solving predictive tracking control of mobile robots by designing a new RNN model which is capable of dealing with nonlinear inequality constraints.

The rest of this paper is organized as follows. In Sect. 2, problem formulations are discussed. In Sect. 3, the neurodynamic optimization approach is delineated. In Sect. 4, simulation results are provided. Finally, Sect. 5 concludes this paper.

2   Problem Formulation

2.1   Dymamic Model of Mobile Robots

Consider a mobile robot with two driving wheels. Denote v_l and v_r as the left wheel speed and the right wheel speed, respectively. The linear velocity and angular velocity is computed as $v = (v_l + v_r)/2$ and $\omega = (v_l - v_r)/b$ where b is the half distance between two wheels. Denote (x, y) and ψ as the position and orientation, respectively. Its kinematic model can be obtained as

$$\dot{\theta} = \begin{bmatrix} \dot{x} \\ \dot{y} \\ \dot{\psi} \end{bmatrix} = \begin{bmatrix} v\cos\psi \\ v\sin\psi \\ \omega \end{bmatrix} = \begin{bmatrix} \cos\psi & 0 \\ \sin\psi & 0 \\ 0 & 1 \end{bmatrix} u, \tag{1}$$

where $\theta = [x; y; \psi]$ is the state vector and $u = [v; \omega]$ is the control input vector. Kinematic control of a mobile robot aims at forcing the state vector θ to track a reference trajectory using a proper design of the input vector u. In practical applications, θ and u are often subject to certain constraints due to safety requirements, performance specifications, and limitations of actuators. These physical constraints can be mathematically denoted as

$$\theta_{\min} \leq \theta \leq \theta_{max}, \; u_{\min} \leq u \leq u_{max}. \tag{2}$$

Denote θ_r and u_r as the reference state and input, and assume that the reference is gemerated by a virtual mobile robot, an error kinematic model can be obtained as follows [1]

$$\dot{x}_e = v_r \cos \psi_e + y_e \omega - v$$
$$\dot{y}_e = v_r \sin \psi_e - x_e \omega$$
$$\dot{\psi}_e = -\omega + \omega_r, \tag{3}$$

where $\theta_e = [x_e; y_e; \psi_e]$ denotes error variables. Based on the kinematic model of error variables, motion control of a mobile robot can be reformulated as a regulation problem where the objective is to steer the state variable θ_e to the origin. However, ω and v are not actual control input variables in practice. In view of the discussions in [12], the relation between velocities and toqrues can be described as follows:

$$\dot{v} = \frac{1}{mr}\tau_r + \frac{1}{mr}\tau_l, \quad \dot{\omega} = \frac{b}{rI_z}\tau_r - \frac{b}{rI_z}\tau_l \tag{4}$$

where m is the mass of the robot, r and the radius of the wheels, and I_z is the moment of inertial of the robot.

In view of (3) and (4), denote $q = [x_e; y_e; \psi_e; v; \omega]$ as a state vector, the dynamic model of the mobile robot can be described as follows:

$$\begin{bmatrix} \dot{x}_e \\ \dot{y}_e \\ \dot{\psi}_e \\ \dot{v} \\ \dot{\omega} \end{bmatrix} = \begin{bmatrix} v_r \cos \psi_e + y_e \omega v \\ v_r \sin \psi_e - x_e \omega \\ -\omega + \omega_r \\ 0 \\ 0 \end{bmatrix} + \begin{bmatrix} 0 & 0 \\ 0 & 0 \\ 0 & 0 \\ 1/mr & 1/mr \\ b/rI_z & -b/rI_z \end{bmatrix} \tau \tag{5}$$

where $\tau = [\tau_r; \tau_l]$ is the control vector to be computed.

2.2 Model Predictive Control

As the sensory and actuary devices on a mobile robot are commonly operate in discrete time, it is desirable to embed a discrete time control system. Based on Euler discretization, (5) is converted to a discrete model as follows:

$$q(k+1) = f(q(k)) + g(q(k))\tau(k) + d(k) \tag{6}$$

where

$$f(q(k)) = \begin{bmatrix} x_e(k) + t_s v_r(k) \cos \psi_e(k) + t_s y_e(k)\omega(k) - t_s v(k) \\ y_e(k) + t_s v_r(k) \sin \psi_e(k) - t_s x_e(k)\omega(k) \\ \psi_e(k) - t_s \omega(k) + t_s \omega_r(k) \\ v(k) \\ \omega(k) \end{bmatrix},$$

t_s is the sampling interval, and $d(k)$ denotes random disturbances. It is assumed that the bound of the disturbances is known in prior, i.e., $d_{\min} \leq d(k)$

$\leq d_{\max} \forall k \geq 0$. In addition, in view of the constraint (2), the system model (6) is required to fulfill the following constraints

$$q_{\min} \leq q(k) \leq q_{\max}, \forall k \geq 0, \quad \tau_{\min} \leq \tau(k) \leq \tau_{\max}, \forall k \geq 0. \tag{7}$$

To explicitly deal with the random disturbances, the min-max strategy is exploited to derive a robust MPC where the worst-case predicted performance is minimized:

$$\min_{\tau(k)} \max_{d(k)} \quad J = \sum_{j=1}^{N} \|r(k+j) - q(k+j)\|_{Q_j}^2 + \sum_{j=0}^{N_u-1} \|\tau(k+j)\|_{R_j}^2 + F(q(k+N))$$

subject to

$$\tau_{\min} \leq \tau(k+j) \leq \tau_{\max}, \ j = 0, 1, \ldots, N_u - 1;$$
$$d_{\min} \leq d(k+j) \leq d_{\max}, \ j = 0, 1, \ldots, N - 1;$$
$$q_{\min} \leq q(k+j) \leq q_{\max}, \ j = 1, 2, \ldots, N;$$
$$q(k+N) \in \mathcal{Q}; \tag{8}$$

where $\tau(k+j)$ denotes the predicted input vector, $q(k+j)$ denotes the predicted state vector, $r(k+j)$ denotes a reference defined as $r(k) = [0; 0; 0; v_r(k); \omega_r(k)]$, N and N_u are respectively prediction horizon $(1 \leq N)$ and control horizon $(0 < N_u \leq N)$, $\| \cdot \|$ denotes the Euclidean norm, Q and R are positive definite weight matrices with compatible dimensions, $q(k + N)$ denotes the predicted terminal state within the prediction horizon, F is a terminal cost, and $\mathcal{Q}$ is a terminal constraint. The predicted values, even in the nominal undisturbed cases, need not be equal to the actual closed-loop values.

In light of the formulation (8), a sufficient condition for the input-to-state stability of the control system (6) can be stated as

1. There exists a local control law $\kappa(q)$ such that $f(q, \kappa(q), d) \in \mathcal{Q}, \forall q \in \mathcal{Q}, d_{\min} \leq d \leq d_{\max}$;
2. There exist $\mathcal{K}_\infty$ functions a_F and b_F such that $a_F(|q|) \leq F(q) \leq b_F(|q|), \forall q \in \mathcal{Q}$;
3. $F(f(q, \kappa(q), d)) - 2F(q) \leq J(q)$.

In this paper, $F(q)$ is defined as a quadratic function, i.e., $F(q) = q^T P q$ where P is a positive definite matrix obtained by solving a discrete-time $\mathcal{H}_\infty$ algebraic Riccati equation offline following the procedures presented in [13].

Define the following vectors

$$\bar{\tau}(k) = [\tau(k); \ldots; \tau(k + N_u - 1)]^T, \bar{d}(k) = [d(k); \ldots; d(k + N - 1)]^T,$$
$$\bar{q}(k) = [q(k+1); \ldots; q(k+N)]^T, \bar{r}(k) = [r(k+1); \ldots; r(k+N)]^T.$$

In view of (6), a vector-valued function $g_q(\cdot)$ can be obtained as the nonlinear mapping between $\bar{q}(k)$ and $\bar{\tau}(k)$, i.e., $\bar{q}(k) = g_q(\bar{\tau}(k), \bar{d}(k))$. Correspondingly, the optimization problem (8) can be written in a compact form as follows:

$$\min_{\bar{\tau}(k)} \max_{\bar{d}(k)} J(\bar{\tau}(k), \bar{d}(k)) = (\bar{r}(k) - g_q)^T Q(\bar{r}(k) - g_q) + \bar{\tau}^T(k) R \bar{\tau}^T(k) + \tilde{g}_q^T P \tilde{g}_q$$

$$\text{s.t.} \bar{\tau}_{min} \leq \bar{\tau}(k) \leq \bar{\tau}_{max}, \ \bar{d}_{min} \leq \bar{d}(k) \leq \bar{d}_{max}, \ \bar{q}_{min} \leq g_q(\bar{\tau}(k)) \leq \bar{q}_{max} \quad (9)$$

By Defining an instrumental variable $z(k) = \max_{\bar{d}(k)} J(\bar{\tau}(k), \bar{d}(k))$, (9) can be equally converted a minimization problem as follows:

$$\min \quad V(\bar{\tau}(k), \bar{d}(k), z(k)) = z(k)$$

$$\text{s.t.} \quad J(\bar{\tau}(k), \bar{d}(k)) \leq z(k), \ \bar{q}_{min} \leq g_q(\bar{\tau}(k)) \leq \bar{q}_{max}$$

$$\bar{\tau}_{min} \leq \bar{\tau}(k) \leq \bar{\tau}_{max}, \ \bar{d}_{min} \leq \bar{d}(k) \leq \bar{d}_{max}. \quad (10)$$

During every sampling interval, the solution to the optimization problem (10) offers a sequence of optimal control signals, and the first control in the sequence will be applied. It is worth noting that (10) is nonconvex due to the nonlinearity of (6).

3 Neurodynamic Optimization

For a nonconvex optimization problem, it is well known that the KKT equations provide the first-order necessary conditions in order for a feasible solution to be optimal. In this section, we first apply a RNN to search for KKT points of the optimization problem (10). As the neurodynamic optimization method proposed in this section applies to any time instant k, we will denote a variable $\xi(k)$ as ξ for simplicity. By letting $\alpha = [z; \bar{\tau}; \bar{d}]$, (10) can be rewritten in a compact form as follows:

$$\min_{\alpha} \quad V(\alpha) = z$$

$$\text{s.t.} \quad c(\alpha) \leq 0 \quad (11)$$

where $c(\alpha)$ denotes the set of inequalilty constraints. The KKT conditions of (11) are stated as follows

$$c(\alpha) \leq 0, \ \mu \geq 0, \ \mu c(\alpha) = 0, \ \nabla V(\alpha) + \nabla c(\alpha)\mu = 0 \quad (12)$$

where μ is a vector of Lagrange multiples, $\nabla V(\alpha)$ is the gradient of V, $\nabla c(\alpha) = (\nabla c_1(\alpha), ..., \nabla c_m(\alpha))$. Based on the projection theorem, the KKT conditions defined in (12) can be equally written as two projection equations

$$\nabla V(\alpha) + \nabla c(\alpha)\mu = 0, \ \mu - (\mu + c(\alpha))^+ = 0 \quad (13)$$

where $\xi^+ = \max(0, \xi)$. To solve the KKT conditions (13), a RNN model can be designed as follows

$$\epsilon \frac{d}{dt} \begin{pmatrix} \alpha \\ \mu \end{pmatrix} = \begin{pmatrix} -\nabla V(\alpha) + \nabla c(\alpha)(\mu + c(\alpha))^+ \\ -\mu + (\mu + c(\alpha))^+ \end{pmatrix} \quad (14)$$

The neural network (14) has a two-layer structure. By comparing (14) with (13), it is straightforward to show that the equilibrium states of the RNN are in one-to-one correspondence with the KKT points of the problem (11).

To further improve the optimality of the canditate solution obtained by (14), we propose apply multiple RNNs in a collective manner. First, a number of RNNs with diverse initial conditions are initialized. Second, each RNN performs constrained local search according to its own dynamics. Next, each RNN evaluates the contemporary solution quality and exchanges information with its peers. According to the exchanged information, each RNN resets its neuronal state and repeats the constrained local search again. These steps are recursively repeated until a termination criterion is met. Let $N_r \geq 2$ be the number of RNNs, N_c be the maximum number of iterations, $\lambda = [\alpha; \mu]$ be the neural state of the ith neural network, and $\bar{\lambda}_i$ be the corresponding equilibrium state. We proceed to describe the procedures of the proposed optimization method.

1. Initialize λ_i, $i = 1, \ldots, N_r$, with a uniformly distributiton in the search space;
2. Perform local search guided by the dynamical Eq. (14) to obtain $\bar{\lambda}_i$;
3. Evaluate the cost function and update the contemporary best solution $\bar{\lambda}^*$ where $V(\bar{\lambda}^*) \leq V(\bar{\lambda}_i), \forall i$;
4. Compute mutual distances between all equilibrium states $d_{ij} = ||\bar{\lambda}_i - \bar{\lambda}_j||$;
5. Select two most distant neural states $\bar{\lambda}_a$, and $\bar{\lambda}_b$;
6. Randomly generate a positive number β lies in $[0, 1]$. If $\beta < CR$, $\lambda_i \leftarrow \bar{\lambda}^* + w_1(\bar{\lambda}_a - \bar{\lambda}_b)$; otherwise, $\lambda_i \leftarrow \bar{\lambda}_i + w_2(\bar{\lambda}^* - \bar{\lambda}_i)$, where CR is a user-defined adaptive crossover probability, w_1 and w_2 are two random vectors whose elements all lie in $[0, 1]$.
7. If the maximum number of iterations N_c is reached, stop; else if $\bar{\lambda}^*$ stops improving, stop; else, go to the Step 3.

4   Simulation Results

In the simulation, it is assumed that the mass of mobile robot is $m = 2\,\text{kg}$, the radium of the wheels are $r = 0.05\,\text{m}$, half distance between the wheels is $b = 0.05\,\text{m}$, the moment of inertia is $I_z = 1\,\text{kg} \cdot \text{m}^2$, and the sampling interval is $t_s = 0.05\,\text{s}$. The trajectory to be tracked consists of a section of straight line where the linear velocity is $1\,\text{m/s}$ along the $x-$axis and a section of a parabolic curve. Besides, the system constraints are $0 \leq v \leq 2$ and $-\frac{\pi}{6} \leq \omega \leq \frac{\pi}{2}$. The initial position of the robot is $(x(0), y(0)) = (0, 2)$. Let $N = 5$, $N_u = 5, Q = \text{diag}(5, 500, 0.1)$, and $R = 0.1I$. In view of the presented MPC approach, the tracking control problem is formulated as a nonconvex optimization problem in form of (10). Figure 1 shows the transient behaviors of a single RNN model with 10 random initial conditions during the first sampling interval. From any initial condition, the RNN converges to an equilibrium state within $1\,\mu s$, though the equilibrium state may not correspond to the global optimal solution. Let $N_r = 3$, $N_c = 20$, Fig. 2 depicts the evolution of the global best solution resulting from the collective RNN approach. It takes the RNN group 5 iterations to information

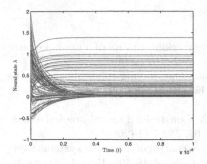

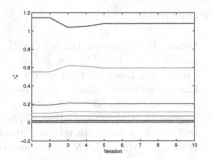

Fig. 1. Transient behaviors of RNN with 10 random initial conditions.

Fig. 2. Evolution of global best solution.

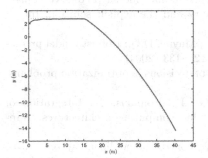

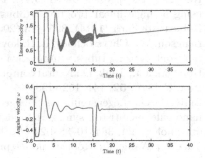

Fig. 3. Trajectory tracking of the mobile robot.

Fig. 4. Linear and angular velocities of the mobile robot.

exchange to find the global optimal solution. In worlds, the proposed RNNs-based optimization approach is capable of computing the optimal control signals in micro-seconds scale in framework of MPC. The overall tracking performances are depicted in Figs. 3 and 4. No constraint is violated during the control process. The results indicate that the proposed MPC approach is superior and highly computationally efficient.

5   Conclusions

In this paper, model predictive tracking control of mob ile robots is investigated where the control problem is formulated as a sequential nonconvex constrained optimization problem. A novel neurodynamic optimization strategy based on multiple cooperative recurrent neural networks is developed for solving the optimization problem. A salient feature of the proposed neural network model is that every equilibrium state corresponds to a KKT point of the optimization problem. The proposed optimization method can be viewed as an emulation of the swarm intelligence in human beings. Simulation results indicate that the proposed control scheme is effective and efficient for real time implementation of MPC of mobile robots.

References

1. Fukao, T., Nakagawa, H., Adachi, N.: Adaptive tracking control of a nonholonomic mobile robot. IEEE Trans. Robot. Autom. **16**, 609–615 (2000)
2. Fierro, R., Lewis, F.L.: Control of a nonholonomic mobile robot: backstepping kinematics into dynamics. In: Proceedings of the 34th IEEE Conference on Decision and Control, pp. 3805–3810 (1995)
3. Mayne, D., Rawlings, J., Rao, C., Scokaert, P.: Constrained model predictive control: stability and optimality. Automatica **36**(6), 789–814 (2000)
4. Fontes, F., Magni, L.: Min-max model predictive control of nonlinear systems using discontinuous feedbacks. IEEE Trans. Autom. Control **48**(10), 1750–1755 (2003)
5. Bemporad, A., Borrelli, F., Morari, M.: Min-max control of constrained uncertain discrete-time linear systems. IEEE Trans. Autom. Control **48**(9), 1600–1606 (2003)
6. Kerrigan, E.C., Maciejowski, J.M.: Feedback min-max model predictive control using a single linear program: Robust stability and the explicit solution. Int. J. Robust Nonlinear **14**(4), 395–413 (2004)
7. Langson, W., Chryssochoos, I., Rakovic, S.V., Mayne, D.Q.: Robust model predictive control using tubes. Automatica **40**(1), 125–133 (2004)
8. Hopfield, J.J., Tank, D.: Neural computation of decisions in optimization problems. Biol. Cybern. **52**, 141–152 (1985)
9. Indiveri, G., Legenstein, R., Deligeorgis, G., Prodromakis, T.: Integration of nanoscale memristor synapses in neuromorphic computing architectures. Nanotechnology **24**, 384010–384022 (2013)
10. Yan, Z., Wang, J.: Robust model predictive control of nonlinear systems with unmodeled dynamics and bounded uncertainties based on neural networks. IEEE Trans. Neural Netw. Learn. Syst. **25**(3), 457–469 (2014)
11. Yan, Z., Wang, J., Li, G.: A collective neurodynamic optimization approach to bound-constrained nonconvex optimization. Neural Netw. **55**, 20–29 (2014)
12. Ryu, J.C., Agrawal, S.K.: Differential flatness-based robust control of mobile robots in the presence of slip. The Int. J. Robot. Res. **30**(4), 463–475 (2011)
13. Raimondo, D.M., Limon, D., Lazar, M., Magni, L., Camacho, E.F.: Min-max model predictive control of nonlinear systems: a unifying overview on stability. Eur. J. Control **15**(1), 5–21 (2009)

Dynamical Analysis of Neural Networks with Time-Varying Delays Using the LMI Approach

Shanmugam Lakshmanan[1]([✉]), C.P. Lim[1], Asim Bhatti[1], David Gao[2], and Saeid Nahavandi[1]

[1] Center for Intelligent Systems Research, Geelong Waurn Ponds Campus, Deakin University, Geelong, Australia
`lakshmanan.shanmugam@deakin.edu.au`
[2] School of Applied and Biomedical Sciences, Federation University, Ballarat, Australia

Abstract. This study is concerned with the delay-range-dependent stability analysis for neural networks with time-varying delay and Markovian jumping parameters. The time-varying delay is assumed to lie in an interval of lower and upper bounds. The Markovian jumping parameters are introduced in delayed neural networks, which are modeled in a continuous-time along with finite-state Markov chain. Moreover, the sufficient condition is derived in terms of linear matrix inequalities based on appropriate Lyapunov-Krasovskii functionals and stochastic stability theory, which guarantees the globally asymptotic stable condition in the mean square. Finally, a numerical example is provided to validate the effectiveness of the proposed conditions.

Keywords: Neural networks · Interval time-varying delay · Stability · Linear matrix inequality

1 Introduction

Neural networks (NNs) constitute an important research topic in field of science and technology because of their extensive applications to various domains such as signal processing, parallel computing, and optimization problems [1,2]. Recently, studies of NNs along with time-delays have become more popular, which make NN models more complicated and interesting. Time-delays are encountered in neural processing and signal transmission, which can destabilize the whole networks, create oscillatory behaviours, and even cause chaos. Therefore, the analysis of NN models with time-delays plays a vital role in directly applying the NN models in real world problems. Indeed, many researchers have conducted dynamical analysis of NN models with time-delays (see, e.g., [3–6]).

Stability analysis plays a significant role in analysing time-delay systems. In the literature, many researchers have conducted stability analysis of time-delay systems using the Lyapunov-Krasowskii methodology with linear matrix inequalities (LMIs) [7,8]. Due to the effects of time-delay, the stability criteria can be

© Springer International Publishing Switzerland 2015
S. Arik et al. (Eds.): ICONIP 2015, Part III, LNCS 9491, pp. 297–305, 2015.
DOI: 10.1007/978-3-319-26555-1_34

classified into two types, i.e., delay-independent stability and delay-dependent stability. The delay-dependent stability criterion is less conservative as compared with the delay-independent one in the case of small time-delays. As such, many researchers have been investigated the delay-dependent stability criterion related to a variety of problems (ref [4–8]). As an example, the authors in [4], discussed the delay-dependent condition for cellular NNs with constant time-delays. Further, the authors in [5–8] argued that delay is varying with respect to time, and derived the stability conditions for NNs with time-varying delays. Based on the proposed results, many researchers have studied the stability criteria of time-varying delays that lie between 0 and their upper bounds, i.e., $0 \leq \tau(t) \leq h$. In practice, a time-delay typically exists in an interval. In other words, a time-delay varies in an interval for which the lower bound is not restricted to 0. For this particular reason, the stability criteria pertaining to the time-delay range has great significance for delayed NN models (see e.g., [9–12]). In [9], the authors initially investigated the stability problem of NNs based on interval time-varying delays by constructing an appropriate Lyapunov-Krasovskii function (LKF) and utilizing the free weight matrix approach. They showed that the proposed results were less conservative as compared with the existing results of NNs with interval time-varying delays.

Recently, NNs with Markovian jumping parameters have been investigated widely due to their random changes of structure. A NN has finite modes, and it may jump from one to another at different times. It has been pointed out in [13] that jumping between different NNs modes can be governed by a Markovian chain. Therefore, many researchers have been investigated NNs with Markovian jumping parameters (see, e.g.,[13–15]). As an example, the authors in [14] investigated the delayed uncertain Hopfield NN models with Markovian parameters, and the problem of state estimation was studied in [15] for jumping recurrent NN models with discrete and distributed delays. However, to the best of the authors' knowledge, the lower bound of time-varying delay is not restricted to 0 in this paper. In addition, not many results pertaining to stability analysis of NNs with Markovian jumping parameters by using convex combination techniques based on the delay interval have been established, which has motivated the present study.

Inspired by the above account, we aim to analyze the delay dependent stability criteria for NN models with interval time-varying delays by constructing suitable LKF and utilizing the free-weighting matrix approach, convex combination technique in this study. The sufficient condition is derived in terms of LMIs [16] for the considered problem with Markovian jumping parameters. The obtained formulae can be determined by using the Matlab LMI control toolbox. A numerical example is provided to illustrate the effectiveness of the proposed results.

Notation: $\mathbb{R}^n$ and $\mathbb{R}^{n \times n}$ represent the n-dimensional Euclidean space and the set of all $n \times n$ real matrices, respectively. For a given matrix, A^{-1} and A^T, denote its inverse and transpose, $X \geq Y$ (similarly, $X > Y$), where X and Y are symmetric matrices, i.e., $X - Y$ is positive semi-definite (similarly, positive definite). $\| \cdot \|$ is the Euclidean norm in $\mathbb{R}^n$. diag$\{\cdots\}$ stands for a block diagonal matrix. The notation $*$ always denotes the symmetric block in a symmetric matrix. $(\Omega, \mathcal{F}, \{\mathcal{F}_t\}_{t \geq 0}, \mathcal{P})$ indicates a complete probability space with a filtration $\{\mathcal{F}_t\}_{t\geq 0}$ satisfying the usual conditions and $\mathcal{E}$ stands for the mathematical expectation. $L^2_{\mathcal{F}_0}([-h_2,0],\mathbb{R}^n)$ denotes the family of all bounded $\mathcal{F}_0$- measurable, $C([-h_2,0],\mathbb{R}^n)$ -valued random variables $\xi = \{\xi(\theta) : -h_2 \leq \theta \leq 0\}$ such that $\int_{-h_2}^{0} |\mathcal{E}(s)|^2 ds < \infty$.

2 Problem Description and Preliminaries

Consider the following NN model with time-varying delays:

$$\dot{y}_i(t) = -a_i y_i(t) + \sum_{j=1}^{n} b_{ij} g_j(y_j(t)) + \sum_{j=1}^{n} c_{ij} g_j(y_j(t - \tau(t))) + I_i, \; i = 1,\dots,n \quad (1)$$

where n denotes the number of neurons in the NN, $y_i(t)$ denotes the state of the ith neuron at time t. $g_j(y_j(t))$ is the activation function of the jth neuron at time t. Parameters b_{ij} and c_{ij} represent, respectively, the connection weights and the delayed connection weights, from the jth neuron to the i neuron. I_i is the external bias on the ith neuron, $a_i > 0$ denotes the rate with which the ith neuron resets its potential to the resting state in isolation when it is disconnected from the network and external inputs. The time-varying delay $\tau(t)$ satisfies the following conditions.

$$0 \leq h_1 \leq \tau(t) \leq h_2, \quad \dot{\tau}(t) \leq \mu, \quad (2)$$

where h_1, h_2, and μ are constants. The NN model defined in (1) can be expressed in the matrix-vector form as follows.

$$\dot{y}(t) = -Ay(t) + Bg(y(t)) + Cg(y(t - \tau(t))) + I, \quad (3)$$

where $y(\cdot) = [y_1(\cdot), y_2(\cdot),\dots,y_n(\cdot)]^T \in \mathbb{R}^n$, $A = diag\{a_1,\dots,a_n\} > 0$, $B = (b_{ij})_{n\times n}$, $C = (c_{ij})_{n\times n}$, $I = [I_1,\dots,I_n]$ and $g(y(\cdot)) = [g_1(y_1(\cdot)),\dots,g_n(y_n(\cdot))]^T$.

Assumption 1: $g_i(\cdot)$ in (1) satisfies

$$l_i^- \leq \frac{g_i(x_1) - g_i(x_2)}{x_1 - x_2} \leq l_i^+, \; \forall x_1, x_2 \in \mathbb{R}, \; x_1 \neq x_2, \; i = 1,\dots,n, \quad (4)$$

where l_i^-, l_i^+ are known constants.

Taking the Markov jumping parameters into account, the delayed NN model defined in (3) becomes

$$\dot{y}(t) = -A(\eta(t))y(t) + B(\eta(t))g(y(t)) + C(\eta(t))g(y(t - \tau(t))) + I \quad (5)$$

where $\eta(t)$ $(t \geq 0)$ is a right-continuous Markov chain on the complete probability space $(\Omega, \mathcal{F}, \{\mathcal{F}_t\}_{t\geq 0}, \mathcal{P})$ taking values in a finite state space $\mathcal{S} = \{1,2,\dots,N\}$ with generator $\Gamma = (\gamma_{ij})_{N\times N}$ and transition probability from the ith mode at t to the jth mode, at $t + \Delta t$ $(i, j \in \mathcal{S})$

$$P\{\eta(t + \Delta t) = j | \eta(t) = i\} = \begin{cases} \gamma_{ij}\Delta t + o(\Delta t), & i \neq j, \\ 1 + \gamma_{ii}\Delta t + o(\Delta t), & i = j, \end{cases}$$

where $\Delta t > 0$ and $\lim_{\Delta t \to 0} \frac{o(\Delta t)}{\Delta t} = 0$, $\gamma_{ij} \geq 0$ is the transition rate from i to j, if $i \neq j$; while $\gamma_{ii} = -\sum_{j=1,\, j\neq i}^{N} \gamma_{ij}$. If we shift the equilibrium point y^* in (5) to the origin by letting $x(t) = y(t) - y^*$, system (5) can be transformed into:

$$\dot{x}(t) = -A_i x(t) + B_i f(x(t)) + C_i f(x(t - \tau(t))), \tag{6}$$

where $x(t) = [x_1(t), \ldots, x_n(t)]^T$ is the state vector of the transformed system, and $f_i(x(t)) = g_i(x_i(t) + y_i^*) - g_i(y_i^*)$, $i = 1, 2, \ldots, n$.
From Assumption 1, $f_i(x(t))$ satisfies

$$l_i^- \leq \frac{f_i(x_1) - f_i(x_2)}{x_1 - x_2} \leq l_i^+, \quad \forall x_1, x_2 \in \mathbb{R}, \; x_1 \neq x_2, \; i = 1, \ldots, n. \tag{7}$$

Let $x(t, \phi)$ be the state trajectories of system (6) with the initial condition $\phi \in L^2_{\mathcal{F}_0}([-h_2, 0], \mathbb{R}^n)$. It can be seen that system (6) admits a trivial solution $x(t, 0) \equiv 0$ corresponding to the initial condition $\phi = 0$.

3   Main Results

For convenience, the following notations are used:

$$L_1 = \mathrm{diag}\{l_1^-, l_2^-, \cdots, l_n^-\}, \quad L_2 = \mathrm{diag}\{l_1^+, l_2^+, \cdots, l_n^+\} \text{ and}$$
$$\xi^T(t) = \begin{bmatrix} x^T(t) & x^T(t - h_1) & x^T(t - \tau(t)) & x^T(t - h_2) & f^T(x(t)) & f^T(x(t - \tau(t))) & \dot{x}^T(t) \end{bmatrix}.$$

We derive a range-dependent time-delay stability condition for delayed NNs (6) with Markovian jumping parameters in the following theorem.

Theorem 1. *Given scalars $h_2 > h_1 \geq 0$ and $\mu \geq 0$, the delayed NN model in (6) is globally asymptotically stable in the mean square if symmetric matrices $P_i > 0$, $Q_l > 0, R_1 > 0, R_2 > 0$ $(l = 1, 2, 3)$, positive diagonal matrices $W, \Delta, \Gamma_1, \Gamma_2$ and real matrices $N_a, M_a, X_a, Y_a, Z_a (a = 1, 2)$ of appropriate dimensions exist, such that the following LMIs hold:*

$$\begin{bmatrix} \Xi^i & \sqrt{h_1}\, \bar{N} & \sqrt{h_2 - h_1}\, \bar{Y} \\ * & -R_1 & 0 \\ * & * & -R_2 \end{bmatrix} < 0, \tag{8}$$

$$\begin{bmatrix} \Xi^i & \sqrt{h_1}\, \bar{N} & \sqrt{h_2 - h_1}\, \bar{X} \\ * & -R_1 & 0 \\ * & * & -R_2 \end{bmatrix} < 0 \tag{9}$$

where $\Xi^i_{7 \times 7}$ with entries:

$$\Xi_{1,1} = -Z_1 A_i - A_i^T Z_1^T + Q_1 + Q_2 + Q_3 + N_1 + N_1^T - 2L_1 \Gamma_1 L_2 + \sum_{j=1}^{N} \gamma_{ij} P_j,$$

$\Xi_{1,2} = -N_1 + N_2^T$, $\Xi_{1,5} = Z_1 B_\imath + \Gamma_1(L_1 + L_2)$, $\Xi_{1,6} = Z_1 C_\imath$,

$\Xi_{1,7} = P - L_1^T W + L_2^T \Delta - A_\imath^T Z_2^T - Z_1$, $\Xi_{2,2} = -Q_1 - N_2 - N_2^T + X_1 + X_1^T$,

$\Xi_{2,3} = -X_1 + X_2^T$, $\Xi_{3,3} = -(1-\mu)Q_3 - X_2 - X_2^T + Y_1 + Y_1^T - 2L_1\Gamma_2 L_2$,

$\Xi_{3,4} = -Y_1 + Y_2^T$, $\Xi_{3,6} = \Gamma_2(L_1 + L_2)$, $\Xi_{4,4} = -Q_2 - Y_2 - Y_2^T$,

$\Xi_{5,5} = -2\Gamma_1$, $\Xi_{5,7} = W - \Delta + B_\imath^T Z_2^T$, $\Xi_{6,6} = -2\Gamma_2$, $\Xi_{6,7} = C_\imath^T Z_2^T$,

$\Xi_{7,7} = h_1 R_1 + (h_2 - h_1)R_2 - Z_2 - Z_2^T$, $\bar{N} = [N_1^T \ N_2^T \ 0 \ 0 \ 0 \ 0 \ 0]^T$,

$\bar{X} = [0 \ X_1^T \ X_2^T \ 0 \ 0 \ 0 \ 0]^T$, $\bar{Y} = [0 \ 0 \ Y_1^T \ Y_2^T \ 0 \ 0 \ 0]^T$.

Proof. Choose the following LKF for the delayed NN with Markovian jumping in (6),

$$
V(x(t), t, \eta(t) = \imath) = x^T(t)P_\imath x(t) + 2\sum_{j=1}^{n}\left(w_j \int_0^{x_j(t)}(f_j(s) - l_\imath^- s)ds + \delta_j \int_0^{x_j(t)}(l_\imath^+ s - f_j(s))ds\right)
$$
$$
+ \int_{t-h_1}^{t} x^T(s)Q_1 x(s)ds + \int_{t-h_2}^{t} x^T(s)Q_2 x(s)ds + \int_{t-\tau(t)}^{t} x^T(s)Q_3 x(s)ds
$$
$$
+ \int_{-h_1}^{0}\int_{t+\theta}^{t} \dot{x}^T(s)R_1\dot{x}(s)dsd\theta + \int_{-h_2}^{-h_1}\int_{t+\theta}^{t} \dot{x}^T(s)R_2\dot{x}(s)dsd\theta. \tag{10}
$$

Let $V(x(t), t, \eta(t) = \imath, t > 0)$ $\mathbb{L}V(t)$ be the stochastic positive LKF. The weak infinitesimal operator is defined as

$$
\mathbb{L}V(x(t), t, \eta(t) = \imath) = \lim_{\Delta t \to 0} \frac{1}{\Delta t}\Big[\mathcal{E}\{Vx((t + \Delta t), r(t + \Delta t), t + \Delta t)|x(t), \eta(t) = \imath\}
$$
$$
- V(x(t), \eta(t) = \imath, t)\Big]
$$
$$
= \frac{\partial V}{\partial t} + \dot{x}^T(t)\frac{\partial V}{\partial x}\Big|_{\eta(t)=\imath} + \sum_{j=1}^{N}\gamma_{\imath j}V(x(t), t, \imath, \jmath).
$$

Now take $\mathbb{L}V(t)$ along a given trajectory of the delayed NN in (6) as follows

$$
\mathbb{L}V(x(t), t, \eta(t) = \imath) \le 2x^T(t)P_\imath\dot{x}(t) + \sum_{j=1}^{N}\gamma_{\imath j}x^T(t)P_\jmath x(t) + 2[f(x(t)) - L_1 x(t)]^T W\dot{x}(t)
$$
$$
+ 2[L_2 x(t) - f(x(t))]^T\Delta\dot{x}(t) + x^T(t)Q_1 x(t) - x^T(t-h_1)Q_1 x(t-h_1)
$$
$$
+ x^T(t)Q_2 x(t) - x(t-h_2)Q_2 x(t-h_2) + x^T(t)Q_3 x(t)
$$
$$
- (1-\mu)x^T(t-\tau(t))Q_3 x(t-\tau(t)) + h_1\dot{x}(t)R_1\dot{x}(t)
$$
$$
- \int_{t-h_1}^{t} \dot{x}^T(s)R_1\dot{x}(s)ds + (h_2 - h_1)\dot{x}^T(t)R_2\dot{x}(t)
$$
$$
- \int_{t-h_2}^{t-h_1} \dot{x}^T(s)R_2\dot{x}(s)ds. \tag{11}
$$

where $W = \text{diag}\{w_1, ..., w_n\}$ and $\Delta = \text{diag}\{\delta_1, ..., \delta_n\}$. It should be noted that

$$-\int_{t-h_2}^{t-h_1} \dot{x}^T(s)R_2\dot{x}(s)ds = -\int_{t-\tau(t)}^{t-h_1} \dot{x}^T(s)R_2\dot{x}(s)ds - \int_{t-h_2}^{t-\tau(t)} \dot{x}^T(s)R_2\dot{x}(s)ds.$$

By Lemma 2.1 in [17], it follows that

$$-\int_{t-h_1}^{t} \dot{x}^T(s)R_1\dot{x}(s)ds \leq h_1\xi^T(t)\bar{N}R_1^{-1}\bar{N}^T\xi(t) + 2\xi^T(t)\bar{N}[x(t) - x(t - h_1)] \quad (12)$$

$$\int_{t-\tau(t)}^{t-h_1} \dot{x}^T(s)R_2\dot{x}(s)ds \leq (\tau(t) - h_1)\xi^T(t)\bar{X}R_2^{-1}\bar{X}^T\xi(t)$$
$$+2\xi^T(t)\bar{X}[x(t - h_1) - x(t - \tau(t))] \quad (13)$$

$$\int_{t-h_2}^{t-\tau(t)} \dot{x}^T(s)R_2\dot{x}(s)ds \leq (h_2 - \tau(t))\xi^T(t)\bar{Y}R_2\bar{Y}^T\xi(t)$$
$$+2\xi^T(t)\bar{Y}[x(t - \tau(t)) - x(t - h_2)]. \quad (14)$$

Further, we add the following zero equation with any chosen matrices of Z_1 and Z_2

$$2[x^T(t)Z_1 + \dot{x}^T(t)Z_2][-A_\imath x(t) + B_\imath f(x(t)) + C_\imath f(x(t - \tau(t)) - \dot{x}(t)] = 0. (15)$$

Noting that for positive diagonal matrices Γ_1, Γ_2 and Assumption 1, one has

$$-f^T(x(t))\Gamma_1 f(x(t)) + 2x^T(t)\Gamma_1(L_1 + L_2)f(x(t)) - 2x^T(t)L_1\Gamma_1 L_2 x(t) \geq 0. (16)$$
$$-f^T(x(t - \tau(t)))\Gamma_2 f(x(t - \tau(t))) + 2x^T(t - \tau(t))\Gamma_2(L_1 + L_2)f(x(t - \tau(t)))$$
$$-2x^T(t - \tau(t))L_1\Gamma_2 L_2 x(t - \tau(t)) \geq 0. \quad (17)$$

Substituting (12)–(14) into (11) and adding (15)–(17) into (11), yields

$$\mathbb{L}V(x(t), t, \eta(t) = \imath) \leq \xi^T(t)\left[\Xi_{\tau(t)}^\imath\right]\xi(t) \quad (18)$$

where $\Xi_{\tau(t)} = \Xi + h_1\bar{N}R_1^{-1}\bar{N}^T + (\tau(t) - h_1)\bar{X}R_2^{-1}\bar{X}^T + (h_2 - \tau(t))\bar{Y}R_2^{-1}\bar{Y}^T$. Taking the mathematical expectation $\mathbb{E}$ on both sides of (18) and from LMIs (8)–(9), we can obtain

$$\mathbb{E}\left\{\mathbb{L}V(x(t), t, \eta(t) = \imath)\right\} \leq \mathbb{E}\left\{\xi^T(t)\left[\Xi_{\tau(t)}^\imath\right]\xi(t)\right\} \leq -\lambda\{\mathbb{E}\{|x(t, \phi, \imath_0)\|^2\}\},$$

where $\lambda = \lambda_{min}\left(-\Xi_{\tau(t)}^\imath\right)$. This implies that system (8) is globally asymptotically stable in the mean square. Notice that $(\tau(t) - h_1)\bar{X}R_2^{-1}\bar{X}^T + (h_2 - \tau(t))\bar{Y}R_2^{-1}\bar{Y}^T$ is a convex combination of matrices $\bar{X}R_2^{-1}\bar{X}^T$ and $\bar{Y}R_2^{-1}\bar{Y}^T$ on $\tau(t) \in [h_1, h_2]$. Therefore, by following the convex analysis approach, $\Xi_{\tau(t)} < 0$ if and only if

$$\Xi_{\tau(t)}^\imath\Big|_{\tau(t)=h_1} < 0, \quad (19)$$

$$\Xi_{\tau(t)}^\imath\Big|_{\tau(t)=h_2} < 0. \quad (20)$$

Using Schur complement, (19)–(20) are equivalent to (8)–(9), respectively. This completes the proof.

Remark 1. A range-dependent time-delay stability criterion has been proposed for a delayed NN model with Marakovian jumping parameters. The sufficient condition has more information of the lower and upper bounds of time-varying delays. Moreover, we have introduced few free weight matrices, and expressed the derived sufficient condition in two LMIs (8) and (9) by using the convex combination technique, which is based on $\tau(t) \in [h_1, h_2]$. Here, it should be mentioned that the restrictive condition of $\mu \leq 1$ has been removed in Theorm 3.1, and we can easily derive the corresponding results in the non-differentiable case of time-varying delays when $Q_3 = 0$ in LKF (10).

4 Numerical Example

A numerical example is presented to illustrate the potential benefits and effectiveness of the developed method for delayed NNs with Markovian jumping parameters. Consider a three-order delayed NN of (6) with mode $\imath = 2$ and the following parameters

$$A_1 = \begin{bmatrix} 1.8 & 0 & 0 \\ 0 & 1.2 & 0 \\ 0 & 0 & 1.4 \end{bmatrix}, \quad A_2 = \begin{bmatrix} 2 & 0 & 0 \\ 0 & 1.3 & 0 \\ 0 & 0 & 1.8 \end{bmatrix}, \quad B_1 = \begin{bmatrix} 0.4 & 1.5 & 0.1 \\ 0.56 & 0 & -1.4 \\ 0.1 & 1 & 1.2 \end{bmatrix},$$

$$B_2 = \begin{bmatrix} -0.5 & 1.2 & 0 \\ -0.5 & 0 & 1 \\ 0.45 & 1.25 & 0.3 \end{bmatrix}. \quad C_1 = \begin{bmatrix} 1.5 & 1 & 0.8 \\ 0 & 1.5 & 0 \\ 0.25 & 1.2 & 0.5 \end{bmatrix}, \quad C_2 = \begin{bmatrix} 1.8 & 0 & 0.24 \\ 0 & 0.8 & 0 \\ 0.54 & 1.2 & 0.9 \end{bmatrix}.$$

In this example, the activation function is assumed to satisfy Assumption 1 with $l_1^- = l_2^- = l_3^- = 0$, $l_1^+ = 0.2$, $l_2^+ = 0.2$ and $l_3^+ = 0.2$. Then the transition probability matrix is assumed to be $\Gamma = \begin{bmatrix} -7 & 7 \\ 6 & -6 \end{bmatrix}$. Let $h_1 = 0.5, h_2 = 1, \mu = 1.1$. Using the Matlab LMI control toolbox to solve LMIs (8)–(9) in Theorem 1, we obtain the following feasible matrices:

$$P_1 = \begin{bmatrix} 6.1996 & -1.5709 & -3.7824 \\ -1.5709 & 5.2842 & -0.2011 \\ -3.7824 & -0.2011 & 7.4683 \end{bmatrix}, \quad P_2 = \begin{bmatrix} 6.0627 & -1.5764 & -3.7717 \\ -1.5764 & 5.1513 & -0.0142 \\ -3.7717 & -0.0142 & 7.1369 \end{bmatrix},$$

$$Q_1 = \begin{bmatrix} 3.5372 & -0.9294 & -1.6140 \\ -0.9294 & 2.7700 & 0.0282 \\ -1.6140 & 0.0282 & 3.5646 \end{bmatrix}, \quad Q_2 = \begin{bmatrix} 3.7535 & -1.0086 & -1.6783 \\ -1.0086 & 2.8251 & -0.0018 \\ -1.6783 & -0.0018 & 3.7860 \end{bmatrix}.$$

$$Q_3 = \begin{bmatrix} 1.3690 & -0.4091 & -0.7585 \\ -0.4091 & 0.8596 & 0.0282 \\ -0.7585 & 0.0282 & 1.3981 \end{bmatrix}, \quad R_1 = \begin{bmatrix} 2.5604 & -0.5296 & -1.3084 \\ -0.5296 & 3.3156 & 0.0930 \\ -1.3084 & 0.0930 & 3.4785 \end{bmatrix},$$

$$R_2 = \begin{bmatrix} 2.6831 & -0.5575 & -1.3459 \\ -0.5575 & 3.3690 & 0.1050 \\ -1.3459 & 0.1050 & 3.5392 \end{bmatrix}, \quad W = \text{diag}\{6.1985, 6.7736, 6.3091\},$$

$$\Delta = \text{diag}\{6.6897, 7.0455, 7.8224\}.$$

Therefore, the proposed NN model with time-varying delays is globally asymptotically stable. In addition, Fig. 1(a) shows the convergence of the state trajectories of the delayed NN model to the zero equilibrium point with different initial conditions. The response of the Markovian jumping modes are shown in Fig. 1(b).

5 Conclusions

In this paper, we have studied the range-dependent time-delay stability criteria for delayed NN models with Markovian jumping parameters. Based on suitable

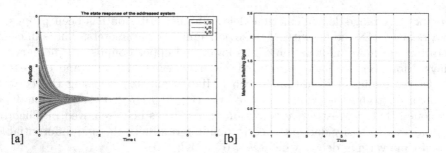

Fig. 1. (a). State trajectories of delayed NNs with different initial conditions. (b). The response of Markovian jumping signal when mode $\imath = 2$.

LKF, integral inequalities, LMI framework, and convex combination technique, the conditions for delay-dependent stability criteria are derived. From the numerical example, it is evident that the proposed method is effective, and is able to provide less conservative results.

References

1. Haykin, S.: Neural Networks: A Comprehensive Foundation. Prentice Hall, NJ (1998)
2. Cichocki, A., Unbehauen, R.: Neural Networks for Optimization and Signal Processing. Wiley, Chichester (1993)
3. Roska, T., Chua, L.O.: Cellular neural networks with nonlinear and delay-type template. Int. J. Circuit Theory Appl. **20**, 469–481 (1992)
4. Xu, S., Lam, J., Ho, D.W.C., Zou, Y.: Novel global asymptotic stability criteria for delayed cellular neural networks. IEEE Trans. Circuits Syst. II, Exp. Briefs **52**, 349–353 (2005)
5. He, Y., Wang, Q.G., Wu, M.: LMI-based stability criteria for neural networks with multiple time-varying delays. Physica D **212**, 126–136 (2005)
6. He, Y., Wu, M., She, J.H.: Delay-dependent exponential stability for delayed neural networks with time-varying delay. IEEE Trans. Circuits Syst. II, Exp. Briefs **53**, 230–234 (2006)
7. Hua, C.C., Long, C.N., Guan, X.P.: New results on stability analysis of neural networks with time-varying delays. Phys. Lett. A **352**, 335–340 (2006)
8. He, Y., Liu, G., Rees, D.: New delay-dependent stability criteria for neural networks with time-varying delay. IEEE Trans. Neural Netw. **18**, 310–314 (2007)
9. He, Y., Liu, G., Rees, D., Wu, M.: Stability analysis for neural networks with time-varying interval delay. IEEE Trans. Neural Netw. **18**, 1850–1854 (2007)
10. Li, C.D., Feng, G.: Delay-interval-dependent stability of recurrent neural networks with time-varying delay. Neurocomputing **72**, 1179–1183 (2009)
11. Singh, V.: A new criterion for global robust stability of interval delayed neural networks. J. Comput. Appl. Math. **221**, 219–225 (2008)
12. Balasubramaniam, P., Lakshmanan, S., Theesar, S.: State estimation for Markovian jumping recurrent neural networks with interval time varying delays. Nonlinear Dyn. **60**, 661–675 (2009)

13. Tino, P., Cernansky, M., Benuskova, L.: Markovian architectural bias of recurrent neural netwoks. IEEE Trans. Neural Netw. **15**, 6–15 (2004)
14. Li, H., Chen, B., Zhou, Q., Lin, C.: Robust exponential stability for delayed uncertain Hopfield neural networks with Markovian jumping parameters. Phys. Lett. A **372**, 4996–5003 (2008)
15. Wang, Z., Liu, Y., Liu, X.: State estimation for jumping recurrent neural networks with discrete and distributed delays. Neural Netw. **22**, 41–48 (2009)
16. Boyd, S., El Ghaouli, L., Feron, E., Balakrishnan, V.: Linear Matrix Inequalities in System and Control Theory. SIAM, Philadelphia (1994)
17. Zhang, D., Yu, L.: H_∞ filtering for linear neutral systems with mixed time-varying delays and nonlinear perturbations. J. Franklin Inst. **347**, 1374–1390 (2010)

Modeling Astrocyte-Neuron Interactions

Soukeina Ben Chikha[(⊠)], Kirmene Marzouki[(⊠)],
and Samir Ben Ahmed

LISI, Laboratoire d'Informatique pour les Systèmes Industriels, Tunis, Tunisia
sokyana@yahoo.fr, kirmene@marzouki.tn,
samirbenahmed@fst.rnu.tn

Abstract. The involvement of astrocytes in information processing in the brain has recently been demonstrated. In this paper, we investigate, using computational models (SOM and MLP), one of the observed astrocyte-neuron interactions for information processing: the neural modulation represented by the observed calcium waves. We apply it to solve classification problems. The results of the performed tests confirmed that the proposed approach improved artificial neural network performance, especially learning time acceleration.

Keywords: Artificial Neural Networks (ANNs) · Multi Layers Perceptron (MLP) · Self Organizing Maps (SOM), Astrocyte-neuron interactions · Calcium waves · Neural modulation

1 Introduction

Biologists have demonstrated the involvement of astrocytes in information processing in the brain. They defined several behavioral observations of astrocytes such as regulating the synaptic transmission through observed calcium waves. In ANN field, few approaches have been proposed to involve astrocytes features in the learning process.

In this paper, we propose an approach modeling the calcium waves created by astrocytes to regulate the synaptic transmission using only neural network paradigms the Self Organizing Maps and the Multi Layers Perceptron.

2 Astrocyte-Neuron Interactions

Currently, biologists concentrate on the study of interactions between glial cells and neurons. Particularly, the interactions between neurons and astrocytes are made by direct connections involving junctions [1] and via neurotransmitters [2, 3]. Consequently, the astrocytes are considered as cellular elements involved in information processing in the nervous system [4]: they have an important implication in the regulation of signal transmission.

In this context, many complex properties of the neuron-astrocyte intracellular signaling were observed. The main observed property of glial cells, in which we are interested, is the neural modulation and is described below.

© Springer International Publishing Switzerland 2015
S. Arik et al. (Eds.): ICONIP 2015, Part III, LNCS 9491, pp. 306–314, 2015.
DOI: 10.1007/978-3-319-26555-1_35

3 Neuronal Modulation: The Calcium Waves

Astrocytic calcium waves in response to glutamate [5] were observed, in the early 90 s. By the measurement of changes of intracellular calcium concentration, it was shown that astrocytes, non-excitable cells, are able to receive and integrate information from neurons. Also, astrocytes communicate with each other and with other cells, particularly neurons and endothelial cells [6, 7].

Astrocytes are endowed with a form of cellular excitability based on changes of intracellular calcium levels. Here, biologists observed intracellular and intercellular calcium waves [8].

The propagation of calcium waves in astrocytes is the basis of the communication between neuronal and glial networks. Actually, the neuronal activity induces calcium responses in astrocytes and, conversely, the increase of the concentration of intracellular calcium in astrocytes causes changes of the neuron's activity [6]. The increase in astrocytic calcium is responsible for the release by astrocytes of active substances (such as glutamate), which could directly affect neuronal activity. This behavior is called neural modulation [9].

Particularly, the astrocyte Ca^{2+} signal shows a complex non-linear relationship with the synaptic activity [8, 10]. Thus, the astrocytes discriminate between the activities of different synapses and respond selectively to different axon pathway. Also, the astrocyte Ca^{2+} signal is modulated by the simultaneous activity of different synaptic inputs. This Ca^{2+} signal modulation depends on cellular intrinsic properties of the astrocytes and depends on the neurotransmitters involved. This modulation is also bi-directionally regulated by the level of synaptic activity, and controls the spatial extension of the intracellular Ca^{2+} signal. Consequently, astrocytic intracellular Ca^{2+} is controlled by the synaptic activity.

4 Biological Inspiration

As the propagation of calcium waves in astrocytes constitutes an essential element for communication between neuronal and glial networks, we were particularly interested to investigate the modulation of synaptic transmission realized by astrocytes.

We remind that, it has been demonstrated that the neuronal activity induced calcium responses in astrocytes and, conversely, the increase of the intracellular concentration of calcium in astrocytes changes the neuronal activity [6]. In addition, the heterogeneity of the distribution of gap junctions produces an asymmetric propagation of calcium waves similar to preferential circuits involving only some cells [8, 10].

In our approach, we model the neural modulation, particularly the calcium waves and the determination of preferential circuits using SOM [11] and MLP [12].

First, we define the preferred circuit, using the concept of Best Matching Unit (BMU) as defined by Kohonen in Self Organizing Map [11] and we introduce the Best BMU (BBMU) as the unit who was selected the maximum times as BMU during the test step. This BBMU unit will represent the preferred chosen path in the artificial network through the hidden layers of the network. In the biological field, the neural modulation is regulated by the level of synaptic activity [8]; in our hypothesis, the definition of the preferred circuit is influenced by the network input. Then, we model the calcium waves using a two dimensional Gaussian function.

Finally, in the learning step, we introduce a new learning rule combing SOM and MLP learning rules: the unit's weight update is influenced by the Gaussian function whose center is the BBMU. The approach details are defined in Sect. 6.

5   Related Work

Several approaches in the ANN field tried to involve astrocytes influence in the learning process. Actually, we observe two teams that published results. They work on classification and prediction problems. It is important to point out that we are more interested in the biological inspiration as a modeling problem.

The first work group proposed in [13] an approach called Chaos Glial Network Connected to Multi-Layer Perceptron. The biological inspiration is based on the fact that in the biological neural network the glial cells affect to the neighbor neurons over a wide range by propagating in the network [6]. In [14] the authors investigate parameter dependency of their approach.

Then, in [15–21] the authors propose a modification of their approach proposed in [13] and investigate the performance and the features of this approach. These modifications are inspired from glial cells observed properties. They modified the weight update rule or on the way how to apply it. They propose an explanation of the obtained results. For computer simulation they applied their approaches to learn two kinds of chaotic time series or to solve the Two-Spiral Problem. They compared their obtained results with conventional MLP and concluded that the simulation result shows that the proposed networks possess better learning performance than the conventional MLP.

Here, we note that there is actually one proposed approach [13] and the others are only variants. Each time, the first approach is modified to fit a new context or a new biological observation.

The second work group presented an approach called Neuron Glia Network (NGN). In [22] the authors investigated the consequence of including artificial astrocytes on ANN performance. They used connectionist systems and evolutionary algorithm. They tested their approach on classification problem using more conventional datasets as Heart Disease (HD), Breast Cancer (BC), Iris Flower (IF), Ionosphere (IS) and Multiplexor (MUX). In NGN the authors introduced a hybrid learning method for training NGN. The authors designed the artificial astrocytes to resemble the signaling properties of biological astrocytes, which respond to neurotransmitters released under high synaptic activity and regulate neurotransmission in a larger temporal scale.

They conclude that the degree of success of NGN is superior to NN on the different used datasets. In [23], the authors detail Artificial Neuron Glia Network (ANGN) architecture and algorithm. They propose and test six variants of the algorithm. The six algorithms were different in two aspects: the management of the activity counter and the respect or not of a weight limit. They tested their six algorithms on MUX and IF and compare their results with ANN trained only by using Genetic Algorithms (GAs). They conclude that all these implemented algorithms tried to emulate the potentiation of synaptic connections caused by astrocytes due to high synaptic activity. Also, they conclude that the six algorithms improved the conventional ANN.

In conclusion, our major observation is that there is not yet a true modeling of glial cells, there is no glial network; only the learning rule is modified and the way it is applied. In fact, biologists demonstrated the involvement of astrocytes in the learning process of human brain. They made several observations describing their influence on neurons. In ANN field, each proposed approach is inspired by one or more astrocytes features.

6 Proposed Approach

In our proposed approach we model the astrocytes modulation of neuronal activity using SOM and MLP. For the network architecture, we use MLP structure where each hidden layer is a SOM map (Fig. 1).

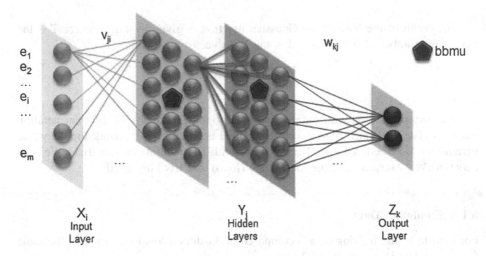

Fig. 1. Our proposed model architecture with preferential circuit (pentagon units)

For the learning algorithm, we use MLP algorithm with some modifications. First, in the test step and at each time epoch, for every input vector we find the BMU as defined by Kohonen using the Euclidian distance and we record that information. Then, at the end of each test step we determine the Best BMU as the unit which was the most selected unit as BMU. In the leaning step of the learning process, the weight of the units is updated using the new learning rule as follow.

We remind SOM learning rule (Eq. 1) and MLP learning rule (Eq. 2)

$$V_{ji}(t + 1) = V_{ji}(t) + \alpha(t) . [x_i - V_{ji}(t)] \tag{1}$$

$$V_{ji}(t+1) = V_{ji}(t) + \boxed{\eta . x_i . F_j} \qquad (2)$$

Where, V_{ji} is the weight vector of unit j when Xi is the input vector.

Our introduced learning rule (Eq. 3) is the combination of Eqs. 1, 2 with a two dimensional Gaussian function ρ (Eq. 4). The center of the Gaussian function is the BBMU and it changes along the learning process.

$$V_{ji}(t+1) = V_{ji}(t) + \boxed{\eta . x_i . F_j} + \rho(i,j,xbbmu,ybbmu,t) \boxed{.a(t).[x_i - V_{ji}(t)]} \qquad (3)$$

Where, η is the learning rate according to MLP and t the time epoch. The learning rate according to SOM is given by Eq. 5 where α_0 represents the initial learning rate.

$$\rho(i,j,xbbmu,ybbmu,t) = F_j . e^{-a(t)\left((i-xbbmu)^2 + (j-ybbmu)^2\right)} \qquad (4)$$

$$\alpha(t) = \alpha_0 . (1 - \frac{t}{T}) \qquad (5)$$

The width of the base of the Gaussian function is given by Eq. 6 where T is the maximum number of iterations and m the input vector dimension.

$$a(t) = \frac{1}{m} . (\frac{3.t}{T} + 1) \qquad (6)$$

We performed tests with different network topologies and with different iteration numbers. The illustrated results were performed according the network topology and parameters described in Tables 1 and 2 below with 10 randomly chosen trials. We used benchmarked datasets [24], the Ionosphere (Iono) and Iris Flower (IF).

6.1   Simulation Data

For Iono tests, the training data is composed of 34-dimensional vectors with the same form. The data is composed of 2 classes

For IF, the training data is composed of 4-dimensional vectors with the same form. The dataset is composed of 3 classes.

Table 1. Simulation parameters description

Parameter	IONO	IF
Training dataset	280	120
Input vector dimension	34	4
Output vector	1	1
α_0	0.8	0.8
η	0.1	0.1
Network architecture	$34 - (4 \times 4) - 1$	$4 - (4 \times 4) - 1$
T: Maximum iteration	5000	10000
Number of classes	2	3

6.2   Learning Quality

To describe simulation results, we compare the obtained results with the conventional MLP. To illustrate our contributions on the improvement of the learning process we investigate the mean squared error (MSE), and we define and describe learning acceleration.

Mean Squared Error. In each simulation, we observe the minimum, the maximum; we calculate the average error and the standard deviation from 10 trials.

Table 2. Iono MSE

	Proposed approach	MLP
Minimum	**0.0015670542255663484**	0.016410360885813807
Maximum	1.002971374796798	0.9987229926268643
Average	0.4026716099113526	0.42168547124279127
Standard deviation	0.1889781337249296	0.07715916125015955

Table 3. IF MSE

	Proposed approach	MLP
Minimum	**0.19390555097770026**	0.377695768753889
Maximum	0.3801785988636056	0.3792897529786541
Average	0.32351789649714396	0.3783518968468071
Standard deviation	0.002930568119670308	1.1899585868428873E-7

Observing the Tables 2 and 3 below of the MSE, the minimum value is given by our approach.

In Fig. 2 we show the general shape obtained along the different simulations. Our model gives better performance than MLP.

Learning Acceleration. Our major contribution is that the learning process occurs fast (Fig. 2).

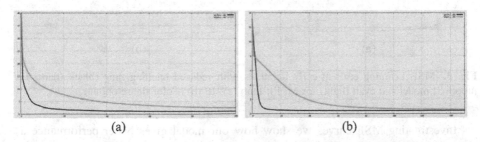

(a)   (b)

Fig. 2. MSE Learning error at early iterations (black (dark) for proposed model and cyan (light) for MLP): Iono (a), IF (b) (Color figure online)

To explore this we note:

- er* the maximum value between the final error obtained by MLP and our model (for one simulation).
- t_{*MLP} 1st time MLP reaches er*.
- $t_{*OurApproach}$ 1st time our approach reaches er*.

We define the learning acceleration as:

$$Learning\_Acc = \frac{t_{*MLP}}{t_{*OurApproach}} \tag{7}$$

For Example, if Learning_Acc equals 6, that means that the learning process using our model is 6 times faster than using MLP.

Table 4. Learning acceleration

	Iono	IF
Minimum	0.0602	0.6069
Maximum	11.26126126126126	34.48275862068966
Average	**5.211151883976925**	**6.220366349525155**
Standard deviation	14.837340799214203	42.430790904383606

As showed in Table 4, for Iono dataset, our approach is about 5 times faster than MLP and can reach 11 times faster. For IF data our approach is also about 6 times faster than MLP and can also reach 34 times faster; which reflect our major contribution.

To confirm the learning acceleration we repeated the simulations with the same conditions and with the same initial states, we only reduced the duration of the learning process to 500.

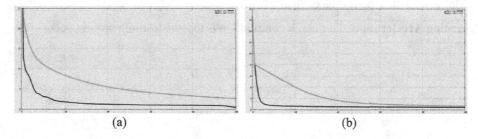

(a)   (b)

Fig. 3. MSE Learning error at early iterations with reduced learning time (black (dark) for proposed model and cyan (light) for MLP): Iono (a), IF (b) (Color figure online)

Investigating MSE curves we show how our model gives better performance as shown in Fig. 3 and yet accelerating the learning time.

7 Conclusion

With the demonstration of the involvement of astrocytes in the information process in the brain, we tried to investigate one of the astrocyte-neuron interactions. In this article, we proposed an approach with a new network architecture, a new learning algorithm and a new learning rule using ANN paradigms. Computer simulations confirmed that the proposed model showed better learning performance than the conventional networks and considerably decreases learning time about 5 times in average (according to the dataset). However, we consider that our major contribution is to attempt to model the modulation of the neuronal activity by astrocytes in ANN field. Actually, the exact role of astrocytes is not yet defined. New biological observations will have an important impact on the ANN.

References

1. Verkhratsky, A., Kirchhoff, F.: Glutamate-mediated neuronal-glial transmission. J. Anat. **210**, 651–660 (2007)
2. Fields, R.D., Stevens, B.: ATP: an extracellular signaling molecule between neurons and glia. Trends Neurosci. **23**, 625–633 (2000)
3. Fields, R.D., Burnstock, G.: Purinergic signalling in neuron-glia interactions. Nat. Rev. Neurosci. **7**, 423–436 (2006)
4. Wigley, R., Hamilton, N., Nishiyama, A., Kirchhoff, F., Butt, A.M.: Morphological and physiological interactions of NG2-glia with astrocytes and neurons. J. Anat. **210**, 661–670 (2007)
5. Cornell-Bell, A.H., Finkbeiner, S.M., Cooper, M.S., Smith, S.J.: Glutamate induces calcium waves in cultured astrocytes: long-range glial signaling. Science **247**, 470–473 (1990)
6. Haydon, P.J.: Glia: listening and talking to the synapse. Nat. Rev. Neurosci. **2**, 185–193 (2001)
7. Zonta, M., Angulo, M.C., Gobbo, S., et al.: Neuron-to-astrocyte signaling is central to the dynamic control of brain microcirculation. Nat. Neurosci. **6**, 43–50 (2003)
8. Perea, G., Araque, A.: Synaptic information processing by astrocytes. J. Physiol. **99**, 92–97 (2006)
9. Nedergaard, M., Ransom, B., Goldman, S.A.: New role for astrocytes: redefining the functional architecture of the brain. Rev. Trends Neurosci. **26**, 523–530 (2003)
10. Venance, L., Stella, N., Glowinski, J., Giaume, C.: Mechanism of initiation and propagation of receptor induced intercellular calcium signaling in cultured at astrocytes. J. Neurosci. **17**, 1981–1992 (1997)
11. Kohonen, T.: The self-organizing map. Proc. IEEE **78**, 1464–1480 (1990)
12. Rumelhart, D.E., Hinton, G.E., Williams, R.J.: Learning representations by back-propagating errors. Nature **323–9**, 533–536 (1986)
13. Ikuta, C., Uwate, Y., Nishio, Y.: Chaos glial network connected to multi-layer perceptron for solving two-spiral problem. In: Proceedings of the ISCAS 2010, pp. 1360–1363 (2010)
14. Ikuta, C., Uwate, Y., Nishio, Y.: Parameter dependency of chaos glial network connected to multi-layer perceptron. In: Proceedings of the NCSP 2010, pp. 596–599 (2010)
15. Ikuta, C., Uwate, Y., Nishio, Y.: Multi-layer perceptron with impulse glial network. In: Proceedings of the NCN 2010, pp. 9–11 (2010)

16. Ikuta, C., Uwate, Y., Nishio, Y.: Multi-layer perceptron with glial network influenced by local external stimulus. In: Proceedings of the NCSP 2011, China, pp. 199–202 (2011)
17. Ikuta, C., Uwate, Y., Nishio, Y.: Performance and features of multi-layer perceptron with impulse glial network. In: Proceedings of the IJCNN 2011, pp. 2536–2541 (2011)
18. Ikuta, C., Uwate, Y., Nishio, Y.: Multi-layer perceptron with pulse glial chain. In: International Symposium on Nonlinear Theory and its Applications, NOLTA 2011, Japan, pp. 435–438 (2011)
19. Ikuta, C., Uwate, Y., Nishio, Y.: Multi-layer perceptron with pulse glial chain for solving two-spiral problem. In: IEEE Workshop on Nonlinear Circuit Networks, pp. 51–53 (2011)
20. Ikuta, C., Uwate, Y., Nishio, Y.: Investigation of Multi-layer perceptron with propagation of glial pulse to two directions. In: Proceedings of the ISCAS 2012, pp. 2099–2102 (2012)
21. Ikuta, C., Uwate, Y., Nishio, Y.: Multi-layer perceptron with positive and negative pulse glial chain for solving two-spiral problem. In: IEEE World Congress on Computational Intelligence, WCCI 2012, Australia, pp. 2590–2595 (2012)
22. Porto-Pazos, A.B., Veiguela, N., Mesejo, P., et al.: Artificial astrocytes improve neural network performance. PLoS ONE 6(4), e19109 (2011)
23. Alvarellos-González, A., Pazos, A., Porto-Pazos, A.B.: Computational models of neuron-astrocyte interactions lead to improved efficacy in the performance of neural networks. Comput. Math. Methods Med., Article ID 476324, 10 (2012)
24. UCI machine learning repository. http://www.ics.uci.edu/~mlearn/MLRepository.html

Growing Greedy Search and Its Application to Hysteresis Neural Networks

Kei Yamaoka and Toshimichi Saito[✉]

Hosei University, Koganei, Tokyo 184-8584, Japan
tsaito@hosei.ac.jp

Abstract. This paper presents the growing greedy search algorithm and its application to associative memories of hysteresis neural networks in which storage of desired memories are guaranteed. In the algorithm, individuals correspond to cross-connection parameters, the cost function evaluates the number of spurious memories, and the set of individuals can grow depending on the global best. Performing basic numerical experiments, the algorithm efficiency is investigated.

Keywords: Greedy search · Hysteresis neural nets · Associative memories

1 Introduction

The hysteresis neural network (HNN [1]) is a continuous-time recurrent-type network characterized by binary hysteresis activation function and ternary connection parameters. Depending on parameters, the HNN can exhibits various phenomena: co-existing equilibrium points, synchronization, chaos, and bifurcation [2]. The dynamics is described by a piecewise linear differential equation and he phenomena can be analyzed precisely. The HNN has been applied to associative memories, analog-to-digital converters, and combinatorial optimization problem solvers [1–5]. In the associative memories, we have several theoretical results for storage and stability of desired memories [1]. However, it is hard to suppress spurious memories.

This paper presents a simple evolutionary algorithm based on the greedy search [6,7] and applies it to the suppression of the spurious memories. In the greedy based algorithm, individuals correspond to cross-connection parameters of the HNN and the cost function evaluates the number of spurious memories. The initial individual is given by the correlation based learning [1] that guarantees storage of desired memories. Bit-inversion and elite strategy are applied and the results are evaluated. Depending on the evaluation, the number of individuals can increase. This growing structure is the major difference from the classic greedy search. We refer to this algorithm as growing greedy search (GGS). Performing numerical experiments for basic examples of associative memories,

---

T. Saito—This work is supported in part by JSPS KAKENHI#15K00350.

S. Arik et al. (Eds.): ICONIP 2015, Part III, LNCS 9491, pp. 315–322, 2015.
DOI: 10.1007/978-3-319-26555-1_36

we have confirmed that the GGS operates effective to suppress the spurious memories. Note that this is the first paper of the GGS. Although the GGS is applied to the HNN in this paper, the GGS will be applied to various systems [8–10].

2 Hysteresis Neural Networks

The dynamics of hysteresis neural network (HNN) is described by

$$\dot{x}_i = -x_i + \sum_{l=1}^{N} w_{ij}y_i + d_i \equiv -(x_i - p_i), \ i = 1 \sim N$$
$$y_i = h(x_i) = \begin{cases} +1 \text{ for } x_i > -Th \\ -1 \text{ for } x_i < Th \end{cases} \tag{1}$$

where $x \equiv (x_1, \cdots, x_N)$ is the inner state vector and $y \equiv (y_1, \cdots, y_N)$, $y_i \in \{-1, 1\}$, is the binary output vector. $h(x)$ is the hysteresis activation function as shown in Fig. 1: $h(x)$ is switched from -1 to 1 (respectively, 1 to -1) if x reaches the right threshold $Th > 0$ (respectively, the left threshold $-Th$). w_{ij} is connection parameters and d_i is offset parameters. For simplicity, we assume that the cross connection parameters are ternary and symmetric: $w_{ij} = w_{ji} \in \{-1, 0, 1\}$. We also assume that $d_i = 0$ and $Th = 0$. An equilibrium point of the HNN is given by

$$p(y) \equiv (p_1, \cdots, p_N), \ p_i = \sum_{l=1}^{N} w_{ij}y_i \tag{2}$$

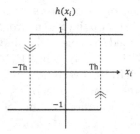

Fig. 1. Binary hysteresis activation function

For stability of the equilibrium point $p(y)$, we have

$$p(y) \text{ is stable } \quad \text{if } p_iy_i > -Th \text{ for any } i$$
$$p(y) \text{ is unstable if } p_iy_i \leq -Th \text{ for some } i \tag{3}$$

In the case where the HNN has equilibrium point, almost all solutions converges to either equilibrium point if

$$w_{ii} + Th > 0 \text{ for } i = 1 \sim N \tag{4}$$

Proofs of Eqs. (2)–(4) can be found in [1]. Hereafter, we consider an application to the associative memory, First, we define the desired memories

$$s^1, \cdots, s^M, \ s^l \equiv (s_1^l, \cdots, s_N^l), \ l = 1 \sim M$$

If we can determine parameters such that the equilibrium point of some desired memory is stable, the desired memory is said to be stored into the HNN. Storage of all the desired memories is guaranteed in the CL-based learning defined as the following [1]. First, the cross connection parameters are given by ternalizeing correlation matrix:

$$w_{ij} = \begin{cases} -1 \text{ for } c_{ij} \leq 0 \\ +1 \text{ for } c_{ij} > 0 \end{cases}, \ c_{ij} = \sum_{l=1}^{M} s_i^l s_j^l, \ i \neq j \tag{5}$$

The auto connection parameter w_{ii} is goven by the minimum integer that satisfies

$$0 < w_{ii} + Th + Q_i, \ Q_i = \min_k \sum_{i \neq j} w_{ij} s_i^k s_j^k \tag{6}$$

If equilibrium point of some output vector is stable and is not any desired memory, the output vector is said to be a spurious memory. Even if all the desired memories are stored, there usually exist spurious memories. Suppression of the spurious memories is an important problem in the associative memories.

3   Growing Greedy Search Algorithm

We present a novel evolutionary algorithm: the growing greedy search method (GGS). First, we give basic definitions. Let t denote evolution steps and let $P^i(t) \equiv (p_1^i(t), \cdots p_M^i(t))$ be the i-th individual at step t where M is the dimension of the individual and $i = 1 \sim k(t)$. Note that the number of individuals $k(t)$ can vary. An individual corresponds to half of the cross-connection parameters:

$$P^i(t) \equiv \{w_{ij}\}, \ i > j, \ i, j = 1 \sim N$$

where $M = N(N-1)/2$. The other half cross-connection parameters are given by the symmetricity, $w_{ji} = w_{ij}$. The cost function is defined by

$$F(P^i(t)) = \text{ the number of spurious memories}$$

Substituting $P^i(t)$ into w_{ij} and substituting the output vectors into Eqs. (2)–(4), this cost function can be calculated where w_{ii} is given by Eq. (6). The GGS consists of the following two sub-programs.

3.1   GGS1: Bit Inversion

Step 1: Initialization. Let $t = 0$ and let $k(t) = 1$. The initial individual $P^i(t)$, $i = 1 \sim k(t)$ is given by the CL-based learning of Eq. (5). The cost function is also initialized, $F(P^i(t))$. Let Gb denote the global best and let $Gb(t) = F(P^i(t))$

Step 2: One bit inversion. For $P^i(t)$, $i = 1 \sim k(t)$, one of the elements is inversed. After the inversion, we obtain $N \times k(t)$ individuals $Q^j(t)$, $j = 1 \sim N \times k(t)$.

Step 3: Evaluation. $\{Q^j(t)\}$ are evaluated by the cost function F. Let $F(Q_b(t))$ be the best value of $F(Q^j(t))$ for $j = 1 \sim N \times k(t)$. The global best is updated:

$$Gb(t) \leftarrow \begin{cases} F(Q^j(t)) & \text{if } F(Q_b(t)) < Gb(t) \\ Gb(t) & \text{otherwise} \end{cases} \qquad (7)$$

If plural individuals $Q^j(t)$ have the same best value of F, then the individuals are preserved in the next step and the number of individuals varies. If the number of the best individuals exceeds the number limit M_a then M_a of the best individuals are selected randomly. For example, if $Q^1(t)$, $Q^7(t)$, and $Q^9(t)$ have the same best value, these three individuals are preserved as $P^1(t)$, $P^2(t)$, and $P^3(t)$, respectively; and let $k(t) \leftarrow 3$.

Step 4: Update check of Gb. If the $Gb(t)$ has not been updated M_b times then the GGS1 is terminated and is switched to GGS2.

Step 5: Let $t \leftarrow t + 1$, go to Step 2, and repeat until the maximum time limit t_{max1}.

3.2   GGS2: Zero Insertion

Step 1: Initialization. Let $t \leftarrow t_1 + 1$ where t_1 is the last step in GGS1. Let $k(t) = 0$. $P^i(t)$, $i = 1 \sim k(t)$ is given by the best individuals after GGS1. If plural best individuals exist, one of them is selected randomly. The cost function and Gb are initialized by the $P^i(t)$.

Step 2: Zero insertion. For $P^i(t)$, $i = 1 \sim k(t)$, zero is inserted into one of the elements. After the insertion, we obtain $N \times k(t)$ individuals $R^j(t)$, $j = 1 \sim N \times k(t)$.

Step 3: Evaluation. $\{R^j(t)\}$ are evaluated by the cost function F. Let $F(R_b(t))$ be the best value of $F(R^j(t))$ for $j = 1 \sim N \times k(t)$. The global best is updated:

$$Gb(t) \leftarrow \begin{cases} F(R^j(t)) & \text{if } F(R_b(t)) < Gb(t) \\ Gb(t) & \text{otherwise} \end{cases} \qquad (8)$$

If plural individuals $Q^j(t)$ have the same best value of F, then the individuals are preserved in the next step and the number of individuals varies. If the number of the best individuals exceeds the number limit M_a then M_a of the best individuals are selected randomly.

Step 4: Update check of Gb. If the $Gb(t)$ has not been updated M_b times then the search is terminated.

Step 5: Let $t \leftarrow t + 1$, go to Step 2, and repeat until the maximum time limit t_{max2}.

Table 1. Parameters after the CL-based learning (1)

j	1	2	3	4	5	6	7	8
w_{1j}	+3	−1	−1	−1	+1	+1	+1	−1
w_{2j}	−1	+1	−1	−1	−1	−1	−1	+1
w_{3j}	+1	−1	+3	−1	−1	−1	+1	−1
w_{4j}	−1	−1	−1	+1	−1	−1	−1	+1
w_{5j}	+1	−1	−1	−1	+3	+1	−1	−1
w_{6j}	+1	−1	−1	−1	+1	+1	−1	+1
w_{7j}	+1	−1	+1	−1	−1	−1	+3	−1
w_{8j}	−1	+1	−1	+1	−1	+1	−1	+1

Table 2. Parameters after the CL-based learning (2)

j	1	2	3	4	5	6	7	8
w_{1j}	+1	+1	+1	−1	−1	−1	+1	−1
w_{2j}	+1	+3	+1	−1	−1	−1	−1	−1
w_{3j}	+1	+1	+1	+1	−1	−1	−1	−1
w_{4j}	−1	−1	+1	+1	+1	+1	−1	−1
w_{5j}	−1	−1	−1	+1	+3	+1	−1	−1
w_{6j}	−1	−1	−1	+1	+1	+1	−1	+1
w_{7j}	−1	−1	−1	−1	−1	−1	+3	+1
w_{8j}	+1	−1	−1	−1	−1	+1	+1	+3

Table 3. Parameters after GGS1 (1)

j	1	2	3	4	5	6	7	8
w_{1j}	+4	+1	+1	−1	−1	−1	−1	−1
w_{2j}	+1	+6	+1	−1	−1	−1	+1	−1
w_{3j}	+1	+1	+2	+1	−1	−1	−1	−1
w_{4j}	−1	−1	+1	0	−1	−1	−1	+1
w_{5j}	−1	−1	−1	−1	0	+1	−1	−1
w_{6j}	−1	−1	−1	−1	+1	0	−1	+1
w_{7j}	−1	−1	−1	−1	+1	−1	+4	−1
w_{8j}	−1	−1	−1	+1	−1	+1	−1	0

Table 4. Parameters after GGS1 (2)

j	1	2	3	4	5	6	7	8
w_{1j}	0	+1	+1	−1	−1	+1	+1	−1
w_{2j}	+1	+2	−1	−1	−1	−1	−1	−1
w_{3j}	+1	−1	0	+1	−1	−1	−1	−1
w_{4j}	−1	−1	+1	0	−1	+1	−1	−1
w_{5j}	−1	−1	−1	−1	+2	+1	−1	−1
w_{6j}	+1	−1	−1	+1	+1	0	−1	+1
w_{7j}	+1	−1	−1	−1	−1	−1	+4	+1
w_{8j}	−1	−1	−1	−1	−1	+1	+1	+4

Table 5. Parameters after GGS2 (1)

j	1	2	3	4	5	6	7	8
w_{1j}	+3	0	+1	0	−1	−1	0	−1
w_{2j}	0	+4	0	−1	−1	−1	+1	−1
w_{3j}	+1	0	+2	0	−1	−1	+1	−1
w_{4j}	0	−1	0	0	−1	−1	−1	+1
w_{5j}	−1	−1	−1	−1	+1	+1	0	−1
w_{6j}	−1	−1	−1	−1	+1	0	−1	+1
w_{7j}	0	+1	+1	−1	0	−1	+2	−1
w_{8j}	−1	−1	−1	+1	−1	+1	+1	0

Table 6. Parameters after GGS2 (2)

j	1	2	3	4	5	6	7	8
w_{1j}	+1	+1	0	−1	−1	0	+1	0
w_{2j}	+1	+1	0	0	−1	−1	−1	0
w_{3j}	0	0	+1	+1	0	−1	−1	−1
w_{4j}	−1	0	+1	+1	0	0	−1	−1
w_{5j}	−1	−1	0	0	+1	+1	0	−1
w_{6j}	0	−1	−1	0	+1	+1	0	+1
w_{7j}	+1	−1	−1	−1	0	0	+2	+1
w_{8j}	0	0	−1	−1	−1	+1	+1	+2

4 Numerical Experiments

In order to investigate the algorithm efficiency, we have performed basic numerical experiments. We consider the following two examples of 14 desired memories.
Example 1 (s^8–s^{14} are inverse patterns of s^1–s^7, respectively.)

$$s^1 \equiv (-1, 1, -1, -1, 1, 1, -1, 1) = -s^8$$
$$s^2 \equiv (1, -1, 1, -1, 1, -1, 1, -1) = -s^9$$
$$s^3 \equiv (1, 1, -1, -1, 1, 1, -1, -1) = -s^{10}$$
$$s^4 \equiv (1, 1, 1, -1, -1, -1, 1, -1) = -s^{11}$$
$$s^5 \equiv (-1, 1, 1, -1, 1, 1, -1, -1) = -s^{12}$$
$$s^6 \equiv (1, -1, 1, -1, 1, 1, -1, -1) = -s^{13}$$
$$s^7 \equiv (-1, 1, 1, 1, -1, -1, -1, 1) = -s^{14}$$

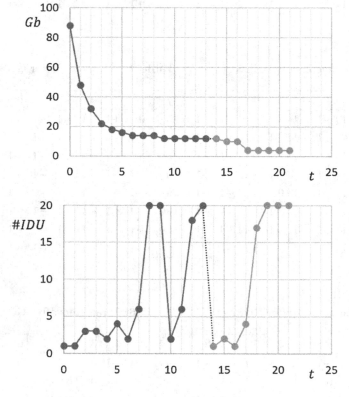

Fig. 2. Evolution process of Gb and #IDU (1)

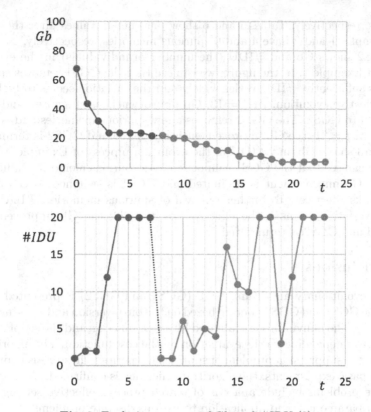

Fig. 3. Evolution process of Gb and #IDU (2)

Example 2 (s^8–s^{14} are inverse patterns of s^1–s^7, respectively.)

$$s^1 \equiv (-1, -1, -1, -1, 1, 1, 1, 1) = -s^8$$
$$s^2 \equiv (1, -1, -1, -1, -1, 1, 1, 1) = -s^9$$
$$s^3 \equiv (1, 1, -1, -1, -1, -1, 1, 1) = -s^{10}$$
$$s^4 \equiv (-1, -1, 1, 1, -1, -1, 1, 1) = -s^{11}$$
$$s^5 \equiv (-1, 1, -1, 1, -1, 1, -1, 1) = -s^{12}$$
$$s^6 \equiv (-1, -1, -1, 1, 1, 1, -1, 1) = -s^{13}$$
$$s^7 \equiv (-1, -1, -1, 1, 1, 1, 1, -1) = -s^{14}$$

Applying the CL-based learning, these desired memories can be stored into 8-dimensional HNN ($N = 8$). Tables 1 and 2 show parameters after the CL-based learning. However, the Examples 1 and 2 have 88 and 68 spurious memories, respectively. In order to suppress the spurious memories, we have applied the GGS1 (bit-inversion) and GGS2 (zero insertion). After trial-and-errors, the algorithm parameters are set as the following.

$$t_{max1} = t_{max2} = 25, M_a = 20, M_b = 5$$

The GGS1 and GGS2 can remuve the spurious memories. Tables 3 and 4 show parameters values after the GGS1. The Examples 1 and 2 have 12 and 22 spurious memories, respectively. Tables 5 and 6 show parameters values after the GGS2. The Examples 1 and 2 have 4 and 6 spurious memories, respectively.

Figure 2 shows Gb and #IDU (the number of individuals) in the evolution process of Example 1. In the figure, we can see that the GGS1 removes spurious memories sufficiently. #IDU varies widely and the variation seems to be helpful for the effective evolution. At $t = 13$, Gb improvement is stagnated and GGS1 is switched to GGS2. The GGS2 removes a few spurious memories and is not so effective. At $t = 21$, the Gb improvement is stagnated and GGS2 is terminated.

Figure 3 shows Gb and #IDU in the evolution process of Example 2. In the figure, we can see that the GGS1 cannot removes spurious memories sufficiently. At $t = 7$, Gb improvement is stagnated and GGS1 is switched to GGS2. The GGS2 works effectively to further removal of spurious memories. #IDU varies and the variation seems to be effective. At $t = 22$, the Gb improvement is stagnated and GGS2 is terminated.

5   Conclusions

A simple evolutionary algorithm, GGS (GGS1 and GGS2) is presented in this paper. The GGS1 and GGS2 operate based on the bit-inversion and zero-insertion, respectively. The individuals is evaluated by a cost function and the set of individuals can grow depending on the evaluation by the cost function. The algorithm is applied to the suppression problem of spurious memories in hysteresis neural networks. In basic experiments. the algorithm efficiency is confirmed.

Future problems include analysis of search process, effective setting of the algorithm parameters, and application to various discrete problems.

References

1. Jin'no, K., Saito, T.: Analysis and synthesis of a continuous-time hysteresis neural network. In: Proceedings of IEEE/ISCAS, pp. 471–474 (1992)
2. Jin'no, K., Nakamura, T., Saito, T.: Analysis of bifurcation phenomena in a 3 cells hysteresis neural network. IEEE Trans. Circuit Syst. I 46(7), 851–857 (1999)
3. Jin'no, K., Tanaka, M.: Hysteresis quantizer, In: Proceedings of IEEE/ISCAS, pp. 661–664 (1997)
4. Nakaguchi, T., Jin'no, K., Tanaka, M.: Hysteresis neural networks for n-queens problems. IEICE Trans. Fund. E82–A(9), 1851–1859 (1999)
5. Nakaguchi, T., Isome, S., Jin'no, K., Tanaka, M.: Box puzzling problem solver by hysteresis neural networks. IEICE Trans. Fund. E84–A(9), 2173–2181 (2001)
6. Couvreur, C., Bresler, Y.: On the optimality of the backward greedy algorithm for the subset selection problem. SIAM J. Matrix Anal. Appl. 21(3), 797–808 (2000)
7. Ertel, W.: Introduction to Artificial Intelligence. Springer, London (2009)
8. Hopfield, J.J.: Neural networks and physical systems with emergent collective computation abilities. Proc. Natl. Acad. Sci. 79, 2554–2558 (1982)
9. Müezzinoglu, M.K., Güzelis, G.: A Boolean Hebb rule for binary associative memory design. IEEE Trans. Neural Netw. 15(1), 195–202 (2004)
10. Tarzan-Lorente, M., Gutierrez-Galvez, Dominique Martinez, D., Marco, S.: A Biologically inspired associative memory for artificial olfaction. In: Proceedings of IJCNN, pp. 25–30 (2010)

Automated Detection of Galaxy Groups Through Probabilistic Hough Transform

Rafee T. Ibrahem[1]([⊠]), Peter Tino[1], Richard J. Pearson[1],
Trevor J. Ponman[1], and Arif Babul[2]

[1] University of Birmingham, Birmingham B15 2TT, UK
{rti273,p.tino}@cs.bham.ac.uk, {richard,tjp}@star.sr.bham.ac.uk
[2] University of Victoria, Victoria, BC V8P 5C2, Canada
babul@uvic.ca

Abstract. Galaxy groups play a significant role in explaining the evolution of the universe. Given the amounts of available survey data, automated discovery of galaxy groups is of utmost interest. We introduce a novel methodology, based on probabilistic Hough transform, for finding galaxy groups embedded in a rich background. The model takes advantage of a typical signature pattern of galaxy groups known as "fingers-of-God". It also allows us to include prior astrophysical knowledge as an inherent part of the method. The proposed method is first tested in large scale controlled experiments with 2-D patterns and then verified on 3-D realistic mock data (comparing with the well-known friends-of-friends method used in astrophysics). The experiments suggest that our methodology is a promising new candidate for galaxy group finders developed within a machine learning framework.

Keywords: Pattern Recognition · Probabilistic Hough transform · Galaxy group finder

1 Introduction

In general, galaxies tend to expand away from one another. However, in certain regions of space there can be an overdensity of galaxies. This results in sufficiently strong gravitational field so that nearby galaxies cannot escape from one another and remain bound together. Galaxy groups play a significant role in explaining the evolution of the universe and measuring its baryonic content. They can also signify gravitational lenses and contribute to the estimation of cosmological parameters [11]. Last but not least, galaxy groups act as laboratories to study different types of galaxy group evolution [18]. Many big galaxy redshift surveys have been conducted to identify galaxy positions in the sky and the recession (line-of-sight) velocities. Given the amounts of available survey data, automated discovery of galaxy groups is of utmost interest to astrophysicists.

One of the common galaxy group finders (with redshift information) is the Friends-of-Friends algorithm (FOF) [7]. The groups are located based on spatial information, linking particles (galaxies) within a pre-specified linking size.

© Springer International Publishing Switzerland 2015
S. Arik et al. (Eds.): ICONIP 2015, Part III, LNCS 9491, pp. 323–331, 2015.
DOI: 10.1007/978-3-319-26555-1_37

Close-by linked pairs (friends) are further aggregated into groups (friends of friends). The linking size is specified according to typical overdensity of galaxies within groups. Several approaches have been proposed to determine the linking size and to measure local galaxy densities [3,14]. Other approaches to galaxy group finding have been based on probabilistic formulations, extending FOF, or including model-based analysis [4,8]. The probabilistic framework enables one to deal consistently with issues such as redshift distortion.

Galaxy groups exhibit a characteristic "fingers of God" (FOG) shape in the angular-Z (redshift) plots - a prolonged dense structure centered at the group position and oriented along the line-of-sight (LOS). We propose to take advantage of such group signatures and develop a dedicated form of model-based probabilistic Hough transform (PHTM). In general, the existing galaxy group finders have many free parameters that need to be carefully set before applying the analysis. This raises issues regarding generality of the results and stability of the calibration process. Hough transform based models have been shown effective in the detection of patterns of interest in cluttered scenes [10]. Probabilistic Hough transform formulation enables us to include explicitly prior expectations on the shape of interest (FOG) through the likelihood model and to treat the background noise consistently.

Hough transform ideas have already been used in astronomy in other contexts, e.g. detection of circular or arc-like forms typically indicative of gravitational lensing [6], identification of continuous gravitational wave signals [2], detection of radial structures on the solar corona [9], or cleaning of the Super-COSMOS Sky Survey (SSS) from the foreground/background noise [16].

The paper has the following organization: After briefly describing the nature of the problem and the data in Sect. 2, our Probabilistic Hough Transform Method (PHTM) is introduced in Sect. 3. We present experimental results in Sect. 4 and conclude in Sect. 5.

2   The Problem of Galaxy Group Identification

The observer on the Earth surveys the universe on a certain patch of the sky identified through two angles - Right Ascension (RA) and Declination (Dec). Besides the spatial position on the sky (RA, Dec), the velocity of the object along the LOS can be deduced from the redshift Z. A typical example of the form of a galaxy survey is shown (as a 2-D slice) in Fig. 1 [1]. Some of the FOG prolonged features (patterns) along the LOS signifying the presence of galaxy groups are clearly visible (marked by red ellipses), some are masked by the background. The challenge is to detect patterns corresponding to the real galaxy groups (true positives), while reducing the detection of similar patterns formed by the fore/background and chance superposition (false positives).

Due to lack of space, we will only briefly outline generation of realistic data involving galaxy groups used in our experiments. The generation process is rather

---

[1] constructed based on a figure from [1].

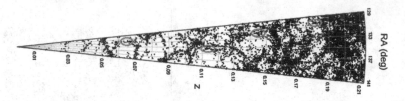

Fig. 1. 2-D slice from a volume of GAMA mock data: RA vs Z. Red elipses signify FOG features [1] (Color figure online).

involved and will be described in detail elsewhere[2]. To test our approach for close redshifts $Z \leq 0.1$, 3-D realistic data consists of two parts - galaxy groups themselves and fore/background galaxies. The key element of the data generation is generation of the individual galaxy groups (while carefully controlling for the extent of galaxy groups of given magnitudes at given Z). The groups are generated from a joint distribution consisting of a Gaussian distribution of dispersed projected velocities along the LOS and the radial distribution in the orthogonal complement of LOS formulated using the Navarro, Frenk and White (NFW) density profiles [12].

Before the methodology is demonstrated on realistic 3-D data, we will first test our method in a large set of controlled experiments in 2-D, where we control for the amount of background noise. In the 2-D setting the LOS direction is the y-axis and the galaxy groups are represented by points (galaxies) generated from Gaussian distributions elongated along the y-axis. In each group we generate 10–25 points from such Gaussian distributions. The background is generated from uniform distribution. The number of background points is determined as $T \cdot N_g$, where N_g is the number of galaxies in galaxy groups and T is a multiplicative factor in the range 5–30. In each setting there are 6 galaxy groups at fixed positions shown in Fig. 2a. A sample of 2D test data obtained with $T = 25$ is presented in 2b.

3 Probabilistic Hough Transform Galaxy Finder

Inspired by a probabilistic formulation of Hough Transform for co-expressed gene detection in 3-color cDNA arrays [17], we have developed a dedicated probabilistic Hough Transform method (PHTM) for galaxy group detection. We first introduce a simplified 2-D model to demonstrate the robustness of the PHTM approach and then introduce the full methodology operating in the 3-D realistic data, taking into account flux limit effects (more distant galaxies of the same intensity are less likely to be observed than closer ones).

3.1 Basic 2-D PHTM Group Finder

The search space is covered by a regular structure of G grid points. On each grid point, we position a noise model representing a possible galaxy group and

---

[2] codes will be available from www.cs.bham.ac.uk/~pxt/my.publ.html.

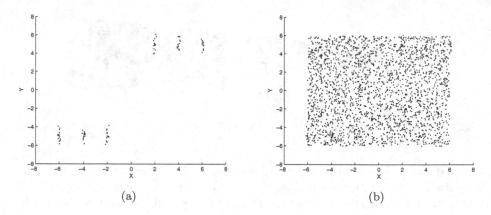

Fig. 2. 2D mock data: (a) Six galaxy groups at fixed positions; (b) galaxy groups with a background density equivalent to 25 times the number of galaxies in galaxy groups.

ask all observed galaxies to ascertain whether they are likely to have come from that group. Formally, for the i-th grid point (x_i, y_i) we have a Gaussian noise model centered at $\mu_i = (x_i, y_i)$ with axis-aligned (diagonal) covariance matrix $C = diag(k \cdot \sigma^2, \sigma^2)$. with variance along the y-axis σ^2 and variance along the x-axis $1/k$ times smaller, i.e. $k \cdot \sigma^2$ (we used $\sigma = 0.5$ and $k = 0.025$). The likelihood model for the i-th grid point is thus a multivariate Gaussian with mean μ_i and covariance C, $p(g|(x_i, y_i), C) = \mathcal{N}(\mu_i, C)$.

Given a galaxy g_q, $q = 1, 2, ..., N$, the degree to which it belongs to the possible group centered at the i-th grid point μ_i is quantified through posterior

$$P(i|g_q) = \frac{p(g_q|\mu_i, C)\cdot)P(i)}{\sum_{j=1}^{G} P(g_q|\mu_j, C)) \cdot P(j)}. \tag{1}$$

We assume no preferred positions for galaxy groups, i.e. flat prior $P(i) = 1/G$. The posterior can be interpreted as a 'soft' vote of the q-th galaxy for the possible galaxy group at position μ_i. The overall vote for the presence of galaxy group at μ_i is then obtained as a flat mixture of posteriors given by the observed galaxies:

$$H(x_i, y_j) = \frac{1}{N} \sum_{q=1}^{N} P(i|g_q). \tag{2}$$

Given a detection threshold $\Theta > 0$, the possible galaxy groups are detected as peaks above Θ in the $H(x_i, y_j)$ landscape. Note that high values of Θ will produce over-cautious conservative detections with a significant number of undetected true galaxy groups (false negatives). On the other hand, low Θ will lead to insignificant low peaks declared as group candidates (false positives).

3.2 Full 3-D PHTM Group Finder in Observational Cone

There are two principal modifications to be made to transform the fundamental model of Sect. 3.1 to the realistic case 3-D mock data (θ, β, Z), where θ and β

denote the RA and Dec, respectively. First, the noise model representing the idea of a galaxy group will be a 3-D Gaussian formulated in the corresponding Cartesian coordinate system (x, y, z) and elongated along the LOS (original axis Z in the cone). Second, in reality, due to the limited sensitivity of observational devices, more distant galaxies are less likely detected than the comparable ones at closer redshift. In what follows we explain how the original model has been adjusted to account for both factors.

After translating from the spherical system (θ, β, Z) to the Cartesian one $x(\theta, \beta, Z), y(\theta, \beta, Z), z(\beta, Z)$, the noise model at the i-th grid point takes the form $p(g|(x_i, y_i, z_i), C) = \mathcal{N}(\mu_i, C)$. To align the prolonged axis-aligned covariance matrix $\tilde{C} = diag(k \cdot \sigma^2, k \cdot \sigma^2, \sigma^2)$ along the LOS, we employ the corresponding rotation matrix R: $C = R\tilde{C}R^T$.

Given the LOS direction $v = (v_x, v_y, v_z)$ in the Cartesian system, the rotation matrix R can be derived by considering the local frame $u = (u_x, u_y, u_z)$, $s = (s_x, s_y, s_z)$ and v. We impose: $u \perp v$, $s \perp v$ and $u \perp s$. In other words, the dot products $v^T u$, $v^T s$ and $u^T s$ vanish. This leads to an undetermined system. By imposing $u = (0, v_z, -v_y)$ we automatically satisfy $v^T u = 0$. Substituting u in $u^T s = 0$, we obtain

$$v_z s_y - v_y s_z = 0, \quad \frac{v_y s_z}{v_z} = s_y. \tag{3}$$

Using $v^T s = 0$, we get

$$v_x s_x + \frac{v_y^2}{v_z} s_z + v_z s_z = 0, \tag{4}$$

yielding

$$s_x = \frac{-s_z(v_y^2 + v_z^2)}{v_x v_z}. \tag{5}$$

We are left with one free parameter, s_z, that can be assigned arbitrary value (we used $s_z = 1$). After normalization of u, s and v into unit vectors, the rotation matrix is formed as $R = [u, s, v]$ (u, s, v form columns of R).

The model developed so far will not work in the real cosmology since it does not account for the flux limit effect. We are more likely to observe galaxies of the same magnitude close by (at smaller Z) than at high Z. Intuitively, a vote from a galaxy of magnitude M observed at high Z should have higher weight than a vote from a closer galaxy of the same magnitude. Galaxies at large Z are harder to observe than those at smaller Z, and there will be more missing votes from undetected galaxies at large Z. In the probabilistic Hough accumulator (2) each observed galaxy has equal weight $1/N$ when voting for galaxy group positions. A principled treatment of this issue in our model formulation is to replace the weight $1/N$ with a redshift and magnitude specific weight. The modified Hough accumulator thus reads

$$H(x_i, y_j, z_i) = \sum_{q=1}^{N} w(q) \cdot P(i|g_q), \tag{6}$$

where $w(q)$ is the weight given to the q-th galaxy based on its redshift Z_q and absolute magnitude M_q. The weights need to sum to 1 and should be inversely related to the luminosity Schechter function $S(M, Z)$, which for a given absolute magnitude M, gives the density of galaxies of that magnitude at redshift Z [15]:

$$S(M, Z) = \frac{\ln(10)}{2.5} \cdot \phi^* \cdot \left(10^{\frac{M^*-M}{2.5}}\right)^{(\alpha+1)} \cdot \exp\left\{-10^{\frac{M^*-M}{2.5}}\right\}, \qquad (7)$$

where $\phi^* = 0.0149 \cdot h^3 \, mpc^{-3}$ is the number density, h is the Hubble parameter, $M^* = -21.35 + 5\log_{10} h$ is the characteristic magnitude and $\alpha = -1.3$ is the faint-end-slope.

The SDSS survey is complete to an apparent Petrosian magnitude limit of $m \approx 17.77$; however, this can vary somewhat across the sky. Following [3], we adopt a more conservative r-band magnitude limit of $m = 17.5$ to simulate SDSS survey [13]. For each galaxy $q = 1, 2, ..., N$, we estimated its absolute magnitude M_q based on m and the redshift Z_q [15]. We propose the following formulation for the weights $w(q)$ that respects both requirements:

$$w(q) = \frac{S(M_q, Z_q)^{-\gamma}}{\sum_{j=1}^{N} S(M_j, Z_j)^{-\gamma}}. \qquad (8)$$

In our experiments, we found $\gamma = 0.3$ to work robustly on the mock data.

4   Experimental Results

We have used the Precision $(TP/(TP + FP))$ versus Recall $(TP/(TP + FN))$ curves in evaluating the group finders, where TP is the number of true positives (correctly detected true groups), FP is the number of false positives (incorrectly detected groups) and FN is the number of false negatives (missed true groups). The precision vs. recall (PvR) curves in Fig. 3 are averages over 10 realizations of background noise in the 2-D data and were obtained by varying the detection threshold Θ. The PHTM method is robust with respect to potentially large amounts of fore/background noise (up to $T = 30$). Note that more direct approaches, such as mixture modeling would end up being swamped with non-group data, even for moderate amounts of background noise (Fig. 2a, b).

For the 3-D realistic data we investigated two settings for the ground truth galaxy groups that needed to be detected: (+5) groups containing at least 5 galaxies (including small, harder to detect groups) and (+10) larger groups containing at least 10 galaxies.

As an example, Fig. 4a–d shows PvR curves of PHTM on two stripes from two different mock data cones consisting of 34 and 26 galaxy groups respectively. The PHTM is compared with FOF method [5]. Note that while it is very natural to create PvR curves from PHTM (by varying Θ), this turned out to be cumbersome for FOF (modifying free parameters can lead to abrupt changes in performance). Therefore, we report a single value (red star) of (precision, recall) obtained with the parameter setting recommended in [5].

To further compare PHTM with FOF, we identify the closest precision value of PHTM to that of FOF and ask if the corresponding recall by PHTM is similar

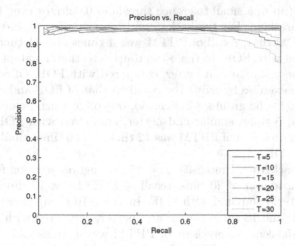

Fig. 3. Average (over 10 realizations of background noise) Precision vs. Recall curves of PHTM on 2D flat mock data

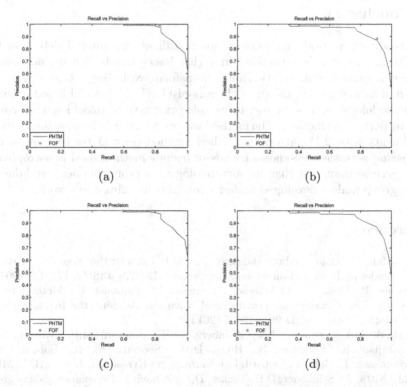

Fig. 4. Precision vs. recall curves of PHTM on realistic mock data when detecting galaxy groups with: (a) +5 galaxies cone-1; (b) +5 galaxies cone-2; (c) +10 galaxies cone-1; (d) +10 galaxies cone-2

to that of FOF (up to a small tolerance threshold 0.025), or even better beyond the tolerance threshold. For the background intensity $T = 5$, in the groups +5 scenario, out of 20 runs, recall of PHTM was 4 times and 11 times similar and better, compared with FOF. In the +10 group detection, recall of PHTM was 2 times and 15 times similar and better, compared with FOF. Of course, one can repeat the same exercise by fixing the recall to that of FOF and comparing the precision values. In the groups +5 scenario, out of 20 runs, precision of PHTM was 8 times and 5 times similar and better, compared with FOF. In the +10 group detection, precision of PHTM was 12 times and 0 times similar and better, compared with FOF.

For higher background intensity $T = 15$ the figures were as follows: In the groups +5 scenario, out of 20 runs, recall of PHTM was 17 times and 0 times similar and better, compared with FOF. In the +10 group detection, recall of PHTM was once and 18 times similar and better, compared with FOF. Finally, in the groups +5 scenario, precision of PHTM was 4 times and 2 times similar and better, compared with FOF. In the +10 group detection, precision of PHTM was 2 times and 3 times similar and better, compared with FOF.

5   Conclusion

We introduced a novel methodology for a difficult automated detection task - finding galaxy groups embedded in a rich background. The methodology is based on a form of probabilistic Hough transform exploiting a typical signature pattern of galaxy groups known as "fingers-of-God". The model based nature of our methodology enables the user to include prior astrophysical knowledge as an inherent part of the method. The method was first tested in large scale controlled experiments with 2-D patterns and then verified on 3-D realistic mock data (comparing with the well-known friends-of-friends method used in astrophysics. The experiments suggest that our methodology is a promising new candidate for galaxy group finders developed within a machine learning framework.

References

1. Alpaslan, M., et al.: Galaxy and mass assembly (gama): the large scale structure of galaxies and comparison to mock universes. MNRAS **438**(1), 177–194 (2014)
2. Astone, P., Colla, A., D'Antonio, S., Frasca, S., Palomba, C.: Method for all-sky searches of continuous gravitational wave signals using the frequency-hough transform. Phys. Rev. D **90**, 042002 (2014)
3. Berlind, A.A., Frieman, J.A., Weinberg, D.H., Blanton, M.R., Warren, M.S., Abazajian, K., Scranton, R., Hogg, D.W., Scoccimarro, R., Bahcall, N.A., Brinkmann, J., Gott, J., Richard, I., Kleinman, S., Krzesinski, J., Lee, B.C., Miller, C.J., Nitta, A., Schneider, D.P., Tucker, D.L., Zehavi, I.: Percolation galaxy groups and clusters in the sdss redshift survey: identification, catalogs, and the multiplicity function. A.J.S **167**, 1–25 (2006)
4. Duarte, M., Mamon, G.A.: Maggie: models and algorithms for galaxy groups, interlopers and environment (2014). arXiv:1412.3364

5. Eke, V.R., et al.: Galaxy groups in the 2dFGRS: the group - finding algorithm and the 2PIGG catalog. MNRAS **348**, 866 (2004)
6. Hollitt, C., Johnston-Hollitt, M.: Feature detection in radio astronomy using the circle hough transform. PASA **29**, 309–317 (2012)
7. Huchra, J.P., Geller, M.J.: Groups of galaxies. I - nearby groups. APJ **257**, 423–437 (1982)
8. Liu, H.B., Hsieh, B., Ho, P.T., Lin, L., Yan, R.: A new galaxy group finding algorithm: probability friends-of-friends. APJ **681**(2), 1046 (2008)
9. Llebaria, A., Lamy, P.: Time domain analysis of solar coronal structures through hough transform techniques. In: Mehringer, D., Plante, R., Roberts, D. (eds.) Astronomical Data Analysis Software and Systems VIII. Astronomical Society of the Pacific Conference Series, vol. 172, p. 46 (1999)
10. Mukhopadhyay, P., Chaudhuri, B.B.: A survey of hough transform. Pattern Recogn. **48**(3), 993–1010 (2015)
11. Mushotzky, R.: Clusters of galaxies: an x-ray perspective. Clusters of Galaxies: Probes of Cosmological Structure and Galaxy Evolution, p. 123 (2004)
12. Navarro, J., Frenk, C.S., White, S.: The structure of cold dark matter halos. APJ **462**, 563 (1996)
13. Pearson, R.J., Ponman, T.J., Norberg, P., Robotham, A.S.G., Farr, W.M.: On optical mass estimation methods for galaxy groups. MNRAS **449**(3), 3082–3106 (2015)
14. Ramella, M., Geller, M.J., Huchra, J.P.: Groups of galaxies in the center for astrophysics redshift survey. APJ **344**, 57–74 (1989)
15. Schechter, P.: An analytic expression for the luminosity function for galaxies. APJ **203**, 297–306 (1976)
16. Storkey, A.J., Hambly, N.C., Williams, C.K.I., Mann, R.G.: Cleaning sky survey data bases using hough transform and renewal string approaches. MNRAS **347**(1), 36–51 (2004)
17. Tino, P., Zhao, H., Yan, H.: Searching for coexpressed genes in three-color cdna microarray data using a probabilistic model-based hough transform. IEEE/ACM Trans. Comput. Biol. Bioinform. **8**(4), 1093–1107 (2011)
18. Tyson, J.A., Valdes, F., Jarvis, J.F., Mills, A.P.: Galaxy mass-distribution from gravitational light deflection. APJ **281**(2), L59–L62 (1984)

A Feature-Based Comparison of Evolutionary Computing Techniques for Constrained Continuous Optimisation

Shayan Poursoltan[(✉)] and Frank Neumann[(✉)]

Optimisation and Logistics, School of Computer Science, University of Adelaide,
Adelaide, SA 5005, Australia
{shayan.poursoltan,frank.neumann}@adelaide.edu.au

Abstract. Evolutionary algorithms have been frequently applied to constrained continuous optimisation problems. We carry out feature based comparisons of different types of evolutionary algorithms such as evolution strategies, differential evolution and particle swarm optimisation for constrained continuous optimisation. In our study, we examine how sets of constraints influence the difficulty of obtaining close to optimal solutions. Using a multi-objective approach, we evolve constrained continuous problems having a set of linear and/or quadratic constraints where the different evolutionary approaches show a significant difference in performance. Afterwards, we discuss the features of the constraints that exhibit a difference in performance of the different evolutionary approaches under consideration.

1 Introduction

There have been many algorithmic approaches proposed to solve complex optimisation problems, including constrained optimisation problems (COP). Several approaches have been proposed to tackle the constraints in constrained problems. Most of the research has been focused on introducing differential evolution (DE) [12], particle swarm optimisation (PSO) [2] and evolutionary strategies (ES) [11] to solve numerical optimisation problems. In order to deal with these constrained problems, there have been techniques that applied to these algorithms such as penalty functions, special operators (separating the constraint and objective function treatment) and decoder based methods. We refer the reader for a survey of constraint handling techniques in evolutionary computing methods to [7].

In order to compare and evaluate the evolutionary algorithms many approaches have been used. One is finding which algorithm performs better on a set of continuous problems using benchmarks sets [3,5]. Recently, there has been an increasing interest to analyse the problem features that make it hard to solve. Initial studies have been carried out in the field of continuous optimisation in [6]. Furthermore, there have been techniques that generate a variation of problem instances from easy to hard. Then, the features of this problem instances are analysed in order to find which of them make the problems hard or easy to solve. Generating the variety

S. Arik et al. (Eds.): ICONIP 2015, Part III, LNCS 9491, pp. 332–343, 2015.
DOI: 10.1007/978-3-319-26555-1_38

of problem instances from easy to hard ensures that the knowledge obtained from analysis is reliable.

Although there is not only a standalone feature that makes a problem hard to solve, but it is assumed that constraints are very important in constrained continuous problems. The evolving approach that has been used to analyse the constraint features and their effects on COP's difficulty is discussed in [8,9]. The idea is to evolve constrained problem instances (by using an evolutionary algorithm) in order to identify the constraint features with more contribution to problem difficulty.

In this paper, by using a single-objective evolutionary algorithm, we generate hard and easy COP instances for DE, ES and PSO algorithms. Later, we solve the generated instances using one algorithm by the other algorithms. The results show that the hardest generated instances using one algorithm are still hard for the other ones. To get better insight, we use multi-objective evolving approach to generate instances that are hard for one algorithm but still easy for the others. By analysing how an algorithm fails in conditions where the rest perform well, we can derive its strengths and weaknesses over constraint features. Our study shows the effectiveness of constraint features that make the problems hard for one and easy for the other algorithms. It can be translated as over which features of constraints, they make the problems hard for a certain algorithm but still easy for the others.

The remainder of this paper is as follows: In Sect. 2 we introduce the concept of COPs. Then we discuss the evolver (single and multi-objective evolutionary approach) and the solver algorithms (DE, ES and PSO) we use in our experiments. In Sect. 3 we analyse the performance of various algorithms on each others hard and easy instances (using the single-objective evolver). Section 4 includes the multi-objective approach that generates hard instances for one but easy for the other algorithms. Furthermore, we carry out the analysis of linear and quadratic constraint features that make the problem hard for one and still easy for the rest. Finally, we conclude with some remarks.

2 Preliminaries

2.1 Constrained Continuous Optimisation Problems

In this study, constrained continuous optimisation problems with inequality and equality constraints are investigated. These problems are optimisation problems where a function $f(x)$ should be optimised with respect to a given set of constraints.

Single-objective functions $f \colon S \to \mathbb{R}$ with $S \subseteq \mathbb{R}^n$ are considered in this research. The constraints impose a feasible subset $F \subseteq S$ of the search space S and the aim is finding $x \in S \cap F$ which minimises f. Formally, we state the problems as follows:

$$
\begin{aligned}
\text{minimize} \quad & f(x), \quad x = (x_1, \ldots, x_n) \in \mathbb{R}^n \\
\text{subject to} \quad & g_i(x) \leq 0 \quad \forall i \in \{1, \ldots, q\} \\
& h_j(x) = 0 \quad \forall j \in \{q+1, \ldots, p\}
\end{aligned}
\tag{1}
$$

where $x = (x_1, x_2, \ldots, x_n)$ is an n dimensional vector and $x \in S \cap F$. The $g_i(x)$ (inequality) and $h_j(x)$ (equality) constraints could be linear/nonlinear. Also, the equality constraints are usually replaced by $|h_j(x)| \leq \varepsilon$ where $\varepsilon = 10e^{-4}$ [5]. The feasible region $F \subseteq S$ of the search space S is defined by

$$l_i \leq x_i \leq u_i, \qquad 1 \leq i \leq n \tag{2}$$

where l_i and u_i denote lower and upper bounds respectively for the ith variable in which $1 \leq i \leq n$. In this paper, we focus on the ability of constraints (linear, quadratic) to make a problem hard or easy. The features of these constraints and their effect on problem difficulty is discussed. The constraints are of the following form:

$$\text{linear constraint} \quad g(x) = b + a_1 x_1 + \ldots + a_n x_n \tag{3}$$

$$\text{quadratic constraint} \quad g(x) = b + a_1 x_1^2 + a_2 x_1 \ldots + a_{2n-1} x_n^2 + a_{2n} x_n \tag{4}$$

or a combination of them, where $x_1, x_2 \ldots, x_n$ are values from Eq. 1 and $a_1, a_2, \ldots, a_n$ are coefficients within lower (l_i) and upper bounds (u_i). We assume univariant quadratic function to analyse each x_n (with exponent 2) independently. Also, unvivarient quadratic constraints are more popular in recent benchmarks [5]. In order to include the optimum of objective function in feasible area, we set $b \leq 0$ (we assume the objective function optimum is zero).

2.2   Algorithms

We now introduce the algorithms for constrained continuous optimisation that are subject to our investigation.

One of the most prominent evolutionary algorithms for COPs is ε-constrained differential evolution with an archive and gradient-based mutation (εDEag). The algorithm is the winner of 2010 CEC competition for continuous COPs [5]. The εDEag uses ε-constrained method to transform algorithms for unconstrained problems to constrained ones. It adopts ε-level comparison to order the possible solutions. In other words, the lexicographic order is used in which constraint violation ($\phi(x)$) has more priority and proceeds the function value ($f(x)$). For more details we refer the reader to [13].

The second algorithm we use in this paper is a $(1+1)$ CMA-ES for constrained optimisation [1]. The $(1+1)$ CMA-ES in [4] is a variant of $(1+1)$-ES which adapts the covariance matrix of its offspring distribution in addition to its global step size. The idea behind the constraint handling approach of this algorithm is to obtain approximations to the normal vectors directions in the vicinity of the current solutions locations by low-pass filtering steps which violates the respective constraints and reducing the variance of the offspring distribution in these directions. Incorporating this constraint handling approach with $(1+1)$ CMA-ES makes an algorithm which is significantly more efficient than other approaches for constrained evolutionary algorithms. Also, the selected algorithm is not sensitive to the rotation of the problem search space. We refer the reader to [1] for more details and implementation.

The third algorithm that is used in our investigation is a particle swarm optimisation. This algorithm (HMPSO) applies a method that uses parallel search operator in which it divides the current swarm into various sub-swarms and locates the solution between them. In each sub-swarm, all particles follow the local best (fittest particle) which improves them to be more fitter. Also, since all sub-swarms are located around different optima (in parallel), then it is more possible to locate multiple optima which improves the diversity of algorithm. Dividing the swarms into sub-swarms improves the diversity of the algorithm. Also, choosing the local best in each sub-swarm can attract the other particles to fitter positions. We refer the reader to [15] for detailed algorithm and implementation.

2.3   Features of Constraints

In this paper we analyse the constraint features of generated problem instances. These features are constraint coefficients relationships such as standard derivation, angle between constraint hyperplanes, feasibility ratio in vicinity of optimum, number of constraints, shortest distance of constraint hyperplane to optimum. The details of these features are discussed in [8].

3   Single-Objective Investigations

We first consider different algorithms and compare their relative performance on each other's generated hard and easy instances. We use single-objective evolver to evolve and generate hard and easy instances for all types of algorithms. The detailed procedure and results for DE instances are discussed in [8]). For this experiment, we perform 30 independent runs generating easy and hard instances for PSO and ES solvers. It means, the single-objective evolver only generates instances that are hard/easy for one type of algorithm (PSO, ES and DE). The required function evaluation number (FEN) for solving these instances (PSO, ES and DE) is used as fitness value for single-objective evolver. The parameters for solvers are identical to [1,13,15]. Also, we run our experiments on Sphere function (bowl shaped) [3]. We now have three groups of easy and hard instances generated for DE, ES and PSO algorithms. We then compare the DE, ES and PSO algorithms by applying them on each other's easy and hard instances. The analysis is done by comparing the required FEN for an algorithm to solve the other's generated problem instances. Then, it is possible to derive strengths and weaknesses of the considered algorithms by observing how well one algorithm performs in conditions where the other algorithms fail (or it is difficult for them). Tables 1 and 2 show different algorithms performance on Sphere objective functions with linear/quadratic constraints (1 to 5 constraints). Considering the required FEN to solve each instances, it is observed that hard instances are still the hardest for their own algorithms and hard for the others. It implies

Table 1. The comparison of algorithms performance on each other's easy and hard instances based on required FEN for Sphere objective function. DE Easy (1 c) means instances that are easy for DE and with 1 **linear constraint**.

Instances	DE algorithm	ES algorithm	PSO algorithm
DE Easy (1 c)	25.6K	28.2K	33.2K
ES Easy (1 c)	26.3K	27.1K	33.9K
PSO Easy (1 c)	24.9K	29.1K	72.5K
DE Easy (2 c)	28.9K	21.9K	32.1K
ES Easy (2 c)	25.2K	24.3K	29.4K
PSO Easy (2 c)	24.2K	25.2K	33.5K
DE Easy (3 c)	32.4K	31.2K	33.9K
ES Easy (3 c)	31.8K	29.1K	33.2K
PSO Easy (3 c)	35.1K	28.6K	35.1K
DE Easy (4 c)	34.2K	29.8K	38.2K
ES Easy (4 c)	32.1K	31.5K	36.1K
PSO Easy (4 c)	35.7K	28.9K	39.5K
DE Easy (5 c)	35.3K	42.1K	46.4K
ES Easy (5 c)	31.2K	45.2K	38.2K
PSO Easy (5 c)	35.3K	44.9K	41.2K
DE Hard (1 c)	91.2K	78.3K	76.4K
ES Hard (1 c)	81.3K	86.4K	78.8K
PSO Hard (1 c)	82.5K	72.5K	85.4K
DE Hard (2 c)	93.4K	81.3K	81.4K
ES Hard (2 c)	84.3K	92.6K	79.4K
PSO Hard (2 c)	85.7K	84.1K	89.4K
DE Hard (3 c)	98.3K	93.8K	78.9K
ES Hard (3 c)	91.4K	108.6K	81.2K
PSO Hard (3 c)	89.1K	98.2K	91.6K
DE Hard (4 c)	104.2K	89.4K	82.5K
ES Hard (4 c)	89.4K	115.1K	78.4K
PSO Hard (4 c)	92.9K	93.5K	115.3K
DE Hard (5 c)	123.2K	111.4K	98.4K
ES Hard (5 c)	98.2K	133.2K	94.9K
PSO Hard (5 c)	101.3K	109.2K	118.3K

Table 2. The comparison of algorithms performance on each other's easy and hard instances based on required FEN for Sphere objective function. DE Easy (1 c) means instances that are easy for DE and with 1 **quadratic constraint**.

Instances	DE algorithm	ES algorithm	PSO algorithm
DE Easy (1 c)	24.2K	23.6K	24.9K
ES Easy (1 c)	24.8K	24.2K	25.4K
PSO Easy (1 c)	26.4K	25.4K	26.4K
DE Easy (2 c)	25.3K	28.1K	26.4K
ES Easy (2 c)	24.1K	27.2K	27.4K
PSO Easy (2 c)	23.5K	29.3K	271.K
DE Easy (3 c)	27.9K	31.9K	35.5K
ES Easy (3 c)	29.4K	32.1K	28.5K
PSO Easy (3 c)	28.1K	28.7K	29.4K
DE Easy (4 c)	34.1K	28.9K	36.4K
ES Easy (4 c)	35.2K	35.3K	31.6K
PSO Easy (4 c)	31.8K	29.5K	33.2K
DE Easy (5 c)	38.7K	29.2K	37.2K
ES Easy (5 c)	35.6K	28.2K	39.5K
PSO Easy (5 c)	36.3K	31.5K	36.2K
DE Hard (1 c)	129.3K	102.7K	105.3K
ES Hard (1 c)	104.3K	121.2K	108.2K
PSO Hard (1 c)	108.2K	104.2K	119.8K
DE Hard (2 c)	132.6K	114.2K	114.9K
ES Hard (2 c)	111.2K	127.1K	112.4K
PSO Hard (2 c)	109.4K	112.4K	125.3K
DE Hard (3 c)	136.2K	116.3K	112.4K
ES Hard (3 c)	117.2K	132.1K	109.9K
PSO Hard (3 c)	119.8K	119.2K	132.6K
DE Hard (4 c)	141.2K	119.9K	119.6K
ES Hard (4 c)	113.8K	131.2K	121.9K
PSO Hard (4 c)	115.4K	121.4K	138.9K
DE Hard (5 c)	149.3K	129.7K	122.9K
ES Hard (5 c)	124.4K	149.6K	126.4K
PSO Hard (5 c)	123.9K	124.2K	148.3K

that the hard instances share some common features to make it difficult to solve for all solvers. However, the obtained knowledge is not enough to compare the algorithm capabilities to solve hard problem instances.

4   Multi-objective Investigations

Based on the experiment results in previous section, hard instances for each algorithm are still hard for the others. In order to extract more useful knowledge about the strengths or weaknesses of certain algorithms on constraint algorithms, we need problem instances that are hard for one and easy for the others. Analysing the features of these instances helps us extracting knowledge regarding the strengths and weaknesses of algorithms by examining why an algorithm performs better on some groups of features while the others fails. This will help us developing more efficient prediction model for automated algorithm selection.

To do this, we use a multi-objective DE algorithm (DEMO) described in [10] to minimise the FEN for one algorithm and maximise it for the others. In other words, the FEN for generated problem instances is higher (harder) for a certain algorithm and lower (easier) for the others. In order to find instances that are hard for one algorithm type and easy for the others, we need to find solution as diverse as possible. Also, the solutions need to be close to pareto front. Satisfying these two aims makes us to use multi-objective evolutionary algorithm to generate problem instances. Hence, we use differential evolution for multi-objective optimisation (DEMO) proposed by Robic in [10]. Based on results in [10], the DEMO achieves efficiently the above two goals. In DEMO, the candidate solution replaces parent when it dominates it and if the parent dominates it, the candidate is discarded. Otherwise, if the candidate and parent cannot dominate each other, the candidate is added to the population. The major difference between DEMO and other multi-objective evolutionary algorithms is that the newly generated good candidates are immediately used in creation of the subsequent candidates. This improves fast convergence to the true pareto front, while the use of non-dominated sorting and crowding distance metric in truncation of the extended population promotes the uniform spread of solutions. We refer the reader to [10] for further details and implementation.

In the following, we discuss the results for algorithms performances comparison. We carry out 30 independent runs for each number of constraints that are hard for one algorithm but still easy for the others. We set the evolving algorithm (DEMO) generation number to 5000 and the other parameters of evolving algorithm are set to pop size $= 40$, CR $= 0.5$, scaling factor $= 0.9$ and FEN_{max} is $300K$. Values for these parameters have been obtained by optimising the performance of the evolving algorithm in order to achieve the more easier and harder problem instances. For each of three algorithms, their best parameters are chosen [3,13,15]. First, the (εDEag) algorithm parameters are considered as: generation number $= 1500$, pop size $= 100$, CR $= 0.5$, scaling factor $= 0.5$. Also, the parameters for e-constraint method are described in [8]. Moreover, for evolutionary strategy we perform $(1,7)$-ES algorithm with 1500 generation using $P_f = 0.4$ with tendency to focus on feasible solution. In HMPSO algorithm, the swarm size N is set to 60, each sub-swarm size (N_s) is 8 and all the PSO parameters are considered as Krohling and Coelho's PSO [14]. In order to solve generated COPs, HMPSO generation number is set to 1500. We need to say the parameters for the solvers are identical to those given in [1,10,13,15]

In our all experiments, we generate set of problem instances that are hard to one algorithm and easy to the other ones. Tables 3, 4, 5, 6, 7 and 8 show the function evaluation number (FEN) required for each algorithm to solve DE/ES/PSO hard instances for Sphere, Ackley and Rosenbrock objective functions (with 1 to 5 linear/quadratic constraints). As it is observed, there is more difference between the required FEN of instances generated by multi-objective algorithm evolver than the single-objective one. For instance, the required FEN for solving DE hard instances are higher for DE algorithm than solving it by ES and PSO algorithm. It means the DE hard instances are only hard for DE algorithm and

Table 3. The FEN required for each algorithm to solve DE/ES/PSO hard instances (Sphere for 1 to 5 **linear constraints**)

Instances	DE algorithm	ES algorithm	PSO algorithm
DE hard (1 c)	**86.3K**	41.5K	43.2K
ES hard (1 c)	45.7K	**84.2K**	48.3K
PSO hard (1 c)	37.2K	41.8K	**80.1K**
DE hard (2 c)	**88.8K**	43.9K	44.2K
ES hard (2 c)	45.9K	**85.4K**	46.3K
PSO hard (2 c)	43.2K	42.5K	**82.9K**
DE hard (3 c)	**91.4K**	44.6K	45.3K
ES hard (3 c)	49.2K	**87.8K**	48.1K
PSO hard (3 c)	46.2K	47.7K	**85.5K**
DE hard (4 c)	**94.2K**	47.5K	47.8K
ES hard (4 c)	51.7K	**89.1K**	50.1K
PSO hard (4 c)	48.7K	49.9K	**87.3K**
DE hard (5 c)	**96.2K**	48.2K	49.5K
ES hard (5 c)	52.4K	**90.4K**	53.5K
PSO hard (5 c)	49.6K	51.4K	**91.6K**

Table 4. The FEN required for each algorithm to solve DE/ES/PSO hard instances (Sphere for 1 to 5 **quadratic constraints**)

Instances	DE algorithm	ES algorithm	PSO algorithm
DE hard (1 c)	**92.3K**	50.2K	51.9K
ES hard (1 c)	48.8K	**91.3K**	49.3K
PSO hard (1 c)	44.5K	46.8K	**93.1K**
DE hard (2 c)	**93.5K**	52.9K	54.2K
ES hard (2 c)	50.9K	**95.9K**	51.2K
PSO hard (2 c)	50.2K	53.2K	**96.3K**
DE hard (3 c)	**95.9K**	54.3K	55.3K
ES hard (3 c)	53.9K	**97.4K**	52.4K
PSO hard (3 c)	57.3K	56.3K	**98.9K**
DE hard (4 c)	**98.3K**	56.4K	57.3K
ES hard (4 c)	56.3K	**102.3K**	52.1K
PSO hard (4 c)	59.2K	58.2K	**101.6K**
DE hard (5 c)	**102.1K**	58.3K	59.4K
ES hard (5 c)	59.2K	**103.2K**	60.2K
PSO hard (5 c)	62.6K	63.8K	**105.2K**

Table 5. The FEN required for each algorithm to solve DE/ES/PSO hard instances (Ackley for 1 to 5 **linear constraints**)

Instances	DE algorithm	ES algorithm	PSO algorithm
DE hard (1 c)	**102.3K**	46.1K	51.4K
ES hard (1 c)	51.2K	**104.7K**	50.2K
PSO hard (1 c)	47.4K	49.8K	**107.4K**
DE hard (2 c)	**112.1K**	56.1K	54.1K
ES hard (2 c)	53.9K	**115.9K**	48.6K
PSO hard (2 c)	55.5K	55.3K	**117.2K**
DE hard (3 c)	**126.1K**	63.7K	65.2K
ES hard (3 c)	59.1K	**128.3K**	58.7K
PSO hard (3 c)	61.7K	62.8K	**134.2K**
DE hard (4 c)	**124.9K**	68.4K	63.1K
ES hard (4 c)	64.1K	**129.8K**	59.2K
PSO hard (4 c)	67.5K	69.2K	**135.2K**
DE hard (5 c)	**138.8K**	75.2K	74.1K
ES hard (5 c)	71.2K	**137.1K**	76.7K
PSO hard (5 c)	73.1K	74.1K	**141.2K**

Table 6. The FEN required for each algorithm to solve DE/ES/PSO hard instances (Ackley for 1 to 5 **quadratic constraints**)

Instances	DE algorithm	ES algorithm	PSO algorithm
DE hard (1 c)	**142.5K**	60.1K	62.5K
ES hard (1 c)	58.5K	**148.2K**	61.4K
PSO hard (1 c)	53.2K	53.9K	**147.7K**
DE hard (2 c)	**153.3K**	58.1K	58.1K
ES hard (2 c)	59.2K	**155.5K**	59.2K
PSO hard (2 c)	57.8K	56.3K	**157.2K**
DE hard (3 c)	**167.3K**	65.2K	68.1K
ES hard (3 c)	63.2K	**169.2K**	69.8K
PSO hard (3 c)	65.7K	67.9K	**167.6K**
DE hard (4 c)	**174.8K**	71.2K	75.1K
ES hard (4 c)	66.8K	**169.1K**	72.9K
PSO hard (4 c)	69.1K	68.3K	**172.9K**
DE hard (5 c)	**179.5K**	75.1K	76.1K
ES hard (5 c)	72.8K	**174.9K**	77.4.2K
PSO hard (5 c)	75.1K	74.9K	**175.9K**

easy for the others. In the following we start analysing constraint features of instances that are hard for one and easy for others.

4.1   Analysis for Linear Constraints

We run our experiments on Sphere, Ackley and Rosenbrock objective functions. The linear constraints are considered as in Eq. 3 with all coefficients a_ns that are in the range of $[-5, 5]$. Also, the problem dimension is set to 30. As it mentioned

Table 7. The FEN required for each algorithm to solve DE/ES/PSO hard instances (Rosenbrock for 1 to 5 **linear constraints**)

Instances	DE algorithm	ES algorithm	PSO algorithm
DE hard (1 c)	**103.1K**	48.2K	53.9K
ES hard (1 c)	53.1K	**107.2K**	52.7K
PSO hard (1 c)	45.7K	48.1K	**109.2K**
DE hard (2 c)	**115.1K**	57.8K	55.7K
ES hard (2 c)	54.7K	**113.4K**	46.1K
PSO hard (2 c)	54.8K	54.9K	**119.5K**
DE hard (3 c)	**124.3K**	65.2K	62.1K
ES hard (3 c)	58.8K	**127.1K**	59.7K
PSO hard (3 c)	62.3K	65.5K	**136.1K**
DE hard (4 c)	**125.5K**	69.1K	65.2K
ES hard (4 c)	67.5K	**128.1K**	58.7K
PSO hard (4 c)	64.9K	70.6K	**137.1K**
DE hard (5 c)	**135.1K**	74.1K	74.7K
ES hard (5 c)	73.7K	**135.8K**	75.1K
PSO hard (5 c)	72.3K	76.5K	**140.9K**

Table 8. The FEN required for each algorithm to solve DE/ES/PSO hard instances (Rosenbrock for 1 to 5 **quadratic constraints**)

Instances	DE algorithm	ES algorithm	PSO algorithm
DE hard (1 c)	**143.1K**	61.4K	63.7K
ES hard (1 c)	59.6K	**149.7K**	62.5K
PSO hard (1 c)	54.3K	54.2K	**143.8K**
DE hard (2 c)	**155.2K**	59.2K	59.2K
ES hard (2 c)	61.3K	**154.8K**	57.9K
PSO hard (2 c)	59.2K	57.1K	**158.0K**
DE hard (3 c)	**168.9K**	63.7K	66.8K
ES hard (3 c)	65.2K	**170.1K**	68.1K
PSO hard (3 c)	63.7K	68.1K	**168.9K**
DE hard (4 c)	**175.1K**	73.8K	76.9K
ES hard (4 c)	68.2K	**172.7K**	75.2K
PSO hard (4 c)	67.7K	69.1K	**176.2K**
DE hard (5 c)	**180.2K**	74.2K	77.8K
ES hard (5 c)	74.2K	**175.1K**	79.2K
PSO hard (5 c)	73.6K	74.4K	**179.4K**

before, to analyse and discuss some features such as shortest distance, we assume that the optimum is zero ($b \leq 0$). We use three types of problem instances. DE hard denotes problem instances that are hard for DE algorithm but still easy for PSO and ES algorithms. Also, ES hard instances are easy for DE and PSO algorithm in this section. PSO hard means the instances that are hard for PSO but easy for the rest. Each constraint is generated using multi-objective evolver to generate instances that are hard for one algorithm but easy for others. In the following we discuss the features of linear constraints.

Figure 1 represents some evidence of linear constraint coefficient relationship (standard deviation). It is shown that standard deviation of (1 to 5) linear constraints are higher for DE hard instances than ES and PSO hard ones. This result is similar for all Sphere, Ackley and Rosenbrock objective functions. This means, the instances that are hard for DE algorithm but easy for ES and PSO have higher standard deviation for their constraints coefficients. In other words, this constraint feature has influence on problem difficulty. This improves the prediction ability for algorithm selection framework.

Box plots shown in Fig. 2 represent the shortest distance from optimum feature for hard instances. Based on the experiments, hard instances for ES algorithm have higher value (closer to optimum) shortest distance than the other algorithms. It is noteworthy that lower value in Fig. 2 means the constraint hyperplane is further from optimum. In other words, the constraints hyperplanes are closer to the optimum in ES hard instances. This relationship holds the pattern for all objective functions in linear constraints. We also study the feasibility ratio in vicinity of the optimum. As observed in Table 10, hard DE instances have lower feasibility ratio comparing to PSO and ES hard instances. This follows the same pattern for all experimented objective functions. Also increasing the number of constraints decreases the problem optimum-local feasibility for all algorithm problem instances. The angle between linear constraints feature is analysed for linear constraints. As it is observed in Table 9, ES hard instances have lower angle values for all Sphere, Ackley and Rosenbrock objective functions. This means, instances that are hard for ES have less angle value between their constraint hyperplanes. Interestingly, all objective function that we use in this experiment follow the same relationship.

As it is observed, to compare the instances, DE hard instances have higher linear constraint coefficient standard deviation. It can be translated as DE algorithm has more difficulty to coefficients standard deviation feature than PSO and ES algorithms. Also, the local-optimum feasibility ratio value is higher in ES and PSO hard instances than DE hard ones. This means, ES and PSO algorithms are more effective to problems with higher optimum feasibility ratio feature. The shortest distance and angle features for ES is less than DE and PSO hard instances. Interestingly, this features are similar for all used objective functions. The linear constraint feature based analysis gives us helpful knowledge to implement algorithm selection framework.

4.2   Analysis for Quadratic Constraints

In this section, we carry out our experiments on Sphere, Ackley and Rosenbrock objective function with quadratic constraints (see Eq. 4) using same setup as previous section. In the following we do feature based analysis of constraints in hard DE, PSO and ES instances (that are easy for the other algorithms).

Figure 1 shows some evidence of quadratic constraint coefficients relationship. Based on our experiments, in each constraint, the quadratic coefficient has more ability than linear coefficients to make problem harder to solve. In other words, in Eq. 4, a_1 is more contributing than a_2 to problem difficulty. As it is shown in the box plots, the standard deviation of 1 to 5 quadratic constraints in DE hard instances are higher comparing the other two algorithm hard instances. In contrast, our results show no systematic relationship between problem difficulty and linear coefficients in each quadratic constraints and quadratic coefficients have more contribution in problem difficulty.

As it is observed in Fig. 2, the shortest distance feature for DE, PSO and ES hard instances are compared. In instances that are hard for ES and easy for the other algorithms, the quadratic constraint hyperplanes are closer to optimum (zero). This applies to all experimented objective functions. Also, calculating the angle feature for quadratic constraint does not show any systematic relationship to problem difficulty. The feasibility ratio near the optimum is analysed for DE, ES and PSO hard instances. As it is shown in Table 11, the feasibility ratio in DE hard instances are lower than the other algorithms hard instances. All objective functions have the same pattern. Also, the number of constraint has a systematic relationship with feasibility ratio.

Based on the results, to compare COP instances with quadratic constraints, DE hard instances have higher coefficient standard deviation value than the other

Table 9. The angle feature for Sphere objective function for linear constraints

	Cons 1,2	Cons 1,3	Cons 1,4	Cons 1,5	Cons 2,3	Cons 2,4	Cons 2,5	Cons 3,4	Cons 3,5	Cons 4,5
DE Hard	74	64	63	58	74	71	68	59	62	86
ES Hard	33	21	37	24	44	46	39	46	48	51
PSO Hard	75	63	82	68	71	73	72	69	81	86

Table 10. Optimum-local feasibility ratio of search space near the optimum for 1, 2, 3, 4 and 5 linear constraint

	1 cons	2 cons	3 cons	4 cons	5 cons
DE Hard	6 %	5 %	3 %	3 %	2 %
ES Hard	16 %	11 %	10 %	6 %	5 %
PSO Hard	17 %	12 %	11 %	8 %	5 %

Table 11. Optimum-local feasibility ratio of search space near the optimum for 1, 2, 3, 4 and 5 quadratic constraint

	1 cons	2 cons	3 cons	4 cons	5 cons
DE Hard	4 %	4 %	3 %	2 %	2 %
ES Hard	14 %	10 %	8 %	7 %	5 %
PSO Hard	15 %	10 %	9 %	8 %	7 %

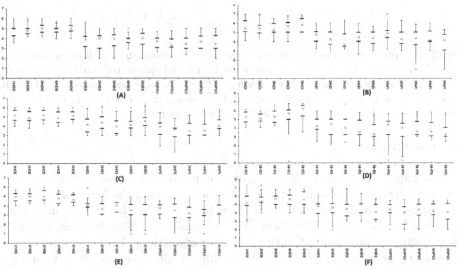

Fig. 1. Box plot for standard deviation of coefficients in linear constraints with objective functions: Sphere (A), Ackley (C) and Rosenbrock (E) and quadratic constraints Sphere (B), Ackley (D) and Rosenbrock (F). Each sub figure includes hard instances (H) with 1 to 5 constraints using algorithms (a/b/c denotes a: number of constraints, b: hard instances and c: hard instances for DE/ES/PSO algorithm).

algorithm hard ones. It is translated as the DE algorithm has more difficulty solving instances with higher standard deviation value for their quadratic constraints than ES and PSO. Also, the quadratic constraints are closer to optimum in ES instances than the other experimented algorithms. In other words, ES algorithm is more influenced by constraint with closer to optimum instances. Moreover, the optimum feasibility ratio in DE instances are lower than PSO and ES.

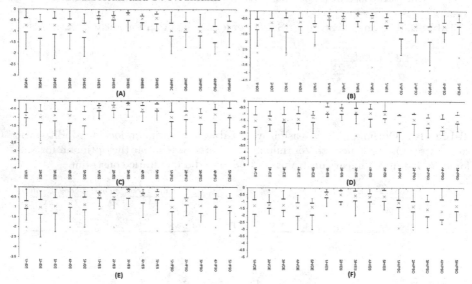

Fig. 2. Box plot for shortest distance feature in linear constraints with objective functions: Sphere (A), Ackley (C) and Rosenbrock (E) and quadratic constraints Sphere (B), Ackley (D) and Rosenbrock (F). Each sub figure includes hard instances (H) with 1 to 5 constraints using algorithms (a/b/c denotes a: number of constraints, b: hard instances and c: hard instances for ES/PSO/DE algorithm).

5   Conclusion

In this paper, we carried out an algorithm performance comparison on each others constrained problem instances. We then analysed the features and characteristics of constraints that make them hard to solve for certain algorithm but easy for the others. It is observed that some constraint features are more contributing to problem difficulty for certain algorithms. In linear constraints, some features such as coefficient relationship, angle, local-optimum feasibility ratio and shortest distance play an important role in problem difficulty to DE and ES algorithms. Considering quadratic instances, angle does not show any relationship to problem difficulty.

By analysing how well one algorithm performs in conditions where other algorithms fail, we can derive its strengths and weaknesses over constrained problems. These results can help us to improve the efficiency of algorithm prediction model.

Acknowledgments. Frank Neumann has been supported by ARC grants DP130104395 and DP140103400.

References

1. Arnold, D.V., Hansen, N.: A (1+1)-CMA-ES for constrained optimisation. In: Proceedings of the 14th Annual Conference on Genetic and Evolutionary Computation, pp. 297–304. ACM (2012)
2. Eberhart, R., Kennedy, J.: A new optimizer using particle swarm theory. In: 1995 Proceedings of the Sixth International Symposium on Micro Machine and Human Science, MHS 1995, pp. 39–43. IEEE (1995)
3. Hansen, N., Auger, A., Finck, S., Ros, R.: Real-parameter black-box optimization benchmarking 2010: experimental setup (2010)
4. Igel, C., Suttorp, T., Hansen, N.: A computational efficient covariance matrix update and a (1+1)-CMA for evolution strategies. In: Proceedings of the 8th Annual Conference on Genetic and Evolutionary Computation, pp. 453–460. ACM (2006)
5. Mallipeddi, R., Suganthan, P.N.: Problem definitions and evaluation criteria for the cec 2010 competition on constrained real-parameter optimization. Nanyang Technological University, Singapore (2010)
6. Mersmann, O., Preuss, M., Trautmann, H.: Benchmarking evolutionary algorithms: towards exploratory landscape analysis. In: Schaefer, R., Cotta, C., Kołodziej, J., Rudolph, G. (eds.) PPSN XI. LNCS, vol. 6238, pp. 73–82. Springer, Heidelberg (2010)
7. Mezura-Montes, E., Coello Coello, C.A.: Constraint-handling in nature-inspired numerical optimization: past, present and future. Swarm Evol. Comput. 1(4), 173–194 (2011)
8. Poursoltan, S., Neumann, F.: A feature-based analysis on the impact of set of constraints for e-constrained differential evolution. CoRR, abs/1506.06848 (2015)
9. Poursoltan, S., Neumann, F.: A feature-based analysis on the impact of linear constraints for ε-constrained differential evolution. In: 2014 IEEE Congress on Evolutionary Computation (CEC), pp. 3088–3095. IEEE (2014)
10. Robič, T., Filipič, B.: DEMO: differential evolution for multiobjective optimization. In: Coello Coello, C.A., Hernández Aguirre, A., Zitzler, E. (eds.) EMO 2005. LNCS, vol. 3410, pp. 520–533. Springer, Heidelberg (2005)
11. Schwefel, H.-P.P.: Evolution and Optimum Seeking: The Sixth Generation. Wiley, New York (1993)
12. Storn, R., Price, K.: Differential evolution-a simple and efficient heuristic for global optimization over continuous spaces. J. Glob. Optim. 11(4), 341–359 (1997)
13. Takahama, T., Sakai, S.: Constrained optimization by the ε constrained differential evolution with an archive and gradient-based mutation. In: 2010 IEEE Congress on Evolutionary Computation (CEC), pp. 1–9. IEEE (2010)
14. Krohling, R., dos Santos Coelho, L., et al.: Coevolutionary particle swarm optimization using Gaussian distribution for solving constrained optimization problems. IEEE Trans. Syst. Man Cybern. Part B Cybern. 36, 1407–1416 (2006)
15. Wang, Y., Cai, Z.: A hybrid multi-swarm particle swarm optimization to solve constrained optimization problems. Frontiers Comput. Sci. China 3(1), 38–52 (2009)

A Feature-Based Analysis on the Impact of Set of Constraints for ε-Constrained Differential Evolution

Shayan Poursoltan[✉] and Frank Neumann[✉]

Optimisation and Logistics, School of Computer Science, University of Adelaide,
Adelaide, SA 5005, Australia
{shayan.poursoltan,frank.neumann}@adelaide.edu.au

Abstract. Different types of evolutionary algorithms have been developed for constrained continuous optimisation. We carry out a feature-based analysis of evolved constrained continuous optimisation instances to understand the characteristics of constraints that make problems hard for evolutionary algorithm. In our study, we examine how various sets of constraints can influence the behaviour of ε-Constrained Differential Evolution. Investigating the evolved instances, we obtain knowledge of what type of constraints and their features make a problem difficult for the examined algorithm.

1 Introduction

Constrained optimisation problems (COPs), specially non-linear ones, are very important and widespread in real world applications [1]. This has motivated introducing various algorithms to solve COPs. The focus of these algorithms is to handle the involved constraints. In order to deal with the constraints, various mechanisms have been adopted by evolutionary algorithms. These techniques include penalty function, decoder-based methods and special operators that separate the treatment of constraints and objective functions. For an overview of different types of methods we refer the reader to Mezura-Montes and Coello Coello [6].

With the increasing number of evolutionary algorithms, it is hard to predict which algorithm performs better for a newly given COP. Various benchmark sets such as CEC'10 [3] and BBOB'10 [2] have been proposed to evaluate the algorithm performances on continuous optimisation problems. The aim of these benchmarks is to find out which algorithm is good on which classes of problems. For constrained continuous optimisation problems, there has been an increasing interest to understanding problem features from a theoretical perspective [9]. The feature-based analysis of hardness for certain classes of algorithms is a relatively new research area. Such studies classify problems as hard or easy for a given algorithm based on the features of given instances. Initial studies in the context of continuous optimisation have recently been carried out in [4,5]. Having enough knowledge on problem properties that make it hard or easy,

© Springer International Publishing Switzerland 2015
S. Arik et al. (Eds.): ICONIP 2015, Part III, LNCS 9491, pp. 344–355, 2015.
DOI: 10.1007/978-3-319-26555-1_39

we may choose the most suited algorithm to solve it. To do this, two steps approach has been proposed by Mersmann et al. [4]. First, one has to extract the important features from a group of investigated problems. Second, in order to build a prediction model, it is necessary to analyse the performance of various algorithms on these features. Feature-based analysis has also been used to gain new insights in algorithm performance for discrete optimisation problems [7,10].

In this paper, we carry out a feature-based analysis for constrained continuous optimisation and generate a variety of problem instances from easy to hard ones by evolving constraints. This ensures that the knowledge obtained by analysing problem features covers a wide range of problem instances that are of particular interest. Although what makes a problem hard to solve is not a standalone feature, it is assumed that constraints are certainly important in COPs. Evolving constraints is a new technique to generate hard and easy instances. So far, the influence of one linear constraint has been studied [8]. However, real world problems have more than one linear constraint (such as linear, quadratic and their combination). Hence, our study is to generate COP instances to investigate which features of the linear and quadratic constraints make the COP hard to solve. To provide this knowledge, we need to use a common suitable evolutionary algorithm that handles the constraints. The ε-constrained differential evolution with an archive and gradient-based mutation (εDEag) [12] is used. The εDEag (winner of CEC 10 special session for constrained problems) is applied to generate hard and easy instances to analyse the impact of set of constraints on it.

Our results provide evidence on the capability of constraints (linear, quadratic or their set of combination) features to classify problem instances to easy and hard ones. Feature analysis by solving the generated instances with εDEag enables us to obtain the knowledge of influence of constraints on problem hardness which could later could be used to design a successful prediction model for algorithm selection.

The rest of the paper is organised as follows. In Sect. 2, we introduce the constrained optimisation problems. Then, we discuss εDEag algorithm that we use to solve the generated problem instances. Section 3 includes our approach to evolve and generate problem instances. Furthermore, the constraint features are discussed. In Sect. 4, we carry out the analysis of the linear and quadratic constraint features. Finally, we conclude with some remarks.

2   Preliminaries

2.1   Constrained Continuous Optimisation Problems

Constrained continuous optimisation problems are optimisation problems where a function $f(x)$ on real-valued variables should be optimised with respect to a given set of constraints. Constraints are usually given by a set of inequalities and/or equalities. Without loss of generality, we present our approach for minimization problems.

Formally, we consider single-objective functions $f: S \to \mathbb{R}$, with $S \subseteq \mathbb{R}^n$. The constraints impose a feasible subset $F \subseteq S$ of the search space S and the goal is to find an element $x \in S \cap F$ that minimizes f.

We consider problems of the following form:

$$\text{minimize} \quad f(x), \quad x = (x_1, \ldots, x_n) \in \mathbb{R}^n$$
$$\text{subject to} \quad g_i(x) \leq 0 \quad \forall i \in \{1, \ldots, q\} \tag{1}$$
$$h_j(x) = 0 \quad \forall j \in \{q+1, \ldots, p\}$$

where $x = (x_1, x_2, \ldots, x_n)$ is an n dimensional vector and $x \in S \cap F$. Also $g_i(x)$ and $h_j(x)$ are inequality and equality constraints respectively. Both inequality and equality constraints could be linear or nonlinear. To handle equality constraints, they are usually transformed into inequality constraints as $|h_j(x)| \leq \varepsilon$, where $\varepsilon = 10e^{-4}$ (used in [3]). Also, the feasible region $F \subseteq S$ of the search space S is defined by

$$l_i \leq x_i \leq u_i, \quad 1 \leq i \leq n \tag{2}$$

where both l_i and u_i denote lower and upper bounds for the ith variable and $1 \leq i \leq n$ respectively.

2.2 εDEag Algorithm

One of the most prominent evolutionary algorithms for COPs is ε-constrained differential evolution with an archive and gradient-based mutation (εDEag). The algorithm is the winner of 2010 CEC competition for constrained continuous problems [3]. The εDEag uses ε-constrained method to transform algorithms for unconstrained problems to constrained ones. It adopts ε-level comparison instead of ordinary ones to order the possible solutions. In other words, the lexicographic order is performed in which constraint violation ($\phi(x)$) has more priority and proceeds the function value ($f(x)$). This means feasibility is more important. Let f_1, f_2 and ϕ_1, ϕ_2 are objective function values and constraint violation at x_1, x_2 respectively. Hence, for all $\varepsilon \geq 0$, the ε-level comparison of two candidates (f_1, ϕ_1) and (f_2, ϕ_2) is defined as the follows:

$$(f_1, \phi_1) <_\varepsilon (f_2, \phi_2) \iff \begin{cases} f_1 < f_2, & \text{if} \quad \phi_1, \phi_2 \leq \varepsilon \\ f_1 < f_2, & \text{if} \quad \phi_1 = \phi_2 \\ \phi_1 < \phi_2, & \text{otherwise} \end{cases}$$

In order to improve the usability, efficiency and stability of the algorithm, an archive has been applied. Using it improves the diversity of individuals. For a detailed presentation of the algorithm, we refer the reader to [12].

3 Evolving Constraints

It is assumed that the role of constraints in problem difficulty is certainly important for COP. Hence, it is necessary to analyse various effects that constraint can impose on a constrained problem. Evolving constraints is a novel methodology to generate hard and easy instances based on the performance of the problem solver (optimisation algorithm).

3.1   Algorithm

In order to analyse the effects of constraints, the variety of them needs to be studied over a fixed objective function. First, constraint coefficients are randomly chosen to construct problem instances. Second, the generated COP is solved by a solver algorithm (εDEag). Then, the required function evaluation number (FEN) to solve this instance is considered as the fitness value for evolving algorithm. This process is repeated until hard and easy instances of constraint problem are generated (see Fig. 1).

To generate hard and easy instances, we use the approach outlined in [8]. It uses fast and robust differential evolution (DE) proposed in [11] to evolve through the problem instances (by generating various constraint coefficients). It is necessary to note that the aim is to optimise (maximise/minimise) the FEN that is required by a solver to solve the generated problem. Also, to solve this generated problem instance and find the required FEN we use εDEag as a solver. The termination condition of this algorithm (evolver) is set to reaching FENmax number of function evaluations or finding a solution close enough to the feasible optimum solution as follows:

$$|f(x_{optimum}) - f(x_{best})| \leq e^{-12} \tag{3}$$

This process generates harder and easier problem instances until it reaches the certain number of generation for the DE algorithm (evolver). Once two distinct sets of easy and hard instances are ready, we start analysing various features of the constraints for these two categories. This could give us the knowledge to understand which features of constraints have more contribution to problem difficulty.

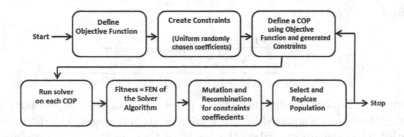

Fig. 1. Evolving constraints process

3.2   Evolving a Set of Inequality Constraints

We focus on analysing the effects of constraints (linear, quadratic and their combination) on the problem and algorithm difficulty. We extract features of constraints and analyse their effect on problem difficulty. The experimented constraints are linear and quadratic as the form of:

$$\text{linear constraint}\quad g(x) = b + a_1 x_1 + \ldots + a_n x_n \tag{4}$$

$$\text{quadratic constraint}\quad g(x) = b + a_1 x_1^2 + a_2 x_1 \ldots + a_{2n-1} x_n^2 + a_{2n} x_n \tag{5}$$

or combination of them. We also consider various numbers of these constraints in this study. Here, $x_1, x_2 \ldots, x_n$ are the variables from Eq. 1 and $a_1, a_2 \ldots, a_n$ are coefficients within the lower and upper bounds (l_c, u_c). We construct COPs where the optimum of the experimented unconstrained problem is feasible. We use quadratic functions of the form of Eq. 5 (univariate) since it is more popular in recent constrained problem benchmarks. Also, the influence of each x_ns can be analysed independently (exponent 2). The optimum of the investigated problems is $x^* = (0, \ldots, 0)$ and we ensure that this point is feasible by requiring $b \leq 0$, when evolving the constraints.

3.3   Constraints Features

We study a set of statistic based features that leads to generating hard and easy problem instances. These features are discussed as follows:

- **Constraint Coefficients Relationship:** It is likely that the statistics such as standard deviation, population standard deviation and variance of the constraints coefficients can represent the constraints influences to problem difficulty. These constraint coefficients are $(b, a_1, a_2, \ldots, a_n)$ in Eqs. 4 and 5.
- **Shortest Distance:** This feature is related to the shortest distance between the objective function optimum and constraint hyperplane. In this paper, the shortest distance to the known optimum from each constraint and their relations to each other is discussed. To find the shortest distance of optimum point $(x_{01}, x_{02}, \ldots, x_{0n})$ to the linear constraint hyperplane $(a_1 x_1 + a_2 x_2 + \cdots + a_n x_n + b = 0)$ we use Eq. 6. Also, for quadratic constraint hyperplane $(a_1 x_1^2 + a_2 x_1 \ldots + a_{(2n-1)} x_n^2 + a_{2n} x_n + b = 0)$ we need to find the minimum of Eq. 7.

$$d_\perp = \frac{a_1 x_{01} + a_2 x_{02} + \ldots a_n x_{0n} + b}{\sqrt{a_1{}^2 + a_2{}^2 + \cdots + a_n{}^2}} \tag{6}$$

$$d_\perp = \sqrt{(x_1 - x_{01})^2 + (x_2 - x_{02})^2 + \cdots + (x_n - x_{0n})^2} \tag{7}$$

where $d_\perp$ in Eq. 7 is the distance from a point to a quadratic hyperplane. Minimising the distance squared $(d_\perp^2)$ is equivalent to minimising the distance $d_\perp$.
- **Angle:** This feature describes the angle of the constraints hyperplanes to each other. It is assumed that the angle between the constraints can influence problem difficulty. To calculate the angle between two linear hyperplanes, we need to find their normal vectors and angle between them using the following equation:

$$\theta = \arccos \frac{n_1 \cdot n_1}{|n_1||n_2|} \tag{8}$$

where n_1, n_2 are normal vectors for two hyperplanes. Also, the angle between two quadratic constraints is the angle between two tangent hyperplanes of their intersection. Then, the angle between these tangent hyperplanes can be found by Eq. 8.

– **Number of Constraints:** Number of constraints plays an important role in problem difficulty. The number of constraints and their effects to make easy and hard problem instances is analysed.

– **Optimum-local Feasibility Ratio:** Although the global feasibility ratio is important to find the initial feasible point, it should not affect the convergence rate during solving the problem. So, the feasibility ratio of generated COP is calculated by choosing random points within the vicinity of the optimum in search space and the ratio of feasible points to all chosen ones is reported. In our experiment, the vicinity of optimum is equivalent to 1/10 of boundaries from optimum for each dimension.

4   Experimental Analysis

We now analyse the features of constraints (linear, quadratic and their combination) for easy and hard instances. We generate these instances for (εDEag) algorithm using well known objective functions. In our experiments, we generate two sets of hard and easy problem instances. Due to stochastic nature of evolutionary algorithms, for each number of constraints we perform 30 independent runs for evolving easy and hard instances. We set the evolving algorithm (DE) generation number to 5000 for obtaining the proper easy and hard instances. The other parameters of evolving algorithm are set to pop size $= 40$, CR $= 0.5$, scaling factor $= 0.9$ and FEN_{max} is $300,000$. Values for these parameters have been obtained by optimising the performance of the evolving algorithm in order to achieve the more easier and harder problem instances. For (εDEag) algorithm, its best parameters are chosen based on [12]. These parameters are: generation number $= 1500$, pop size $= 40$, CR $= 0.5$, scaling factor $= 0.9$. Also, the parameters for ε-constraint method are set to control generation (Tc) $= 1000$, initial e level (q) $= 0.9$, archive size $= 100n$ (n is dimension number), gradient-based mutation rate (Pg) $= 0.2$ and number of repeating the mutation (Rg) $= 3$.

4.1   Analysis for Linear Constraints

In order to focus only on constraints, we carry out our experiments on various well-known objective functions. These functions are: Sphere (bowl shaped), Ackley (many local optima), Rosenbrock (valley shaped) and Schaffer (many local minima) (see [2]). The linear constraint is as the form of Eq. 4 with dimension (n) as 30 and all coefficients are within the range of $[-5, 5]$. Also, number of constraints is considered as 1 to 5. To discuss and study some features such as shortest distance to optimum, we assume that zero is optimum (all bs should be negative). We used (εDEag) algorithm as solver to generate more easy and hard instances. In the following we will present our findings based on various features for linear constraints (for each dimension).

Table 1. The angle feature for Sphere objective function

	Cons 1,2	Cons 1,3	Cons 1,4	Cons 1,5	Cons 2,3	Cons 2,4	Cons 2,5	Cons 3,4	Cons 3,5	Cons 4,5
DE Easy	15	17	25	21	32	27	41	47	45	43
DE Hard	45	51	63	59	62	73	76	69	79	86

Figure 2 shows some evidence about linear constraints coefficients relationship such as standard deviation. It is obvious that there is a systematic relationship between the standard deviation of linear constraint coefficients and problem difficulty. The box plot (see Fig. 2) represents the results for easy and hard instances using all objective function for (εDEag) algorithm (solver). As it is observed, the standard deviation for coefficients in each constraint (1 to 5) for easy instances are lower than hard ones. Both these coefficient values can be a significant role to make a problem harder or easier to solve. Interestingly, all different objective functions follow the same pattern.

Figure 3 represents variation of shortest distance to optimum feature for easy and hard instances using (εDEag) algorithm. Lower value means a higher distance from the optimum. This means, the linear hyperplanes in easy instances are further from optimum. Based on results, there is a strong relationship between problem hardness and shortest distance of constraint hyperplanes to optimum. In other word, this feature is contributing to problem difficulty. As expected, all objective functions follow the same systematic relationship between their feature and problem difficulty. This means, this feature can be used as a proper source of knowledge for predicting problem difficulty.

The angle between linear constraint hyperplanes feature shows relationship between the angle and problem difficulty. The angle between constraints in easier instances are less than higher ones (see Table 1). So, this feature is contributing in problem difficulty. Table 2 explains the variation of number of constraints feature group. It is shown that the problem difficulty (required FEN for easy and hard instances) has a strong systematic relationship with number of constraints for the experimented algorithm. To calculate the optimum-local feasibility ratio, 10^6 points are generated within the vicinity of optimum (zero in our problems). Later, the ratios of feasible points to all generated points are investigated for easy and hard instances. Results point out that increasing number of linear constraints, decreases the feasibility ratio for experimented algorithms (see Table 4).

In summary the variation of feature values over the problem difficulty is more prominent in some of them than the other groups. Features such as, coefficients standard deviation, shortest distance, angle, number of constraints and feasibility ratio exhibit a relationship to problem hardness. This relationship is stronger for some features.

4.2   Analysis for Quadratic Constraints

In this section, we carry out our experiments on quadratic constraints. We use objective functions, dimension and coefficient range similar to linear analysis.

Table 2. The FEN for linear constraints

Constraint - Function	DE Easy	DE Hard
1 c Sphere	25.6K	91.2K
2 c Sphere	28.9K	93.4K
3 c Sphere	32.4K	98.3K
4 c Sphere	34.2K	104.2K
5 c Sphere	35.5K	123.2K
1 c Ackley	65.2K	232.1K
2 c Ackley	69.3K	243.7K
3 c Ackley	74.2K	265.4K
4 c Ackley	86.4K	271.3K
5 c Ackley	92.3K	277.2K
1 c Rosenbrock	32.8K	145.2K
2 c Rosenbrock	35.9K	153.3K
3 c Rosenbrock	34.5K	167.9K
4 c Rosenbrock	42.2K	172.4K
5 c Rosenbrock	48.3K	176.8K
1 c Schaffer	84.8K	247.1K
2 c Schaffer	87.9K	259.1K
3 c Schaffer	93.5K	280.3K
4 c Schaffer	103.2K	293.8K
5 c Schaffer	112.4K	297.4K

Table 3. The FEN for quadratic constraints

Constraint - Function	DE Easy	DE Hard
1 c Sphere	24.2K	129.3K
2 c Sphere	25.3K	132.6K
3 c Sphere	27.9K	136.2K
4 c Sphere	34.1K	141.2K
5 c Sphere	38.7K	149.3K
1 c Ackley	68.4K	228.3
2 c Ackley	72.9K	232.5K
3 c Ackley	84.5K	239.6K
4 c Ackley	95.3K	247.9K
5 c Ackley	98.1K	251.9K
1 c Rosenbrock	31.4K	173.2K
2 c Rosenbrock	32.45K	182.3K
3 c Rosenbrock	42.5K	190.6K
4 c Rosenbrock	52.7K	192.8K
5 c Rosenbrock	71.1K	213.4K
1 c Schaffer	91.3K	278.9K
2 c Schaffer	94.9K	283.1K
3 c Schaffer	103.7K	289.3K
4 c Schaffer	114.1K	296.1K
5 c Schaffer	123.4	300k

In the following the group of features are studied for easy and hard instances using quadratic constraints.

Observing the Fig. 2, we can identify the relationship of quadratic coefficients and their ability to make problem hard or easy. Based on the experiments, quadratic coefficients has the ability to make problems harder or easier for algorithms. In other words, in each constraint, the quadratic coefficients (within the quadratic constraint) are more contributing to problem difficulty than linear coefficients (see Eq. 5). Figure 2 shows the standard deviation of quadratic coefficients for easy and hard COPs. As shown, the standard deviation of quadratic coefficient in 1 to 5 constraints in easy instances are less than harder one. In contrast to quadratic coefficients, our experiments show there is no systematic relationship between the linear coefficient in quadratic constraints and problem hardness. In other words, quadratic coefficients (a_{2n-1}) are more contributing than linear ones (a_{2n}) in the same quadratic constraint (see Eq. 5).

Box plots shown in Fig. 3 represent the shortest distance of a quadratic constraint hyperplanes to optimum. As it is observed, harder instances have constraint hyperplanes closer to optimum than easier ones. Calculating the angles between constraints do not follow any systematic pattern and there is no relationship between angle feature and problem difficulty for quadratic constraints. We also study the number of quadratic constraints feature. As it is shown in

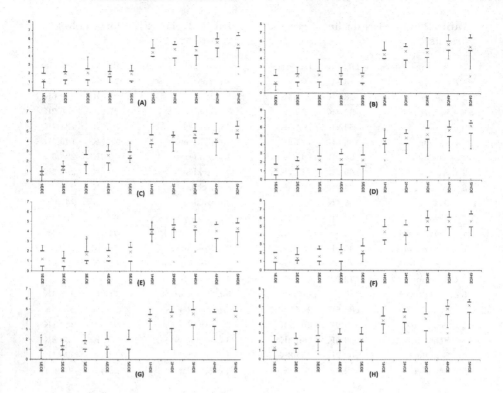

Fig. 2. Box plot for standard deviation of coefficients in linear (A,C,E,G) and quadratic (B,D,F,H) constraints for Sphere (A,B), Ackley (C,D), Rosenbrok (E,F) and Schaffer (G,H). Each sub figure includes 2 sets of hard (H) and Easy (E) instances with 1 to 5 constraints using algorithms (a/b/c denotes a: constraint number, b: easy/hard instances and c:algorithm)

Table 3, the number of quadratic constraints is contributing to problem difficulty. It is obvious that increasing the number of quadratic constraints makes a problem harder to solve (increases FEN). As observed in Table 5, our investigations on the feasibility ratio show that increasing the number of constraints decreases the problem optimum-local feasibility ratio for easy and hard instances respectively. As it is observed, some groups of features are more contributing to problem difficulty than others. It is shown that the angle feature does not follow any systematic relationship with problem hardness for the considered algorithm in the case of quadratic constraints. On the other hand, the standard deviation, feasibility ratio and number of constraints are more influencing the performance of εDEag.

4.3 Analysis for Combined Constraints

In this section, we consider the combination of linear and quadratic constraints. The generated COPs have different numbers of linear and quadratic constraints

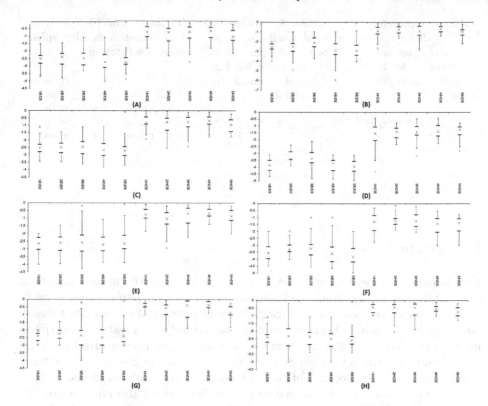

Fig. 3. Box plot for the shortest distance to optimum of linear (A,C,E,G) and quadratic (B,D,F,H) constraints for Sphere (A,B), Ackley (C,D), Rosenbrok (E,F) and Schaffer (G,H). Each sub figure includes 2 sets of hard (H) and Easy (E) instances with 1 to 5 constraints using DE algorithm (a/b/c denotes a: constraint number, b: easy/hard instances and c:algorithm)

(up to 5 constraints). The obtained results show the higher effectiveness of quadratic constraints than linear constraints. In other words, these constraints are more contributing to problem difficulty than linear ones. By analysing the various number of constraints (See Table 6) we can conclude that the required FEN for sets of constraints with more quadratic ones is higher than sets with more linear constraints. This relationship holds the pattern for both easy and hard instances.

In summary, it is observed that the variation of linear and quadratic constraint coefficients over the problem difficulty is more contributing for some group of features. Considering quadratic constraints only, it is obvious that some features such as angle do not provide useful knowledge for problem difficulty. In general, this experiments point out the relationship of the various constraint features of easy and hard instances with the problem difficulty while moving from easy to hard ones. This improves the understanding of the constraint structures and their ability to make a problem hard or easy for a specific group of evolutionary algorithms.

Table 4. Optimum-local feasibility ratio of search space near the optimum for 1,2, 3,4 and 5 linear constraint

	DE Easy	DE Hard
1 cons	42 %	7 %
2 cons	32 %	6 %
3 cons	22 %	4 %
4 cons	17 %	3 %
5 cons	11 %	2 %

Table 5. Optimum-local feasibility ratio of search space near the optimum for 1,2, 3,4 and 5 quadratic constraint

	DE Easy	DE Hard
1 cons	36 %	11 %
2 cons	27 %	7 %
3 cons	12 %	4 %
4 cons	11 %	3 %
5 cons	8 %	2 %

Table 6. The FEN for combined constraints using Sphere objective function

	DE Easy	DE Hard
1 Lin 4 Quad	22.4K	97.5K
2 Lin 3 Quad	17.5K	95.1K
3 Lin 2 Quad	16.5K	94.2K
4 Lin 1 Quad	14.1K	91.4K

5   Conclusions

In this paper, we performed a feature-based analysis on the impact of sets of constraints (linear, quadratic and their combination) on performance of well-known evolutionary algorithm (εDEag). Various features of constraints for easy and hard instances have been analysed to understand which features contribute more to problem difficulty. The sets of constraints have been evolved using an evolutionary algorithm to generate hard and easy problem instances for εDEag. Furthermore, the relationship of the features with the problem difficulty have been examined while moving from easy to hard instances. Later on, these results can be used to design an algorithm prediction model.

Acknowledgments. Frank Neumann has been supported by ARC grants DP130104395 and DP140103400.

References

1. Floudas, C.A., Pardalos, P.M.: A Collection of Test Problems for Constrained Global Optimization Algorithms. LNCS, vol. 455. Springer, Heidelberg (1990)
2. Hansen, N., Auger, A., Finck, S., Ros, R.: Real-parameter black-box optimization benchmarking 2010: Experimental setup (2010)
3. Mallipeddi, R., Suganthan, P.N.: Problem definitions and evaluation criteria for the CEC 2010 competition on constrained real-parameter optimization. Nanyang Technological University, Singapore (2010)
4. Mersmann, O., Bischl, B., Trautmann, H., Preuss, M., Weihs, C., Rudolph, G.: Exploratory landscape analysis. In: Proceedings of the 13th Annual Conference on Genetic and Evolutionary Computation, pp. 829–836. ACM (2011)
5. Mersmann, O., Preuss, M., Trautmann, H.: Benchmarking evolutionary algorithms: towards exploratory landscape analysis. In: Schaefer, R., Cotta, C., Kołodziej, J., Rudolph, G. (eds.) PPSN XI. LNCS, vol. 6238, pp. 73–82. Springer, Heidelberg (2010)

6. Mezura-Montes, E., Coello Coello, C.A.: Constraint-handling in nature-inspired numerical optimization: past, present and future. Swarm Evol. Comput. **1**(4), 173–194 (2011)
7. Nallaperuma, S., Wagner, M., Neumann, F., Bischl, B., Mersmann, O., Trautmann, H.: A feature-based comparison of local search and the christofides algorithm for the travelling salesperson problem. In: Proceedings of the Twelfth Workshop on Foundations of Genetic Algorithms XII, pp. 147–160. ACM (2013)
8. Poursoltan, S., Neumann, F.: A feature-based analysis on the impact of linear constraints for ε-constrained differential evolution. In: 2014 IEEE Congress on Evolutionary Computation (CEC), pp. 3088–3095. IEEE (2014)
9. Poursoltan, S., Neumann, F.: Ruggedness quantifying for constrained continuous fitness landscapes. In: Datta, R., Deb, K. (eds.) Evolutionary Constrained Optimization, pp. 29–50. Springer, Heidelberg (2015)
10. Smith-Miles, K., van Hemert, J., Lim, X.Y.: Understanding TSP difficulty by learning from evolved instances. In: Blum, C., Battiti, R. (eds.) LION 4. LNCS, vol. 6073, pp. 266–280. Springer, Heidelberg (2010)
11. Storn, R., Price, K.: Differential evolution-a simple and efficient heuristic for global optimization over continuous spaces. J. Glob. Optim. **11**(4), 341–359 (1997)
12. Takahama, T., Sakai, S.: Constrained optimization by the ε constrained differential evolution with an archive and gradient-based mutation. In: 2010 IEEE Congress on Evolutionary Computation (CEC), pp. 1–9. IEEE (2010)

Convolutional Associative Memory: FIR Filter Model of Synapse

Rama Murthy Garimella[1]([✉]), Sai Dileep Munugoti[2], and Anil Rayala[1]

[1] International Institute of Information Technology, Hyderabad, India
rammurthy@iiit.ac.in, anil.rayala@students.iiit.ac.in
[2] Indian Institute of Technology, Guwahati, India
d.munugoti@iitg.ernet.in

Abstract. In this research paper, a novel Convolutional Associative Memory is proposed. In the proposed model, Synapse of each neuron is modeled as a Linear FIR filter. The dynamics of Convolutional Associative Memory is discussed. A new method called Sub-sampling is given. Proof of convergence theorem is discussed. An example depicting the convergence is shown. Some potential applications of the proposed model are also proposed.

Keywords: Convolutional Associative Memory · FIR filter · Sub-sampling matrix · Hankel matrix

1 Introduction

Artificial Neural Networks (ANNs) such as Multi-Layer Perceptron (MLP), Hopfield Neural Network (HNN) are based on synaptic weights that are scalars. Researchers conceived of ANNs based on dynamic synapse i.e. the synapse is modeled as a linear filter e.g. Finite Impulse Response (FIR) filter [1–3]. One of the motivations of such a model of synapse is the ability to filter noise corrupting the signals.

In recent years, ANNs based on deep learning paradigm are able to provide excellent results in many applications (with the ability to learn features from the data automatically). Specifically Convolutional Neural Networks (CVNNs) pioneered hand writing recognition systems and other related Artificial Intelligence (AI) systems.

One of the main goals of the research in AI is to build a system which has Multi-Modal Intelligence i.e. a system combining and reasoning with different inputs like Vision, Sound, Smell and Touch which is similar to how the human brain works. It is biologically more appealing for such system to have a memory which is similar to that of human memory.

Human memory is based on associations with the memories it contains. For example a part of a well-known tune is enough to bring the whole song back to mind. Associative memories can be used for tasks like

© Springer International Publishing Switzerland 2015
S. Arik et al. (Eds.): ICONIP 2015, Part III, LNCS 9491, pp. 356–364, 2015.
DOI: 10.1007/978-3-319-26555-1_40

- Completing information if some part is missing
- Denoising if noisy input is given
- To guess information that means if a pattern is presented, the most similar stored pattern is determined etc.

Hopfield Neural Network is widely used as an Associative memory.

In this research paper, we propose and study a novel associative memory based on FIR filter model of synapse. Several other researches proposed certain types of Convolutional Associative Memories [5], but, the approach of Sub-Sampling performed in our method is very different from any other contribution. We expect our Convolutional Associative Memory to find many applications.

This research paper is organized as follows: In Sect. 2, the previous known literature of Hopfield Network is reviewed. In Sect. 3, the dynamics of the proposed novel Convolutional Associative Memory is studied. In Sect. 4, examples of the proposed model of convolutional associative memory are given. In Sect. 5, some applications of the proposed Convolutional Associative Memory are discussed. The paper concludes in Sect. 6.

2   Review of Literature

In 1982, John Hopfield proposed a form of neural network which potentially acts as an Associative memory called Hopfield Neural Network. A Hopfield Neural network is a form of recurrent neural network which is a non-linear dynamical system based on weighted, undirected graph. In a Hopfield Network, each node of the graph represent an artificial neuron which assumes a binary value $+1$ or -1 and the edge weights correspond to synaptic weights. The order of the network corresponds to number of neurons. If all the synaptic weights of the network are represented in a synaptic weight matrix $\mathbf{M}$, a Hopfield neural network can be represented by $\mathbf{M}$ and threshold vector $\mathbf{T}$.

Since each neuron has a state of $+1$ or -1, the state space of an $\mathbf{N}$ neuron Hopfield network can be given by an N dimensional unit -hypercube. Let the state of the non-linear dynamical system is represented by the $N \times 1$ vector S(t) and let $S_i(t) \in \{1, -1\}$ represent the state of ith neuron at time instant t. The state update of ith neuron is done by

$$S_i(t+1) = Sign(\sum_{j=1}^{N} M_{ij}S_j(t) - T_i)$$

Depending on the number of neurons at which state is updated at a given time instant, the network operation can be classified.

Two main types of network operation are:

(1) **Serial Mode updating:** Here, at any time instant, only state of one node of the network is updated.
(2) **Fully Parallel Mode updating:** Here, at every time instant, state of all the nodes are updated.

The network is said to be stable/converged if and only if

$$S(t+1) = S(t)$$

Thus, if the network has reached stable state/converged, then irrespective of the mode of operation, the network remains in that state forever. The following convergence theorem summarizes the dynamics of Hopfield Network.

Theorem 1: Let the pair $N = (M, T)$ specify a Hopfield neural network, then the following hold true:

(1) If the network is operating in a serial mode and the elements of the diagonal of M are non-negative, the network will always converge to a stable state (i.e. there are no cycles in the state space).
(2) If the network is operating in the fully parallel mode, the network will always converge to a stable state or to a cycle of length 2 (i.e. the cycles in the state space are of length almost 2).

Proof of the above theorem can be found in [4].

3   Dynamics of Convolutional Associative Memory

Before we begin the dynamics of Convolutional Associative Memory, some important innovative concepts are discussed:

3.1   State as a Sequence

In the convolutional Associative Memory, each neuron is modeled to have a state which is a sequence of binary values $\{+1, -1\}$ of length **L** rather than a single value $\{+1 \text{ or } -1\}$ as in the case of Hopfield network. All the neurons are interconnected and it is assumed that there are no self connections. For practical considerations it is taken that the sequence starts at time n = 0.

Notation:
S represents a Matrix of size $(L \times N)$ where **N** correspond to the number of neurons and S_i(i^{th} column of S) correspond to the state of ith neuron.

3.2   Synapse as a Linear FIR Filter: Sub-sampling

Based on biological motivation, the author for the first time proposed the idea of modeling the synapse as a linear filter in [1]. The synapse of each connection is taken to be a Linear Discrete time Finite Impulse Response Filter of length **F** rather than a single constant as in the case of Hopfield Network. From practical and theoretical considerations, this type of model is considered to be more realistic. It is assumed that, the sequence starts at time n = 0. Since output of each synapse is convolution of state sequence of a neuron and a filter, the output of the synapse is a sequence of length $\mathbf{L} + \mathbf{F} - 1$ and since this sequence is used to update the state of a neuron, it needs to be compressed to a sequence of length L, this compression is called as Sub-Sampling. The compression can be done in infinite ways, it can observed that the compression is a linear operator and can be realized by a matrix of size $(L \times L + F - 1)$. Let $\mathbf{K}$ represent the sub-sampling matrix which does the task of compression. In digital signal processing, it is very well known that windows such as Hamming, Hanning, Blackman window are utilized. We invoke such results in the design of synaptic FIR filter.

3.3   Convolution as Matrix Multiplication

In the convolutional associative memory, the convolution of the synaptic filter sequence (between the neurons i, j) and state sequence can be realized by multiplying a convolution matrix $H_{i,j}$ $(L + F - 1 \times L)$ with the state sequence of length $\mathbf{L}$.

The $H_{i,j}$ matrix can be observed to be a Toeplitz matrix and the first column can be given as a vector of length $\mathbf{L} + \mathbf{F} - 1$ in which only first F elements are non-zero, and the first row is a vector of length $\mathbf{L}$ with only first element as non-zero. For L = 4, F = 3: $H_{i,j}$ matrix for a synapse can be given as:

It can be seen that, multiplying $H_{i,j}$ with a vector of length L is same as convolving a sequence of length L (state sequence) of each neuron and the impulse response of length F, $\{h(0), h(1), h(2)\}$.

$$\begin{bmatrix} h(0) & 0 & 0 & 0 \\ h(1) & h(0) & 0 & 0 \\ h(2) & h(1) & h(0) & 0 \\ 0 & h(2) & h(1) & h(0) \\ 0 & 0 & h(2) & h(1) \\ 0 & 0 & 0 & h(2) \end{bmatrix}$$

Notation: Let H denote an $N \times N$ Cell with each element as a Matrix. $H_{i,j}$ represents the synaptic filter convolution matrix between the neurons i an j. It is assumed that $H_{i,j}$ and $H_{j,i}$ are same for all i and j. Since it is also assumed that there are no self-connections (Sect. 3.1) i.e. $H_{i,i}$ is a null matrix for all i.

3.4   Serial Mode Updation

In serial mode updating, the state sequence of ith neuron is updated by

$$S_i(t + 1) = sign(\sum_j K * H_{i,j} * S_j(t)) \tag{1}$$

$S_i(t + 1)$ in the above equation is the updated state sequence (length L) of ith neuron. $S_j(t)$ is the state sequence (length L) of jth neuron. $H_{i,j}$ is the convolution Matrix of the synapse between neuron i and j. Matrix multiplication is denoted by *.

3.5   Energy Function of the Network

$$E = -\sum_{i,j} S_i(t)^T * K * H_{i,j} * S_j(t) \tag{2}$$

It can be observed that for an **N** neuron network and for given $H_{i,j}$, the above defined energy function is bounded below.

In Eq. (2), $S_i(t)^T$ denotes the transpose of $S_i(t)$.

3.6   Proof of Convergence of Energy in Serial Mode

If the state of ith neuron is changed from $S_i(t)$ to $S_i(t+1)$ the change in energy can be given by

$$E(S_i(t+1)) - E(S_i(t)) = -1 * \{\sum_j ((S_i(t+1) - S_i(t))^T * K * H_{i,j} * S_j(t)$$

$$+ \sum_j S_j(t)^T * K * H_{j,i} * (S_i(t+1) - S_i(t))\}$$

Note: If the energy of the network has to converge, the change in energy should be either negative or zero for all time instants.

We know that, $S_i(t+1) - S_i(t)$ (element wise) and $\sum_j K * H_{i,j} * S_j(t)$ have same sign since $S_i(t+1) = sign\sum_j K * H_{i,j} * S_j(t)$. Hence we can say that

$-\sum_j (S_i(t+1) - S_i(t))^T * K * H_{i,j} * S_j(t)$ is always negative or zero.

So the first term of the change in energy is always negative or zero and if the second term is equal to the first term then we can say that the total change energy of the network is always negative or zero.

If $K * H_{j,i}$ is a symmetrical Matrix, then $\sum_j S_j(t)^T * K * H_{j,i} * (S_i(t+1) - S_i(t))$

and $\sum_j (S_i(t+1) - S_i(t))^T * K * H_{i,j} * S_j(t))$ are equal, which can be easily seen by taking transpose of either of them and also using the fact that $H_{i,j} = H_{j,i}$ from Sect. 3.3.

Therefore, $E(S_i(t+1)) - E(S_i(t))$ is always either negative or zero if $K * H_{j,i}$ is a symmetric matrix. Since the energy is a lower bounded function and since it always decreases or remains constant we can say that the energy of the system converges. We know that, $H_{j,i}$ is in the form as suggested in Sect. 3.3 for all i and j, if K is a Hankel matrix, then $K * H_{j,i}$ will be a symmetric matrix.

3.7   Proposed Form of K

K has to be a Hankel Matrix. For $L = 4$; $F = 3$, the K is:

$$\begin{bmatrix} b(1) & b(2) & b(3) & b(4) & b(5) & b(6) \\ b(2) & b(3) & b(4) & b(5) & b(6) & b(7) \\ b(3) & b(4) & b(5) & b(6) & b(7) & b(8) \\ b(4) & b(5) & b(6) & b(7) & b(8) & b(9) \end{bmatrix}$$

The above proposed form is one of many forms of sub-sampling for which the networks energy converges and it should not be mistaken as the only form.

3.8   Convergence of the Network to a Stable State

If the energy of the system has converged, then there are two possible cases

(1) $S_i(t+1) - S_i(t) = 0$ and change in energy is zero.
(2) All the elements of $S_i(t+1) - S_i(t)$ might not be zero but the corresponding elements in $\sum_j K * H_{i,j} * S_j(t)$ might be zero hence making change in energy zero. This kind of change is only possible when one or many elements of S_i changes from -1 to $+1$.

Hence once the energy in the network has converged it is clear from the foregoing facts that the network will reach a stable state after at most $L \times N^2$ time intervals.

It is well known that convergence proof in parallel mode can be reduced to that in serial mode [6].

In view of the above proof, we have the following convergence theorem.

Theorem 2: Let the pair R = (S, H) represent the Convolutional Associative Memory where S in the form as discussed in Sect. 3.1 and H in the form as discussed in Sect. 3.3. Considering K to be a Hankel Matrix, then the following hold true:

(1) If the network is operating in a serial mode, the network will always converge to a stable state (i.e. There are no cycles in the state space).
(2) If the network is operating in the fully parallel mode, the network will always converge to a stable state or to a cycle of maximum length 2.

The proof for parallel mode can be done by converting parallel mode into serial mode as proposed in [6].

4   Example

Consider the case where,
Number of neurons $= 4$
Pattern length $= 3$
Filter length $= 2$
Initial State Matrix is

$$\begin{bmatrix} -1 & -1 & 1 & 1 \\ -1 & 1 & 1 & -1 \\ -1 & 1 & 1 & 1 \end{bmatrix}$$

where each column denotes state of a neuron Let $H\{i,j\}$ corresponds to filter coefficients for a synapse between i^{th} and j^{th} neuron and take

$$H\{1,2\} = H\{2,1\} = \{4,1\}$$
$$H\{1,3\} = H\{3,1\} = \{-2,0\}$$
$$H\{1,4\} = H\{4,1\} = \{1,3\}$$
$$H\{2,3\} = H\{3,2\} = \{-5,0\}$$
$$H\{2,4\} = H\{4,2\} = \{-1,1\}$$
$$H\{3,4\} = H\{4,3\} = \{0,-3\}$$

let us take sub-sampling matrix

$$K = \begin{bmatrix} -6 & 2 & -4 & 4 \\ 2 & -4 & 4 & -6 \\ -4 & 4 & -6 & 2 \end{bmatrix}$$

which is in the required form.

Initial energy of the system E in $= 512$.

Now let us update the first neuron, the state matrix of network becomes

$$\begin{bmatrix} 1 & -1 & 1 & 1 \\ -1 & 1 & 1 & -1 \\ 1 & -1 & 1 & 1 \end{bmatrix}$$

Updated energy of the system $E_{upd} = 0$.

Now let us update the second neuron, the state matrix of network becomes

$$\begin{bmatrix} 1 & 1 & 1 & 1 \\ -1 & -1 & 1 & -1 \\ 1 & 1 & 1 & 1 \end{bmatrix}$$

Updated energy of the system $E_{upd} = -320$.

Now let us update the third neuron, the state matrix of network becomes

$$\begin{bmatrix} 1 & 1 & 1 & 1 \\ -1 & -1 & -1 & -1 \\ 1 & 1 & 1 & 1 \end{bmatrix}$$

Updated energy of the system $E_{upd} = -432$.

Now let us update the fourth neuron, the state matrix of network becomes

$$\begin{bmatrix} 1 & 1 & 1 & 1 \\ -1 & -1 & -1 & -1 \\ 1 & 1 & 1 & 1 \end{bmatrix}$$

Updated energy of the system $E_{upd} = -432$.

Again updating the states of all the neurons doesnt alter the states of neurons which implies the system has converged.

Hence the example we took converged.

5   Applications: Multi Modal Intelligence

(1) One of the major challenges to the research in Artificial Intelligence has been Multi modal intelligence i.e. A system combining and reasoning with different inputs e.g. Vision, Sound, Smell, Touch. This is akin to how we humans perceive the world around us - our sight works in conjunction with our sense of hearing, touch etc. AI technologies that leverage multi-modal data are likely to be more accurate. The problem of Multi-Model Intelligence can be solved by using a Convolutional Associative Memory. Since each neuron has a state which is a vector of length L, different elements of the vector can correspond to different inputs, i.e. the last element of the vector at each neuron can correspond to speech the above one can correspond to visual data etc.

Example:

- A video camera might be able to accurately "recognize" a human if it recognizes human voices and touch.
- In Many security systems, only one input from a user is taken to know if he is authorized to enter or not. Using convolutional Associative memory proposed in this paper many inputs can be taken from user i.e. voice, fingerprint, visual etc. to have a better security system.

(2) The proposed convolutional associative memory can be used in robust speech recognition, since each synapse is modeled as a linear FIR filter it is more likely that such a system filters out the noise and produces more accurate results than that of existing systems.

6   Conclusion

In this research paper, the synapse is modeled as a discrete time linear FIR filter of length **F** and state of each neuron is modeled as a sequence of length **L**. It is proved that such a system converges in serial mode operation for the proposed type of sub-sampling matrix. It is expected that such an artificial neural network will find many applications.

References

1. Rama Murthy, G.: Some novel real/complex-valued neural network models. In: Proceedings of 9th Fuzzy Days (International Conference on Computational Intelligence), Dortmund, Germany, September 18–20 2006. Advances in Soft Computing, Springer Series, Computational Intelligence, Theory and Applications (2006)
2. Rama Murthy, G.: Finite impulse response (FIR) filter model of synapses: associated neural networks. In: International Conference on Natural Computation (2008)
3. Rama Murthy, G.: Multi-dimensional Neural Networks: Unified Theory. New Age International Publishers, New Delhi (2007)

4. Hopfield, J.J.: Neural networks and physical systems with emergent collective computational abilities. Proc. Natl. Acad. Sci. USA **79**, 2554–2558 (1982)
5. Karbasi, A., Salavati, A.H., Shokrollahi, A.: Convolutional Neural Associative Memories: Massive Capacity with Noise Tolerance. arXiv.org arXiv:1407.6513
6. Bruck, J., Goodman, J.W.: A generalized convergence theorem for neural networks. IEEE Trans. Inf. Theor. **34**(5), 1089–1092 (1988)

Exploiting Latent Relations Between Users and Items for Collaborative Filtering

Yingmin Zhou, Binheng Song, and Hai-Tao Zheng$^{(\boxtimes)}$

Graduate School at Shenzhen, Tsinghua University, Building H, Tsinghua Campus,
University Town, Shenzhen, China
zhou-ym14@mails.tsinghua.edu.cn,
{songbinheng,zheng.haitao}@sz.tsinghua.edu.cn

Abstract. As one of the most important techniques in recommender systems, collaborative filtering (CF) generates the recommendations or predictions based on the observed preferences. Most traditional recommender systems fail to discover the latent associations between the same or similar items with different names, which is called synonymy problem. With the rapid increasing number of users and items, the user-item rating data is extremely sparse. Based on the limited number of user ratings, we cannot capture enough information from the user history using the traditional CF techniques, which could reduce the effectiveness of the recommender systems.

In this paper, we propose a novel model User-Relation-Item Model (URIM) for CF, which exploits the latent relationship between different user interest domains and item types. By introducing a component named user-item-relation matrix, which reflects the latent major association patterns behind users and items, URIM tackles the synonymy problem, and therefore achieves a significant performance improvement. We compared our method with several state-of-the-art recommendation algorithms on two real-world datasets. Experimental results validate the effectiveness of our model in terms of prediction accuracy (RMSE) and top-N recommendation quality (Recall and Precision). More specifically, URIM reduces the RMSE by nearly 10 % and 5 % on the two datasets, respectively.

Keywords: Recommender system · Collaborative filtering · Batch gradient descent · Top-N recommendation

1   Introduction

In the past few years, people have been inundated with massive information. Recommender system has become a useful tool for users to filter out useless information, which can help users locate in the information they really need. There are two major categories in personalized recommender systems [15]: content-based filtering (CBF) [1] and collaborative filtering (CF). The former method generates recommendations based on the item or user features extracted from

© Springer International Publishing Switzerland 2015
S. Arik et al. (Eds.): ICONIP 2015, Part III, LNCS 9491, pp. 365–374, 2015.
DOI: 10.1007/978-3-319-26555-1_41

domain knowledge in advance. Obviously, it needs to pre-process the items to obtain its features and relies on special domain knowledge. It cannot find out the items that hold the users' potential interest while he is not so familiar.

The two more successful approaches of CF are memory-based [18], and model-based algorithms. Memory-based model is centered on capturing the relationships between users or items [14]. However, there are two main typical drawbacks of memory-based model. First, it's hard to decide which similarity measurement to choose on different domains. Second, the similarities are based on common rated items and therefore are unreliable when the data is sparse, which leads to the poor performance in the sparse datasets. To overcome the shortcomings of memory-based techniques, model-based CF has been investigated. Model-based approaches use the observed rating data to estimate and learn a parameter model to make predictions [2].

However, there are several limitations for most conventional factor-based CF models. First, the rating matrix is usually factorized into two matrices, denoting the latent factors of users and items respectively, and they may not reflect the associative pattern between user's interest and item's type. As a result, they may fail to capture some implicit information behind the limited amount of rating data. Besides, it's hard to decide the number of latent factors. A trade-off should be made between prediction performance and complexity.

To address these problems, we propose a simple but effective model: User-Relation-Item Model (URIM) for CF. Our contributions are summarized as follows:

- We investigated the latent relationship between user's interest domain and item's type for CF-based recommendation. To the best of our knowledge, this is the first work which denotes users' and items' latent factors in different dimensions.
- By introducing a component named user-item-relation matrix, which reflects the latent major associations between user interest and item types, we can alleviate the data sparse and synonymy problems.
- We conducted a series of experiments on the real-world datasets. Experimental results show that URIM provides good predictive accuracy in terms of RMSE as well as in top-N recommendations.

The rest of this paper is organized as follows: Sect. 2 will introduce the details of URIM, including the model description and training algorithms. We report experimental settings, results and our analysis in Sect. 3. And we'll make a conclusion in Sect. 4.

2   User-Relation-Item Model

2.1   Description

Recall that, conventional CF algorithms consider the users and items in the same dimension, which may fail to capture the latent association between users' interest and items' types. To match the user's interest and item's types, we assume that there exists a fixed rating pattern for the certain user interest domain and item type. Intuitively, we introduce a correlation matrix which correlates the user interest and item types to exploit the latent relationship between users and items. We factorize the rating matrix R into three small matrices as follows:

$$R = UPV \tag{1}$$

Here, U is a $M \times K$ matrix, denoting that there are M users and each user has K interest domains. P is a $K \times S$ matrix, which intermediates the relations between user interest and item types. And similarly, V is a $S \times N$ matrix, representing S item types and N items. Each user i is associated with a vector u_i, and each item j is associated with a vector v_j. For a given user i, the elements of u_i measure the extent to which the user is interested in this domain. Consistently, for a given item j, the elements of v_j measure the extent to which the item possesses those types. And P_{ks} captures the latent association between k_{th} user interest s_{th} item types. We use $\hat{r}_{ij}$ to denote the predicted rates for user i on the item j, leading to the estimate:

$$\hat{r}_{ij} = u_i \cdot P \cdot v_j \tag{2}$$

Here, for a realistic setting, we set $u_i \in [0,1]$, $v_i \in [0,1]$ and $P \in [-H/2, H/2]$, where H denotes the rating range in the corresponding dataset.

The whole framework is demonstrated on Fig. 1. We summarized the major steps as follows: URIM processes the user rating data in the training set and initialize the parameters (e.g. user interest number, item type number, learning rate α, regression parameter λ and threshold θ) with the input values. During the training phrase, we train the proposed model using batch gradient descent iteratively, to obtain the optimal values for user interest matrix, user-item correlation matrix as well as item type matrix respectively. We'll describe the details of the algorithms in the next section.

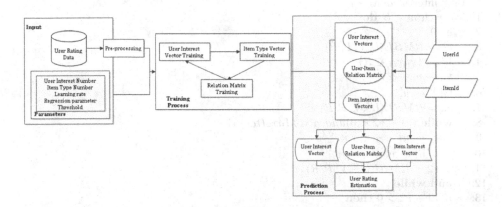

Fig. 1. Framework of URIM

2.2   Algorithms

To eliminate the influence of other factors such as a single user's bias on the prediction accuracy, we pre-process the rating. Assuming that the rating data in the training set ranges from 0 to H, for a given user, we map the minimum rate (r_{min}) into $H/2$, the average rate (r_{avg}) into 0, and the maximum rate (r_{max}) into $H/2$. In this way, we linearize other ratings(r) from two sides.

To get the user interest matrix U, user-item correlation matrix P and item type matrix V, we use the batch gradient descent to train the model and get the local optimum. To avoid overfitting, we choose the technique which explicitly penalizes overly complex models. As the physical meaning of u denotes the ratio of user's interest domains, while v represents the composition of item's types, the sum of the vectors are supposed to draw near 1. To regularize the learned parameters, we define the improved function as follows:

$$J = \min_{U,P,V} (\frac{1}{2} \times \sum_{r_{ij} \in \tau} (r_{ij} - U_{ik}P_{ks}V_{sj})^2 + \frac{1}{2} \times \lambda \times (\sum_{i}(||U_i||^2-1)^2 + \sum_{j}(||V_j||^2-1)^2)))$$

(3)

Here, $||U_i|| = \sqrt{U_{i1}^2 + ... + U_{ik}^2}$ and $||v_j|| = \sqrt{V_{j1}^2 + ... + V_{js}^2}$.

As U, P and V are unknown at the first, the equation is not convex. However, if we fix two of the unknown matrices, the optimization problem becomes quadratic and can be solved optimally.

Algorithm 1. User Interest Matrix Training Process

Input:
   A collection of rating $r_{ij} \in R$ given by the users $u_i \in U$ for the items $v_j \in V$, learning rating α, maximum number of Iteration Max_Iter

Output:
   User interest matrix U

1: **for each** $u_i \in U$ **do**
2:    n=0
3:    $r1 = RMSE(U, V, P, R)$
4:    $V' = \{v_j | r_{ij} \in R\}$
5:    $u'_i \leftarrow u_i - \alpha \times \frac{\partial J}{\partial u_i}$
6:    $U[i]' \leftarrow u'_i$
7:    $r2 = RMSE(U', V, P, R)$
8:    **while** $(r1 - r2 < 0) and (n < Max\_Iter)$ **do**
9:       $u'_i \leftarrow u_i - \frac{1}{2} \times \alpha \times \frac{\partial J}{\partial u_i}$
10:      $U'[i] \leftarrow u'_i$
11:      $r2 = RMSE(U', V, P, R)$
12:   **end while**
13:   **if** $r1 - r2 > 0$ **then**
14:      $U[i] \leftarrow u'_i$
15:   **end if**
16: **end for**

Algorithm 1 describes the process of training user interest matrix. Note that $RMSE(U, P, V, R)$ is a function which compute the RMSE value on current parameters. It's an iterative process as a whole. During each iteration, for each user u, we can get a set of items vectors V' rated by u. And then update the user vector u using the update function denoted in Eq. 7, which will be inferred later in this section. Figure out whether the RMSE value decreased, if so, update the user vector u and come to next user. Otherwise, decrease the gradient value by half each time automatically until the iteration number comes to maximum iteration number or the update vector is suitable, which contributes to save time without re-computing. The training process of item type matrix and relation matrix is similar. The only difference is that: for each rating pairs in the training set, we update each element in the P matrix separately and decide whether to update or not depending on the change of RMSE automatically. By repeating the training process described above over and over again until the loss of two iterations is less than the threshold, we can obtain optimal values fitting our proposed models.

To get the gradient value of these variables, we infer the derivation equations of U, P and V respectively as follows (given that two other matrices is fixed):

$$\frac{\partial J}{\partial U_{ik}} = \sum_{r_{ij} \in \tau} (r_{ij} - \sum_s U_{ik} P_{ks} V_{sj}) \cdot (-\sum_s P_{ns} V_{sj}) + \lambda \cdot (||u_i||^2 - 1) \cdot u_{ik} \quad (4)$$

$$\frac{\partial J}{\partial V_{sj}} = \sum_{r_{ij} \in \tau} (r_{ij} - \sum_k U_{ik} P_{ks} V_{sj}) \cdot (-\sum_k U_{ik} P_{ks}) + \lambda \cdot (||v_j||^2 - 1) \cdot v_{sj} \quad (5)$$

$$\frac{\partial J}{\partial P_{ks}} = \sum_{r_{ij} \in \tau} (r_{ij} - \sum_s U_{ik} P_{ks} V_{sj}) \cdot (-U_{ik} V_{sj}) \quad (6)$$

To reduce the computation complexity, we define a user vector and item vector as u and v respectively, Eqs. 4 and 5 can be written as:

$$\frac{\partial J}{\partial u} = (r - uPV) \cdot (-PV)^T + \lambda \cdot (||u||^2 - 1) \cdot u \quad (7)$$

$$\frac{\partial J}{\partial v} = (r - UPv) \cdot (-UP) + \lambda \cdot (||v||^2 - 1) \cdot v \quad (8)$$

Thus, the update function can be described as:

$$u \leftarrow u - \alpha \times \frac{\partial J}{\partial u}, v \leftarrow v - \alpha \times \frac{\partial J}{\partial v}, P_{ks} \leftarrow P_{ks} - \alpha \times \frac{\partial J}{\partial P_{ks}} \quad (9)$$

After obtaining the three well-trained matrices, we can compute the estimated rating for i_{th} user on the j_{th} item by the following equation:

$$r = U_i P V_j \quad (10)$$

As we scale the ratings into -H/2 and H/2 at the beginning, we scale it onto 0 and H.

3 Experimental Study

3.1 Experimental Settings

Dataset. We use the Movielens [4] and Netflix [3]. As for the training set, we randomly selected 80 % of the items every user has rated as the training data and the remaining 20 % ratings as the test data set.

Evaluation Metrics. We only consider the quality of the prediction accuracy and top-N recommendation. For prediction accuracy, we choose Root Mean Square Error ($RMSE = \sqrt{\dfrac{\sum\limits_{r_{ij} \in \tau} (r_{ij} - \hat{r}_{ij})^2}{|\tau|}}$) as our indicator, where lower values denotes better performance. While for top-N recommendation evaluation, we use the testing methodology similar to [5]. We use $recall(N) = \frac{\#hits}{|T|}$ and $recall(N) = \frac{\#hits}{|T|}$ to measure the performance, where $|T|$ denotes the number of ratings in the test set. By definition, recall for this single test can assume whether 0 (the case of miss) or 1 (the case of hit).

Compared Methods

- ItemAvg: this method proceeds a simple prediction rule: we use the average rating of the movie as the predict rating r'_{ij}.
- ItemBasedCF: as referred to [12], we selected k = 100 in the Movielens and k = 150 in the Netflix, which performs best in our experiment.
- PureSVD: we consider all missing values in the user rating matrix as mean values to reduce the errors. The user rating matrix R is estimated by the factorization refered to [5].
- SVD++: we adopt the integrated method [8], which take the temporal effects into account, and therefore performs better in SVD++ methods. As referred to [8], we used the following values for the parameters: $\gamma_1 = \gamma_2 = 0.007, \gamma_3 = 0.001, \lambda_1 = 0.005, \lambda_2 = \lambda_3 = 0.015$.

3.2 Experimental Result

Prediction Accuracy. For Movielens and Netflix dataset, we have conducted a series of experiments. Note that the training parameters can automatically altered in our framework, the values of learning rate α and regression parameter λ only has effect on the coverage speed, without affecting the final result. And threshold θ controls the coverage degree, it can be decided based on realistic settings. Hence, we set the $\alpha = 0.05$, $\lambda = 0.01$ and $\theta = 0.0001$. Besides, PureSVD50 means the SVD model with 50 latent factors, while PureSVD200 with 200 latent factors. Consistently, SVD++50 denotes 50 factors, and SVD++200 represents 200 factors. And the experimental results on the two datasets have been shown in Table 1.

Table 1. The RMSE values on Movielens and Netflix

User Interest	2	2	2	2	3	3	3	3
Item Types	2	3	5	10	2	3	5	10
MovieLens	0.797	0.797	0.802	0.797	0.797	**0.796**	0.798	0.799
Netflix	0.864	0.859	0.861	0.855	0.863	0.869	0.857	0.859
User Interest	5	5	5	5	10	10	10	10
Item Types	2	3	5	10	2	3	5	10
MovieLens	0.801	**0.796**	0.800	0.806	0.797	0.799	0.801	0.808
Netflix	0.857	0.868	0.869	0.849	0.855	0.863	**0.843**	0.865

Table 2. The Comparision of RMSE values: lower values indicate better performace

Dataset	ItemAvg	ItemBasedCF	PureSVD 50	PureSVD 200	SVD++ 50	SVD++200	URIM
Movielens	1.041	0.921	0.895	0.893	0.909	0.906	**0.795**
Netflix	1.053	0.952	0.918	0.912	0.895	0.893	**0.843**

In Movielens, we can find that the RMSE values range from 0.796 to 0.808. And when the number of user interest is 5 and the item type is 3, URIM performs best: RMSE comes down to nearly 0.796, which is a very promising value. While in Netflix, a more sparse dataset, it's easy for us to find that URIM performs best with 10 user interest domains and 5 item types, where the RMSE is 0.843. An interesting finding is that the RMSE values varied a little with the change of the number of user's interest domains and items' types. The reason for this phenomenon is that we formulate the prediction model at the full space without reducing dimensions. We can capture most of the useful explicit information in this full-reserved model. Hence, the choice of user interest domain and item type numbers don't have too many effects on the final performance.

Table 2 reports the comparison of different alogrithms in terms of RMSE. It's apparent that both in Movielens and Netflix datasets, URIM outperforms other methods a lot. Especially in Movielens, the RMSE decreased from about 0.90 to about 0.79, which makes a significant improvement. There are several reasons. On the one hand, facing with the few number of user rating history, the neighbor information is not so valid in ItemBasedCF. However, by introducing the user-item-relation component, URIM corporates the latent associations between users and items, which helps exploit more information within the limited data. Besides, URIM reserved most of the explicit indications in the three components, while PureSVD and SVD++ may filter out some useful information during reducing the dimensions.

Top-N Recommendation. Figures 2 and 3 present the performance of the algorithms on the Movielens dataset over the full test set. It is apparent that the five methods have great performance differences in terms of top-N accuracy. It's not surprising for us to find that ItemAvg method performs worst, as it's a non-personalized method, and didn't take any personal information into consideration. It's a little strange for us to find that for PureSVD performs better when the number of latent factor is 50 other than 200. The reason is that, PureSVD may filter out more noises when we only concentrated on the top-N recommendation. In addition, as we can see, our proposed model-URIM performs best in terms of recall. The recall of URIM comes to 63 % at N = 20. This means that about 63 % of the 5-star rated items will be recommended to the user within a top-20 recommendation list using URIM.

Precision-Recall curves measure the proportion of preferred items that are actually recommended. Figure 3 confirms that URIM outperforms than other algorithms in terms of precision metrics, as it dominates other algorithms all the time. Consistent to Movielens, it's apparent that URIM outperforms other algorithms in terms of recall and precision. When the N = 20, the recall of URIM comes to 61 %. It means URIM has a probability of 61 % recommending the appealing items to uses in the top-20 list. Similarly, the strange and somehow unexpected result of PureSVD is that algorithm with fewer later factors performs better. However, as for precision-recall curve, PureSVD and SVD++ intersect with each other all the time. Anyway, with all different selections of recall values, URIM performs best on the Netflix dataset in terms of precision (Figs. 4 and 5).

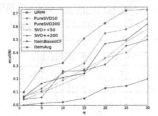

Fig. 2. Recall on Movielens

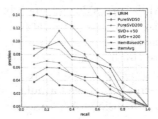

Fig. 3. Pre VS Rec on Movielens

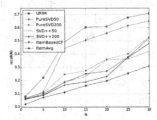

Fig. 4. Recall on Netflix

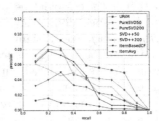

Fig. 5. Pre VS Rec on Netflix

Over both Movielens and Netflix datasets, URIM is the most significant algorithm in terms of recall as well as precision, outperforming other compared methods. As we expected before, given its naive assumption and simple model design, URIM makes a significant improve to other state-of-the-art model. In summary, URIM has multiple advantages. First, the number of user interest domains and item types has little effect on the final result, as URIM reserved the information fully in the three well-defined components. Secondly, URIM has the convenience of representing the users as a vector of user interest, and the items as a vector of item types, offering designers flexibility in handling new users, new ratings by existing users and explaining the reasoning behind the generated recommendations. In addition, we take the user and the item in the different dimensions, exploiting the major association patterns between user interest domains and item types, which fits the realistic settings more naturally and reasonably. Finally, we consider all possible user-item pairs (regardless of rating availability) in the training phase while other methods only consider the known ratings in the training set.

3.3 Complexity Analysis

In case of URIM, the offline part requires more time compared to the correlation-based algorithm, but less than SVD-based algorithms. Assume that N_I represent the number of iteration, r denotes the number of ratings, K denotes the number of user interest domains and S denotes the number of item types, we can get the running time complexity by the following equation:

$$N_I * (r + r + K * S * r) = N_I * r * (2 + K * S) \tag{11}$$

Usually, k, s and N_I are small, but r is large. Luckily, in case of URIM recommendation generation, the model training process can be done offline. Offline computation is not very critical to the performance of the recommender system. As for storage, however, URIM is more efficient, we need to store just user, item and relation matrices of size $M \times K$, $K \times S$ and $S \times N$ respectively, a total of $O(M + N)$, as K and S are small constants. But in case of the correlation-based CF algorithm, an $M \times M$ all-to-all correlation table must be stored requiring $O(M^2)$ storage, which can be substantially large with millions of users and items.

4 Conclusion and Future Work

In this paper, we proposed a simple but effective model for CF-based recommendation: User-Relation-Item Model (URIM). By introducing a component named user-item-relation matrix, which reflects the latent major associations between user interest domains and item types, URIM alleviates the synonymy problem and the data sparse problem efficiently. Most importantly, experimental results of URIM show a significant increase in predictive accuracy in terms of RMSE. As for top-N recommendation, URIM shows its high quality in terms of recall and precision compared with other state-of-the-art typical CF algorithms. And we'll investigate reasonable improvement to our model in reducing its complexity in the future.

References

1. Adomavicius, G., Tuzhilin, A.: Toward the next generation of recommender systems: a survey of the state-of-the-art and possible extensions. IEEE Trans. Knowl. Data Eng. **17**(6), 734–749 (2005)
2. Bell, R., Koren, Y., Volinsky, C.: Modeling relationships at multiple scales to improve accuracy of large recommender systems, KDD 2007, pp. 95–104. ACM, New York (2007)
3. Bennett, J., Lanning, S., Netflix, N.: The netflix prize. In: KDD Cup and Workshop in Conjunction with KDD (2007)
4. Cantador, I., Brusilovsky, P., Kuflik, T.: 2nd workshop on information heterogeneity and fusion in recommender systems (hetrec 2011). In: Proceedings of the 5th ACM Conference on Recommender Systems, RecSys 2011. ACM, New York (2011)
5. Cremonesi, P., Koren, Y., Turrin, R.: Performance of recommender algorithms on top-n recommendation tasks. In: Proceedings of the Fourth ACM Conference on Recommender Systems, RecSys 2010, pp. 39–46. ACM, New York (2010)
6. Deshpande, M., Karypis, G.: Item-based top-n recommendation algorithms. ACM Trans. Inf. Syst. **22**, 143–177 (2004)
7. Koren, Y., Bell, R., Volinsky, C.: Matrix factorization techniques for recommender systems. Computer **42**(8), 30–37 (2009)
8. Koren, Y.: Factorization meets the neighborhood: a multifaceted collaborative filtering model. In: Proceedings of the 14th ACM SIGKDD International Conference on Knowledge Discovery and Data Mining, KDD 2008, pp. 426–434. ACM, New York (2008)
9. Koren, Y.: Factor in the neighbors: scalable and accurate collaborative filtering. ACM Trans. Knowl. Discov. Data **4**(1), 1:1–1:24 (2010)
10. Lee, J., Sun, M., Lebanon, G.: A comparative study of collaborative filtering algorithms. CoRR abs/1205.3193 (2012)
11. McLaughlin, M.R., Herlocker, J.L.: A collaborative filtering algorithm and evaluation metric that accurately model the user experience, SIGIR 2004, pp. 329–336. ACM, New York (2004)
12. Papagelis, M., Plexousakis, D.: Qualitative analysis of user-based and item-based prediction algorithms for recommendation agents. Eng. Appl. Artif. Intell. **18**(7), 781–789 (2005)
13. Piotte, M., Chabbert, M.: The pragmatic theory solution to the netflix grand prize. In: Netflix Prize Documentation (2009)
14. Sarwar, B., Karypis, G., Konstan, J., Riedl, J.: Item-based collaborative filtering recommendation algorithms. In: Proceedings of the 10th International Conference on World Wide Web, WWW 2001, pp. 285–295. ACM, New York (2001)
15. Shani, G., Gunawardana, A.: Evaluating recommendation systems. In: Ricci, F., Rokach, L., Shapira, B., Kantor, P.B. (eds.) Recommender Systems Handbook, pp. 257–297. Springer, New York (2011)
16. Shi, Y., Larson, M., Hanjalic, A.: Collaborative filtering beyond the user-item matrix: a survey of the state of the art and future challenges. ACM Comput. Surv. **47**(1), 3:1–3:45 (2014)
17. Su, X., Khoshgoftaar, T.M.: A survey of collaborative filtering techniques. Adv. Artif. Intell. **2009**, 4:2 (2009)
18. Takács, G., Pilászy, I., Németh, B., Tikk, D.: Major components of the gravity recommendation system. SIGKDD Explor. Newsl. **9**(2), 80–83 (2007)

An Efficient Incremental Collaborative Filtering System

Aghiles Salah[(✉)], Nicoleta Rogovschi, and Mohamed Nadif

LIPADE, University of Paris Descartes, 45, Rue des Saints-Peres, Paris, France
{Aghiles.Salah,Nicoleta.Rogovschi,Mohamed.Nadif}@parisdescartes.fr

Abstract. Collaborative filtering (CF) systems aim at recommending a set of personalized items for an active user, according to the preferences of other similar users. Many methods have been developed and some, such those based on *Similarity* and *Matrix Factorization (MF)* can achieve very good recommendation accuracy, but unfortunately they are computationally prohibitive. Thus, applying such approaches to real-world applications in which available information evolves frequently, is a non-trivial task. To address this problem, we propose a novel efficient incremental CF system, based on a weighted clustering approach. Our system is able to provide a high quality of recommendations with a very low computation cost. Experimental results on several real-world datasets, confirm the efficiency and the effectiveness of our method by demonstrating that it is significantly better than existing incremental CF methods in terms of both scalability and recommendation quality.

Keywords: Collaborative filtering · Recommender systems · Clustering

1 Introduction

In order to filter large amounts of information, recommender systems (RSs) have been adopted by many real-world applications such as Amazon [8] and Netflix [7]. Collaborative filtering (CF) is the most often used approach in RSs. It consists in predicting the items that an active user will enjoy, based on the items that the people who are most similar to this user have enjoyed. Among CF systems two distinct types of approach are to be found:

Memory-based CF. These approaches are based on computing similarities [1]. User-based collaborative filtering looks for similarities between the active user $\mathbf{u}_a$ and all other users and tries to predict the preference of $\mathbf{u}_a$ for a set of new items, according to the preferences of the K most similar users to $\mathbf{u}_a$. Item-based collaborative filtering consists in finding the K nearest neighbors of each item and making recommendations according to the neighborhood of items enjoyed by the user $\mathbf{u}_a$.

Model-based CF. These approaches begin by suggesting a model that will learn from the user/item rating matrix $\mathbf{U}$ in order to capture the hidden features

© Springer International Publishing Switzerland 2015
S. Arik et al. (Eds.): ICONIP 2015, Part III, LNCS 9491, pp. 375–383, 2015.
DOI: 10.1007/978-3-319-26555-1_42

of the users and items. They then predict the missing ratings according to this model. Many model-based CF techniques have been proposed, the most popular being those based on clustering [12], co-clustering [4], and matrix factorization MF [2,11].

Traditional CF approaches, such as matrix factorization (MF) and memory-based methods, can achieve good prediction accuracy, but their computation time rises steeply as the number of users and items increases. Further, these methods need to be performed periodically (i.e., offline) in order to take into account new ratings, new users and items. However, with this strategy, new information which appear between two offline computations are not considered. As a result, applying traditional CF techniques to real-world applications such as Netflix in which the sets of users, items and ratings are frequently updated, remains therefore a challenge.

To overcome the problem of computation time, incremental CF systems have been proposed. The most popular are incremental CF based on MF approaches [5,10], incremental CF based on co-clustering [4,6], and incremental memory-based CF, including user [9] and item [13] based approaches. All these efforts have demonstrated the effectiveness of developing incremental models to provide scalable collaborative filtering systems. But often, these will significantly reduce the quality of recommendations. Further, most of these approaches (except memory-based CF) do not handle all possible dynamic scenarios (i.e., submission of new ratings, update of existing ratings, appearance of new users and new items). For instance incremental CF based on singular value decomposition [10], do not treat the two first scenarios.

In this paper we focus on the problem of computation time in CF systems. In order to overcome this drawback we propose a novel incremental CF approach, which is based on a weighted version of the online spherical k-means algorithm: OSK-means [14]. Our method is able to handle in a very short time the frequent changes in CF data; including the submission of new ratings, the update of existing ratings, the appearance of new users and items. Below, we summarize the key contributions we make in this paper.

- We derive a novel efficient CF system, based on a weighted clustering approach.
- In order to handle frequent changes in CF data, we design incremental updates, which allow to efficiently treat submissions of new ratings, updates of existing ratings, and occurrence of new users and items.

Numerical experiments validate our approach. The results on several real datasets show that our method outperforms significantly state-of-the-art incremental methods in terms of scalability and recommendation quality.

The rest of this paper is organized as follows. Section 2 introduces the formalism of traditional OSK-means. Section 3 provides details about the weighted version of OSK-means and our CF system: training, prediction and incremental training steps. Section 4 presents the results obtained on real-world datasets, in terms of recommendation quality and computation time. Finally, the conclusion summarizes the advantages of our contribution.

2   Online Spherical K-Means

In this paper matrices are denoted with boldface uppercase letters and vectors with boldface lowercase letters. Given a matrix $\mathbf{U} = (u_{ij})$ of size $n \times p$, the ith row (user) of this matrix is represented by a vector $\mathbf{u}_i = (u_{i1}, \ldots, u_{ip})^T$, where T denotes the transpose. The column j corresponds to the jth item. The partition of the set of rows into K clusters can be represented by a classification matrix $\mathbf{z}$ of elements z_{ik} in $\{0,1\}^K$ satisfying $\sum_{k=1}^{K} z_{ik} = 1$. The notation $\mathbf{z} = (z_1, \ldots, z_n)^T$, where $z_i \in \{1, \ldots, K\}$ represents the cluster of i, will be also used.

Before describing OSK-means, we will first introduce spherical K-means (SK-means). The SK-means [3] algorithm is a K-means algorithm in which the objects (users) $\mathbf{u}_1, \ldots, \mathbf{u}_n$ are assumed to lie on a hypersphere. SK-means originally maximizes the sum of the dot product between the elements of the data points and the K means directions characterizing the clusters. This is equivalent to maximizing the sum of the cosine similarity of the normalized data. The algorithm then maximizes the following objective function:

$$L = \sum_{i=1}^{n} \sum_{k=1}^{K} z_{ik} \cos(\mathbf{u}_i, \boldsymbol{\mu}_k) = \sum_{i=1}^{n} \sum_{k=1}^{K} z_{ik} \mathbf{u}_i^T \boldsymbol{\mu}_k, \tag{1}$$

where $z_{ik} \in \{0,1\}$; $z_{ik} = 1$ if $\mathbf{u}_i \in k^{th}$ cluster, $z_{ik} = 0$ otherwise. SK-means repeats the following two steps:

- For $i = 1, \ldots, n$, assign $\mathbf{u}_i$ to the kth cluster, where $z_i = \arg\max_k \left(\mathbf{u}_i^T \boldsymbol{\mu}_k \right)$, $k = 1, \ldots, K$.
- Calculate $\boldsymbol{\mu}_k = \frac{\sum_{i,k} z_{ik} \mathbf{u}_i}{\| \sum_{i,k} z_{ik} \mathbf{u}_i \|}$.

OSK-means [14] uses competitive learning (Winner-Takes-All strategy) to minimize the objective function (1), which leads to

$$\boldsymbol{\mu}_k^{new} = \frac{\boldsymbol{\mu}_k + \eta \frac{\partial L_i}{\partial \boldsymbol{\mu}_k}}{\|\boldsymbol{\mu}_k + \eta \frac{\partial L_i}{\partial \boldsymbol{\mu}_k}\|} = \frac{\boldsymbol{\mu}_k + \eta \mathbf{u}_i}{\|\boldsymbol{\mu}_k + \eta \mathbf{u}_i\|},$$

where η is the learning rate, $\boldsymbol{\mu}_k$ is the closest centroid to the object $\mathbf{u}_i$, and L_i denotes $\sum_{k=1}^{K} z_{ik} \mathbf{u}_i^T \boldsymbol{\mu}_k$.

In the OSK-means method, each centroid is updated incrementally with a learning rate η. Zhong [14] proposed an exponentially decreasing learning rate $\eta^t = \eta_0 (\frac{\eta_f}{\eta_0})^{\frac{t}{n \times B}}$, where $\eta_0 = 1.0$, $\eta_f = 0.01$, B is the number of batch iterations and $t, (0 \leq t \leq n \times B)$ is the current iteration.

3   Efficient Incremental Collaborative Filtering System (EICF)

In this section we describe our collaborative filtering system EICF, designed to provide a high quality of recommendations with a very low computation cost. This system can be divided into three main steps: training, prediction, and incremental training. The different steps are described as follows:

3.1   Training Step

This step, consists in clustering the users into K groups. Unfortunately the traditional OSK-means which has been proposed in the context of documents clustering, is not adapted for CF data. Unlike text data, the sparsity in CF is caused by unknown ratings, which requires a different handling than if the sparsity is caused by entries of "zero". To address this problem, we propose novel version of OSK-means which is more suitable for CF. It consists in introducing user weights, in order to tackle the sparsity problem; by giving more importance for users who provided many ratings. Thereby, the resulting clusters will be highly influenced by the most useful users (i.e., users with high weights). Below we give more details about this weighted version of OSK-means. Let w_i denote the weight of the ith user, the weighted objective function of the SK-means is given by:

$$L^w = \sum_{i=1}^{n} L_i^w, \text{ where } L_i^w = \sum_{k=1}^{K} w_i z_{ik} \mathbf{u}_i^T \boldsymbol{\mu}_k, \tag{2}$$

Thus, the corresponding update centroid for the weighted OSK-means is given by:

$$\boldsymbol{\mu}_k^{new} = \frac{\boldsymbol{\mu}_k + \eta \frac{\partial L_i^w}{\partial \boldsymbol{\mu}_k}}{||\boldsymbol{\mu}_k + \eta \frac{\partial L_i^w}{\partial \boldsymbol{\mu}_k}||} = \frac{\boldsymbol{\mu}_k + \eta w_i \mathbf{u}_i}{||\boldsymbol{\mu}_k + \eta w_i \mathbf{u}_i||}, \tag{3}$$

We now give an intuitive formulation of user weights. Let $\mathbf{M} = (m_{ij})$ be an $(n \times p)$ binary matrix, such that $m_{ij} = 1$ if the rating u_{ij} is available, and $m_{ij} = 0$ otherwise. Its ith row corresponds to a vector $\mathbf{m}_i = (m_{i1}, \ldots, m_{ip})^T$ indicating which items have been rated by the ith user. Thus, we define the weight of the ith user to be proportional to the number of his available ratings as follows:

$$w_i = (\mathbf{m}_i^T \mathbb{1}) \times \sigma(\mathbf{u}_i) \tag{4}$$

where $\mathbb{1}$ is the vector of the appropriate dimension which all its values are 1, and $\sigma(\mathbf{u}_i)$ denotes the standard deviation of ratings provided by $\mathbf{u}_i$. We consider the standard deviation in order to give less importance for users who provide only low ratings or similarly, only high ratings (i.e., users who expressed the same preference for all items they have rated). Algorithm algotrain describes in more details our training step.

3.2   Prediction Step

In this step, unknown ratings are predicted according to the clustering results. However, it is difficult to make consistent predictions, even when the best clustering results are achieved, because there are so many unknown ratings in $\mathbf{U}$. To overcome this difficulty we propose to estimate unknown ratings by a weighted average of observed ratings, as follows:

$$u_{aj} = \frac{\sum_{i=1}^{n} w_i z_{ik} \mathbf{u}_i^T \boldsymbol{\mu}_k \times u_{ij}}{\sum_{i=1}^{n} w_i z_{ik} \mathbf{u}_i^T \boldsymbol{\mu}_k}, \tag{5}$$

Algorithm 1. EICF training.

Input: n users $\mathbf{u}_i$ in $\mathbb{R}^p$, K is the number of clusters;

Output: K centers $\boldsymbol{\mu}_k$ in $\mathbb{R}^p$, and $\mathbf{z} = (z_1, \ldots, z_n)$;

Steps:

1. Normalize the users (i.e., $\mathbf{u}_i = \frac{\mathbf{u}_i}{\|\mathbf{u}_i\|}$, $\forall i$)
2. Compute user weights: $w_i = (\mathbf{m}_i^T \mathbb{1}) \times \sigma(\mathbf{u}_i)$;
3. Initialization: random initialization of the partition $\mathbf{z}$, $t = 1$;
4. Estimation of the initial centroids:

$$\mu_{kj} = \frac{\sum_i m_{ij} w_i z_{ik} u_{ij}}{\| \sum_i w_i z_{ik} \mathbf{u}_i \|}$$

for $b = 1$ **to** B **do**
   for each $\mathbf{u}_i$ in **U do**
      5. User assignment: compute $z_i = \arg\max_k(w_i \mathbf{u}_i^T \boldsymbol{\mu}_k)$;
      6. Centroid update: compute the winner centroid by $\hat{\boldsymbol{\mu}}_{z_i} = \frac{\boldsymbol{\mu}_{z_i} + \eta w_i \mathbf{u}_i}{\|\boldsymbol{\mu}_{z_i} + \eta w_i \mathbf{u}_i\|}$;
      $t = t + 1$;
   end for
end for

Let $\mathbf{u}_a$ denote the active user, $k = z_a$. The key idea behind this strategy is to weight the available ratings u_{ij} according to the similarity between each user $\mathbf{u}_i$ and its corresponding centroid $\boldsymbol{\mu}_k$, and to weight by w_i, in order to give greater importance for users closest to their centroid, and respectively to give more importance for ratings provided by most important users.

The prediction Eq. (5) is attractive because it depends only on the clustering results, which means that it can be performed offline and stored in a $(K \times p)$ matrix $\mathbf{P}$, which leads to very short prediction times.

3.3 Incremental Training Step

In the sequel, we design incremental updates, in order to handle the frequent changes in CF data. Thus, we can distinguish four main situations: (1) submission of new ratings, (2) update existing ratings, (3) appearance of new users, (4) appearance of new items. In the following, we give the update formulas for each situation.

Submission of a new rating. Let $\mathbf{u}_a$ denote an active user who submits a new rating for an item j. The equations below, give the different incremental updates to perform in this case.

- Update the norm of $\mathbf{u}_a$: $\|\mathbf{u}_a^+\| = \sqrt{\|\mathbf{u}_a\|^2 + u_{aj}^2}$
- For each k, update the similarity between $\mathbf{u}_a$ and $\boldsymbol{\mu}_k$:

$$cos(\mathbf{u}_a^+, \boldsymbol{\mu}_k) = \frac{1}{\|\mathbf{u}_a^+\|}[\|\mathbf{u}_a\| \times \mathbf{u}_a^T \boldsymbol{\mu}_k + u_{aj}\mu_{kj}],$$

- Update the weight of the active user: $\hat{w}_a = (\frac{w_a}{\sigma(\mathbf{u}_a)} + 1) \times \sigma(\mathbf{u}_a^+)$
- Update the assignment of $\mathbf{u}_a$: $\hat{z}_a = \arg\max_k cos(\mathbf{u}_a^+, \boldsymbol{\mu}_k)$.
- Update the corresponding centroid $\boldsymbol{\mu}_{\hat{z}_a}$, by using formula (3) where

$$\sigma(\mathbf{u}_a^+)^2 = \frac{N_a \times (\sigma(\mathbf{u}_a)^2 + \bar{u}_a^2) + u_{aj}^2}{N_a + 1} - \left(\frac{N_a \bar{u}_a + u_{aj}}{N_a + 1}\right)^2,$$

thanks to König-Huygens formula, i.e., $\sigma(\mathbf{u}_a) = \sqrt{\frac{1}{N_a} \sum_j u_{aj}^2 - \bar{u}_a^2}$. The notation $\mathbf{u}_a^+$ denotes the active user $\mathbf{u}_a$ with the new rating u_{ij} available, N_a and $\bar{u}_a$ denote respectively, the number of ratings and the average rating of $\mathbf{u}_a$ before evaluating item j. Note that, as the centroids are stable at the end of training, the two latter incremental updates concerning the assignment of $\mathbf{u}_a$, do not need to be performed after each one new rating.

Update an existing rating. In this case, the active user updates an existing rating for an item j. As for the submission of a new rating, the main updates are summarized below.

- Update the norm of $\mathbf{u}_a$: $\|\mathbf{u}_a^+\| = \sqrt{\|\mathbf{u}_a\|^2 - u_{aj}^2 + \hat{u}_{aj}^2}$
- For each k, update the similarity between $\mathbf{u}_a$ and $\boldsymbol{\mu}_k$:

$$cos(\mathbf{u}_a^+, \boldsymbol{\mu}_k) = \frac{1}{\|\mathbf{u}_a^+\|}[\|\mathbf{u}_a\| \times \mathbf{u}_a^T \boldsymbol{\mu}_k - u_{aj}\mu_{kj} + \hat{u}_{aj}\mu_{kj}]$$

- Update the weight of the active user: $\hat{w}_a = \frac{w_a}{\sigma(\mathbf{u}_a)} \times \sigma(\mathbf{u}_a^+)$
- Update the assignment of $\mathbf{u}_a$: $\hat{z}_a = \arg\max_k cos(\mathbf{u}_a^+, \boldsymbol{\mu}_k)$.
- Update the corresponding centroid $\hat{z}_a$, by using Eq. (3) where

$$\sigma(\mathbf{u}_a^+)^2 = \left(\sigma(\mathbf{u}_a)^2 + \bar{u}_a^2 + \frac{\hat{u}_{aj}^2 - u_{aj}^2}{N_a}\right) - \left(\bar{u}_a + \frac{\hat{u}_{aj} - u_{aj}}{N_a}\right)^2,$$

$\hat{u}_{aj}$ denotes the new value substituted for the existing rating u_{aj}, and the notation $\mathbf{u}_a^+$ represents the active user after updating the known rating u_{aj}.

Appearance of new user. In this situation, a new user is incorporated into the model in real time. Let $\hat{\mathbf{u}}_a$ denote a new user. The model is incremented as follows:

- Compute the weight of $\hat{\mathbf{u}}_a$, by using Eq. (4).
- Assign $\hat{\mathbf{u}}_a$ to k^{th} cluster where $k = \arg\max_{1 \leq k \leq K} (\frac{\hat{\mathbf{u}}_a^T \boldsymbol{\mu}_k}{\|\hat{\mathbf{u}}_a\|})$.
- Update the corresponding centroid: $\hat{\boldsymbol{\mu}}_k = \frac{\boldsymbol{\mu}_k + \eta w_a \frac{\hat{\mathbf{u}}_a}{\|\hat{\mathbf{u}}_a\|}}{\|\boldsymbol{\mu}_k + \eta w_a \frac{\hat{\mathbf{u}}_a}{\|\hat{\mathbf{u}}_a\|}\|}$.

Appearance of new item. When a new item appears, it has no ratings, so there nothing to change in the model. When a new item starts receiving ratings, handling new item, reduces to handling the submission of new ratings.

4   Experimental Results

Hereafter, we propose to evaluate the performances of our CF system on three real-world datasets. The first is *MovieLens*[1] *(ML-1M)*, consisting of 1,000,209 ratings provided by 6040 users for 3952 movies (only 4.2 % of ratings are observed). The second is *MovieLens (ML-100k)*, containing 100,000 ratings given by 943 users for 1664 movies. The proportion of observed ratings in this dataset is 6.4 %. The last dataset is Epinions[2], with 664,824 ratings from 49,290 users on 139,738 items (movies, musics, electronic products, books, ...). The Epinion dataset is more than 99 % sparse.

We compare EICF with several popular methods, namely: incremental user-based CF IUCF [9], incremental item-based CF IICF [13], and incremental CF based on co-clustering COCLUST [4]. All the evaluations are made under the same machine (OS: ubuntu 14.04 LTS 64-bit, Memory: 16 GiB, Processor: Intel® Core™ i7-3770 CPU @ 3.40 GHz × 8). To evaluate our CF system we focus on the quality of the recommendations and computational time. Thus, we chose the F-measure F1 [1] as the evaluation metric. Unlike accuracy metrics such as mean average error and root means square error, the F-measure allows to evaluate the quality of the set of recommendations [1], which is more relevant in the context of CF. The results reported in Table 1 are obtained as follows: (1) We generate ten random training-test (80–20 %) sets from each dataset. (2) Users in test sets are considered as new ones, and are incorporated incrementally. (3) Finally, we report the average F-measure for each method, over different recommendation lists (i.e., containing 10, 25 and 40 items). We also report the average computation time required by each method, for incorporating and generating recommendations, for users from the test sets. Note that, in terms of computation time, IICF is favoured in this comparison; unlike the other methods, incorporating new users is not the most expensive computation for this approach.

From Table 1, we note that our method provides a high quality of recommendations, thanks to our strategy for alleviating the sparsity problem; by introducing user weights. In fact, our CF system EICF exhibits the best quality of recommendations, over all datasets. Moreover, from Table 1 we observe that EICF requires much less time for handling new information and generating recommendations, than the other incremental methods, including IICF although it is favoured. This performance rises significantly as the volume of data increases. In fact, contrary to the other methods, the complexity of EICF does not depend on the number of users and items, as reported in Table 2. Therefore, EICF is more suitable than the other incremental methods, for real world-applications involving large databases in which users, items and ratings are frequently updated. Note that, the computation time of COCLUST reported in Table 1 is high, even

---

[1] http://grouplens.org/datasets/movielens/.
[2] http://www.epinions.com.

Table 1. Comparison of several CF systems in terms of F1 and computation time

Datasets		Recom. lists	CF methods			
			COCLUST	IICF	IUCF	EICF
ML-100k	F1	10	0.02	0.10	0.16	**0.21**
		25	0.06	0.16	0.22	**0.30**
		40	0.05	0.10	0.24	**0.33**
	Comp. time (s)		0.97	0.88	1.78	**0.51**
ML-1M	F1	10	0.2e-4	0.04	0.10	**0.14**
		25	0.02	0.08	0.15	**0.21**
		40	0.05	0.11	0.17	**0.25**
	Comp. time (s)		11.85	7.96	138.1	**2.86**
Epinion	F1	10	0.45e-4	0.008	0.022	**0.038**
		25	0.38e-4	0.011	0.029	**0.043**
		40	0.73e-4	0.012	0.032	**0.046**
	Comp. time (s)		212.71	927.22	4041.01	**149.20**

Table 2. Comparison of computational times (in the worst case) in various situations. W^* denotes the number of observed ratings in $\mathbf{U}$. K and L are the number of row and column clusters, K also denotes the number of neighbours for memory CF (IUCF, IICF). p^* is the number of observed ratings for a new user, n^* denotes the number of available ratings for a new item. Finally, B denotes the number of iterations

Algorithm	Static training	Prediction	Inc. train
IUCF	$O(nW^*)$	$O(K)$	$O(np^*)$
EICF	$O(BKW^*)$	$O(1)$	$O(Kp^*)$
COCLUST	$O(BW^* + nBKL + pBKL)$	$O(1)$	$O(p^*)$
IICF	$O(pW^*)$	$O(p^*)$	$O(pn^*)$

if its complexity in the dynamic situation (i.e., inc. train: $O(p^*)$) might appear attractive. The reason is that, this approach provides only partial updates, and the co-clustering is performed periodically to completely incorporate new information.

5   Conclusion

We presented EICF, a novel efficient and effective incremental CF system, which is based on a weighted clustering approach. To achieve high quality of recommendations, we introduced user weights into the clustering process, to lessen the effect of users who provided only few ratings. In order to address the computational time problem, we designed incremental updates, which allows our system to handle in a very short time, the frequent changes in CF data; such as submissions of new ratings, appearance of new users and items. Numerical experiments on real-world datasets demonstrate the efficiency and the effectiveness of our method which provides better quality of recommendations than existing incremental CF systems, while requiring less computation time. Thus, our CF system is more suitable than existing incremental approaches, for real-world applications involving huge databases, in which available information (i.e., users, items and ratings) changes frequently. For future work, we will investigate other strategies for handling the sparsity problem in CF, and try to develop a parallel version of EICF, that can support distributed computations.

References

1. Bobadilla, J., Ortega, F., Hernando, A., Gutiérrez, A.: Recommender systems survey. Knowl. Based Syst. **46**, 109–132 (2013)
2. Delporte, J., Karatzoglou, A., Matuszczyk, T., Canu, S.: Socially enabled preference learning from implicit feedback data. In: Blockeel, H., Kersting, K., Nijssen, S., Železný, F. (eds.) ECML PKDD 2013, Part II. LNCS, vol. 8189, pp. 145–160. Springer, Heidelberg (2013)
3. Dhillon, I.S., Modha, D.S.: Concept decompositions for large sparse text data using clustering. Mach. Learn. **42**(1–2), 143–175 (2001)
4. George, T., Merugu, S.: A scalable collaborative filtering framework based on co-clustering. In: Fifth IEEE International Conference on Data Mining, pp. 625–628 (2005)
5. Han, S., Yang, Y., Liu, W.: Incremental learning for dynamic collaborative filtering. J. Softw. **6**, 969–976 (2011)
6. Khoshneshin, M., Street, W.N.: Incremental collaborative filtering via evolutionary co-clustering. In: RecSys, pp. 325–328 (2010)
7. Koren, Y.: The Bellkor solution to the Netflix grand prize (2009)
8. Linden, G., Smith, B., York, J.: Amazon.com recommendations: item-to-item collaborative filtering. IEEE Internet Comput. **7**, 76–80 (2003)
9. Papagelis, M., Rousidis, I., Plexousakis, D., Theoharopoulos, E.: Incremental collaborative filtering for highly-scalable recommendation algorithms. In: Hacid, M.-S., Murray, N.V., Raś, Z.W., Tsumoto, S. (eds.) ISMIS 2005. LNCS (LNAI), vol. 3488, pp. 553–561. Springer, Heidelberg (2005)
10. Sarwar, B., Karypis, G., Konstan, J., Riedl, J.: Incremental singular value decomposition algorithms for highly scalable recommender systems. In: ICIS, pp. 27–28 (2002)
11. Sarwar, B.M., Karypis, G., Konstan, J.A., Riedl, J.T.: Application of dimensionality reduction in recommender system - a case study. In: ACM WEBKDD Workshop (2000)
12. Ungar, L.H., Foster, D.P.: Clustering methods for collaborative filtering. In: AAAI Workshop on Recommendation Systems, vol. 1, pp. 114–129 (1998)
13. Yang, X., Zhang, Z., Wang, K.: Scalable collaborative filtering using incremental update and local link prediction. In: CIKM, pp. 2371–2374 (2012)
14. Zhong, S.: Efficient online spherical k-means clustering. In: IJCNN, pp. 3180–3185 (2005)

MonkeyDroid: Detecting Unreasonable Privacy Leakages of Android Applications

Kai Ma[1]([⊠]), Mengyang Liu[1]([⊠]), Shanqing Guo[1]([⊠]), and Tao Ban[2]([⊠])

[1] Department of Computer Science and Technology, Shandong University,
Jinan, China
{makaisdu,mengyang.liu}@gmail.com, guoshanqing@sdu.edu.cn
[2] NICT, Tokyo, Japan
bantao@nict.go.jp

Abstract. Static and dynamic taint-analysis approaches have been developed to detect the processing of sensitive information. Unfortunately, faced with the result of analysis about operations of sensitive information, people have no idea of which operation is legitimate operation and which is stealthy malicious behavior. In this paper, we present Monkeydroid to pinpoint automatically whether the android application would leak sensitive information of users by distinguishing the reasonable and unreasonable operation of sensitive information on the basis of information provided by developer and market provider. We evaluated Monkeydroid over the top 500 apps on the Google play and experiments show that our tool can effectively distinguish malicious operations of sensitive information from legitimate ones.

Keywords: Android security · Privacy leakage detection · Static taint analysis · Natural language processing

1 Introduction

According to the statistical data from Strategy Analytics [1], Android operating system accounted for about 84.6 % of the mobile terminal market in 2014. Because of the importance of mobile devices in our daily life such as storing a large amounts of sensitive personal information and the openness of the Android platform, the android platform has become an ideal land of the wanton growth of malicious software. As more and more apps are used for private and privileged tasks, concerns are also rising about the consequences of failure to protect or respect users privacy.

As a result, many approaches have been proposed to automatically reveal privacy disclosures in Android apps and can be grouped into two major categories: static analysis (e.g..ScAndroid [2], Androidleaks [3], Chex [4]) and dynamic analysis(e.g..Taintdroid [5], Appsplayground [6]). Although these work could be successful to reveal the operations of sensitive information in the Android apps. However,there is no an approach to judge the legitimacy of the detected flows of sensitive information(e.g..location, device identifier).

© Springer International Publishing Switzerland 2015
S. Arik et al. (Eds.): ICONIP 2015, Part III, LNCS 9491, pp. 384–391, 2015.
DOI: 10.1007/978-3-319-26555-1_43

In this paper, we propose Monkeydroid, an approach to automatically pinpoint whether the android application will leak android users sensitive information in a mobile device. With the help of application description and Android API document, this approach leads to a significant reduction in false positives by distinguishing the reasonable and unreasonable behaviors of android application about sensitive information. We evaluated Monkeydroid over the top 500 apps on the Google play. Our experiments show that our tool can effectively distinguish malicious operations of sensitive information from legitimate ones.

The contributions of this paper are as follows:

- We propose an approach to judge the legitimacy of operations of sensitive information in Android applications based on natural language processing of application description and static taint analysis.
- We implement a system named Monkeydroid, which is a prototype of our approach to detect privacy leakages of Android applications. We evaluate Monkeydroid on the top 500 apps from the Google Play store. The results show Monkeydroid has a high accuracy and high performance in distinguishing malicious operations of sensitive information from legitimate ones.

The rest of this paper is organized as following. In Sect. 2, we introduce the problem statement of our work. Section 3 gives the overview of Monkeydroid and implementation of Monkeydroid. In Sect. 4, we evaluate Monkeydroid. Section 5 presents the limitation of Monkeydroid. We draw the conclusions in Sect. 6.

2 Problem Statement

Let us consider a weather application (com.devexpert.weather.apk) that the users location need to be sent to the server when the user utilizes the app to obtain the local weather information. Using current detection systems to analyze the app, we will get a report which contains a privacy disclosure about users location, regardless of the fact that the apps legitimate function depends on users location. In fact,we lack an effective tool to inform the user which are legitimate and malicious privacy disclosures reported by static or dynamic analysis.

Based on the above facts, we propose one solution to this problem, where we try to use application description to judge the legitimacy of operations of sensitive information in an application.

Question 1. Why do we try to use market description to help us analyze the privacy disclosure of Android apps?

The programmer of the application utilizes code to realize his intention in the app and makes use of natural language to express his intention in the application description if he is honest. If the intention of code can match the intention of application description, we think that the user can know how the apps deal with his privacy by reading the application description and we can believe that privacy disclosure does not exist in the app. If the application description does not contain some kinds of sensitive information but the code of the application contains them, we can identify benign and malicious operations of sensitive information through our approach.

Question 2. How to bridge the semantic gap between apk and its description?

Although users can easily understand the meaning of a natural language document, it's difficult for computers to perform like human beings to understand the meaning of natural language sentences. Therefore, an effective intermediate expression is in need to help the computer automatically breaks the semantic gap between application and its description. Considering that the meaning of a well constructed natural language sentence that describes operations can be regarded as a statement, it is reasonable to convert these sentences to logic expressions. Recent research has shown the adequacy of using First-Order-Logic expression for NLP(Natural Language Processing) related analysis tasks [7]. So we construct our intermediate-representation, semantic graph, based on the First-Order-Logic expression.

3   Monkeydroid

In this section, we present the architecture of Monkeydroid (in Fig. 1) and give the details of design and implementation.

3.1   Extracting Sensitive-Information Behaviors

Monkeydroid unzips an apk file and decomposes the Dalvik bytecode executable file into Jimple representation, which is a typed-3 address intermediate representation suitable for analysis and optimization on the Soot framework. Leveraging layout files and manifest file, Monkeydroid constructs precise call graph and ICFG(Inter-procedural Control-Flow Graph) [8]. Monkeydroid marks sensitive sources and outgoing channels(sinks) provided by SuSi [10] and makes use of static taint analysis to reveal flows from sources to outgoing channels(sinks).

After static taint analysis, Monkeydroid reduced sensitive information flows to source-to-sink form. Then, Monkeydroid constructs semantic graphs(in Fig. 3)

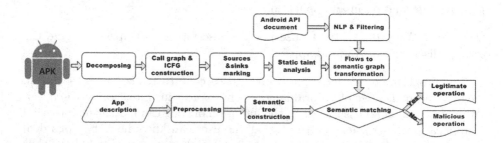

Fig. 1. The architecture of Monkeydroid

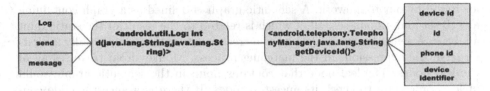

Fig. 2. Android API document

Fig. 3. The semantic graph of this flow

of the simplified flows on the basis of knowledge base. The knowledge base is defined as a collection of verbs or nouns associated with particular API name. Knowledge atoms can be represented as <API name - verb or noun set>. To ensure the accuracy of the semantic graph, Monkeydroid uses Android API document to generate a knowledge base containing semantic information of sensitive resource related API. In the Android API document, each API name (e.g. public double getLatitude()) is followed by several nature language sentences (e.g. "Get the latitude, in degrees.") describing the function of this API. For the remaining sensitive API, Monkeydroid uses the Stanford parser to identify the part of speech (POS, such as nouns and verbs) a particular word in a sentence in the description of an API (in Fig. 2). After POS tagging, Monkeydroid traverses the tagged sentences and extracts the nouns or verbs respectively for API in the source or the sink list. The extracted verbs or nouns are organized in the form of knowledge atoms mentioned above.

3.2 Application Description Analysis

This subsection presents the approach of the transformation between application description and corresponding semantic tree used in Monkeydroid. This part in Monkeydroid is implemented as the following two units:

1. Preprocessor. The preprocessor is designed for segmenting target application description. It accepts the natural language description of target application and generates a list of separated sentences which contains the keyword of the source generated by the program analyser. The separated sentences are put into an automation to check if they contains a keyword.
2. Semantic-tree construction. The semantic-tree construction is designed for the generation of a semantic-tree of a sentence which contains the keyword of the source. It accepts the sentence list generated by the preprocessor and provide a semantic-tree for the checker. The semantic-tree is represented by the FOL representation. We implement the semantic-tree construction with the help of Stanford NLP tools [9].

3.3 Semantic Graphs of APK and Application Description Matching

This subsection is about to find the contradiction between program context and application description. After the construction of the semantic graph and the semantic tree, Monkeydroid tries to justify the rationality of the sensitive behavior found by our framework. A semantic graph is defined as a graph containing a verb set and a noun set, all of which is related to the result of the static taint analysis. A semantic tree is defined as a tree structure that is essentially a First-Order-Logic expression. In the matching process, Monkeydroid traverses all of the leaf nodes. The leaf node that contains noun in the semantic graph would yield a procedure to check its ancestor nodes. If there is a ancestor node containing a verb in the corresponding semantic graph, the semantic graph matches the semantic tree, otherwise the semantic graph does not match the semantic tree. The algorithm(in Algorithm 1) shows the process in pseudo-code.

Input: *SemGraph*, *SemTree*
Output: *inDesc*
1 *inDesc = False*;
2 **for** *node in SemTree.leafNodes* **do**
3   **if** *node.content in SemGraph.nounSet* **then**
4   *iter = node*;
5   **while** *iter.hasParent* **do**
6   *iter = iter.parent*;
7   **if** *iter.content in SemGraph.verbSet* **then**
8   *inDesc = True*;
9   return *inDesc*;
10   **end**
11   **end**
12   **end**
13 **end**
14 return *inDesc*;

Algorithm 1. Match semantic graph with semantic tree

4 Evaluation

4.1 Study Subjects

In the experiment, we evaluate the effectiveness of Monkeydroid over the top 500 apps downloaded from the Google play store during March, 2014. Although there are more than one million applications on the Google play, most smartphone users only concentrate on the top applications and it's rational for us to choose the top 500 apps as the study subjects. By testing these apps with Monkeydroid, we evaluate whether Monkeydroid can precisely identify benign operation on sensitive information, particularly those delivering sensitive information to the network.

4.2   Experimental Environment

We setup our evaluation on a sever with 12×2.00 GHz processors and 32.0 GB physical memory. Each analysis task (for analyzing an app) is assigned to 8×2.00 GHz processors and 16 GB, which runs JDK 1.7.0 21. Because of either insufficient memory or failure of type resolving, Monkeydroid cannot analyze some apps in the evaluation and we share same problem with other flowDroid-dependent tools [8]. Thus, below we show our experimental results over the remaining apps.

4.3   Results

We analyze the top 500 apps, among which 164 apps fail to go through and 256 apps contain behaviors of the sensitive information(e.g..location, IMEI, contact). Monkeydroid identifies that behaviors of the sensitive information are legitimate in 65 apps and unreasonable in 227 apps by matching apk with app description.

We show a part of the summarized result in Table 1 for these apps. From left to right of the table, the table shows apps package name, sensitive information and sink. If the sensitive information is sent to the outside through the sink, the flow could be recorded in the table. The fourth column of the table is "In description" and we fill "Yes" or "No" in the blank if the sensitive information mentioned in the market description or not. If the operation described in the market description, we take it as the rational operation and use a dot to mark in the table. To check the result of Monkeydroid, we also utilize the VirusTotal to test the apps and record how many anti-virus engines in VirusTotal identify the app as malicious.

From this table, our observations and findings are as follows. First, we can see that the most frequently leaked sensitive information is device ID (IMEI). The related research work such as Taintdroid [5] can verify this result. Second, if the sensitive information mentioned in the market description, this operation of app is likely to be rational. Third, the applications developed by the mature company like Facebook, Twitter are usual to be reinforced and we can't analyze these apps completely.

5   Limitation

There are several limitations with Monkeydroid. As other static analysis systems on Android, Monkeydroid cannot detect the disclosures caused by Java reflection, code encryption, JNI calls, or dynamical code loading. As mentioned in Sect. 3.1, knowledge base only from Android API is not complete and this can produce false positives in matching semantic graphs. Faced with multiple sentences about one kind of sensitive information, Monkeydroid cannot take accurate analysis and this is the major problem we will solve in the future work.

Table 1. Partial experimental results

Package name	Sensitive information	Sink	In description	Rational	Not rational	Virus total
Vitamio Plugin.apk	IMEI	Log	Yes	–	•	0/57
	IMSI	Log	No	–	•	
	SimSerialNumber	Log	No	–	•	
	SimCountryIso	Log	No	–	•	
com.whatsapp.wallpaper.apk	location	Internet	No	–	•	0/57
com.gameloft.android.apk	location	Internet	No	–	•	0/57
com.photo.ghost.prank.apk	location	Internet	No	–	•	0/57
AccuWeather 3.2.14.1	IMEI	Log	No	–	•	0/57
	IMEI	Internet	No	–	•	
	location	Log	No	–	•	
	location	Internet	Yes	•	–	
Amazon 2.9.7	location	Internet	No	–	•	0/57
Backgrounds HD Wallpapers 2.0.1	IMEI	Log	No	–	•	0/57
	IMEI	Internet	No	–	•	
	Contact	Log	Yes	•	–	
Chase Mobile 3.16	Contact	Internet	Yes	•	–	0/57
ebay 2.6.0.98	Location	Internet	Yes	•	–	0/57
HotPads 3.1	Location	Internet	Yes	•	–	0/55

6 Conclusion

In this paper, we propose an approach to judge the legitimacy of operations of sensitive information in Android Applications based on natural language processing of application description and static taint analysis. We implement a system named Monkeydroid, which is a prototype of our approach. Compared with previous work, Monkeydroid effectively rules out the legitimate and malicious privacy disclosures, which exposes those privacy leakages that cannot be associated with apps functions. As a result, Monkeydroid can greatly increase the detection rate of threatening privacy leaks, and at the same time, considerably prompt developer to pay attention to the description of apps and to be honest.

Acknowledgments. This work is partially supported by National Natural Science Foundation of China (61173068, 61173139), Program for New Century Excellent Talents in University of the Ministry of Education, the Key Science Technology Project of Shandong Province (2014GGD01063), the Independent Innovation Foundation of Shandong Province(2014CGZH1106) and the Shandong Provincial Natural Science Foundation (ZR2014FM020, ZR2014FM031).

References

1. https://www.strategyanalytics.com/access-services/devices/mobile-phones
2. Fuchs, A.P., Chaudhuri, A., Foster, J.S.: Scandroid: automated security certification of android applications. Manuscript, University of Maryland **2**(3), (2009). http://www.cs.umd.edu/avik/projects/scandroidascaa

3. Gibler, C., Crussell, J., Erickson, J., Chen, H.: AndroidLeaks: automatically detecting potential privacy leaks in android applications on a large scale. In: Katzenbeisser, S., Weippl, E., Camp, L.J., Volkamer, M., Reiter, M., Zhang, X. (eds.) Trust 2012. LNCS, vol. 7344, pp. 291–307. Springer, Heidelberg (2012)
4. Lu, L., Li, Z., Wu, Z., et al.: Chex: statically vetting android apps for component hijacking vulnerabilities. In: Proceedings of the 2012 ACM Conference on Computer and Communications Security, pp. 229–240. ACM (2012)
5. Enck, W., Gilbert, P., Han, S., et al.: TaintDroid: an information-flow tracking system for realtime privacy monitoring on smartphones. ACM Trans. Comput. Syst. (TOCS) **32**(2), 5 (2014)
6. Rastogi, V., Chen, Y., Enck, W.: AppsPlayground: automatic security analysis of smartphone applications. In: Proceedings of the third ACM Conference on Data and Application Security and Privacy, pp. 209–220. ACM (2013)
7. Pandita, R., Xiao, X., Zhong, H., et al.: Inferring method specifications from natural language API descriptions. In: Proceedings of the 34th International Conference on Software Engineering, pp. 815–825. IEEE Press (2012)
8. Arzt, S., Rasthofer, S., Fritz, C., et al.: Flowdroid: precise context, flow, field, object-sensitive and lifecycle-aware taint analysis for android apps. ACM SIG-PLAN Not. **49**(6), 259–269 (2014). ACM
9. Pandita, R., Xiao, X., Yang, W., et al.: WHYPER: towards automating risk assessment of mobile applications. In: USENIX Security 2013 (2013)
10. Rasthofer, S., Arzt, S., Bodden, E.: A machine-learning approach for classifying and categorizing android sources and sinks. In: 2014 Network and Distributed System Security Symposium (NDSS), February 2014, to appear. http://www.bodden.de/pubs/rab14classifying.pdf

Statistical Prior Based Deformable Models for People Detection and Tracking

Amira Soudani$^{(\boxtimes)}$ and Ezzeddine Zagrouba

Equipe de Recherche SIIVA, Laboratoire RIADI, Institut Supérieur d'Informatique,
Université de Tunis El Manar, 2 Rue Abou Rayhane Bayrouni, 2080 Ariana, Tunisia
`amira.soudani@gmail.com`, `ezzeddine.zagrouba@fsm.rnu.tn`

Abstract. This paper presents a new approach to segment and track
people in video. The basic idea is the use of deformable model with
incorporation of statistical prior. We propose an hybrid energy model
that incorporates a global and a statistical based energy terms in order
to improve the tracking task even under occlusion conditions. Target
models are initialized at the first frame, then predictions are constructed
based on motion vectors. Therefore, we apply an hybrid active contour
model in order to segment tracked people. Experiments show the ability
of the proposed algorithm to detect, segment and track people well.

Keywords: Tracking · Segmentation · Deformable models · Multiple
targets · Active contours · Occlusion

1 Introduction

Object tracking is the process of locating moving targets throughout the frames
of video sequences. In this aim several works were proposed and divided into
three categories [1] based on the tracked feature: Point, Kernel and Silhouette
based tracking methods. In this paper, we propose to classify methods of the
state of art based on the representation of the tracked targets. In fact, a target object is represented by a model that includes information about its shape
and appearance. The shape of the object of interest can be approximated by a
basic, articulated or deformable representation. In [2], authors propose a tracking
method based on basic representation and particle filtering. Indeed, each target
is initialized in the first frame then it is divided into seven parts. Multi-part
RGB kernel histograms are computed for each part. The particle filter is used
to look for candidates in the current image based on their previous states. For
each particle provided by the particle filtering process, a multi-parts representation is computed as same as the models of the observations. Thereafter, they
choose the optimal candidate through l_1 regularized least square approach and
the appearance model is updated. A tracking method based on articulated representation is presented by [3]. They propose a method for detecting articulated
people and estimating their pose from static images based on a new representation of deformable part models. Several tracking methods based on deformable

© Springer International Publishing Switzerland 2015
S. Arik et al. (Eds.): ICONIP 2015, Part III, LNCS 9491, pp. 392–401, 2015.
DOI: 10.1007/978-3-319-26555-1_44

representation are proposed. Authors in [4] use the active contour to model the human body and propose a novel method for tracking the body's contour by combining the colour and the depth cues adaptively. Their method consists on two mainly stages. First, they evolve the active contour to the object's boundary by integrating the edge and the region cues of the depth image and the region cue of the colour image in level set framework. In the second stage, they refine the tracking result provided by the level set method using the two properties of the body surface in depth image. In [5], authors propose an approach to segment and track multiple persons in a video sequence via graph-cuts optimization technique. They extract initial silhouettes that will be modeled by ellipses. Then, a prediction step based on optical flow vectors is used to detect if an occlusion will handle. Hence, they identify the occluding persons by the use of the Chi-squared similarity metric based on the intensity histogram and they update the objects models of the interacting persons. Finally, a segmentation based on graph-cuts optimization is performed based on the predicted models.

In this paper, we address the problem of tracking people based on deformable models. In fact, we focus on region based active contour methods in order to segment and track people. The novelty of our approach is the incorporation of a color constraint in the energy functional which is composed of a local and global region based energy terms. The rest of the paper is organized as follows. Section 2 presents the proposed approach. Section 3 outlines the algorithm followed by experimental results in Sect. 4. Finally, conclusion and perspectives are presented in Sect. 5.

2   Hybrid Active Contour Based Person Segmentation and Tracking

2.1   Shape Description and Initialization

At first, we proceed to a background substraction using a reference frame in order to extract N initial targets to be tracked in the video sequence. We associate to each target $P^{(i,t_0)}, i \in \{1 \ldots N\}$ a closed contour $\zeta^{(i,t_0)}$ and its corresponding convex hull $\Psi^{(i,t_0)}$ (Fig. 1).

Fig. 1. Shape initialization

2.2 Contour Prediction

Once the people are detected, we will compute a prediction of the contour $\zeta_{pred}^{(i,t)}$ for each object i based on the average of optical flow vectors $d^{(i,t-1)}$ of its previous displacements. Each predicted contour $\zeta_{pred}^{(i,t)}$ is represented by a corresponding convex hull $\Psi_{pred}^{(i,t)}$. These predictions will be useful in the tracking process. In fact, they will be incorporated in the energy functional in order to deduce the final contour of each tracked person.

$$\zeta_{pred}^{(i,t)} = \{x + d^{(i,t-1)} \,|\, x \in \zeta^{(i,t-1)}\}. \tag{1}$$

2.3 Active Contour Model with Statistical Prior

In this paper, we propose an energy functional based on statistical prior. In fact, for each target i at time t, we will minimize an energy function as follows:

$$E(\zeta^{(i,t)}) = \alpha E_{glob}(\zeta^{(i,t)}) + (1 - \alpha)E_{stat}(\zeta^{(i,t)}). \tag{2}$$

Where α is positive user fixed constants. Therefore we add a regularization term in order to keep the curve smooth Eq. 3.

$$E(\zeta^{(i,t)}) = \alpha E_{glob}(\zeta^{(i,t)}) + (1 - \alpha)E_{stat}(\zeta^{(i,t)}) + \gamma|\zeta^{(i,t)}|. \tag{3}$$

Where γ is a positive user fixed constant and $|\zeta^{(i,t)}|$ is the contour's length.

Global Energy. Let's denote by $I : \Omega \to \Re$ a given image function where $I(x)$ is the intensity of pixel $x \in \Omega$. The first term E_{glob} is based on the uniform modeling energy proposed by Chan and Vese [6]. The energy is based on the means intensities inside and outside a closed contour ζ.

$$E_{glob}(\zeta^{(i,t)}) = \int_{\Omega_{in}} (I(x) - u_{in})^2 dx + \int_{\Omega_{out}} (I(x) - u_{out})^2 dx \tag{4}$$

Where u_{in} and u_{out} are respectively the means intensities inside and outside $\zeta^{(i,t)}$, Ω_{in} and Ω_{out} are respectively the areas inside and outside $\zeta^{(i,t)}$. The minimization problem is solved by the use of level set representation. Therefore, the closed contour $\zeta^{(i,t)}$ associated to target i at time t is represented as the zero level set of the signed function $\Phi^{(i,t)}$ such that:

$$\begin{cases} \zeta^{(i,t)} = \{x \in \Omega | \Phi^{(i,t)}(x) = 0\}; \\ inside(\zeta^{(i,t)}) = \{x \in \Omega | \Phi^{(i,t)}(x) > 0\}; \\ outside(\zeta^{(i,t)}) = \{x \in \Omega | \Phi^{(i,t)}(x) < 0\}. \end{cases}$$

Thus the energy functional is reformulated in terms of level set function:

$$E_{glob}(\Phi^{(i,t)}) = \int_{\Omega} H(\Phi^{(i,t)}(x))(I(x) - u_{in})^2 dx +$$

$$\int_{\Omega} (1 - H(\Phi^{(i,t)}(x)))(I(x) - u_{out})^2 dx. \tag{5}$$

Where $H(\Phi^{(i,t)}(x))$ is an approximation of the Heaviside function whose value is 1 if $\Phi^{(i,t)}(x)$ is positive and null otherwise while u_{in} and u_{out} are defined as follows:

$$u_{in} = \frac{\int_\Omega I(x)H(\Phi^{(i,t)}(x))dx}{\int_\Omega H(\Phi^{(i,t)}(x))dx}, u_{out} = \frac{\int_\Omega I(x)(1-H(\Phi^{(i,t)}(x)))dx}{\int_\Omega (1-H(\Phi^{(i,t)}(x)))dx}. \quad (6)$$

Statistical Constraint. In order to improve the tracking task, we will incorporate a statistical prior to constrain the segmentation. We compute the color distribution associated to the background and the foreground which is a gaussian mixture model adjusted respectively to the set of pixels outside and inside the contour $\zeta^{(i,t)}$ of target i at time t. Hence, the statistical term is defined as follows:

$$E_{stat}(\zeta^{(i,t)}) = -\int_{\Omega_{in}} logP(I(x)|\psi_{in})dx - \int_{\Omega_{out}} logP(I(x)|\psi_{out})dx \quad (7)$$

where $P(I(x)|.)$ denotes the Generalized Gaussian distributions assumed to represent the likelihood that a pixel of intensity $I(x)$ belongs to Ω_{in} or Ω_{out} while ψ_{in} and ψ_{out} are the parameters of the distribution respectively inside and outside $\zeta^{(i,t)}$.

Then, we rewrite E_{stat} in terms of level set function

$$E_{stat}(\Phi^{(i,t)}) = -\int_\Omega H(\Phi^{(i,t)}(x)) logP(I(x)|\psi_{in}) dx$$

$$-\int_\Omega (1-H(\Phi^{(i,t)}(x))) logP(I(x)|\psi_{out}) dx. \quad (8)$$

Final Model

$$E(\Phi^{(i,t)}) = \alpha E_{glob}(\Phi^{(i,t)}) + (1-\alpha)E_{stat}(\Phi^{(i,t)}) +$$

$$\gamma \int_\Omega \delta(\Phi^{(i,t)}(x))|\nabla\Phi^{(i,t)}(x)|dx. \quad (9)$$

We use the Euler-Lagrange equations to solve the minimization problem. Then the level set function can be updated by gradient descent method and the evolution equation is expressed by

$$\frac{\partial\Phi^{(i,t)}}{\partial t}(x) = \alpha\delta(\Phi^{(i,t)}(x))\left((I(x)-u_{in})^2 - (I(x)-u_{out})^2\right) +$$

$$(1-\alpha)\,\delta(\Phi^{(i,t)}(x))\left(-\frac{log(p(I(x|\psi_{in})))}{log(p(I(x|\psi_{out})))}\right) + \gamma\,\delta(\Phi^{(i,t)}(x))\left(div\left(\frac{\nabla\Phi^{(i,t)}(x)}{|\nabla\Phi^{(i,t)}(x)|}\right)\right). \quad (10)$$

2.4   Occlusion Handling

A major challenge of tracking multiple objets is the occlusion handling. We inspired form the work of Brox and Weickert [7] who proposed an algorithm for multiple region segmentation based on the idea of competing regions. We consider that the evolution equation is composed of a retreat and an advance term denoted U and V respectively. The retreat term is always negative and aims to move the curve inward whereas the advance component is always positive and tries to move the curve outward along its normal. Then, our goal is to allow multiple contours to compete with each other. To do this, we retain the notion of competition between advance and retreat forces and combine them in a different way. These two terms can be expressed for each tracked object i at time t as follows (Fig. 2):

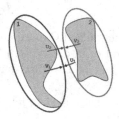

Fig. 2. Advance and retreat components

$$U^{i,t} = \int_{\Omega_{out}} (I(x) - u_{out})^2 dx - \int_{\Omega_{out}} logP(I(x)|\psi_{out})dx + \frac{\gamma}{2}|\zeta^{(i,t)}|. \quad (11)$$

$$V^{i,t} = \int_{\Omega_{in}} (I(x) - u_{in})^2 dx - \int_{\Omega_{in}} logP(I(x)|\psi_{in})dx + \frac{\gamma}{2}|\zeta^{(i,t)}|. \quad (12)$$

Given two adjacent objects $\{i, j \in 1..N \mid i \neq j\}$, then the corresponding U and V terms are defined:

$$\begin{cases} if\, U^{i,t} < -V^{j,t} then\, U^{i,t} = 0; \\ if\, V^{i,t} < -U^{j,t}\, then\, V^{i,t} = 0. \end{cases}$$

This allows us to better deal with occlusion cases. In fact, this allows a contour to not converge to the adjacent region.

3 Algorithm

The proposed algorithm can be summarized as follow:

Algorithm 1. Overview of the proposed method

1.. Initialization of targets
2. Process at time $t > t_0$
   (a) Compute Optical flow vectors
   (b) Predict Targets
   (c) Energy minimization eq. 10
   (d) Update predictions
   set t=t+1 and return to 2

---

4 Experiments

The main goal behind the proposed algorithm is to track persons in video sequence. In order to evaluate the effectiveness and performance of the proposed method, we will test it on several datasets for single and multiple objects tracking (Skater, PETS 2009). Experiments are implemented under Matlab 7.4 in a personal computer with a processor Intel Core i5-3210M, 250 GHZ, 600 GB RAM, Windows 7. The parameters are set as follows: $\alpha = 0.5$, $\beta = 0.5$, $\gamma = 0.5$ and the number of components of the GMM components is set to 5.

4.1 Qualitative Evaluation

Tracking Single Object. First, we test our method on video sequences including a single object. We present on Figs. 3 and 4 results of the proposed method

(a) 12   (b) 18   (c) 24   (d) 38

(e) 54   (f) 76   (g) 84   (h) 92

Fig. 3. Female Skater results

(a) 30   (b) 64   (c) 102   (d) 114

(e) 124   (f) 132   (g) 144   (h) 158

Fig. 4. Male Skater results

(a) 224   (b) 231   (c) 237

(d) 240   (e) 255   (f) 261

Fig. 5. Results on PETS 2009 sequence

on single object sequences. The obtained results show good contour detection and people tracking. In fact, the contours of the objects are well detected despite the shape deformations and fast motion of the tracked targets.

Tracking Multiple Objects. In this paragraph, we show results of the proposed method on PETS2009 dataset (View 001) (Figs. 5 and 6). People on the sequence are well detected and tracked. We notice the entrance of a new pedestrian in Fig. 5(b) and the proposed method succeeds to detect and to track it along the frames. In Fig. 6, we show occlusion cases in Fig. 6(a, f, h). We deduce

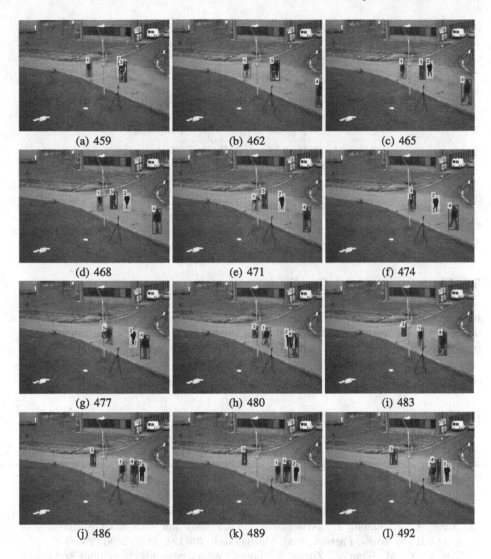

(a) 459　　　　　　　　　(b) 462　　　　　　　　　(c) 465

(d) 468　　　　　　　　　(e) 471　　　　　　　　　(f) 474

(g) 477　　　　　　　　　(h) 480　　　　　　　　　(i) 483

(j) 486　　　　　　　　　(k) 489　　　　　　　　　(l) 492

Fig. 6. Results on PETS 2009 sequence

that the proposed algorithm succeeds to detect and track pedestrians even under severe occlusion cases. On the other hand, in Fig. 6(b), a new target appears on the sequence and is well detected and tracked.

4.2   Quantitative Evaluation

We use a selection of metrics proposed in the Video Analysis and Content Extraction (VACE) protocol [8]. In Table 1, we evaluate the overall accuracy (MOTA) and precision (MOTP) of the tracking algorithm. We compare the tracking results on PETS 2009 sequence (View 001) of the proposed method with the works of [5,9,10]. As shown on Table 1, the proposed method gives better tracking and accuracy scores.

Table 1. Quantitative results on the PETS 2009 database (S2.L1).

	LP [9]	SDP [9]	[10]	[5]	Proposed
MOTA	0.82	0.11	0.89	0.67	0.904
MOTP	0.56	0.1	0.562	0.61	0.605
MODA	0.85	0.11	0.908	0.7	0.921
MODP	0.57	0.12	0.573	0.57	0.603

5   Conclusion and Outlines

In this paper we presented a multi-persons tracking approach based on deformable models. We incorporate a statistical prior in the energy term. Experiments were performed on several video sequences including single and multiple objects and provided us good tracking results. As future directions, we aim to improve the segmentation task by incorporating graph-cuts optimization.

References

1. Yilmaz, A., Li, X., Shah, M.: Contour based object tracking with occlusion handling in video acquired using mobile cameras. IEEE Trans. Pattern Anal. Mach. Intell. **26**, 1531–1536 (2004)
2. Shao, J., Dong, N., Tong, M.: Multi-part sparse representation in random crowded scenes tracking. Pattern Recogn. Lett. **34**(7), 780–788 (2013)
3. Yang, Y., Ramanan, D.: Articulated human detection with flexible mixtures of parts. IEEE Trans. Pattern Anal. Mach. Intell. **35**(12), 2878–2890 (2013)
4. Xu, Y., Ye, M., Tian, Z., Zhang, X.: Locally adaptive combining colour and depth for human body contour tracking using level set method. IET Comput. Vision **8**(4), 316–328 (2014)
5. Soudani, A., Zagrouba, E.: People tracking based on predictions and graph-cuts segmentation. In: Bebis, G., Boyle, R., Parvin, B., Koracin, D., Li, B., Porikli, F., Zordan, V., Klosowski, J., Coquillart, S., Luo, X., Chen, M., Gotz, D. (eds.) ISVC 2013, Part II. LNCS, vol. 8034, pp. 158–167. Springer, Heidelberg (2013)
6. Chan, T.F., Vese, L.A.: Active contours without edges. Trans. Img. Proc. **10**(2), 266–277 (2001)
7. Brox, T., Weickert, J.: Level set segmentation with multiple regions. IEEE Trans. Image Process. **15**(10), 3213–3218 (2006)
8. Kasturi, R., et al.: Framework for performance evaluation of face, text, and vehicle detection and tracking in video: data, metrics, and protocol. IEEE Trans. Pattern Anal. Mach. Intell. **31**(2), 319–336 (2009)

9. Berclaz, J., Fleuret, F., Fua, P.: Multiple object tracking using flow linear programming. In: The 12th IEEE International Workshop on Performance Evaluation of Tracking and Surveillance, pp. 1–8 (2009)
10. Andriyenko, A., Schindler, K., Roth, S.: Discrete-continuous optimization for multi-target tracking. In: IEEE Conference on Computer Vision and Pattern Recognition, pp. 1926–1933 (2012)

Visual and Dynamic Change Detection
for Data Streams

Lydia Boudjeloud-Assala[1]([⊠]), Philippe Pinheiro[2], Alexandre Blansché[1],
Thomas Tamisier[2], and Benoît Otjaques[2]

[1] Université de Lorraine, Laboratoire d'Informatique Théorique et Appliquée - EA
3097, Ile du Saulcy, 57045 Metz Cedex 1, France
lydia.boudjeloud-assala@univ-lorraine.fr
[2] LIST - Luxembourg Institute of Science and Technology,
Esch-sur-Alzette, Luxembourg

Abstract. We propose in this paper a new approach to detect and visualize the change in a streaming clustering. This approach can be used to
explore visually the data streams. We assume that the data stream structure can be different during the time. Our objective is to alert the user
on the structure change during the time period. A common approach
to deal with data streams is to observe and process it in a window. The
principle of the proposed approach is to apply a data exploration method
on each window. We then propose to visualize the change between all
windows for each extracted cluster. The user can investigate more precisely the change between the two windows through a visual projection
for each extracted cluster.

Keywords: Visual exploration · Dynamic clustering · Change
detection · Data stream

1 Introduction

A data stream is an ordered sequence of data items arriving continuously, generally time-stamped, in a database [13]. In recent years, data streams have received
considerable attention in various applications, such as Internet traffic [3], financial tickers [20] and scientific literature analysis [21]. When dealing with data
streams, one important task is the indication of the stream change. Change
detection is the process of identifying differences in the state of an object or
phenomenon by observing it at different times or different locations in space.
In the streaming context, it is the process of segmenting a data stream into
different segments by identifying the points where the stream dynamics changes
[18]. The design of general, scalable and statistically relevant change detection
methods is a great challenge. Change detection consists in comparing the underlying distribution of tuples observed at different times. Two temporal windows
are defined: the *reference* and the *current*. An overview of the main change
detection approaches is given by Dries [11]. Change detection in the distribution

© Springer International Publishing Switzerland 2015
S. Arik et al. (Eds.): ICONIP 2015, Part III, LNCS 9491, pp. 402–410, 2015.
DOI: 10.1007/978-3-319-26555-1_45

of tuples can be considered as a statistical hypothesis test which involves two samples of multidimensional tuples. Such problems are studied in the statistical literature. Approaches based on nearest-neighbor analyses [14] or distance between density estimates [4] have been developed. To detect change in a data stream, the algorithm proposed by Kifer et al. [15] compares the distributional distance between two sliding windows with a given threshold. Their approach requires no prior assumptions on the nature of the data distribution. They can compare two sliding windows of different sizes because the distance used by their algorithm is the Kolmogorov-Smirnov statistical distance. In Bondu and Boullé [8] the change detection problem is turned into a supervised learning task. Authors chose to exploit the supervised discretization method of a continuous variable and estimate the conditional distribution of classes owing to a piecewise constant estimator. We assume in our work, that the two consecutive windows of data are given. Many studies have been devoted to developing strategies of choosing, sampling, splitting, growing, and shrinking the windows for optimal change detection [1,5,12]. In terms of visualization, the only relevant work, to our knowledge, is the work of Kranen et al. [16]. The visualization component allows to visualize the stream as well as the clustering results (the CluStream algorithm [2]), and compare experiments with different settings in parallel (F1-measure, Precision, Recall, or other evaluation measures [6]).

This paper proposes a new approach to explore visually the data stream. A data stream evolves over time and a clustering result and visualization are desired at each time step. We want to alert the user on the structure change during the time period, assuming that the data stream structure can be different in time. Since a data stream is infinite in nature, a common approach is observing and processing it in a window. The principle of the proposed approach is to apply a clustering method on each window. The clustering algorithm is not applying to find the clustering results but is applied to explore the data and find optimal structure in the current time. We then study a data structure in each window in a static way, and a stream structure can change between two adjacent (consecutive) windows. We propose to visualize the change between all windows as a stream. If the data structure changes, it is indicated through a visual streaming change and then the user can investigate more precisely the change between the two windows through a visual projection. As a multidimensional data visualization methods, we use a two dimensional reduction method Stochastic Neighbor Embedding (t-SNE) [19]. The approach for the data exploration in different windows, uses a cluster extraction method based on a cluster limit detection method for clusters defined by their center. To extract clusters we use a deterministic approach, which simplifies the change detection. Our objective is not to search or to find the real classes in the data, even if our method finds them with a certain accuracy rate [9]. Our objective is to use a clustering deterministic approach to explore a data stream through different windows, and alert the user if change is detected between windows. The clusters are extracted according to a specific order and this order is fix as long as the data structure does not change. We can then evaluate the stability between two clustering structures. The first originality of our approach is the possibility to detect the change for a data stream, through a deterministic clustering approach, which is impossible to achieve with other clustering algorithms such as k-means, because the different algorithms are generally stochastic and the solutions are generally instable. The second originality is to allow this change detection through a visual heat map of the streaming change and the window data 2D projection.

The remainder of this paper is organized as follows: Sect. 2 gives the details of the proposed deterministic clustering algorithm for data exploration with evaluations. Section 3 describes the change detection approach and evaluation comments. Section 4 presents the proposed tools for visual streaming change detection and comments of the experimental results. Finally, we draw the conclusions and future work.

2   Data Stream Exploration

We propose a generic iterative approach that extract clusters one after the other, based on an optimization method. The method extracts the best cluster at each iteration according to the evaluation criterion. The extracted cluster represents a subset of the most homogeneous and separate data from other objects in each iteration. The iterative process is performed by a deterministic optimization method finding at each iteration the best cluster based on a particular evaluation criterion. For our application on data stream exploration and change detection, we fix the number of iterations and in the same time the total number of extracted clusters at the beginning. These extracted clusters summarize the information of the data structure. We assume that the clusters are defined by their center. In order to obtain the set of objects contained in the cluster, we select the ones close enough to the center. We choose the threshold by computing the distance between all the objects and the cluster center and ordering them from the closest object to the farthest. We then try to find an abrupt increasing of distance that will indicate the limit of the cluster [9]. Most of the evaluation criteria for clustering evaluate the complete clustering and not each cluster separately. Only two criteria exist, the Wemmert-Gançarski compactness criterion [10] gives an evaluation for each cluster, but this evaluation is based on the position of other cluster centers. As we only define a cluster by its center and its limit, we cannot use such a criterion. The intra-class inertia also gives an evaluation for each cluster, but is biased by the cluster size and the variance on the dimensions. We propose two evaluation criteria for single clusters. The first one is the inertia ratio IR, the ratio of the intra-class inertia by the total inertia of the data, normalized by the number of objects in the extracted cluster and the whole data set:

$$IR\left(C_k\right) = \frac{Card(D) \sum\limits_{o \in C_k} d\left(o, c_k\right)^2}{Card(C_k) \sum\limits_{o \in D} d\left(o, g\right)^2} \tag{1}$$

The second proposed evaluation criterion, the cluster limit ratio CLR, the ratio between the distance of the last object of the extracted cluster and the first object outside the cluster.

$$CLR\left(C_k\right) = \frac{\max\limits_{o \in C_k}\left(d\left(o, c_k\right)\right)}{\min\limits_{o \notin C_k}\left(d\left(o, c_k\right)\right)} \tag{2}$$

Each cluster is extracted independently. The two proposed criteria provide an evaluation of the compactness of one cluster. However, as each cluster is extracted independently, we need to make sure that the extracted clusters are different (different subset of objects). To ensure that, we propose to add a penalty for overlapping cluster OP, defined as follows:

$$OP\left(C_k\right) = \lambda\frac{\sum\limits_{o \in C_k} nb_o}{Card(C_k)} \tag{3}$$

where nb_o is the number of clusters to which the object o has been previously assigned, and $\lambda \geq 0$ is a weight chosen for the penalty depending on the significance one gives to overlapping clusters. This penalty is simply added to the criterion IR or CLR.

To extract the cluster, we propose an optimization method. For our data stream application through windows, we use an exhaustive search, which tests each data objects as center, in different windows. The extracted clusters have then a deterministic order. For each cluster we obtain a center ID and a threshold (radius, cluster limit) according to the two criterion IR and CLR.

3   Change Detection Method

A common approach to deal with data streams is a window processing. Based on the window width, a window can be a fixed-size window or a variable-size window, it can be also a time-driven model or tuple-driven model. For our application we choose a sliding fixed-size and tuple-driven window model to deal with a data stream. Sliding window model is useful when we need to make decisions based on the recent observations such as stock data stream or sensor data streams. Let window $w[t_c, t_c + N - 1]$ be a sliding window, where t_c is the current time. The window starts at the time t_c and ends at the current time $t_c + N - 1$, the window size is N (in terms of data objects). An object is expired if its time stamp t is less than the time t_c. An active object is defined as an item whose time stamp t lies in the range $[t_c, t_c + N - 1]$. Data objects are inserted into the sliding window as they arrive, and the oldest objects are removed from it. We consider two possibilities of sliding windows:

- sliding windows with overlapping windows, with the overlapping size set to $N/2$ between two adjacent windows,
- sliding windows without overlapping.

We study data in each window off-line, and a stream structure can change between two adjacent (consecutive) windows. We assume that the data stream structure can be different during the time. Our objective is to alert the user on the structure change during the time period. We describe in the following the proposed method for the visual change detection.

The key idea underlying the window-based change detection methods is to compare two samples extracted from two windows: the reference window and the current window. Let w_1 and w_2 denote two basic windows of size N (in term of data objects). The problem of change detection in data streams is to define a change. Given a distance function which measures the dissimilarity of two sliding windows $d(w_1, w_2)$, and ω a distance-based threshold used to decide whether a change occurs. A change occurs if the dissimilarity measure between two windows exceeds a given threshold. Window-based methods for detecting changes may reduce the memory and time required to execute a change detector. For our approach, we firstly propose to compute a distance between all adjacent windows $(d(w_1, w_2), d(w_2, w_3), \ldots, d(w_{t_c-N+1}, w_{t_c}))$, and to compute the distance between a referenced w_{ref} window with others $(d(w_{ref}, w_1), d(w_{ref}, w_2), \ldots, d(w_{ref}, w_{t_c-N+1}, d(w_{ref}, w_{t_c}))$. As we summarize the information of the data structure in the extracted clusters according to different windows, we do not apply a change detection method on two extracted samples, but we apply the distance function on the extracted centers that summarise the data structure information. We then obtain a distance vector $(d(w_{t_c-N+1}, w_{t_c}))$ associated to each cluster. A new cell value is added to this vector at each studied new window. This distance vector contains the streaming change information of each cluster.

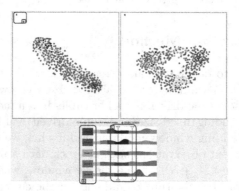

Fig. 1. The tool interface: (a) a window number, (b) cluster identification, (c) change detection - user can chose the windows (between window 4 and window 6 as seen in the t-SNE projection) (Color figure online)

We propose to visualize the change for each cluster between all windows as a stream. If the data structure changes, it is indicated through a visual streaming change and then the user can investigate more precisely the change between the two windows through a visual projection. The approach for the exploration of data in different windows uses a cluster extraction method. This method is based on a cluster limit detection for clusters defined by their center. For the change detection method, we summarize streaming change information of each cluster through a distance vector. We then use a time series with variation color for each cluster to project the streaming change of the vector distance change, where the color represent the distance measure between two all adjacent windows or with a referenced window. The ω threshold, is then automatically determined from a variation of the normalized vector distance (in the range $[0,1]$). As we can see in the Fig. 1, the change detection between two consecutive or with referenced windows are indicated through a heat map information where the smaller values ($\simeq 0$) are indicated in light color, the larger ones ($\simeq 1$) are indicated in dark color. Each cluster is represented separately one after other, we can see an example with 5 clusters in the Fig. 1b. The color variation represents the difference between two adjacent windows ($\Delta = d(w_{t_c-N+1}, w_{t_c})$), which represents the distance between the cluster centers at each step of the stream. If the cluster center change is not important, it is indicated in light color, otherwise, is indicated in dark color. The variations in height of the time series indicate the importance of the extracted cluster in term of elements.

4   Experimental Results

In order to evaluate the change detection method through different windows on stream data set, we provide some results on real data stream. We firstly demonstrate the effectiveness of the proposed method with WaveForm data set from University of California at Irvine (UCI) machine learning benchmark repository [7]. WaveFrom data set is composed on 5000 objects with 21 dimensions (attributes), all of which include noise. WaveFrom data set is composed by 3 classes of waves. Each class is generated from a combination of 2 or 3 "base" waves and each instance is generated with added noise (mean 0, variance 1) in each dimension. The projected form of this data set is a triangle form. We apply our tools on unsupervised exploration process, so the classes are not visualized. We decompose the data set on 10 windows, in each window we have 500 objects sampled from the whole data set, and they have also a triangle form in the 2D projection. We insert between window 3 and 5, a sample data set composed by 500 objects, taken from one axis of the whole data set, that we call the *intrusive set* (Fig. 1). The objective is to find as good as possible the change that occurs by the window 4. The data stream analysis proceeds up to the window 9, the Figs. 2 and 3 present two possibilities for the user to see the variations. From the user selected windows (Fig. 3), here, the user select with the cursor the 4th window and the indicated windows are projected (Fig. 1). The variation for all two consecutive windows (Fig. 2), here, the cursor indicate the projected windows (Fig. 1).

We also test our method on Sea Concepts [17] data set from Ubiquitous Knowledge Discovery and Data Analysis repository. Sea Concepts Data set is composed by 60000 objects, 3 dimensions and 3 classes. Dimensions are numeric

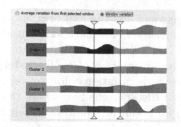

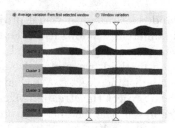

Fig. 2. Visual change detection with two consecutive windows variation for WaveForm data set.

Fig. 3. Visual change detection with average variation from the user selected windows for WaveForm data set.

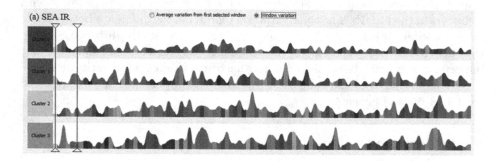

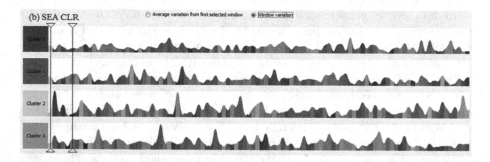

Fig. 4. Visual change detection with two consecutive windows variation for Sea Concepts data set: (a) With IR cluster evaluation, (b) With CLR cluster evaluation (Color figure online).

between 0 and 10. Data set has about 10 % of noise. We decompose the data sets on 100 windows respectively, in each window we fix the size of 600 as a number of tuple entering in the window. This data set is generally used for data stream mining problems. In the Fig. 4, we can see the obtained visual time series with IR and CLR cluster evaluation. The Sea data set has 3 clusters but we iterate our approach at 4 iteration, because the first extracted cluster has very small size compared with other clusters (the amplitude of the time series). In term of color change, the first and the third clusters are rather stable (Fig. 4), the second and the forth are less. In the last third clusters we can see 4 dark zone (Fig. 4a), that indicate change in the stream. The user can then, use this tools to explore and interpret more easily the obtained results. The number of the clusters and the number of projected windows are fixed by the user. The different experiments and visualizations on synthetic and real data sets show the effectiveness of the proposed approach.

5   Conclusions

We propose in this paper a new approach to explore visually data streams. We assume that the data stream structure can be different during the time. We want to alert the user on the structure change during the time period. The first originality of our approach is the possibility to detect the change for a data stream, through a clustering deterministic approach. The second originality is to allow this change detection through a visual projection of the streaming change and the 2D projection of the windows data. Experiments and visualizations on data sets show the effectiveness of the proposed approach. We can visually detect the change and explore the visual data set projections to explain and interpret this change. In a visual and interactive point of view, there are certainly more things to do, such as improving the interaction. In fact, it may be interesting to see the change for a specific cluster trough the stream. It is also, interesting to test all other possibilities of visual interaction, to thereby improve the visual data stream mining.

References

1. Adams, R., MacKay, D.: Bayesian online changepoint detection. Technical report, University of Cambridge (2007)
2. Aggarwal, C.C., Watson, T.J., Ctr, R., Han, J., Wang, J., Yu, P.S.: A framework for clustering evolving data streams. In: VLDB, pp. 81–92 (2003)
3. Aizen, J., Huttenlocher, D., Kleinberg, J., Novak, A.: Traffic-based feedback on the web. Proc. Natl. Acad.Sci. 101(Suppl. 1), 5254–5260 (2004)
4. Anderson, N.H., Hall, P., Titterington, D.M.: Two-sample test statistics for measuring discrepancies between two multivariate probability density functions using kernel-based density estimates. J. Multivar. Anal. 50(1), 41–54 (1994)
5. Bifet, A., Gavalda, R.: Learning from time-changing data with adaptive windowing. In: Seventh SIAM International Conference Data Mining, pp. 443–448 (2007)
6. Bifet, A., Holmes, G., Pfahringer, B., Kranen, P., Kremer, H., Jansen, T., Seidl, T.: Moa: massive online analysis, a framework for stream classification and clustering. In: WAPA, pp. 44–50 (2010)
7. Blake, C., Merz, C.: UCI repository of machine learning databses. Technical report, University of California, Irvine, Department of Information and Computer Sciences (1998). http://archive.ics.uci.edu/ml/datasets.html. Accessed January 2014
8. Bondu, A., Boulle, M.: A supervised approach for change detection in data streams. In: The 2011 International Joint Conference on Neural Networks (IJCNN), pp. 519–526 (2011)

9. Boudjeloud-Assala, L., Blansché, A.: Iterative evolutionary subspace clustering. In: International Conference on Neural Information Processing, vol. 1, pp. 424–431 (2012)

10. Wemmert, C., Gançarski, P., Korczak, J.: A collaborative approach to combine multiple learning methods. Int. J. Artif. Intell. Tools **9**(1), 59–78 (2000)

11. Dries, A., Rückert, U.: Adaptive concept drift detection. Stat. Anal. Data Min. **2**(5–6), 311–327 (2009)

12. Gama, J., Medas, P., Castillo, G., Rodrigues, P.: Learning with drift detection. In: Bazzan, A.L.C., Labidi, S. (eds.) SBIA 2004. LNCS (LNAI), vol. 3171, pp. 286–295. Springer, Heidelberg (2004)

13. Golab, L., Özsu, M.T.: Issues in data stream management. SIGMOD Rec. **32**(2), 5–14 (2003)

14. Hall, P.: Permutation tests for equality of distributions in high-dimensional settings. Biometrika **89**(2), 359–374 (2002)

15. Kifer, D., Ben-David, S., Gehrke, J.: Detecting change in data streams. In: Endowment, V. (ed.) Proceedings of the Thirtieth International Conference on Very Large Data Bases, vol. 30, p. 191 (2004)

16. Kranen, P., Kremer, H., Jansen, T., Seidl, T., Bifet, A., Holmes, G., Pfahringer, B.: Clustering performance on evolving data streams: assessing algorithms and evaluation measures within MOA. In: ICDM Workshops, pp. 1400–1403 (2010)

17. Street, W.N., Kim, Y.: A streaming ensemble algorithm SEA for large-scale classification. ACM Press, pp. 377–382 (2001). http://www.liaad.up.pt/kdus/products/datasets-for-concept-drift Accessed April 2014

18. Tran, D.H.: Change detection in streaming mining. Ph.D. thesis, Faculty of computer science and automation, Ilmenau University of Technology (2013)

19. van der Maaten, L., Hinton, G.: Visualizing high-dimensional data using t-SNE. J. Mach. Learn. Res. **9**, 2579–2605 (2008). Nov

20. Vlachos, M., Wu, K.-L., Chen, S.-K., Yu, P.S.: Fast burst correlation of financial data. In: Jorge, A.M., Torgo, L., Brazdil, P.B., Camacho, R., Gama, J. (eds.) PKDD 2005. LNCS (LNAI), vol. 3721, pp. 368–379. Springer, Heidelberg (2005)

21. Wei, C.-P., Chang, Y.-H.: Discovering event evolution patterns from document sequences. IEEE Trans. Syst. Man Cybern. A Syst. Hum. **37**(2), 273–283 (2007)

Adaptive Location for Multiple Salient Objects Detection

Shaoyong Jia[1], Yuding Liang[1], Xianyang Chen[1], Yun Gu[1], Jie Yang[1],
Nikola Kasabov[2], and Yu Qiao[1(✉)]

[1] Institue of Image Processing and Pattern Recognition,
Shanghai Jiao Tong University, Shanghai, China
{jiashaoyong,liangyuding,st_tommy,geron762,jieyang,qiaoyu}@sjtu.edu.cn
[2] Knowledge Engineering and Discovery Research Institute,
Auckland University of Technology, Auckland, New Zealand
nkasabov@aut.ac.nz

Abstract. Salient objects detection aims to locate objects that capture
human attention within images. Recent progresses in saliency detection
have exploited the center prior, to combine with other cues such as background information, object size or region contrast, achieving competitive
results. However, previous approaches of center prior supposing salient
object locates nearly at image center is very simple, fragile, especially
not suitable for multiple salient objects detection, but the assumption is
mostly heuristic. In this paper, we present an adaptive location method
based on geodesic filtering framework to address these issues. First, we
detect salient points by the adjustive color Harris algorithm. Second, we
involve the Affinity Propagation (AP) method to automatically cluster
the salient points for a coarse objects location. Then, we utilize geodesic
filtering framework for a final saliency map by multiplying objects location and size. Experimental results on two more challenging databases of
off-center and multiple salient objects demonstrate our approach is more
robust to the location variations of salient objects, against state-of-the-
art methods for saliency detection.

Keywords: Saliency detection · Adaptive location · AP cluster method

1 Introduction

Recent years have seen greatly increasing interest in salient object detection [3].
It is motivated by the importance of saliency detection in applications such as
object detection and recognition [10], image segmentation [6], image and video
compression [5] and visual tracking [11]. Because of the loss of high level knowledge, all bottom up methods depend on assumptions about the properties of
objects and backgrounds. Among them, some researches usually add a Gaussian
map to models for center prior to enhance saliency computation in [7,17], which
suppose that salient objects locate closely at the image center.

© Springer International Publishing Switzerland 2015
S. Arik et al. (Eds.): ICONIP 2015, Part III, LNCS 9491, pp. 411–418, 2015.
DOI: 10.1007/978-3-319-26555-1_46

We observed two issues about previous assumptions of center prior. The first is to simply treat the image center as potential salient object location which ignores the fact that multiple salient objects are off-center at different levels. This is fragile and may fail on more challenging databases, such as the SED2 [2] of which the images contain multiple off-centered salient objects. In this case, image center assumption may become the bottleneck when it is integrated with other cues for saliency detection. Secondly, while some methods re-estimate the mean and radius of the gaussian map from an initial saliency map, this strategy is still not suitable for multiple off-center salient objects.

In this paper, we present an adaptive location method to address the above two problems. Our main contribution is a novel and reliable salient objects location measure, called *adaptive location*. Instead of simply assuming salient object locating at the image center, the proposed method aims to automatically detect the salient objects location. Our method is more robust as it characterizes the spatial layout of salient objects. In detail, we firstly detect salient points and cluster them by AP algorithm [4]. Then we utilize the *geodesic filtering* framework and "soft" region size computing method proposed in [17] for a final saliency map.

We enhance the baseline proposed in [17]. Since images in SED2 [2] have only two salient objects and simple background, we select the images of multiple objects against more complex background from DUT-OMRON [16] and

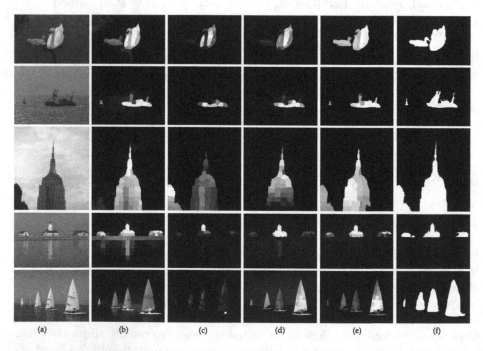

Fig. 1. Saliency detection results on challenging examples. (a) Input images; (b) GS [13]; (c) SF [9]; (d) Base [17]; (e) Ours; (f) ground truth.

PASCAL-S [8], called dataset MDUT-SAL, to further validate our algorithm. Experimental comparisons show that our approach outperforms Base [17], especially on MDUT-SAL. The examples in Fig. 1 show that comparisons against other methods of different difficulties: background interference, small salient object, background touching, three and four saliency objects. More importantly, the performance of all previous methods are further improved with our results combined than Base [17], and new state-of-the-art results are achieved.

2   Geodesic Filtering Framework

The papers [17,18] both proposed geodesic filtering framework based on a regular superpixel image representation, which encodes the information of image segmentation in an implicit and soft manner.

Firstly, an image is converted into CIELab color space and decomposed into N superpixels representation by SLIC algorithm [1]. Then an undirected weighted graph is constructed by connecting spatially adjacent superpixels. The Euclidean distance between superpixels i and j is denoted as the edge weight $w_{i,j}$ according to average colors of superpixels. The *geodesic distance* between any two superpixels G_d is computed as:

$$G_d(i,j) = \min_{i=v_1,v_2,\ldots,v_n=j} \sum_{k=1}^{n-1} w_{v_k,v_{k+1}} \tag{1}$$

where $v_1, v_2, \ldots, v_n$ is a shortest path in the graph linking nodes i and j, and $G_d(i,i)$ is set to 0. Then the *geodesic connectivity* is defined as:

$$G_c(i,j) = exp(-\frac{G_d^2(i,j)}{2\sigma^2}) \tag{2}$$

Secondly, the *geodesic filtering* framework is defined to measure the properties of image regions from superpixels representation. Suppose $I(j)$ is the property value of superpixel j to be filtered, the geodesic filtering computes the property of the region that superpixels j belongs to as:

$$GF(I,j) = \frac{\sum_{j=1}^{N} G_c(i,j) \times I(j)}{\sum_{j=1}^{N} G_c i,j} \tag{3}$$

It aggregates and smoothes the property values within the same homogeneous region. After filtering, all superpixels in the same region have similar property values of that region. As proposed in [17], Eq. (3) is used to estimate salient object centerness by replacing $I(j)$ with a gaussian map $M(j)$, which is too simple and weak for multiple salient objects detection. We propose our approach in Sect. 3 to alleviate this problem. And an un-normalized version of GF by removing the denominator is used to estimate the object size.

3   Our Approach

Many saliency methods are biased to assign image center regions with higher saliency. However, previous methods simply use a gaussian fall-off map with mean at the image center and a fixed radius, or re-estimate the mean and radius of the gaussian map from an initial saliency map which highly depends on the quality of the initial saliency map. These strategies are problematic for multiple salient objects.

We propose a method which can detect the salient objects location automatically, which characterizes the spatial layout of salient objects. We follow below five steps to implement our algorithm with enough motivation in detail.

Image smoothing: some image background or noise may be so complex that they affect subsequent salient points detection. We smooth images firstly via L0 gradient minimization [15] which can remove low-amplitude structures and globally preserve and enhance salient edges. The salient points detection and clustering results before and after smoothing are shown in Fig. 2(d) and (e). We can see that the salient points coming from background are eliminated and the cluster center locates at the object center basically after smoothing the original input images.

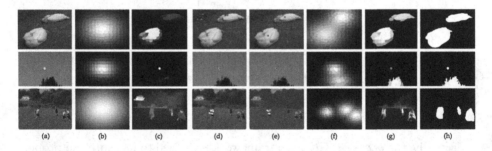

Fig. 2. Illustration of our approach. (a) Input images; (b) gaussian maps [17]; (c) final saliency maps [17]; (d) salient points without image smoothing and corresponding cluster center (red); (e) salient points with image smoothing and corresponding center (red); (f) our gaussian maps; (g) our final results; (h) ground truth (Color figure online).

Salient points detection: traditional luminance-based saliency detection methods incline to completely ignore the color information and thus are very sensitive to the background noises. In [12], they applied the boosting color saliency theory to Harris detector and show that the resulting saliency points are much more informative than the luminance-based Harris points.

In this paper, we adopt the color boosting Harris points [12] as salient points to catch the corners or marginal points of visual salient region in color image and eliminate the salient points near image boundary. Then the saliency points provide us a coarse location of the salient areas even if there are multiple salient objects. As the color boosting Harris points usually gather around the saliency region, the salient points center usually locates at the object center. We denote the salient points as SP_k, $k = 1, 2, ..., K$, where K is the number of salient points. Besides, it is good for subsequent clustering to locate salient objects adaptively. Note that even though few salient points from background noises do not make an obvious negative effect on cluster center even the final saliency map.

Adaptive location: In [14], they proposed the concept of convex hull derived from salient points and adopted k-means method to group superpixels inside and outside the convex hull for eliminating the effect of the noisy region included in the convex hull based on Bayesian model. However, they are simply used for single salient object, which is quite different with ours.

We adopt the AP method to cluster K salient points into l clusters, which is basically consistent with the number of salient objects, with m salient points respectively, represented as $SP_j^i = \{X_j^i, Y_j^i\}$, where $j = 1, 2, ..., m$, $i = 1, 2, ..., l$, namely:

$$K = \sum_{i=1}^{i=l} SP_m^i. \tag{4}$$

Then we calculate the center of each cluster, namely *adaptive location*, the average of spacial positions following below formula:

$$C_i = \frac{1}{m} \sum_{j=1}^{j=m} SP_j^i = \frac{1}{m} \sum_{j=1}^{j=m} \{X_j^i, Y_j^i\}. \tag{5}$$

And we define the cluster radius R_i as the average Euclidean distance between each salient point and corresponding cluster center:

$$R_i = \frac{1}{m} \sum_{j=1}^{j=m} \|SP_j^i - C_i\|^2. \tag{6}$$

Then we get a gaussian fall-off map G by combining R_i, as shown in Fig. 2(f), with mean at cluster center and standard deviation equals to its cluster radius for each cluster. Note that we add a small constant value to cluster radius to avoid the degenerate case when they are equal to 0.

Final saliency map: we replace I with G in Eq. 3 to acquire a saliency map based on our *adaptive location*. Then we completely follow the background prior and approximate computation of region size in [17] for final saliency map, shown in Fig. 2(g), which are much better than the Base [17] results in Fig. 2(c). This fully shows that we further optimize the proposed method in [17].

4 Experiments

For experimental comparison, we use a standard benchmark dataset SED2 [2] which contains 100 images of two salient objects with largely different sizes and locations while background is relatively simple, and our more challenging MDUT-SAL, consisting of 220 images with multiple salient objects and complex background by combining most examples in DUT-OMRON [16] and PASCAL-S [8]. We follow [17] to compute the standard precision-recall curves (PR curves) and F-measures evaluation metrics. As complementary, we also introduce the mean absolute error (MAE) into the evaluation which measures how close a saliency map is to the ground truth.

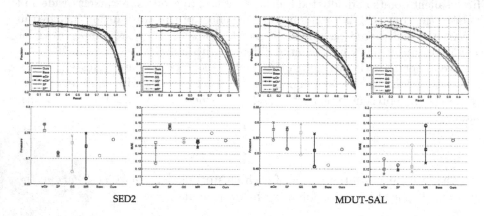

SED2   MDUT-SAL

Fig. 3. (Better viewed in color) Precision-recall curves, F-measure and MAE of various methods on SED2 [2] (left) and MDUT-SAL. In the PR curves, results of dotted lines and (*) are obtained by combining our results. In the F-measure and MAE, the circle marker is the value of some state-of-the-art methods, square and cross markers are the results after combing with Base and ours, respectively (Color figure online).

We compare against the most recent state-of-the-art saliency methods, including saliency filter (SF) [9], manifold ranking (MR) [16], geodesic saliency (GS) [13], and saliency optimization (wCtr) [18]. All of them implemented algorithms based on SLIC [1] superpixels and achieved competitive results in recent years. Example results of recent state-of-the-art original results, after combining Base [17] and our approach are shown in Fig. 4.

4.1   Comparison with State-of-the-Art

Figure 3 reports the PR curves, F-measures and MAE of all methods on two databases, before and after combining with our approach. We can make several obviously observations. Firstly, our approach outperforms Base [17] in terms of three evaluation metrics especially on dataset MDUT-SAL, which demonstrates that our method is more robust and general for multiple salient objects detection. Secondly, all previous methods are higher improved after combination with our method on dataset MDUT-SAL. We consider that this is because SED2 [2] is relatively simple and other complex algorithms are possibly overfitted to SED2 dataset and do not generalize well to MDUT-SAL. Specifically, it is more obvious that wCtr [18] which acquires the best result on both two databases, and improved results are best on multiple salient objects detection up to now. The motivation for combination has been fully proven in [17]. Finally, the performance gaps between previous methods are much smaller after combination as shown in Fig. 3 in sight of three metrics. Thus, the approach we proposed is an enhanced baseling for state-of-the-art methods.

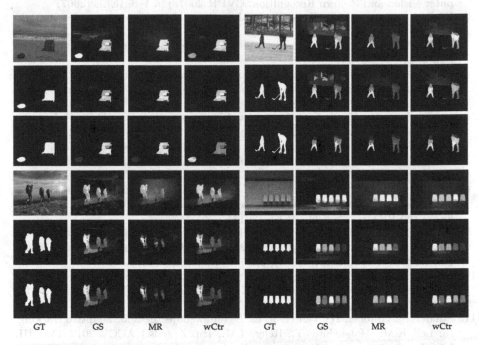

GT   GS   MR   wCtr   GT   GS   MR   wCtr

Fig. 4. Example results of three recent state-of-the-art methods. For each image, the first row shows the input image and related original results. The second row shows the ground truth and related improved results by combining Base [17]. The last row shows the ground truth and related enhanced results after combining our method.

5 Conclusion

We present an adaptive location for multiple salient objects detection based on *geodesic filtering* framework. It mainly introduces the salient points detection algorithm and Affinity Propagation (AP) clustering method to acquire a coarse salient objects location, called *adaptive location*. Then we use the *geodesic filtering* framework for a final fine saliency map. By comparing against the state-of-the-art methods, we find that our approach outperforms Base and improves other state-of-the-art methods after combination. For further validating our method, we propose a more challenging database MDUT-SAL than SED2. We hope our work and dataset can enhance the understanding of multiple salient objects detection in future.

References

1. Achanta, R., Shaji, A., Smith, K., Lucchi, A., Fua, P., Susstrunk, S.: Slic superpixels compared to state-of-the-art superpixel methods. IEEE Trans. Pattern Anal. Mach. Intell. **34**(11), 2274–2282 (2012)
2. Alpert, S., Galun, M., Basri, R., Brandt, A.: Image segmentation by probabilistic bottom-up aggregation and cue integration. In: 2007 IEEE Conference on Computer Vision and Pattern Recognition, CVPR 2007, pp. 1–8. IEEE (2007)
3. Borji, A., Cheng, M.M., Jiang, H., Li, J.: Salient object detection: a benchmark. ArXiv e-prints (2015)
4. Frey, B.J., Dueck, D.: Clustering by passing messages between data points. Science **315**(5814), 972–976 (2007)
5. Guo, C., Zhang, L.: A novel multiresolution spatiotemporal saliency detection model and its applications in image and video compression. IEEE Trans. Image Process. **19**(1), 185–198 (2010)
6. Li, Q., Zhou, Y., Yang, J.: Saliency based image segmentation. In: 2011 International Conference on Multimedia Technology (ICMT), pp. 5068–5071. IEEE (2011)
7. Li, Y., Fu, K., Zhou, L., Qiao, Y., Yang, J.: Saliency detection via foreground rendering and background exclusion. In: 2014 IEEE International Conference on Image Processing (ICIP), pp. 3263–3267. IEEE (2014)
8. Li, Y., Hou, X., Koch, C., Rehg, J.M., Yuille, A.L.: The secrets of salient object segmentation. In: 2014 IEEE Conference on Computer Vision and Pattern Recognition (CVPR), pp. 280–287. IEEE (2014)
9. Perazzi, F., Krahenbuhl, P., Pritch, Y., Hornung, A.: Saliency filters: contrast based filtering for salient region detection. In: 2012 IEEE Conference on Computer Vision and Pattern Recognition (CVPR), pp. 733–740. IEEE (2012)
10. Ren, Z., Gao, S., Chia, L.T., Tsang, I.H.: Region-based saliency detection and its application in object recognition. IEEE Trans. Circ. Syst. Video Technol. **24**(5), 769–779 (2014)
11. Stalder, S., Grabner, H., Van Gool, L.: Dynamic objectness for adaptive tracking. In: Lee, K.M., Matsushita, Y., Rehg, J.M., Hu, Z. (eds.) ACCV 2012, Part III. LNCS, vol. 7726, pp. 43–56. Springer, Heidelberg (2013)
12. Van De Weijer, J., Gevers, T., Bagdanov, A.D.: Boosting color saliency in image feature detection. IEEE Trans. Pattern Anal. Mach. Intell. **28**(1), 150–156 (2006)
13. Wei, Y., Wen, F., Zhu, W., Sun, J.: Geodesic saliency using background priors. In: Fitzgibbon, A., Lazebnik, S., Perona, P., Sato, Y., Schmid, C. (eds.) ECCV 2012, Part III. LNCS, vol. 7574, pp. 29–42. Springer, Heidelberg (2012)
14. Xie, Y., Lu, H.: Visual saliency detection based on bayesian model. In: 2011 18th IEEE International Conference on Image Processing (ICIP), pp. 645–648. IEEE (2011)
15. Xu, L., Lu, C., Xu, Y., Jia, J.: Image smoothing via l 0 gradient minimization. ACM Trans. Graph. (TOG) **30**, 174 (2011). ACM
16. Yang, C., Zhang, L., Lu, H., Ruan, X., Yang, M.H.: Saliency detection via graph-based manifold ranking. In: 2013 IEEE Conference on Computer Vision and Pattern Recognition (CVPR), pp. 3166–3173. IEEE (2013)
17. Zhao, L., Liang, S., Wei, Y., Jia, J.: Size and location matter: a new baseline for salient object detection. In: Cremers, D., Reid, I., Saito, H., Yang, M.-H. (eds.) ACCV 2014. LNCS, vol. 9005, pp. 578–592. Springer, Heidelberg (2015)
18. Zhu, W., Liang, S., Wei, Y., Sun, J.: Saliency optimization from robust background detection. In: 2014 IEEE Conference on Computer Vision and Pattern Recognition (CVPR), pp. 2814–2821. IEEE (2014)

Robust Detection of Anomalies via Sparse Methods

Zoltán Á. Milacski[1], Marvin Ludersdorfer[3], András Lőrincz[1],
and Patrick van der Smagt[2,3](✉)

[1] Faculty of Informatics, Eötvös Loránd University, Budapest, Hungary
srph25@gmail.com, lorincz@inf.elte.hu
[2] Department for Informatics, Technische Universität München, Munich, Germany
smagt@brml.org
[3] fortiss, An-Institut Technische Universität München, Munich, Germany
ludersdorfer@fortiss.org

Abstract. The problem of *anomaly detection* is a critical topic across application domains and is the subject of extensive research. Applications include finding frauds and intrusions, warning on robot safety, and many others. Standard approaches in this field exploit simple or complex system models, created by experts using detailed domain knowledge.

In this paper, we put forth a statistics-based anomaly detector motivated by the fact that anomalies are sparse by their very nature. Powerful sparsity directed algorithms—namely Robust Principal Component Analysis and the Group Fused LASSO—form the basis of the methodology. Our novel unsupervised single-step solution imposes a convex optimisation task on the vector time series data of the monitored system by employing group-structured, switching and robust regularisation techniques.

We evaluated our method on data generated by using a Baxter robot arm that was disturbed randomly by a human operator. Our procedure was able to outperform two baseline schemes in terms of F_1 score. Generalisations to more complex dynamical scenarios are desired.

1 Introduction

The standard approach to describing system behaviour over time is by devising intricate dynamical models—typically in terms of first- or second-order differential equations—which are based on detailed knowledge about the system. Such models can then help to both control as well as predict the temporal evolution of the system and thus serve as a basis to detect faults.

We investigate the common case where models are either difficult to obtain or not rich enough to describe the dynamic behaviour of the nonlinear plant. This can happen when the plant has many degrees of freedom or high-dimensional sensors, and when it is embedded into a complex environment. Such is typically true for robotic systems, intelligent vehicles, or manufacturing sites; i.e., for a typical modern actor–sensor system that we depend on.

© Springer International Publishing Switzerland 2015
S. Arik et al. (Eds.): ICONIP 2015, Part III, LNCS 9491, pp. 419–426, 2015.
DOI: 10.1007/978-3-319-26555-1_47

In such cases, the quality of fault detection deteriorates: too many false positives (i.e., false alarms) make the fault detection useless, while too many false negatives (i.e., unobserved faults) may harm the system or its environment. Rather than fully trusting incomplete models, we put forth a methodology which creates a *probabilistic vector time series model* of the system from the recorded data and detects outliers with respect to this learned model. Following standard procedures [4], detecting such outliers will be called *anomaly detection*.

This type of detection is notoriously difficult as it is an ill-posed problem. First, the notion of anomaly strongly depends on the domain. Then, the boundary between "normal" and "anomalous" might not be precise and might evolve over time. Also, anomalies might appear normal or be obscured by noise. Finally, collecting anomalous data is very difficult, and labelling them is even more so [4]. Two observations are important to make: (i) anomalies are sparse by their very nature, and (ii) in a high-dimensional real-world scenario it will not be possible to rigorously define "normal" and "anomalous" regions of the data space. We therefore focus on an unsupervised approach: in a single step, a probabilistic vector time series model of the system's data is created, and (patterns of) samples that do not fit in the model are conjectured to be the sought anomalies.

In this paper we introduce a new two-factor convex optimisation problem using (i) group-structured, (ii) switching, and (iii) robust regularisation techniques; aiming to discover anomalies in stochastic dynamical systems. We assume that there is a *family of behaviours* between which the system switches randomly. We further assume that the time of the switching is stochastic, and that the system is coupled, i.e., both switching points and anomalies span across dimensions. Given that the general behaviour between the switches can be approximated by a random parameter set defining a *family of dynamics*, we are interested in detecting rare anomalies that may occur with respect to the "normal" behaviour (hence there are two factors). To the best of our knowledge, the combination of the techniques (i)–(iii) is novel.

To test our methods and demonstrate our results, we generated data with a Baxter robot. This system serves as a realistic place holder for a general system with complex dynamics in a high-dimensional space, in which the data cannot be easily mapped to a lower-dimensional plane, while the sensory data are not trivial. We generate realistic anomalies by having the robot perform predefined movements, with random physical disturbances from a human.

2   Theoretical Background

2.1   LASSO and Group LASSO

LASSO [9] is an ℓ_1-regularised least-squares problem defined as follows. Let $D < N$ and let us denote an input vector by $\mathbf{x} \in \mathbb{R}^D$, an overcomplete vector system by $\mathbf{D} \in \mathbb{R}^{D \times N}$, a representation of $\mathbf{x}$ in system $\mathbf{D}$ by $\mathbf{a} \in \mathbb{R}^N$, a tradeoff of penalties by λ. Then LASSO tries to find

$$\min_{\mathbf{a}} \frac{1}{2} \|\mathbf{x} - \mathbf{Da}\|_2^2 + \lambda \|\mathbf{a}\|_1, \tag{1}$$

which—for a sufficiently large value of λ—will result in a gross-but-sparse representation for vector $\mathbf{a}$: only a small subset of components $\mathbf{a}_i$ will be non-zero (but large) while the corresponding columns $\mathbf{D}_{.,i}$ will still span $\mathbf{x}$ closely. Model complexity (sparsity) is implicitly controlled by λ. LASSO is the best convex approximation of the NP-hard ℓ_0 version of the same problem, since convexity ensures a unique global optimum value and polynomial-time algorithms grant one to find a global solution [3,5].

A useful extension to LASSO is the so-called Group LASSO [11]: instead of selecting individual components, groups of variables are chosen. This is done by defining a disjoint group structure on $\mathbf{a}$ (or equivalently on the columns of $\mathbf{D}$):

$$G_i \subseteq \{1,\ldots,N\}, \quad |G_i| = N_i, \quad i = 1,\ldots,L, \tag{2}$$

$$G_i \cap G_j = \emptyset, \quad \forall i \neq j, \tag{3}$$

$$\bigcup_{i=1}^{L} G_i = \{1,\ldots,N\}. \tag{4}$$

Then one can impose a mixed ℓ_1/ℓ_2-regularisation task:

$$\min_{\mathbf{a}} \frac{1}{2}\|\mathbf{x} - \mathbf{D}\mathbf{a}\|_2^2 + \lambda \sum_{i=1}^{L} \|\mathbf{a}_{G_i}\|_2, \tag{5}$$

where $\mathbf{a}_{G_i}$ denotes the subset of components in $\mathbf{a}$ with indices contained by set G_i. The last term is often referred to as the $\ell_{1,2}$ norm of $\mathbf{a}$. Thus the elements from the same group are either forced to vanish together or form a dense representation within the selected groups, resulting in so-called *group sparsity*.

2.2 Robust Principal Component Analysis

Similarly to the LASSO problem, one can define an ℓ_1 norm-based optimisation for compressing the data into a small subspace in the presence of a few outliers. Let $\|\cdot\|_*$ and $\|\text{vec}(\cdot)\|_1$ denote the singular valuewise ℓ_1 norm (nuclear norm) and the elementwise ℓ_1 norm of a matrix, respectively. Then the Robust Principal Component Analysis (RPCA) [2] task for given $\mathbf{X} \in \mathbb{R}^{D \times T}$ and unknowns $\mathbf{U}, \mathbf{S} \in \mathbb{R}^{D \times T}$ is as follows:

$$\min_{\mathbf{U},\mathbf{S}} \frac{1}{2}\|\mathbf{X} - \mathbf{U} - \mathbf{S}\|_F^2 + \lambda\|\mathbf{U}\|_* + \mu\|\text{vec}(\mathbf{S})\|_1, \tag{6}$$

where $\mathbf{U}$ is a robust low-rank approximation to $\mathbf{X}$, gross-but-sparse (non-Gaussian) anomalous errors are collected in $\mathbf{S}$, and $\|\cdot\|_F$ denotes the Frobenius norm. Then as a post-processing step, singular value decomposition can be performed on the outlier-free component $\mathbf{U} = \mathbf{W}\mathbf{\Sigma}\mathbf{V}^T$, resulting in a robust estimate of rank $k \in \mathbb{N}$. Note that the classical Principal Component Analysis (PCA) of the input matrix $\mathbf{X}$ would be vulnerable to outliers.

2.3   Fused LASSO and Group Fused LASSO

LASSO can also be modified to impose sparsity for linearly transformed ($\mathbf{Q} \in \mathbb{R}^{T \times M}$ for arbitrary M) components of $\mathbf{v} \in \mathbb{R}^T$:

$$\min_{\mathbf{v}} \frac{1}{2} \|\mathbf{y} - \mathbf{v}\|_2^2 + \lambda \|\mathbf{Q}^T \mathbf{v}\|_1. \tag{7}$$

In the special case when $\mathbf{y} \in \mathbb{R}^T$ is a time series and $\mathbf{Q}$ is a finite differencing operator of order p, this technique is called Fused LASSO [10] and yields a piecewise polynomial approximation of $\mathbf{v}$ of degree $(p-1)$. One can also consider the version when $\mathbf{y}$ is replaced with a multivariate time series $\mathbf{X}, \mathbf{V} \in \mathbb{R}^{D \times T}$:

$$\min_{\mathbf{V}} \frac{1}{2} \|\mathbf{X} - \mathbf{V}\|_F^2 + \lambda \|\text{vec}(\mathbf{V}\mathbf{Q})\|_1. \tag{8}$$

Change points can be localised in the different components and by assuming a coupled system, change points may be co-localised in time. The formulation that allows for such group-sparsity is the $\ell_{1,2}$ norm of Group LASSO:

$$\min_{\mathbf{V}} \frac{1}{2} \|\mathbf{X} - \mathbf{V}\|_F^2 + \lambda \sum_{t=1}^{T-p} \|\mathbf{V}\mathbf{Q}_{.,t}\|_2 \tag{9}$$

for order of differentiation p (with $M = T - p$) [1].

3   Methods

3.1   Problem Formulation

We assume that there is a piecewise polynomial trajectory in a multi-dimensional space, like the motion of a robotic arm in configuration space. We also assume that the robot executes certain actions one-by-one, e.g., it is displacing objects giving rise to sharp changes in the trajectory. However, we are not aware of the plan and we have no additional information and thus, we do not know the points in time or space when the trajectory would switch. These points will be called *switching points* or *change points*. Yet we know that such change points of different configuration components are co-localised in time (i.e., the plan spans across dimensions).

We also assume that at random times an anomaly disturbs the motion, but the system compensates, reverts back to the trajectory, and executes the task.

We propose to use the following $\ell_{1,2}, \ell_{1,2}$ convex optimisation problem for the detection of the above kind of anomalies. With known $\mathbf{X} \in \mathbb{R}^{D \times T}$ and variables $\mathbf{V}, \mathbf{S} \in \mathbb{R}^{D \times T}$, solve

$$\min_{\mathbf{V}, \mathbf{S}} \frac{1}{2} \|\mathbf{X} - \mathbf{V} - \mathbf{S}\|_F^2 + \lambda \sum_{t=1}^{T-p} \|\mathbf{V}\mathbf{Q}_{.,t}\|_2 + \mu \sum_{t=1}^{T} \|\mathbf{S}_{.,t}\|_2, \tag{10}$$

where $\mathbf{Q}$ is a finite differencing operator. For the sake of simplicity, we assume that order of $\mathbf{Q}$ is $p = 2$:

$$\mathbf{Q}_{d,t} = \begin{cases} 1, & \text{if } d = t \text{ or } d = t + 2, \\ -2, & \text{if } d = t + 1, \\ 0, & \text{otherwise.} \end{cases} \tag{11}$$

This is a combination of the Robust PCA and the Group Fused LASSO tasks: it tries to find a robust piecewise linear approximation $\mathbf{V}$ and additive error term $\mathbf{S}$ for $\mathbf{X}$ with group-sparse switches *and* group-sparse anomalies that contaminate the unknown plan. The second term says that we are searching for a fit, which has group-sparse second-order finite differences. The third term ($\mathbf{S}$) corresponds to gross-but-group-sparse *additive* anomalies not represented by the switching model.

We also use the corresponding ℓ_1, ℓ_1 variant as a comparison:

$$\min_{\mathbf{V}, \mathbf{S}} \frac{1}{2} \|\mathbf{X} - \mathbf{V} - \mathbf{S}\|_F^2 + \lambda \|\text{vec}(\mathbf{V}\mathbf{Q})\|_1 + \mu \|\text{vec}(\mathbf{S})\|_1, \tag{12}$$

which allows change points and anomalies to occur independently in the components (via ordinary sparsity instead of group-sparsity).

We used Matlab R2014b with CVX 3.0 beta [6,7] for minimisations, capable of transforming cost functions to equivalent forms that suit fast solvers, e.g., the Splitting Conic Solvers (SCS) 1.0 [8] and used the sparse direct linear system option.

Dependencies on the (λ, μ) tuple were searched on the whole data set within the domain of $\{2^{-9}, 2^{-7}, \ldots, 2^{15}\} \times \{2^{-16}, 2^{-14}, \ldots, 2^8\}$. The best values were selected according to the F_1 score, the harmonic mean of precision and sensitivity: $F_1 = \frac{2\,\text{TP}}{(2\,\text{TP}+\text{FP}+\text{FN})}$, where TP, FP, FN are the number of true positives, false positives and false negatives for anomalous segments, respectively. Predicted momentary values (i.e., norms of the gross-but-group-sparse additive error term: $\|\mathbf{S}_{.,t}\|_2$, $t = 1, \ldots, T$) were thresholded into momentary binary labels with respect to 0.01. Note that these labels have many spikes and gaps between them, thus a morphological closing operator was used to fill-in such gaps up to 1 s length during post-processing, resulting in the segmented labels of the F_1 calculations. A baseline algorithm using internal torque information—not available for our methods—together with some differencing heuristics and parameter optimisation was also added for a rough orientation about performance.

3.2   Data Set

We used 300 trials of 7-dimensional joint configurations of one arm of a Baxter research robot[1] as our data set. The configuration was characterised by 2 shoulder, 2 elbow and 3 wrist angles. The transitions between prescribed configurations were realised as an approximately piecewise linear time series with common

---

[1] http://www.rethinkrobotics.com/baxter-research-robot/.

Fig. 1. Frame series of the operator hitting the robot arm.

change points (due to some minor irregularities of the position controller). Anomalies were generated by manual intrusion into the process: an assistant for the experiment kept hitting the robot arm when asked, as depicted in Fig. 1. The controller reverted back to the original trajectory as soon as possible. Timestamps of 822 collision commands were logged and served as ground-truth labels for the anomalies. The actual impacts were delayed by up to 5 s in the data because of human reaction time. We took this uncertain delay into consideration when computing the F_1 performance metric: we tried to pair detected anomalous segments with the timestamps provided with respect to the 5 s threshold. The data were recorded at 800 Hz frequency and interpolated uniformly to 50 Hz.

4   Results

Parameter dependencies with respect to the mean F_1 score for the $\ell_{1,2}, \ell_{1,2}$ (10) and the ℓ_1, ℓ_1 (12) methods are shown in Fig. 2(a) and (b), respectively. The best parameter combination for both schemes was $(\lambda, \mu) = (2^{-1}, 2^{-6})$. With these settings, the former algorithm achieved 709 true positive, 94 false positive and 113 false negative anomalous segments. Figure 2(c) includes the best achieved mean F_1 scores for all approaches, including the heuristic baseline procedure.

An example highlights prediction scenarios for the $\ell_{1,2}, \ell_{1,2}$ method in Fig. 3: (a) shows the approximately piecewise linear input time series, with markers indicating common change points on each curve, as well as 5 anomalies pointed out by red rectangles; (b) shows the output gross-but-group-sparse norm values of our procedure, the 0.01 threshold level and the logged green ground-truth anomaly labels; while (c) provides close-up views of the actual and predicted anomalous segments. Anomalies 3 to 5 (around time points 700, 1160 and 1440) are true positives, as they are paired with ground-truth labels and are above the threshold. Anomaly 2 (around 420) is marked false negative, as it is improperly

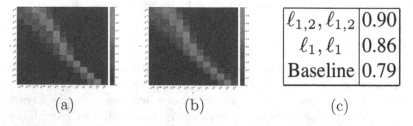

$\ell_{1,2}, \ell_{1,2}$	0.90
ℓ_1, ℓ_1	0.86
Baseline	0.79

(a)   (b)   (c)

Fig. 2. Mean F_1 scores with different (λ, μ) parameters for (a): $\ell_{1,2}, \ell_{1,2}$ (10), (b): ℓ_1, ℓ_1 (12) algorithms. (c): Best achieved mean F_1 scores including baseline heuristics.

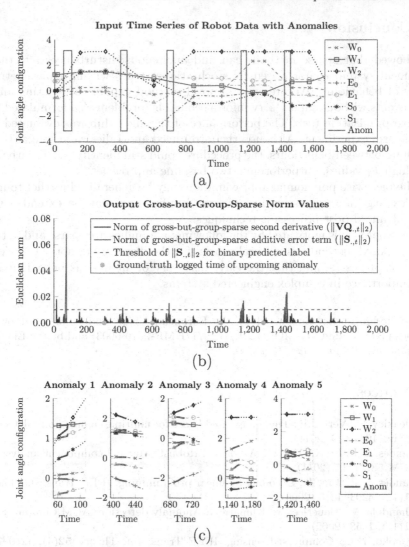

Fig. 3. Prediction scenarios for the $\ell_{1,2}, \ell_{1,2}$ method. (a): Input time series of joint angles. W: wrist, E: elbow, S: shoulder; Red rectangles: anomalies. (b): Blue (red): estimated change points (anomalies); red dashed line: optimal threshold level; green dots: ground-truth anomaly labels. (c): Close-ups for actual and detected (red) anomalous segments: Anomaly 1/2: false positive/negative, 3/4/5: true positives (Color figure online).

thresholded (there are 71 examples for this in the entire data set). Anomaly 1 (around 80) is stigmatised false positive, as it lacks the ground-truth label, while an anomaly is certainly present in the segment due to controller imprecision (the total number of such cases is 17). Mean F_1 scores would be somewhat higher with more sophisticated thresholding and controller policies.

5 Conclusion

We showed that sparse methods can find anomalous disturbances affecting a dynamical system characterised by stochastic switches within a parametrised family of behaviours. We presented a novel method capable of locating anomalous events without supervisory information. The problem was formulated as a convex optimisation task. The performance of the algorithm was evaluated in a robotic arm experiment. Although the introduced anomalies were similar to the switching between behaviours, the procedure could still identify the disturbances with high F_1 values, outperforming two baseline approaches.

The piecewise polynomial approximation may be generalised further to more complex, e.g., autoregressive behavioural families and may be extended with more advanced post-processing techniques.

Our approach is motivated by the following: anomalies are sparse and in turn, sparsity based methods are natural choices for the discovery of not yet modelled events or processes; and that the early discovery of anomalous behaviour is of high importance in complex engineered systems.

Acknowledgements. Supported by the European Union and co-financed by the European Social Fund (TÁMOP 4.2.1./B-09/1/KMR-2010-0003) and by the EIT Digital grant on *CPS for Smart Factories*.

References

1. Bleakley, K., Vert, J.P.: The group fused Lasso for multiple change-point detection. arXiv:1106.4199 (2011)
2. Candès, E.J., Li, X., Ma, Y., Wright, J.: Robust principal component analysis? J. ACM **58**(3), 11 (2011)
3. Candès, E.J., Tao, T.: Decoding by linear programming. IEEE Trans. Inf. Theory **51**(12), 4203–4215 (2005)
4. Chandola, V., Banerjee, A., Kumar, V.: Anomaly detection. ACM Comput. Surv. **41**(15), 1–58 (2009)
5. Donoho, D.L.: Compressed sensing. IEEE Trans. Inf. Theory **52**(4), 1289–1306 (2006)
6. Grant, M., Boyd, S., Ye, Y.: CVX. Recent Adv. Learn. Cont. pp. 95–110 (2012)
7. Grant, M.C., Boyd, S.P.: Graph implementations for nonsmooth convex programs. In: Blondel, V.D., Boyd, S.P., Kimura, H. (eds.) Recent Advances in Learning and Control, pp. 95–110. Springer, London (2008)
8. O'Donoghue, B., Chu, E., Parikh, N., Boyd, S.: Operator splitting for conic optimization via homogeneous self-dual embedding. arXiv:1312.3039 (2013)
9. Tibshirani, R.: Regression shrinkage and selection via the lasso. J. Roy. Stat. Soc. B **58**, 267–288 (1996)
10. Tibshirani, R., Saunders, M., Rosset, S., Zhu, J., Knight, K.: Sparsity and smoothness via the fused lasso. J. Roy. Stat. Soc. B **67**(1), 91–108 (2005)
11. Yuan, M., Lin, Y.: Model selection and estimation in regression with grouped variables. J. Roy. Stat. Soc. B **68**(1), 49–67 (2006)

Vehicle Detection Using Appearance and Shape Constrained Active Basis Model

Sai Liu[✉] and Mingtao Pei

Beijing Laboratory of Intelligent Information Technology,
School of Computer Science, Beijing Institute of Technology,
Beijing 100081, People's Republic of China
{liusai,peimt}@bit.edu.cn

Abstract. In this paper, we propose an Appearance and Shape Constrained Active Basis Model (ASC-ABM) to detect vehicles in image. ASC-ABM effectively incorporates the appearance and shape prior of vehicles in the active basis model. Therefore, compared with the original ABM, it can effectively remove the false positives caused by the clutter background and traffic lines. Experiment results demonstrate the effectiveness of the proposed method.

Keywords: Active basis model · Vehicle detection · Appearance information · Shape constraint

1 Introduction and Related Work

Vision-based vehicle detection is essential to intelligent transportation systems. It is the prerequisite for traffic surveillance systems and intelligent vehicles.

Many detection methods apply motion information to detect vehicles [5–7]. However, motion detection is not reliable as it is easily affected by the environment such as illumination and shadow. In addition to the motion based methods, many other methods are based on the appearance information of vehicles [2, 3]. Chen et al. [2] present a new symmetrical SURF descriptor to detect vehicle. Tian B et al. [3] treat vehicle as multiple salient parts, which are represented by their distinctive colors, textures, and region features. Ramirez et al. [6] propose a framework for detecting vehicle by utilizing the motion information in addition to the appearance information. Similar to motion, the appearance of vehicles is not stable in complex illumination environment. When the illumination changes, the appearance of vehicle will change, and it is difficult to model all of the possible appearance.

Edges between the vehicle and road are stable to the illumination changes compared to the appearance. Therefore, edge information is a promising feature for vehicle detection. Lin et al. [5] extract part-based appearance and edge features with multiple models to detect vehicles in the blind-spot area. Cheng et al. [10] use color features and edge features with dynamic Bayesian network to classify vehicles for aerial surveillance. Their method can't handle clutter background very well for their method only concentrate on local information, lacking of incorporating the global shape prior.

© Springer International Publishing Switzerland 2015
S. Arik et al. (Eds.): ICONIP 2015, Part III, LNCS 9491, pp. 427–434, 2015.
DOI: 10.1007/978-3-319-26555-1_48

Active basis model [1] is a framework for detecting generic objects by learning a deformable template, with its elements being Gabor functions [9], capturing edge feature and allowing elements of template to shift in the detection process. However, when applied for vehicle detection in complex traffic environment, the ABM will produce many false positives as the traffic lines and pedestrians in the environment have strong responds to the Gabor functions. We observe that usually the vehicle and the road have quite different appearance. When the illumination changes, the appearances of vehicle and road also change, but they are still different from each other. Another observation is that the vehicles are rigid objects. The shape of vehicles are different when the vehicle is on different positions in the scene, but the shape change should be small. Based on the above observations, we propose an appearance and shape constrained active basis model (ASC-ABM) to detect vehicles. The appearance constraint is that the appearance of the regions on the two sides of the ABM boundary should be different (the appearance of vehicle and road should be different). The shape constraint is that the elements of the ABM can shift with a penalty to capture the shape changes of vehicle, and prevent large shape change as vehicle is rigid. As the ASC-ABM effectively incorporates the appearance and shape prior of vehicles in the active basis model. It can effectively remove the false positives caused by the clutter background and traffic lines.

The rest of the paper is organized as follows: a brief introduction of Active Basis Model and its deficiencies in vehicle detection are given in Sect. 2. Section 3 describes our Appearance and Shape Constrained Active Basis Model. Section 4 presents the experimental results. Finally, conclusion is given in Sect. 5.

2   Active Basis Model

2.1   Overview of ABM

Active basis model provides a computational architecture for representing, learning and recognizing deformable object. It consists of a set of Gabor wavelet elements at selected locations and orientations. Gabor wavelet elements are selected from a dictionary of Gabor wavelets $\{B_{x,y,s,\alpha}, \forall(x, y, s, \alpha)\}$. An image I can be represented as,

$$\mathbf{I} = \sum_{i=1}^{n} c_i B_i + U \tag{1}$$

where n is the number of selected elements, B_i is the i'th element of the template, c_i is the coefficient, and U is the residual image. In the training process, Gabor wavelet elements are chosen from dictionary, given M training images $\{I_m, m = 1...M\}$, I_m can be represented as,

$$\mathbf{I}_m = \sum_{i=1}^{n} c_{m,i} B_{m,i} + U_m \tag{2}$$

where $B_{m,i}$ is the shifted version of B_i. Details of the training process can be found in [1]. In the detection process, the log-likelihood ratio can be used to score the template matching,

$$\log \frac{p(\mathrm{I}|B)}{q(\mathrm{I})} = \sum_{i-1}^{n} \left[\widehat{\lambda}_i h \left(|\langle \mathrm{I}, \hat{B}_i \rangle|^2 \right) - \log z \left(\widehat{\lambda}_i \right) \right] \tag{3}$$

where $\widehat{\lambda}_i (i = 1 \ldots n)$ are computed in the training process, I is a test image, B is the learned template. B_i corresponds to the i'th template element, $\widehat{B}_i$ is the final shifted version of B_i. $\left| \mathrm{I}, \widehat{B}_i \right|$ represents the operation of filtering image with Gabor functions. $h \left(|\langle \mathrm{I}, \hat{B}_i \rangle|^2 \right) = \max_{r \subset \mathcal{R}} \left(|\langle \mathrm{I}, \hat{B}_{i,r} \rangle|^2 \right)$ represents the shifting process, where $\mathcal{R}$ is the given shift ranges, we search shift r form $\mathcal{R}$ to find the maximum Gabor filtered value, $B_{i,r}$ represents shifting B_i with respects to r.

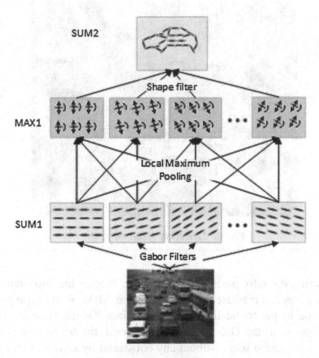

Fig. 1. The scheme of sum-max maps

The detection process can be accomplished by a computation scheme of sum-max maps. Figure 1 illustrates the scheme.

Test image is filtered by Gabor functions with different orientations, the filtered images are called SUM1 maps as Gabor filter is a local summation operation. MAX1 maps are produced by applying local maximization operator to the SUM1 maps. Finding local maximum, with respect to horizontal direction, vertical direction, and orientation in the given ranges, is equivalent to shift. $h \left(|\langle \mathrm{I}, \hat{B}_i \rangle|^2 \right)$ in Eq. (3) means the i'th element value in MAX1 maps after shifting. The values of MAX1 maps are local maximum of SUM1 maps, Then MAX1 maps are filtered by the learned template using the log-likelihood ratio (3), producing SUM2 map. Values in SUM2 map are template matching scores. Vehicle candidates can be obtained bases on SUM2 map.

2.2 Deficiencies of ABM in Vehicle Detection

As template matching scores depend on the response of local Gabor filter operation, the whole detection process only makes use of edge feature. Many false positives will be detected over traffic lines and pedestrians which with strong edges, as Fig. 2 shows.

Fig. 2. Detection result of ABM

After analyzing the false positives, we observe that: as the appearance differences between the vehicles and road are not utilized in the ABM, many false positive results are produced due to the traffic lines and pedestrians, which have strong edges and obtain large response to the Gabor functions. Second, the template matching process seeks to increase matching score without any constraint by shifting in the given ranges to find the local maximum, which does not use the prior that the vehicles are rigid objects and their shape should not change much in the scene. Based on the above observations, we incorporate the appearance information and shape constraint in the ABM to eliminate the false positives. Figure 3 illustrates the workflow.

3 Appearance and Shape Constraint ABM

3.1 Appearance Constraint

To make background less clutter, test images are firstly smoothed via L0 gradient minimization [4]. Based on the observation that the appearance of the regions on the

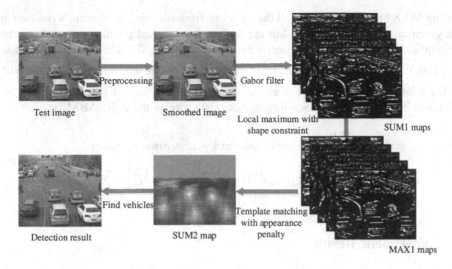

Fig. 3. Workflow of the ASC-ABM

two sides of the ABM boundary should be different (the appearance of vehicle and road should be different), we add a penalty term to Eq. (3),

$$
\begin{aligned}
score &= \log \frac{p(I|B)}{q(I)} - \textit{appearance penalty} \\
&= \sum_{i=1}^{n} \left[\widehat{\lambda}_i h\left(\left| I, \widehat{B}_i \right|^2 \right) - \log Z\left(\widehat{\lambda}_i \right) \right] - \sum_{i=1}^{n} \partial_i \, (\partial_i \geq 0)
\end{aligned}
\tag{4}
$$

$$
\partial_i = \begin{cases} \alpha, & B_i \textit{ is boundary element and regions on the two sides of it are similar} \\ 0, & \textit{otherwise} \end{cases}
$$

Two 125 dimensions color histograms h_1 and h_2 are extracted in the regions on the opposite sides of the boundary element and their intersection distance id are computed [8],

$$
id = \sum_{i=1}^{n} min(h_1(i), h_2(i)), \quad n = 125;
\tag{5}
$$

if id is less than a predefine threshold, we consider the two regions are similar and add penalty α to this element.

3.2 Shape Constraint

As shift increase the matching score under no constraint by finding the largest value in the SUM1 maps, the final shape of the template after shifting may change much. Due to vehicle is rigid object, template elements should not shift dramatically. We add a penalty term to scores aiming to preserving the shape. Specially, we change the way of getting MAX1 maps. We search all the positions from the original element's position in the given ranges as ABM does, but the farther the searched positions away from the original element's position, the greater penalty it will take. We add shape constraint as $h\left(|\langle I, \hat{B}_i \rangle|^2\right) = \max_{r \subset \mathcal{R}}\left(\left(|\langle I, B_{i,r} \rangle|^2\right) - \kappa \delta_{i,r}\right)$. $\delta_{i,r}$ is the absolute shift value with shift r relative to the original i'th element's position. κ is weight parameter.

Combining with appearance information, the score of the ASC-ABM is,

$$\begin{aligned} score &= max\left(log\frac{p(I|B)}{q(I)} - shape\ constraint\right) - appearance\ penalty \\ &= \sum_{i=1}^{n}\left[\widehat{\lambda}_i \max_{r \subset \mathcal{R}}\left(\left(|I, B_{i,r}|^2\right) - \kappa \delta_{i,r}\right) - \log z\left(\widehat{\lambda}_i\right)\right] - \sum_{i=1}^{n} \partial_i(\kappa \geq 0) \end{aligned} \tag{6}$$

4 Experiment Result

To verify the effectiveness of the ASC-ABM, we build a vehicle dataset. There are 658 images with size of 524 × 616, which are taken from a camera mounted on a portal frame. 4780 vehicle positions are labeled manually. One sample image is shown in Fig. 4.

Fig. 4. Sample image of our vehicle dataset

To evaluate the performance of ASC-ABM, the recall-rate and false-alarm rate are used. Recall-rate is the ratio of the number of true positives to the number of ground truths. False-alarm rate is the ratio of the number of false positives to the total detection. The two measures are defined as,

$$Recall\_rate = C_{TP}/N \text{ and } False\_alarm = C_{FP}/C \tag{7}$$

where N is the number of ground truths, C_{TP} is the number of true positives, C_{FP} is the number of false positives and $C = C_{TP} + C_{FP}$ is the number of total detection. Figure 5 shows some detection results. Quantitative results are shown in Table 1. The original ABM is used as baseline, which has the highest false-alarm rate. By adding appearance constraint in the ABM, the recall-rate increases 0.05 %, and false-alarm rate drops 5.62 %. When adding both the appearance and shape constraint, we get encouraging result, with the recall-rate increases 2.1 % and false-alarm rate drops 19.49 %.

(a). ABM   (b). ASC-ABM

Fig. 5. Detection result

Table 1. Result of the experiments

Method	Detection number	True positive	False positive	Missing	Recall rate	False alarm
ABM	6577	4319	2258	461	90.35 %	34.33 %
AC-ABM	6061	4321	1740	459	90.40 %	28.71 %
ASC-ABM	5189	4419	770	361	92.45 %	14.84 %

5 Conclusion

In the paper, we have presented an Appearance and Shape Constrained Active Basis Model (ASC-ABM) for vehicle detection. First, we take appearance information into account beside Gabor feature. Second, we preserve the vehicle shape by adding shape constraint to the template, preventing elements of the template shifting much, as vehicle is rigid. Experiments result show the efficiency of our method.

References

1. Wu, Y.N., Si, Z., Gong, H., et al.: Learning active basis model for object detection and recognition. Int. J. Comput. Vis. **90**(2), 198–235 (2010)
2. Chen, L.C., Hsieh, J.W., Yan, Y., et al.: Vehicle make and model recognition using sparse representation and symmetrical SURFs. Pattern Recogn. **48**, 1979–1998 (2015)
3. Tian, B., Li, Y., Li, B., et al.: Rear-view vehicle detection and tracking by combining multiple parts for complex urban surveillance. IEEE Trans. Intell. Transp. Syst. **15**(2), 597–606 (2014)
4. Xu, L., Lu, C., Xu, Y., et al.: Image smoothing via L 0 gradient minimization. ACM Trans. Graph. (TOG) **30**(6), 174 (2011)
5. Lin, B.F., Chan, Y.M., Fu, L.C., et al.: Integrating appearance and edge features for sedan vehicle detection in the blind-spot area. IEEE Trans. Intell. Transp. Syst. **13**(2), 737–747 (2012)
6. Ramirez, A., Ohn-Bar, E., Trivedi, M.M.: Go with the flow: improving multi-view vehicle detection with motion cues. In: 2014 22nd International Conference on Pattern Recognition (ICPR), pp. 4140–4145. IEEE (2014)
7. Jazayeri, A., Cai, H., Zheng, J.Y., et al.: Vehicle detection and tracking in car video based on motion model. IEEE Trans. Intell. Transp. Syst. **12**(2), 583–595 (2011)
8. Freytag, A., Rodner, E., Bodesheim, P., Denzler, J.: Rapid uncertainty computation with gaussian processes and histogram intersection kernels. In: Lee, K.M., Matsushita, Y., Rehg, J.M., Hu, Z. (eds.) ACCV 2012, Part II. LNCS, vol. 7725, pp. 511–524. Springer, Heidelberg (2013)
9. Daugman, J.G.: Uncertainty relation for resolution in space, spatial frequency, and orientation optimized by two-dimensional visual cortical filters. JOSA A **2**(7), 1160–1169 (1985)
10. Cheng, H.Y., Weng, C.C., Chen, Y.Y.: Vehicle detection in aerial surveillance using dynamic bayesian networks. IEEE Trans. Image Process. **21**(4), 2152–2159 (2012)

Denoising Cluster Analysis

Ruqi Zhang, Zhirong Yang$^{(\boxtimes)}$, and Jukka Corander

Helsinki Institute for Information Technology HIIT,
University of Helsinki, Helsinki, Finland
zhirong.yang@helsinki.fi

Abstract. Clustering or cluster analysis is an important and common task in data mining and analysis, with applications in many fields. However, most existing clustering methods are sensitive in the presence of limited amounts of data per cluster in real-world applications. Here we propose a new method called denoising cluster analysis to improve the accuracy. We first construct base clusterings with artificially corrupted data samples and later learn their ensemble based on mutual information. We develop multiplicative updates for learning the aggregated cluster assignment probabilities. Experiments on real-world data sets show that our method unequivocally improves cluster purity over several other clustering approaches.

1 Introduction

Cluster analysis or clustering is an exploratory data analysis tool which aims at dividing data objects into groups such that objects in the same group are more similar than those in other groups. As one of the major tools for modern data mining and analysis, clustering research has found a wide variety of applications in many domains of science and technology.

Most clustering methods are built upon statistical laws, assuming a wealth of samples are available per cluster. With a limited amount of data points, many existing cluster analysis approaches can only achieve mediocre performance. Especially, methods that employ non-convex objectives are prone to yield poor local optima, which demands more complicated pre-training or initialization.

In this paper we propose a new clustering technique called denoising cluster analysis (DECLU). We first manually incorporate a small amount of noise among the data points. This is equivalent to sampling from the underlying smoothed data distribution, which potentially can generate infinite amounts of training data. We first build a base clustering for each noisy version of the data set. Next we aggregate the basal partitions into a single final clustering by using an information theoretic measure based on mutual information.

As an algorithmic contribution, we develop a new clustering ensemble method based on nonnegative learning, without construction of dense and expensive consensus relationship graph. We derive the multiplicative update rule for right stochastic matrices, which result in probabilistic cluster assignments.

We test the new method on various real-world data sets and compare it with several other clustering and ensemble clustering methods. The experimental

© Springer International Publishing Switzerland 2015
S. Arik et al. (Eds.): ICONIP 2015, Part III, LNCS 9491, pp. 435–442, 2015.
DOI: 10.1007/978-3-319-26555-1_49

results indicate that our method is more advantageous in terms of clustering accuracy.

The remaining paper is organized as follows. Section 2 briefly reviews mutual information and its application in comparing clusterings. Next we present our method in Sect. 3, with the base clustering generation, the ensemble clustering objective, and its optimization using multiplicative updates. In Sect. 4, we report the experiment setting and results. Section 5 summarizes the paper and discusses some possibilities for future work.

In what follows, a clustering is represented by a cluster indicator matrix, for example, denoted by U where $U_{ik} = 1$ if the ith sample is in the kth cluster, and $U_{ik} = 0$ elsewhere.

2   Preliminary: Mutual Information

In probability theory and information theory, the mutual information (MI) of two random variables is a measure of the mutual dependence between variables. The mutual information of two discrete random variables X and Y can be defined as

$$I(X;Y) = \sum_{y \in Y} \sum_{x \in X} P(x,y) \log \frac{P(x,y)}{P(x)P(y)}. \tag{1}$$

Mutual information measures the information that X and Y share. If X and Y are independent, then knowing X does not give any information about Y and vice versa, so their mutual information is zero. At the other extreme, if X is a deterministic function of Y and Y is a deterministic function of X then all information conveyed by X is shared with Y. In this case the mutual information is the same as the uncertainty (entropy) contained in Y (or X) alone.

Mutual information (MI) can be used to compare two clusterings U and V (see e.g., [12]). Let n_{ij} be the number of objects that are common to the ith cluster in U and the jth cluster in V, $a_i = \sum_j n_{ij}$, $b_j = \sum_i n_{ij}$, and $\sum_{ij} n_{ij} = N$. Then

$$I(U;V) = \sum_i \sum_j \frac{n_{ij}}{N} \log \frac{n_{ij}/N}{a_i b_j / N^2}. \tag{2}$$

A larger MI indicates that the two clusterings are closer up to a certain cluster permutation. Various normalizations can be applied to fix the MI range in $[0, 1]$. See [12] for a summary.

3   Denoising Clustering

Learning with artificially corrupted data, represented by training samples with manually incorporated noise, is a well-known trick in many machine learning settings, for example, generating additional training examples for Support Vector

Machine classifier to improve generalization performance (see e.g., [2,4]), and reconstructing input data from artificial corruption in Denoising Auto-Encoder for learning useful representations of data (see e.g., [10,11]). In this work, we apply a similar trick to cluster analysis. This encompasses two steps: we first construct base clusterings for noisy versions of the data, and then aggregate them into a single final clustering.

3.1 Generating Base Clusterings

A good clustering should respect the data distribution that underlies finite amount of sample data points. Kernel smoothing or Parzen window method [6] is a common approach to estimate the distribution that underlies the given data points.

In this work, instead of explicitly run kernel density estimation, which is usually expensive, we employ an implicit way to achieve a similar regularization for the clustering task for better accuracy. In the following, we implicitly use the Parzen window with Gaussian radial kernel, but the same technique can be extended to other kernels in a straightforward manner.

We add white noise to each data point x_i:

$$\tilde{x}_i = x_i + \epsilon_i, \tag{3}$$

where $\epsilon_i \sim \mathcal{N}(0, \sigma)$. Next we apply a relatively simple clustering method to partition the noisy data $\tilde{X} = \{\tilde{x}_1, \ldots, \tilde{x}_N\}$. We choose Normalized Cut [8] for the base clusterings because it is less sensitive to the initializations and performs better for data in curved manifolds.

3.2 Clustering Ensemble Using Mutual Information

Next we find the consensus clustering of the basal partitions based on noisy versions of the original data. Classical clustering ensemble methods require construction of a pairwise relationship graph, which is quadratic to the number of samples and thus prohibitive for large-scale data sets. Here we propose a new method that directly learns the cluster assignment probabilities of size $N \times r$ for N data points and r clusters. This significantly reduces the computational cost.

Let $U^{(m)}$ denote the mth base clustering indicator matrix, where $m = 1, \ldots, M$. We seek a probabilistic cluster ensemble W where W_{ik} is the probability of the kth cluster given the ith sample. The ensemble minimizes the total difference to the base clusterings, measured by the "D_{sum}" distance based on mutual information [12]:

$$\underset{W \geq 0}{\text{minimize}} \quad \mathcal{J}(W) = \sum_m H(W) + H(U^{(m)}) - 2I(W; U^{(m)}) \tag{4}$$

$$\text{subject to} \quad \sum_{k=1}^{r} W_{ik} = 1, \ i = 1, \ldots, N, \tag{5}$$

where H is the entropy of cluster assignment. We choose this objective because it is a valid metric, bounded in $[0, M \log N]$, and with relatively simple gradient for optimization.

Writing out the objective, we have

$$
\begin{aligned}
\mathcal{J}(W) = & \sum_m \Bigg[-\sum_k \frac{1}{N} \sum_i W_{ik} \log \frac{1}{N} \sum_i W_{ik} \\
& -\sum_k \frac{1}{N} \sum_i U_{ik}^{(m)} \log \frac{1}{N} \sum_i U_{ik}^{(m)} \\
& -2\sum_{kl} \frac{1}{N} \sum_i W_{ik} U_{il}^{(m)} \log \frac{\frac{1}{N} \sum_i W_{ik} U_{il}^{(m)}}{\frac{1}{N} \sum_i W_{ik} \frac{1}{N} \sum_i U_{il}^{(m)}} \Bigg] \\
= & \sum_m \Bigg[\frac{1}{N} \sum_k \sum_i W_{ik} \log \sum_i W_{ik} \\
& -\frac{2}{N} \sum_{kl} \sum_i W_{ik} U_{il}^{(m)} \log \sum_i W_{ik} U_{il}^{(m)} \\
& +\frac{2}{N} \sum_i \left(\sum_k W_{ik} \right) \sum_l U_{il}^{(m)} \log \sum_j U_{jl}^{(m)} \Bigg] + \text{constant}. \quad (6)
\end{aligned}
$$

Using Lagrangian multipliers $\lambda = [\lambda_1, \ldots, \lambda_N]$ for the sum-to-one constraints, the relaxed objective function is

$$
\widetilde{\mathcal{J}}(W, \lambda) = \mathcal{J}(W) - \sum_i \lambda_i \left(\sum_k W_{ik} - 1 \right). \quad (7)
$$

Its gradient w.r.t. W is $\dfrac{\partial \widetilde{\mathcal{J}}(W)}{\partial W_{ik}} = \nabla_{ik}^+ - \nabla_{ik}^- - \lambda_i$, where ∇^+ and ∇^- are the positive and (unsigned) negative parts of the $\frac{\partial \mathcal{J}(W)}{\partial W}$

$$
\nabla_{ik}^+ = -\frac{2}{N} \sum_m \sum_t \left(\log \frac{\sum_i W_{ik} U_{it}^{(m)}}{N} \right) U_{it}^{(m)} + \frac{M}{N} (1 + \log N) \quad (8)
$$

$$
\nabla_{ik}^- = -\frac{M}{N} \log \frac{\sum_i W_{ik}}{N} - \frac{2}{N} \sum_m \sum_l U_{il}^{(m)} \log \frac{\sum_j U_{jl}^{(m)}}{N}. \quad (9)
$$

This suggests the preliminary multiplicative update rule $W_{ik}' = W_{ik} \dfrac{\nabla_{ik}^- + \lambda_i}{\nabla_{ik}^+}$.

Imposing the constraints $\sum_k W_{ik}' = 1$, we have $\sum_k W_{ik} \dfrac{\nabla_{ik}^-}{\nabla_{ik}^+} + \lambda_i \sum_k \dfrac{W_{ik}}{\nabla_{ik}^+} = 1$. Solving the equation we obtain

$$
\lambda_i = \frac{1 - \sum_k W_{ik} \nabla_{ik}^- / \nabla_{ik}^+}{\sum_k W_{ik} / \nabla_{ik}^+} \quad (10)
$$

Putting them back to the preliminary rule, we have $W'_{ik} = W_{ik} \frac{\nabla^-_{ik} A_{ik} + 1 - B_{ik}}{\nabla^+_{ik} A_{ik}}$, where

$$A_{ik} = \sum_k \frac{W_{ik}}{\nabla^+_{ik}} \text{ and } B_{ik} = \sum_k W_{ik} \frac{\nabla^-_{ik}}{\nabla^+_{ik}}. \tag{11}$$

There is a negative term $-B_{ik}$ in the numerator, which may cause negative entries in the updated W. To overcome this, we apply the "moving term" trick [15–18,21] to resettle B_{ik} to the denominator, giving the final update rule

$$W^{\text{new}}_{ik} = W_{ik} \frac{\nabla^-_{ik} A_{ik} + 1}{\nabla^+_{ik} A_{ik} + B_{ik}}. \tag{12}$$

Our ensemble algorithm simply iterates the above update rule until W converges. The updates have the following guarantee

Theorem 1. $\tilde{\mathcal{J}}(W^{new}, \lambda) \leq \tilde{\mathcal{J}}(W, \lambda)$, with λ given in Eq. 10.

The proof is given in the Appendix. The theorem shows that the algorithm jointly reduces $\mathcal{J}(W)$ while steering W rows to the probability simplex. The tradeoff between these two forces is adaptively adjusted by A_{ik}.

4   Experiments

We have tested our new method on six real-world data sets from the UCI repository[1]. Their statistical characteristics are given in Table 1 and below a brief verbal description of each data set is given:

- ECOLI, the UCI *Ecoli* data set, containing protein localization sites, originally with 8 attributes.
- WINE, the UCI *Wine* data set, which is a result of a chemical analysis of wines grown in the same region in Italy but derived from three different cultivars, originally with 13 dimensional features;
- MFEAT, the UCI *Multiple Features* data set, which consists of features of handwritten numerals; the digits are represented in terms of 649 features from six aspects;
- SEGMENT, the UCI *Image Segmentation* data set, image patches from 7 outdoor images, originally with 19 high-level features;
- OPTDIGITS, the UCI *optical recognition of handwritten digits* data set, originally with 64 dimensions;
- PENDIGITS, the UCI *pen-based recognition of handwritten digits* data set, originally with 16 dimensions.

---

[1] http://archive.ics.uci.edu/ml/.

Table 1. Statistics of the data sets.

Data set	Samples	Dimensions	Classes
ECOLI	327	7	5
WINE	178	13	3
MFEAT	2000	649	10
SEGMENT	2310	19	7
OPTDIGITS	5620	64	10
PENDIGITS	10992	16	10

We have compared DECLU with five other clustering approaches, three single clustering methods and two clustering ensemble methods:

- *Normalized Cut* (Ncut) [8], a spectral clustering method that projects the tailing eigenvectors of symmetric normalized Laplacian of the similarity matrix to the closest cluster indicator matrix;
- *Probabilistic Latent Semantic Indexing* (PLSI) [5] which factorize the similarity matrix $P(x_i, x_j) \approx \sum_k P(x_i|C = k)P(x_j|C = k)P(C = k)$, where C is the cluster variable; the cluster assignment $P(C = k|x_i)$ can then be obtained with $P(x_i|C = k)$ and $P(C = k)$ through Bayes rule;
- *Left Stochastic Decomposition* (LSD) [1] which factorizes the similarity matrix into two left-stochastic matrices;
- *Cluster-based similarity partitioning algorithm* (CSPA) [9], the similarity between two data-points is defined to be directly proportional to number of constituent clusterings of the ensemble in which they are clustered together;
- *Meta-clustering algorithm* (MCLA) [9], which is based on clustering clusters; first, it tries to solve the cluster correspondence problem and then uses voting to place data-points into the final consensus clusters.

We use the default setting in the above methods. The number of clusters is set to the number of known classes in each data set. We followed the convention that uses kmeans for generating base clustering for CSPA and MCLA. For DECLU,

Table 2. Clustering purities for the compared methods. Boldface numbers indicate the best for each data set.

Data set	Ncut	PLSI	LSD	CSPA	MCLA	DECLU
ECOLI	0.79	0.80	0.68	0.75	0.79	**0.82**
WINE	0.72	0.72	0.72	0.69	0.70	**0.73**
MFEAT	0.76	0.75	**0.78**	0.57	0.66	**0.78**
SEGMENT	0.62	**0.63**	0.59	0.51	0.60	**0.63**
OPTDIGITS	0.84	0.85	0.81	0.77	0.80	**0.90**
PENDIGITS	0.74	0.77	0.84	0.65	0.68	**0.85**

we have used $\sigma = 0.02$ in noise generation and $K = 5$ in K-Nearest-Neighbor similarity graph. For all cluster ensemble methods, we have used $M = 10$ base clusterings.

The clustering performance is evaluated by cluster purity, calculated by

purity $= \dfrac{1}{N} \sum_{k=1}^{r} \max_{1 \le l \le r} n_k^l$, where n_k^l is the number of data samples in the cluster

k that belong to ground-truth class l. A larger purity in general corresponds to a better clustering result.

The resulting cluster purities are shown in Table 2. We can see that DECLU yields top performance for all data sets, with a tie with a different method on two (MFEAT and SEGMENT) out of the six data sets in total.

5 Conclusion

In this paper we have proposed a new clustering method which consists of two steps. First, we repeatedly incorporate a small amount of noise to the data to generate multiple base partitions of the data. Next, we have developed a new ensemble method using information theoretic metric and its multiplicative optimization algorithm. Empirical studies on the new method indicate that it outperforms several other existing clustering approaches in terms of clustering accuracy.

In this work we used a fixed amount of Gaussian noise. Other types of noise would be interesting to study in the future. Similarly, it would be valuable to investigate how to automatically adjust the noise level for generating better base clusterings. Moreover, we aim at examining other information divergence (e.g., [3,20]) or mutual information variants besides D_{sum} to improve the ensemble performance (e.g., [7,12,19]). Other types of base clustering generation methods (e.g., [13,14]) could further improve the accuracy. In summary, the consistently satisfactory performance of DECLU and its computational scalability suggest considerable potential for further development of denoising based clustering methods.

Appendix: Proof of Theorem 1

Proof. We use W for current estimate, $\widetilde{W}$ for variable, and W^{new} for the new estimate, respectively. The objective function $\widetilde{\mathcal{J}}$ fulfills the theorem conditions in [16]. Therefore, we can construct the majorization function

$$G(\widetilde{W}, W) = \sum_{ik} \left[\nabla_{ik}^+ \widetilde{W}_{ik} - \nabla_{ik}^- W_{ik} \log \widetilde{W}_{ik} + \frac{B_{ik}}{A_{ik}} W_{ik} - \frac{W_{ik}}{A_{ik}} \log \widetilde{W}_{ik} \right] + \text{constant}$$

such that $G(\widetilde{W}, W) \ge \widetilde{\mathcal{J}}(\widetilde{W}, \lambda)$ and $G(W, W) = \widetilde{\mathcal{J}}(W, \lambda)$. Let W^{new} be the minimum of $G(\widetilde{W}, W)$, which is implemented by zeroing $\partial G / \partial \widetilde{W}$ and yields Eq. 12. Therefore $\widetilde{\mathcal{J}}(W^{\text{new}}, \lambda) \le G(W^{\text{new}}, W) \le G(W, W) = \widetilde{\mathcal{J}}(W, \lambda)$.

References

1. Arora, R., Gupta, M., Kapila, A., Fazel, M.: Clustering by left-stochastic matrix factorization. In: ICML (2011)
2. Bishop, C.: Training with noise is equivalent to Tikhonov regularization. Neural Comput. **7**(1), 108–116 (1995)
3. Dikmen, O., Yang, Z., Oja, E.: Learning the information divergence. IEEE Trans. Pattern Anal. Mach. Intell. **37**(7), 1442–1454 (2015)
4. Herbrich, R., Graepel, T.: Invariant pattern recognition by semidefinite programming machines. In: NIPS (2004)
5. Hofmann, T.: Probabilistic latent semantic indexing. In: SIGIR, pp. 50–57 (1999)
6. Parzen, E.: On estimation of a probability density function and mode. Ann. Math. Stat. **33**(3), 1065–1076 (1962)
7. Romano, S., Bailey, J., Nguyen, V., Verspoor, K.: Standardized mutual information for clustering comparisons: one step further in adjustment for chance. In: ICML (2014)
8. Shi, J., Malik, J.: Normalized cuts and image segmentation. IEEE Trans. Pattern Anal. Mach. Intell. **22**(8), 888–905 (2000)
9. Strehl, A., Ghosh, J.: Cluster ensembles - a knowledge reuse framework for combining multiple partitions. J. Mach. Learn. Res. **3**, 583–617 (2002)
10. Vincent, P., Larochelle, H., Bengio, Y., Manzagol, P.: Extracting and composing robust features with denoising autoencoders. In: ICML (2008)
11. Vincent, P., Larochelle, H., Lajoie, I., Bengio, Y., Manzagol, P.A.: Stacked denoising autoencoders: learning useful representations in a deep network with a local denoising criterion. J. Mach. Learn. Res. **11**, 3371–3408 (2010)
12. Vinh, N.X., Epps, J., Bailey, J.: Information theoretic measures for clusterings comparison: variants, properties, normalization and correction for chance. J. Mach. Learn. Res. **11**, 2837–2854 (2010)
13. Yang, Z., Hao, T., Dikmen, O., Chen, X., Oja, E.: Clustering by nonnegative matrix factorization using graph random walk. In: NIPS (2012)
14. Yang, Z., Laaksonen, J.: Multiplicative updates for non-negative projections. Neurocomputing **71**(1–3), 363–373 (2007)
15. Yang, Z., Oja, E.: Linear and nonlinear projective nonnegative matrix factorization. IEEE Trans. Neural Netw. **21**(5), 734–749 (2010)
16. Yang, Z., Oja, E.: Unified development of multiplicative algorithms for linear and quadratic nonnegative matrix factorization. IEEE Trans. Neural Netw. **22**(12), 1878–1891 (2011)
17. Yang, Z., Oja, E.: Clustering by low-rank doubly stochastic matrix decomposition. In: ICML (2012)
18. Yang, Z., Oja, E.: Quadratic nonnegative matrix factorization. Pattern Recogn. **45**(4), 1500–1510 (2012)
19. Yang, Z., Peltonen, J., Kaski, S.: Optimization equivalence of divergences improves neighbor embedding. In: ICML (2014)
20. Yang, Z., Zhang, H., Yuan, Z., Oja, E.: Kullback-leibler divergence for nonnegative matrix factorization. In: Honkela, T. (ed.) ICANN 2011, Part I. LNCS, vol. 6791, pp. 250–257. Springer, Heidelberg (2011)
21. Zhu, Z., Yang, Z., Oja, E.: Multiplicative updates for learning with stochastic matrices. In: Kämäräinen, J.-K., Koskela, M. (eds.) SCIA 2013. LNCS, vol. 7944, pp. 143–152. Springer, Heidelberg (2013)

Novel Information Processing for Image De-noising Based on Sparse Basis

Sheikh Md. Rabiul Islam[1]([✉]), Xu Huang[1], Keng Liang Ou[2],
Raul Fernandez Rojas[1], and Hongyan Cui[3]

[1] Faculty of ESTeM, University of Canberra, Canberra, Australia
{Sheikh.Md.RabiulIslam,Xu.Huang,Raul.Fernandez.Rojas}@canberra.edu.au
[2] College of Oral Medicine, Taipei Medical University, Taipei, Taiwan
klou@tmu.edu.tw
[3] Bejing University of Posts and Telecommunication, Beijing, China
cuihy@bupt.edu.cn

Abstract. Image de-noising is one of the important information processing technologies and a fundamental image processing step for improving the overall quality of medical images. Conventional de-noising methods, however, tend to over-suppress high-frequency details. To overcome this problem, in this paper we present a novel compressive sensing (CS) based noise removing algorithm using proposed sparse basis on CDF9/7 wavelet transform. The measurement matrix is applied to the transform coefficients of the noisy image for compressive sampling. The orthogonal matching pursuit (OMP) and Basis Pursuit (BP) are applied to reconstruct image from noisy sparse image. In the reconstruction process, the proposed threshold with Bayeshrink thresholding strategies is used. Experimental results demonstrate that the proposed method removes noise much better than existing state-of-the-art methods in the sense image quality evaluation indexes.

Keywords: ATVD · BP · CS · OMP · Sparse

1 Introduction

From the compressive sensing (CS) theory, we know that a sparse signal can be reconstructed from far fewer samples than the samples required by Nyquist rate. That is why CS has been widely used in medical image and remote sensing. In a CS-based image processing system, de-noising is a classical problem to be improved in the quality of images using a reduced amount of data. In recent years, many researchers have worked on advanced methods for image de-noising, including total variation image regularization [6], texture preserving variational de-noising using adaptive fidelity (ATVD), which is a modified version of TVD algorithm [4], basis pursuit de-noising [3]. In addition to mentioned above methods for image de-noising, we shall next focus on compressive sensing as a new method for image de-noising. This paper is to develop a novel image de-noising algorithm based on compressive sensing, which is faster, simpler and also can

© Springer International Publishing Switzerland 2015
S. Arik et al. (Eds.): ICONIP 2015, Part III, LNCS 9491, pp. 443–451, 2015.
DOI: 10.1007/978-3-319-26555-1_50

keep strong edge preservation. In this novel work, we have applied proposed sparse basis on CDF9/7 wavelet transform and a convex optimization technique for reconstruction of image is used, such as which is also called Basis Pursuit (BP) [3] and greedy pursuit such as Orthogonal Matching Pursuit (OMP) algorithm [9]. Even though after algorithm processing, there is still some noise owing to random can be observed. Hence, we take our proposed threshold just before applied inverse CDF9/7 wavelet transform and make a filter operation. To find the best de-noised image, we used TVD algorithm [6] and TV with adaptive fidelity (ATVD) [4]. For making a good comparison, we have used sparse basis DWT and DCT and used image quality assessment scheme [5] to assessed images de-noised.

This paper is organized as follows. Section 2 gives a review of the compressive sensing (CS) problem statement. Section 3 describes sparse image representation by wavelet lifting scheme. Section 4 describes compressive image de-noising method with an algorithm. Experimental results are reported in Sect. 5. The paper completed with a brief conclusion.

2   About Compressed Sensing

The concept of compressive sensing (CS) is to acquire significant data directly without sampling the signal. Thus, it is shown that if the signal is 'sparse' or compressible, then the acquired data is sufficient to reconstruct the original signal from sparse signal with a high probability [1]. Sparsity is mainly defined by the appropriate basis such as DCT or WT and others transformation for that signal. According to compressive sensing theory, CS can be further described as below:

First let following the notations of [1] and we have $f = f_1, \ldots\ldots, f_N$ be N real-valued samples of a signal, which can be represented by the transform coefficients, x. That is,

$$f = \Psi x = \sum_{i=1}^{N} x_i \psi_i \qquad (1)$$

where $\Psi = [\psi_1, \psi_2, \ldots\ldots, \psi_N]$ is an $N \times N$ transform basis matrix, which determines the domain in which the signal is sparse and $x = [x_1, x_2, \ldots\ldots, x_N]$ is an N-dimension vector of coefficients with $x_i = < x, \psi_i >$. We assume that x is S-sparse, meaning that there are only significant elements in x with $S \ll N$.

Suppose a general linear measurement process computes inner product with $M < N$, between f and a collection of vectors, ϕ_j, giving $y_j = < f, \phi_j >$; $j = 1 \ldots\ldots M$. If Φ denotes the $M \times N$ matrix with ϕ_j as row vectors, then the measurements $y = [y_1, y_2, \ldots\ldots, y_M]$ are given by:

$$y = \Phi f = \Phi \Psi x = \Theta x \qquad (2)$$

where y is M-dimensional observation vector, Φ is the $M \times N$ random measurement matrix, and $\Theta = \Phi \Psi$ is called sensing matrix. For reconstruction ability x is S-sparse, if Θ satisfies the Restricted Isometric Property (RIP) [2].

The signal reconstruction is an ill-conditioned problem involves using y to reconstruct the N-length signal, x, that is S-sparse, given Φ and Ψ. Many researchers have developed a number of different algorithms to solve these three CS reconstruction problems. In this paper we have used Basis Pursuit (BP) algorithm [3] to solve the l_1 norm and gradient-based algorithms as OMP algorithm [9] to solve the l_2 norm for proposed image de-noising applications.

3 Proposed Sparse Representation of Image

Currently, multi-resolution pyramid decomposition and synthesis algorithm, namely the Mallat algorithm, is the most commonly used in wavelet research area. Cohen-Daubechies-Feauveau 9/7 (CDF 9/7) Wavelet Transform (WT) [8] is a lifting scheme based wavelet transform that can reduce the computational complexity. The principle of lifting scheme is described by considering an input image x fed in parallel into a $\widehat{h}$ (low pass filter) and $\widehat{g}$ (high pass filter). The outputs of the two filters are then sub sampled by 2 ($\downarrow$ 2) to obtain low-pass subband y_L and high-pass subband y_H as shown in Fig. 1. The original signal can be reconstructed by synthesis filters h (low pass) and g (high pass), which take the up-sampled by 2 ($\uparrow$ 2) for y_L and y_H as inputs. An analysis and synthesis system has the perfect reconstruction property if and only if $x' = x$.

The mathematical representations of y_L and y_H can be defined as

$$\begin{cases} y_L(n) = \sum_{i=o}^{N_L-1}\widehat{h}(i)x(2n-i), \\ y_H(n) = \sum_{i=o}^{N_H-1}\widehat{g}(i)x(2n-i) \end{cases} \tag{3}$$

where N_L and N_H are the lengths of $\widehat{h}$ and $\widehat{g}$ respectively.

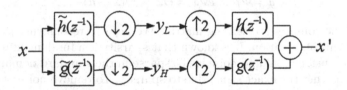

Fig. 1. Discrete wavelet transform (or subband transform) analysis and synthesis system [8].

This paper develops multi-layer lifting scheme WT with sparse basis, which decomposes an image x into 4 parts for each layer: $LL1$, $HL1$, $LH1$, and $HH1$ and each layer multiply by sparse matrix. This concept is also applied to the second and third level decomposition based on the principle of multi-resolution analysis. For example the $LL1$ subband is decomposed into four smaller subbands: $LL2$, $LH2$, $HL2$, and $HH2$. The three layer subbands are sparse so that

Orthogonal Matching Pursuit (OMP) algorithm [9] or Basis Pursuit (BP)[3] can be adopted to rebuild these parts directly.

4 Proposed Image De-noising Method Based on Compressive Sensing

For image de-noising application, we propose another novel image de-noising based on compressive sensing framework. The objective of novel compressive sensing de-noising process is to estimate the original image x with dimension $N \times N$ pixels by discarding the corrupted together with three popular different noises: Gaussian Noise, Poisson noise, and impulse (salt & pepper) noise n from the function f:

$$f = x + n \tag{4}$$

The real value sample of signal f which can be represented by transform coefficients x. That is

$$f = \Psi x = \sum_{i=1}^{N} (x + n)_i \psi_i \tag{5}$$

where $\Psi = [\psi_1, \psi_2, \ldots\ldots, \psi_N]$ is the transform basis matrix using by sparsity CDF 9/7 wavelet transform and $s = [(x + n)_1, (x + n)_2, \ldots, (x + n)_N]$ is an N-vector of coefficients with $x_i = \langle x, \psi_i \rangle$ and there are only with $S \ll N$ significant elements in x. Most natural signals can be transformed to sparse domain by several conventional transforms.

We sample from f by mixing matrix or measurement matrix Φ that is stable and incoherence with matrix transform Ψ:

$$y = \Phi f = \Phi \Psi x = \Theta x = \Theta(x + n) \tag{6}$$

where Θ is the compressive sensing matrix. We need to reconstruct the original signal from this observation. It is known that sparsity is a fundamental principle in fidelity reconstruction, and noise is not sparse in the standard domain. Hence, we can reconstruct the exact signal due to sparsity. To remove noise, we develop a novel CS image de-noising Algorithm 1 and its flow chart is shown in Fig. 2.

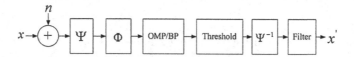

Fig. 2. Proposed image de-noising framework based on compressive sensing.

Algorithm 1. Proposed Image De-noising Algorithm with CS

---

1   Introduce noise to the image f.

2   Perform sparse domain CDF 9/7 wavelet transform to signal f into several sparse directions to CDF 9/7 wavelet sub-band $f_{j,l}$, where j is the decomposition level and l the number of direction level at each scale are expressed as

$$\widehat{f}^1 = W^1 f = \left[\widehat{f}^1_{LL} \widehat{f}^1_{LH} \widehat{f}^1_{HL} \widehat{f}^1_{HH} \right]^T$$

$$\widehat{f}^2 = W^2 \widehat{f}^1_{LL} = \left[\widehat{f}^2_{LL} \widehat{f}^2_{LH} \widehat{f}^2_{HL} \widehat{f}^2_{HH} \right]^T$$

$$\widehat{f}^3 = W^3 \widehat{f}^2_{LL} = \left[\widehat{f}^3_{LL} \widehat{f}^3_{LH} \widehat{f}^3_{HL} \widehat{f}^3_{HH} \right]^T$$

where $W^j, j \in 1, 2, 3$ represents the 2D CDF9/7 wavelet transform matrix of level j.

3   Apply CS scheme to each direction and decomposition level as follows:

$$y = \Phi_{j,l} \Psi x_{j,l}$$

4   Apply $l_1 - norm$ approach BP or OMP to reconstruct the signal x from y using step 3.

5   Apply proposed threshold (T) with the aid of Bayeshrink thresholding strategies via proposed threshold and is given by

$$T(j,l) = \beta \frac{\sigma^2_{j,l}}{\sigma^2_{w,j,l}}$$

$$\sigma^2_{j,l} = \left(\frac{median(|HP_{j,l}|)}{0.6745} \right)^2$$

$$\widehat{\sigma}^2_{w,j,l} = max(\widehat{\sigma}^2_y - \widehat{\sigma}^2, 0)$$

where $\widehat{\sigma}^2_y = \frac{1}{M \times N} \sum_{j,l=1}^{M,N} y^2_{j,l}$. β is the parameter define by threshold. With proposed threshold, a de-noised OMP/BP coefficient $\widehat{x}_{T,j,l}$ is calculated as follows:

$$\widehat{x}_{T,j,l} = \begin{cases} \frac{x_{r,j,l}[x_{r,j,l} - T^\alpha]}{|x_{r,j,l}|} & \text{for } |x_{r,j,l}| \geq T(j,l) \\ 0 & \text{for } |x_{r,j,l}| < T(j,l) \end{cases}$$

where $x_{r,j,l}$ is the coefficients value after image reconstruction by OMP/BP. α is smooth signal parameter and we have chosen $\alpha = 20$ and β is 0.3 for proposed threshold.

6   After threshold, we apply inverse CDF 9/7 wavelet transform to recovered image.

7   Finally recovered image is applied to filter operation methods (TVD algorithm [6], or ATVD algorithm [4]) to obtain de-noised image x'.

8   Sparse basis DWT or DCT is used to decompose the image into feature and non-feature regions in steps 2 and apply inverse DWT or DCT in step 6 for evaluation of de-noised image.

---

5   Experimental Results

For evaluating the performance of proposed de-noising algorithm on images
were selected from open source databases [7]. Consider, the noise is a combination of three popular noises including fixed Poisson noise at $\lambda = 0.9686$,
100 % impulse noise density, and white Gaussian noise with a fixed deviation
level $\sigma \in [5, 10, 15, 20, 25]$ in the effect of low-light noise distribution in digital camera. This low-light noise is used as camera sensor matrix and camera
performance. For comparison, besides the proposed de-noising frame based on
proposed sparse basis CDF 9/7 WT with BP or OMP including TVD/ATVD
and other de-noising techniques with sparse basis DWT and DCT by OMP or
BP including TVD/ATVD algorithm are also used. This paper compares the
evaluation of quality indexes (EIQ) for de-noised images on several measurements $M > N(N = 256)$. It is noted that the more measurements of M, the
better will be the quality of de-noised images. Figure 3 shows the quality indexes
of de-noised images obtained with different de-noising methods performed on
the X-ray images. From these figures, UIQI and Q(Kurtosis) values of proposed
sparse basis CDF9/7 WT with Gaussian measurements, OMP and proposed
threshold show higher values than others. However, there is some noise inside
de-noised image, so de-noising method such as TVD/ATVD has been applied
there. The experimental results clearly show that the proposed method based
on proposed sparse basis CDF 9/7 WT and proposed threshold by OMP or BP
algorithm including TVD, ATVD out-performed all other four proposed methods for all values of the noise deviation in the range [5, 25]. Figure 4 shows the
visual comparison of de-noised X-ray images for $\sigma = 10$ and for the necessity
of filtering TVD or ATVD for the proposed de-noising algorithm. The comparison is clear with data plotted in Fig. 3(a)–(b), which shows the relationship
between the UIQI and Q(Kurtosis) for different σ performed on an X-ray image.
Similarly the visual comparison of de-noised brain (MR)-1 image in Fig. 6 for
$\sigma = 5$ with data plotted in Fig. 5(a) and (b). The proposed sparse basis CDF
9/7 WT with BP and proposed threshold including TVD/ATVD show higher

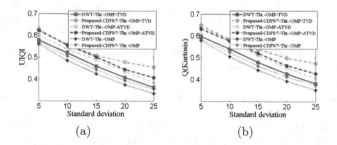

(a)   (b)

Fig. 3. Plots of image quality indexes for X-ray image versus noise standard deviations
(σ) in the range [5, 25] for six different proposed de-noising approach.

quality of de-noised brain (MR)-1 image as shown in Fig. 6(a) and (c) as compared with sparse basis DCT for Fig. 6(b) and (d) in the range $\sigma = [5, 25]$. So, the TVD and ATVD is more important for this proposed de-noising algorithm. We also observes from Fig. 7 the computational complexity for novel de-noising framework based on CS.

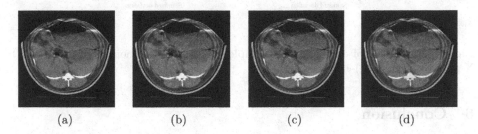

(a)   (b)   (c)   (d)

Fig. 4. De-noising of X-ray image (a) Sparse CDF9/7 WT with BP algorithm and Thr., (b) DWT with BP and Thr., (e) Sparse CDF 9/7 WT with BP, Thr., and TVD, (f) DWT with BP, Thr., and TVD.

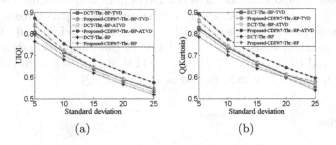

(a)   (b)

Fig. 5. Plots of image quality indexes versus noise standard deviations (σ) in the range [5, 25] for six different proposed de-noising approach for brain (MR)-1 image.

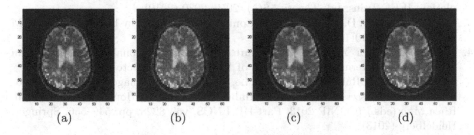

(a)   (b)   (c)   (d)

Fig. 6. De-noising of brain (MR)-1 image with (a) CDF9/7 WT with BP and Thr., (b) DCT with BP and Thr., (c) CDF 9/7 WT with BP, Thr., and ATVD, (d) DCT with BP, Thr., and ATVD.

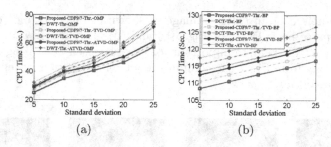

Fig. 7. Comparison of computational complexity for novel de-noising scheme for (a) X-ray, (b) brain (MR)-1 image.

6   Conclusion

In this paper, we proposed sparse domain CDF 9/7 wavelet transform for image compression based on CS. We also proposed an image de-noising algorithm based on compressive sensing framework that addresses the simultaneous removal of undesired noise components and preservation of high-frequency details in images. Experimental results demonstrate that the proposed sparse basis CDF 9/7 wavelet transform and proposed threshold are more efficient and effective for removing noise than sparse DWT and DCT in the image reconstruction algorithm OMP and BP including TVD and ADTV of proposed approach in terms of proposed image quality indexes and others like UIQI.

References

1. Baig, M.Y., Lai, E.M.K., Punchihewa, A.: Compressed sensing-based distributed image compression. J. Appl. Sci. **4**(2), 128–147 (2014)
2. Candes, E.J., Tao, T.: The power of convex relaxation: near-optimal matrix completion. IEEE Trans. Inf. Theory **56**(5), 2053–2080 (2010)
3. Chen, S., Donoho, D., Saunders, M.: Atomic decomposition by basis pursuit. SIAM J. Sci. Comput. **20**(1), 33–61 (1998)
4. Gilboa, G., Sochen, Y.Y., Nir, Z.: Texture preserving variational denoising using an adaptive fidelity term. In: Proceedings of VLSM, pp. 1–8, October 2003
5. Islam, Sheikh MdRabiul, Huang, Xu, Le, Kim: A Novel Image Quality Index for Image Quality Assessment. In: Lee, Minho, Hirose, Akira, Hou, Zeng-Guang, Kil, Rhee Man (eds.) ICONIP 2013, Part III. LNCS, vol. 8228, pp. 549–556. Springer, Heidelberg (2013)
6. Rudin, L.I., Osher, S., Fatemi, E.: Nonlinear total variation based noise removal algorithms. J. Phys. D Nonlinear Phenom. **60**(14), 259–268 (1992)
7. St. Xavier's: TEST IMAGES. http://decsai.ugr.es/cvg/dbimagenes/index.php (2010)

8. Sweldens, W.: The lifting scheme: a new philosophy in biorthogonal wavelet constructions. Proceeding of SPIE in Wavelet Applications in Signal and Image Processing III, 68–79 (September 1995)
9. Tropp, J., Gilbert, A.: Signal recovery from random measurements via orthogonal matching pursuit. IEEE Trans. Inf. Theory 53(12), 4655–4666 (2007)

Trajectory Abstracting with Group-Based Signal Denoising

Xiaoxiao Luo, Qing Xu$^{(\boxtimes)}$, Yuejun Guo, Hao Wei, and Yimin Lv

School of Computer Science and Technology, Tianjin University, Tianjin, China
qingxu@tju.edu.cn
http://cs.tju.edu.cn/szdw/jsfjs/xuqing/

Abstract. Trajectory abstracting is to compendiously summarize the substance of a lot of information delivered by the trajectory data. In this paper, to cope with complex trajectory data, we propose a novel framework for abstracting trajectories from the perspective of signal processing. That is, trajectories are designated as signals, manifesting the copious information that varies with time and space, and denoising is exploited to concisely communicate the trajectory data. Resampling of trajectory data is firstly performed, based on achieving the minimum *Jensen-Shannon divergence* of the trajectories before and after being resampled. The resampled trajectories are matched into groups according to their similarity and, a *non-local* denoising approach based on wavelet transformation is developed to produce summaries of trajectory groups. Our new framework can not only offer multi-granularity abstractions of trajectory data, but also identify outlier trajectories. Extensive experimental studies have shown that the proposed framework achieves very potential results in trajectory summarization, in terms of both objective evaluation metrics and subjective visual effects. To the best of our knowledge, this is the first to deploy the group-based signal denoising technique in the context of summarizing the trajectory data.

Keywords: Trajectory abstracting · Multi-granularity abstractions · Signal processing · *non-local* denoising · Wavelet transformation

1 Introduction

Trajectory data is very useful for a lot of practical fields such as intelligent transportation and so on [13]. The processing of the trajectory data is fundamentally based on clustering, which is basically one of the most powerful techniques to obtain the patterns and knowledge of these data for the better abstraction and full employment of them [8]. Unfortunately, the performance of clustering may degrade when dealing with complex trajectory data. Here the meaning of complexity is at least threefold. First, different trajectories can have largely diverse numbers of sample points. Second, similarity between trajectories happening in one area of a scene can significantly differ from that in another area. Third, outliers occur together with common clusters of trajectories. An example in Fig. 4(f)

© Springer International Publishing Switzerland 2015
S. Arik et al. (Eds.): ICONIP 2015, Part III, LNCS 9491, pp. 452–461, 2015.
DOI: 10.1007/978-3-319-26555-1_51

shows the difficulty of the clustering technique for treating complex trajectory data.

In this paper, in order to effectively analyze and understand complex trajectories, we propose a totally new approach, called trajectory abstracting, which is significantly different from the clustering scheme. That is, considering that trajectories are in fact signals handling the information that change with time and space, the abstraction of trajectory data is performed from the perspective of signal denoising. At first, resampling of trajectories is taken for combatting the situation that numbers of trajectory sample points are largely different. Then, variations of trajectory similarities occurring in different scene areas are tackled by iterative group-based "*non-local*" denoising. Our denoising scheme provides trajectory details and summarizations with multiple granularities, leading to better analysis and understanding of the complexity of data.

The remainder of this paper is organized as follow. The next Section covers the related work. The developed trajectory abstracting framework is described in Sect. 3. Section 4 introduces two metrics to evaluate the proposed framework for trajectory abstraction. Experimental results are presented in Sect. 5. The final Section concludes the paper.

2 Related Work

The use of clusters (with their centroids) may the most popular way to represent the patterns of trajectory data [11]. A lot of good clustering techniques used for trajectory data have been developed [11,13]. For example, classical algorithms include k-means [9], BIRCH [17], DBSCAN [6], OPTICS [4], and STING [16]. Among them, k-means and DBSCAN may be the mostly applied in practice due to their easy use and efficiency. However, as indicated in Sect. 1, the clustering may not benefit for the reasonable handling of the complex trajectory data. For the sake of treating complex trajectory data, data filtering is a good preprocess. For instance, Keogh et al. apply the wavelet transform to represent a single trajectory and do clustering by k-means [15]. Notice that our proposed approach is largely different from that by Keogh et al., we perform the wavelet denoising across all the trajectories in a similarity group.

Block-Matching and 3D Filtering (*BM3D*) may be the best state-of-the-art filter for noisy image/video data [5], which utilize coherence among pixel patches to do denoising very effectively. The core of *BM3D* inspires us, but we significantly extend *BM3D* to cope with the trajectory data that are greatly different from the image/video data.

3 The Framework for Trajectory Abstraction

Our proposed framework is composed of two main components, resampling and group-based *non-local* denoising. An abstraction in one granularity is obtained by an iteration of denoising. Our method iteratively outputs trajectory abstractions with multi-granularities and outliers.

3.1 Resampling

In general, complex trajectories have various numbers of sample points. Also, trajectories are usually corrupted with noise. We propose to make use of a resampling technique to smooth out noise and to make resampled trajectories have a same number of sample points, leading to a better trajectory abstraction. Each trajectory is resampled to have a equal distance interval. For the sake of obtaining the minimum bias introduced by resampling, we find an optimal number n_{opt} of sample points to preserve information of the original trajectory as much as possible. Since the trajectory shape is very important, we obtain n_{opt} by minimizing the widely used *Jensen-Shannon divergence* (*JSD*) between the trajectory shapes before and after being re-sampled. Actually, the trajectory shape can be characterized by the distribution of the angles at the sample points.

Suppose a trajectory $T = \{p_1, p_2, \ldots p_n\}$, is with n sample points, here $p_i = (x_i, y_i)$ is the 2-D coordinates in a ground plane. We obtain its angles $\{\theta_j\}, j = 1, 2, \ldots, n-2$ (Fig. 1), to build up a probability distribution, $\mathbf{P}^{ori} = \{\mathbf{P}_1, \mathbf{P}_2, \ldots, \mathbf{P}_M\}$, using the angle histogram with M bins ($M = 10$ in this paper). For the resampled version T', we obtain its probability distribution $\mathbf{P}^{res}$ similarly, and then the *JSD* distance between T and T' is computed by

$$JSD(T; T') = H(\frac{\mathbf{P}^{ori} + \mathbf{P}^{res}}{2}) - \frac{H(\mathbf{P}^{ori}) + H(\mathbf{P}^{res})}{2} \tag{1}$$

where $H(.)$ denotes the *Shannon* entropy. In this paper, the *JSD* value between a dataset before and after being resampled is defined as the mean of *JSD*s for the trajectories of this dataset. Apparently, the lower *JSD* indicates a higher similarity between the original and the resampled trajectory data. Thus, n_{opt} is located at the minimum in all the *JSD* values resulted from the trajectories with different numbers of resampled points. For example, Fig. 2 presents such a plot for the *Pedestrian* dataset (Sect. 5) and undoubtedly, $n_{opt} = 21$ corresponds to the minimum *JSD*.

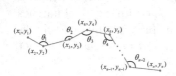

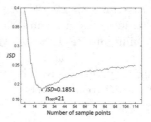

Fig. 1. A trajectory with its angles

Fig. 2. Determination of n_{opt} based on the *JSD* value

3.2 *non-local* Denoising

With all the trajectories after being resampled, we perform a procedure called "*non-local*" denoising, involving three phases, *Matching, Thresholding* and *Combining*, in multiple iterations for obtaining the multi-granularity abstractions. In fact, the summaries in different abstracted levels give a more clear and better understanding about the trajectory data. It is worthy to point out that outliers can be detected in the process of *non-local* denoising. The outliers, picked out in an iteration, are not be included for the later iterative operations.

For easy understanding, we define $\mathbf{TR}_k = \{T_{k,1}, T_{k,2}, \ldots, T_{k,l}\}$ ($k \geq 1$) as the k-th iterative output with l abstracted trajectories (and also as the input of the $(k+1)$-th iteration). Here $\mathbf{TR}_0$ is the product by the trajectory resampling, as the input of the 1-st iteration. The iterative execution is terminated when the output abstractions between two consecutive iterations do not change anymore.

Matching. In this phase, similar trajectories are matched to establish groups. Notice that, simultaneously, outlier trajectories may be identified by the similarity matching. In concrete, each input trajectory is considered as a "reference", and the nearby trajectories similar to this reference are matched to form a group. Assume that there exists a reference trajectory $T_{k,r} \in \mathbf{TR}_k$, its similarity group $\mathbf{TG}_{k,r}$ is

$$\mathbf{TG}_{k,r} = \{T_{k,j} \in \mathbf{TR}_k | Diff(T_{k,r}, T_{k,j}) < \tau \ (j \neq r)\} \tag{2}$$

where $Diff(.)$ is a distance function and the simple and widely used Euclidean distance is adopted here. τ is a threshold, selected adaptively (Sect. 3.3), to determine whether two trajectories are distant or not.

Notice that, a trajectory can be used in more than one matching-based groups. Some outliers with few similar trajectories can be distinguished in this phase. That is, a trajectory is defined as outlier if the number of trajectories in its similarity group is less than η ($\eta = 3$ is used, for simplicity).

Thresholding. The wavelet threshold technique [7] is operated for the trajectories in the whole group, rather than just for a single trajectory, to obtain a compacted and summarized representation of these trajectories.

Suppose a reference trajectory $T_{k,r}$, having g number of neighboring trajectories $T_{k,j}$ ($j = 1, 2, ..., g; j \neq r$), is with a similarity group which is now denoted as a matrix

$$\mathbf{TG}_{k,r} = \begin{bmatrix} T_{k,1}^r \\ T_{k,2}^r \\ ... \\ T_{k,g}^r \\ T_{k,r}^r \end{bmatrix} = \begin{bmatrix} p_1^1 \ p_2^1 \ p_3^1 \ ... \\ p_1^2 \ p_2^2 \ p_3^2 \ ... \\ ... \\ p_1^g \ p_2^g \ p_3^g \ ... \\ p_1^r \ p_2^r \ p_3^r \ ... \end{bmatrix} = \begin{bmatrix} s_1 \ s_2 \ s_3 \ ... \end{bmatrix}. \tag{3}$$

Here $s_i = \left[p_i^1, p_i^2, ...p_i^g, p_i^r\right]^{\mathbf{T}}$ is designated as the i-th signal of this similarity group, which is actually the collection of the i-th sample points of the different trajectories in the group. The wavelet threshold is used for the noise filtering for s_i. At first, wavelet transform is performed on s_i to obtain the corresponding coefficients. Then the high frequency components are completely suppressed if and only if their values are smaller than a adaptively determined ϵ (Sect. 3.3). Finally, the denoised signal $\tilde{s}_i$ is given by using inverse wavelet transform

$$\tilde{s}_i = F^{-1}(\Upsilon(F(s_i))), \tag{4}$$

where Υ, F and F^{-1} denote the wavelet thresholding, wavelet transform and its inverse, respectively. In this paper Haar wavelet is employed and the denoised similarity group is as follows

$$\widetilde{\mathbf{TG}}_{k,r} = [\tilde{s}_1, \tilde{s}_2, ...] = \begin{bmatrix} \widetilde{T^r_{k,1}} \\ \widetilde{T^r_{k,2}} \\ ... \\ \widetilde{T^r_{k,g}} \\ \widetilde{T^r_{k,r}} \end{bmatrix} \tag{5}$$

Combining. Basically, for a resampled trajectory, more than one denoised versions can be obtained if it belongs to several similarity groups. In this case, we further combine these versions by averaging them to obtain a better abstraction with richer *non-local* similarity information.

3.3 Parameter Selection for Group-Based Denoising

All the important parameters used for grouping and denoising are adaptively selected, achieving the effectiveness and robustness of the proposed technique.

The distance threshold τ for the matching phase is determined based on the statistically averaged Euclidean distance between two trajectories in the dataset under consideration. Suppose there exists n trajectories in a dataset. For each trajectory T, $n-1$ Euclidean distances between T and all the other trajectories are sorted ascendingly. The average of x $(1 \leqslant x < n)$ smallest distances, denoted by Y, can be calculated. A plot of Y versus x is then provided (Fig. 3(a)). The Y value corresponding to the maximum of the second derivative of this plot (Fig. 3(b)) can be found. All the Y values resulted from n trajectories are averaged to obtain τ. We have observed that ϵ used for the thresholding phase is closely related with τ, $\epsilon = 2.5 \times \tau$ can do very well.

4   Evaluation

In this paper, we develop two quantitative criteria, *Integrality* (*INT*) and *Fidelity* (*FID*), to systematically and objectively evaluate the performance of our proposed framework for trajectory abstracting.

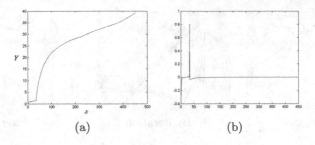

Fig. 3. (a) The plot of Y versus x of the trajectory T. (b) The second derivative of (a). The maximum point in (b) with red color corresponds to the red point in (a) (Colour figure online)

4.1   Fidelity (*FID*)

FID is to measure the similarity between the trajectory datasets before and after being abstracted, and this is a quality evaluation on the trajectory summarization. Given a original trajectory dataset without detected outliers $\mathbf{TR} = \{T_1, T_2, ..., T_n\}$ and its abstract of the last interation (or the cluster centroids) $\mathbf{TR'} = \{T'_1, T'_2, ..., T'_m\}$, we define *FID* by

$$FID(\mathbf{TR}, \mathbf{TR'}) = 1 - \frac{\max_i \{\min_j Diff(T_i, T'_j)\}}{MaxDiff}, \tag{6}$$

Here *Diff*() is the hausdorff distance, and *MaxDiff* is the global maximum distance between $\mathbf{TR}$ and $\mathbf{TR'}$. Apparently, a higher *FID* means the trajectory abstraction can express the dataset more accurately.

4.2   Integrality (*INT*)

INT measures the degree of coverage by abstracted trajectories for the trajectory dataset. Suppose $\mathbf{TR'} = \{T'_1, T'_2, ..., T'_m\}$ is the abstracted output by the last interation (or the cluster centroids), Its original trajectory without detected outliers is $\mathbf{TR} = \{T_1, T_2, ..., T_n\}$. The *INT* between original data and the abstraction is defined as

$$INT(\mathbf{TR}, \mathbf{TR'}) = \frac{1}{|\mathbf{TR}|} \sum_i Diff(T_i, T'_i) \tag{7}$$

Here *Diff*() is the hausdorff distance. A low *INT* indicates a completely coverage of the original.

5   Experiments

We have done extensive tests to evaluate the performance of our proposed framework, including 7 public databases listed in Table 1. Considering that trajectory abstracting and cluster are both aimed at pattern mining, two typical and effective cluster methods widely used in practical applications, DBSCAN and k-means, are compared with our technique.

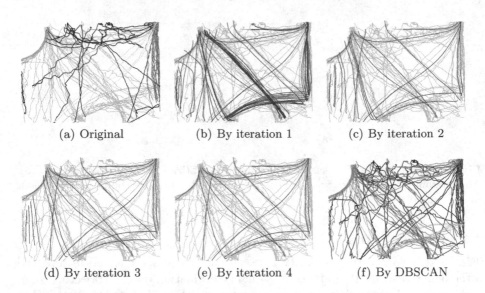

(a) Original   (b) By iteration 1   (c) By iteration 2

(d) By iteration 3   (e) By iteration 4   (f) By DBSCAN

Fig. 4. Comparision of different methods on *Pedestrian* (Colour figure online)

5.1 Results Analysis of *Pedestrian* Dataset

Figure 4(a) is the original trajectories of *Pedestrian*, where twenty trajectories in black color are outliers. The numbers of sample points in this dataset are largely diverse, varying between 33 and 422. The abstracted trajectories obtained in four iterations are respectively shown in Fig. 4(b)–(e). Ten clusters and outliers given by DBSCAN are displayed in Fig. 4(f) with different colors. Overall, the proposed technique compresses and summarizes very messy data into neat abstractions with multiple granularities. Fourteen most abstracted patterns, obtained in the last iteration, in reality exactly exhibit the fourteen pedestrian paths in this scenario (Fig. 4(e)).

Complex trajectory data can be represented by the multi-granularity abstractions very well. However, DBSCAN may give out unsatisfactory outputs. For example, the trajectories in red color (Fig. 4(a)) are messy and, they have two opposite directions. Iteration 1 makes them become neat and close, and then Iteration 2 outputs two abstracted patterns presenting trajectories with two opposite directions. As a result, the products by both two iterations indicate the processing of abstracting and, more importantly, jointly present trajectory details and summarizations. In contrast, DBSCAN falsely merge these trajectories two opposite directions into a single cluster.

As for the complex trajectory data with diverse trajectory similarities in different scene areas, abstractions by the proposed technique can give a concise representation of the original data. But the clusters by DBSCAN may be mistaken in this case. For instance, four trajectories in yellow color (Fig. 4(a)) are distant, compared with the trajectories in other areas. DBSCAN wrongly iden-

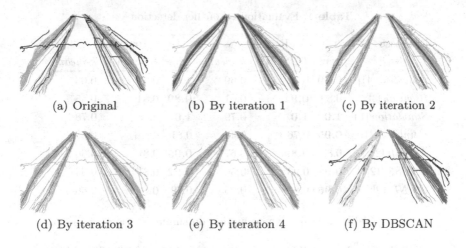

(a) Original   (b) By iteration 1   (c) By iteration 2

(d) By iteration 3   (e) By iteration 4   (f) By DBSCAN

Fig. 5. Comparision of different methods on *highway* (Colour figure online)

tify these four as outliers. Notably, our approach generates a condensed pattern, depicting the data satisfactorily.

5.2 Results Analysis of *Highway* Dataset

Original trajectories of *Highway* are shown in Fig. 5(a), and eighteen black trajectories are outliers. The trajectory abstractions at multiple granularities clearly expose the four traffic lanes. The lengths of the trajectories in the leftmost lane are diverse. Most of them are longer than the four trajectories rendered in red (Fig. 5(a)). These four short trajectories are compressed into one abstracted pattern which differs from their adjacent trajectories in the same lane. DBSCAN uses a single cluster for all the trajectories belonging to this traffic lane. But here a single cluster is not enough to feature the trajectories in largely different lengths.

5.3 Objective Evaluation

The recall and precision values by several methods for anomaly detection are listed in Table 1. It is obvious that our method performs the best to detect outliers. This is due to that *non-local* denoising can emphasize outliers in the process of trajectory summarizing.

The *FID* and *INT* scores listed in Table 2 indicate that the proposed technique achieves the best, consistent with the visual results discussed above.

6   Conclusions

In this paper, for the purpose of abstracting the complex trajectory data, we have established a novel effective technique in which a *non-local* signal denoising approach is exploited to obtain the summaries of the trajectory groups.

Table 1. Evaluation on outlier detection

Database	Recall			Precision		
	ours	DBSCAN	k-means	ours	DBSCAN	k-means
Pedestrian [1]	**0.90**	0.87	0.69	**1.0**	0.95	0.63
Highway [3]	0.89	**0.94**	0.61	**0.89**	0.84	0.55
Simulation [14]	**1.0**	**1.0**	0.79	**1.0**	0.95	0.78
Edinburgh [1]	**0.95**	0.76	0.48	0.81	**0.86**	0.55
Aircraft [2]	0.88	**0.89**	0.67	**0.95**	0.81	0.60
CROSS [12]	**0.98**	0.93	0.50	**0.92**	0.81	0.64
OMNI [12]	**0.96**	0.84	0.55	0.79	**0.85**	0.66

Table 2. Objective evaluation on summarization

	Pedestrian		Highway		Simulation		Edinburgh		Aircraft		CROSS		OMNI	
	FID	INT	FID	INT	FID	INT	FID	INT	FID	INT	FID	INT	FID	INT
ours	**0.59**	**323**	**0.58**	**147**	0.79	**159**	**0.45**	871	**0.88**	**101**	0.73	**87**	**0.63**	338
DBSCAN	0.53	434	0.56	181	**0.81**	191	0.38	920	0.61	155	0.69	109	0.55	395
k-means	0.56	451	0.51	252	0.75	171	0.43	**833**	0.68	116	**0.80**	101	0.61	**296**

The widely used *JSD* is investigated to do the resampling of the trajectories with varied lengths. The group-based denoising is iterated to obtain the multi-granularity condensed and summarized trajectory representations, which in the meantime may include some outliers of the data. We have also proposed two metrics to quantitatively evaluate the new framework for trajectory abstraction. A lot of experiments have clearly shown that the proposed technique is very helpful for understanding and utilizing the complex trajectory data in practice. To our knowledge this is the first deployment of the mighty group-based signal denoising technique for trajectory abstracting.

Several improvements for the proposed trajectory summarizing will be performed in our future work. In order to have some emphasis on the local features of long trajectories, a long trajectory would be segmented and, our proposed framework would be extended to deal with the trajectory segments. Furthermore, the powerful and general *visual analytics* technique [10] would be used to develop visual interactions to improve the trajectory abstraction.

Acknowledgments. This work has been funded by Natural Science Foundation of China (61471261, 61179067, U1333110), and by grants TIN2013-47276-C6-1-R from Spanish Government and 2014-SGR-1232 from Catalan Government (Spain).

References

1. http://homepages.inf.ed.ac.uk/rbf/forumtracking/
2. https://c3.nasa.gov/dashlink/resources/132/ (2011)
3. Anjum, N., Cavallaro, A.: Multifeature object trajectory clustering for video analysis. IEEE Trans. Circ. Syst. Video Technol. **18**(11), 1555–1564 (2008)
4. Ankerst, M., Breunig, M.M., Kriegel, H.P., Sander, J.: Optics: ordering points to identify the clustering structure. In: ACM Sigmod Record, vol. 28, pp. 49–60. ACM (1999)
5. Dabov, K., Foi, A., Katkovnik, V., Egiazarian, K.: Image denoising by sparse 3-d transform-domain collaborative filtering. IEEE Trans. Image Proces. **16**(8), 2080–2095 (2007)
6. Ester, M., Kriegel, H.P., Sander, J., Xu, X.: A density-based algorithm for discovering clusters in large spatial databases with noise. In: Kdd, vol. 96, pp. 226–231 (1996)
7. Johnstone, I.M., Silverman, B.W.: Wavelet threshold estimators for data with correlated noise. J. Royal Stat. Soc.: Ser. B (Stat. Methodol.) **59**(2), 319–351 (1997)
8. Laxhammar, R., Falkman, G.: Online learning and sequential anomaly detection in trajectories. IEEE Trans. Pattern Anal. Mach. Intell. **36**(6), 1158–1173 (2014)
9. Lloyd, S.: Least squares quantization in pcm. IEEE Trans. Inf. Theory **28**(2), 129–137 (1982)
10. May, R., Hanrahan, P., Keim, D.A., Shneiderman, B., Card, S.: The state of visual analytics: views on what visual analytics is and where it is going. In: 2010 IEEE Symposium on Visual Analytics Science and Technology (VAST), pp. 257–259. IEEE (2010)
11. Morris, B.T., Trivedi, M.M.: A survey of vision-based trajectory learning and analysis for surveillance. IEEE Trans. Circ. Syst. Video Technol. **18**(8), 1114–1127 (2008)
12. Morris, B.T., Trivedi, M.M.: Trajectory learning for activity understanding: unsupervised, multilevel, and long-term adaptive approach. IEEE Trans. Pattern Anal. Mach. Intell. **33**(11), 2287–2301 (2011)
13. Morris, B.T., Trivedi, M.M.: Understanding vehicular traffic behavior from video: a survey of unsupervised approaches. J. Electron. Imaging **22**(4), 041113 (2013)
14. Piciarelli, C., Micheloni, C., Foresti, G.L.: Trajectory-based anomalous event detection. IEEE Trans. Circ. Syst. Video Technol. **18**(11), 1544–1554 (2008)
15. Vlachos, M., Lin, J., Keogh, E., Gunopulos, D.: A wavelet-based anytime algorithm for k-means clustering of time series. In: Proceedings of the Workshop on Clustering High Dimensionality Data and Its Applications, Citeseer (2003)
16. Wang, W., Yang, J., Muntz, R., et al.: Sting: a statistical information grid approach to spatial data mining. VLDB **97**, 186–195 (1997)
17. Zhang, T., Ramakrishnan, R., Livny, M.: Birch: an efficient data clustering method for very large databases. In: ACM SIGMOD Record, vol. 25, pp. 103–114. ACM (1996)

Multi-scale Fractional-Order Sparse Representation for Image Denoising

Leilei Geng[1(✉)], Quansen Sun[1], Peng Fu[1], and Yunhao Yuan[2]

[1] Nanjing University of Science and Technology,
Nanjing 210094, China
leileigeng_njust@163.com, qssun@126.com,
njust_fupeng@sina.com
[2] Jiangnan University, Wuxi 214122, China
yyhzbh@163.com

Abstract. Sparse representation models code image patches as a linear combination of a few atoms selected from a given dictionary. Sparse representation-based image denoising (SRID) models, learning an adaptive dictionary directly from the noisy image itself, has shown promising results for image denoising. However, due to the noise of the observed image, these conventional models cannot obtain good estimations of sparse coefficients and the dictionary. To improve the performance of SRID models, we propose a multi-scale fractional-order sparse representation (MFSR) model for image denoising. Firstly, a novel sample space is re-estimated by respectively correcting singular values with the non-linear fractional-order technique in wavelet domain. Then, the denoised image can be reconstructed with the accurate sparse coefficients and optimal dictionary in the novel sample space. Compared with the conventional SRID models and other state-of-the-art image denoising algorithms, the experimental results show that the performances of our proposed MFSR model are much better in terms of the accuracy, efficiency and robustness.

Keywords: Multi-scale · Fractional-order · Sparse representation-based image denoising

1 Introduction

Sparse representation models are very effective to describe the inner-structure of the image [1]. In this model, the image x is consider as a linear combination by a small number of atoms from a given dictionary, $x \approx D\alpha$. The D is an over-complete dictionary, and most entries of the sparse coefficients α are zero or close to zero. The conventional SRID models are iterative algorithms, which alternate between sparse coding and dictionary updating [2]. In the sparse coding stage, the sparse coefficients of samples are estimated using Orthogonal Matching Pursuit (OMP) algorithm [3, 4]. In the dictionary updating stage, each dictionary atom is updated given the current sparse coefficients of samples [5]. In the past few years, much relevant works were invested for the theoretical analysis and practical use of the SRID models [6–9]. Elad [6] fused together the training and the denoising steps into one coherent and iterated process.

© Springer International Publishing Switzerland 2015
S. Arik et al. (Eds.): ICONIP 2015, Part III, LNCS 9491, pp. 462–470, 2015.
DOI: 10.1007/978-3-319-26555-1_52

Moreover, Dong [7] sought for good estimation of sparse coefficients by exploiting nonlocal self-similarity in the observed image.

As mentioned above, most conventional SRID models concentrated on learning the adaptive dictionary directly from the noisy image itself [6], instead of that from the high-quality image database [9, 10]. The main advantage of these SRID models is that the self-structure information of the noisy image is exploited and the trained dictionary is adaptive to the images of interest. Hence, these SRID models are more effective than those learning dictionary in an external training database. However, the training samples, constructed directly from the observed image, are unavoidable to be corrupted by the noise. In this way, the accurate sparse coefficients and optimal dictionary, used for reconstructing denoised image, cannot be obtained fast. Consequently, the performance and the efficiency of conventional SRID models should be improved.

In this paper, we propose a multi-scale fractional-order sparse representation (MFSR) model to obtain better estimation of both sparse coefficients and the dictionary. There are two main contributions in our work. One is to decompose the training sample space, directly constructed from the noisy image, into different frequency sub-spaces through the discrete Harr wavelet transform (DHWT). In wavelet sub-spaces, the extrinsic factor (noise) is isolated from the intrinsic factor (texture). The other one is to re-estimate the frequency sub-spaces by respectively correcting singular values with the non-linear fractional-order technique. In the new sub-spaces, the noise is suppressed with preserving image textures. In this way, the accurate sparse coefficients and optimal dictionary, used for reconstructing denoised image, can be obtained fast. Compared with the conventional SRID models [6] and the BM3D image denoising with shape-adaptive PCA (BM3D-SAPCA) algorithm [11], our proposed MFSR model performs much better in terms of the accuracy and efficiency.

2 Multi-scale Fractional-Order Sparse Representation (MFSR)

For a large image x, $x_{ij} = \mathbf{R}_{ij}x$ denotes the (i,j)-patch, where $\mathbf{R}_{ij}$ is an extracting matrix. Given a dictionary $\mathbf{D}_x$, each patch can be represented as $x_{ij} \approx \mathbf{D}_x\alpha_{x,ij}$, and the α_x and $\mathbf{D}_x$ can be obtained by Eq. (1). Then the large image x can be constructed by $x \approx \mathbf{D}_x\alpha_x$.

$$\left\{\alpha_{x,ij}, \mathbf{D}_x\right\} = \arg\min_{\alpha_{x,ij}, \mathbf{D}_x} \sum_{i,j} \left\|\mathbf{D}_x\alpha_{x,ij} - \mathbf{R}_{ij}x\right\|_2^2 + \sum_{i,j} \mu_{ij}\left\|\alpha_{x,ij}\right\|_1 \tag{1}$$

In the scenario of image denoising, the noisy image is modeled as $y = x+n$, where n is an additive Gaussian white noise with zero-mean and standard deviation σ. The conventional SRID models recover x from noisy image y by solving Eq. (2). Then the denoised image $\hat{x}$ is then reconstructed by Eq. (3).

$$\left\{\alpha_{y,ij}, \mathbf{D}_y\right\} = \arg\min_{\alpha_{y,ij}, \mathbf{D}_y} \lambda \left\|x - y\right\|_2^2 + \sum_{i,j} \mu_{ij}\left\|\alpha_{y,ij}\right\|_1 + \sum_{i,j} \left\|\mathbf{D}_y\alpha_{y,ij} - \mathbf{R}_{ij}x\right\|_2^2 \tag{2}$$

$$\hat{x} = \mathbf{D}_y \alpha_y = \left(\lambda \mathbf{I} + \sum_{i,j} \mathbf{R}_{ij}^T \mathbf{R}_{ij} \right)^{-1} \left(\lambda y + \sum_{i,j} \mathbf{R}_{ij}^T \mathbf{D}_y \alpha_{y,ij} \right) \tag{3}$$

2.1 Motivation

In the conventional SRID models, the dictionary $\mathbf{D}_y$ and the sparse coefficients α_y are considered as the estimation of the true ones. In order to improve the quality of the denoised image, the $\mathbf{D}_y$ and α_y are expected to be as close as possible to their true ones [7]. However, due to the noise of the observed image, the $\mathbf{D}_y$ and α_y will deviate from their true ones. In this paper, we define the deviated degrees of the dictionary $\mathbf{D}_y$ as Dev $(\mathbf{D}_x, \mathbf{D}_y)$, and define the deviated degree of the sparse coefficients α_y as $Dev(\alpha_x, \alpha_y)$:

$$Dev(\mathbf{D}_x, \mathbf{D}_y) = 1 - |\langle \mathbf{D}_x, \mathbf{D}_y \rangle| / (\|\mathbf{D}_x\| \cdot \|\mathbf{D}_y\|) \tag{4}$$

$$Dev(\alpha_x, \alpha_y) = \|\alpha_y - \alpha_x\|_F / \|\alpha_x\|_F \tag{5}$$

To show the deviated degrees, we conduct some experiments with the Lena image. We add Gaussian white noise to the original image x to get the noisy image y (the standard deviation $\sigma = 20$). The $Dev(\alpha_x, \alpha_y)$ is obtained using fixed DCT bases. In Fig. 1 and in Fig. 2, we plot the distribution of $Dev(\mathbf{D}_x, \mathbf{D}_y)$ and $Dev(\alpha_x, \alpha_y)$, respectively. It can be seen that the deviated degrees are high, and the accurate α_y and optimal $\mathbf{D}_y$ cannot be obtained from the noisy image by the conventional SRID models. This observation motivates us to seek for better estimation of the α_y and $\mathbf{D}_y$ to improve the quality of the denoised image $\hat{x}$ by Eq. (3).

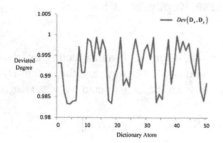
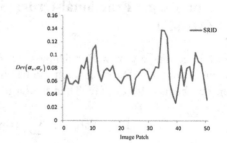

Fig. 1. Deviated degree of dictionary   **Fig. 2.** Deviated degree of sparse coefficients

2.2 Multi-scale Re-Estimation of the Novel Sample Space

The definition of $Dev(\alpha_x, \alpha_y)$ and $Dev(\mathbf{D}_x, \mathbf{D}_y)$ indicates that we could further improve the quality of the denoised image by improving the accuracy of the α_y and $\mathbf{D}_y$. However, it is difficult to directly improve the accuracy of the α_y and $\mathbf{D}_y$. To deal with the weakness, we re-estimate a novel sample space by respectively correcting singular values with the non-linear fractional-order technique in wavelet domain. The singular values are obtained by the singular value decomposition (SVD). They can be inter-preted as the nonlocal variance estimation and exposes the geometric structure of an image. Hence, the singular values represent the intrinsic and algebraic image proper-ties. However, due to the noise, the singular values deviate from their true ones. We define the deviated degree of the singular values as:

$$Dev\left(\mathbf{S}_x, \mathbf{S}_y\right) = \left\|\mathbf{S}_y - \mathbf{S}_x\right\|_F \Big/ \left\|\mathbf{S}_x\right\|_F \tag{6}$$

In order to suppress $Dev(\mathbf{S}_x, \mathbf{S}_y)$, we employ the idea of the non-linear fractional-order technique to respectively correct the singular values. In recent years, fractional-order techniques have been used in multi-scale texture enhancement [12], image registration [13], and face representation [14]. This method has been demon-strated to have the better capability of non-linearly enhancing complex texture details than traditional integral-based algorithms.

Therefore, we propose a novel sample space though multi-scale fractional-order estimation, as defined in Eq. (7), including three stages: (1) The noisy sample space is decomposed into different frequency sub-spaces by the DHWT with 2 levels. In the frequency sub-spaces, the extrinsic factor (noise) is isolated from the intrinsic factor (texture). (2) The novel sub-spaces are estimated in the Harr wavelet domain. The singular values of each sub-space are respectively corrected though the non-linear fractional-order technique. In this way, the extrinsic disturbance (noise) is restrained and complex texture details avoid to be interrupted. (3) The novel sample space is constructed though the inverse DHWT.

$$\mathbf{x}^H = \sum_{i=1}\sum_{j=1}\mathbf{x}_{ij}^H = \sum_{i=1}\sum_{j=1}\mathbf{u}_{ij}\Lambda_{ij}^H\mathbf{v}_{ij}^T \quad \Lambda_{ij}^H = diag\left(s_1^{\eta_1}, s_2^{\eta_2}, \cdots, s_r^{\eta_r}\right) \tag{7}$$

where $\mathbf{x}_{ij}^H$ is the sample in novel sample space $\mathbf{x}^H$; $\mathbf{u}_{ij}$ and $\mathbf{v}_{ij}^T$ are respectively the left and the right singular vector matrices of $\mathbf{x}_{ij}$; Λ_{ij} is the singular value matrices; $s_1 \geq s_2 \geq \cdots \geq s_r$ are the descending order singular values; r is equal to the rank of the matrix $\mathbf{x}_{ij}$; the fractional-order vector $\mathbf{\eta} = (\eta_1, \cdots, \eta_r)$ is the parameter of our proposed MFSR model. Each fractional-order η_i is determined by a series of experiments on synthetic data (images), as defined in Eq. (8):

$$\eta_i = \arg\min_{\eta_i} Dev\left(\mathbf{S}_{x,i}, \mathbf{S}_{y,i}\right) = \arg\min_{\eta_i} \left\|\left(\mathbf{S}_{y,i}\right)^{\eta_i} - \mathbf{S}_{x,i}\right\|_F \Big/ \left\|\mathbf{S}_{x,i}\right\|_F \tag{8}$$

where the $\mathbf{S}_{x,i}$ and $\mathbf{S}_{y,i}$ are the ith dimension of singular value vectors composed from original samples (image patches) and noisy samples, respectively.

2.3   Modeling of MFSR

Our proposed MFSR model is an iterative algorithm and each iteration includes four stages: (1) estimate a novel sample space in wavelet domain as described in Sect. 2.2;

(2) gain the sparse coefficients α_y^H of the samples using OMP algorithm; (3) update the dictionary $\mathbf{D}_y^H$; (4) calculate the denoised image. Therefore, our proposed MFSR model can be defined as the outcome of a minimization problem:

$$\left\{\alpha_{y,ij}^H, \mathbf{D}_y^H, \hat{x}\right\} = \underset{\alpha_{y,ij}, \mathbf{D}_y}{\arg\min} \lambda \left\|x^H - y\right\|_2^2 + \sum_{i,j} \mu_{ij} \left\|\alpha_{y,ij}^H\right\|_1 + \sum_{i,j} \left\|\mathbf{D}_y^H \alpha_{y,ij}^H - \mathbf{R}_{ij} x^H\right\|_2^2 \quad (9)$$

where $\alpha_{y,ij}^H$ represents the sparse coefficients for the novel (i, j)-patch x_{ij}^H based on the dictionary $\mathbf{D}_y^H$. Our proposed MFSR model is presented as following.

Algorithm 1. Our proposed MFSR model

Task: Reconstruct an noisy image y

Parameters: n-block size, k-dictionary size, J-number of iterations

Initialization: Set $\hat{x}^{(0)} = y$, $\mathbf{D}_y^{H(0)}$ =DCT dictionary

Repeat J times:
1. Multi-scale re-estimation of the sample space:
    (1) Construct the frequency sub-spaces using DWT
    (2) Estimate the new sub-spaces by Eq. (7)
    (3) Construct the novel space using inverse DWT
2. Sparse Coding: Calculate $\alpha_{y,ij}^H$ for each patch x_{ij}^H
3. Dictionary Updating: Update atom d_l, $l=1,2,...,k$ by
    (1) Find patches where $\alpha_{y,ij}^H$ is equal nonzero and calculate error matrix
    (2) Update the dictionary atom and the coefficient values
4. Calculate the mth denoised image by Eq. (3)

---

3 Experimental Results

In this section, we verify the performance of our proposed MFSR model for image denoising. The basic parameters of our proposed MFSR model are empirically set as follows: (1) The image patch size is 8×8; (2) The dictionary size is 64×256; (3) The Lagrange multiplier is $\lambda = 30/\sigma$; (4) The parameter $c = 1.15$ is the stopping criteria for the sparse coding stage. The set of original images include 5 typical images and each noisy image is tested for 10 times. We compare our proposed MFSR model with two recently developed state-of-the-art image denoising algorithms, including the conventional SRID models [6], and the BM3D-SAPCA algorithm [11].

3.1 Necessity of Multi-scale Decomposition

When the singular values are corrected directly in the original sample space, the unexpected perturbations are involved. As illustrated in Fig. 3, the noisy image is added by white Gaussian noise with zero-mean and the standard deviation $\sigma = 50$. And the right image is reconstructed directly using the corresponding singular values obtained from the original image. As shown in the reconstructed image, the noise is suppressed, but other new disturbances are involved. To avoid involving the unexpected perturbations, the novel sample space should be re-estimated in wavelet domain.

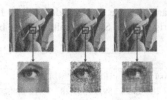

Fig. 3. Comparison of experimental results: clean image (left), noisy image (middle), reconstructed image (right).

We compare our proposed MFSR model with two models, including the conventional SRID model [6] and the fractional-order sparse representation (FSR) model in original sample space. To verify the necessity of the multi-scale decomposition, we evaluate the quality of denoised images using the popular Peak Signal to Noise Ratio (PSNR). The PSNR results of noisy images with various noise standard deviations are calculated, as reported in Table 1. It can be seen that our proposed MFSR model achieves highly competitive denoising performance. In term of the average and the standard deviation (*std*) PSNR results, there is little difference of denoising performance between the FSR model and the conventional SRID model. Therefore, it can be concluded that it is necessary to correct singular values with the non-linear fractional-order technique in wavelet domain.

Table 1. PSNR results with various noise standard deviations. In each cell, the top left is the FSR result; the top right is the SRID result; and the bottom is our proposed MFSR result.

σ	Iteration										Iteration			
	Lena		Barbara		Boat		House		Pepper		*average*		*std*	
5	38.33	38.04	38.10	38.10	36.70	35.75	39.33	39.48	36.88	38.29	37.87	37.69	1.09	1.42
	39.53		**39.45**		**38.39**		**40.59**		**38.79**		**39.35**		**0.84**	
10	35.35	35.40	34.26	34.26	33.53	33.21	35.70	36.08	34.60	35.21	34.70	34.79	0.89	1.15
	36.90		**35.77**		**34.66**		**37.15**		**35.46**		**35.99**		**1.03**	
20	32.27	32.15	30.45	30.45	30.26	30.29	33.14	30.41	32.24	32.34	31.67	31.69	1.26	1.21
	34.01		**32.16**		**31.41**		**34.35**		**32.64**		**32.91**		**1.24**	
35	29.52	29.54	26.31	26.31	27.61	27.59	30.37	28.13	29.88	29.62	28.74	28.92	1.72	1.33
	30.89		**30.24**		**28.72**		**31.83**		**30.37**		**30.41**		**1.13**	
50	27.68	27.88	23.98	23.98	25.84	25.87	27.82	33.12	27.99	27.83	26.66	27.00	1.73	1.32
	29.54		**27.57**		**27.14**		**29.83**		**28.30**		**28.48**		**1.18**	

3.2 Effectiveness of Denoising

We compare our proposed MFSR model with two recently developed state-of-the-art image denoising algorithms, including the conventional SRID models [6] and the BM3D-SAPCA algorithm [11]. The PSNR results of the three algorithms are reported in Table 2. From Table 2, we can conclude that our proposed MFSR model performs

Table 2. PSNR results with various noise standard deviations. In each cell, the top left is the BM3D-SAPCA result; the top right is the SRID result; and the bottom is our proposed MFSR result.

σ	Iteration										Iteration			
	Lena		Barbara		Boat		House		Pepper		*average*		*std*	
5	38.72	38.04	38.31	36.91	37.28	35.75	39.83	39.48	38.12	38.29	38.45	37.69	0.93	1.42
	39.53		**39.45**		**38.39**		**40.59**		**38.79**		**39.35**		**0.84**	
10	35.93	35.40	34.98	34.05	33.92	33.21	36.71	36.08	34.68	35.21	35.24	34.79	1.09	1.15
	36.90		**35.77**		**34.66**		**37.15**		**35.46**		**35.99**		**1.03**	
20	33.05	32.15	33.05	30.57	33.05	30.29	33.05	33.12	33.05	32.34	33.05	31.69	33.05	1.21
	34.01		**32.16**		**31.41**		**34.35**		**32.64**		**32.91**		**1.24**	
35	30.56	29.54	28.98	27.43	28.43	27.59	31.38	30.41	28.52	29.62	29.57	28.92	1.32	1.33
	30.89		**30.24**		**28.72**		**31.83**		**30.37**		**30.41**		**1.13**	
50	29.05	27.88	27.23	25.27	26.78	25.87	29.69	28.13	26.68	27.83	27.89	27.00	1.39	1.32
	29.54		**27.57**		**27.14**		**29.83**		**28.30**		**28.48**		**1.18**	

much better than other algorithms. In term of the *std* results, our proposed MFSR model presents the best robustness for simple textures and complex textures.

3.3   Computational Efficiency

As the conventional SRID models and our proposed MFSR model are the iterative algorithms, we evaluate the computational efficiency using the iterations. The stop-ping criterion is that the change of the PSNR continues being less than 0.001 for three times. The iteration results are reported in Table 3. In order to better visualize the results, the Fig. 4 presents the average of iterations (as drawn by the full lines) and the standard deviation of iterations (as described by the error bars). It can be seen that our proposed MFSR model can fast converge with half iterations of the conventional SRID models. In term of the std, our proposed MFSR model is much more stable than the conventional SRID models. Therefore, our proposed MFSR model exhibits the higher computational efficiency and better robustness than the SRID models.

Table 3. Iteration results with various noise standard deviations. In each cell, the left is the conventional SRID result, the right is our proposed MFSR result.

σ	Iteration										Iteration			
	Lena		Barbara		Boat		House		Pepper		*average*		*std*	
5	87.20	**23.40**	71.20	**23.20**	92.40	**23.20**	98.20	**24.20**	55.00	**23.20**	80.80	**23.44**	17.58	**0.43**
10	54.80	**19.20**	56.00	**21.20**	72.00	**20.40**	75.00	**21.40**	47.20	**21.40**	61.00	**20.72**	11.95	**0.94**
20	39.00	**12.00**	31.60	**12.20**	38.20	**14.20**	38.20	**12.20**	29.00	**13.20**	35.20	**12.76**	4.58	**0.93**
35	21.20	**10.40**	18.80	**10.40**	23.20	**11.40**	17.00	**11.00**	17.80	**11.00**	19.60	**10.84**	2.56	**0.43**
50	16.00	**8.20**	13.00	**7.00**	16.00	**8.20**	16.00	**8.40**	13.00	**8.40**	14.80	**8.04**	1.64	**0.59**

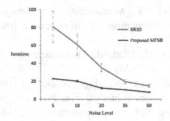

Fig. 4. The iteration results with various noise standard deviations

4 Conclusion

In this paper, we have presented a novel MFSR model for image denoising. Due to the noise, the accurate sparse coefficients and optimal dictionary, directly constructed from the noisy sample space, cannot be obtained fast. To deal with this weakness, we re-estimate a novel sample space by respectively correcting singular values with the non-linear fractional-order technique in wavelet domain. And the fractional-order techniques are introduced into the domain of sparsity-based image denoising for the first time. Compared with the BM3D-SAPCA algorithm and the conventional SRID models, our proposed MFSR model can achieve highly competitive performance in terms of the accuracy, efficiency and robustness.

Acknowledgements. This work was supported in part by the National Nature Science Foundation of China under Grant No. 61273251, No. 61402203 and No. 61401209

References

1. Bruckstein, A.M., Donoho, D.L., Elad, M.: From sparse solutions of systems of equations to sparse modeling of signals and images. SIAM Rev. **51**(1), 34–81 (2009)
2. Aharon, M., Bruckstein, A.M.: K-SVD: An algorithm for designing over complete dictionaries for sparse representation. Trans. Sig. Process. **54**(11), 4311–4322 (2006)
3. Rubinstein, R., Peleg, T., Elad, M.: Analysis K-SVD: a dictionary-learning algorithm for the analysis sparse model. Trans. Sig. Process. **61**(3), 661–677 (2013)
4. Yang, J.C., Wang, Z.W., Lin, Z.: Coupled dictionary training for image super resolution. Trans. Image Process. **21**(8), 3467–3478 (2012)
5. Elad, M., Aharon, M.: Image denoising via learned dictionaries and sparse representation. In: 22th IEEE International Conference on Computer Vision and Pattern Recognition, pp. 895–900. IEEE Press, New York (2006)
6. Romano, Y. Elad. M.: Improving K-SVD denoising by post-processing its method-noise. In: 20th IEEE International Conference on Image Processing, pp. 435–439. IEEE Press, Melbourne (2013)
7. Dong, W.S., Zhang, L., Shi, G.M., Li, X.: Nonlocally centralized sparse representation for image restoration. Transactions on Image Processing **22**(4), 1620–1630 (2013)
8. Sulam, J., Ophir, B., Elad, M.: Image denoising though multi-scale dictionary learning. In: 21th IEEE International Conference on Image Processing, pp. 808–812. IEEE Press, Pairs (2014)

9. Mairal, J., Sapiro, G., Elad, M.: Multi-scale sparse image representation with learned dictionaries. In: 13th IEEE International Conference on Image Processing, pp. 105–108. IEEE Press, Atlanta (2006)
10. Rubinstein, R., Bruckstein, A.M., Elad, M.: Dictionaries for sparse representation modeling. Process. IEEE **98**(6), 1045–1057 (2010)
11. Dabov, K., Foi, A., Katkovnik, V., Egiazarian, K.: BM3D image denoising with shape-adaptive principal component analysis. In: Proceedings of the 2th Workshop Signal Process with Adaptive Sparse Struct Representations, pp. 1–6. Springer, Saint-Malo (2009)
12. Pu, Y.F., Zhou, J.L., Yuan, X.: Fractional differential mask: a fractional differential-based approach for multi-scale texture enhancement. Trans. Image Process. **19**(2), 491–511 (2010)
13. Pan, W., Qin, K., Chen, Y.: An adaptable-multilayer fractional Fourier transform approach for image registration. Trans. Pattern Anal. Mach. Intell. **31**(3), 400–413 (2009)
14. Yuan, Y.H., Sun, Q.S., Ge, H.W.: Fractional-order embedding canonical correlation analysis and its applications to multi-view dimensionality reduction and recognition. Pattern Recogn. **47**(3), 1411–1424 (2014)

Linear Hyperbolic Diffusion-Based Image Denoising Technique

Tudor Barbu[✉]

Institute of Computer Science of the Romanian Academy, Iaşi, Romania
`tudor.barbu@iit.academiaromana-is.ro`

Abstract. A novel PDE-based image restoration approach is proposed in this article. The provided PDE model is based on a linear second-order hyperbolic diffusion equation. The well-posedness of the proposed differential model and some nonlinear PDE schemes derived from it are also discussed. A consistent and fast-converging numerical approximation scheme using finite differences is then constructed for the continuous hyperbolic PDE model. Some image restoration experiments using this approach and several method comparisons are also described.

Keywords: Image denoising · Linear second-order PDE model · Hyperbolic diffusion equation · Edge-preservation · Finite-difference based discretization scheme

1 Introduction

The partial differential equation (PDE) - based techniques have been increasingly used in the image processing and analysis fields in the last 25 years, since the conventional image processing approaches have numerous drawbacks and cannot solve a lot of issues related to this domain [1]. One of these issues is the preservation of details (edges and other structures) during the image denoising process.

The nonlinear PDE-based restoration models, such as Perona-Malik anisotropic diffusion scheme [2], TV Denoising [3] and other diffusion-based and variational approaches derived from these influential algorithms [4], clearly outperform the classic 2D image filters [1], removing the blurring effect and preserving the essential image features. Unfortunately, the second-order nonlinear diffusion based methods often generate the undesired staircasing, or blocky, effect.

Numerous improved second-order nonlinear PDE techniques trying to alleviate this staircase effect, such as Adaptive TV denoising [5], anisotropic HDTV regularizer [6], TV-L1 model [7] and TV minimization with Split Bregman [8], have been developed in recent years. We have also proposed many nonlinear anisotropic diffusion and PDE variational image restoration models that remove successfully the Gausian noise, preserve the image features and reduce the blocky effect, in this period [9–14]. Although

© Springer International Publishing Switzerland 2015
S. Arik et al. (Eds.): ICONIP 2015, Part III, LNCS 9491, pp. 471–478, 2015.
DOI: 10.1007/978-3-319-26555-1_53

these improved nonlinear PDE schemes reduce that effect, they cannot remove it completely. They may also have other disadvantages, such as the high computational cost and execution time. So, we have decided to take another look at the linear PDE-based denoising solutions in this paper.

The linear diffusion models have long been considered the simplest PDE–based image denoising solutions, since 2D Gaussian filtering process is equivalent to the linear diffusion given by the heat equation [4]. The main drawbacks of the most linear PDE restoration models are the image blurring effect, which affects edges and other details, and the absence of the localization property. Also, they may dislocate the boundaries when moving from finer to coarser scales [15].

So, we consider here an improved linear PDE-based image restoration technique that overcome these drawbacks and also executes fast and avoid completely the staircasing effect. The proposed image denoising approach is based on a second-order hyperbolic diffusion equation. The linear PDE hyperbolic model described in the next section provides effective image restoration results, reduces the blurring effect and has the localization property [15]. That means the solution of the second-order hyperbolic equation propagates with finite speed. Another reason we consider a linear PDE for image noise removal is that our proposed hyperbolic scheme may lead to some very performant nonlinear PDE denoising models.

A robust and consistent numerical approximation scheme is constructed for the continuous smoothing model and described in the third section. The obtained iterative discretization algorithm is based on finite differences and converges fast to that solution, representing the denoised image. The performed noise removal experiments and method comparison that are discussed in the fifth section prove the effectiveness of the proposed PDE-based method and its advantages comparing to other restoration schemes. Our article finalizes with a section of conclusions and a list of bibliographical references.

2   Novel Linear Second-Order PDE-Based Restoration Model

We consider a robust diffusion-based image restoration model. It is composed of a linear second-order hyperbolic partial differential equation and a set of boundary conditions. Thus, we have:

$$
\begin{cases}
\lambda \frac{\partial^2 u}{\partial t^2} + \gamma^2 \frac{\partial u}{\partial t} - \alpha \nabla^2 u + \zeta \left(u - u_0 \right) = 0 \\
u\left(0, x, y\right) = u_0\left(x, y\right) \\
\frac{\partial u}{\partial t}\left(0, x, y\right) = u_1\left(x, y\right) \\
u\left(t, x, y\right) = 0, \ \forall t \geq 0, \ \left(x, y\right) \in \partial \Omega
\end{cases}
, \left(x, y\right) \in \Omega \qquad (1)
$$

where the domain $\Omega \subseteq (0, \infty) \times R^2$, $\lambda, \gamma, \alpha \in (0, 3]$, $\zeta \in (0, 0.5]$ and u_0 constitutes the initial image that is affected by Gaussian noise.

The 2nd – order hyperbolic PDE framework given by (1) represents a non-Fourier model for the heat propagation. This model is well-posed, admitting a unique and weak solution u, which propagates with finite speed [16]. Therefore, our well-posed PDE model (1) has also the localization property [15]. So, while the Fourier model of heat propagation, represented as $\frac{\partial u}{\partial t} = \Delta u$, where $(x, y) \in R^2$, $t > 0$ and $u(0, x, y) = u_0(x, y)$, has no localization property, its solution propagating with infinite speed, the non-Fourier heat propagating scheme proposed here behaves considerably better in this aspect.

This unique solution of model (1) will be approximated numerically by applying a consistent finite-difference based discretization scheme that is described in the next section. Our iterative numerical approximation algorithm will converges fast to the PDE solution. The component $\zeta(u - u_0)$ from the hyperbolic equation has been introduced to stabilize this optimal denoising solution, preventing the further degradation of the smoothed image.

The proposed linear PDE-based denoising model is important not only because it executes successfully, produce satisfactory image noise reduction results and outperform other filtering approaches, but also because some more effective nonlinear PDE-based image enhancement schemes can be derived from it. So, we will investigate both second-order and fourth-order nonlinear versions of our linear hyperbolic diffusion based technique. Second-order nonlinear hyperbolic PDE models that may provide better deblurring results could be achieved by transforming the equation from (1) into:

$$\lambda \frac{\partial^2 u}{\partial t^2} + \gamma^2 \frac{\partial u}{\partial t} - \alpha(\Delta u)\nabla^2 u + \zeta(u - u_0) = 0 \tag{2}$$

where α becomes a function of current image's Laplacian $(\Delta u = \nabla^2 u)$, or into the following nonlinear anisotropic diffusion:

$$\lambda \frac{\partial^2 u}{\partial t^2} + \gamma^2 \frac{\partial u}{\partial t} - div(\alpha(|\nabla u|)\nabla u) + \zeta(u - u_0) = 0 \tag{3}$$

where parameter α becomes a function of the evolving image's gradient. Equation (2) may be further transformed into the following nonlinear fourth-order hyperbolic PDE:

$$\lambda \frac{\partial^2 u}{\partial t^2} + \gamma^2 \frac{\partial u}{\partial t} - \nabla^2(\alpha(\Delta u)\nabla^2 u) + \zeta(u - u_0) = 0 \tag{4}$$

A 4th - order hyperbolic diffusion model based on (4) may overcome more successfully the staircasing effect during the denoising process. The noise removal techniques based on (2) – (4) will be developed as part of our future research.

3   Finite-Difference Based Numerical Approximation Scheme

A consistent numerical approximation scheme is proposed for the continuous PDE model (1). The discretization of our hyperbolic model is based on the finite-difference method [17].

So, we consider a space grid size of h and a time step Δt. The space and time coordinates are quantized as following:

$$x = ih, y = jh, t = n\Delta t, \forall i \in \{0, 1, \dots, I\}, j \in \{0, 1, \dots, J\}, n \in \{0, 1, \dots, N\} \tag{5}$$

The hyperbolic diffusion equation from (1) is equivalent to the following second-order PDE:

$$\lambda \frac{\partial^2 u}{\partial t^2} + \gamma^2 \frac{\partial u}{\partial t} - \alpha \left(\frac{\partial^2 u}{\partial x^2} + \frac{\partial^2 u}{\partial y^2} \right) + \zeta \left(u - u_0 \right) = 0 \tag{6}$$

that will be discretized, by using the finite differences [17], as following:

$$\begin{aligned} &\lambda \frac{u^{n+\Delta t}(i,j) + u^{n-\Delta t}(i,j) - 2u^n(i,j)}{\Delta t^2} + \gamma^2 \frac{u^{n+\Delta t}(i,j) - u^{n-\Delta t}(i,j)}{2\Delta t} \\ &-\alpha \frac{u^n(i+h,j) + u^n(i-h,j) + u^n(i,j+h) + u^n(i,j-h) - 4u^n(i,j)}{h^2} \\ &+\zeta \left(u^n(i,j) - u^0(i,j) \right) = 0 \end{aligned} \tag{7}$$

If one considers $h = 1$ and $\Delta t = 1$, then (7) leads to the next explicit numerical approximation scheme of the PDE model (1):

$$\begin{aligned} u^{n+1}(i,j) &= \frac{2\zeta - 4\lambda}{2\lambda + \gamma^2} u^n(i,j) + \frac{\gamma^2 - 2\lambda}{2\lambda + \gamma^2} u^{n-1}(i,j) + 2\zeta u^0(i,j) \\ &+ 2\alpha \left(u^n(i+1,j) + u^n(i-1,j) + u^n(i,j+1) + u^n(i,j-1) - 4u^n(i,j) \right) \end{aligned} \tag{8}$$

where $u^0(i,j) = u_0(i,j)$ and $n > 0$.

The developed iterative restoration algorithm receives the initial $[I \times J]$ image affected by noise and applies repeatedly the procedure provided by (8), for each $n = 1$, $2, \ldots, N$. The number of iterations of our numerical approximating scheme, N, is quite low, since this discretization algorithm converges very fast to the solution representing the optimal image restoration. That means the approximation scheme (8) is also consistent to the PDE model (1).

4   Experiments and Method Comparison

We have successfully tested the proposed linear hyperbolic PDE-based denoising technique, by applying our iterative scheme (8) on hundreds of images affected by various levels of Gaussian noise.

The described image restoration model achieves satisfactory smoothing results while preserving the edges and other details quite well, although the blurring is not completely avoided. As one can see in the next figure, our approach overcomes successfully other unintended effects, such as staircasing [18] and speckle. We have identified the following set of parameters of this PDE model that provide the optimal denoising results:

$$\lambda = 2.4, \gamma = 1.5, \alpha = 1.8, \zeta = 0.15, \Delta t = 1, h = 1, N = 12 \tag{9}$$

Therefore, the optimal smoothing is achieved after a low number of iterations, $N = 12$. That means our restoration technique executes very fast, its running time being less than 1 s.

From the performed method comparison we have found that our linear hyperbolic diffusion technique outperforms not only the two-dimensional conventional filters, but

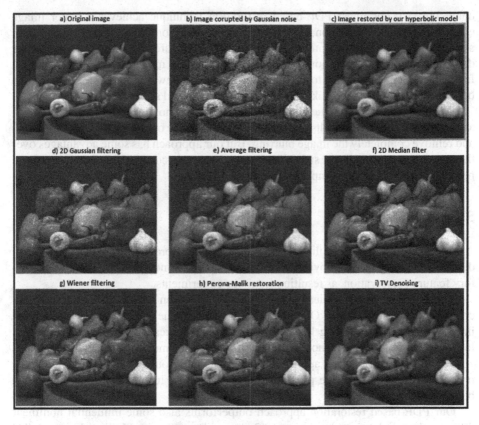

Fig. 1. Method comparison: image restored by various denoising techniques

also many linear and nonlinear PDE-based denoising methods. The restoration perform-ance of the proposed hyperbolic approach is assesed by using the Peak Signal-to-Noise Ratio (PSNR) measure [1]. Our denoising scheme obtains higher PSNR values than well-known 2D classic filters, such as Gaussian, Average, Median and Wiener [1], and even influential nonlinear diffusion-based models, like Perona-Malik scheme [2] and TV Denoising [3], as one can see in Table 1.

Table 1. PSNR values for several noise removal techniques

This filter	Gaussian	Average	Median	Wiener	P-M	TV
27.11(dB)	22.31(dB)	23.28(dB)	24.18(dB)	24.5(dB)	25.79(dB)	24.27(dB)

The smoothing results achieved by these methods are displayed in Fig. 1, whose images represent: (a) the original [512 × 512] *Peppers* image; (b) that image corrupted by Gaussian noise characterized by $\mu = 0.21$ and $var = 0.023$; (c) the image restored by our hyperbolic scheme; (d) – (g) denoising results obtained by the [3 × 3] 2D filter kernels (Gaussian, average, median and Wiener); (h) Perona-Malik smoothing; TV denoising. So, our linear PDE-based restoration technique achieves a better image enhancement than conventional and classic linear diffusion filters, removing a greater amount of Gaussian noise and reducing the blurring effect. It also provides better denoising results than some nonlinear PDE and variational models [2–4], executes much faster than them (converging to the optimal denoising in fewer iterations) and overcome better the undesired effects, like the blocky effect [18] and speckling.

We have found that some improved second-order nonlinear PDE models [5–8] achieve better denoising results (higher PSNR values) than our linear hyperbolic method and remove completely the image blurring. But our approach has some advantages over them, too. It avoids totally the staircase effect and executes considerably faster than those methods, given its lower computational cost.

5   Conclusions

An effective linear hyperbolic PDE-based image restoration framework has been described in this paper. It provides a successful Gaussian noise removal and a satisfactory feature preservation, as resulting from our experiments.

The proposed second-order linear hyperbolic diffusion model represents the main contribution of this article. Our novel denoising scheme represents a considerable improvement of the existing linear diffusion models, because, unlike them, it has a localization property and do not dislocate the edges when moving to coarser scales. The conventional image filtering approaches are also clearly outperformed by it, since the proposed restoration scheme provides better smoothing results and overcome much better the blurring effect.

Our PDE-based restoration approach outperforms also some influential nonlinear diffusion-based denoising techniques [2–4], such as Perona-Malik scheme or TV Denoising. Although some nonlinear PDE denoising models, like several improved versions of TV denoising [5–7] outperform our hyperbolic model in some ways, getting higher PSNR values and providing better deblurring results, unlike our restoration scheme, they may still suffer from staircasing effect [18] and generate speckle noise. Also, the nonlinear anisotropic diffusion based techniques have a much higher computational cost and consequently execute slower than our denoising approach that provides a fast image smoothing.

The low running time of our restoration scheme is due to the proposed fast-converging finite-difference based numerical approximation algorithm, which is consistent to the continuous PDE-based model and represents another contribution of this paper. A low execution time could be an important advantage of a denoising method, allowing it to be used for restoration of images from large databases or as a restoration component of an image analysis system.

Also, it is important that the drawbacks of the proposed linear PDE model can be corrected by improving our hyperbolic filtering scheme, such that to become a nonlinear diffusion-based smoothing approach. The second and fourth order nonlinear PDE models that could be derived from it, discussed in the second section, may represent much better smoothing solutions that remove all the unintended effects and outperform the state-of-the-art nonlinear diffusion techniques. The considered nonlinear hyperbolic models will be developed as part of our future research in the PDE-based image restoration domain.

Acknowledgments. The research of this work was mainly supported by the project PN-II-ID-PCE-2011-3-0027-160/5.10.2011, financed by UEFSCDI Romania. It was supported also by the Institute of Computer Science of the Romanian Academy, Iaşi, Romania.

References

1. Jain, A.K.: Fundamentals of Digital Image Processing. Prentice Hall, Upper Saddle River (1989)
2. Perona, P., Malik, J.: Scale-space and edge detection using anisotropic diffusion. In: Proceedings of IEEE Computer Society Workshop on Computer Vision, pp. 16–22, November 1987
3. Rudin, L., Osher, S., Fatemi, E.: Nonlinear total variation based noise removal algorithms. Physica D: Nonlinear Phenom. **60**(1), 259–268 (1992)
4. Weickert, J.: Anisotropic diffusion in image processing. European Consortium for Mathematics in Industry. B. G. Teubner, Stuttgart (1998)
5. Chen, Q., Montesinos, P., Sun, Q., Heng, P., Xia, D.: Adaptive total variation denoising based on difference curvature. Image Vis. Comput. **28**(3), 298–306 (2010)
6. Hu, Y., Jacob, M.: Higher degree total variation (HDTV) regularization for image recovery. IEEE Trans. Image Process. **21**(5), 2559–2571 (2012)
7. Micchelli, C.A., Shen, L., Xu, Y., Zeng, X.: Proximity algorithms for image models II: L1/TV denosing. Advances in Computational Mathematics, online version available (2011)
8. Cai, J.F., Osher, S., Shen, Z.: Split Bregman methods and frame based image restoration. Multiscale Model. Sim. **8**(2), 337–369 (2009)
9. Barbu, T., Barbu, V., Biga, V., Coca, D.: A PDE variational approach to image denoising and restorations. Nonlinear Anal. RWA **10**, 1351–1361 (2009)
10. Barbu, T.: Variational image denoising approach with diffusion porous media flow. Abstract and Applied Analysis, vol. 2013, Article ID 856876, 8 p. Hindawi Publishing Corporation (2013)
11. Barbu, T.: A Novel Variational PDE Technique for Image Denoising. In: Lee, M., Hirose, A., Hou, Z.-G., Kil, R.M. (eds.) ICONIP 2013, Part III. LNCS, vol. 8228, pp. 501–508. Springer, Heidelberg (2013)
12. Barbu, T.: Robust anisotropic diffusion scheme for image noise removal. In: Proceedings of 18th International Conference in Knowledge Based and Intelligent Information & Engineering Systems, KES 2014, September 15–17, Gdynia, Poland, Procedia Computer Science, vol. 35, pp. 522–530. Elsevier (2014)
13. Barbu, T.: Nonlinear diffusion-based image restoration model. In: Proceedings of the 5th International Conference on Circuits, Systems, Control, Signals, CSCS 2014, Salerno, Italy, June 3–5, Latest Trends in Circuits, Systems, Signal Processing and Automatic Control, pp. 122–125 (2014)
14. Barbu, T., Favini, A.: Rigorous mathematical investigation of a nonlinear anisotropic diffusion-based image restoration model. Electron. J. Differ. Equ. **129**, 1–9 (2014)
15. Witkin, A. P.: Scale-space filtering. In: Proceedings of the Eighth International Joint Conference on Artificial Intelligence (IJCAI 1983, Karlsruhe, Aug. 8–12), vol. 2, pp. 1019–1022 (1983)

16. Barbu, V.: Nonlinear Semigroups and Differential Equations in Banach Spaces. Noordhoff International Publishing, Leyden (1976)
17. Johnson, P.: Finite Difference for PDEs. School of Mathematics, University of Manchester, Semester I, (2008)
18. Buades, A., Coll, B., Morel, J.-M.: The staircasing effect in neighborhood filters and its solution. IEEE Trans. Image Process. **15**(6), 1499–1505 (2006)

Noise on Gradient Systems with Forgetting

Chang Su[1], John Sum[1(✉)], Chi-Sing Leung[2], and Kevin I.-J. Ho[3]

[1] ITM, National Chung Hsing University, Taichung, Taiwan
pfsum@nchu.edu.tw
[2] EE, City University of Hong Kong, Kowloon Tong, Hong Kong
eeleungc@cityu.edu.hk
[3] CSCE, Providence University, Taichung, Taiwan
ho@pu.edu.tw

Abstract. In this paper, we study the effect of noise on a gradient system with forgetting. The noise include multiplicative noise, additive noise and chaotic noise. For multiplicative or additive noise, the noise is a mean zero Gaussian noise. It is added to the state vector of the system. For chaotic noise, it is added to the gradient vector. Let $\mathbf{x}$ be the state vector of a system, S_b be the variance of the Gaussian noise, κ' is average noise level of the chaotic noise, λ is a positive constant, $V(\mathbf{x})$ be the energy function of the original gradient system, $V_\otimes(\mathbf{x})$, $V_\oplus(\mathbf{x})$ and $V_\odot(\mathbf{x})$ be the energy functions of the gradient systems, if multiplicative, additive and chaotic noises are introduced. Suppose $V(\mathbf{x}) = F(\mathbf{x}) + \lambda\|\mathbf{x}\|_2^2$. It is shown that $V_\otimes(\mathbf{x}) = V(\mathbf{x}) + (S_b/2)\sum_{j=1}^n (\partial^2 F(\mathbf{x})/\partial x_j^2)x_j^2 - S_b \sum_{j=1}^n \int x_j(\partial^2 F(\mathbf{x})/\partial x_j^2)dx_j$, $V_\oplus(\mathbf{x}) = V(\mathbf{x}) + (S_b/2)\sum_{j=1}^n \partial^2 F(\mathbf{x})/\partial x_j^2$, and $V_\odot(\mathbf{x}) = V(\mathbf{x}) + \kappa'\sum_{i=1}^n x_i$. The first two results imply that multiplicative or additive noise has no effect on the system if $F(\mathbf{x})$ is quadratic. While the third result implies that adding chaotic noise can have no effect on the system if κ' is zero. As many learning algorithms are developed based on the method of gradient descent, these results can be applied in analyzing the effect of noise on those algorithms.

1 Introduction

Research on the effect of noise on neural networks has been conducted for almost two decades. From the earlier 90 s to the mid 90 s, researchers investigated the effect of noise on the performance of a multilayer perceptron (MLP)/recurrent neural networks (RNN) [6,11–13,15] and the associative networks [2,19]. Later, from the mid 90 s to the late 90 s, researchers started to analyze the effects of *additive input noise* (AIN) [5,7,8,14] and additive weight noise (AWN) [1] on back-propagation learning. The objective functions for these noise injection-based learning algorithms were revealed. In the 2000s, researchers investigated the effect of *chaotic noise* (CN) on MLP [3,4].

In recent years, the effects of AWN and *multiplicative weight noise* (MWN) on the RBF and MLP learning algorithms have been investigated [9,10,16,17]. It is shown that the objective function of the RBF learning algorithm with adding AWN or MWN is identical to the original RBF learning algorithm [9].

© Springer International Publishing Switzerland 2015
S. Arik et al. (Eds.): ICONIP 2015, Part III, LNCS 9491, pp. 479–487, 2015.
DOI: 10.1007/978-3-319-26555-1_54

Hence, adding AWN or MWN during RBF learning cannot improve the generalization ability of an RBF. Adding AWN during MLP learning can improve the generalization ability of an MLP. Adding MWN during MLP learning might not be [10, 17]. These results clarify a common missconception that adding noise during learning is able to improve the generalization ability of a neural network.

Now, we would like to investigate another question. *Would similar results be obtained for other learning algorithms?* To do so, one obvious approach is to investigate the effect of noise on a gradient system as many learning algorithms are developed by the method of gradient descent. Understanding the effect of noise on gradient systems can aid in the understanding of the effect of noise on these learning algorithms. Therefore, the objective of the paper is to investigate the effects of three types of noise (multiplicative noise, additive noise and chaotic noise) on a gradient system with forgetting. The energy functions of the corresponding gradient systems are revealed.

In the next section, the gradient systems with noise are introduced. The energy functions of these gradient systems with noise will be analyzed in Sect. 3. Effect of noise on the gradient systems will be elucidated in Sect. 4. Finally. Section 5 gives the conclusion of the paper.

2   Models

Let $\mathbf{x}(t) \in R^n$ and $F(\mathbf{x}) \in R$ is a bounded smooth function of $\mathbf{x}$. The energy function is given by $V(\mathbf{x}) = F(\mathbf{x}) + \lambda\|\mathbf{x}\|_2^2$, where λ is a small positive number called forgetting factor. The gradient system is defined as follows:

$$\mathbf{x}(t+1) = \mathbf{x}(t) - \mu\left(\frac{\partial F(\mathbf{x}(t))}{\partial \mathbf{x}} + \lambda\mathbf{x}(t)\right), \tag{1}$$

where μ is the learning step and it is a small positive number, and $\partial F(\mathbf{x}(t))/\partial\mathbf{x} = \partial F(\mathbf{x})/\partial\mathbf{x}|_{\mathbf{x}=\mathbf{x}(t)}$.

2.1   Multiplicative/Additive Noise

With multiplicative noise, the vector $\mathbf{x}(t)$ in (1) is replaced by $\tilde{\mathbf{x}}(t)$, where

$$\tilde{\mathbf{x}}(t) = \mathbf{x}(t) + \mathbf{b}(t) \otimes \mathbf{x}(t). \tag{2}$$
$$\mathbf{b}(t) \otimes \mathbf{x}(t) = (b_1(t)x_1(t), b_2(t)x_2(t), \cdots, b_n(t)x_n(t))^T.$$

With additive noise,

$$\tilde{\mathbf{x}}(t) = \mathbf{x}(t) + \mathbf{b}(t). \tag{3}$$

In (2) and (3), $\mathbf{b}(t) \in R^n$ is a Gaussian random vector with mean $\mathbf{0}$ and covariance matrix $S_b\mathbf{I}_{n\times n}$. Moreover, $E[b_i(t)] = 0$ for all $i = 1, \cdots, n$ and $t \geq 0$. $E[b_i^2(t)]$ equals to S_b and $E[b_i(t)b_j(t)]$ equals zero if $i \neq j$. $E[b_i(t_1)b_i(t_2)] = 0$ if $t_1 \neq t_2$. The gradient system with noise is given as follows:

$$\mathbf{x}(t+1) = \tilde{\mathbf{x}}(t) - \mu\left(\frac{\partial F(\tilde{\mathbf{x}}(t))}{\partial\mathbf{x}} + \lambda\tilde{\mathbf{x}}(t)\right). \tag{4}$$

2.2   Chaotic Noise

With chaotic noise injection, the noise is added to the gradient vector as follows [3, 4, 20]:

$$\mathbf{x}(t+1) = \mathbf{x}(t) - \mu \left(\frac{\partial F(\mathbf{x}(t))}{\partial \mathbf{x}} + \lambda \mathbf{x}(t) + \kappa n(t)\mathbf{e} \right), \tag{5}$$

where $\mathbf{e}$ is a constant vector of all 1s, κ is a positive constant and $n(t)$ is a deterministic noise generated by

$$n(t+1) = \alpha n(t)(1 - n(t)), \quad 3.6 < \alpha < 4. \tag{6}$$

3   Energy Functions

In this section, the energy functions of these gradient systems with noise are revealed. The effect of noise on the gradient systems will be discussed in the next section.

3.1   Multiplicative/Additive Noise

Given $\mathbf{x}(t)$, the mean update of (4) can be written as follows:

$$E[\mathbf{x}(t+1)|\mathbf{x}(t)] = E[\tilde{\mathbf{x}}(t)|\mathbf{x}(t)] - \mu E \left[\frac{\partial F(\tilde{\mathbf{x}}(t))}{\partial \mathbf{x}} + \lambda \tilde{\mathbf{x}}(t) \middle| \mathbf{x}(t) \right]. \tag{7}$$

In (7), the expectation is taken over the probability space of $\tilde{\mathbf{x}}(t)$. Since $E[\mathbf{b}(t)] = \mathbf{0}$, by (2) we get that $E[\tilde{\mathbf{x}}(t)|\mathbf{x}(t)] = \mathbf{x}(t)$. Equation (7) can be rewritten as follows:

$$E[\mathbf{x}(t+1)|\mathbf{x}(t)] = \mathbf{x}(t) - \mu \left(E \left[\frac{\partial F(\tilde{\mathbf{x}})}{\partial \mathbf{x}} \middle| \mathbf{x}(t) \right] + \lambda \mathbf{x}(t) \right). \tag{8}$$

Next, we let $V_\otimes(\mathbf{x})$ be a scalar function such that

$$E[\mathbf{x}(t+1)|\mathbf{x}(t)] = \mathbf{x}(t) - \mu \frac{\partial V_\otimes(\mathbf{x}(t))}{\partial \mathbf{x}}. \tag{9}$$

The energy function is stated in the following theorem.

Theorem 1. *For a gradient system defined as (1) and $\mathbf{x}(t)$ is corrupted by multiplicative noise as stated in (2),*

$$E[F(\tilde{\mathbf{x}})|\mathbf{x}] = F(\mathbf{x}) + \frac{S_b}{2} \sum_{j=1}^{n} \frac{\partial^2 F(\mathbf{x})}{\partial x_j \partial x_j} x_j^2 \tag{10}$$

and

$$V_\otimes(\mathbf{x}) = F(\mathbf{x}) + \frac{\lambda}{2}\|\mathbf{x}\|_2^2 + \frac{S_b}{2} \sum_{j=1}^{n} \frac{\partial^2 F(\mathbf{x})}{\partial x_j \partial x_j} x_j^2 - S_b \int \mathbf{x} \otimes \mathbf{diag}\,\{\mathbf{H}(\mathbf{x})\} \cdot d\mathbf{x}. \tag{11}$$

where $\int$ is the line integral, $\mathbf{H}(\mathbf{x})$ is the Hessian matrix of $F(\mathbf{x})$, i.e. $\mathbf{H}(\mathbf{x}) = \nabla\nabla_{\mathbf{x}}F(\mathbf{x})$ and

$$\text{diag}\{\mathbf{H}(\mathbf{x})\} = \left(\frac{\partial^2 F(\mathbf{x})}{\partial x_1^2}, \frac{\partial^2 F(\mathbf{x})}{\partial x_2^2}, \cdots, \frac{\partial^2 F(\mathbf{x})}{\partial x_n^2}\right)^T.$$

Proof: Consider (8) and let $\partial F(\mathbf{x})/\partial x_i$ be the ith element of $\partial F(\mathbf{x})/\partial\mathbf{x}$.

$$\frac{\partial F(\tilde{\mathbf{x}})}{\partial x_i} = \frac{\partial F(\mathbf{x})}{\partial x_i} + \sum_{j=1}^{n}\frac{\partial^2 F(\mathbf{x})}{\partial x_j \partial x_i}(b_j x_j) + \frac{1}{2}\sum_{k=1}^{n}\sum_{j=1}^{n}\frac{\partial^3 F(\mathbf{x})}{\partial x_k \partial x_j \partial x_i}b_k b_j x_k x_j. \quad (12)$$

Therefore,

$$E\left[\frac{\partial F(\tilde{\mathbf{x}})}{\partial x_i}\bigg|\mathbf{x}\right] = \frac{\partial F(\mathbf{x})}{\partial x_i} + \frac{S_b}{2}\sum_{j=1}^{n}\frac{\partial^3 F(\mathbf{x})}{\partial x_j \partial x_j \partial x_i}x_j^2. \quad (13)$$

On the other hand, by expanding $F(\tilde{\mathbf{x}})$ about $\mathbf{x}$, we get that

$$F(\tilde{\mathbf{x}}) = F(\mathbf{x}) + \sum_{i=1}^{n}\frac{\partial F(\mathbf{x})}{\partial x_i}b_i x_i + \frac{1}{2}\sum_{j=1}^{n}\sum_{i=1}^{n}\frac{\partial^2 F(\mathbf{x})}{\partial x_j \partial x_i}b_j b_i x_j x_i$$

and hence

$$E[F(\tilde{\mathbf{x}})|\mathbf{x}] = F(\mathbf{x}) + \frac{S_b}{2}\sum_{j=1}^{n}\frac{\partial^2 F(\mathbf{x})}{\partial x_j \partial x_j}x_j^2. \quad (14)$$

Differentiate both side of (14) with respect to x_i, we get that

$$\frac{\partial}{\partial x_i}E[F(\tilde{\mathbf{x}})|\mathbf{x}] = \frac{\partial F(\mathbf{x})}{\partial x_i} + \frac{S_b}{2}\sum_{j=1}^{n}\frac{\partial^3 F(\mathbf{x})}{\partial x_i \partial x_j \partial x_j}x_j^2 + S_b\frac{\partial^2 F(\mathbf{x})}{\partial x_i \partial x_i}x_i. \quad (15)$$

As $F(\mathbf{x})$ is smooth, $\partial^3 F(\mathbf{x})/\partial x_j \partial x_j \partial x_i = \partial^3 F(\mathbf{x})/\partial x_i \partial x_j \partial x_j$. Compare (13) and (15), we get that

$$E\left[\frac{\partial F(\tilde{\mathbf{x}})}{\partial x_i}\bigg|\mathbf{x}\right] = \frac{\partial E[F(\tilde{\mathbf{x}})|\mathbf{x}]}{\partial x_i} - S_b\frac{\partial^2 F(\mathbf{x})}{\partial x_i \partial x_i}x_i.$$

Further by (8) and (9), we get that

$$V_{\otimes}(\mathbf{x}) = E[F(\tilde{\mathbf{x}})|\mathbf{x}] - S_b\int \mathbf{x}\otimes\text{diag}\{\mathbf{H}(\mathbf{x})\}\cdot dx + \frac{\lambda}{2}\|x\|_2^2. \quad (16)$$

Putting (10) in (16) and rearranging the terms, we can get the energy function as given in (11) and the proof is completed.   **Q.E.D.**

Similarly, the energy function of the gradient system with additive noise is stated in the following theorem.

Theorem 2. *For a gradient system defined as (1) and* $\mathbf{x}(t)$ *is corrupted by additive noise as stated in (3),*

$$V_{\oplus}(\mathbf{x}) = F(\mathbf{x}) + \frac{\lambda}{2}\|\mathbf{x}\|_2^2 + \frac{S_b}{2}\sum_{j=1}^{n}\frac{\partial^2 F(\mathbf{x})}{\partial x_j \partial x_j}. \tag{17}$$

Proof: For additive noise, the noisy $\tilde{\mathbf{x}}$ in (4) is given by $\tilde{\mathbf{x}} = \mathbf{x} + \mathbf{b}$. Similarly, we consider (8) and let $\partial F(\mathbf{x})/\partial x_i$ be the ith element of $\partial F(\mathbf{x})/\partial \mathbf{x}$.

$$\frac{\partial F(\tilde{\mathbf{x}})}{\partial x_i} = \frac{\partial F(\mathbf{x})}{\partial x_i} + \sum_{j=1}^{n}\frac{\partial^2 F(\mathbf{x})}{\partial x_j \partial x_i}b_j + \frac{1}{2}\sum_{k=1}^{n}\sum_{j=1}^{n}\frac{\partial^3 F(\mathbf{x})}{\partial x_k \partial x_j \partial x_i}b_k b_j.$$

Therefore,

$$E\left[\frac{\partial F(\tilde{\mathbf{x}})}{\partial x_i}\bigg|\mathbf{x}\right] = \frac{\partial F(\mathbf{x})}{\partial x_i} + \frac{S_b}{2}\sum_{j=1}^{n}\frac{\partial^3 F(\mathbf{x})}{\partial x_j \partial x_j \partial x_i}. \tag{18}$$

Using similar technique as in multiplicative noise, we can get that

$$E[F(\tilde{\mathbf{x}})|\mathbf{x}] = F(\mathbf{x}) + \frac{S_b}{2}\sum_{j=1}^{n}\frac{\partial^2 F(\mathbf{x})}{\partial x_j \partial x_j} \tag{19}$$

and

$$\frac{\partial}{\partial x_i}E[F(\tilde{\mathbf{x}})|\mathbf{x}] = \frac{\partial F(\mathbf{x})}{\partial x_i} + \frac{S_b}{2}\sum_{j=1}^{n}\frac{\partial^3 F(\mathbf{x})}{\partial x_i \partial x_j \partial x_j}. \tag{20}$$

Compare (18) with (20), we get that $E\left[\partial F(\tilde{\mathbf{x}})/\partial x_i|\mathbf{x}\right] = \partial E[F(\tilde{\mathbf{x}})|\mathbf{x}]/\partial x_i$ and thus

$$\frac{\partial V_{\otimes}(\mathbf{x}(t))}{\partial \mathbf{x}} = \frac{\partial E[F(\tilde{\mathbf{x}})|\mathbf{x}]}{\partial \mathbf{x}} + \lambda \mathbf{x} \tag{21}$$

By (19), (20) and (21), the energy function as stated in (17) can be obtained and the proof is completed.   **Q.E.D.**

3.2   Chaotic Noise

For the system with chaotic noise injection, all elements in $\mathbf{x}$ suffered the same amount of noise $\kappa n(t)$ in the tth step. As observed from Fig. 1 which plots the value $\sum_{\tau=0}^{T-1} n(t+\tau)/T$ for $t = 1, \cdots, 2000$ and for different values of T, it is reasonable to assume that $\sum_{\tau=0}^{T-1} n(t+\tau)/T$ is a constant for all t if $T \gg 1$. Then, we can get the following theorem on the energy function of a gradient system with chaotic noise.

Theorem 3. *For a gradient system defined as (5) and* $\mu T \to 0$,

$$V_{\odot}(\mathbf{x}) = F(\mathbf{x}) + \frac{\lambda}{2}\|\mathbf{x}\|_2^2 + \kappa' \sum_{i=1}^{n} x_i, \tag{22}$$

where κ' *is a constant.*

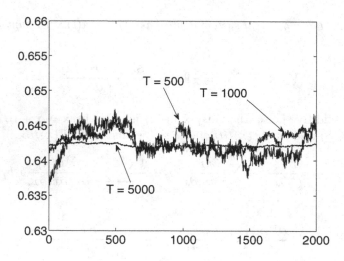

Fig. 1. $\sum_{\tau=0}^{T-1} n(t+\tau)/T$ against t, for $t = 1, \cdots, 2\,K$; $T = 500, 1\,K, 5\,K$. $\alpha = 3.8$

Proof: Suppose $\mu T \to 0$ for all t, we could assume that

$$\mathbf{x}(t+\tau) = \mathbf{x}(t), \quad \frac{\partial F(\mathbf{x}(t+\tau))}{\partial \mathbf{x}} = \frac{\partial F(\mathbf{x}(t))}{\partial \mathbf{x}}$$

for $\tau = 0, 1, \cdots, T - 1$. In such case, we can get from (5) that

$$\mathbf{x}(t+T) = \mathbf{x}(t) - \mu'\left(\frac{\partial F(\mathbf{x}(t))}{\partial \mathbf{x}} + \lambda\mathbf{x}(t) + \frac{\kappa}{T}\sum_{\tau=0}^{T-1} n(t+\tau)\mathbf{e}\right)$$

$$= \mathbf{x}(t) - \mu'\left(\frac{\partial F(\mathbf{x}(t))}{\partial \mathbf{x}} + \lambda\mathbf{x}(t) + \kappa\bar{n}\mathbf{e}\right), \tag{23}$$

where $\mu' = \mu T$ and $\bar{n} = \lim_{T\to\infty}\sum_{t=1}^{T} n(t)/T$. Clearly, the energy function is given by (22) and $\kappa' = \kappa\lim_{T\to\infty}\sum_{t=1}^{T} n(t)/T$. **Q.E.D.**

4 Effect of Noise

For multiplicative noise, let us rewrite that $V_\otimes(\mathbf{x}) = F(\mathbf{x}) + \frac{\lambda}{2}\|\mathbf{x}\|_2^2 + S_b\mathcal{R}(\mathbf{x})$, where $\mathcal{R}(\mathbf{x})$ corresponds to a regularizer. From (11), it is given by

$$\mathcal{R}(\mathbf{x}) = \frac{1}{2}\sum_{j=1}^{n} \frac{\partial^2 F(\mathbf{x})}{\partial x_j \partial x_j} x_j^2 - \int \mathbf{x} \otimes \mathbf{diag}\{\mathbf{H}(\mathbf{x})\} \cdot d\mathbf{x}. \tag{24}$$

The effect of the first term is to bring $\mathbf{x}$ closer to the zero vector while the second term is to push it away. Therefore, the existence of multiplicative noise in a gradient system would lead to two opposite effects. It should also be noted that $\mathbf{H}(\mathbf{x})$ is a constant matrix (say $\bar{\mathbf{H}}$) and $\mathcal{R}(\mathbf{x}) = 0$ if $F(\mathbf{x})$ is quadratic. Existence of multiplicative noise has no effect on the gradient system.

For additive noise, the additional term $(S_b/2) \sum_{j=1}^{n} \partial^2 F(\mathbf{x})/\partial x_j \partial x_j$ has the effect that brings the solution closer to the zeros vector. This term reduces to a constant if $F(\mathbf{x})$ is quadratic. The different between $V_{\oplus}(\mathbf{x})$ and $V(\mathbf{x})$ is just a constant. So, existence of additive noise has no effect on a gradient system if $F(\mathbf{x})$ is quadratic.

For chaotic noise, from (22), the additional term is $\kappa' \sum_{i=1}^{n} x_i$. Its effect is to let $\mathbf{x}$ slide along the direction $-[1\ 1\ \cdots\ 1]^T$. If all the x_is are positive, the additional term will bring them move slightly towards the zero vector. If all the x_i are negative, it will move $\mathbf{x}$ slightly further away from the zero vector. The effect of noise on the gradient system will depend on the minimum point of the $V(\mathbf{x})$ and it has no effect if $\lim_{T \to \infty} \sum_{t=1}^{T} n(t)/T = 0$.

5 Conclusion

In this paper, we have introduced the models of the gradient systems with three different type of noise, namely multiplicative, additive and chaotic noise. The energy functions of the corresponding gradient systems with noise have been revealed. By investigating the additional term in the energy functions as compared with the original energy function, it is found that only additive noise has a clear effect on the gradient system. It enforces the state vector moving slightly towards the zero vector. With multiplicative noise, two opposite effects exists, moving towards and away. With chaotic noise, the effect will be depended on the location of the minimum point of $V(\mathbf{x})$. It could be enforced to move towards or away from the zero vector. Moreover, if $F(\mathbf{x})$ is quadratic, either multiplicative or additive noise will have no effect on the gradient system.

Treating (i) the state vector as the weight vector of a neural network, (ii) $F(\cdot)$ as the mean square errors and (iii) moving toward the zero vector as improving generalization, our results imply that (a) injecting AWN during MLP learning can improve generalization, (b) injecting MWN or CN during MLP learning might not be and (c) injecting AWN or MWN during RBF learning cannot improve generalization. Results (a) and (b) are equally applied to other nonlinear neural networks. Treating x as the state variable in the stochastic Wang's kWTA model [18] or $\mathbf{x}$ as the neuronal outputs of Hopfield network, the effect of noise on these models can thus be analyzed by the same technique. Due to page limit, those results will be presented elsewhere.

Acknowledgments. The work presented in this paper is supported in part by research grants from Taiwan National Science Council numbering 100-2221-E-126-015 and 101-2221-E-126-016.

References

1. An, G.: The effects of adding noise during backpropagation training on a generalization performance. Neural Comput. **8**, 643–674 (1996)
2. Asai, H., Onodera, K., Kamio, T., Ninomiya, H.: A study of Hopfield neural networks with external noises. In: Proceedings IEEE International Conference on Neural Networks, vol. 4, pp. 1584–1589 (1995)
3. Azamimi, A., Uwate, Y., Nishio, Y.: An analysis of chaotic noise injected to backpropagation algorithm in feedforward neural network. In: Proceedings of IWVCC08, pp. 70–73 (2008)
4. Azamimi, A., Uwate, Y., Nishio, Y.: Effect of chaos noise on the learning ability of back propagation algorithm in feed forward neural network. In: Proceedings of the 6th Internationa Colloquium on Signal Processing and Its Applications (CSPA) (2010)
5. Bishop, C.M.: Training with noise is equivalent to Tikhonov regularization. Neural Comput. **7**, 108–116 (1995)
6. Bolt, G.: Fault tolerant in multi-layer Perceptrons. Ph.D. Thesis, University of York, UK (1992)
7. Grandvalet, Y., Canu, S.: A comment on noise injection into inputs in backpropagation learning. IEEE Trans. Syst. Man Cybern. **25**, 678–681 (1995)
8. Grandvalet, Y., Canu, S., Boucheron, S.: Noise injection: theoretical prospects. Neural Comput. **9**, 1093–1108 (1997)
9. Ho, K., Leung, C.S., Sum, J.: Convergence and objective functions of some fault/noise injection-based online learning algorithms for RBF networks. IEEE Trans. Neural Netw. **21**(6), 938–947 (2010)
10. Ho, K., Leung, C.S., Sum, J.: Objective functions of the online weight noise injection training algorithms for MLP. IEEE Trans. Neural Netw. **22**(2), 317–323 (2011)
11. Jim, K.C., Giles, C.L., Horne, B.G.: An analysis of noise in recurrent neural networks: convergence and generalization. IEEE Trans. Neural Netw. **7**, 1424–1438 (1996)
12. Murray, A.F., Edwards, P.J.: Synaptic weight noise during multilayer perceptron training: fault tolerance and training improvements. IEEE Trans. Neural Netw. **4**(4), 722–725 (1993)
13. Murray, A.F., Edwards, P.J.: Enhanced MLP performance and fault tolerance resulting from synaptic weight noise during training. IEEE Trans. Neural Netw. **5**(5), 792–802 (1994)
14. Reed, R., Marks II, R.J., Oh, S.: Similarities of error regularization, sigmoid gain scaling, target smoothing, and training with jitter. IEEE Trans. Neural Netw. **6**(3), 529–538 (1995)
15. Sequin, C.H., Clay, R.D.: Fault tolerance in feedforward artificial neural networks. Neural Netw. **4**, 111–141 (1991)
16. Sum, J., Leung, C.S., Ho, K.: Convergence analysis of on-line node fault injection-based training algorithms for MLP networks. IEEE Trans. Neural Netw. Learn. Syst. **23**(2), 211–222 (2012)
17. Sum, J., Leung, C.S., Ho, K.: Convergence analyses on on-line weight noise injection-based training algorithms for MLPs. IEEE Trans. Neural Netw. Learn. Syst. **23**(11), 1827–1840 (2012)
18. Sum, John, Leung, Chi-sing, Ho, Kevin: Effect of input noise and output node stochastic on Wang's kWTA. IEEE Trans. Neural Netw. Learn. Syst. **24**(9), 1472–1478 (2013)

19. Wang, L.: Noise injection into inputs in sparsely connected Hopfield and winner-take-all neural networks. IEEE Trans. Syst. Man Cybern. Part B: Cybern. **27**(5), 868–870 (1997)

20. Zhang, H., Zhang, Y., Xu, D., Liu, X.: Deterministic convergence of chaos injection-based gradient method for training feedforward neural networks, to appear in Cognitive Neurodynamic

User Recommendation Based on Network Structure in Social Networks

Yi Chen, Xiaolong Wang, Buzhou Tang$^{(\boxtimes)}$, Junzhao Bu,
and Xin Xiang

Key Laboratory of Network Oriented Intelligent Computation,
Shenzhen Graduate School, Harbin Institute of Technology,
Shenzhen 518055, China
{chenyi,bujunzhao,xiangxin}@hitsz.edu.cn,
wangxl@insun.hit.edu.cn, tangbuzhou@gmail.com

Abstract. Advances in Web 2.0 technology has led to the popularity of social networking sites. One fundamental task for social networking sites is to recommend appropriate new friends for users. In recent years, network structure has been used for user recommendation. Most existing network structure-based recommendation methods either need to pre-specify the group number and structure type or fail to improve performance. In this paper, we propose a novel network structure-based user recommendation method, called Bayesian nonparametric mixture matrix factorization (BNPM-MF). The BNPM-MF model first employs a Bayesian nonparametric model to automatically determine the group number and the network structure in networks and then applies a matrix factorization method on each structure to user recommendation for improvement. Experiments conducted on a number of real networks demonstrate that the BNPM-MF model is competitive with other state-of-the-art methods.

Keywords: User recommendation · Network structure · Matrix factorization · Bayesian nonparametric model · Social network

1 Introduction

The development of Web 2.0 technology allows people to access many social networking sites (e.g. Twitter, Facebook, Slashdot and YouTube). People in social networking sites can construct their social circles via sharing their favorites, commenting on others and making friends. One fundamental task for social network analysis is to recommend appropriate new friends and interesting information for users. Existing content-based recommendation approaches face a challenge in Twitter-style social network [1], and work in [2] found that user relationship play a dominant role in making friends. A large number of relation-based recommendation methods have been proposed during the last several years such as matrix factorization methods [3, 4], and a detailed survey about them was presented by Hasan [5].

Recently, network structure (e.g., community structure and bipartite structure) has attracted a great deal of attention in user recommendation, and several network structure-based recommendation methods have been proposed. The existing network

© Springer International Publishing Switzerland 2015
S. Arik et al. (Eds.): ICONIP 2015, Part III, LNCS 9491, pp. 488–496, 2015.
DOI: 10.1007/978-3-319-26555-1_55

structure-based methods may fall into two categories: two-phase methods [1, 6] and probabilistic graphical methods [7, 8]. The two-phase methods first employ a structure detection method to determine the structure in networks and then apply user recommendation methods on the structure. For example, Xu et al. first proposed an effective method to find meaningful subgroups and then extended the traditional collaborative filtering algorithm on the subgroups to obtain top-N user recommendations [6]. Zhao et al. first applied an LDA-based method to discover communities and then adopted the matrix factorization method for user recommendation on each community [1]. The limitation of the two-phase methods lies in that most structure detection methods used in the first phase need to pre-specify the group number and the structure type in networks. In most real world networks, however, neither the group number nor the structure type is available in advance. The probabilistic graphical methods introduce Bayesian nonparametric models [9] to determine the group number and structure type in networks and use the model parameters for link recommendation. Kemp et al. proposed an infinite relational model to discover structure in relational datasets [7]. Chen et al. proposed a Bayesian nonparametric mixture model to determine group number and group partition [8]. Although the probabilistic graphical methods recommend links without pre-specifying the group number and structure type, they fail to apply user recommendation methods on the structures for improving performance.

In this paper, we propose a novel structure-based user recommendation method, called the Bayesian nonparametric mixture matrix factorization (BNPM-MF). The BNPM-MF takes advantages of both two phase methods and probabilistic graphical models. It first employs a Bayesian nonparametric model to automatically determine the group number and the network structure in networks and then applies a matrix factorization method on each structure to user recommendation for improvement. Experiments conducted on a number of real networks demonstrate that our BNPM-MF model is competitive with other state-of-the-art methods.

The remainder of this paper is organized as follows. Section 2 introduces the BNPM-MF model. Experiments are presented in Sect. 3. Section 4 draws conclusions.

2 Bayesian Nonparametric Mixture Matrix Factorization (BNPM-MF) Model

The BNPM-MF is a two-phase method. It first employs a Bayesian nonparametric mixture (BNPM) model [8] to determine the group number and the network structure in networks and then applies matrix factorization on the network structure for user recommendation.

2.1 Network Structure Detection

The BNPM model is a Bayesian nonparametric model that allows the group number of a network K to approach infinity, but only a finite number of groups are used to generate the observed network. It has ability to determine the group number and the group partition in networks automatically. Before detailedly describing the BNPM model, we summarize the symbols in Table 1.

Table 1. Meanings of BNPM symbols

Symbol	Meaning
N	Node set
i	Node index
L_i	The out neighbor edges of node i
z_i	The group assignment of node i
α	Dirichlet process prior parameter for group multinomials
β	Dirichlet prior parameter for node multinomials
π	Parameter for group multinomials
θ_k	Parameter for node multinomials

In the BNPM model, a Chinese Restaurant Process (CRP) [10] is used to simulate the nonparametric distribution of latent groups, and a network with N nodes is generated in the following way (see Fig. 1):

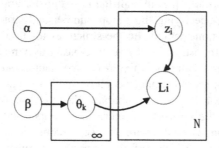

Fig. 1. Overview of the BNPM

1. Draw π from $CRP(\alpha)$;
2. For each group k in ∞ groups;

   Draw θ_k from $Dirichlet(\beta)$;

3. For each new node i:
   (a) Draw a latent group z_i from π;
   (b) For each edge in L_i:
       Draw an end node j from θ_{z_i}.

In a social network, let U be the set of users and E be the set of directed edges connecting users. An edge $e_{ij} \in E$ denotes a directed edge from user i to user j. Let H and T be the set of head users who have out edges and the set of tail users who have incoming edges respectively, defined as

$$H = \{i | i \in U \wedge \exists j \in U \wedge \exists e_{ij} \in E\}$$
$$T = \{i | i \in U \wedge \exists j \in U \wedge \exists e_{ji} \in E\} \tag{1}$$

For user i, use L_i to denote the set of out edges of it:

$$L_i = \{e_{ij} | j \in U \wedge \exists e_{ij} \in E\} \tag{2}$$

Therefore the edge set E can be given by

$$E = \bigcup_{i \in H} L_i \tag{3}$$

After the group number K and the group partition Z of a network are automatically determined by the BNPM, all edges fall into one of $K \times K$ blocks, forming different types of network structures. For each block, a user-user matrix $m_{gh}(1 \leq g \leq K, 1 \leq h \leq K)$ with head users, tail users and edges can be denoted by

$$
\begin{aligned}
m_{gh}.H &= \{i | i \in H \wedge z_i = g\} \\
m_{gh}.T &= \{i | i \in T \wedge z_i = h\} \\
m_{gh}.E &= \{e_{ij} | e_{ij} \in E \wedge i \in m_{gh}.H \wedge j \in m_{gh}.T\},
\end{aligned}
\tag{4}
$$

where each entry $m_{gh}[i,j] = 1$ if there is an edge $e_{ij} \in m_{gh}.E$, otherwise $m_{gh}[i,j] = 0$. Thus, we can obtain $K \times K$ user-user matrices from the output of the BNPM model.

Algorithm 1: BNPM-MF algorithm

Input   : Set of user-user relationships E; Number of latent factors c

Output : Ranked recommendation list

1    $H \leftarrow \{i | \exists e_{ij} \in E\}$

2    $T \leftarrow \{j | \exists e_{ij} \in E\}$

3    $L_i \leftarrow \{j | i \in U \wedge j \in U \wedge \exists e_{ij} \in E\}$

4    $L = \bigcup_{i \in H} L_i$

5    $K, Z \leftarrow BNPM(L)$

6   for g=1 to K do

7   for h=1 to K do

8            $m_{gh}.H = \{i | i \in H \wedge z_i = g\}$

9            $m_{gh}.T = \{i | i \in T \wedge z_i = h\}$

10            $m_{gh}.E = \{e_{ij} | e_{ij} \in E \wedge i \in m_{gh}.H \wedge j \in m_{gh}.T\}$

11   construct matrix m_{gh}

12           $c = \min(c, |m_{gh}.H|, |m_{gh}.T|)$

13           $P^{|m_{gh}.H| \times c}, Q^{|m_{gh}.T| \times c} \leftarrow IF - MF(m_{gh}, c)$

14   Endfor

15   Endfor

16   foreach *pair*(i, j) do

17   compute *score*(i, j) according to equation (5)

17   Endfor

19   Return the ranked lists of users

2.2   User Recommendation

After the network structure of a network is detected, matrix factorization is applied on each user-user matrix. Several matrix factorization methods have been proposed for recommender systems, such as the implicit feedback matrix factorization (IF-MF) [3] and the Bayesian Personalized Ranking matrix factorization (BPR-MF) [4]. In this paper, we try the IF-MF method.

Given a matrix m and the number of latent factors c, the IF-MF can obtain two matrices, namely $P^{|m.H \times c|}$ and $Q^{|m.T \times c|}$, which denote the mappings of head and tail users in the reduced latent space respectively. Correspondingly, each head user i is associated with a vector $p_i \in P^{|m.H \times c|}$ and each tail user j is associated with a vector $q_j \in Q^{|m.T \times c|}$.

With these matrix mappings, the score used to measure how strong user i chooses user j as a friend can be the inner product of p_i and q_j

$$score(i,j) = \langle p_i, q_j \rangle, (p_i \in P^{|m_{z_iz_j}.H \times c|}, q_j \in Q^{|m_{z_iz_j}.T \times c|}) \tag{5}$$

For each user i, we sort the scores of unseen edges starting from it and recommend the tail users of the ranked edges to it. Algorithm 1 summarizes the BNPM-MF model. Given the edge set E of a social network and the number of latent factors c, a ranked list of recommended users is returned by the BNPM-MF model.

3   Experiments

To investigate the effectiveness of the BNPM-MF model for user recommendation, we test it on six real world networks, including two networks with gold network structure (Cora and Citeseer) and four networks without gold structure (Epinion, Slashdot08, Twitter and Facebook). Cora and Citeseer are two pre-processed scientific citation networks; Epinion, Slashdot08, Twitter and Facebook are four online social networks. Following [1], we pre-process the Epinion, Slashdot08, Twitter and Facebook datasets and improve the density by removing nodes which have less 10 edges. Table 2 lists the detailed statistics of these networks and Fig. 2 shows the distribution of users that have the same number of edges in each network.

We select five state-of-the-art models for comparison. They are two matrix factorization-based methods (i.e., IF-MF [3] and BPR-MF [4]), a two-phase method (i.e., community-based matrix factorization, CB-MF [1]) and two probabilistic graphical methods (i.e., Infinite relational model (IRM) [7] and BNPM [8]). As the source codes of the IF-MF and BPR-MF are provided in [11] and the source codes of the IRM and BNPM are provided by their authors, we use the release codes and implement the CB-MF by ourselves. The CB-MF model needs to pre-specify the group number, we set the gold for the Cora and Citeseer and 10 for the Epinion, Slashdot08, Twitter and Facebook as [1].

For each dataset, we randomly choose 5 % edges as testing data and keep the rest as training data. The performance of all models is measured by two evaluation metrics:

Table 2. Detailed information of six networks in our study.

NID	Name	N	E	Sparity
1	Cora[a]	2,708	10,564	99.85 %
2	Citeseer[a]	3,312	9,330	99.91 %
3	Epinion[b]	11,005	568,262	99.53 %
4	Slashdot08[b]	27,278	843,862	99.89 %
5	Twitter[b]	52,184	2,471,060	99.91 %
6	Facebook[c]	29,440	1,384,590	99.84 %

[a]http://linqs.cs.umd.edu/projects/projects/lbc/
[b]https://snap.stanford.edu/data/
[c]http://konect.uni-koblenz.de/networks/facebook-wosn-links

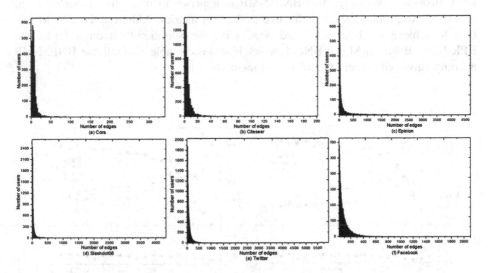

Fig. 2. Characteristics of the six datasets

Mean Average Precision (MAP) [12] and Normalized Discounted Cumulative Gain (NDCG) [13]. The MAP is computed by

$$MAP = \frac{\sum_{i}^{N} prec(i)}{N}, \tag{6}$$

where N denotes the number of actual nodes and $prec(i)$ is the percentage of the top i nodes in predicted ranking position. The NDCG is computed as

$$NDCG_k = \frac{1}{IDCG_k} \times \sum_{i=1}^{k} \frac{2^{b_i-1}}{\log(i+1)}, \tag{7}$$

where $b_i = 1$ if the predicted node at position i hit actual node and $b_i = 0$ otherwise, $IDCG_k$ is the maximum $NDCG_k$ that corresponds to the optimal ranking list. For MAP and NDCG, a higher result means a good recommendation. Particularly, perfect MAP or NDCG can be 1.

Before the comparison, we analyze the latent factor c of matrix factorization (MF) algorithm. Figure 3 illustrates the results of IF-MF, BPR-MF, CB-MF and BNPM-MF on the Cora network with c gradually changing from 5 to 100. As can be seen, the IF-MF and BPR-MF first increase with c and then reach a stationary state; the CB-MF and BNPM-MF first increase with c and then decrease. The reason is that the MF functions on the whole network for IF-MF and BPR-MF but on each structure for CB-MF and BNPM-MF. For the following experiments, we fix $c = 65$ for IF-MF, BPR-MF and $c = 10$ for CB-MF, BNPM-MF.

Tables 3 and 4 show the MAPs and NDCGs identified by different models on the six networks respectively. The BNPM-MF is superior to other five models on all networks except the Slashdot08 for two evaluation metrics. Although the BNPM-MF fails to achieve the best MAP and NDCG on the Slashdot08, it outperforms the BPR-MF, CB-MF, IRM and BNPM, which is still acceptable. Overall, the BNPM-MF is competitive with other state-of-the-art methods.

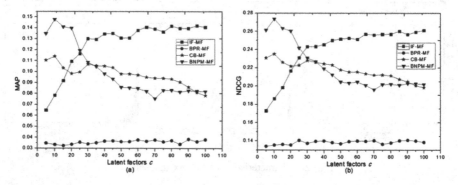

Fig. 3. Four matrix factorization based methods with different latent factors.

Table 3. MAPs identified by six models on the six datasets.

NID	IF-MF	BPR-MF	CB-MF	IRM	BNPM	BNPM-MF
1	0.1407	0.0364	0.1215	0.0362	0.0936	**0.1478**
2	0.1067	0.0251	0.0803	0.04	0.0542	**0.1158**
3	0.0645	0.0625	0.0568	0.0309	0.0379	**0.0827**
4	**0.049**	0.0175	0.0344	0.02	0.0202	0.0431
5	0.1068	0.2136	0.1962	0.0508	0.1128	**0.2597**
6	0.0636	0.0944	0.0589	0.0149	0.0149	**0.1201**

Table 4. NDCGs identified by six models on the six datasets.

NID	IF-MF	BPR-MF	CB-MF	IRM	BNPM	BNPM-MF
1	0.2572	0.1396	0.2454	0.1504	0.2254	**0.2736**
2	0.2066	0.1112	0.1987	0.1555	0.1741	**0.2308**
3	0.2406	0.2345	0.2327	0.1958	0.2017	**0.2517**
4	**0.1811**	0.1436	0.1678	0.1504	0.1473	0.1691
5	0.284	0.414	0.3866	0.2216	0.2931	**0.4505**
6	0.2337	0.279	0.2287	0.1658	0.1772	**0.2977**

4 Conclusion

In this paper, we propose a novel structure-based user recommendation method, called the Bayesian nonparametric mixture matrix factorization (BNPM-MF). The BNPM-MF model first employs a Bayesian nonparametric model to determine the group number and the network structure in networks and then applies a matrix factorization method on structures to user recommendation for improvement. Experiments conducted on a number of real networks demonstrate that the BNPM-MF model is competitive with other state-of-the-art methods.

Acknowledgements. This paper is supported in part by grants: National 863 Program of China (2015AA015405), NSFCs (National Natural Science Foundation of China) (61402128, 61473101, 61173075 and 61272383), Strategic Emerging Industry Development Special Funds of Shenzhen (JCYJ20140508161040764 and JCYJ20140417172417105) and Scientific Research Foundation in Shenzhen (Grant No. JCYJ20140627163809422).

References

1. Zhao, G., Lee, M.L., Hsu, W., Chen, W., Hu, H.: Community-based user recommendation in uni-directional social networks. In: Proceedings of the 22nd ACM International Conference on Information & Knowledge Management, San Francisco, CA, USA, pp. 189–198 (2013)
2. Hannon, J., Bennett, M., Bennett, M.: Recommending twitter users to follow using content and collaborative filtering approaches. In: Proceedings of the Fourth ACM Conference on Recommender Systems, Barcelona, Spain, pp. 199–206 (2010)
3. Hu, Y., Koren, Y., Volinsky, C.: Collaborative filtering for implicit feedback datasets. In: IEEE International Conference on Data Mining, Las vegas Nevada, USA, pp. 263–272 (2008)
4. Rendle, S., Freudenthaler, C., Gantner, Z., Schmidt-Thieme, L.: BPR: Bayesian personalized ranking from implicit feedback. In: Proceedings of the Twenty-Fifth Conference on Uncertainty in Artificial Intelligence, Arlington, Virginia, USA, pp. 452–461 (2009)
5. Hasan, M.A., Zaki, M.J.: A survey of link prediction in social networks. In: Aggarwal, C.C. (ed.) Social Network Data Analytics, pp. 243–275. Springer, Heidelberg (2011)
6. Xu, B., Bu, J., Chen, C., Cai, D.: An exploration of improving collaborative recommender systems via user-item subgroups. In: Proceedings of the 21st International Conference on World Wide Web, Lyon, France, pp. 21–30 (2012)

7. Kemp, C., Tenenbaum, J.B., Griffiths, T.L., Yamada, T., Ueda, N.: Learning systems of concepts with an infinite relational model. In: Proceedings of the 21st National Conference on Artificial intelligence, Boston, USA, pp. 381–388 (2006)

8. Chen, Y., Wang, X.L., Xiang, X., Tang, B.Z., Chen, Q.C., Yuan, B., Bu, J.Z.: Automatic exploration of structural regularities in networks http://arxiv.org/abs/1403.0466

9. Gershman, S.J., Blei, D.M.: A tutorial on Bayesian nonparametric models. J. Math. Psychol. **56**, 1–12 (2012)

10. Pitman, J.: Combinatorial stochastic processes, pp. 56–60. Springer, Heidelberg (2002)

11. Gantner, Z., Rendle, S., Freudenthaler, C., Schmidt-Thieme, L.: MyMediaLite: a free recommender system library. In: Proceedings of the Fifth ACM Conference on Recommender Systems, Chicago, Illinois USA, pp. 305–308 (2011)

12. Yue, Y., Finley, T., Radlinski, F., Joachims, T.: A support vector method for optimizing average precision. In: Proceedings of the 30th Annual International ACM SIGIR Conference on Research and Development in Information Retrieval, Amsterdam, Dutch, pp. 271–278 (2007)

13. Jarvelin, K., Kekalainen, J.: Cumulated gain-based evalution for ir techniques. ACM Trans. Inf. Syst. (TOIS) **20**(4), 422–446 (2002)

Decoupled Modeling of Gene Regulatory Networks Using Michaelis-Menten Kinetics

Ahammed Sherief Kizhakkethil Youseph[1,3]($\boxtimes$), Madhu Chetty[2,3], and Gour Karmakar[2]

[1] Faculty of Information Technology, Monash University, Melbourne, Australia
[2] Faculty of Science and Technology, Federation University Australia, Gippsland, Australia
{madhu.chetty,gour.karmakar}@federation.edu.au
[3] National Information and Communication Technology Australia (NICTA), Melbourne, Australia
ahammed.youseph@monash.edu, {ahammed.youseph,madhu.chetty}@nicta.com.au

Abstract. A set of genes and their regulatory interactions are represented in a gene regulatory network (GRN). Since GRNs play a major role in maintaining the cellular activities, inferring these networks is significant for understanding biological processes. Among the models available for GRN reconstruction, our recently developed nonlinear model [1] using Michaelis-Menten kinetics is considered to be more biologically relevant. However, the model remains coupled in the current form making the process computationally expensive, especially for large GRNs. In this paper, we enhance the existing model leading to a decoupled form which not only speeds up the computation, but also makes the model more realistic by representing the strength of each regulatory arc by a distinct Michaelis-Menten constant. The parameter estimation is carried out using differential evolution algorithm. The model is validated by inferring two synthetic networks. Results show that while the accuracy of reconstruction is similar to the coupled model, they are achieved at a faster speed.

Keywords: Gene regulatory network · Michaelis-Menten kinetics · Differential evolution · Computational complexity

1 Introduction

Cells are the basic building blocks of any living organism. The activities within a cell are governed by the regulatory interactions occurring at genetic level. Therefore, inferring the regulatory interactions is key to understanding the behavior of biological processes.

The microarray data, usually available for inferring gene regulatory networks (GRN), gives the expression levels of different genes at different instant of time. Various models are proposed for reconstruction of GRNs from microarray data. For example, Boolean network models use binary variables to represent gene expression level, 1(ON) if expressed and 0(OFF) if not [2,3]. However, the gene

© Springer International Publishing Switzerland 2015
S. Arik et al. (Eds.): ICONIP 2015, Part III, LNCS 9491, pp. 497–505, 2015.
DOI: 10.1007/978-3-319-26555-1_56

expressions are continuously varying and these Boolean approximations are not realistic. Bayesian network models use conditional probability to represent the causal relationship between genes [4,5]. These models are linear models and not suitable to represent biological systems which are nonlinear in nature. Other linear models used in GRN inference include linear differential equation models [6,7], state space models [8,9], etc. A popular nonlinear GRN model is the S-system model [10,11], which suffers from high computational cost.

Michaelis-Menten kinetics are widely used in modeling biochemical reactions. However, as the kinetic equations can be written only when the interactions are known, it has not been used for reverse engineering GRNs. In our work reported recently [1], we developed a GRN model based on Michaelis-Menten kinetics which is biologically relevant and having fewer parameters compared to the S-system model. However, the model in its current form with coupled equations is not suitable for inferring large networks because all the parameters for each gene cannot be estimated without considering other genes. In this paper, we enhance the model leading to a decoupled set of equations allowing us to perform the identification process with a reduced computational complexity. We used data from an *in silico* network and an *in vivo* network. The parameters were estimated through optimization using differential evolution algorithm.

Rest of the paper is organized as follows: Sect. 2 details the background of the model. The proposed model, its identification and computational complexity analyses are described in Sect. 3. In Sect. 4, we discuss the experimental results. The conclusions and future scope of the work are presented in Sect. 5.

2   Background

Michaelis-Menten kinetics is widely used in modeling biochemical reactions. They have been usually used in modeling genetic regulations when the regulations are known a-priori. For the reverse engineering of a GRN, we developed a model based on Michaelis-Menten kinetics in [1]. For N genes, the model was defined as:

$$\dot{x}_i = \alpha_i \prod_{j=1}^{N} \left(\frac{g_{ij}x_j + h_{ij}K_j}{K_j + x_j} \right) - \beta_i x_i \ . \tag{1}$$

where, α_i is the maximum rate of expression of gene-i, β_i is the self decay rate of mRNA expressed by gene-i, K_j is Michaelis-Menten constant which is a measure of the binding affinity of the regulatory protein from gene-j; the smaller the value of K_j, the larger is the binding affinity and $h_{ij}, g_{ij} \in \{0,1\}, h_{ij} + g_{ij} \geq 1 \ \forall \ i,j = 1,\ldots,N$; $h_{ij} = 1$ and $g_{ij} = 1$, implies that there is no regulation from gene-j to gene-i, $h_{ij} = 0$ and $g_{ij} = 1$ implies that gene-j activates gene-i and $h_{ij} = 1$ and $g_{ij} = 0$ implies that gene-j inhibits gene-i.

It may be noted that in this model, for each regulatory gene, the value of K_j is fixed irrespective of the gene being regulated, i.e., the binding affinity of the regulatory protein from gene-j is considered to be same for all the genes it regulates. For example, if gene-1 regulates gene-2 and gene-3, the binding affinity of the protein from gene-1 towards gene-2 and gene-3 are assumed to be the same. This is correct provided one gene regulates only one gene. But, this is not the case in reality, especially in large scale networks, where a gene is more likely to regulate multiple genes.

The identification of systems described by nonlinear ordinary differential equation (ODE) models is computationally expensive due to the high cost involved in numerical integration. The parameters involved in such models (e.g. S-system) used in GRN inference, are usually estimated in a step by step method [10]. The ODE describing the rate of change of expression of one gene is considered at a time while the expression values of other genes are given as input to the estimation algorithm. In such a decoupled system, the parameters associated with the rate of change of expression of each gene are estimated one by one.

For this reason, in this paper, we propose a modification in our coupled ODE model [1] which will resolve both the above mentioned problems in dealing with the model in [1]. This is presented in the next section.

3 The Method

The enhancement of the model, the parameter estimation using the decoupled set of equations and the comparison of the computational complexities of the two methods are described in this section.

3.1 The Model

Our coupled model presented in [1] is modified to allow differing values for the Michaelis-Menten constant for each regulation, irrespective of the regulatory gene. If a gene-i is regulating gene-j and gene-k, two different constants K_{ji} and K_{ki} are used to differentiate the binding of protein from gene-i to the promoter region of gene-j from its binding to the promoter region of gene-k. The model defined in (1) is modified as:

$$\dot{x}_i = \alpha_i \prod_{j=1}^{N} \left(\frac{g_{ij} x_j + h_{ij} K_{ij}}{K_{ij} + x_j} \right) - \beta_i x_i \ . \tag{2}$$

where, K_{ij} is Michaelis-Menten constant for the regulation of gene i by gene j if the regulation exists and other constants are same as in (1).

h_{ij} and g_{ij} both cannot be zero for any i and j [1]. As with our previous model [1], instead of h_{ij} and g_{ij}, we can estimate a new variable $\delta_{ij} \in \{-1, 0, 1\}$ to reduce the computational cost as below:

$$\delta_{ij} = \begin{cases} 0, & \text{if } g_{ij} = 1 \text{ and } h_{ij} = 1 \ . \\ +1, & \text{if } g_{ij} = 1 \text{ and } h_{ij} = 0 \ . \\ -1, & \text{if } g_{ij} = 0 \text{ and } h_{ij} = 1 \ . \end{cases} \tag{3}$$

Substituting g and h in (2) with functions of δ gives the following:

$$\frac{dx_i}{dt} = \alpha_i \prod_{j=1}^{N} \left(\frac{\frac{\delta_{ij}+1}{|\delta_{ij}|+1}x_j + \frac{1-\delta_{ij}}{|\delta_{ij}|+1}K_{ij}}{K_{ij} + x_j} \right) - \beta_i x_i \ . \tag{4}$$

3.2   Parameter Estimation

In our previous model [1], the parameters to be estimated for each gene are shown in Fig. 1(a). For all genes, in each equation the parameters are different except $K_1, K_2, \ldots, K_N$ that are the same for all genes, i.e., $\forall i = 1, \ldots, N$ in (1). The total number of parameters to be estimated in this coupled model is $N(2+N) + N = N^2 + 3N$ and they have to be estimated all together.

$$
\begin{matrix}
1 \\ 2 \\ - \\ i \\ - \\ N
\end{matrix}
\begin{pmatrix}
\alpha_1 & \beta_1 & \delta_{11} & - & \delta_{1N} & K_1 & - & K_N \\
\alpha_2 & \beta_2 & \delta_{21} & - & \delta_{2N} & K_1 & - & K_N \\
- & - & - & & - & - & & - \\
\alpha_i & \beta_i & \delta_{i1} & - & \delta_{iN} & K_1 & - & K_N \\
- & - & - & & - & - & & - \\
\alpha_N & \beta_N & \delta_{N1} & - & \delta_{NN} & K_1 & - & K_N
\end{pmatrix}
\qquad
\begin{matrix}
1 \\ 2 \\ - \\ i \\ - \\ N
\end{matrix}
\begin{pmatrix}
\alpha_1 & \beta_1 & \delta_{11} & - & \delta_{1N} & K_{11} & - & K_{1N} \\
\alpha_2 & \beta_2 & \delta_{21} & - & \delta_{2N} & K_{21} & - & K_{2N} \\
- & - & - & & - & - & & - \\
\alpha_i & \beta_i & \delta_{i1} & - & \delta_{iN} & K_{i1} & - & K_{iN} \\
- & - & - & & - & - & & - \\
\alpha_N & \beta_N & \delta_{N1} & - & \delta_{NN} & K_{N1} & - & K_{NN}
\end{pmatrix}
$$

(a) Coupled method defined in (1)   (b) Decoupled method defined in (2)

Fig. 1. The parameters to be estimated for each gene

Now, as we have modified the model, the parameters for each gene in our new model defined in (2) are shown in Fig. 1(b). None of the parameters are coupled for different genes and hence the model defined in (2) can be decoupled, thereby allowing estimation of parameters to be carried out for one gene at a time. The decoupled model can finally be expressed as follows:

$$\frac{dx_i}{dt} = \alpha_i \left(\frac{g_{ii}x_i + h_{ii}K_{ii}}{K_{ii} + x_i} \right) \prod_{\substack{j=1 \\ j \neq i}}^{N} \left(\frac{g_{ij}x_j^* + h_{ij}K_{ij}}{K_{ij} + x_j^*} \right) - \beta_i x_i \ . \tag{5}$$

where, x_j^* are the expression levels of genes other than gene-i. They can be determined using any numerical interpolation method before the parameter estimation of gene-i is calculated. Figure 1(b) shows that each gene has $2 + N + N = 2(N+1)$ parameters. Therefore, total number of parameters that need to be estimated for N genes is $2N(N+1)$.

Fitness Function. The parameters are estimated using an optimization algorithm by which an objective function is minimized. The fitness function usually considered in reverse-engineering GRNs is a sum of the deviation of the model prediction from the experimental values and a penalty on the network complexity. We used a function analogous to [12]. The fitness function for gene-i is the following:

$$f_i = \sum_{t=1}^{T} \left\{ \frac{x_i(t)^{cal} - x_i(t)^{exp}}{x_i(t)^{exp}} \right\}^2 + C_i \frac{N}{Z_i} \qquad (6)$$

where, $x_i(t)^{cal}$ and $x_i(t)^{exp}$ are the expression levels of gene-i at t^{th} time instant as calculated by the model and as in the microarray data, respectively, N is the total number of genes, Z_i is the total number of non-regulations, $Z_i = N - r_i$, where, r_i = total number of regulations for gene-i = $\sum_{j=1}^{N} | \delta_{ij} |$ and C_i is a constant as defined in [12].

3.3 Computational Complexity

In this section we evaluate and compare the computational complexity of the proposed decoupled model and the coupled model presented in [1]. For this, we first present the algorithm being proposed. Next, the algorithm with coupled equation is briefly described to highlight the reduction in complexity achieved.

Enhanced Model with the Decoupled Equation Defined in (5). The total number of parameters to be estimated is $2N(N + 1)$ and the problem is divided into N subproblems where in each subproblem, $2(N + 1)$ parameters of one gene are estimated by providing the expression profiles of the other genes. The profiles are calculated through interpolation and hence can be considered constant.

Step A. Interpolate the gene expression data using cubic spline for the necessary samples. These interpolated values are required while the numerical integration is carried out to calculate the expression profile of a gene from the model. The interpolation is to be carried out for $2N_t$ samples where, N_t is the number of steps used in numerical integration, $N_t = (T_f - T_i)/h$, T_f and T_i are the time for the last and the first samples, respectively in the microarray data and h is the step size for the fourth order Runge Kutta (RK4) method used for numerical integration.

Step B. For each gene estimate the parameters individually.

1 Initialize NP members of the population, where, NP is usually taken in proportion to the number of parameters, $(2 + 2N)$ [13].

2 Generate NP new members through mutation, do cross over on each element of the member and ensure bound-constraint satisfaction.

3 For each new member, evaluate the fitness function.

    3.1 Calculate the network complexity, the second term on RHS of (6).

    3.2 For each time instance from the second time sample, get the gene expression calculated by the model defined in (5) using RK4.

        3.2.1 Given the gene expression level at a given time, t, calculate the four slopes of RK4 method using the model and interpolated data to get the gene expression level at time $t + \Delta t$.

3.3 Find the modeling error, the first term on RHS of (6) for all samples.

3.4 Sum of both the terms in steps 3.1 and 3.3 gives the fitness function.

4 Selection of next generation from the two populations based on the fitness function values.

5 Go to step 2 if the termination criteria is not met.

6 If the termination criteria is met, the best member of the current population is the candidate solution.

Computational complexity of finding spline coefficients is $O(N)$ and of interpolation is $O(logN)$. In our case, we provide microarray data having the expression levels of N genes at N_s time instances to interpolate the data for $2N_t$ samples. So the complexity for step A is $O(N \times N_s + log(N \times N_t))$.

The complexity of step 3.1 is $O(1)$. The complexity of step 3.2.1, RK4 using the function in (5), is $O(N)$. Step 3.2.1 needs to be repeated N_t times to calculate the expression levels at all the time instances. Therefore, the complexity of step 3.2 is $O(N_t N)$. The steps 3.3 and 3.4 are of $O(N_s)$, $O(1)$, respectively. Since the calculation is to be carried out for NP members, the complexity of step 3 is $O(NP \times N_t \times N) = O(N^2 N_t)$ as $NP + O(N)$.

Computational complexity of step 4 is $O(NP) = O(N)$. So, the complexity of each iteration is $O(N^2 N_t)$, neglecting the lower order terms. If the termination criteria is budget-oriented, the steps are to be repeated G_{max} times, where G_{max} is the maximum number of generations set. As the whole estimation is to be carried out for N genes, the total complexity is $O(G_{max} N^3 N_t)$.

Original Model with Coupled Equation Presented in (1). The total number of parameters to be estimated is $3N + N^2$. The parameter estimation algorithm is very similar to the decoupled one, except that all the parameters are estimated through a single optimization process. The step A of the decoupled model is not required here. The RK4 for the function evaluation in step 3.2.1 is $O(N^2)$ as the slopes are vectors of size N. As the error is to be calculated for all genes and samples, the complexity of function evaluation (step 3) is $O(N_t N \times N^2) = O(N^3 N_t)$. As the population size in this method is $O(N^2)$, the total computational complexity will be $O(G_{max} N^2 \times N^3 N_t) = O(G_{max} N^5 N_t)$.

The computational complexity of our decoupled method is two-order less than that of our coupled method. This exhibits a huge improvement in the computational complexity of our decoupled method and indicates that it will be faster compared to its coupled counter part for any dataset irrespective of its size.

4   Results and Discussion

We used the standard performance metrics, sensitivity, specificity and F-score, to evaluate the simulation outcomes. We also conducted a statistical test, namely paired t-test to determine whether the results are significantly different.

4.1 *In silico* Network

For experimental studies, we generated the ten datasets using a 5-gene-synthetic network for ten time instances by the coupled model [1]. The experimental results are presented in Table 1. The simulation time taken for 800 iterations and nine steps of RK4 per function evaluation for both models are also included in the table. Investigations of other synthetic networks, showed similar results, but are not presented here due to space constraint. We can observe that the proposed model, while maintaining the reconstruction accuracy (shown by S_n, S_p and F), shows a significant improvement in time for simulation. The values of all performance metrics are same as that of the coupled method except for a noise level of 10 % and it is comparable with the coupled method with a p-value 0.61, derived from a statistical test, t-test. This p-value shows that the reconstruction accuracies are not statistically different, i.e. the results are not significantly different.

Table 1. The best of the experimental results for *in silico* network

Input data	Coupled [1]				Proposed			
	S_n	S_p	F	Simulation time (s)	S_n	S_p	F	Simulation time (s)
0 % Noise	1.0	1.0	1.0	851.2	1.0	1.0	1.0	805.4
5 % Noise	1.0	1.0	1.0	851.6	1.0	1.0	1.0	806.8
7.5 % Noise	1.0	0.94	0.94	860.4	1.0	0.94	0.94	810.1
10 % Noise	1.0	0.82	0.84	862.3	0.86	0.89	0.80	812.2

4.2 *In vivo* Network - IRMA

Irene et al. [14] reported a 5-gene synthetic network in yeast, (i.e. *Saccharomyces cerevisiae*) for "In vivo Reverse engineering and Modeling Assessment (IRMA)". The five genes are CBF1, GAL4, SWI5, GAL80 and ASH1. There are six transcriptional regulations, namely CBF1 activates GAL4, GAL4 activate SWI5, SWI5 activates CBF1, GAL80 and ASH1, ASH1 inhibits CBF1. The network has two protein-protein interactions, Gal4p and Gal80p inhibiting each other. The original network consists of all eight regulatory arcs and a simplified network excludes the two protein-protein interactions.The data sets of this network [14] have been used for evaluating the proposed model.

Table 2 compares the performance of the proposed decoupled model with other methods that have been used earlier to reconstruct IRMA network from both, ON and OFF datasets. With the exception of coupled model, the results of the proposed decoupled model for ON data are superior to all other methods. For OFF data, the prediction accuracy of our decoupled method is better than all other methods. For the simplified network, decoupled model outperforms the coupled, while for the original network, the results are comparable, with a p-value 0.74 which also indicates that there is no significant difference among

Table 2. The experimental results for IRMA network, reconstructed from ON and OFF datasets. The results for other methods were reported in [15].

Data	ON Data						OFF Data					
Network	Original			Simplified			Original			Simplified		
Method	S_n	S_p	F	S_n	S_p	F	S_n	S_p	F	S_n	S_p	F
Proposed (best)	0.71	0.89	0.71	1.00	0.82	0.83	1.00	0.84	0.80	1.00	0.82	0.83
Coupled [1]	0.71	0.89	0.71	1.00	0.82	0.83	1.00	0.83	0.82	1.00	0.73	0.77
REGARD [12]	0.69	0.83	0.64	0.70	0.75	0.61	0.77	0.76	0.63	0.80	0.79	0.70
TD-ARACNE [16]	0.63	0.88	0.67	0.67	0.90	0.73	0.60	-	0.46	0.75	-	0.60
BANJO [17]	0.24	0.76	0.29	0.50	0.70	0.50	0.38	0.88	0.46	0.33	0.90	0.44
ALG [18]	0.77	0.27	0.40	0.80	0.42	0.50	0.76	0.56	0.57	0.80	0.75	0.67

the results. However, as observed in Sect. 4.1 above with the 5-gene network, the similar level of performance is achieved with significantly less computational time. For 800 iterations and 2×10^5 time steps of RK4 per function evaluation, the coupled method required 106 hours. This performance was achieved in 68 hours by the proposed decoupled method. In other words, decoupled method is 55.88 % faster than our coupled method [1].

5 Conclusions

Reverse-engineering GRNs is a challenging problem in computational biology. Our previously proposed GRN model [1], is biologically relevant model as it is based on well formulated Michaelis-Menten kinetics. However, this model in its current coupled form is computationally expensive. Further, in the coupled model, the binding coefficients can only have equal values which is not realistic. In this paper, we have enhanced the coupled model thereby allowing the binding coefficient to have different values for different gene interaction. This improvement also enables the underlying differential equations describing the model to become decoupled thereby opening a different way of parameter estimation resulting in speeding up the algorithm. Since the binding coefficients can acquire different values, the modification adds biological relevance to the model. We are currently evaluating the method by its application to larger network.

Acknowledgments. This work was partly supported by Australian Federal and Victoria State Governments and the Australian Research Council through the ICT Centre of Excellence program, National ICT Australia (NICTA).

References

1. Youseph, A.S.K., Chetty, M., Karmakar, G.: Gene regulatory network inference using Michaelis-Menten kinetics. In: IEEE Congress on Evolutionary Computation (CEC), pp. 2392–2397 (2015)

2. Kauffman, S.A.: Metabolic stability and epigenesis in randomly constructed genetic nets. J. Theor. Biol. **22**(3), 437–467 (1969)
3. Akutsu, T., Miyano, S., Kuhara, S.: Identification of genetic networks from a small number of gene expression patterns under the boolean network model. In: Pacific Symposium on Biocomputing, vol. 4, pp. 17–28 (1999)
4. Ram, R., Chetty, M.: A markov-blanket-based model for gene regulatory network inference. IEEE/ACM Trans. Comput. Biol. Bioinform. **8**(2), 353–367 (2011)
5. Xuan, N., Chetty, M., Coppel, R., Wangikar, P.: Gene regulatory network modeling via global optimization of high-order dynamic Bayesian network. BMC Bioinform. **13**(1), 131 (2012)
6. Kabir, M., Noman, N., Iba, H.: Reverse engineering gene regulatory network from microarray data using linear time-variant model. BMC Bioinform. **11**(Suppl. 1), S56 (2010)
7. Wu, F.X., Liu, L.Z., Xia, Z.H.: Identification of gene regulatory networks from time course gene expression data. In: Annual International Conference of the IEEE Engineering in Medicine and Biology Society (EMBC), pp. 795–798 (2010)
8. Hirose, O., Yoshida, R., Imoto, S., Yamaguchi, R., Higuchi, T., Charnock-Jones, D.S., Print, C., Miyano, S.: Statistical inference of transcriptional module-based gene networks from time course gene expression profiles by using state space models. Bioinformatics **24**(7), 932–942 (2008)
9. Tamada, Y., Yamaguchi, R., Imoto, S., Hirose, O., Yoshida, R., Nagasaki, M., Miyano, S.: Sign-ssm: open source parallel software for estimating gene networks with state space models. Bioinformatics **27**(8), 1172–1173 (2011)
10. Maki, Y., Ueda, T., Okamoto, M., Uematsu, N., Inamura, K., Uchida, K., Takahashi, Y., Eguchi, Y.: Inference of genetic network using the expression profile time course data of mouse p19 cells. Genome Inf. **13**, 382–383 (2002)
11. Chowdhury, A., Chetty, M.: An improved method to infer gene regulatory network using S-system. In: IEEE Congress on Evolutionary Computation (CEC), pp. 1012–1019 (2011)
12. Chowdhury, A., Chetty, M., Vinh, N.X.: Adaptive regulatory genes cardinality for reconstructing genetic networks. In: IEEE Congress on Evolutionary Computation (CEC), pp. 1–8 (2012)
13. Storn, R., Price, K.: Differential evolution - a simple and efficient adaptive scheme for global optimization over continuous spaces. Technical report. TR-95-012, ICSI, March 1995. http://www1.icsi.berkeley.edu/~storn/litera.html
14. Cantone, I., Marucci, L., Iorio, F., Ricci, M.A., Belcastro, V., Bansal, M., Santini, S., di Bernardo, M., di Bernardo, D., Cosma, M.P.: A yeast synthetic network for in vivo assessment of reverse-engineering and modeling approaches. Cell **137**(1), 172–181 (2009)
15. Chowdhury, A., Chetty, M., Vinh, N.: Incorporating time-delays in S-system model for reverse engineering genetic networks. BMC Bioinform. **14**(1), 196 (2013)
16. Zoppoli, P., Morganella, S., Ceccarelli, M.: Timedelay-aracne: reverse engineering of gene networks from time-course data by an information theoretic approach. BMC Bioinform. **11**(1), 154 (2010)
17. Yu, J., Smith, V.A., Wang, P.P., Hartemink, A.J., Jarvis, E.D.: Advances to Bayesian network inference for generating causal networks from observational biological data. Bioinformatics **20**(18), 3594–3603 (2004)
18. Noman, N., Iba, H.: Inferring gene regulatory networks using differential evolution with local search heuristics. IEEE/ACM Trans. Comput. Biol. Bioinf. **4**(4), 634–647 (2007)

Neural Networks with Marginalized Corrupted Hidden Layer

Yanjun Li[1], Xin Xin[1(✉)], and Ping Guo[1,2]

[1] School of Computer Science and Technology, Beijing Institute of Technology,
Beijing 100081, China
liyanjunzgz@126.com, xxin@bit.edu.cn, pguo@ieee.org
[2] Image Processing and Pattern Recognition Laboratory,
Beijing Normal University, Beijing 100875, China

Abstract. Overfitting is an important problem in neural networks (NNs) training. When the number of samples in the training set is limited, explicitly extending the training set with artificially generated samples is an effective solution. However, this method has the problem of high computational costs. In this paper we propose a new learning scheme to train single-hidden layer feedforward neural networks (SLFNs) with implicitly extended training set. The training set is extended by corrupting the hidden layer outputs of training samples with noise from exponential family distribution. When the number of corruption approaches infinity, in objective function explicitly generated samples can be expressed as the form of expectation. Our method, called marginalized corrupted hidden layer (MCHL), trains SLFNs by minimizing the loss function expected values under the corrupting distribution. In this way MCHL is trained with infinite samples. Experimental results on multiple data sets show that MCHL can be trained efficiently, and generalizes better to test data.

Keywords: Neural network · Overfitting · Classification

1 Introduction

Overfitting is an important problem in NNs training [1,2]. Overfitting occurs when the network has too many free parameters relative to the number of training samples. In this situation the network adapts to the particular details of the training set and leads to poor generalization performance.

A typical method of solving overfitting is to extend the training set. Because overfitting can become less severe as the size of the training set increases. When the number of samples in the training set is limited, explicitly extending the training set with artificially generated samples is an effective solution. However, this method has the problem of high computational costs. Because the training time increases with the number of samples.

To solve overfitting, in this paper we propose a new learning scheme to train single-hidden layer feedforward neural networks (SLFNs) with implicitly

© Springer International Publishing Switzerland 2015
S. Arik et al. (Eds.): ICONIP 2015, Part III, LNCS 9491, pp. 506–514, 2015.
DOI: 10.1007/978-3-319-26555-1_57

extended training set. The training set is extended by corrupting the hidden layer outputs of training samples with noise from exponential family distribution. When the number of corruption approaches infinity, in objective function explicitly generated samples can be expressed as the form of expectation. Our method, called marginalized corrupted hidden layer (MCHL), trains SLFNs by minimizing the loss function expected values under the corrupting distribution. In this way MCHL is trained with infinite samples.

Parameter optimization of NNs is a big challenge. NNs are normally optimized with backpropagation (BP) algorithm [3]. As a first order gradient descent parameter optimization method, BP algorithm has the problems of local minimum and slow convergence.

To optimize MCHL efficiently, we propose to optimize the parameters of MCHL by pseudo inverse operation. Our optimization method is inspired by the work of Guo and Michael [4]. Different from BP algorithm, for MCHL the previously trained weights in the network are not changed. This makes the training of MCHL more efficient. In addition, the model parameters have analytical solution, so MCHL tends to achieve global minima.

Experimental results on multiple data sets show that MCHL can be trained efficiently, and generalizes better to test data. In summary, we make the following contributions: (1) to solve overfitting, we propose a new learning scheme to train SLFNs with implicitly extended training set; (2) for MCHL, we propose an efficient parameter optimization method; (3) on several data sets, we show that MCHL can be trained efficiently, and generalizes better to test data.

2   Related Works

In this subsection we brief the related works about corrupting samples during training. Burges and Scholkopf [5] first propose to improve the generalization ability of predictors by explicitly corrupting training data. Hinton et al. [1] propose a method to reduce overfitting by randomly omitting half of the feature detectors on each training case. Vincent et al. [6] propose a unsupervised representation learning method, which corrupts the input data and keeps the desired output unchanged. Their approach is commonly used to train autoencoders, and these denoising autoencoders can be further used to initialize deep architectures. Chen et al. [7] propose marginalized denoising autoencoders for domain adaptation which are linear denoising autoencoders. Maaten et al. [8] propose to extend the training set with infinitely many artificial samples by corrupting the original training data. Our method differs from Maaten et al. work in that we corrupt the hidden layer outputs instead of the original training samples. Except corrupting features, there is another research direction (corrupting labels). Chen et al. [9] propose a fast image annotation method based on labels corruption. Lawrence and Schölkopf [10] propose an algorithm for constructing a kernel Fisher discriminant from training examples with noisy labels.

3   MCHL

In this section we first introduce MCHL learning scheme, then analyze how to marginalize the noise introduced in the hidden layer outputs analytically by minimizing the loss function expected values under the corrupting distribution, i.e., solve the weights of hidden layer to output layer.

3.1   Learning Scheme

Given a training set $D = \{(\mathbf{x}_i, \mathbf{y}_i) | \mathbf{x}_i \in \mathbf{R}^d, \mathbf{y}_i \in \mathbf{R}^k, i = 1, \cdots, M\}$, let L denotes the number of hidden nodes, $h(\mathbf{x})$ denotes the feature mapping function and $\mathbf{h}(\mathbf{x})$ denotes mapping result of data $\mathbf{x}$. In MCHL different hidden neurons can use different feature mapping functions. In real applications $h(\mathbf{x})$ can be defined as

$$h(\mathbf{x}) = F(\mathbf{a}, b, \mathbf{x}), \mathbf{a} \in \mathbf{R}^d, b \in R, \tag{1}$$

where $(\mathbf{a}, b)$ are hidden node parameters. $F(\mathbf{a}, b, \mathbf{x})$ can be any piecewise continuous function which meets universal approximation capability theorem [11]. Typically used feature mapping functions $F(\mathbf{a}, b, \mathbf{x})$ include Sigmoid ($\frac{1}{1+\exp(-(\mathbf{a}\cdot\mathbf{x}+b))}$), Gaussian ($\exp(-b\|\mathbf{x} - \mathbf{a}\|)$), Hyperbolic tangent ($\frac{1-\exp(-(\mathbf{a}\cdot\mathbf{x}+b))}{1+\exp(-(\mathbf{a}\cdot\mathbf{x}+b))}$) and Cosine ($\cos(\mathbf{a} \cdot \mathbf{x} + b)$).

We use pseudo inverse operation to learn the parameters of MCHL. MCHL trains a single-hidden layer feedforward neural network (SLFN) by two stages: (1) map training data into a new space (called MCHL space); (2) solve the parameters of hidden layer to output layer in MCHL space. We first introduce how to solve parameters of hidden layer to output layer (weight $\mathbf{W}_2$), then depict how to figure out the parameters of input layer to hidden layer (weight $\mathbf{W}_1$, i.e. parameters $\mathbf{a}$ and b).

MCHL solves weight $\mathbf{W}_2$ of hidden layer to output layer in MCHL feature space by minimizing the training error

$$\min_{\mathbf{W}_2} \|\mathbf{H}\mathbf{W}_2 - \mathbf{Y}\|^2, \tag{2}$$

where $\mathbf{H}$ is the matrix of hidden layer outputs. The smallest norm least squares solution of optimization problem (2) is $\mathbf{W}_2^* = \mathbf{H}^\dagger \mathbf{Y}$, where $\mathbf{H}^\dagger$ is the pseudo inverse of matrix $\mathbf{H}$.

We hope the outputs of hidden layer are irrelevant. To achieve this objective we first randomly generate a $M \times L$ full rank matrix $\mathbf{P}$. Huang et al. [11] have proved that SLFNs with arbitrarily assigned input weights and hidden layer biases and with almost any nonzero activation function can universally approximate any continuous functions on any compact input sets. Solve equation

$$\mathbf{X}\mathbf{W}_1 = \mathbf{P}, \tag{3}$$

we get $\mathbf{W}_1^* = \mathbf{X}^\dagger \mathbf{P}$. Different from conventional NNs trained with BP algorithm, parameters in MCHL have analytical solutions, hence MCHL can be trained efficiently.

3.2 Marginalizing the Noise

Intuitively we can improve the generalization ability of SLFNs by extending
the training set in the MCHL feature space. This can be achieved by explicitly
corrupting each training sample in the MCHL feature space.

Given the training set D, let $\mathbf{t}_i$ denotes the mapping result of sample $\mathbf{x}_i$ in
MCHL feature space, i.e. $\mathbf{t}_i = \mathbf{h}(\mathbf{x}_i)$. We can corrupt each sample in MCHL
feature space N times according to a fixed noise distribution to generate a new
data set $\widetilde{D}$ with MN samples. For each sample $\mathbf{t}_m$ in MCHL feature space,
corruption corresponds to generate new samples $\widetilde{\mathbf{t}}_{mn}$ (with $n = 1, \cdots, N$). For
convenience, we take binary classification, $y \in \{-1, +1\}$, for example. The newly
generated data set $\widetilde{D}$ can be used for training by minimizing

$$\mathcal{L}(\widetilde{D}; \Theta) = \sum_{m=1}^{M} \frac{1}{N} \sum_{n=1}^{N} L(\widetilde{\mathbf{t}}_{mn}, y_m; \Theta), \tag{4}$$

where $\widetilde{\mathbf{t}}_{mn} \sim p(\widetilde{\mathbf{t}}_{mn}|\mathbf{t}_m), \Theta$ is the set of model parameters, and $L(\widetilde{\mathbf{t}}_{mn}, y_m; \Theta)$
is the loss function of the model. The binary classification can be extended to
multiclass (with k classes) by replace label y with label vector $\mathbf{y} = \{-1, 1\}^k$.

Explicit corruption is effective, but it has the problem of high computational
costs. The computational complexity of the minimization of $\mathcal{L}(\widetilde{D}; \Theta)$ scales
linearly in the number of corrupted samples. Here, we consider the limiting
case, i.e. $N \to \infty$. By applying the weak law of large numbers, we can rewrite
$\frac{1}{N} \sum_{n=1}^{N} L(\widetilde{\mathbf{t}}_{mn}, y_m; \Theta)$ as its expectation, i.e.,

$$\mathcal{L}(\widetilde{D}; \Theta) = \sum_{m=1}^{M} E[L(\widetilde{\mathbf{t}}_m, y_m; \Theta)]_{p(\widetilde{\mathbf{t}}_m|\mathbf{t}_m)}, \tag{5}$$

We assume: (1) corruption distribution is a member of the natural exponential family and the corruption of each dimension of $\mathbf{t}$ is independent; (2)
corruption distribution is unbiased, that is to say $E[\widetilde{\mathbf{t}}_m]_{p(\widetilde{\mathbf{t}}_m|\mathbf{t}_m)} = \mathbf{t}_m$. Here, we
use $\mathbf{w}$ to denote weights of hidden layer to output layer. When loss function is
quadratic loss function, $\mathbf{w}$ can be achieved by minimizing the loss function:

$$\mathcal{L}(\widetilde{D}; \mathbf{w}) = \sum_{m=1}^{M} E[(\mathbf{w}^T \widetilde{\mathbf{t}}_m - y_m)^2]_{p(\widetilde{\mathbf{t}}_m|\mathbf{t}_m)}$$

$$= \mathbf{w}^T (\sum_{m=1}^{M} E[\widetilde{\mathbf{t}}_m] E[\widetilde{\mathbf{t}}_m]^T + V[\widetilde{\mathbf{t}}_m]) \mathbf{w} - 2(\sum_{m=1}^{M} y_m E[\widetilde{\mathbf{t}}_m])^T \mathbf{w} + M \tag{6}$$

where $V[\mathbf{t}]$ is the covariance of $\mathbf{t}$, and all expectation and covariance are under
$p(\widetilde{\mathbf{t}}_m|\mathbf{t}_m)$. According to assumption (1), we can show that $V[\mathbf{t}]$ is a diagonal
matrix which stores the variances of $\mathbf{t}$. Set the derivatives of $\mathcal{L}(\widetilde{D}; \mathbf{w})$ with
respect to $\mathbf{w}$ equal to zero, we obtain the optimal solution

$$\mathbf{w}^* = (\sum_{m=1}^{M} E[\widetilde{\mathbf{t}}_m] E[\widetilde{\mathbf{t}}_m]^T + V[\widetilde{\mathbf{t}}_m])^\dagger (\sum_{m=1}^{M} y_m E[\widetilde{\mathbf{t}}_m]) \tag{7}$$

Probability density function (PDF), mean and variance of typically used corrupting distributions are listed in Table 1.

Table 1. PDF, mean and variance of typically used corrupting distributions.

Noise distribution	PDF	$E[\widetilde{t}]_{p(\widetilde{t}|t)}$	$V[\widetilde{t}]_{p(\widetilde{t}|t)}$
Blankout	$p(\widetilde{t}=0)=q,\ p(\widetilde{t}=\frac{1}{1-q}t)=1-q$	t	$\frac{q}{1-q}t^2$
Gaussian	$p(\widetilde{t}|t)=N(\widetilde{t}|t,\sigma^2)$	t	σ^2
Laplace	$p(\widetilde{t}|t)=Laplace(\widetilde{t}|t,\lambda)$	t	$2\lambda^2$
Poisson	$p(\widetilde{t}|t)=Poisson(\widetilde{t}|t)$	t	t

In summary, training process of MCHL can be summarized as follow:
Given a training set $D = \{(\mathbf{x}_i,\mathbf{y}_i)|\mathbf{x}_i \in \mathbf{R}^d, \mathbf{y}_i \in \mathbf{R}^k, i = 1,\cdots,M\}$,
feature mapping function $F(\mathbf{a},b,\mathbf{x})$, and hidden neuron number L,
    step1: Randomly generate a $M \times L$ full rank matrix $\mathbf{P}$.
    step2: Evaluate the hidden node parameters $(\mathbf{a},b)$ by solving Eq. (3).
    step3: Calculate the hidden layer output matrix $\mathbf{H}$.
    step4: Calculate the output weight $\mathbf{w}^*$ according to formula (7).

4   Experiments

Experiments include three parts: (1) analyze the influence of blankout corruption level q to classification performance of MCHL (We use blankout noise corruption and assume same noise level q for each dimension of feature.); (2) analyze the influence of hidden nodes number to classification performance of MCHL; (3) analyze the classification performance of MCHL.

Feature mapping function uses sigmoid function. l_2 regularizer is added to the weights calculation of hidden layer to output layer. Wide type of data sets are used in this section, most of the data sets are taken from UCI Machine Learning Repository [13]. We consider binary classification and multiclass classification two cases. Binary classification data sets include: Colon [12], Diabete [13], SPECTF [13], Heart [14], Madelon [15], Australian Credit [13] and Dimdata [16]. Multiclass classification data sets include: Iris [13], Glass [13], Win [13], Ecoli [13], Segment [13], Vehicle [13] and Letter [13]. The corresponding categories number are 3, 6, 3, 8, 7, 4 and 26, separately. Detailed information about the data sets are listed in Table 2.

4.1   Influence of Blankout Corruption Level q

We explore the classification performance of MCHL as a function of the blankout corruption level q. Blankout corruption level $q = 0$ means that MCHL do not corrupt the hidden layer outputs. We set the hidden nodes number to the sample

Table 2. Basic statistics of data sets.

Binary	Colon	Diabete	SPECTF	Heart	Madelon	Australian	Dimdata
# total	62	768	267	270	2600	690	4192
# features	2000	8	44	13	500	6	14
Multiclass	Iris	Glass	Win	Ecoli	Segment	Vehicle	Letter
# total	150	214	178	336	2310	846	20000
# features	4	9	13	7	19	18	16

features number and regularization parameter C to 10^{-4}. Four data sets are used in this subsection. The training data number for each data set are 150 (Heart), 1400 (Madelon), 40 (Colon) and 150 (SPECTF). Experimental results are listed in Table 3.

Table 3. Relation between classification results and blankout corruption level q.

	$q=0$	$q=0.1$	$q=0.2$	$q=0.3$	$q=0.4$	$q=0.5$	$q=0.6$	$q=0.7$	$q=0.8$	$q=0.9$
Heart	83.33	**85.00**	80.83	77.50	70.83	67.50	65.83	63.33	60.00	59.17
SPECTF	80.34	**81.20**	81.20	81.20	81.20	81.20	81.20	81.20	81.20	81.20
Colon	81.82	81.82	**86.36**	86.36	86.36	86.36	86.36	86.36	86.36	86.36
Madelon	56.67	57.83	**58.50**	58.17	58.00	58.00	58.00	57.83	57.83	58.33

From Table 3 we can find that: (1) Marginalizing the noise introduced in the hidden layer outputs can improve the classification results (1.67 % on Heart, 0.86 % on SPECTF, 4.54 % on Colon and 1.83 % on Madelon). (2) On SPECTF, Colon and Madelon data sets MCHL consistently improves the classification results on all blankout corruption level; (3) On the whole the best performance tends to be obtained by MCHL with low corruption levels, i.e. the order of q is around 0.2.

4.2 Influence of Hidden Nodes Number

As a kind of SLFNs, hidden layer of MCHL can nonlinear map training data into a high dimensional feature space. In this subsection we analysis the impact of hidden nodes number to the classification performance of MCHL. Heart and Madelon data sets are used in this subsection, and the training data number are 150 and 1400, respectively. Regularization parameter C is set to 10^{-4}. We analyze two cases, $q = 0$ (does not have corruption) and $q = 0.1$ (has corruption). Experimental results are listed in Tables 4 and 5. First column of Tables 5 and 6 corresponds to the primary feature dimension of the data.

From Tables 4 and 5, we can find that an appropriate increase in the number of hidden nodes can improve the classification performance of MCHL. Nonlinear

Table 4. Classification accuracies of different hidden nodes number on Heart data set.

	L = 13	L = 50	L = 100	L = 200	L = 400	L = 800	L = 1000	L = 1500	L = 2000
q = 0	78.33	**82.50**	73.33	70.00	69.17	70.83	72.50	70.83	72.00
q = 0.1	81.67	84.17	83.33	**85.00**	84.17	83.33	83.33	80.83	83.33

Table 5. Classification accuracies of different hidden nodes number on Madelon data set.

	L = 500	L = 50	L = 100	L = 200	L = 400	L = 800	L = 1000	L = 1500	L = 2000
q = 0	55.00	49.00	53.83	55.50	54.33	**56.83**	55.67	52.67	54.33
q = 0.1	56.50	50.67	52.33	55.17	54.50	58.17	57.67	**58.50**	58.33

feature mapping in MCHL has a similar effect of kernel function used in support vector machine (SVM).

4.3   Classification Performance

This subsection we make detailed experiments to analyze the classification performance of MCHL. SVM is used as baseline. All of data sets are used in the subsection. The training data number for each data set are: 150 (Hear), 1400 (Madelon), 40 (Colon), 150 (SPECTF), 510 (Diabete), 460 (Australian),1000 (Dimdata), 100 (Iris), 140 (Glass), 120 (Win), 220 (Ecoli), 1540 (Segment), 560 (Vehicle) and 13333 (Letter).

SVM uses popular RBF Kernel ($k(\mathbf{x}_i, \mathbf{x}_j) = \exp(-\gamma \|\mathbf{x}_i - \mathbf{x}_j\|^2)$). Experimental parameters are selected by cross-validation. Parameters C and γ are searched on grid $\{2^{-16}, 2^{-14}, 2^{-12}, \cdots, 2^{12}, 2^{14}, 2^{16}\}$. The number of hidden layer nodes is selected on grid $\{50, 100, 200, 400, 800, 1000, 1500, 2000\}$. Blankout noise corruption level q is searched on grid $\{0, 0.1, 0.2, \cdots, 0.9\}$. Experiments on each dataset are repeated ten times with randomly selected training and test data. The mean and standard deviation of classification accuracy are recorded. Experimental results are shown in Tables 6 and 7. From Tables 6 and 7, we can find that the classification performance of MCHL is slightly better than SVM. In addition the

Table 6. Binary class classification performance compare.

Data set	Colon	Diabete	SPECTF	Heart	Madelon	Australian	Dimdata
SVM	**89.55 ± 5.77**	79.22 ± 1.92	**82.22 ± 3.19**	**86.33 ± 2.01**	58.12 ± 1.04	86.43 ± 1.50	95.68 ± 0.24
MCHL	88.64 ± 6.51	**79.38 ± 1.56**	81.54 ± 3.16	85.75 ± 2.75	**59.15 ± 0.96**	**87.65 ± 1.89**	**95.74 ± 0.29**

Table 7. Multiclass classification performance compare.

Data set	Iris	Glass	Win	Ecoli	Segment	Vehicle	Letter
SVM	97.80 ± 1.89	**68.24 ± 4.57**	98.10 ± 1.21	87.07 ± 3.27	**96.43 ± 0.63**	**84.09 ± 1.71**	92.85 ± 0.28
MCHL	**98.20 ± 1.66**	67.84 ± 4.79	**98.45 ± 1.43**	**87.59 ± 2.62**	96.14 ± 0.65	83.57 ± 1.57	**93.50 ± 0.22**

parameters of MCHL have analytical solutions, this makes the training efficiency of MCHL higher than SVM.

5   Conclusions

Generalization ability of NNs is limited by the number of training samples. Explicitly extending the training set with artificially generated samples by corrupting hidden layer outputs can improve the generalization ability of NNs. But it has the problem of high computation costs. We propose MCHL which improves the generalization ability of SLFNs by marginalizing the noise introduced in the hidden layer outputs. In this way MCHL is trained with infinite samples. Experimental results on multiple data sets show that MCHL can be trained efficiently, and generalizes better to test data.

Acknowledgments. The work described in this paper was mainly supported by National Natural Science Foundation of China (No. 61300076 and 61375045), Ph.D. Programs Foundation of Ministry of Education of China (No. 20131101120035), Beijing Natural Science Foundation(4142030), Excellent young scholars Research Fund of Beijing Institute of Technology.

References

1. Hinton, G.E., Srivastava, N., Krizhevsky, A., Sutskever, I., Salakhutdinov, R.R.: Improving neural networks by preventing co-adaptation of feature detectors (2012)
2. Guo, P., Michael, R.L., Chen, C.L.P.: Regularization parameter estimation for feedforward neural networks. IEEE Trans. Syst. Man Cybern. Part B **33**(1), 35–44 (2003)
3. Rumelhart, D.E., Hinton, G.E., Williams, R.J.: Learning representations by back-propagating errors. Nature **323**(9), 533–536 (1986)
4. Guo, P., Michael, R.L.: A pseudoinverse learning algorithm for feedforward neural networks with stacked generalization application to software reliability growth data. Neurocomputing **56**(1), 101–121 (2004)
5. Burges, C.J.C., Scholkopf, B.: Improving the accuracy and speed of support vector machines. In: Advances in Neural Information Processing Systems, vol. 9, pp. 375–381 (1997)
6. Vincent, P., Larochelle, H., Lajoie, I., Bengio, Y., Manzagol, P.A.: Stacked denoising autoencoders: learning useful representations in a deep network with a local denoising criterion. J. Mach. Learn. Res. **11**, 3371–3408 (2010)
7. Chen, M., Xu, Z., Weinberger, K.Q., Sha, F.: Marginalized denoising autoencoders for domain adaptation. In: International Conference on Machine Learning, pp. 767–774 (2012)
8. Maaten, L., Chen, M., Tyree, S., et al.: Learning with marginalized corrupted features. In: International Conference on Machine Learning, pp. 410–418 (2013)
9. Chen, M., Zheng, A., Weinberger, K.: Fast image tagging. In: International Conference on Machine Learning, pp. 1274–1282 (2013)
10. Lawrence, N.D., Schölkopf, B.: Estimating a kernel fisher discriminant in the presence of label noise. In: International Conference on Machine Learning, pp. 306–313 (2001)

11. Huang, G.B., Chen, L., Siew, C.K.: Universal approximation using incremental constructive feedforward networks with arbitrary input weights. Technical report ICIS/46/2003
12. Li, J., Liu, H.: Kent Ridge Bio-Medical Data Set Repository (2004). http://levis.tongji.edu.cn/gzli/data/mirror-kentridge.html
13. Bache, K., Lichman, M.: UCI Machine Learning Repository (2013). http://archive.ics.uci.edu/ml
14. Mike, M.: Statistical Datasets (1989). http://lib.stat.cmu.edu/datasets/
15. Guyon, I.: Design of experiments of the NIPS 2003 variable selection benchmark (2003). http://www.nipsfsc.ecs.soton.ac.uk/papers/Datasets.pdf
16. Odewahn, S.C., Stockwell, E.B., Pennington, R.L., Humphreys, R.M., Zumach, W.A.: Automated star/galaxy discrimination with neural networks. Astronomical 103(1), 318–331 (1992)

An Incremental Network
with Local Experts Ensemble

Shaofeng Shen[1], Qiang Gan[1], Furao Shen[1(✉)], Chaomin Luo[2], and Jinxi Zhao[1]

[1] National Key Laboratory for Novel Software Technology,
Department of Computer Science and Technology,
Nanjing University, Nanjing, China
{shaofeng2014,njucsgq}@gmail.com, {frshen,jxzhao}@nju.edu.cn
[2] Department of Electrical and Computer Engineering,
University of Detroit Mercy, Detroit, USA
luoch@udmercy.edu

Abstract. Ensemble learning algorithms aim to train a group of classifiers to enhance the generalization ability. However, vast of those algorithms are learning in batches and the base classifiers (e.g. number, type) must be predetermined. In this paper, we propose an ensemble algorithm called INLEX (Incremental Network with Local EXperts ensemble) to learn suitable number of linear classifiers in an online incremental mode. Specifically, it incrementally learns the representational nodes of the input space. In the incremental process, INLEX finds nodes in the decision boundary area (boundary nodes) based on the theory of entropy: boundary nodes are considered to be disordered. In this paper, boundary nodes are activated as experts, each of which is a local linear classifier. Combination of these linear experts with dynamical weights will constitute a decision boundary to solve nonlinear classification tasks. Experimental results show that INLEX obtains promising performance on real-world classification benchmarks.

1 Introduction

Ensemble learning algorithms have attracted much attention for decades. Bagging [1] and AdaBoost [2] are well-known ensemble learning algorithms. They enhance the generalizing ability by training a group of classifiers with different distribution. The combinational strategies are voting (e.g. Bagging) or weighted voting (e.g. AdaBoost). Another well-known ensemble model is Mixture of experts (ME) [3] in which the divide-and-conquer strategy is used. In ME, several local experts are trained to partition the input space, the weights of experts are computed dynamically based on the input.

Online incremental learning is another important topic in machine learning. Online learning algorithms process the data one-by-one and are suitable to the applications with continuous online data or insufficient memory space (e.g. visual tracking, embedded system). Incremental learning addresses the ability

© Springer International Publishing Switzerland 2015
S. Arik et al. (Eds.): ICONIP 2015, Part III, LNCS 9491, pp. 515–522, 2015.
DOI: 10.1007/978-3-319-26555-1_58

of repeatedly training a system using new data without destroying the old prototype patterns [4]. Online incremental learning algorithms often adapt their models to the input well.

In the ensemble learning literature, some online ensemble algorithms have been proposed [5–7]. However, in those methods, the base classifier or expert must be predetermined. That means we must determine both suitable number and type of base classifiers before applying those algorithms. In an online learning system, each data arrives one-by-one. Therefore, some artificial methods (e.g. brute searching, cross-validation) cannot be used. As a result, we can not obtain some prior knowledge such as suitable number and type of the base classifier. We need to get those knowledge in the learning process. That inspires the study of online incremental ensemble learning. Specifically, this issue can be that: (1) the base classifiers are generated in the online incremental process, (2) the base classifiers are efficient and simplify. In another aspect, as far as Minsky concerns that, the human solve the classification problem just by perceptrons. Minsky thinks that the brain is not a unified entity but a society of elements that both complement and complete with one another. For efficiency and simplicity, linear classifier is mostly suitable to be the base classifier in an online ensemble system.

In this paper, we propose an ensemble algorithm called INLEX (Incremental Network with Local EXperts ensemble) to learn suitable number of linear classifiers in an online incremental mode. INLEX incrementally learns the representational nodes of the training data based on competitive Hebbian rule [8]. In the incremental process, it finds nodes in the decision boundary area (boundary nodes) based on the theory of entropy: boundary nodes are considered to be disordered. In addition, boundary nodes will be activated as experts, each of which is a local linear classifier. In total, INLEX aims to find and train the experts in the decision boundary area. As a result, dynamical combination of them can constitute a decision boundary to solve classification tasks. Experimental results show that INLEX obtains promising performance on real-world classification benchmarks.

2   Proposed Method

In Fig. 1, we show the motivation of INLEX. Illustrated in Fig. 1, we define the area between class A and B as decision boundary area. In step 1, suitable number of nodes are learned. These nodes are able to represent the distribution of the train data. In step 2, we find the nodes in the decision boundary area (boundary nodes) and activate them as experts. Each expert is a local linear classifier. In step 3, the experts are trained in a supervised way. As a result, these local experts can solve the classification task competitively and complementally. Namely, combination of them can constitute a decision boundary to solve this classification task.

In this paper, we implement that motivation in an online incremental mode in INLEX which can also solve multi-classification tasks. Nodes are generated

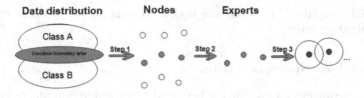

Fig. 1. Motivation of INLEX

incrementally in a self-organized manner. We do not need some prior knowledge such as number of nodes. In the learning process, new experts will be activated if new boundary nodes have generated. Meanwhile, experts in INLEX online learns each labeled data.

To realize these targets, we emphasize the key aspects of nodes growing procedure, boundary nodes detection, experts activation and training.

2.1 Nodes Growing Procedure

To learn the representational nodes of the training data incrementally, competitive Hebbian rule and similarity threshold criterion are used in INLEX. The competitive Hebbian rule can be described as: for each input signal, find its two closest nodes (measured by Euclidean distance) and connect with an edge between them. The similarity threshold criterion is defined in Definition 1. The input signal is a new node if the distance between the signal and the nearest node (or second nearest node) is greater than a threshold T.

Definition 1. *Similarity Threshold Criterion: If node i has neighbors (there are some nodes connected to i), its similarity threshold is defined by the largest Euclidean distance between i and its neighbors. Otherwise, its similarity threshold is defined by the smallest Euclidean distance between i and other nodes.*

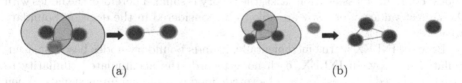

Fig. 2. Two situations of nodes growing procedure. The green nodes are the input signals and other nodes are nodes in INLEX. In Fig. 2(a), an input lies within the local region of its winners, the connection between two winners will be established and the nearest one will be pulled to the input sample. In Fig. 2(b), the input is far from its two winners and lies in a new region. Therefore, a new node will be inserted to represents the new region (Color figure online).

Shown as Fig. 2, based on the competitive Hebbian rule, INLEX learns the topology of the input space. Based on the similarity threshold, new node will be inserted adaptively. In total, the nodes growing procedure continues in a self-organized manner.

2.2   Boundary Nodes Detection

To solve a classification task, the key is to find the decision boundary. Therefore, in INLEX, only boundary nodes are activated. Herein, we will propose the online boundary nodes detection method.

For an input signal (x, d), where x is the input features vector and d is the label, we can define the similarity between x and node i as:

$$s_i = e^{-||w_{s_i} - x||} \qquad (1)$$

where w_{s_i} represents the position of node i. Then, for each input signal, every node i learns its similarity to class d with s_i. Namely, in the learning process, each node i records its accumulated similarity to every class. For T-classification task, the accumulated similarity vector is defined as:

$$sv_i = (p_{1_i}, p_{2_i}, ..., p_{T_i}) \qquad (2)$$

p_{t_i} is the accumulated similarity between node i and class t. Then, entropy of sv_i can be used to judge the position of node i. The nodes with the largest values of entropy are considered in the decision boundary area.

In thermodynamics, entropy is a measure of the number of specific ways in which a thermodynamic system may be arranged, commonly understood as a measure of disorder. In sv_i, p_{t_i} is the similarity between node i and class t. Consider that in a binary classification task. If node i is in the boundary between class A and class B, the values of its accumulated similarity points p_{A_i} and p_{B_i} are approaching. Namely, it is similar to both class A and class B. Incorporated with the theory of entropy, we can interpret that as: the similarity vectors of the boundary nodes are usually disordered. Their values of entropy are universally larger than other nodes. For multi-classification tasks, the theory is same. Therefore, the nodes with the largest values of entropy are able to be considered in the decision boundary area.

Because INLEX learns incrementally, it finds boundary nodes based on accumulated entropy. In INLEX, each node records the accumulated similarity to each class based on latest learned L_2 (predefined parameter) input signals. When the learning times are multiple of L_2, each node i calculates the entropy E_i of the similarity vector and adds E_i to its accumulated entropy. The nodes which are not experts and with the largest values of accumulated entropy are activated.

2.3   Experts Activation and Training

Each expert is a linear classifier. Therefore, while node i is activated as an expert, two extra parameters will be assigned to it. The first one in expert i is

$\theta_i = \{w_{i_1}, ..., w_{T_1}\}$. To solve T-classification task, each expert i gives its output for class t $(t = 1, ..., T)$ based on w_{t_i}, namely:

$$O_{t_i} = \frac{1}{1 + e^{-w_{t_i}^T x}} \tag{3}$$

The second extra parameter in expert i is w_{g_i} which is used to assign weight. w_{g_i} is initialized by w_{s_i}. For each input data, the weight g_i of expert i is defined by:

$$g_i = \frac{e^{-||w_{g_i} - x||}}{\sum_{i \in E_s} e^{-||w_{g_i} - x||}} \tag{4}$$

where E_s is the experts set. The training for experts aims to update parameter θ_i and slightly adjust the position of expert i. As a result, these experts solve the classification task both competitively and complementally. Therefore, to preserve the topology of nodes, we assign each expert i an extra parameter w_{g_i} which is used in the experts training process.

Next, INLEX will combine the output values of these experts to make prediction. For each data (x, d), the ensemble output for class t and predictive label are:

$$O_t = \sum_{i \in E_s} g_i O_{t_i} \quad y = arg \max_t O_t \tag{5}$$

We define the loss function as:

$$L_e = -log \sum_{i \in E_s} g_i e^{-\frac{1}{2}(1 - O_{d_i})^2} \tag{6}$$

Computing the derivatives of the loss function with respect to w_{d_i} and w_{g_i} of expert i will yield the update strategies for the parameters under the method of gradient descent. Namely,

$$\Delta w_{g_i} = -\eta(h_i - g_i)\frac{x - w_{g_i}}{||x - w_{g_i}||}O_{t_i}(1 - O_{t_i}) \tag{7}$$

$$\Delta w_{d_i} = \eta h_i(1 - O_{d_i})O_{d_i}(1 - O_{d_i}) \tag{8}$$

$$h_i = -\frac{g_i e^{-\frac{1}{2}(1 - O_{d_i})^2}}{\sum_{k \in E_s} g_k e^{-\frac{1}{2}(1 - O_{d_k})^2}} \tag{9}$$

In addition, a penalty to each parameter w_{t_i} $(t \neq d, i \in E_s)$ is provide so that the output value for class t decreases. Namely,

$$\Delta w_{t_i} = \eta h_i(0 - O_{t_i})O_{t_i}(1 - O_{t_i}) \tag{10}$$

In Eqs. (7), (8) and (10), η is the learning rate and calculated by: $\eta = 1/M_j$. M_j is the winning time of node j.

Algorithm 1. Complete algorithm of INLEX

1. Initially, two nodes are initialized randomly. One of them is activated as an expert.
2. Input a labeled data point (x, d).
3. Find the nearest node n_1 and second nearest node n_2 of x.
4. Get the threshold T_{n_1}/T_{n_2} of n_1/n_2 based on the similarity threshold criterion.
5. If $||w_{s_{n_1}} - x|| > T_{n_1}$ or $||w_{s_{n_2}} - x|| > T_{n_2}$, x is a new node. Otherwise, update the positions of n_1 by:

$$\Delta w_{s_{n_1}} = \frac{1}{M_{n_1}}(x - w_{s_{n_1}}) \tag{11}$$

6. If there is no edge connecting n_1 and n_2, connect n_1 and n_2 by an edge. Set the age of this edge to 0. Otherwise, only refresh this age: set its age to 0.
7. Find all edges connected to n_1. Add the ages of those edges by 1. Delete those edges whose ages are larger than a_{max} (predefined parameter).
8. For each node i, calculate the similarity s_i between i and x by Eq. (1), add its accumulated similarity point p_{d_i} by s_i.
9. Each expert updates its parameters according to Eqs. (7), (8) and (10).
10. If the learning times are multiple of L_1 (predefined parameter), go to step 11. Otherwise, go to step 12.
11. Denoising: delete the nodes which are not experts and have no neighboring node.
12. If the learning times are multiple of L_2, go to step 13. Otherwise, goto step 2.
13. Find and activate boundary nodes:
    - Calculated the entropy of each node i in this period:

$$E_i = -\sum_{t=1}^{T} \frac{p_{t_i}}{\sum_{t=1}^{T} p_{t_i}} log(\frac{p_{t_i}}{\sum_{t=1}^{T} p_{t_i}}) \tag{12}$$

    - For each node i, add its accumulated entropy $E_i{}^A$ by E_i. Calculate the maximum/mean accumulated entropy $E_{max}{}^A/E_{mean}{}^A$. The threshold T_e is defined by $(E_{max}{}^A + E_{mean}{}^A)/2$.
    - For each node i, if $E_i{}^A$ is larger than T_e, it is a boundary node. And if boundary node i is not an expert, activate node i as expert.
    - Clear the similarity vector.
14. Goto step 2.

2.4   The Complete Algorithm of INLEX

As a summary, we provide the complete algorithm of INLEX in Algorithm 1. In Algorithm 1, while a new node i is inserted, it will be added into nodes set N_s ($N_s = i \cap N_s$). Its accumulated similarity vector is 0. Its accumulated entropy $E_i{}^A$ is set to 0. Its times as winners (M_i) are set to 1. While node i is activated as expert i, it will be added into experts set E_s ($E_s = i \cap E_s$). When the following input signals arrive, the new expert will be trained.

3   Experiments

This section evaluates INLEX's performance on classification benchmark data sets. The details of the data sets are shown in Table 1. They are all from

UCI machine learning repository [9]. In experiments, each data set is divided into ten parts. We randomly select one part as testing set and the remaining parts as training set. Besides, we compare INLEX with leading online Boosting (OnBoost) algorithm proposed in [6] to show its generalizing performance.

Experiment Setup: Before applying OnBoost on classification tasks, suitable number and type of base classifiers must be predetermined. In the experiment, we set the number of base classifiers of OnBoost to 100, which is same as that in [6]. Then, we test the performance of OnBoost with different type of base classifier. The base classifiers of OnBoost are Perceptron, Naive Bayes and Multi-Layer Perceptrons (MLP), respectively. We notate OnBoost with Perceptron, Naive Bayes and Multi-Layer Perceptrons as OB_P, OB_NB and OB_MLP, respectively.

There are three predefined parameters in INLEX. They are a_{max}, L_1 and L_2. We set them to 100, 200, 600, respectively, in our following experiment. In experiments, we also find that our method achieves stable results with respect to settings of these parameters. In addition, the learning times of INLEX on iris, balance-scale and breast are set to 3000. In the remaining data sets, the learning times are set to be identical to the instances number of each training set. Namely, INLEX learns those data sets with only one iteration.

Comparing Results: The error rates of INLEX and OnBoost are summarized in Table 1. At first, we analyze the results of OnBoost. The performances of OnBoost with different type of base classifier are different. When the base classifier is linear (perceptron), OnBoost's performance is poor. OnBoost also gets unstable results when its base classifier is nonlinear. For example, OB_MLP performs well in iris and nursery but much poorly in balance-scale and breast. OB_NB gets good results in breast and balance-scale but poor results in waveform and waveform-n. As the results demonstrating that, when apply OnBoost to solve classification tasks, we must select suitable type of base classifier.

Then, we compare INLEX with OnBoost. Summarized in Table 1, INLEX performs over OnBoost in all the data sets other than nursery. INLEX incorporates the merits of incremental learning. It usually adapts to each data set well. In nursery, the performance of INLEX is worse than that of OB_NB and OB_MLP. It is notable that one of the critical weakness of MLP is that its converging speed is so slow. In our experiment, ON_MLP learns each data set 50 iterations. However, INLEX only learns the data set of nursery one iteration. Therefore, OB_MLP performs better than INLEX in nursery. Even so, OB_MLP performs poorly on some data sets as well.

Next, the experts numbers of INLEX on those data sets are 14, 29.6, 30.1, 17.4, 39.1 and 59.2 respectively. Because the experts are activated from nodes, the number of experts in INLEX is related to the distribution of each data set. It has relevance to the instances number of each data set.

Table 1. Error rates for INLEX VS OnBoost

Data set	Instances	Classes	Attributes	INLEX	OB_P	OB_NB	OB_MLP
Iris	150	3	4	**0.0400**	0.0733	0.0600	**0.0400**
Balance-scale	625	3	4	**0.0996**	0.1520	0.1253	0.1680
Breast	683	2	9	**0.0293**	0.0322	0.0427	0.0864
Waveform-n	5000	3	40	**0.1378**	0.2240	0.1920	0.1680
Waveform	5000	3	21	**0.1436**	0.2820	0.1860	0.1700
Nursery	12960	5	9	0.1134	0.1427	0.0908	**0.0555**

4   Conclusion

In this paper, an online incremental ensemble algorithm called INLEX is proposed. INLEX incrementally learns the nodes in the decision boundary area and activates them as experts. Each expert is a local linear classifier. Combination of experts will constitute a decision boundary to solve classification tasks. Experimental results show that INLEX has promising generalizing performance.

The future work includes decreasing the experts number. Some strategy should be used to avoid activating similar nodes. Specifically, in the experts activation step, if a new boundary node is found, we should use some strategies to judge if it is necessary to activate this node as expert.

Acknowledgments. This work was supported by the National Natural Science Foundation of China (Grant No. 61375064, 61373001 and 61321491), Foundation of Jiangsu NSF (Grant No. BK20131279).

References

1. Breiman, L.: Bagging predictors. Mach. Learn. **24**(2), 123–140 (1996)
2. Freund, Y., Schapire, R.E.: A decision-theoretic generalization of on-line learning and an application to boosting. J. Comput. Syst. Sci. **55**(1), 119–139 (1997)
3. Jacobs, R.A., Jordan, M.I., Nowlan, S.J., et al.: Adaptive mixtures of local experts. Neural Comput. **3**(1), 79–87 (1991)
4. Shen, F.R., Hasegawa, O.: An incremental network for on-line unsupervised classification and topology learning. Neural Netw. **19**(1), 90–106 (2006)
5. Jordan, M.I., Jacobs, R.A.: Hierarchical mixtures of experts and the EM algorithm. Neural Comput. **6**(2), 181–214 (1994)
6. Oza, N.C.: Online bagging and boosting. In: 2005 IEEE International Conference on Systems, Man and Cybernetics, vol. 3, pp. 2340–2345. IEEE (2005)
7. Chen, S.T., Lin, H.T., Lu, C.J.: An online boosting algorithm with theoretical justifications. In: ICML, pp. 1007–1014 (2012)
8. Martinetz, T., Schulten, K.: Topology representing networks. Neural Netw. **7**(3), 507–522 (1994)
9. Bache, K., Lichman, M.: UCI machine learning repository (2013). http://archive.ics.uci.edu/ml

Nitric Oxide Diffusion and Multi-compartmental Systems: Modeling and Implications

Pablo Fernández López[1], Patricio García Báez[2],
and Carmen Paz Suárez Araujo[1(✉)]

[1] Instituto Universitario de Ciencias y Tecnologías Cibernéticas,
Universidad de Las Palmas de Gran Canaria, Campus Universitario de Tafira,
35017 Las Palmas de Gran Canaria, Spain
{pfernandez,cpsuarez}@dis.ulpgc.es
[2] Departamento de Ingeniería Informática y de Sistemas,
Universidad de La Laguna, 38071 La Laguna, TF, Spain
pgarcia@ull.es

Abstract. The *volume transmission* (VT), a new type of cellular signaling, is based on the diffusion of neuro-active substances such as *Nitric Oxide* (NO) in the *Extracellular Space* (ECS). It is not homogeneous, critically dependent on, and limited by, its structure and physico-chemical properties. We present a different computational model of the NO diffusion based on multi-compartmental systems and transportation phenomena. It allows incorporating these ECS characteristics and the biological features and restrictions of the NO dynamics.

This discrete model will allow to determine the NO dynamics and its capabilities in cellular communication and formation of complex structures in biological and artificial environments.

This paper addresses the design model and its analysis in one-dimensional and three-dimensional environment, over trapezoidal generation and diffusion processes.

Keywords: Multi-compartmental systems · Nitric oxide diffusion · Volume transmission · NO generation/synthesis · Diffusion neighbourhood

1 Introduction

An important mechanism of cellular communication is the *volume transmission* (VT). It coexists with synaptic transmission and it is based on the diffusion of neuro-active substances such as *Nitric Oxide* (NO) in the *Extracellular Space* (ECS), which is not homogeneous, critically dependent on, and limited by, its structure and physico-chemical properties [1]. NO is a free radical gas, highly diffusible in both aqueous and lipid environments. It is considered an *atypical* cellular messenger, since it needs no receptor, does not accumulate in synaptic vesicles and it freely diffuses through membranes affecting neighbouring cells. NO dynamics make up diverse processes: *Generation*, *Diffusion*, *Selfregulation* and *Recombination*, because NO regulates its own production.

© Springer International Publishing Switzerland 2015
S. Arik et al. (Eds.): ICONIP 2015, Part III, LNCS 9491, pp. 523–531, 2015.
DOI: 10.1007/978-3-319-26555-1_59

In the absence of determinant experimental data for understanding how NO functions as a neural signalling molecule, we have developed a computational model of NO diffusion based on Multi-Compartmental Systems [2] and transportation phenomena. It is different from the previously proposed models, [3–7], because of it allows to incorporate the biological features and restrictions of the NO dynamics and of the environment where the NO diffusion processes take place. Our proposal can use whatever morphology of NO generation and to consider the diffusion in anisotropic and non-homogeneous environment. The main objective is to define a formal framework for determining the NO dynamics and its capabilities in cellular signalling and formation of complex structures, in biological and artificial environments.

This paper, addresses the model and its analysis in one-dimensional and three-dimensional environment, over generation and diffusion processes.

2   Multi-compartmental Model of NO Diffusion

The multi-compartmental model of NO diffusion appears to be able to consider a plausible biologically morphology of NO generation/synthesis. This feature is not present in the spontaneous generation, which is used in the analytical model [3]. This is because NO diffuse at the same instant of its synthesis. We consider this biological reality and model the NO diffusion at the molecular level where it is mathematically discrete. We define, as a theoretical abstraction, *the compartment*. It is the minimal volume where the NO diffusion is expressed. Its biological counterpart can be found at any neural level, from the molecule to the circuit. We study the NO behavior in complex systems composed of many interconnected compartments. Its formalization is based on multi-compartmental systems [2,8], which are a subclass of linear dynamic systems, given by an equation set like the one given in Eq. (1).

$$\frac{dq_i}{dt} = -(J_{0i} + \sum_{j \neq i} J_{ji})q_i + \sum_{j \neq i} J_{ji}q_i + E_i \tag{1}$$

Where E_i is the input flow. J_{hk}, the transference coefficients, defined as $J_{hk} = f(\mathbf{q}, t)$, the sustance transferences, $t_{ij} = J_{ij}q_j$ and $t_{ji} = J_{ji}q_i$ and the out flow $f_{0i} = J_{0i}q_i$, depending on the substance, $\mathbf{q}$, inside of the compartments.

On the other hand, the transport phenomena takes place when, due to the gradient of a physical magnitude, another is displaced and invades and affects its setting in a structural and/or functional way. Its behavior can be studied from a phenomenological point of view [9].

The proposed model is sustained on these two pillars, where the expressions that define the NO diffusion dynamics are deduced.

Let us consider a set of compartments in a one-dimensional environment, Fig. 1. Any of these compartments allow possibilities of NO generation, diffusion, self-regulation and recombination. The generation process is defined by the F_i function. F_i defines both, the quantity and the morphology of the generated

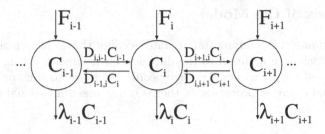

Fig. 1. One-dimensional environment of compartments

NO in the compartment. The transportation phenomena cause the speed of NO flow between two compartments to be proportional to their NO concentration difference. The NO concentration gradient in a compartment takes into account the self-regulation process of NO [9]. This is mathematically expressed by the Eq. (2).

$$\frac{dC_i}{dt} = D_{i,i-1}(C_{i-1} - C_i) + D_{i,i+1}(C_{i+1} - C_i) - \lambda_i C_i + F_i \qquad (2)$$

where $D_{i,i-1}$ and $D_{i,i+1}$ are the environmental coefficients of diffusion between the indicated compartments. λ_i is the self-regulation parameter of NO, which is proportional to the quantity of NO concentration.

The model of the NO diffusion dynamics is defined by a system of first order differential equations, as in Eq. (2), where we can consider specific cyclic contour conditions, Eq. (3). It can be extended in direct form to two and three-dimensional environments.

$$\frac{d\mathbf{C}}{dt} = \mathbf{HC} + \mathbf{F} \qquad (3)$$

Where $\mathbf{C} = (C_1, C_2, ..., C_N)^T$, $\mathbf{F} = (F_1, F_2, ..., F_N)^T$, and $\mathbf{H}$ (matriz tridiagonal), Eq. (4). Needing to consider the semi-discretization factor $1/(\Delta x)^2$ and multiplying by $\mathbf{H}$.

$$\mathbf{H} = \begin{pmatrix} -(D_{1,N} + D_{1,2} + \lambda_1) & D_{1,2} & \cdots \\ D_{2,1} & -(D_{2,1} + D_{2,3} + \lambda_2) & \cdots \\ \vdots & \vdots & \ddots \end{pmatrix} \qquad (4)$$

The model allows considering the non-homogeneity of the environment, and is also able to cause different forms in the processes of generation of NO in distinct regions of the environment, as well as variations in the self-regulation of NO according to the treated region. This is represented in the model by F_i and λ_i functions. In addition, it allows the anisotropy of the environment to be established by means of the use of specific diffusion coefficients for every inter-compartmental environment, where the NO spreads by. These capabilities provide a model which can be categorized as a generalized formal tool with high power to emulate and to study the NO dynamics very close to the NO behavior in the biological environment.

3   Analysis of the Model

The proposed model is analyzed taking into account those characteristics found in the NO dynamic in addition to the diffusion environment, fitting to the biological reality of the dynamic. We pay specific attention to the analysis of the generation and diffusion processes, of the NO, in a one-dimensional case and in the volumetric case.

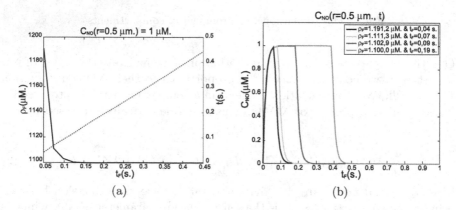

Fig. 2. (a) Behaviour of the multi-compartmental model of NO diffusion for different generation strengths of NO. Single NO generation. (b) Profiles of NO concentration at $r = 0.5\,\mu\mathrm{m}$ as a function of ρ_F and t_F. Single NO generation (Color figure online).

The first part of the study focuses on the generation process of the NO and we call it the F_i Function. It is a trapezoidial function, and has as its source data studies carried out by Aleh Balbatun and other researchers on endothelial cells of rats and rabbits [10].

An analysis of the model behaviour is carried out with respect to strength of the NO generation. We will confirm its consistency, using different values for strength, to those experimentally obtained from Tadeusz Malinski and others [11]. These results determine an induced maximum value of the NO concentration for $1\,\mu\mathrm{m}$ in an endothelial cell membrane of $1\,\mu\mathrm{m}$ in diameter.

We use an environment of 21 linearly arranged compartments, see Fig. 1. Each one represents a space measuring $0.1\,\mu\mathrm{m}$. Given that the NO diffusion is symmetric along the entire sphere which circumscribes the endothelial cell, we analyze the behaviour of the NO concentration which passes through the diameter of this sphere. Given that the endothelial cell is in the center of the environment, the NO generation occurs in compartment 11. We analyze the strength that this generation must have so that the reached value of the NO concentration at $r = 0.5\,\mu\mathrm{m}$ (compartments 6 and 16 of the environment) is $1\,\mu\mathrm{M}$.

Behaviour of the model regarding strength of the NO generation (ρ_F) is shown in Fig. 2(a). We show the relationship between ρ_F and t_F (generation time) (black line), for established biological values [11]. The maximum value is

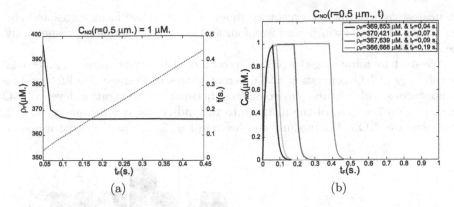

Fig. 3. (a) Behaviour of the multi-compartmental NO diffusion model for different generation strengths of multiple NO generation. (b) NO concentration profiles, for $r = 0.5\,\mu\text{m}$ as a function of ρ_F and t_F. Multiple NO generation.

attained in times very close to the length of the generation process (red line). We observe that there is an exponential increase in the strength of the generation process as the time duration is reduced. In addition, it is noted that the required times for C_6 and C_16 to reach a value of $1\,\mu\text{M}$ is linear with respect to the duration of the generation processes. Similarly, both Figs. 2(b) and 3(b) with NO concentration profiles when $r = 0.5\,\mu\text{m}$ and different values of ρ_F and t_F, reveals that the $1\,\mu\text{m}$ value reached is independent of the duration of the generation process.

When different NO generation processes coexist in the center of the endothelial cell the behaviour of the model is similar to the simple generation, Fig. 3(a). In our study for 3 processes in compartments 10, 11 and 12 values obtained for ρ_F, are approximately one third of the values obtained previously. Presence of self-regulation is not assumed here nor in the single generation, and consequently the relationship of the coexisting sources in the environment and in time is additive.

The second part of our analysis of the model focuses on the NO diffusion, the creation of *diffusion neighborhood* (DNB) and the emergence of complex structures. Data used comes from biological experiments [5]. The diffusion coefficient in an isotropic medium is $D = 3.3 \times 10^3\,\mu\text{m}^2\,\text{s}^{-1}$. The quantity of NO at time $t = 0\,\text{s}$, is $0.24\,\text{nmol cm}^{-3}$, which is the time from the generation of the NO until the detection of the diffusion, specifically $400 \pm 20\,\text{ms}$, with a growth ratio of $1.2 \pm 0.05\,\text{nmol cm}^{-3}\text{s}^{-1}$, maximum concentration, $4.30 \pm 0.15\,\text{nmol cm}^{-3}$ at time of $600 \pm 20\,\text{ms}$ and a mean NO lifetime that ranges from $0.5\,\text{s}$ and $5\,\text{s}$. The used three-dimensional isotropic, non-isotropic, homogeneous and non-homogeneous environment, is a $110 \times 110 \times 110\,\mu\text{m}^3$, which includes $11 \times 11 \times 11$ compartments see Fig. 4(e), where $\Delta x = \Delta y = \Delta z = 110\,\mu\text{m}$ and $\delta_i = 10\,\mu\text{m}$.

The generation of NO occurs in centralized compartments of the aforementioned environment and each compartment has a neighborhood made up of the six nearest compartments, except for those located on the faces, edges and corners of the 3D shape which have five, four and three compartments, respectively, see Fig. 4(a)–(d).

We start by analyzing the NO diffusion model behavior using a trapezoidal morphology of NO generation and different values of average NO lifetime in a homogeneous and isotropic environment. Maximum concentration levels of NO reached, as a function of the distance to the individual compartment, (i, j, k), that generates NO in the medium are shown in Fig. 5(a). The time that is needed

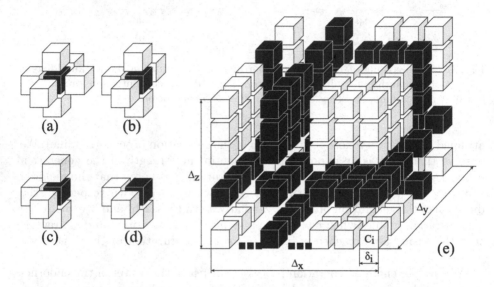

Fig. 4. 3D environment, where the model is analyzed. Black colors represent planes and lines for a visualization of the NO dynamic, (e). Details of neighborhoods for compartment $C_{i,j,k}$ when it is located inside of the 3D shape (a), when it forms part of a face (b), an edge (c), or a corner of the 3D shape (d).

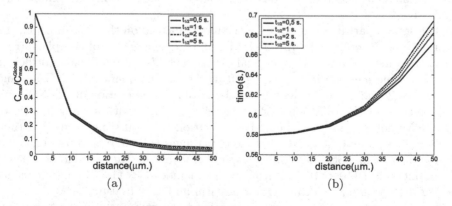

Fig. 5. (a) C_{max}/C_{max}^{global}, as a function of distance. Single generation. (b) Time to reach maximum concentration C_{max} as a function of distance.

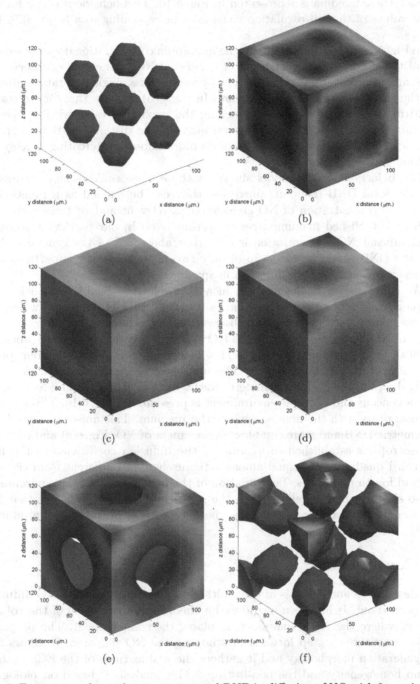

Fig. 6. Emergence of complex structures and DNB in diffusion of NO with 8 coexisting generation processes in an $11 \times 11 \times 11$ compartment environment.

to reach these maxima is represented in Fig. 5(b). This behaviour occur for different values of the self-regulation constant, corresponding to a $t_{1/2} = 0.5$, 1, 2 and 5 s.

In Fig. 5(a) we see that at 30 μm the maximum concentration does not exceed 5 % of the maximum concentration of generated NO. We can also observe the little influence that self-regulation of NO presents for a single generation process and the generated quantity of NO is low. In Fig. 5(b) we show that for a distance of 10 μm there is no delay when reaching the maximum value with respect to the same generation process and as we increase the distance to the generator compartment, the self-regulation becomes more important, creating a delay for reaching such maximum value.

The behaviour of the model shown in these figures allows non symmetric and non local DNB to be obtained and they can be defined as a function of the spread concentration of NO considered as relevant and/or time needed to reach an established maximum concentration level. In our particular study, if the significant NO concentration is only that above 5 % of the generated NO, then the DNB would be made up of only two of the nearest compartments to the generator, one for each direction in space.

We continue our model analysis in a plausible biological situation for NO diffusion, considering the features and restrictions of the environment where the process takes place. We carry out observations for behaviour of volumetric diffusion of NO, including simultaneous NO generation with 8 coexisting processes Fig. 6(a), in an environment of 11 × 11 × 11 compartments with different properties of isotropy and homogeneity. The emergence of complex structures occurs and is shown in Fig. 6. Total symmetry in the formation of DNB and its range in a homogeneous and isotropic environment is presented in Fig. 6(b). This symmetry disappears with the non-isotropy of the medium. This anisotropy produces asymmetric DNB and more complex displacement of NO, Fig. 6(c) and (d). The non-isotropy is established in quadrants, the diffusion coefficients of the first and third quadrants are equal amongst themselves and different from the second and fourth quadrants. The variation of the homogeneity of the environment is also shown, which is related with the self-regulation, and indicates that NO disappears more rapidly in the area where generation processes are produced, see Fig. 6(e) and (f).

4   Conclusions

The description and analysis of the multi-compartmental model of NO diffusion was carried out. It has been introduced a theoretical abstraction of the volume element where the NO dynamics take place, the compartment. The proposed model constitutes a step forward in studying the NO diffusion. It allows any NO generation morphology and it gathers the real features of the ECS such as the no homogeneity and the non-isotropy. This analysis is based on biological data given by Malinski et al. It has been done studying the NO behaviour, with specific attention on the generation and diffusion process, in one-dimensional and three-dimensional environments.

We analysed the NO model with single and multiple generation processes, using trapezoidal morphology. The obtained results fit with biological behaviour of NO. Important aspects regarding generation strength and time have been observed and the no dependency of the time generation for reaching the maximum NO induced concentration. When multiple NO generation processes coexist, DNBs and the formation of complex structures appear. They present high complexity and asymmetries for non-homogenous and asymmetric environments. Therefore, our study shows the possibility of non-local and non-symmetric DNB generation and the influence of the self-regulation process in that dynamic.

Finally, our model is powerful to study and determine the NO dynamics in biological and artificial environments and it represents a generalized formal tool for designing and interpreting biological experiments on NO behaviour.

References

1. Syková, E.: Extrasynaptic volume transmission and diffusion parameters of the extracellular space. Neuroscience **129**, 861–876 (2004)
2. Jacquez, J.A., Simon, C.P.: Qualitative theory of compartmental systems. SIAM Rev. **35**(1), 43–79 (1993)
3. Fernández López, P., García Baez, P., Suárez Araujo, C.P.: Dynamic of nitric oxide diffusion in volume transmission: model and validation. In: Loo, C.K., Yap, K.S., Wong, K.W., Teoh, A., Huang, K. (eds.) ICONIP 2014, Part I. LNCS, vol. 8834, pp. 50–58. Springer, Heidelberg (2014)
4. Lancaster, Jr., J.R.: Simulation of the diffusion and reaction of endogenously produced nitric oxide. Proc. Natl. Acad. Sci. USA **91**, 8137–8141 (1994)
5. Wood, J., Garthwaite, J.: Models of the diffusional spread of nitric oxide: implications for neural nitric oxide signalling and its pharmacological properties. Neuropharmacology **33**(11), 1235–1244 (1994)
6. Krekelberg, B.: Modelling cortical seft-orgnization by volumen learning; Doctoral Dissertation, Kings College London (1997)
7. Philippides, A., et al.: Four-dimensional neuronal signaling by NO: a computational analysis. J. Neurosci. **20**(3), 1199–1207 (2000)
8. Godfrey, K.: Compartmental models and their application. Academic Press, London (1983)
9. Suárez Araujo, C.P.: Study and reflections on the functional and organisational role of neuromessenger Nitric Oxide in learning: an artificial and biological approach. In: Computer Anticipatory Systems, AIP, vol. 517, pp. 296–307 (2000)
10. Balbatun, A., et al.: Dynamics of nitric oxide release in the cardiovascular system. Acta Biochim. Pol. **50**(1), 61–68 (2003)
11. Malinski, T., et al.: Diffusion of nitric oxide in the aorta wall monitored in situ by porphyrinic microsensors. Biochem. Biophys. Res. Commun. **193**(3), 1076–1082 (1993)

Structural Regularity Exploration in Multidimensional Networks

Yi Chen, Xiaolong Wang, Buzhou Tang[✉], Junzhao Bu,
Qingcai Chen, and Xin Xiang

Key Laboratory of Network Oriented Intelligent Computation,
Harbin Institute of Technology Shenzhen Graduate School,
Shenzhen 518055, China
{chenyi,bujunzhao,xiangxin}@hitsz.edu.cn,
wangxl@insun.hit.edu.cn,
{tangbuzhou,qingcai.chen}@gmail.com

Abstract. Multidimensional networks, networks with multiple kinds of relations, widely exist in various fields. Structure exploration (i.e., structural regularity exploration) is one fundamental task of network analysis. Most existing structural regularity exploration methods for multidimensional networks need to pre-assume which type of structure they have, and some methods that do not need to pre-assume the structure type usually perform poorly. To explore structural regularities in multidimensional networks well without pre-assuming which type of structure they have, we propose a novel feature aggregation method based on a mixture model and Bayesian theory, called the multidimensional Bayesian mixture (MBM) model. Experiments conducted on a number of synthetic and real multidimensional networks show that the MBM model achieves better performance than other relative models on most networks.

Keywords: Multidimensional networks · Network structure · Structural regularity exploration · Mixture model · Bayesian theory

1 Introduction

In recent years, with the rapid development of social networking sites (e.g. Twitter, Facebook, Flickr and YouTube) which allow people to build up many kinds of networks, network analysis has attracted considerable attention from more and more researchers in various fields. Most existing studies focus on monodimensional networks composed of nodes with only one type of relationship. In the real world, however, there are also many multidimensional networks where more than one type of relationship (i.e., dimension) exists between nodes [1]. For example, in a social network, various types of relationships such as friend, colleague, schoolmate, romance, neighbor and family may co-exist between people.

One fundamental task of network analysis is to explore structural regularities, that is to explore network structure, such as assortative structure (i.e., community structure), disassortative structure (e.g., bipartite structure) [2] and mixture structure [3]. A great challenge for structural regularity exploration in multidimensional networks is how to

© Springer International Publishing Switzerland 2015
S. Arik et al. (Eds.): ICONIP 2015, Part III, LNCS 9491, pp. 532–540, 2015.
DOI: 10.1007/978-3-319-26555-1_60

aggregate multiple dimensions of the networks. A number of structural regularity exploration methods for multidimensional networks have been proposed recently. They may fall into three categories: network aggregation methods [1, 4], feature aggregation methods [5, 6] and mixture models [7]. The network aggregation methods first aggregate all dimensions into a unified network, and then employ existing structure exploration methods on the unified network. Tang et al. presented a simple united network via calculating the average links among nodes in multidimensional networks [1]. The limitation of this method lies in that all dimensions are treated equivalently. To solve this limitation, Zhu and Li proposed a new unified network to distinguish different dimensions according to their importance and node similarity [4]. The feature aggregation methods first extract features or clusters from each dimension, and then aggregate them. Tang and Liu proposed a principal modularity maximization (PMM) method to handle community detection in multidimensional networks [5]. Mucha et al. extended the modularity to simultaneously model interslice connections and intraslice connections for community detection in multidimensional networks [6]. This method aims at providing a partition for each dimension of multidimensional networks rather than for the whole networks. Both the network aggregation and feature aggregation methods need a pre-assumed structure type such as community structure which is usually unavailable in advance. The mixture models use graphical models to generate multidimensional networks such as marginal product mixture model (MPMM) [7]. Compared with the network aggregation and feature aggregation methods, the mixture models do not need to pre-assume which type of structure a network has, but they usually perform poorly on multidimensional networks.

In order to explore structural regularities in multidimensional networks well without pre-assuming which type of structure they have, in this paper, we propose a novel feature aggregation method based on a mixture model and Bayesian theory, called the multidimensional Bayesian mixture model (MBM). To get the MBM model, we first extend the Newman's mixture model (NMM) [2], a model for structural regularity exploration in monodimensional networks, for multidimensional networks, called the multidimensional Newman's mixture model (MNM), and then further extend the MNM model using Bayesian theory. Experiments conducted on a number of synthetic and real multidimensional networks show that our MBM model achieves better performance than other relative models on most networks.

The remainder of this paper is organized as follows. Section 2 introduces the MBM model. Experiments are presented in Sect. 3. Section 4 draws conclusions.

2 Multidimensional Mixture Models

In this section, we first give the definition of multidimensional networks, and then introduce the MNM model and the MBM model one after another.

A multidimensional network with N nodes and D dimensions can be represented as $A = \{A^{(1)}, \ldots, A^{(d)}, \ldots, A^{(D)}\}$, where $A^{(d)} = (1 \leq d \leq D)$ is the adjacent matrix of the dth dimension. For the dth dimension, $A_{ij}^{(d)} = 1, (1 \leq i, j \leq N)$ if there is a link from node i to node j and 0 otherwise. We use $L_i^{(d)}$ to denote the out links of node i in the dth dimension (i.e., a set of neighbor edges of node i in an undirected network).

2.1 Multidimensional Newman's Mixture (MNM) Model

A monodimensional network with K (a predefined number) groups can be generated by the NMM model with two parameters π_k and θ_{kj}, where π_k denotes the probability of a node in group $k (k \in \{1, 2, \ldots, K\})$ and θ_{kj} denotes the probability of a link from a node in group k connecting to node j, subjected to the normalization constraint $\sum_{j=1}^{N} \theta_{kj} = 1$. The vector θ_k represents the characteristic of nodes in group k linking to other nodes. According to θ_k, nodes connecting to other nodes in similar patterns are grouped together. In the NMM model, a network is generated in the following way: for each node i and its links L_i, (1) Node i falls into a group z_i with probability π_{z_i} ; (2) Each link $A_{ij} \in L_i$ selects node j with probability $\theta_{z_i j}$. The probability of a network A with N nodes can be written as

$$p(A, z \mid \pi, \theta) = p(A, z \mid \theta) \cdot p(z \mid \pi) = \prod_{i=1}^{N} \left(\pi_{z_i} \cdot \prod_{j=1}^{L_i} \theta_{z_i j}^{A_{ij}} \right) \tag{1}$$

where $z_i \in \{1, \ldots, K\}$ is a hidden variable that needs to be inferred. The logarithm of the Eq. (1):

$$\ln p(A, z \mid \pi, \theta) = \sum_{i=1}^{N} \left(\ln \pi_{z_i} + \sum_{j=1}^{L_i} A_{ij} \ln \theta_{z_i j} \right) \tag{2}$$

The hidden variables $z_i, i = 1, \ldots, N$, parameters π_k and θ_{kj} can be estimated using an expectation-maximization (EM) [8] algorithm. In the E step, the expectation of Eq. (2) is:

$$\bar{L} = \sum_{i,k} q_{ik} \left[\ln \pi_k + \ln \sum_j A_{ij} \ln \theta_{kj} \right] \tag{3}$$

with

$$q_{ik} = \frac{\pi_k \prod_j \theta_{kj}^{A_{ij}}}{\sum_k \pi_k \prod_j \theta_{kj}^{A_{ij}}}, \tag{4}$$

where q_{ik} denotes the probability of node i belonging to group k. In the M step, the parameters π and θ can be re-estimated by optimizing Eq. (3) as

$$\pi_k = \frac{\sum_i q_{ik}}{N}, \tag{5}$$

$$\theta_{kj} = \frac{\sum_i A_{ij} q_{ik}}{\sum_j \sum_i A_{ij} q_{ik}}, \tag{6}$$

The θ represents the preference or feature of a group of nodes about which other nodes they connect to. Nodes with similar θ should be grouped together. That is, nodes i and j should be in the same group if $(\theta_{1i}, \ldots, \theta_{ki}, \ldots, \theta_{Ki})$ and $(\theta_{1j}, \ldots, \theta_{kj}, \ldots, \theta_{Kj})$ are similar enough. In particular, in a community structure, if node i is in group k, but node j is not, then θ_{ki} should be much higher than θ_{kj}; in a bipartite structure, if node i is in group k, but node j is not, then θ_{ki} should be much lower than θ_{kj}. In the MNM model, we use

$$M = \begin{bmatrix} \theta_{11} & \cdots & \theta_{k1} & \cdots & \theta_{K1} \\ \cdots & & \cdots & & \cdots \\ \theta_{1j} & \cdots & \theta_{kj} & \cdots & \theta_{Kj} \\ \cdots & & \cdots & & \cdots \\ \theta_{1N} & \cdots & \theta_{kN} & \cdots & \theta_{KN} \end{bmatrix}$$

as the network structural features.

For multidimensional networks, we follow the PMM model [5] to aggregate the structural features of all dimensions as follows

$$\mathbf{M} = \left[M^{(1)}, \ldots, M^{(d)}, \ldots, M^{(D)} \right], \tag{7}$$

and then perform k-means on M to get the final node partition.

In the MNM model, node i may not fall into group k because of $q_{ik} = 0$ in Eq. (4), and will be wrongly assigned to another group. Specially, when $q_{i1} = \ldots = q_{iK} = 0$ on a network, node i may not fall into any group, resulting in that the MNM model fails. To avoid this problem, we extend the MNM model using Bayesian theory.

2.2 Multidimensional Bayesian Mixture (MBM) Model

In a Bayesian setting, the model parameters π and θ_k themselves will be random variables with prior distributions. The prior distributions are usually conjugate functions such as the Dirichlet distribution

$$\begin{aligned} \pi &\sim Dirichlet\,(\alpha) \\ \theta_k &\sim Dirichlet\,(\beta), \end{aligned} \tag{8}$$

where α and β are the Dirichlet hyperparameters.

The generative process for a network is summarized as follows:

1. Draw π from *Dirichlet* (α);
2. For each group k in K groups;

   Draw θ_k from *Dirichlet* (β);

3. For each new node i:
   (a) Draw a latent group z_i from π;
   (b) For each link in L_i: Draw an end node j from θ_{z_i}.

Then the probability of a network A with N nodes is (refer to Eq. (1)):

$$p(A, z \mid \alpha, \beta) = p(A, z \mid \pi, \theta)p(\pi \mid \alpha)(\theta \mid \beta) \tag{9}$$

Due to the conjugacy between the Dirichlet and Multinomial distributions, Eq. (9) can be simplified as:

$$p(A, z \mid \alpha, \beta) = p(A \mid z, \beta)p(z \mid \alpha) \tag{10}$$

with

$$p(z \mid \alpha) = \int p(z \mid \pi)p(\pi \mid \alpha)d\pi$$
$$p(A \mid z, \beta) = \int p(A \mid z, \theta)p(\theta \mid \beta)d\theta \tag{11}$$

The latent variable z of the Bayesian model (Eq. 10) cannot be exactly inferred, but can be approximately inferred by the Monte Carlo approaches. We employ Markov Chain Monte Carlo (MCMC) and follow an iterative procedure to achieve posterior inference over the latent variables. The sampler iterates as follows:

Sampling z We sample each node of z in succession using the Gibbs sampling. For each node i, given the group assignment for all other nodes, the group probability of it choosing group k is as follows:

$$p(z_i = k \mid z_{\neg i}, A)) \propto \prod_{j=1}^{L_i} \frac{m_{k,\neg i}^j + \beta}{m_{k,\neg i} + N\beta + j - 1} \cdot \frac{n_k + \alpha}{N + K\alpha} \tag{14}$$

where n_k denotes the number of nodes belongs to group k, $m_{k,\neg i}$ denotes the number of out links from nodes that belong to k except node i, $m_{k,\neg i}^j$ denotes the number of out links from nodes that belong to k except node i to node j. The $p(z_i = k \mid z_{\neg i}, A)$ has the same meaning of q_{ik} (see Eq. 4), but cannot be zero as $\frac{m_{k,\neg i}^j + \beta}{m_{k,\neg i} + N\beta + j - 1} > \frac{m_{k,\neg i}^j}{m_{k,\neg i} + (N-1)\beta + j - 1} \geq 0$.

Hyperparameters We use slice sampling [9] to determine the optimal hyperparameters α and β in $(0,1)$.

The m_k^j denotes the number of out links from nodes falling into group k to node j. It plays a similar role as θ_{kj} in NMM. So we use

$$
M = \begin{bmatrix}
m_1^1 & \cdots & m_k^1 & \cdots & m_K^1 \\
\cdots & & & & \cdots \\
m_1^j & \cdots & m_k^j & \cdots & m_K^j \\
\cdots & & \cdots & & \cdots \\
m_1^N & \cdots & m_k^N & \cdots & m_K^N
\end{bmatrix}
$$

as the network structural features.

Table 1. Detailed information of nine networks in our study.

Name	N	D	K	Directed	Structure types	Source
Syn-com	350	4	3	No	Community	Synthetic
Ckm[a]	246	3	4	No	Community	Real
Syn-tri	150	4	3	No	Tripartite	Synthetic
DutchElite[b]	4,747	4	2	No	Bipartite	Real
Syn-mix	150	4	5	No	Mixture	Synthetic
Lazega[a]	71	3	2	Yes	Mixture	Real
Gd99[b]	234	4	8	No	Mixture	Real
Bay[b]	125	2	7	Yes	Mixture	Real
Email[c]	4,408	4	13	No	Mixture	Real

[a]http://moreno.ss.uci.edu/data.html
[b]http://vlado.fmf.uni-lj.si/pub/networks/data/
[c]http://www-poleia.lip6.fr/ ~ jacoby/Research/research_en.html

3 Experiments

To investigate the effectiveness of the MBM model for structural regularity exploration in multidimensional networks, we compare it with other related models on nine synthetic and real multidimensional networks, including three synthetic networks and six real networks. The detailed information of all networks is shown in Table 1 and the synthetic networks are illustrated in Fig. 1.

Six models coming from the three categories mentioned in the introduction section are selected for comparison. They are three network aggregation methods (i.e., the Unify-Infomap (UI), Unify-OSLOM (UO), Unify-Louvain (UL)) proposed by [4], a feature aggregation method (i.e., the PMM) proposed by [5], a mixture model (i.e., the MPMM) proposed by [7] and MNM. As the source code of the PMM model has been released by the authors, we use the released source code in this study. For other models, we implement them by ourselves. Among all the models, as some models (i.e., PMM, MNM and MBM) require a pre-defined group number, we adopt the gold standard for them, run them 10 times and report the performance of the best time.

The performance of all models is measured by the Normalized Mutual Information (NMI) [10]:

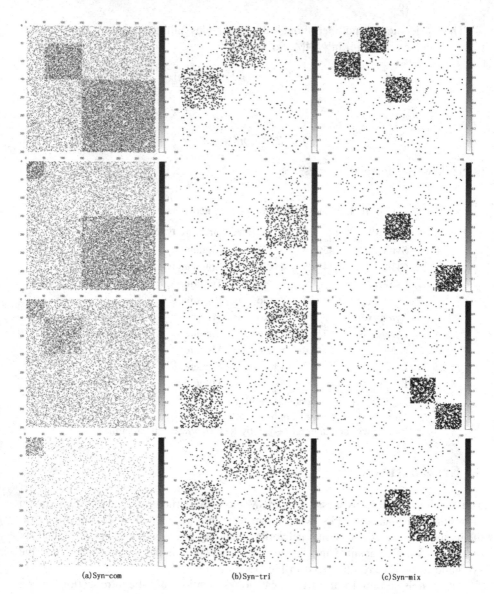

Fig. 1. The adjacency matrix of synthetic multidimensional networks: (a) the Syn-com network, (b) the Syn-tri network, (c) the Syn-mix network.

$$P_{nmi}(G, G') = \frac{2MI(G, G')}{H(G) + H(G')},$$ (15)

where $G = (G_1, G_2, \ldots, G_K)$ are defined groups in a network, $G' = (G'_1, G'_2, \ldots, G'_K)$ are groups detected by an algorithm, $H(G)$ and $H(G')$ are the entropies of G and G', and $MI(G, G')$ is the mutual information between them. A high P_{nmi} means a good detection. Specially, $P_{nmi} = 1$ means a perfect detection.

Table 2. P_{nmi}s identified by seven models on the nine multidimensional networks.

Name	UI	UO	UL	PMM	MPMM	MNM	MBM
Syn-com	0	0.9822	0.1946	0.9665	0.0074	0.9727	**1**
Ckm	**0.9628**	0.6485	0.5668	0.648	0.0944	0.7164	0.9385
Syn-tri	0	0.0451	0.2745	0.0085	0.0027	1	**1**
DutchElite	0.0233	0.0323	0.0228	0.0081	0.0016	0.0182	**0.3467**
Syn-mix	0.8538	0.7252	0.3977	0.8013	0.1104	1	**1**
Lazega	0	0.5294	0.2171	0.0133	0.0002	NA	**0.5699**
Gd99	0.2012	0.1098	**0.2063**	0.1087	0.0541	0.1964	0.1942
Bay	0	0.0419	0.3773	0.4443	0.1146	0.3832	**0.4726**
Email	0.2746	0.2964	0.2257	0.3479	0.0016	NA	**0.4078**

Table 2 shows the P_{nmi}s identified by different models on the nine networks, where NA denotes that a model is inapplicable on a network. For community structure exploration, the MBM model achieves a perfect P_{nmi} of 1 on the Syn-com network, that is, all nodes are partitioned into the correct groups, and a P_{nmi} of 0.9385 on the Ckm network, which is slightly inferior to the UI model but outperforms all other models. For disassortative structure exploration, the MBM model achieves a perfect P_{nmi} of 1 on the Syn-tri network, which is the same as the MNM model and significantly out-performs other models, and a P_{nmi} of 0.3467 on the DutchElite network, which is higher than other six models by at least 0.3144. For mixture structure exploration, the MBM model is applicable on all networks and achieves the best P_{nmi}s on the Syn-mix, Lazega, Bay and Email networks. The P_{nmi} on the Gd99 network is slightly inferior to the UI, UL and MNM networks but superior to other models, which is still acceptable. Overall, the MBM model achieves highest P_{nmi} on seven networks. Compared with the existing five models, the MNM model achieves highest P_{nmi} on three networks (i.e., Syn-com, Syn-tri and Syn-mix), but is inapplicable on two networks (i.e., Lazega and Email) due to the problem of $q_{ik} = 0$, which is avoided by the MBM model.

4   Conclusion

In this paper, we propose a novel feature aggregation method based on a mixture model and Bayesian theory for multidimensional networks without pre-assuming the structure type, called the multidimensional Bayesian mixture (MBM) model. Experiments conducted on a number of synthetic and real multidimensional networks show that the MBM model is able to explore structural regularities in multidimensional networks very well. However, the MBM model suffers from an inherent limitation of Bayesian models, that is, they require a pre-specified group number. As future work, we will further extend the MBM model using Bayesian nonparametric theory to automatically determine the group number.

Acknowledgements. This paper is supported in part by grants: National 863 Program of China (2015AA015405), NSFCs (National Natural Science Foundation of China) (61402128, 61473101, 61173075 and 61272383), Strategic Emerging Industry Development Special Funds of Shenzhen (JCYJ20140508161040764 and JCYJ20140417172417105) and Scientific Research Foundation in Shenzhen (Grant No. JCYJ20140627163809422).

References

1. Tang, L., Wang, X.F., Liu, H.: Community detection in multi-dimensional networks. Technical report, DTIC Document, January 2010
2. Newman, M.E.J., Leicht, E.A.: Mixture models and exploratory analysis in networks. Proc. Natl. Acad. Sci. **104**(23), 9564–9569 (2007)
3. Chai, B.-F., Yu, J., Jia, C.-Y., Yang, T.-B., Jiang, Y.-W.: Combining a popularity-productivity stochastic block model with a discriminative-content model for general structure detection. Phys. Rev. E **88**(1), 012807 (2013)
4. Zhu, G., Li, K.: A unified model for community detection of multiplex networks. In: Benatallah, B., Bestavros, A., Manolopoulos, Y., Vakali, A., Zhang, Y. (eds.) WISE 2014, Part I. LNCS, vol. 8786, pp. 31–46. Springer, Heidelberg (2014)
5. Tang, L., Liu, H.: Uncovering cross-dimension group structures in multi-dimensional networks. In: SDM Workshop on Analysis of Dynamic Networks, Sparks, NV (2009)
6. Mucha, P.J., Richardson, T., Macon, K., Porter, M.A., Onnela, J.-P.: Community structure in time-dependent, multiscale, and multiplex networks. Science **328**(5980), 876–878 (2010)
7. DuBois, C., Smyth, P.: Modeling relational events via latent classes. In: Proceedings of the 16th ACM SIGKDD international conference on Knowledge discovery and data mining, pp. 803–812, Washington, USA (2010)
8. Dempster, A.P., Laird, N.M., Rubin, D.B.: Maximum likelihood from incomplete data via the EM algorithm. J. Roy. Stat. Soc. B **39**, 1–38 (1977)
9. Neal, R.M.: Slice sampling. Ann. Stat. **31**, 705–741 (2003)
10. Ana, L., Jain, A.K.: Robust data clustering. In: Proceedings of the 2003 IEEE Computer Society Conference on Computer Vision and Pattern Recognition, pp. 128–133, Madison, USA (2003)

Proposal of Channel Prediction by Complex-Valued Neural Networks that Deals with Polarization as a Transverse Wave Entity

Tetsuya Murata, Tianben Ding, and Akira Hirose[✉]

Department of Electrical Engineering and Information Systems,
The University of Tokyo, 7-3-1 Hongo, Bunkyo-ku, Tokyo 113-8654, Japan
{murata,ding_tei}@eis.t.u-tokyo.ac.jp, ahirose@ee.t.u-tokyo.ac.jp
http://www.eis.t.u-tokyo.ac.jp/

Abstract. Multipath fading is one of the most serious problems in mobile communications. Various methods to solve or mitigate it have been proposed in time or frequency domain. Previously we proposed a channel prediction method that combines complex-valued neural networks and chirp z-transform that utilizes both the time- and frequency-domain representation, resulting in much higher performance. In this paper, we propose to deal with polarization additionally in its adaptive channel prediction to improve the performance further. A preliminary experiment demonstrates improvement larger than what is expected by a simple diversity gain.

Keywords: Channel prediction · Fading · Complex-valued neural network

1 Introduction

Mobile communications often suffers from various fading phenomena. Diversity, error-correction coding and other techniques are used to mitigate the degradation. Other methods are pre-equalization and transmission-power control. Such adaptive techniques require channel prediction. The channel prediction is useful also in so-called high-capacity spatial division multiple access (HC-SDMA) and multi-user multiple-input multiple-output (MIMO) systems.

There exist various channel prediction methods, namely, those based on linear prediction [1,2], autoregression models [3–6], neural-network-based nonlinear prediction [7,8] and super-resolution methods such as ROOT-MUSIC and ESPRIT. These methods work in time or frequency domain. However, they have problems in its prediction accuracy and/or calculation cost. Time-domain methods mostly present low accuracy, while frequency-domain methods such as ROOT-MUSIC and ESPRIT need a high cost in processing.

The authors previously proposed a high-accuracy and low-cost method by employing chirp z-transform (CZT) [10] based on the Jakes model, and later the

© Springer International Publishing Switzerland 2015
S. Arik et al. (Eds.): ICONIP 2015, Part III, LNCS 9491, pp. 541–549, 2015.
DOI: 10.1007/978-3-319-26555-1_61

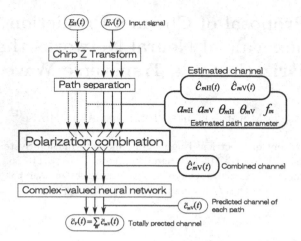

Fig. 1. Processing flow of the CZT-CVNN channel prediction including the polarization combining

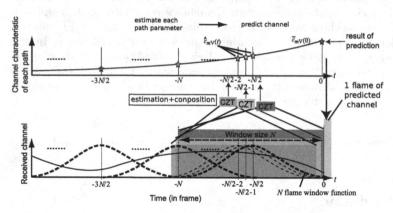

Fig. 2. Parameter estimation through CZT for channel prediction [9]

combination of CZT with linear prediction [11,12] as well as Lagrange extrapolation [13] to demonstrate improved prediction accuracy.

We applied our methods in various propagation situations to find that they sometimes work only insufficiently. Consequently we also proposed a method that combines the CZT path-separation and complex-valued neural networks (CVNN). This CZT-CVNN method shows high performance even in difficult channel situations [9,14,15]. However, it does not deal with polarization that is an important entity of electromagnetic wave.

In present communications, spatial diversity is often used to compensate fading to some extent. Polarization diversity is also available. Such researches suggest that the introduction of polarization into channel prediction brings about further improvement in the prediction. However, polarization combining in the channel prediction is usually difficult to realize.

This paper presents a method of channel prediction incorporating the polarization combining and the CZT-CVNN adaptive processing. The path-separation employing the CZT realizes the estimation of a polarization state in each path from the received electrical components. In other words, we propose a channel prediction that deals with the electromagnetic wave having polarization as a physical entity. Then we can obtain a certain amount of improvement in the prediction accuracy larger than what we expect in a simple polarization energy summation.

2   Combining Polarization Through Path Separation

In most of channel prediction methods including our previous CZT-CVNN method, we dealt with only a single polarization component. However, in actual communication situations, transmitted electromagnetic wave is scattered in its propagation, and changes in the polarization states. Then the multiple-path waves are superposed at the receiving antenna in a polarization-wise way to generate signals corresponding to instantaneous electric field components. In this paper, first we estimate the polarization state of each channel path wave by employing the CZT path separation, which enables us to combine polarization-wise signals appropriately in the following processing. Then we finally predict the future channel state more accurately based on the polarization-combined signals.

Figure 1 is the processing flow in the prediction. First we transform the horizontal and vertical components, E_H and E_V, by CZT respectively as shown in Fig. 2 [9]. A set of obtained path parameters represent a channel state at the time of the CZT-window center. We combine the horizontal and vertical path parameters, and use the results in the following channel prediction.

The detail is explained as follows. Through the path separation, we estimate

$\hat{a}_{mH}$: Estimated amplitude of horizontal component of m-th path.
$\hat{a}_{mV}$: Estimated amplitude of vertical component of m-th path.
$\hat{\theta}_{mH}$: Estimated phase of horizontal component of m-th path.
$\hat{\theta}_{mV}$: Estimated phase of vertical component of m-th path.
$\hat{f}_m$: Estimated Doppler frequency of m-th path.

Figure 3 shows the amplitude and phase spectra obtained for E_H and E_V by setting an N-frame long window to the time-domain signals to apply the CZT [9]. Figure 3(a) and (b) shows the amplitude spectra versus Doppler frequency for E_H and E_V, respectively, whereas Fig. 3(c) and (d) show corresponding phase spectra. In the amplitude spectra in Fig. 3(a) and (b), we find amplitude peaks which show respective paths, and can estimate also the Doppler frequency $\hat{f}_m$. Here, $\hat{a}_{mH}$ denote the peaks in E_H, while $\hat{a}_{mV}$ denote those in E_V. We can estimate $\hat{a}_{mV}$ for a horizontal peak by reading the vertical amplitude at $\hat{f}_m$ obtained in horizontal spectrum, and vice versa. We estimate $\hat{\theta}_{mH}$ and $\hat{\theta}_{mV}$ in

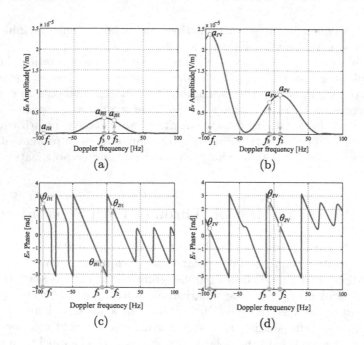

Fig. 3. Peak extraction to estimate channel Doppler frequencies and amplitude parameters in (a) horizontal and (b) vertical amplitude spectra as well as estimation of phase parameters in (c) horizontal and (d) vertical phase spectra

the same manner in Fig. 3(c) and (d). Accordingly we obtain the parameters including the polarization in the respective paths.

We combine the estimated polarization components $\hat{c}_{mH}(t)$ and $\hat{c}_{mV}(t)$ into a merged vertical component $\hat{c}'_{mV}(t)$ as

$$\hat{c}_{mH}(t) = \hat{a}_{mH}(t)e^{j(2\pi\hat{f}_m(t)t+\hat{\theta}_{mH}(t))} \tag{1}$$

$$\hat{c}_{mV}(t) = \hat{a}_{mV}(t)e^{j(2\pi\hat{f}_m(t)t+\hat{\theta}_{mV}(t))} \tag{2}$$

$$\hat{c}'_{mV}(t) = \hat{c}_{mH}(t)e^{j(\hat{\theta}_{mV}(t)-\hat{\theta}_{mH}(t))} + \hat{c}_{mV}(t) \tag{3}$$

The CVNN learns the time-sequential merged channel $\hat{c}'_{mV}(t)$ for prediction of the respective channels. We use a layered CVNN. The details of the CVNN dynamics are explained in Ref. [9]. We sum the neural output signals $\tilde{c}_{mV}(t)$, i.e., respective path prediction, to generate a finally predicted channel $\tilde{c}_V(t)$.

3   Simulation Experiment

3.1   Experimental Setup

We evaluate the performance of our proposal of polarization-combining CVNN channel prediction with polarization estimation through CZT channel separation.

Figure 4 shows the geometrical setup assumed in the simulation experiment which includes two scatterers and a user (mobile terminal) moving with a velocity of $v = 14$ m/s. This sever situation causes such a fast, complicated and heavy fading that it is very difficult to predict the channel. The user receives a direct wave (path p_1) and two scattered waves (p_2, p_3). The carrier frequency is $f_c = 2$ GHz

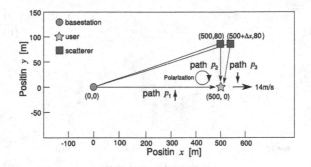

Fig. 4. Geometrical setup of simulation experiment

We assume scattering matrices $\mathbf{S}_2$ and $\mathbf{S}_3$ for the two scatterers in Paths p_2 and p_3, respectively, as

$$\mathbf{S}_2 = \frac{1}{2}\begin{bmatrix} 1 & -j \\ -j & -1 \end{bmatrix}, \quad \mathbf{S}_3 = \begin{bmatrix} -1 & 0 \\ 0 & 1 \end{bmatrix} \tag{4}$$

Path 2 (p_2) presents a circular polarization, resulting in a severer situation. Such scattering occurs, for example, when two 45 deg and −45 deg slant line scatterers exist $\lambda/8$ (λ : wavelength) away from each other. As a communication system, we assume orthogonal frequency division multiplexing (OFDM) as asimulation on a realistic condition shown in Table 1 [9]. We employ quaternary phase shift keying (QPSK) and time-domain duplex (TDD) in which we use the same frequency for uplink and downlink in time sharing. The number of OFDM subcarriers is 52. We predict the channel state in one TDD frame (5 ms) starting at $t = 0$. The CZT window center is put for every frame. For simplicity, the difference in the direction of arrivals is ignored in the present experiment.

3.2 Polarization State Estimation

Figure 5(a),(c),(e) present the instantaneous electrif field, showing the polarization states to be received in the simulation experiment. The horizontal and vertical components, $E_H(t)$ and $E_V(t)$, are shown as a function of time t for (a) p_1, (c)p_2 and (e)p_3, respectively. At the receiver, they are superposed to exhibit instantaneous electric field depicted in Fig. 5(g), often resulting in fading. It is found that the instantaneous electric field changes fast and complicatedly, showing a polarization very different from the vertical one that we expect for a non-muptipath situation.

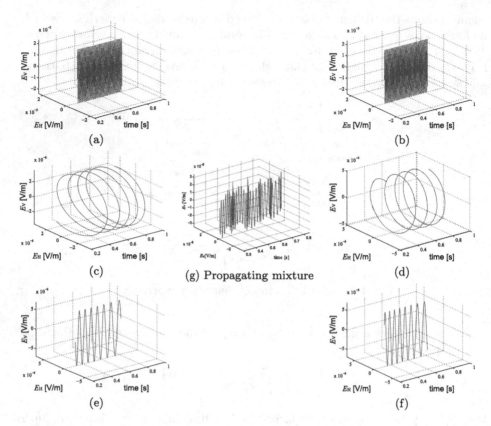

Fig. 5. Electric field versus time to illustrate (a) actual and (b)estimated p_1 polarization, (b) actual and (d)estimated p_2 polarization, (e) actual and (f) estimated p_2 polarization as well as (g) superposition received at the receiver.

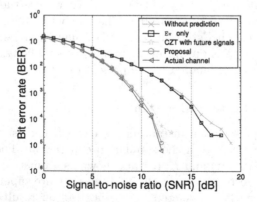

Fig. 6. Bit-error-rate (BER) curves of various methods versus signal-to-noise ratio (SNR).

Table 1. Parameters in the OFDM system and the channel prediction

OFDM Parameter	Value
QPSK symbol number	1612
QPSK symbol rate F	406.25 kHz
Number of carriers K	52
FFT Size	64
Carrier spacing F/K	7812.5 Hz
Spacing between carriers	3.98 MHz
Size of guard interval	16
TDD frame length	5 ms
TDD symbol number in a frame s	2500 symbol
Sampling rate	500 kHz
Channel Prediction Parameter	Value
CZT window length	N frame
Symbol number in a CZT window S	$N \cdot s$ frame·symbol
Iteration of ML-CVNN weight update R_{ML}	10 times

We examined whether the system is capable of estimating the polarization state separate from one another based on the mixed received instantaneous electric field presented in Fig. 5(g). The CZT window length is $N = 32$ frame. A longer window leads to a high-resolution CZT and, in this sense, maybe more accurate estimation. However, it may be not adaptable to fast changing situations. The CZT-separated electric fields are shown in Figs. 5(b), (d) and (f). The figures represent vertical components (p_1, p_3) and circular one (p_2), which are in a good agreement with the actual field before mixture. The polarization state estimation is found successful.

3.3 BER for the Polarization-Combining CZT-CVNN Channel Prediction

Next we evaluate the channel prediction performance in the simulation experiment. The CZT window length is $N = 8$ frame and the distance between the two scatterers is $\Delta x = 20$ m. For comparison, we also obtained results when we receive only the vertical components, "E_V only," and also those when we used only path-separation estimation process, without a prediction, by using $\hat{a}_{m\mathrm{V}}(-N/2)$, $\hat{f}_m(-N/2)$ and $\hat{\theta}_{m\mathrm{V}}(-N/2)$, namely, "Without Prediction."

Figure 6 shows the bit-error rate (BER) obtained with the channel predicted by the above methods. In this figure, "CZT with future signals" means the BER obtained by a set of channel parameters estimated by CZT having its window center at $t = 0$ including future signals ($t > 0$), which is not available in reality. "Actual channel" shows the BER when we know the actual channel exactly. The nearer to "Actual channel" a curve exists, the better the performance is. "Proposal" is very similar to "Actual channel," showing a high accuracy in the prediction. However, "E_V only" shows a 5–6 dB degradation at $BER = 10^{-5}$. That is, the polarization combining resulted in 5–6 dB improvement. Since the power in each channel is almost the same, a simple addition of the circular channel power will make only 1.8 dB (1.5 times in power) improvement. The additional gain is attributed to the improvement in the prediction accuracy.

4 Summary

This paper proposed to deal with the polarization additionally in the CZZT-CVNN channel prediction. We demonstrated a high improvement of the BER performance. By regarding the polarity as one of the wave entities, we can make the CVNN work much better to show a higher prediction performance.

References

1. Maehara, F., Sasamori, F., Takahata, F.: Linear predictive maximal ratio combining transmitter diversity for OFDM-TDMA/TDD systems. IEICE Trans. Commun. **E86-B**, 221–229 (2003)
2. Bui, H., Ogawa, Y., Nishimura, T., Ohgane, T.: Performance evaluation of a multiuser MIMO system with prediction of time-varying indoor channels. IEEE Trans. Antennas Propag. **61**(1), 371–379 (2012)
3. Eyceoz, T., Duel-Hallen, A., Hallen, H.: Prediction of fast fading parameters by resolving the interference pattern. In: Proceedings of the Asilomar Conference on Signals, Systems, and Computers, vol. 1, pp. 167–171 (1997)
4. Eyceoz, T., Duel-Hallen, A., Hallen, H.: Deterministic channel modeling and long range prediction of fast fading mobile radio channels. IEEE Trans. Commun. Lett. **2**(9), 254–256 (1998)
5. Arredondo, A., Dandekar, K., Xu, G.: Vector channel modeling and prediction for the improvement of downlink received power. IEEE Trans. Commun. **50**(7), 1121–1129 (2002)
6. Sternad, M., Aronsson, D.: Channel estimation and prediction for adaptive OFDM downlinks. In: Proceedings of the IEEE Vehicular Technology Conference 2003-Fall, vol. 2, pp. 1283–1287, October 2003
7. Liu, W., Yang, L.-L., Hanzo, L.: Recurrent neural network based narrowband channel prediction. In: Proceedings of the IEEE Vehicular Technology Conference 2006-Spring, vol. 5, pp. 2173–2177, May 2006
8. Potter, C., Venayagmoorthy, G.K., Kosbar, K.: RNN based MIMO channel prediction. Sign. Proces. **90**, 440–450 (2010)
9. Ding, T., Hirose, A.: Fading channel prediction based on combination of complex-valued neural networks and chirp z-transform. IEEE Trans. Neural Networks Learn. Syst. **25**, 1686–1695 (2014)
10. Tan, S., Hirose, A.: Low-calculation-cost fading channel prediction using chirp z-transform. Electron. Lett. **45**(8), 418–420 (2009)
11. Ozawa, S., Tan, S., Hirose, A.: Channel prediction experiment based on linear prediction in frequency domain. In: Asia-Pacific Microwave Conference (APMC, 2010), Yokohama, pp. 1280–1283, December 2010

12. Ozawa, S., Tan, S., Hirose, A.: Errors in channel prediction based on linear prediction in the frequency domain: a combination of frequency-domain and time-domain techniques. URSI Radio Sci. Bull. **337**, 25–29 (2011)
13. Matsui, H., Hirose, A.: Nonlinear prediction of frequency-domain channel parameters for channel prediction in fading and fast doppler-shift change environment. In: International Symposium on Antennas and Propagation (ISAP), pp. 1132–1135 (2012)
14. Ding, T., Hirose, A.: Fading channel prediction based on complex-valued neural networks in frequency domain. In: International Symposium on Electromagnetic Theory (EMTS 2013), Hiroshima, pp. 640–643. IEICE, May 2013
15. Ding, T., Hirose, A.: Fading channel prediction based on self-optimizing neural networks. In: Loo, C.K., Yap, K.S., Wong, K.W., Teoh, A., Huang, K. (eds.) ICONIP 2014, Part I. LNCS, vol. 8834, pp. 175–182. Springer, Heidelberg (2014)

A Scalable and Feasible Matrix Completion Approach Using Random Projection

Xiang Cao[✉]

Department of Computer Science and Engineering,
Shanghai Jiao Tong University, Shanghai 200240, China
xiang.cx@sjtu.edu.cn

Abstract. The low rank matrix completion problem has attracted great attention and been widely studied in collaborative filtering and recommendation systems. The rank minimization problem is NP-hard, so the problem is usually relaxed into a matrix nuclear norm minimization. However, the usage is limited in scability due to the high computational complexity of singular value decomposition (SVD). In this paper we introduce a random projection to handle this limitation. In particular, we use a randomized SVD to accelerate the classical Soft-Impute algorithm for the matrix completion problem. The empirical results show that our approach is more efficient while achieving almost same performance.

1   Introduction

In many machine learning applications such as collaborative filtering and recommendation systems, the data is usually represented as a matrix with many missing entries, yielding the problem of matrix completion. More precisely, matrix completion is to recover the missing entries based on the observed parts. Moreover, such a data matrix has a low-rank property. Thus, matrix completion can be modeled as a low-rank regularization problem [3] and can be relaxed to the minimization of the nuclear norm [2].

Meanwhile, many methods have been successively proposed to solve the nuclear norm regularized minimization problem. A class of representative algorithms are based on a so-called soft-thresholding singular value decomposition (ST-SVD) [9,10,13]. However, these algorithms are limited in applications of large-scale data sets because of the high computation complexity of SVD. To avoid computing SVD of large matrices, Srebro et al. [12] proposed a variant by representing the original matrix as factorization of two low-rank matrices. This variant can be solved by using an alternating iterative procedure [8,15]. But as this variant is formulated as a nonconvex optimization problem, the solution usually relies on initialization and sacrifices.

Additionally, coordinate gradient descent methods were studied in [5,11,14]. More specifically, this method computes the top SVD and generates a rank-one matrix at each iteration, followed by updating the weights of the present matrices. However, it is quite difficult to refine the weights. Large weights make it hard for the algorithms to converge, and smaller ones make the algorithms slow.

© Springer International Publishing Switzerland 2015
S. Arik et al. (Eds.): ICONIP 2015, Part III, LNCS 9491, pp. 550–558, 2015.
DOI: 10.1007/978-3-319-26555-1_62

In this paper we would like to address the computational problem that the conventional ST-SVD suffers for. Our work is motivated by the idea of random projection (RP). Halko et al. [7] proposed a novel randomized SVD(RSVD) method, which projects the original large size data matrix into a small size matrix by using a random Gaussian projection and performs SVD on the small matrix. The randomized SVD is provably efficient and effective and has been widely used and improved [4]. Taking the sparsity of the input matrix into consideration, we choose a sparse matrix instead of the Gaussian matrix, so as to further accelerate the algorithm. Moreover, even for a dense matrix, the computational complexity is lower [4]. We introduce the randomized SVD method to the ST-SVD matrix completion method, giving rise to a fast ST-SVD method.

Our method can significantly reduce computational costs. Compared with the Soft-Impute algorithm [10] where the Lanczos bi-diagonalization with partial reorthogonalization is used to compute the truncated SVD, our method is much more efficient. Compared with the Soft-Impute-ALS algorithm of Hastie et al. [8], our method has more accurate and faster performance.

The remainder of the paper is organized as follows. We introduce some notation, the matrix completion problem, and the related work in Sect. 2. In Sect. 3, we give an algorithm of randomized SVD and apply it into the Soft-Impute algorithm with an analysis of the algorithm. We compare the Soft-Impute-RP with Soft-Impute, Soft-Impute-ALS in numerical experiment in Sect. 4. And the final section is a conclusion of the whole paper.

2 Preliminaries

2.1 Notation

Assume that the data is represented as an $m \times n$ matrix $Y = [Y_{ij}]$. Let $\Omega \subset \{1, \ldots, m\} \times \{1, \ldots, n\}$ denote the set of the indices of the observed entries of Y, and $P_{\Omega}(Y)$ denote an $m \times n$ projection matrix. That is, $[P_{\Omega}(Y)]_{i,j} = Y_{ij}$ if $(i, j) \in \Omega$, and $[P_{\Omega}(Y)]_{i,j} = 0$ otherwise. The complement of $P_{\Omega}(Y)$ is defined as $P_{\Omega}^{\perp}(Y) = Y - P_{\Omega}(Y)$.

Given a matrix X of rank r, the condensed singular value decomposition (SVD) is given as $X = U\Sigma V^T$, where $U \in \mathbb{R}^{m \times r}$ and $V \in \mathbb{R}^{n \times r}$ satisfy $U^T U = I$ and $V^T V = I$, and $\Sigma = \text{diag}(\sigma_1(X), \ldots, \sigma_r(X))$ is the diagonal matrix of the singular values with $\sigma_1(X) \geq \sigma_2(X) \geq \cdots \sigma_r(X) > 0$.

Let $\|X\|_F = \sqrt{\sum_{i,j} X_{ij}^2} = \sqrt{\sum_{i=1}^{r} \sigma_i(X)^2}$ denote the Frobenius norm, $\|X\|_* = \sum_{i=1}^{r} \sigma_i(X)$ denote the nuclear norm, and $S_\lambda(X) = U(\Sigma - \lambda I)_+ V^T$ be the singular value thresholding (SVT) for $\lambda > 0$. Here I is an identity matrix and $(W)_+ = \text{diag}([\sigma_1(X) - \lambda]_+, \ldots, [\sigma_r(X) - \lambda]_+)$ where $[u - \lambda]_+ = u - \lambda$ if $u > \lambda$ and $[u - \lambda]_+ = 0$ if $u \leq \lambda$.

2.2 Problem Formulation

Given an $m \times n$ observed matrix $P_\Omega(Y)$, the matrix complete problem is to find a low rank matrix X to approximate Y. Specifically, it is formulated as the following optimization problem:

$$\min_X \text{rank}(X), \quad s.t. \; P_\Omega(Y) = P_\Omega(X). \tag{1}$$

However, the rank minimization problem in (1) is NP-Hard. As is well known, the nuclear norm is the tightest convex relaxation of the matrix rank [2]. As a result, the following optimization problem is alternatively used:

$$\min_X \|X\|_*, \quad s.t. \; P_\Omega(Y) = P_\Omega(X). \tag{2}$$

Using the nuclear norm as a regularizer [10], the above problem can be equivalently formulated as:

$$\min_X \frac{1}{2}\|P_\Omega(Y) - P_\Omega(X)\|_F^2 + \lambda\|X\|_*. \tag{3}$$

2.3 Related Work

The related work includes the classical SVT [1], Soft-Impute [10] and Soft-Impute-ALS algorithms [8].

Lemma 1. *Assume $X \in \mathbb{R}^{m \times n}$ is a rank-r matrix. The solution to the optimization problem*

$$\arg\min_X \frac{1}{2}\|Y - X\|_F^2 + \lambda\|X\|_* \tag{4}$$

is $\hat{X} = S_\lambda(Y)$.

The lemma was proposed by Cai et al. [1]. They solved the problem iteratively as follows

$$\begin{cases} X^k = S_\lambda(Y^k), \\ Y^k = Y^{k-1} + \delta_k P_\Omega(Y - X^k). \end{cases} \tag{5}$$

The resulting algorithm is called a SVT (singular value thresholding) algorithm.

Mazumder et al. [10] proposed a Soft-Impute algorithm to solve the problem in (3). Let $Y^* = P_\Omega(Y) + P_\Omega^\perp(X)$. Then the problem of (3) is rewritten as:

$$\min_X \frac{1}{2}\|Y^* - X\|_F^2 + \lambda\|X\|_*. \tag{6}$$

Based on Lemma 1, the solution to (6) is $X = S_\lambda(Y^*)$. So in each iteration the Soft-Impute algorithm goes as follows:

$$\begin{cases} Y^* = P_\Omega(Y) + P_\Omega^\perp(\hat{X}), \\ \hat{X} = S_\lambda(Y^*). \end{cases} \tag{7}$$

When updating X, the Soft-Impute algorithm requires to do the SVD of Y^*. Thus, when the matrix is large-scale, the computational cost becomes huge. As a result, the use of the algorithm is limited for large-scale matrix completion problems.

Lemma 2. *Given an $m \times n$ matrix X, then we have*

$$\|X\|_* = \min_{A,B:X=AB^T} \frac{1}{2}(\|A\|_F^2 + \|B\|_F^2). \tag{8}$$

The lemma is given in [12] for the maximum-margin matrix factorization (MMMF) algorithm.

Recently, Hastie et al. [8] used Lemma 2 to address the high computation that the Soft-Impute algorithm suffers for. In particular, they transformed the problem in (6) as

$$\min_{A,B} \frac{1}{2}\|Y^* - AB^T\|_F^2 + \frac{\lambda}{2}\|A\|_F^2 + \frac{\lambda}{2}\|B\|_F^2. \tag{9}$$

Then the current problem is solved by using an alternating minimization algorithms (ALS).

3 Methodology

In this section, we first introduce the randomized SVD (RSVD) and give a brief analysis of RSVD. Furthermore, we present an algorithm by combining the Soft-Impute algorithm and RSVD for the matrix completion problem.

3.1 Randomized SVD (RSVD)

Our main contribution is to accelerate the Soft-Impute algorithm via accelerating SVD by using random projection. As is shown in (7), we need to do SVD for Y^* which is of the same size as the data matrix. Therefore when the size of the data matrix is large, each iteration will be quite slow. Moreover, implementing SVD of a large-scale matrix may result in the problem of numerical instability, which in turn makes the algorithm less effective.

Random projection is a feasible approach to reducing the time complexity of SVD. proposed a randomized SVD method. Clarkson et al. [4] proposed a low rank approximation in input sparsity time based on [7]. The basic idea is to first construct a low-dimensional subspace that captures the action of the matrix and then compute SVD on a smaller matrix. More specifically, the method consists of the two stages.

Stage 1: Compute an approximate basis Q for the range of the original data matrix X, which can be formulated as:

$$X \approx QQ^T X, \tag{10}$$

Algorithm 1. Randomized Singular Value Decomposition (RSVD)

Input: Matrix $X_{m \times n}$, the target number of singular values k, the number of elements in each column vector of the sparse matrix q

Output: an approximate rank-k SVD of X: $[U, D, V]$

1: Initialize l
2: Generate a sparse matrix, of which each column has q elements, $S = GenS(q)$
3: Form $Y = XS^T$
4: $Q = \text{orth}(Y)$
5: Form $B = Q^T X$
6: Compute the SVD of the smaller matrix $B = \tilde{U}DV^T$
7: Set $U = Q\tilde{U}$
8: $U = U(:, 1:k), D = D(1:k, 1:k), V = V(:, 1:k)$

Algorithm 2. Soft-Impute-RP

Input: Matrix $Y_{m \times n}$, parameter λ, target rank k

Output: Matrix $X_{m \times n}$

1: Initiallize $\hat{X} = 0$, $q = 1$(for example).
2: **repeat**
3:    Form $Y^* = P_\Omega(Y) + P_\Omega^\perp(\hat{X})$
4:    Compute the RSVD of Y^* by Algorithm 1, $[U, D, V] = \text{RSVD}(Y^*, k, q)$
5:    Compute $\hat{X} = U(D - \lambda I)_+ V^T$
6: **until** some condition is met

---

where Q has orthonormal columns. Intuitively, the matrix Q may contain as few columns as possible, but it is more important to get an accurate approximation of the input matrix. Therefore, how to form the matrix Q is the key to RSVD. The concrete construction for Q is given in Algorithm 1 and the bounds is given in Theorem 1 to ensure the effectiveness.

Stage 2: Once the matrix Q has been obtained in Stage 1, one projects the original matrix X to a smaller matrix $B = Q^T X$ and compute the standard SVD of B, and then get the approximate SVD of X.

Theorem 1. *Let S be a $k \times n$ sparse matrix, X be an $m \times n$ matrix, Q be the column space of XS^T,*

$$\Pr_Q(\|QQ^T X - X\|_F > \frac{3}{\sqrt{k}}\|X\|_F) \leq \frac{1}{10} \tag{11}$$

3.2   Soft-Impute Based on RSVD

As mentioned in Sect. 2.3, the Soft-Impute algorithm is a classical algorithm for matrix completion. It is used to solve the problem in (3). The Soft-Impute algorithm is based on SVT (see Eq. (5)). With replacing the missing elements with those obtained in soft-threshold SVD, the problem is transformed to (6) and the solution is given by (7).

In each iteration of the Soft-Impute algorithm, there needs doing SVD of the matrix Y^*. In the conventional Soft-Impute, the Lanczos method is used to approximately compute the truncated SVD of Y^*. The Lanczos based SVD (LanSVD) method can reduce computational complexity, but it has low stability because it is a numerical iteration procedure.

In order to reduce the complexity of computing SVD and improve the accuracy, we employ RSVD in Algorithm 1 instead of LanSVD. Accordingly, we have a new fast Soft-Impute algorithm(Soft-Impute-RP), which is displayed in Algorithm 2.

4   Empirical Analysis

We compare our algorithm with the conventional Soft-Impute [10] and the Soft-Impute-ALS [8] in recommendation system, because all these algorithms are based on Soft-Impute. We run experiments in MovieLens10M in Matlab on PC environment with 2.6 GHz Intel Processors and 128 G memory. And the experiments in other data sets are run in Matlab on PC environment with 3.33 GHz Intel Processors and 8G memory.

4.1   Setup

We choose two typical recommendation data sets: skinny data sets Jester [6] and MovieLens. The details of the data sets are presented in Table 1. We treat 80 % of each data set as being observed and the rest 20 % as missing for testing. We randomly repeat the treatment for 5 times to implement our experiments and report average results.

Table 1. Size and rating of the data sets.

data set	Row	Column	Rating
Jester1	24,983	100	10^6
Jester2	23,500	100	10^6
Jester3	24,938	100	6×10^5
MovieLens100k	943	1,682	10^5
MovieLens1M	6,040	3,706	10^6
MovieLens10M	69,878	10,677	10^7

In the experiment, the Root-Mean-Square Error (RMSE) is used to evaluate the performance of the three algorithms. It is defined as

$$RMSE = \sqrt{\sum_{(i,j) \in S_{test}} (X_{ij} - T_{ij})^2 / |S_{test}|}, \qquad (12)$$

where S_{test} represents the test set and $|S_{test}|$ is the cardinality of S_{test}, and X_{ij} is the solution from the algorithms and T_{ij} is the corresponding true value.

Table 2. The RMSE of each algorithm in the data sets

Algorithms	Jester1	Jester2	Jester3	MovieLens100k	MovieLens1M	MovieLens10M
Soft-Impute	4.38	4.47	**5.15**	0.949	**0.879**	**0.895**
Soft-Impute-ALS	4.96	4.78	5.67	1.000	0.954	0.969
Soft-Impute-RP	**4.08**	**4.34**	5.25	**0.934**	0.890	0.904

Table 3. Time measured in seconds of each algorithm in the data sets

Algorithms	Jester1	Jester2	Jester3	MovieLens100k	MovieLens1M	MovieLens10M
Soft-Impute	13.80	10.50	14.70	39	969	7560
Soft-Impute-ALS	6.10	6.54	10.70	12.6	203	2295
Soft-Impute-RP	**4.80**	**5.28**	**4.96**	**4.95**	**184.91**	**2093**

4.2 Results

Combining the data in Tables 2, 3 with Figs. 1, 2, we compare the accuracy and efficiency among the three algorithms.

For the skinny Jester, our Soft-Impute-RP algorithm is the first to get convergence, followed by Soft-Impute-ALS and Soft-Impute. And it takes less than half of the time that Soft-Impute takes, when achieving almost the same or better results. Furthermore, Soft-Impute-RP takes less time but performs better than Soft-Impute-ALS.

As for the results of the algorithms on the different-size MovieLens data sets, our Soft-Impute-RP algorithm performs well as Soft-Impute does and much better than Soft-Impute-ALS. Taking efficiency into consideration, Soft-Impute-RP is the fastest in all the three data sets.

In summary, our algorithm achieves an expected balance between efficiency and effectiveness. Soft-Impute-RP accelerates the convergence speed of Soft-Impute by up to 7 times while achieving almost same or better performance. Also, Soft-Impute-RP performs much better in both accuracy and efficiency than Soft-Impute-ALS in all the data sets.

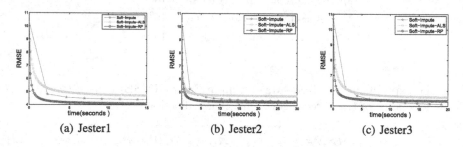

(a) Jester1   (b) Jester2   (c) Jester3

Fig. 1. RMSE-time of Jester

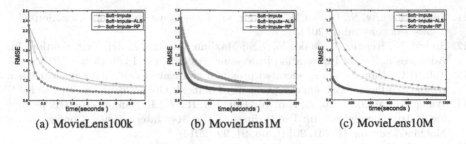

(a) MovieLens100k   (b) MovieLens1M   (c) MovieLens10M

Fig. 2. RMSE-time of MovieLens at early iterations

5 Conclusion

In this paper we have devised a scalable Soft-Impute by using a randomized SVD technique. We have empirically validated that our approach is more efficient than the conventional Soft-Impute algorithm [10] when achieving similar completion accuracy, on different kinds of recommendation data sets. Compared to Soft-Impute-ALS [8] which is another accelerating method based on Soft-Impute, our algorithm enjoys a better trade-off between efficiency and effectiveness, and behaves much better in accuracy and stability.

References

1. Cai, J.F., Candès, E.J., Shen, Z.: A singular value thresholding algorithm for matrix completion. SIAM J. Optim. **20**(4), 1956–1982 (2010)
2. Candès, E.J., Recht, B.: Exact matrix completion via convex optimization. Found. Comput. Math. **9**(6), 717–772 (2009)
3. Candès, E.J., Tao, T.: The power of convex relaxation: near-optimal matrix completion. IEEE Trans. Inf. Theor. **56**(5), 2053–2080 (2010)
4. Clarkson, K.L., Woodruff, D.P.: Low rank approximation and regression in input sparsity time. In: Proceedings of the Forty-Fifth Annual ACM Symposium on Theory of Computing, pp. 81–90. ACM (2013)
5. Dudik, M., Harchaoui, Z., Malick, J., et al.: Lifted coordinate descent for learning with trace-norm regularization. In: AISTATS-Proceedings of the Fifteenth International Conference on Artificial Intelligence and Statistics-2012, vol. 22, pp. 327–336 (2012)
6. Goldberg, K., Roeder, T., Gupta, D., Perkins, C.: Eigentaste: a constant time collaborative filtering algorithm. Inf. Retrieval **4**(2), 133–151 (2001)
7. Halko, N., Martinsson, P.G., Tropp, J.A.: Finding structure with randomness: probabilistic algorithms for constructing approximate matrix decompositions. SIAM Rev. **53**(2), 217–288 (2011)
8. Hastie, T., Mazumder, R., Lee, J., Zadeh, R.: Matrix completion and low-rank svd via fast alternating least squares (2014). arXiv preprint arXiv:1410.2596
9. Ma, S., Goldfarb, D., Chen, L.: Fixed point and bregman iterative methods for matrix rank minimization. Math. Prog. **128**(1–2), 321–353 (2011)
10. Mazumder, R., Hastie, T., Tibshirani, R.: Spectral regularization algorithms for learning large incomplete matrices. J. Mach. Learn. Res. **11**, 2287–2322 (2010)

11. Shalev-Shwartz, S., Gonen, A., Shamir, O.: Large-scale convex minimization with a low-rank constraint (2011)
12. Srebro, N., Rennie, J., Jaakkola, T.S.: Maximum-margin matrix factorization. In: Advances in Neural Information Processing Systems, pp. 1329–1336 (2004)
13. Toh, K.C., Yun, S.: An accelerated proximal gradient algorithm for nuclear norm regularized linear least squares problems. Pacific J. Optim. **6**(615–640), 15 (2010)
14. Wang, Z., Lai, M.J., Lu, Z., Fan, W., Davulcu, H., Ye, J.: Rank-one matrix pursuit for matrix completion. In: Proceedings of the 31st International Conference on Machine Learning (ICML 2014), pp. 91–99 (2014)
15. Wen, Z., Yin, W., Zhang, Y.: Solving a low-rank factorization model for matrix completion by a nonlinear successive over-relaxation algorithm. Math. Prog. Comput. **4**(4), 333–361 (2012)

CuPAN – High Throughput On-chip Interconnection for Neural Networks

Ali Yasoubi[1(✉)], Reza Hojabr[1], Hengameh Takshi[1],
Mehdi Modarressi[1,2], and Masoud Daneshtalab[3]

[1] School of Electrical and Computer Engineering, College of Engineering,
University of Tehran, Tehran, Iran
{a.yasoubi,r.hojabr,h.takshi,modarressi}@ut.ac.ir
[2] School of Computer Science, Institute for Researches
in Fundamental Sciences, Tehran, Iran
[3] Department of Electronic and Embedded Systems,
Royal Institute of Technology, Stockholm, Sweden
masdan@kth.se

Abstract. In this paper, we present a *Cu*stom *P*arallel *A*rchitecture for *N*eural networks (CuPAN). CuPAN consists of streamlined nodes that each node is able to integrate a single or a group of neurons. It relies on a high-throughput and low-cost Clos on-chip interconnection network in order to efficiently handle inter-neuron communication. We show that the similarity between the traffic pattern of neural networks (multicast-based multi-stage traffic) and topological characteristics of multi-stage interconnection networks (MINs) makes neural networks naturally suited to the MINs. The Clos network, as one of the most important classes of MINs, provide scalable low-cost interconnection fabric composed of several stages of switches to connect two groups of nodes and interestingly, can support multicast in an efficient manner. Our evaluation results show that CuPAN can manage the multicast-based traffic of neural networks better than the mesh-based topologies used in many parallel neural network implementations and gives lower average message latency, which directly translates to faster neural processing.

Keywords: Neural networks · Multicast · Clos network · Network-on-chip

1 Introduction

Neural networks are widely used to solve problems in various fields of computer science including vision, speech and pattern recognition, and prediction. In addition to common applications of neural networks, the capability of them in approximate calculation of time-consuming functions of a program has recently been exploited to accelerate program execution [1].

Neural network can be implemented on software (running on general purpose processors and recently, on GPUs) or customized hardware. Hardware implementation of neural networks has long attracted many attentions, as it offers lower power consumption, higher throughput and shorter execution time, in particular for real-time

© Springer International Publishing Switzerland 2015
S. Arik et al. (Eds.): ICONIP 2015, Part III, LNCS 9491, pp. 559–566, 2015.
DOI: 10.1007/978-3-319-26555-1_63

neural processing [2, 3]. These hardware implementations vary from small-scale single-application functional units for embedded systems [4, 5] to large massively parallel architectures that simulate billions of neurons at real-time [1, 2]. Customized hardware for neural networks typically consists of several processing nodes arranged as a multi- or many-core system-on-chip. Multiply-and-accumulate (MAC) units are the heart of these nodes. Based on the mapping policy and required performance, one or multiple neurons may be mapped onto each node. Neuron outputs are sent in the form of packets to other nodes according to the neural network topology (connectivity pattern).

Early examples of hardware implementation of neural networks use a shared bus topology, as a simple and low-cost interconnect, to connect neurons [6, 7]. However, low scalability of bus in terms of power and performance motivated researchers to replace bus with the more scalable network-on-chip (NoC). The superiority of NoCs for neural network chips have been discussed in many prior work [1, 2, 7, 8].

Most existing NoC proposals for neural networks use either the baseline or some variations of the well-known mesh topology. The mesh topology consists of an n-dimensional array of nodes connected by a regular interconnection structure that connect each node to its direct neighbors and hence, benefits from ease of circuit-level implementation. However, as the NoC size increases, average latency of mesh increases considerably due to the lack of direct paths among remotely located nodes in the topology. In addition, the traffic of neural networks is in the form of multicast, so a NoC that supports multicast communication is critical for them.

In this paper, we present CuPAN, a Custom Parallel Architecture for Neural networks that uses the efficient low-cost multi-stage Clos interconnection topology for inter-neuron connections. Multi-stage interconnection networks (MINs) are a class of indirect interconnection networks that adopt multiple layers of switches to connect any pair of input and output ports [9]. MINs are primarily a replacement for crossbar switches for systems with a large number of input and outputs, where the crossbar's cost is prohibitive. Clos is the most favorite family of MINs and has been used in many commercial devices and machines, including Infiniband and Myrinet switches [10] for low-cost switch ingress and egress port connection.

Clos is an instance of high-radix networks that has attracted many attentions in recent years [11–16]. In high-radix networks, routes have more, but narrower, links compared to a low-radix network (like mesh). Due to the more ports (and links) that a high-radix switch has, high-radix networks provide lower diameter and hence, much better scalability in terms of zero-load latency and throughput. However, they suffer from higher serialization delay, if the packet length is much larger than the link bit-width, but it is not a problem in neural networks with intrinsically short packets: each packet in the NoCs used in neural networks should carry neuron output, which is a single floating point number. Thus, the packets contain 32- or 64-bit data plus a few control bits. This narrow bit-width is the first reason that makes high-radix low-bit width Clos an appropriate option for implementing the communication part of neural networks. Fast low-cost input-output connections and efficient support of multicast is the second interesting property of Clos that make it suitable to the multicast-based traffic of neural networks.

Experimental results show that the proposed architecture, CuPAN, can make more appropriate trade-off between cost and performance than state-of-the-art prior work for hardware implementation of neural networks.

2 Clos Topology

Figure 1 shows a 3-stage Clos network that connects $N(= n \times r)$ processors at the input layer to N processors at the output layer. The links shown in the figure are unidirectional from input to output switches. Each layer is a stack of crossbar switches. Generally, $a(m.n.r)$ Clos network has m switches at the middle-stage, r input switches each with n input ports, and r output switches each with n output ports. Each middle stage switch has one input link from every input switch and one output link to every output switch. Consequently, input switches are of size $n \times m$, middle switches are of size $r \times r$, and output switches are of size $m \times n$.

In our proposed Clos, each switch is implemented as an input-queued crossbar switch. A credit-based flow control mechanism is applied on two sides of every link to guarantee that there is always available buffering space in the downstream switch before sending a packet.

The fact that neural network packets are short and single-flit, omits many issues and considerations in multiple-flit packet NoCs that make the arbitration and switching methods complicated and expensive.

For a Clos network with m middle switches, there are m distinct routes from any input to any output, one through each middle stage switch. In our design, every switch in the network has multicast capability: each input link of a switch can be simultaneously connected to any subset of output links of the switch (provided that the output ports are idle).

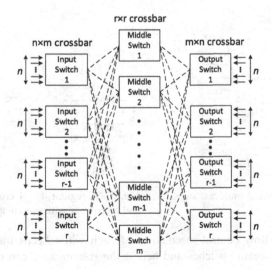

Fig. 1. 3-stage (m, n, r) Clos network

3 Neural Networks on Clos Topology

The proposed architecture consists of a number of processing units connected by a Clos network. The processing elements can be either a general-purpose processor or a customized MAC unit. In either case, the same traffic pattern will be injected into the network. If the neural network has more neurons than the processing elements, multiple neurons are mapped onto a single processing element.

3.1 Broadcast on Clos

Since a neuron in layer n should send its output to all neurons in layer $n + 1$, assuming layer n and layer $n + 1$ are respectively mapped onto the input and output ports of the Clos, data should be broadcasted from its corresponding input port of the Clos network to all outputs. To this end, the neuron data is first packetized and delivered to the first stage switch (stage 1 in Fig. 2). Then, it is sent to a middle-stage and output layer switches with large output speedup (Fig. 3) after the routing and arbitration operations.

First-Stage Routing. At an input switch, once a packet gets to the head of the queue, the routing unit of the switch determines the output port (middle switch) to which the packet should be sent. The routing decision is made in such a way that the queuing time (blocking latency) of the packet in the middle switches is minimized and the load is balanced across the middle switches.

To this end, the queue length of the downstream input port in the middle switches is used as congestion metric. More complex metrics, like average waiting time in middle switches can also be used, but our evaluations show that the average queue length makes an appropriate trade-off between the measurement complexity and network performance. Please note that this is the only stage that makes routing decision. In consequent stages, packets are limited to follow a single path.

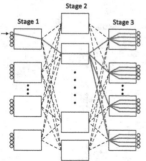

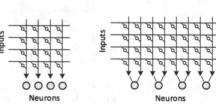

Fig. 2. The proposed multicast steps

Fig. 3. A regular 4 × 4 crossbar switch (left) and a crossbar with output speedup of 2 (right)

A credit-based flow control mechanism of each witch keeps track of the queue length at all downstream switches and hence, the routing unit can obtain this information locally with no cost.

Arbitration. After routing, the packet requests the selected output port. An arbiter associated to each output port grants the port to one of the requesting packets based on the round-robin policy. The winner is then sent to the middle switch, whereas the losers will continue to request in the next cycle.

Multicast. The switches in the middle and output layers of Clos broadcast every received packet. Each switch of these layers selects input ports in a round-robin manner and broadcasts the head-of-line packet of that input port to all output ports. Stages 2 and 3 in Fig. 2 show this stage in the middle and output switches, respectively.

If a packet should be delivered to a subset of Clos outputs, a bit vector is used as a mask to select proper output ports at the middle and output switches. The vectors are stored in routers and are indexed by an identifier in packet header.

Clos Area Reduction. When the Clos topology is used to broadcast data, an imbalance between the logical input and output bisection occurs; at each cycle, input ports can feed $n \times r$ new data to the network, while the network can only deliver a single data to all output ports. To increase throughput, we propose to use crossbars with output speedup of two or more in the output layers. In a crossbar with output speedup of p, each output port of Clos (and hence the target processor connected to it) is connected to p crossbar outputs and can receive p distinct data from the crossbar. We set the number of middle switches to p to balance the output and middle bandwidth. However, the input bandwidth is still larger than the middle and output bandwidth: in a Clos, there are $m \times r$ links between the input and middle switches that are capable to carry a maximum of $m \times r$ packets to middle stage switches, but only a maximum of m (or p) packets can be sent (broadcasted) from the middle stage switches to the output stage. To remove unused bisection width, we limit the output degree of input switches, i.e. the number of connections that each input stage switch has to middle stage switches, from m to a number less than which is 2 in this work.

4 Evaluation

We have implemented the proposed Clos network in a cycle-accurate network-on-chip simulator. We have also implemented the concentrated mesh topology of [8] and the H-NoC topology of EBMRACE [2] in the simulator for the comparison purpose. We use the same broadcast algorithms developed for these topologies [2, 8]. The three considered networks have 128 nodes with 40-bit links (to carry a fixed point number and routing information in a single flit). All input ports are equipped with 8-flit buffers. The only exception is the last level switches of CuPAN that use 2-flit buffers, since the input and output bandwidth of these switches are equal, so there is no need to deep buffers. When there is no contention, all switches forward packets after buffering, routing, and arbitration in three cycles. In the simulations, we used a (4,4,32) Clos with reduced links and output speedup of 4. In H-NoC, 8 neurons (functional units) are connected to a neuron router, 8 neuron routers are connected to a tile router, and two tile routers are connected by a cluster router to form a 128-node network. In the concentrated mesh, 4 nodes are connected to a single router arranged as an 8×4 mesh. In this experiment, we assumed that processing nodes perform the multiply and accumulate operations fast enough to process all the data delivered by network without any queuing delay and in a single cycle.

Figure 4 compares the broadcast performance of the three networks by demonstrating the average broadcast latency of them under the uniform traffic. The broadcast latency of a packet is measured from the packet generation time to the time when it is delivered to the last destination.

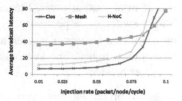   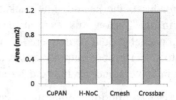

Fig. 4. Average broadcast latency of concentrated mesh, H-NoC, and CuPAN under different injection rates

Fig. 5. Area of full crossbar, concentrated mesh (Cmesh), H-NoC, and CuPAN in mm^2

As the figure shows, the fixed hop count of three in CuPAN (compared to the maximum hop count of 6 of H-NoC and 12 of mesh) leads to shorter latency. However, the more links provided by a concentrated mesh topology leads to slightly better latency in high traffic.

Power consumption evaluation in a selected injection rate (just before the saturation point of H-NoC) shows the power usage of 858 mW for H-NoC, 602 mW for CuPAN and 912 mW for the concentrated mesh. CuPAN consumes less power due to its shorter hop count. Power results just include the communication power and are obtained by an architectural-level power evaluation tool in 45 nm.

We also evaluate the area of the three considered NoCs by synthesizing the VHDL model of them using a commercial synthesis and physical design tool. The figure also shows the area of a 128 × 128 crossbar switch for reference. The results are depicted in Fig. 5 and show that the 128-port Clos with output speedup of 4 for the last level switches is smaller than other considered networks. The next smaller network is H-NoC that is 13 % larger than an equivalent CuPAN. The main reason of the better area of CuAPN is that it needs smaller buffering space and simpler routing unit.

We further evaluate the proposed NoC under some realistic neural networks. The benchmarks and the neural network topology established for them are listed in Table 1. We implement all benchmarks on three considered networks of size 128; this involves mapping multiple neurons onto a single NoC node (Clos input, in our case). The same mapping is considered for all NoCs. Table 1 compares the performance of the three NoCs in terms of the throughput. The neural network throughput is defined as the neural network input arrival rate at which the NoC can still work normally by keeping the latency under 100 cycles. The results are normalized with respect to the results obtained by concentrated mesh. As Table 1 shows, the CuPAN can provide a higher throughput than H-NoC and mesh due to the better latency it exhibits. As the table demonstrates, for small benchmarks, H-NoC shows the same behavior as a 3-stage Clos and can offer similar performance.

In a 3-layer neural network, as mentioned before, a bidirectional Clos is used. Bidirectional CuPAN consumes more area than H-NoC, but this area overhead can be compensated by up to 50 % higher throughput it offers.

Table 1. Normalized neural network throughput under four realistic applications for the considered NoCs. The results are normalized with respect to the Cmesh results

Benchmark	NN topology (in:hidden:out)	Normalized throughput (input/cycle)		
		CuPAN	H-NoC	CMesh
Performance modeling [17]	128:256:128	3.75	2.81	1
Object classification [18]	3072:3000:10	3.08	2.06	1
Census data analysis [19]	14:12:12:2	2.25	2.23	1
Hand writing digit recognition [20]	784:700:10	3.65	2.92	1

5 Conclusion

In this paper, we proposed CuPAN, a customized parallel architecture for neural networks, which utilizes the well-known Clos topology to support heavy inter-neuron communication. The synergy between the multicast-based multilayer communication of neural networks and the topology of Clos makes Clos an appropriate selection for hardware-based implementation of neural networks. In this paper we showed that Clos can offer higher throughput and lower latency than the conventional mesh and the state-of-the-art HNoC topologies. Clos has long been a popular topology for interconnection networks; in particular, several area-efficient layouts have been proposed for Clos that can be used for efficient implementation of the proposed NoC.

Acknowledgement. This work was partially supported by VINNOVA (Swedish Agency for Innovation Systems) within the CUBRIC project.

References

1. Painkras, E., Plana, L.A., Garside, J., et al.: SpiNNaker: a 1-W 18-core system-on-chip for massively-parallel neural network simulation. IEEE J. Solid State Circuits **48**(8), 1943–1953 (2013)
2. Carrillo, S., Harkin, J., McDaid, L.J., et al.: Scalable hierarchical network-on-chiparchitecture for spiking neural network hardware implementations. IEEE Trans. Parallel Distrib. Syst. **24**, 2451–2461 (2013)
3. Zhang, Q., Wang, T., Tian, Y., et al.: ApproxANN: an approximate computing framework for artificial neural network. In: Design Automation and Test in Europe Conference Exhibition (DATE), pp. 701–706 (2015)
4. Venkataramani, S., Ranjan, A., Roy, K., Raghunathan, A.: AxNN. In: Proceedings of the International Symposium on Low power Electronics and Design-ISLPED 2014, pp. 27–32 (2014)

5. Esmaeilzadeh, H., Sampson, A., Ceze, L., Burger, D.: Neural acceleration for general-purpose approximate programs. IEEE Micro **33**, 16–27 (2012)
6. Fakhraie, S.M., Smith, K.C.: VLSI — Compatible Implementations for Artificial Neural Networks. Springer, Heidelberg (1997)
7. Vainbrand, D., Ginosar, R.: Network-on-chip architectures for neural networks. In: 2010 Fourth ACM/IEEE International Symposium on Networks-on-Chip, pp. 135–144. IEEE (2010)
8. Dong, Y., Li, C., Lin, Z., Watanabe, T.: Multiple network-on-chip model for high performance neural network. J. Semicond. Technol. Sci. **10**, 1 (2010)
9. Dally, W.J., Towles, B.: Principles and Practices of Interconnection Networks, 1st edn. Morgan-Kaufmann Publishers, San Francisco (2004)
10. Legacy Myrinet-2000. https://www.myricom.com/hardware/myrinet-2000-switches.html. Accessed July 2015
11. Kao, Y.-H., Alfaraj, N., Yang, M., Chao, H.J.: Design of high-radix clos network-on-chip. In: Fourth ACM/IEEE International Symposium on Networks-on-Chip, pp. 181–188. IEEE (2010)
12. Kim, J., Dally, W.J., Towles, B., Gupta, A.K.: Microarchitecture of a high-radix router. In: 32nd International Symposium on Computer Architecture, pp. 420–431. IEEE (2005)
13. Kao, Y.-H., Yang, M., Artan, N.S., Chao, H.J.: CNoC: high-radix clos network-on-chip. IEEE Trans. Comput. Des. Integr. Circuits Syst. **30**, 1897–1910 (2011)
14. Chen, L., Zhao, L., Wang, R., Pinkston, T.M.: MP3: minimizing performance penalty for power-gating of clos network-on-chip. In: 2014 IEEE 20th International Symposium on High Performance Computer Architecture, pp. 296–307. IEEE (2014)
15. Kamali, M., Petre, L., Sere, K., Daneshtalab, M.: Formal modeling of multicast communication in 3D NoCs. In: 2011 14th Euromicro Conference on Digital System Design, pp. 634–642. IEEE (2011)
16. Ebrahimi, M., Daneshtalab, M., Liljeberg, P., Plosila, J., Flich, J., et al.: Path-based partitioning methods for 3D networks-on-chip with minimal adaptive routing. IEEE Trans. Comput. **63**, 718–733 (2014)
17. Esmaeilzadeh, H., Saeedi, P., Araabi, B.N., et al.: Neural network stream processing core (NnSP) for embedded systems. In: IEEE International Symposium Circuits System (2006)
18. Krizhevsky, A.: Learning multiple layers of features from tiny images. M.S. thesis, University of Toronto (2009)
19. Lichman, M.: UCI machine learning repository. In: University of California; Irvine; School of Information and Computer Science (2013). http://archive.ics.uci.edu/ml
20. Lecun, Y., Bottou, L., Bengio, Y., Haffner, P.: Gradient-based learning applied to document recognition. Proc. IEEE **86**, 2278–2324 (1998)

Forecasting Bike Sharing Demand Using Fuzzy Inference Mechanism

Syed Moshfeq Salaken[✉], Mohammad Anwar Hosen, Abbas Khosravi, and Saeid Nahavandi

Center for Intelligent Systems Research, Deakin University, Geelong, Australia
syed.salaken@gmail.com

Abstract. Forecasting bike sharing demand is of paramount importance for management of fleet in city level. Rapidly changing demand in this service is due to a number of factors including workday, weekend, holiday and weather condition. These nonlinear dependencies make the prediction a difficult task. This work shows that type-1 and type-2 fuzzy inference-based prediction mechanisms can capture this highly variable trend with good accuracy. Wang-Mendel rule generation method is utilized to generate rulebase and then only current information like date related information and weather condition is used to forecast bike share demand at any given point in future. Simulation results reveal that fuzzy inference predictors can potentially outperform traditional feedforward neural network in terms of prediction accuracy.

1 Introduction

Bike sharing demand forecasting in city level granularity is a relatively new problem in research community, with only a few researchers approaching this problem with machine learning techniques [1,2]. In addition, most of the existing works in current literature addresses the problem based on short span of prior data for knowledge discovery. Only [1–3] utilizes approximately 2 years of data for prediction. However, authors in [2] used historical data for bike share demand forecasting ahead of 24 hours, researchers in [1] used this for anomaly detection and authors in [3] used this for forecasting with signal processing techniques. Our work utilizes the same dataset used by [1,2]. Among other related works, researchers in [4] addresses balancing uneven demand among stations with several mover trucks, authors in [5] addresses the same issue with only one mover truck, authors in [6] investigates availability of bike in station within a predefined time span in the city of Dublin and researchers in [7] addresses the near future bike availability prediction in stations. Work in [8] forecasts the number of bikes entering and exiting a particular station with neural network as a part of building fuzzy and neural network coupled relocation algorithm for bike sharing system. Please note, [8] does not use fuzzy logic for bike demand forecasting. Finally, researchers in [9] addresses the bike availability issue with a small 7 week dataset.

© Springer International Publishing Switzerland 2015
S. Arik et al. (Eds.): ICONIP 2015, Part III, LNCS 9491, pp. 567–574, 2015.
DOI: 10.1007/978-3-319-26555-1_64

Among these works, only investigation in [3] forecasts the bike share demand with current variables like weather, holiday and number of users. However, even this work does not attempt to forecast the bike share demand based on only date, holiday and weather without the knowledge of user base volume. Finally, none of the works in existing literature employs a solution which not only makes a decent forecast, but also has the capability to explain the reason behind the forecasted volume.

In this work, we demonstrate the fuzzy inference can predict the bike share demand with good accuracy at any given point of time with the knowledge of calendar information coupled with weather variables like temperature, wind speed and humidity. Once the fuzzy inference model is built with prior knowledge, past data are no longer required to make a prediction. This is the key difference of our work with [7], which is most similar to our work based on a similar dataset from Barcelona. However, work in [7] forecasts only station level demand, not system level demand and therefore, our results are not comparable with theirs. Research in [2] predicts one day ahead demand by introducing several artificial time dependent features and missing samples. This kind of data preprocessing is not necessary for fuzzy inference. In addition, this fuzzy inference mechanism has the ability to explain the causal effect of date and weather situation on the prediction value.

Please note, no new fuzzy inference mechanism is claimed in this work, instead it is demonstrated that existing fuzzy inference mechanism, including type-1 and interval type-2 fuzzy sets, is capable of capturing the highly diverse nature of bike share demand. To the best of our knowledge, this is the first fuzzy inference based prediction study on bike share demand to forecast overall bike sharing demand. It is also shown that fuzzy inference mechanism is better than neural network prediction on the same dataset, without performing any optimization on either FLS parameters or neural network parameters, and therefore provides a better overall accuracy along with easily interpretable prediction mechanism. This, of course, does not imply anyway that a better tuned neural network will not outperform a FLS. The focus of this work lies within the scope of demonstrating that a simple FLS can forecast the highly variable nature of bike share demand while offering an insight to the causal relationship among demand and different contextual factors.

The rest of the paper is organized as follows: Sect. 2 describes the experimental procedure along with preprocessing, Sect. 3 discusses and compares the results, and finally Sect. 4 concludes the work.

2   Experimental Procedure

The data used in this study is taken from Capital Bikeshare system, Washington D.C., USA for year 2011 and 2012, as provided in [1]. It is available in daily and hourly format. Both of them contain index, date, season, year, month, day of the week, hour (only in hourly format) weekend, public holiday, weather situation, temperature, feeling temperature e.g. real feel of the temperature, humidity and wind speed. It also provides number of registered and casual users in addition to total user count.

2.1 Data Preparation

Original data contains few extraordinary environmental scenario which is not helpful to this regression problem, but can be of significant importance for some cases. As these are not in the scope of this work, those data are discarded as part of data cleansing. For example, effects of hurricane Sandy and extra ordinary rain on December 7, 2011 are discarded. After this is done, date information and index are also discarded. This is done to ensure the purpose of this work, i.e. showing fuzzy inference is capable of forecasting highly diverse nature of bike sharing demand, is not biased by gradually increasing demand of the service. Next, the data is shuffled randomly before split into training and testing chunks. This randomization may provide different level of success in forecasting. Therefore, each forecasting computation is run 5 times to make the result statistically meaningful. Data is neither scaled nor normalized in this study. However, other aspects of developing fuzzy inference model for this purpose will be undertaken in future work. Please note that, one out of every six samples is taken into consideration for hourly forecast to reduce the forecasting time. If all data is taken into account, it is very likely to produce even more accurate result.

2.2 Fuzzy Rulebase Construction

This work utilizes the renowned Wang-Mendel rule generation method [10] to generate fuzzy rulebase from numerical data. As different input variables are known from historical data, it is very easy to produce an exhaustive rulebase from them. After which this database is pruned by picking only one rule from a conflict group. Gaussian membership function is utilized throughout the work to ensure whole input domain is covered in order to avoid any abrupt change in control surface. A total of 10 membership functions are used on each of 9 inputs to determine the fuzziness of each input sample. Because the range of output is very large, 35 membership functions with a very large standard deviation are used to capture the output domain.

Two types of fuzzy inference mechanism are used and compared. Please note, the parameters of fuzzy inference system or fuzzy logic system are not optimized as this work only intends to show fuzzy inference mechanism is a good predictor for this kind of data. In future works, tuning will be done to achieve best performance form fuzzy inference process. Parameter values used in this experiment are listed in Table 1. Sample membership functions for both type-1 (T1) and interval type-2 fuzzy inference (IT2 FLS) are shown in Fig. 1a and b. MFs on all input are not shown due to page limitation.

2.3 Fuzzy Inference

Once the fuzzy rulebase is constructed, this is applied on testing data. This is done by taking one sample of data which contains one instance of every input and passing that to fuzzy inference function. Inside the inference mechanism, fuzziness of each input is first done by the means of membership degree computation. Afterwards, the applicability of each rule is calculated and center of corresponding rule output is captured. Finally, they are aggregated and defuzzified using the method described in [10]. For interval type-2 fuzzy inference, first the firing interval is calculated in the same way applicability of rule is computed in type-1 inference and then defuzzification is done using EIASC algorithm [11].

2.4 Rulebase Adaptation

As described in [10], rulebase can be adapted to enhance the performance of fuzzy inference based prediction. A similar approach is taken in this work as well to address the highly variable nature of bike share demand. When one sample of test set is predicted, it is incorporated in fuzzy rulebase with actual output and used to produce another set of exhaustive and pruned rulebase. This effectively increases the exhaustive rulebase size which is used throughout the prediction. In addition, exhaustive rulebase is used in first inference, instead of pruned rulebase, to ensure highest prediction accuracy from the beginning.

2.5 Neural Network Prediction

Since there is no work in literature which tackles this forecasting problem from the perspective we are investigating, a neural network is built, trained and tested for performance evaluation of fuzzy inference mechanism. All inputs and outputs are scaled using their minimum and maximum value prior to neural network training and then final result in scaled back for comparison with actual output. However, no such preprocessing is done for fuzzy inference based prediction. In the neural network, a 3 layer configuration is used with 1 hidden layer and 10 hidden neuron. Bayesian regularization [12,13] is used as training algorithm as it gives more generalization power to neural network [12,14]. Sum of squared error is used as cost function as this is proven to produce most accurate prediction after few trial and error.

Table 1. Parameter values in experiment

Inference type		Input	Output
	Number	9	1
	Range	[0,25]	[22,8714]
T1	Number of MF	10	35
	Standard deviation	0.25	60
IT2	Number of MF	10	35
	Standard deviation	0.15, 0.35	60

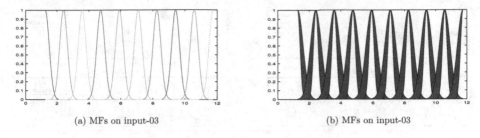

(a) MFs on input-03   (b) MFs on input-03

Fig. 1. Samples of T1 and IT2 membership functions on input

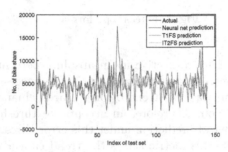

Fig. 2. Comparison of fuzzy inference and neural network prediction (Bayesian regularization, 10 neuron in 1 hidden layer and SSE)

3   Results and Discussion

As RMSE is prone to occasional large error and this dataset contains a wide range of output data (see Table 1), which is neither scaled nor normalized for fuzzy inference, root mean squared logarithmic error (RMSLE) is used as the primary metric for comparison. Because RMSLE is not easy to interpret, other metrics e.g. mean absolute percentage error and percentage of predictions under defined threshold value are also presented. Even though none of these metrics directly shows the level of success in capturing the trend, they establish a benchmark for further study. These values are not comparable with existing literature as this particular problem was never investigated within research community from the same perspective i.e. predicting future demand based on current knowledge only where no previous demand (e.g. demand of last 24 hours etc.) is known. Few related works utilized relative absolute error [1,7], root relative squared error [1], root mean square error for 5 min [6] and mean absolute error on a different, but slightly related, problem [3,9].

As the data provides two sets of datafile for hourly and daily usage, fuzzy inference mechanism is applied on both of them. Result for daily forecasting performance can be seen in Table 2 and Fig. 3a. Hourly forecasting performance is listed in Table 3 and Fig. 3b. From Table 2, note that RMSLE for type-1 (T1) prediction is always lower than interval type-2 (IT2) prediction. This surely indicates that T1 prediction is more accurate than IT2 inference based prediction.

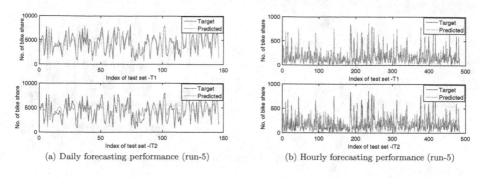

(a) Daily forecasting performance (run-5)   (b) Hourly forecasting performance (run-5)

Fig. 3. Forecast performance

This can be confirmed by respective mean absolute percentage error (MAPE) value as well. However, none of them captures the success of fuzzy inference in capturing the highly variable trend of bike share demand for either T1 prediction or IT2 prediction. Therefore, an acceptance threshold variable is also introduced and number of prediction under this error threshold is calculated in percentage. Note that, this metric shows the trend capturing success partially and preserve the earlier observation of higher T1 prediction success. However, it is neither related with RMSLE nor MAPE value. This can be realized by examining the first and second row of Table 2 which shows a decrease of both RMSLE and MAPE with a decrease of "close point(%)" value whereas an increase is intuitively expected as those error values decreased.

From Fig. 3 and Table 3, observe that fuzzy inference based prediction is even more successful in capturing the changing nature of bike share demand in the hourly forecast. Even in this case, T1 inference based prediction is more accurate than IT2 based prediction. However, please note that no optimization is done on either T1 FLS or IT2 FLS in order to tune their parameter and enhance the performance. It is indeed possible that a better tuned IT2 fuzzy logic system (FLS) may outperform a T1 FLS.

To provide a benchmark, daily forecast prediction of a neural network containing one hidden layer with 10 neurons, which is trained with Bayesian regularization algorithm, is compared with that of fuzzy inference system. The comparison is shown in Fig. 2. Please note, this configuration of neural network is finally chosen after few trial and error with different number of hidden layer neurons and training algorithm as this configuration is found to be producing most accurate prediction. Observe from this figure that neural network prediction is not closely following the target data and occasionally showing very large deviation. Clearly both T1 and IT2 FLS prediction outperforms neural network prediction. Please note, other training algorithms and different number of neurons in hidden layer are also tested and the best result is presented in this figure. However, this result does not imply that a better tuned neural network will not outperform the fuzzy inference based system. Making such a comparison with optimization techniques is beyond the scope of this work and will be investigated in future works.

Table 2. Performance in daily forecast

Run ID	Acceptance threshold	Type-1			Interval type-2		
		RMSLE	MAPE	Close point (%)	RMSLE	MAPE	Close point (%)
1	500	0.3194	24.3735	53.4247	0.3952	36.2882	36.3014
2	500	0.3024	22.3361	48.6301	0.3549	30.626	31.5068
3	500	0.3078	24.5076	47.2603	0.4078	37.2364	34.9315
4	500	0.3694	28.3052	52.0548	0.4431	41.7659	36.3014
5	500	0.3275	24.2118	50	0.3895	35.3197	36.9863

Table 3. Performance in hourly forecast

Run ID	Acceptance threshold	Type-1			Interval type-2		
		RMSLE	MAPE	Close point (%)	RMSLE	MAPE	Close point (%)
1	50	0.4577	36.86	75.4132	0.6596	82.131	57.2314
2	50	0.4922	39.455	72.314	0.6226	75.2026	56.6116
3	50	0.4682	39.2181	71.4876	0.6202	74.5585	53.719
4	50	0.487	39.4747	72.5207	0.6309	77.0695	55.5785
5	50	0.461	40.9236	72.314	0.6351	78.5359	53.0992

Furthermore, it is important to note that results of this study show that T1 fuzzy system outperforms IT2 fuzzy systems in terms of prediction accuracy. This observation complies with existing literature as current literature states that only optimally tuned IT2FLS may outperform another optimally tuned T1FLS [15]. We want to emphasize again that no such optimization is done in this study and footprints of uncertainty for IT2 FS are arbitrarily designed.

4 Conclusion

In this study, it is shown that fuzzy inference mechanism can effectively capture the rapidly changing nature of bike share demand in city level forecasting. Both type-1 and interval type-2 fuzzy inference models are investigated. The result demonstrates that unoptimized fuzzy inference systems can widely outperform unoptimized traditional feed forward neural networks. Thus, simple fuzzy inference based prediction is more effective than costly neural network based prediction approach in the sense of achieving insight among bike sharing demand and contextual factors. Most importantly, this work shows that forecasting of bike sharing demand can be done with fuzzy inference, which has never been done before using a fuzzy logic system in existing literature.

References

1. Fanaee-T, H., Gama, J.: Event labeling combining ensemble detectors and background knowledge. Prog. Artif. Intell. **2**, 1–15 (2013). doi:10.1007/s13748-013-0040-3
2. Giot, R., Cherrier, R.: Predicting bikeshare system usage up to one day ahead. In: 2014 IEEE Symposium on Computational Intelligence in Vehicles and Transportation Systems (CIVTS), pp. 22–29. IEEE (2014)

3. Borgnat, P., Abry, P., Flandrin, P., Robardet, C., Rouquier, J.-B., Fleury, E.: Shared bicycles in a city: a signal processing and data analysis perspective. Adv. Complex Syst. **14**(3), 415–438 (2011)
4. Benchimol, M., Benchimol, P., Chappert, B., De La Taille, A., Laroche, F., Meunier, F., Robinet, L.: Balancing the stations of a self service bike hire system. RAIRO-Oper. Res. **45**(01), 37–61 (2011)
5. Chemla, D., Meunier, F., Calvo, R.W.: Bike sharing systems: solving the static rebalancing problem. Discrete Optim. **10**(2), 120–146 (2013)
6. Yoon, J.W., Pinelli, F., Calabrese, F.: Cityride: a predictive bike sharing journey advisor. In: 2012 IEEE 13th International Conference on Mobile Data Management (MDM), pp. 306–311. IEEE (2012)
7. Froehlich, J., Neumann, J., Oliver, N.: Sensing and predicting the pulse of the city through shared bicycling. In: IJCAI, vol. 9, pp. 1420–1426 (2009)
8. Caggiani, L., Ottomanelli, M.: A modular soft computing based method for vehicles repositioning in bike-sharing systems. In: 2012 Proceedings of EWGT2012 - 15th Meeting of the EURO Working Group on Transportation, Procedia - Social and Behavioral Sciences, vol. 54, pp. 675–684, Paris, September 2012. http://www.sciencedirect.com/science/article/pii/S1877042812042474
9. Kaltenbrunner, A., Meza, R., Grivolla, J., Codina, J., Banchs, R.: Urban cycles and mobility patterns: exploring and predicting trends in a bicycle-based public transport system. Pervasive Mobile Comput. **6**(4), 455–466 (2010)
10. Wang, L.-X., Mendel, J.M.: Generating fuzzy rules by learning from examples. IEEE Trans. Syst. Man Cybern. **22**(6), 1414–1427 (1992)
11. Wu, D., Nie, M.: Comparison and practical implementation of type-reduction algorithms for type-2 fuzzy sets and systems. In: 2011 IEEE International Conference on Fuzzy Systems (FUZZ), pp. 2131–2138. IEEE (2011)
12. MacKay, D.J.: A practical Bayesian framework for backpropagation networks. Neural Comput. **4**(3), 448–472 (1992)
13. Girosi, F., Jones, M., Poggio, T.: Regularization theory and neural networks architectures. Neural Comput. **7**(2), 219–269 (1995)
14. Foresee, F.D., Hagan, M.T.: Gauss-Newton approximation to Bayesian learning. In: Proceedings of the 1997 International Joint Conference on Neural Networks, vol. 3, pp. 1930–1935. IEEE, Piscataway (1997)
15. Wu, D., Mendel, J.: Designing practical interval type-2 fuzzy logic systems made simple. In: 2014 IEEE International Conference on Fuzzy Systems (FUZZ-IEEE), pp. 800–807, July 2014

Prior Image Transformation for Presbyopia Employing Serially-Cascaded Neural Network

Hideaki Kawano[1]([✉]), Kouichirou Hayashi[1], Hideaki Orii[2], and Hiroshi Maeda[1]

[1] Kyushu Insititute of Technology, Kitakyushu 804-8550, Japan
kawano@ecs.kyutech.ac.jp
http://www.ssc.ecs.kyutech.ac.jp/
[2] Fukuoka University, Fukuoka 814-0180, Japan
oriih@fukuoka-u.ac.jp

Abstract. Visual functions of the elderly are gradually changing with age. As one of the changes, aged eyes have a different property in perceiving high-frequency components from younger eyes. In general, the elderly perceives images differently from the younger. To give the same perception for the same image, a different image from an original image needs to be displayed to the elderly. In this paper, a method of generating the inverse characteristic of the image filter for the presbyopia is proposed. To this end, a serially-cascaded neural network model is proposed. The neural network is composed of 4 layers. The 4-layer neural network is divided into 2 blocks on the aspect of function. The upper and the lower layers play a role of simulating the image conversion and obtaining its opposite characteristic, respectively. The performance of the proposed framework is evaluated by the experiments on the image pre-conversion for the presbyopia.

Keywords: Presbyopia · Elderly vision · Image transformation · Serially-cascaded neural networks · Inverse transformation

1 Introduction

Recent popularizations of small IT devices, e.g. cell phones, smart phones, and tablet PCs, are deeply concerned about adverse effects for humans' visual power. If human eyes which adjust focus at the unconscious level are kept focusing on close objects for a long time, an ability of focus adjusting might be deteriorated. And also, this might bring decay of the visual power.

In general, the percentage of low visual power people glows with seniority. This is a so-called presbyopia [1,2]. In addition, as previously mentioned, it is considered that factors raising visual degradations increase even in young people. It is said that a half of recent Japanese people suffers from the visual degradation. Such visual unclearness might bring the presbyopes a lot of inconveniences in daily life. Indeed glasses and contact lens are widely used for dealing with the presbyopia, but those compensators may make the users feel inconvenience from physical restraints and constrictive view. Especially, the contact lens are

© Springer International Publishing Switzerland 2015
S. Arik et al. (Eds.): ICONIP 2015, Part III, LNCS 9491, pp. 575–582, 2015.
DOI: 10.1007/978-3-319-26555-1_65

concerned about the possibility of setting up eye troubles because of a long period of use and adhesions of dirt. On the other hand, there are some medical treatments of visual recovery, e.g. eye muscle trainings, the LASIK (laser in situ keratomileusis) [3], and the Orthokeratology [4]. Though it is expected to recover visibilities, there are some problems: it might take several years before any effect is seen, it might work differently for different people. Therefore, it is difficult to find an effective medical treatment of visual recovery suitable for each person.

Aoki, M. et al. [5] proposed a virtual display to let the presbyope perceive an original image. The virtual display creates a transformed image which can be perceived as non-blurred image. The transformed image can be produced by prior image transformation considering the inverse characterisitics of presbyope's visual perception. The transformation method is based on a Wiener filter. In this framework, the Wiener filter produce a transformed image with the characterisitcs to cancel the presbyope's blurring effects. However, the result of the Wiener filter involves abnormal values which are not falling within an image intensity range. To avoid this result, the contrast of original image should be decreased before the transformation. As a result, the preprocessing brings a deterioration of the transformed result.

In this study, by obtaining the inverse characteristics of the presbyopia vision, we propose a image transformation method for generating an image for which the presbyope perceives the same impression as the clear-sighted people. The proposed method employs a multi-layer perceptron with four layers. In the MLP network, an improved image on visibility can be obtained effectively. To valid the effectiveness of the proposed method, some experiments are addressed with respect to image transformation for the presbyopes.

2   Serially-Cascaded Neural Network

To let the presbyopes perceive an image as the normal vision people do, an appropriately transformed image needs to be presented to the presbyopes. Figure 1 shows a schematic overview of the proposed framework. In the Fig. 1, the transformation model represents a perception model of the presbyopes. In this study, it is assumed that the perception model is given, and the presbyopes perception is modeled by a Gaussian filter as the previous study [5] used. A major issue is to generate an inverse transformation model representing the inverse perception model of the presbyopes. To this end, we construct a serially-cascaded neural network connecting a simulated model of presbyopes perception and an inverse model for the presbyopes visual characteristics. Inputs and outputs of the neural network are intensitiy values of the image, and the neural network is learnt in the manner of back propagation.

2.1   Simulated Model of Presbyopia Vision

At first, a Gaussian filter which simulates blurring effect by the presbyopia is prepared for obtaining the presbyopia simulated image which is described as the

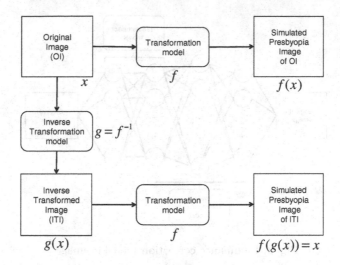

Fig. 1. Overview of the proposed method.

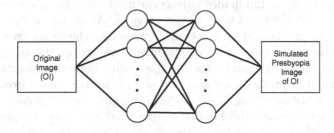

Fig. 2. Simulated model of the presbyopia vision.

transform image in Fig. 1. Next, an original image to let the presbyope perceive and its simulated presbyopia image are fed to a 2-layer MLP neural network for training as inputs and teachers, respectively. As a result of learning for the input-output pairs, a simulated model of presbyopia vision can be obtained as shown in Fig. 2. As shown in Fig. 2, the insensity values are learnt, i.e. let the image size be n by m, the neural network model has n by m inputs and n by m outputs. In other words, a relationship between the luminance values in the original image and the simulated presbyopia image is mapped to the network.

2.2 Luminance Correction Model of Image

Luminance correction model can be obtained by using a 4-layer neural network involving the simulated model of presbyopia vision constructed in the previous process. The parameters of the neural network, i.e. weights and biases, from the third layer to the fourth layer in the 4-layer network are assigned by the use of the simulated model of presbyopia vision. The assigned parameters are not updated during the following training. By applying the original image as both input and

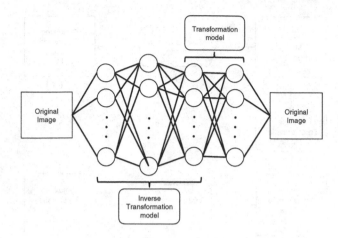

Fig. 3. Luminance correction model of image.

teacher, respectively, a luminance correction model can be generated between the first layer and the third layer of the network. As shown in Fig. 3, training is performed for a whole neural network connecting the luminance correction model with the simulated model of presbyopia vision. In the training phase, the neural network is trained by the same learning method as constructing the simulated model of presbyopia vision. By picking up the values from the third layer, a luminance-corrected image to be presented to the presbyope can be generated. By presenting the luminance-corrected image to the presbyope, it can be expected for the presbyope to perceive the similar appearance as the original image.

3   Experimental Results

To verify the effectiveness of the proposed method, a preliminary experiment is addressed on the comparative to the conventional study [5]. Image size of the image "Penguins" used in the experiment is of 72 by 96 pixels as shown in Figs. 4(a) and 6(a). Image size of the image "Document" used in the experiment is of 80 by 100 pixels as shown in Figs. 5(a) and 7(a).

3.1   Experimental Results of the Conventional Method

Figure 4 shows experimental results for the conventional method. In Fig. 4, the image (b) shows a luminance correction image which is presented to the presbyopes, the image (c) shows a perceived image of original image which is simulated by a Gaussian filtering, and the image (d) shows a perceived image of the luminance correction image. As shown in Fig. 4(d), the contrast of perceived image is quite low. It is confirmed that the results for "Document" image have a similar tendency as shown in Fig. 5.

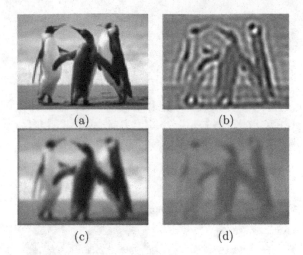

(a) (b)

(c) (d)

Fig. 4. Experimental results for a "Penguins" image by the conventional method [5]. (a) Original image, (b) Luminance correction image, (c) Presbyopia simulation of the original image, (d) Presbyopia simulation of the luminance correction image.

For the quantitative evaluation, the RMSE between the image in Fig. 4(a) and the image in Fig. 4(d) is calculated. As a result, the RMSE was 58.9. And also, the RMSE between the image in Fig. 5(a) and the image in Fig. 5(d) was 44.3.

3.2 Experimental Results of the Proposed Method

The number of units in each layer for the 2-layer neural network is 72 by 96 for input layer and 72 by 96 for output layer, respectively. In addition, the number of units in each layer for the 4-layer neural network connecting the luminance correction model with the presbyopia visual simulated model is 72 by 96 for input layer, 500 for the second layer, 72 by 96 for the third layer, and 72 by 96 for the output layer, respectively. Since the luminance correction model is located in the 1st to 3rd layer of the 4-layer neural network, the number of units of each layer is 72 by 96 for the input layer, 500 for the intermediate layer, and 72 by 96 for the output layer, respectively. Sigmoid functions are used as a threshold function for each model.

Furthermore, the processing time for the learning was approx. 3 min; the processing time for the output of the test image was approx. 3 seconds. As shown in the experimental results, indeed the image obtained by the luminance correction has become like a binary image compared to the original image, but the perceived image in Fig. 6(d) is better than one by the conventional method.

It is confirmed that the results for "Document" image have a similar tendency as shown in Fig. 7.

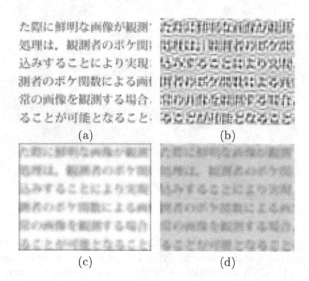

Fig. 5. Experimental results for a "Document" image by the conventional method [5]. (a) Original image, (b) Luminance correction image, (c) Presbyopia simulation of the original image, (d) Presbyopia simulation of the luminance correction image.

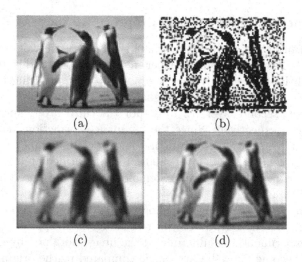

Fig. 6. Experimental results for a "Penguins" image. (a) Original image, (b) Luminance correction image, (c) Presbyopia simulation of the original image, (d) Presbyopia simulation of the luminance correction image.

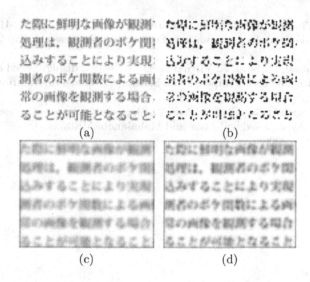

Fig. 7. Experimental results for a "Document" image. (a) Original image, (b) Luminance correction image, (c) Presbyopia simulation of the original image, (d) Presbyopia simulation of the luminance correction image.

For the quantitative evaluation, the RMSE between the image in Fig. 6(a) and the image in Fig. 6(d) is also calculated. As a result, the RMSE was 12.6. And also, the RMSE between the image in Fig. 7(a) and the image in Fig. 7(d) was 30.0.

4   Conclusions

In this study, a method to generate an inverse characteristic of presbyopia simulation model was proposed. The proposed method uses four-layer neural network cascading an image transform model and its inverse model. As a result of experiments, we confirmed the proposed method could successfully obtain an inversely transformed image with respect to the presbyopia transformation characteristics. Future works involve a development of application to other color transformation cases. And also, our proposed method can generalize to restorations of image deteriorations such as image super-resolutions [6].

Acknowledgments. This work is partially supported by JSPS KAKENHI Grant Number 25730138. The authors also gratefully acknowledge the helpful comments and suggestions of the reviewers, which have improved the presentation.

References

1. Ohno, S., Kinoshita, S.: Standard Ophthalmology. Igaku-Shoin, Tokyo (2010)
2. Schachar, R.A.: The Mechanism of Accommodation and Presbyopia. Kugler Publications, Amsterdam (2012)

3. Ihihashi, Y., Toda, I.: LASIK. Jpn. J. Cataract Refract. Surg. **26**(4), 419–426 (2012)
4. Goto, T.: Orthokeratology: evaluation of efficacy and lens care. J. Eye **27**(11), 1513–1518 (2010)
5. Aoki, M., Sakaue, F., Sato, J.: Virtual visual correction display based on pirior blur restoration. In: The 20th Symposium on Sensing via Image Information, IS2-23, pp. 1–7 (2014)
6. Kawano, H., Suetake, N., Cha, B., Aso, T.: Sharpness preserving image enlargement by using self-decomposed codebook and Mahalanobis distance. Image Vis. Comput. **27**(6), 684–693 (2009)

Computational Complexity Reduction for Functional Connectivity Estimation in Large Scale Neural Network

JeongHun Baek[1]([✉]), Shigeyuki Oba[1], Junichiro Yoshimoto[2,3], Kenji Doya[2], and Shin Ishii[1]

[1] Graduate School of Informatics, Kyoto University, Yoshidahonmachi 36-1, Sakyo, Kyoto, Japan
baek-j@sys.i.kyoto-u.ac.jp, oba@i.kyoto-u.ac.jp
[2] Neural Computation Unit, Okinawa Institute of Science and Technology Graduate University, 1919-1 Tancha, Onna-son, Kunigami-gun, Okinawa 904-0495, Japan
doya@oist.jp
[3] Graduate School of Information Science, Nara Institute of Science and Technology, 8916-5 Takayama, Ikoma, Nara, Japan
juniti-y@is.naist.jp

Abstract. Identification of functional connectivity between neurons is an important issue in computational neuroscience. Recently, the number of simultaneously recorded neurons is increasing, and computational complexity to estimate functional connectivity is exploding. In this study, we propose a two-stage algorithm to estimate spike response functions between neurons in a large scale network. We applied the proposed algorithm to various scales of neural networks and showed that the computational complexity is reduced without sacrificing estimation accuracy.

Keywords: Computational complexity reduction · Functional connectivity · Two-stage algorithm · Large scale networks

1 Introduction

In computational neuroscience, estimation of causal interactions between neurons has been considered as an important issue to investigate functions of the brain [1]. Recently, calcium imaging technique [2] is enabling us to record a history of emitted signals of many neurons as a spike train data. Based on the spike train data, causal interactions between neurons are estimated and such estimated interactions are called functional connectivity.

Although there have been a lot of studies to compute the functional connectivity, they are roughly classified into two categories: One is based on descriptive statistics such as partial correlation (e.g. [3]) and generalized transfer entropy (e.g. [4]); and another is based on generative models (e.g. [5–7]). The generalized linear model (GLM) is a representative model for the latter category and has an advantage that it can simulate a dynamic behavior of the neural population in

© Springer International Publishing Switzerland 2015
S. Arik et al. (Eds.): ICONIP 2015, Part III, LNCS 9491, pp. 583–591, 2015.
DOI: 10.1007/978-3-319-26555-1_66

addition to the functional connectivity estimation once we determine the model parameters from the data. On the other hand, computational cost to determine GLM's parameters is much heavier than that to compute descriptive statistics, and the difference becomes more critical as the number of the neurons increases. To relieve the issue, we aim to reduce computational complexity for parameter estimation of the GLM without sacrificing the estimation accuracy.

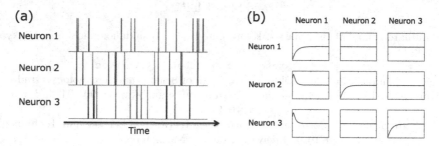

Fig. 1. Estimation of spike response function between neurons. (a) An example of spike train data for neurons 1, 2, and 3. (b) An example of the estimated spike response functions for any pairs of the three neurons. Rows and columns denote post- and pre-synaptic (target and trigger) neurons. Horizontal and vertical axes of each panel denote time-lag and intensity of the spike response, respectively.

A response function represents a strength of functional connectivity between a pair of pre- and post-synaptic neurons as a function of time-lag (Fig. 1). For example, the response function from neuron 1 to neuron 2, (row 2 column 1 in Fig. 1(b)) means that spike probability of neurons 2 is enhanced immediately after a spike signal of neuron 1 and the enhancement to the probability gradually decreases at larger time-lag. Likewise, the detail function shape of the response function brings rich information about the functional connectivity and is expected to contribute to understanding and simulating neural network's activity.

Recently, the number of neurons that are simultaneously observed and recorded is rapidly growing, owing to the advance of variety of imaging technology. Consequently, increasing amount of computational complexity is required to calculate response functions [8]. Currently, it takes about several hours to estimate functional connectivity between 1,000 neurons [3]. In the future, functional connectivity analyses among 3,000 or 10,000 neurons are prospected to take several weeks or several months, respectively. We, then, have to reduce computational complexity without sacrificing estimation accuracy.

In this study, we propose a two-stage algorithm with a particular interest in reducing computational complexity of spike response function estimation. The density of functional connections in typical neural networks is around 1 % or less [9]. Thus, if we separate the estimation into a rough estimation of binary presence of functional connectivity and a detailed estimation of the shape of response function, the latter detailed estimation can omit around 99 % of the cost for fully estimating all the response function shapes.

2 Spike Response Model and Its Estimation

2.1 Original Model

We consider spike train data of a form $\{N_i(t) \in \{0,1\}, i = 1, ..., C, t = 1, ..., T\}$ such that $N_i(t) = 1$ denotes that a neuronal spike is observed on neuron i at time t, where C is the number of neurons and T is the observation time length. According to GLM, the probability to observe a spike on neuron i at time t is represented by

$$P(N_i(t) = 1|Z_i(t)) = f(Z_i(t)) \tag{1}$$

where $f(x) = 1/(1 + \exp(-x))$ is a logistic function and

$$Z_i(t) = \text{bias}_i + \sum_{c=1}^{C} \sum_{s=1}^{S} R_{ic}(s)N_c(t - s) \tag{2}$$

is a linear model of the internal potential that determines the spike probability. bias_i is a scalar value and S is the maximum length of time-lag. $R_{ic}(s)$ is a linear coefficient corresponding to $N_c(t - s)$, i.e., when there is a spike of neuron c at time $t - s$, spike probability of neuron i at time t is enhanced by the value of $R_{ic}(s)$. Thus, $R_{ic}(s)$, as a function of time-lag s, is called a spike response function of a pair of pre- and post-synaptic neurons, c and i.

Main purpose to consider the GLM is to estimate the response functions $R_{ic}(s)$ of all pairs of neurons $i = 1, ..., C, c = 1, ..., C$ because the estimated response function $R_{ic}(s)$ provides existence, polarity, strength, delay, and other features of functional connection between the pair of neurons.

In order to represent smoothness and to reduce the effective number of parameters to estimate, the response function is represented as a linear sum of basis functions:

$$R_{ic}(s) = \sum_{k=1}^{K} W_{ick}B_k(s) \tag{3}$$

where W_{ick} is a weight coefficient and $B_k(s)$ is the k-th basis function. The basis set $B_k(s), k = 1, ..., K$ should represent possible variations of response functions. Specifically, we use a set of gamma distributions (Fig. 2) with $K = 10$.

The GLM has been applied to response function estimation in several works [5,10]. When basis functions are introduced to GLM, such a modified GLM is called generalized functional additive model (GFAM) [10].

2.2 Logistic Regression with L1 Regularization

The estimation of the response function, or equivalently the weight coefficient W in GLM or GFAM is a logistic regression problem. In particular, we applied logistic regression with $L1$ regularization because functional connections within real

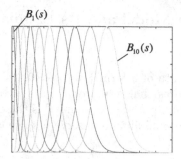

Fig. 2. A set of gamma distributions utilized as basis functions. $B_k(s)$ represents the k-th basis function. Each basis function is normalized between 0 to 1 and obtained by the formula; $B_k(s) = s^{0.5(k+1)^2-1} \exp(-s)/|B_k(s)|$

neural networks are often sparse, allowing response function $R_{ic}(s)$ to be exactly zero for many pairs of neurons (i, c). We used the dual augmented Lagrangian (DAL) method to solve the logistic regression problem [11].

Note that, to match the response model to a standard form of logistic regression model, the convolution $X_{ck}(t) = \sum_{s=1}^{S} B_k(s)N_c(t-s)$ is performed.

$$Z_i(t) = \text{bias}_i + \sum_{c=1}^{C}\sum_{k=1}^{K} W_{ick}X_{ck}(t) \tag{4}$$

2.3   Computational Complexity

Computational complexity of logistic regression is proportional to the number of unknown variables and data length. Thus, introduction of basis functions is effective for reducing the complexity; namely, the complexity proportional to the number of maximum time-lag S was reduced to those proportional to the number of bases K, typically $S \approx 100$ to $K \approx 10$. The total computational complexity is then represented as $O(C^2TK)$; thus, it is proportional to the numbers of pre- and post-synaptic neuron pairs C, observed data length T, and the number of basis functions K. We usually need a certain number of bases, because too small number of bases may lead to insufficiency for representing the true response.

2.4   Two-Stage Algorithm

We propose an algorithm consisting of two stages. We control the number of basis functions for each stage to reduce the total computational complexity without sacrificing the accuracy.

First Stage. In the first stage, we roughly determine a binary label denoting the presence/absence of the functional connection for each pair of neurons. This rough estimation is performed by the same algorithm as described in Sect. 2.2, but the algorithm settings are modified; namely, the number of bases K is reduced into a small one enough to reduce computational cost, and the threshold to judge the presence/absence of each connection is set large enough. Specifically, the number of basis functions was set at $K' = 1$. The absolute value of the area of the spike response function is used for discrimination. Figure 3(a) illustrates this rough estimation, in which gray color represents the presence of the corresponding connection.

Second Stage. In the second stage, we calculated a detailed shape of the response function corresponding to each connection that had been determined as present in the first stage. The number of basis functions K was set at the original, hence sufficiently large, number. Figure 3(b) illustrates this second stage. Thus, the computational complexity is proportional to the number of connections that have been determined as present in the first stage. Consequently, the total amount of computational complexity of the first and the second stages is expected to be far smaller than the original when the first stage effectively cuts off a large portion of the whole set of neuron pairs.

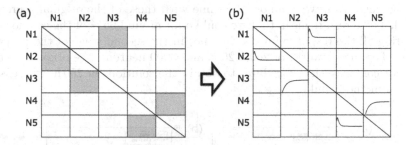

Fig. 3. (a) The result of the first stage. (b) The result of the second stage. The response function is estimated in detail, for each neuron pair whose connection has been determined as present in the first stage (a).

3   Result

3.1   Numerical Experiment Setting

For numerical experiment, we simulated the neural network of 3000 neurons. Before the application, we performed the selection of neurons. We sorted 3000 neurons by the number of spikes and selected 500 neurons, and then applied the proposed method to the sub-network composed of the 500 selected neurons. The reason why we selected neurons is that neurons which have few spikes don't have enough statistical information to estimate functional connectivity, and they obviously caused inaccurate estimation. To compare the original and proposed methods, we needed to perform credible estimation, and thus we should have selected neurons which have enough spikes. Similarly, we applied the proposed method to sub-networks of the 1000, 1500, 2000, 2500 and 3000 neurons.

For each case, spike train data of time length 10,000 (ms) was generated by an artificially designed GLM model. Each model was designed to have a sparse structure whose connection probability was $0.3 \sim 1.3\%$. For each connected pair of neurons, we set a response function at one of physiologically reasonable options. We employed the original method (described in Sect. 2.2) and our proposed method to estimate response functions of all pairs of neurons.

3.2 Estimated Connectivity Proportion

We applied a common threshold to determine the presence/absence of all possible functional connections, where the common threshold was determined by inspecting the receiver operating characteristic (ROC) curve [12]; in particular, we selected such a threshold that expected false positive rate was less than 5%. Figure 4(a) shows that 5% of false positive rates were used as the criterion to set the threshold in the first stage. Figure 4(b) shows estimated connectivity proportion of each case. In the connectivity proportion analysis, the results of the two-stage algorithm were almost the same with those of the original regression method. These results mean that we could omit around 95% of the cost for fully estimating all the response function shapes in the second stage of the proposed method. For the networks of 2000, 2500 and 3000 neurons, the original method could not complete the procedures to obtain the solutions in 250 h, whereas the proposed method could.

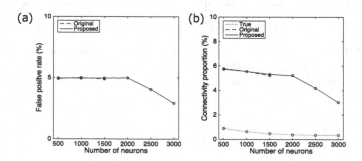

Fig. 4. (a) False positive rates, (b) Connectivity proportion of the true network and of the functional connectivity estimation by the original method and the proposed method, denoted by dotted, dashed and solid lines, respectively.

3.3 Comparison of the Elapsed Time

Elapsed computation times were compared (Fig. 5). The original method took over four times more than the proposed method. These results mean that the proposed method was effective in reducing the computational complexity and hence in applicability to large-scale neural networks. Notice that the second stage required more time than the first stage in Fig. 5. Corresponding to the truly low connectivity probability (Fig. 4(b)), more than 90 % of possible connections were filtered out in the first stage. Figure 5(b), however, suggests that the second stage grew the bottleneck as the network size became large. Since there were certain number of false positives, which were determined as present but were in fact absent, both by our method and the original method (Fig. 4(b)), we could have performed further cut-off in the first stage. Thus, there might be some room for improving the first stage of the proposed method. In this study, we used an Intel Xeon X5690 CPU (3.46 GHz) and MATLAB as a programming language.

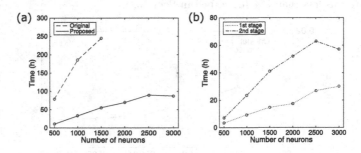

Fig. 5. (a) Elapsed time comparison between the original method and the proposed method. (b) Elapsed time at the 1st and 2nd stages in the proposed method.

3.4   Comparison of the Estimation Accuracy

Comparison of ROC-AUC. Binary discrimination of functional connectivity was compared (Fig. 6). We determined the existence of functional connection for each pair of neurons by applying an arbitrary threshold to the peak absolute value of the response function. The determination error was measured by area under the curve (AUC) of receiver operating characteristic (ROC).

The ROC-AUC values were smaller for larger network size. This tendency occurred partly because larger network in our simulation included larger number of such neurons that emitted small number of spikes and thus functional connectivity detection was statistically difficult with the lack of information. Larger false negative rate with larger network size (Fig. 6(b)) also supported this reason.

In this case, the proposed method performed slightly better than the original method excepting when the number of neurons was 500 (Fig. 6(a) and (b)). This result was not expected but is understandable because the true response functions used in the artificial neural networks were so simple that the GLM with a small base number K in the first stage fitted better than that with a large number K in the second stage. From these results, we can say that the accuracy of estimating the presence/absence in the functional connectivity analysis was kept, and sometimes even better than that by the original method.

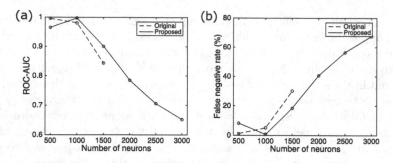

Fig. 6. (a) ROC-AUC values, (b) False negative rates of the functional connectivity estimation. Notice that false negative rates are obtained at the threshold where false positive rate was less than 5 % (described in Sect. 3.2).

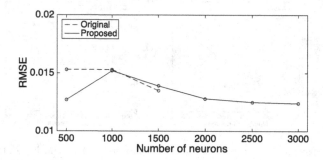

Fig. 7. Comparison of RMSE to estimate response functions.

Comparison of RMSE. The estimation accuracy of the response function was compared (Fig. 7). Root mean square error (RMSE) between the true and estimated response functions was calculated for each pair of neurons and then averaged over all the pairs for which there was truly functional connection. The results of the two-stage algorithm were almost the same with those of the original method excepting when the number of neurons was 500. Notice that the difference in RMSE between the two methods was slight, actually less than 0.005. The proposed method would have produced a smaller RMSE than the original method, partly because the data size per a model parameter was larger in the proposed method. Although we need some more investigation to support this speculation, we can say that the estimation accuracy of the response function was found to be similar to that of the original method.

4   Summary

We proposed the two-stage algorithm for functional connectivity analysis, which successfully reduced computational complexity whereas maintaining the accuracy of functional connectivity analysis, and tested it using artificial neural networks of variety of size. The new algorithm reduced the computational complexity, less than half of the original method, without sacrificing estimation accuracy.

We also found that the estimation accuracy using the two stage algorithm was better than expected both in terms of determining presence/absence of functional connectivity and of RMSE to identify detailed shape of spike response functions. Future work would be able to clarify the reason for the accuracy improvement, and moreover, we can expect further advantage of our method in applications to even large-scale networks.

References

1. Brown, E.N., Kass, R.E., Mitra, P.P.: Multiple neural spike train data analysis: state-of-the-art and future challenges. Nat. Neurosci. **7**(5), 456–461 (2004)
2. Stosiek, C., Garaschuk, O., Holthoff, K., Konnerth, A.: In vivo two-photon calcium imaging of neuronal networks. Proc. Natl. Acad. Sci. **100**(12), 7319–7324 (2003)
3. Sutera, A., Joly, A., François-Lavet, V., Qiu, Z.A., Louppe, G., Ernst, D., Geurts, P.: Simple connectome inference from partial correlation statistics in calcium imaging (2014). arXiv:1406.7865
4. Stetter, O., Battaglia, D., Soriano, J., Geisel, T.: Model-free reconstruction of excitatory neuronal connectivity from calcium imaging signals. PLoS Comput. Biol. **8**(8), e1002653 (2012)
5. Kim, S., Putrino, D., Ghosh, S., Brown, E.N.: A Granger causality measure for point process models of ensemble neural spiking activity. PLoS Comput. Biol. **7**(3), e1001110 (2011)
6. Stevenson, I.H., Rebesco, J.M., Miller, L.E., Kording, K.P.: Inferring functional connections between neurons. Curr. Opin. Neurobiol. **18**(6), 582–588 (2008). Dec
7. Fletcher, A.K., Rangan, S.: Scalable inference for neuronal connectivity from calcium imaging. In: Ghahramani, Z., Welling, M., Cortes, C., Lawrence, N.D., Weinberger, K.Q. (eds.) Advances in Neural Information Processing Systems 27, pp. 2843–2851. Curran Associates Inc, Red Hook (2014)
8. Stevenson, I.H., Kording, K.P.: How advances in neural recording affect data analysis. Nat. Neurosci. **14**(2), 139–142 (2011)
9. Stuart, G., Spruston, N., Häusser, M.: Dendrites. Oxford University Press, New York (2008)
10. Song, D., Wang, H., Tu, C.Y., Marmarelis, V.Z., Hampson, R.E., Deadwyler, S.A., Berger, T.W.: Identification of sparse neural functional connectivity using penalized likelihood estimation and basis functions. J. Comput. Neurosci. **35**(3), 335–357 (2013)
11. Tomioka, R., Suzuki, T., Sugiyama, M.: Super-linear convergence of dual augmented Lagrangian algorithm for sparsity regularized estimation. J. Mach. Learn. Res. **12**, 1537–1586 (2011)
12. Hanley, J.A., McNeil, B.J.: The meaning and use of the area under a receiver operating characteristic (ROC) curve. Radiology **143**(1), 29–36 (1982)

Matrix-Completion-Based Method for Cold-Start of Distributed Recommender Systems

Bo Pan[✉] and Shu-Tao Xia

Tsinghua-Southampton Web Science Laboratory, Graduate School at Shenzhen,
Tsinghua University, Shenzhen, China
einstein246@vip.qq.com, xiast@sz.tsinghua.edu.cn

Abstract. Recommender systems has been wildly used in many websites. These perform much better on users for which they have more information. Satisfying the needs of users new to a system has become an important problem. It is even more accurate considering that some of these hard to describe new users try out the system which unfamiliar to them by their ability to immediately provide them with satisfying recommendations, and may quickly abandon the system when disappointed. Quickly determining user preferences often through a boot process to achieve, it guides users to provide their opinions on certain carefully chosen items or categories. In particular, we advocate a matrix completion solution as the most appropriate tool for this task. We focus on online and offline algorithms that use data compression algorithm and the decision tree which has been built to do real-time recommendation. We merge the three algorithms : distributed matrix completion, cluster-based and decision-tree-based, We chose different algorithms based on different scenarios. The experimental study delivered encouraging results, with the matrix completion bootstrapping process significantly outperforming previous approaches. *abstract* environment.

Keywords: Matrix completion · Recommender system · Collaborative filtering · Decision tree · Spectral clustering

1 Introduction

With the rise of "big data," low-rank matrix completion algorithm has recently been widely used in the field of data mining. Matrix completion algorithm has been widely used in personalized recommender system, such as Netflix, Pandora [18]. The latest research results show that low-rank matrix factorization to solve the matrix completion problems which is filled with noise is very successful [10]. At its heart, according to the observed data, it calculate the unobserved data. Matrix completion can be used in recommender system, video surveillance [11,12], graphical model selection [13], so it has been know as a "top-10" algorithm [14].

© Springer International Publishing Switzerland 2015
S. Arik et al. (Eds.): ICONIP 2015, Part III, LNCS 9491, pp. 592–599, 2015.
DOI: 10.1007/978-3-319-26555-1_67

Now, a particular challenge that the recommender system facing is to recommend to new users, this is known as the *user cold start problem*. The quality of recommendations strongly depends on the amount of information collect from the user, so the users new to the system hardly receive reasonable recommendations [3, 16, 17]. In order to quantify this problem, Fig. 1 shows how the error on Netflix test data decreases as users provide more ratings. The users that have many ratings can enjoy error rates around 0.85 compare to those users with just a few known ratings get the error rates around 1. However, providing new users with an accurate recommendation is essential to growing the user base of the system. The essence of bootstrapping a recommender system is to promote the interaction, users that will receive accurate recommendation just pay a low-effort to interact with the system [4, 6].

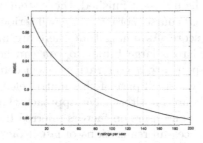

Fig. 1. The test error rate of Netflix data. Lower y-axis values represent more accurate predictions. The x-axis describes the exact number of ratings taken for each user. Results are computed by the factorized item-item model [1].

In this paper, we study new users cold start problem for recommender systems. First, we use the $TFOCS$ [15] to fill the rating data over a *MapReduce* cluster. Second, we use spectral clustering algorithm to extract feature matrix. Third, we improved a decision tree method to build a decision tree. Forth, we calculate the recommendation of the previous three steps respectively. Finally, according to the recommendation accuracy we choose the appropriate algorithm.

To sum things up, the main contributions of the paper are:

1. We divide the recommender system into two parts: the offline and online. Offline recommender system provide the result which is used to be the training set, online provide the real time recommendation to new users. We transplant the $TFOCS$ algorithm into the *hadoop* environment.

2. We proposed a method of extracting data, we use small data contain enough information to recommend to users rather than big data.

3. We improved the decision tree algorithm, we use the matrix completion algorithm increases the ratings data. Using this data to build a new decision tree.

4. We compare the current method for solving the problem of cold-start to confirm the correctness of our approach.

2   Related Work

In this section, we provide a brief introduction to explain why the recommendation problem can be abstracted into a matrix completion problem. Then, we introduce a seed sets select method to generate the seed sets. Last, We introduce an existing solutions to solve this problem using the decision tree.

2.1   Matrix Completion Problem

To understand the relationship between the matrix completion and recommender system, it is necessary to introduce the *Netflix Prize* [14]. Its purpose is to predict the unknown user data based on existing user data. We start from a simple model, describing how to make this recommender system. Such as, *netflix* has a Critics System, after the user rent finished, the system will the system will let users rate these movies. Of course, not everyone will seriously score, in fact, only a handful of users will score. Assuming that there are m users and n movies, we put all ratings fitted to a large table, so you can get the matrix. In the matrix, each line corresponds to a user ratings, each column corresponds to a film ratings. Clearly, only a small part of this matrix element is known.

Latent factor models are an alternative approach that explains ratings by characterizing both items and users on factors inferred from the pattern of ratings. One of the most successful realizations of latent factor models is based on *matrix factorization*, e.g., [2,8,9].

2.2   Decision Tree

A relative work by the group [5] contemplates using decision trees for solving the cold start problem. They propose to first cluster users into groups, and then to build a classical decision tree classifier such as $ID3$ that will closely recover the known user clusters based on user ratings of specific items. However, this method is not practically feasible. Another relative work by group [7] propose a decision tree method such as $C4.5$. They set the ratings to be three part, they take four and five star ratings for a movie as a "like" indication. The one-,two- and three-star ratings indicate "dislike". Finally, unrated items are taken as "unknown".

Each tree node [7], whether internal or a leaf, represents a group of users. Consequently, each node of the tree predicts item ratings by taking the mean rating by it's corresponding users. The mean prediction minimizes the error associated with each node, which is defined as the *squared deviation* between the predicted ratings at the node and the true ratings for the users represented by the node.This method have the best performance compare to the HELF, Entropy0 and GreedyExtend algorithm.

3   Matrix-Completion-Based Method for Distributed Recommender System

In this section, we discuss how to give a cold start recommendation. We use three steps to do this. First, we use distributed matrix completion algorithm to file the rating data. Second, we use spectral cluster algorithm to extract the feature matrix, then use this matrix to solve the cold start problem. Third, we use the improved decision tree algorithm to recommend to new users. Last, we combine these algorithm together, choose different algorithms based on different scenarios. Figure 2 shows the structure of the system, the direction of the arrow indicates the processing flow of the system. Details are described in the Fig. 2 below.

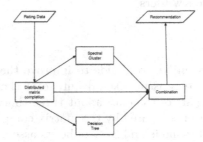

Fig. 2. Using distributed matrix completion algorithm file the rating data, putting these result into spectral cluster algorithm and decision tree respectively, combining the result of distributed matrix completion, spectral cluster and decision tree according to the certain users select different results to users.

3.1   Distributed Matrix Completion

TFOCS [15] is a general framework for solving a variety of convex cone problems that frequently arise in signal processing, machine learning, statistics, and other fields. We transplant the *TFOCS* algorithm to adjust the need of big data matrix computation. There are three nodes we built in the *hadoop* cluster. One namenode, two datanodes is set in the cluster.

The approach works as follows:

1. determine a conic formulation of the problem
2. determine its dual
3. apply smoothing
4. solve using an optimal first-order method

In the first step, we use the TFOCS [15] algorithm to complete the data set V, this process will take several hours. The new data set V_* will be saved in database. To ensure the sparsity of data, we just put the observed data V into the distributed algorithm, the completed rating result will be used in the follow steps.

3.2 Spectral Cluster

In order to deal with the cold start problem, we should extract the feature matrix from the result of first step. We use spectral cluster algorithm to do users-cluster and items-cluster respectively. we extract the big matrix into a small one. But it does not meet the matrix completion requirements of low rank. Thus, we should do something with this matrix, in order to make it into a low rank matrix. There is a simple solution to do this. We set the matrix elements in odd rows even position to 0, and the elements in even rows odd position to 0. So the matrix become a low rank matrix.

When the new users come to our system, we guide them to rate a few items. Then we insert a new row of the vector which just has been rated by the new users in the last row of the matrix. We use the stand-alone mode algorithm recommended results to new users.

3.3 Decision Tree

As previously mentioned in [7], we decide to improve the experiment result of that method. A key strategy is to take advantage of characteristics of matrix completion, we extract part of the data about three times of the original from the matrix which has been computed by the matrix completion algorithm.

Formally, let t be a tree node and $S_t \subseteq U$ be its associated set of users. The value predicted by the node t to item i is $\mu(t)_i = |S_t \bigcap R(i)|^{-1} \sum_{u \in S_t \bigcap R(i)} r_{ui}$. The squared error associated with node t and item i is: $e^2(t)_i = \sum_{u \in S_t \bigcap R(i)} (r_{ui} - \mu(t)_i)^2$. The overall squared error at node t is $e^2(t) = \sum_i e^2(t)_i$.

Construction of the decision tree follows a common top-down practice. For each internal node we choose the best splitting item, namely the item that partitions the users in to three sets such that the total squared prediction error is minimized. Accordingly, user population is partitioned among subtrees, and then the process continues recursively with each of the subtrees.

3.4 Combination

After the first three steps of the calculation, we can make recommendations for new users. We consolidate results calculated before, depending on the system requirements, select a different result, recommend for new users.

$$recommendation \leftarrow \begin{cases} Spectral\ Cluster & rating\ number \leq \\ Decision\ Tree & 5 \geq rating\ number \leq 8 \\ Spectral\ Cluster & 8 \geq rating\ number \leq k \\ Distributed\ Matrix & rating\ number \geq k \end{cases}$$

We choose appropriate algorithm result according to the number of user ratings. When the ratings are less than 5, we select the spectral-cluster-based algorithm, e.g. The number k is determined by the algorithm of distributed TFOCS and spectral cluster.

4 Experiment

We compare the spectral cluster based algorithm with HELF, Entropy0, and GreedyExtend algorithm.

From the left of Fig. 3 we can see our method get the best performance. This is because the other three methods are static method, the seed set they select from the total set will not be changed, they can not dynamically adapt the changes from users ratings. However, our method can recommend different recommendation dynamically with the ratings. And the category of the recommendation mostly are different.

Next, we compare the decision tree based algorithm with HEFL, Entropy0, GreedyExtend, and decision tree.

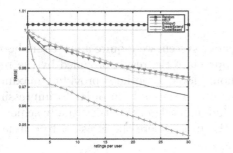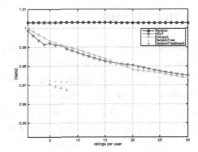

Fig. 3. The x-axis represents the number of user ratings, the y-axis represents the RMSE error.

From the right of Fig. 3 we can see our decision tree based method get the lowest RMSE error. The decision tree method is a dynamic algorithm, it is better than HELF, Entropy0, and GreedyExtend. But our improved decision tree method is better than the original decision tree method, because we get three times users rating data than it, due to matrix completion algorithm.

We compare the number of ratings per user when they get the same RMSE error.

From the left Fig. 4 we can see that our algorithm need least ratings compare with other algorithm and the netflix test set result when we achieve the same RMSE error.

Last, we compare all our method to determine how to select a proper method.

From the right Fig. 4 we can see that the decision tree based get the best result when the ratings between 5 and 8. When the ratings more than 30, the distributed matrix completion algorithm is the best method.In other bound, we select the spectral cluster based algorithm to give recommendation to new users. Since the decision tree levels are between 5 and 8, when the ratings beyond 8, the recommendation results are almost fixed, so we also must select the other two methods to solve the new users cold start problem.

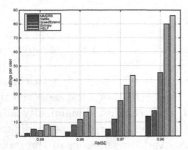 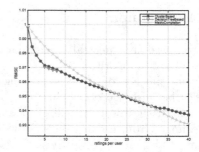

Fig. 4. The left fig :x-axis represents the RMSE error, the y-axis represents the ratings per user;The right fig :x-axis represents the number of user ratings, the y-axis represents the RMSE error.

5   Conclusions

Introducing a new user into a recommender system we require a low-effort bootstrapping process, then the user can appreciate the result provided by the recommender system. System designers should look at methods providing maximal prediction accuracy at minimal distraction to the user, in order to impressing their new users, who might be the most judgmental ones. This is usually achieved by conducting a short interview with the user, where she is asked to evaluate certain products or categories.

We have shown a major accuracy improvement when replacing the decision tree bootstrapping process. Adaptive bootstrapping is achieved by a hybrid algorithm recommender, which is consisted of three parts distributed matrix completion, Spectral Cluster, and decision tree. We compared the results with the major methods such as HELF, Entropy0, GreedyExtend, and Decision Tree, our method is more effective. We combine the three method of us together, propose the scenarios that can be used, the combined method is not limited by other conditions,and it can be used in most actual environment.

References

1. Koren, Y.: Factor in the neighbors: scalable and accurate collaborative filtering. ACM Trans. Knowl. Disc. Data (TKDD) **4**(1), 1 (2010)
2. Koren, Y., Bell, R., Volinsky, C.: Matrix factorization techniques for recommender systems. Computer **8**, 30–37 (2009)
3. Rashid, A.M., Albert, I., Cosley, D., et al.: Getting to know you: learning new user preferences in recommender systems. In: Proceedings of the 7th International Conference on Intelligent User Interfaces, pp. 127–134. ACM (2002)
4. Kohrs, A., Merialdo, B.: Improving collaborative filtering for new users by smart object selection. In: Proceedings of International Conference on Media Features (ICMF) (2001)
5. Rashid, A.M., Karypis, G., Riedl, J.: Learning preferences of new users in recommender systems: an information theoretic approach. ACM SIGKDD Explor. Newslett. **10**(2), 90–100 (2008)

6. Golbandi, N., Koren, Y., Lempel, R.: On bootstrapping recommender systems. In: Proceedings of the 19th ACM International Conference on Information and Knowledge Management, pp. 1805–1808. ACM (2010)
7. Golbandi, N., Koren, Y., Lempel, R.: Adaptive bootstrapping of recommender systems using decision trees. In: Proceedings of the Fourth ACM International Conference on Web Search and Data Mining, pp. 595–604. ACM (2011)
8. Candes, E., Tao, T.: The Dantzig selector: statistical estimation when p is much larger than n. Ann. Stat. **35**(6), 2313–2351 (2007)
9. James, G.M., Radchenko, P., Lv, J.: DASSO: connections between the Dantzig selector and lasso. J. Roy. Stat. Soc. B (Statistical Methodology) **71**(1), 127–142 (2009)
10. Keshavan, R., Montanari, A., Oh, S.: Matrix completion from noisy entries. In: Advances in Neural Information Processing Systems, pp. 952–960 (2009)
11. Cands, E.J., Li, X., Ma, Y., et al.: Robust principal component analysis? J. ACM (JACM) **58**(3), 11 (2011)
12. Mackey, L.W., Jordan, M.I., Talwalkar, A.: Divide-and-conquer matrix factorization. In: Advances in Neural Information Processing Systems, pp. 1134–1142 (2011)
13. Mnih, A., Salakhutdinov, R.: Probabilistic matrix factorization. In: Advances in Neural Information Processing Systems, pp. 1257–1264 (2007)
14. Bennett, J., Lanning, S.: The netflix prize. In: Proceedings of KDD Cup and Workshop 2007, p. 35 (2007)
15. Becker, S.R., Cands, E.J., Grant, M.C.: Templates for convex cone problems with applications to sparse signal recovery. Math. Program. Comput. **3**(3), 165–218 (2011)
16. Chen, P.L.: A linear ensemble of individual and blended models for music rating prediction
17. Zhou, Y., Wilkinson, D., Schreiber, R., Pan, R.: Large-scale parallel collaborative filtering for the netflix prize. In: Fleischer, R., Xu, J. (eds.) AAIM 2008. LNCS, vol. 5034, pp. 337–348. Springer, Heidelberg (2008)
18. Gemulla, R., Nijkamp, E., Haas, P.J., et al.: Large-scale matrix factorization with distributed stochastic gradient descent. In: Proceedings of the 17th ACM SIGKDD International Conference on Knowledge Discovery and Data Mining, pp. 69–77 ACM (2011)

Weighted Joint Sparse Representation Based Visual Tracking

Xiping Duan[1,2(✉)], Jiafeng Liu[1], and Xianglong Tang[1]

[1] School of Computer Science and Technology,
Harbin Institute of Technology, Harbin, China
xpduan_1999@126.com, {jefferyliu, tangxl}@hit.edu.cn
[2] College of Computer Science and Information Engineering,
Harbin Normal University, Harbin, China

Abstract. Aiming at various tracking environments, a weighted joint sparse representation based tracker is proposed. Specifically, each object template is weighted according to its similarity to each candidate. Then all candidates are represented sparsely and jointly, and the sparse coefficients are used to compute the observation probabilities of candidates. The candidate with the maximum observation probability is determined as the object. The object function is solved by a modified accelerated proximal gradient (APG) algorithm. Experiments on several representative image sequences show that the proposed tracking method performs better than the other trackers in the scenarios of illumination variation, occlusion, pose change and rotation.

Keywords: Computer vision · Visual tracking · Kernel sparse representation

1 Introduction

Visual tracking is a hot topic of computer vision, and extensively applied into such fields as intelligent monitoring, car navigation, advanced human-computer interaction, and so forth. However, due to all kinds of internal and external factors, such as noises, occlusion, illumination and viewpoint variations, pose change, rotation, and so on, realizing the robust visual tracking is still a challenging task. The existing visual tracking methods can be categorized into the generative and the discriminative. The generative methods treat the visual tracking as a matching problem. For example, both Eigentracker [1] and meanshift tracker [2], which are proposed by Black and Comaniciu respectively, have desired real-time performance, but lack the necessary updating of target templates. IVT [3] proposed by Ross, adapts to the appearance change by incrementally learning a low dimensional domain. VTD [4] proposed by Kwon extends the traditional particle filter tracker with multiple motion models and multiple appearance models. The discriminative methods are regarded as a binary classification problem, and need to learn and update a classifier by sampling positive and negative samples. The representative methods are: Avidan et al. [5] proposed a SVM based tracker. Collins et al. [6] discriminate the target from background by online feature selection. Grabner et al. [7] proposed a online semi-supervised boosting method to avoid drifting away from the target and improve tracking performance. Babenko et al. [8] avoid bias and

© Springer International Publishing Switzerland 2015
S. Arik et al. (Eds.): ICONIP 2015, Part III, LNCS 9491, pp. 600–609, 2015.
DOI: 10.1007/978-3-319-26555-1_68

drift by multi-instance learning. Zhang et al. [9] formulate the tracking task as a binary classification problem in the compressed domain and obtain the real-time performance.

Recently, sparse representation based visual tracking methods are very popular. For instance, in [10–12], each candidate is represented sparsely as the linear combination of templates, with excellent tracking results. Due to the small sampling radium, candidates are similar to each other. [13] mines the similarity between candidates by joint sparse representation, improving the representation accuracy of candidates and the robustness of the whole tracking system. However, the aforementioned sparse representation based tracking methods assume that different templates have the same possibilities to represent each candidate. According to [14], the template which is similar to a candidate should have larger probability to represent the candidate.

In this paper, a weighted joint sparse representation based visual tracking method is proposed to robustly track the target in the complex environment. Specifically, to represent a candidate, all the templates are weighted according to their similarity to the candidate. Then, all the candidates are represented by the joint sparse representation to reflect the similarity among them. The sparse coefficient is used to compute the observation probability of the corresponding candidate. The candidate with the maximal observation probability is determined as the target. In this model, a template similar to a candidate has the larger possibility to represent the candidate. To adapt to the illumination change, a locally normalized feature is used for appearance representation, which is an illumination invariant feature. Experiments show that the proposed tracker can well adapt to the possible illumination variation, occlusion, pose change, rotation, and the influence of comprehensive conditions.

2 Weighted Joint Sparse Representation Based Tracker

2.1 Weighted Joint Sparse Representation

Let $T = \{T_1, T_2, \ldots, T_M\}$ and $\{y_1, y_2, \ldots, y_N\}$ denote the target templates set and candidates of current frame respectively. To locate the target accurately, Zhang et al. proposed a multi-task learning based tracking method, which takes into account the similarity relation among candidates and represents all the candidates by solving the $\ell_{2,1}$ mixed norm regularized minimization problem [13].

$$\hat{c} = \arg\min_c \frac{1}{2} \sum_{i=1}^{N} \|\mathbf{y}_i - \mathbf{T}c_i\|_2^2 + \lambda \sum_{j=1}^{J} \|c^j\|_2 \tag{1}$$

where

$$c = [c_1, c_2, \ldots, c_N] = [c^1; c^2; \ldots, c^J] = \begin{pmatrix} c_1^1 & \cdots & c_N^1 \\ \vdots & \ddots & \vdots \\ c_1^J & \cdots & c_N^J \end{pmatrix}$$

is the sparse representation coefficient matrix. Specifically, the i th column c_i of c corresponds to the representation coefficient of the candidate $\mathbf{y}_i$, the j row c^j are the coefficients of all the candidates corresponding to the target template T_j, and c_i^j is the j th representation coefficient of the candidate $\mathbf{y}_i$ corresponding to the target template T_j. The regularization term of (2) is $\ell_{2,1}$ mixed norm, which can cause the joint sparse representation of candidates. By mining the similarity relation among candidates, [13] improves the representation accuracy and the tracking robustness compared with other sparse representation based trackers [10–12]. However, all the aforementioned trackers [10–13] assume that all the target templates have the same possibilities to represent a candidate. According to [14], the target template more similar to a candidate has larger possibility to be chosen to represent the candidate. So in this paper, before representing a candidate, all the target templates are weighted using their distance to the candidate. Then, considering the similarity relation among candidates, all the candidates are represented using joint sparse representation.

$$\hat{c} = \arg\min_c \frac{1}{2} \sum_{i=1}^{N} \|\mathbf{y}_i - \mathbf{T}c_i\|_2^2 + \lambda \sum_{j=1}^{J} \|w^j \odot c^j\|_2 \tag{2}$$

Compare with (1), the weight matrix

$$w = [w_1, w_2, \ldots, w_N] = [w^1; w^2; \ldots, w^J] = \begin{pmatrix} w_1^1 & \cdots & w_N^1 \\ \vdots & \ddots & \vdots \\ w_1^J & \cdots & w_N^J \end{pmatrix}$$

is introduced into (2), where w_i^j is the weight of the target template T_j representing the candidate $\mathbf{y}_i$, and calculated as the distance between them

$$w_i^j = \exp\left(\frac{\|y_i - T_j\|_2}{\sigma}\right)$$

The internal $\|y_i - T_j\|_2$ is the Euclidean distance between T_j and $\mathbf{y}_i$.

The sparse representation coefficients $\{c_i\}_{i=1}^{N}$ of candidates $\{y_i\}_{i=1}^{N}$ are obtained by solving (2), and used to calculate the reconstruction errors $\{r_i\}_{i=1}^{N}$ of candidates.

$$r_i = \|\mathbf{y}_i - T \cdot c_i\|_2 \tag{3}$$

r_i is further used for calculating the observation probability of the candidate $\mathbf{y}_i$.

$$P(\mathbf{y}_i|o) \propto \mu \cdot \exp(-r_i) \tag{4}$$

where μ is the normalization constant. Among a group of given candidates $\{y_1, y_2, \ldots, y_N\}$, the candidate y_i with minimum observation probability

$$i = \arg\min_j P(\mathbf{y}_j|o) \tag{5}$$

is determined as the target.

2.2 Solving of Objective Function

The key for the aforementioned weighted joint sparse representation is to solve (2), the corresponding objective function is denoted as

$$f(c) = \frac{1}{2}\sum_{i=1}^{N} \|\mathbf{y}_i - \mathbf{T}c_i\|_2^2 + \lambda \sum_{j=1}^{J} \|w^j \odot c^j\|_2 \tag{6}$$

To solve (6), an auxiliary function is constructed

$$g(c) = \frac{1}{2}\sum_{i=1}^{N} \|\mathbf{y}_i - \mathbf{T} * (w_i)^{-1} \odot c_i\|_2^2 + \lambda \sum_{j=1}^{J} \|c^j\|_2 \tag{7}$$

where $(w_i)^{-1} = \left[\frac{1}{w_i^1}; \frac{1}{w_i^2}; \ldots; \frac{1}{w_i^{p+n}}\right]$, and $\odot$ is the element-wise multiplication operator, that is, the multiplication is conducted among corresponding elements of two matrices or two vectors. The same as (1), $g(c)$ is a multi-task joint sparse representation problem, and can be solved by a modified APG algorithm [15]. Suppose the optimal solutions of $g(c)$ and $f(c)$ are $\hat{c}' = \arg\min_c g(c)$ and $\hat{c} = \arg\min_c f(c)$ respectively, then similar to [13], there is the following correspondence

$$\hat{c} = w^{-1} \odot \hat{c}'$$

where

$$
\begin{aligned}
w^{-1} &= \left[(w_1)^{-1}, (w_2)^{-1}, \ldots, (w_N)^{-1}\right] \\
&= \left[(w^1)^{-1}; (w^2)^{-1}; \ldots; (w^{p+n})^{-1}\right] \\
&= \begin{pmatrix} \frac{1}{w_1^1} & \cdots & \frac{1}{w_N^1} \\ \vdots & \ddots & \vdots \\ \frac{1}{w_1^{p+n}} & \cdots & \frac{1}{w_N^{p+n}} \end{pmatrix}
\end{aligned} \tag{8}
$$

2.3 Template Updating

A simple template updating strategy is adopted. (1) The target of the 1st frame is added into the target template set and not allowed be updated. (2) Determine whether the target appearance of the current frame greatly changes or not. If the amount of change is greater than a given threshold, update target templates set. Otherwise, not update. Specifically, if the similarity degree of the target y_i and the target template T_j is smaller than a given threshold τ, then the target y_i is used to update the target template T_k, where

$$i = \arg\min_j P(\mathbf{y}_j|o),$$

$$j = \arg\min_l \|T_l - \mathbf{y}_i\|_2,$$

$$k = \arg\max_l \|T_l - \mathbf{y}_i\|_2.$$

2.4 Tracking Algorithm

In this paper, the locally normalized feature is adopted for appearance representation, as shown in Fig. 1. In particular, the candidate or target template is firstly divided into several regions. Secondly the vector representation of each region is obtained by concatenating all the columns sequentially, and normalizing to the ℓ_2 unit length. Further the normalized vectors are concatenated into a longer vector as the feature vector of the candidate or target template.

The detailed algorithm of this paper is shown in Algorithm 1.

Algorithm 1 Weighted joint sparse representation based visual tracking algorithm

(1) Initialization: The target state x_0 of the 1st frame, initial target templates set $T = \{T_1, T_2, ..., T_M\}$;

(2) From the 2nd frame, execute the following steps continuously and alternately, till to the final frame.

a. Sample N candidates $\{x_1, x_2, ..., x_N\}$ within a circular region of radium r away from the target location of the previous frame, and obtain their feature vectors $\{y_1, y_2, ..., y_N\}$ as shown in Fig.1.

b. Obtain the sparse representation coefficient matrix $\hat{c}$ by solving (2), with the i th column c_i corresponding to the candidate y_i.

c. Compute the reconstruction error r_i and observation probability $P(\mathbf{y}_i|o)$ of the candidate x_i by (3) and (4) respectively.

d. Determine the candidate x_i satisfying $i = \arg\max_j P(y_j|o)$ as the target of current frame.

e. Adaptive template updating: Compute the similarity degree between the tracked target $\mathbf{y}_i$ and the most similar target template T_j ($j = \arg\min_l \|T_l - \mathbf{y}_i\|_2$). If the similarity degree is less than the threshold τ, the target $\mathbf{y}_i$ is used to update target template set.

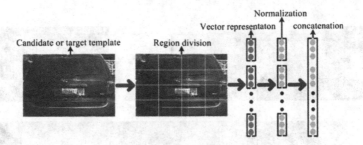

Fig. 1. Appearance representation

3 Experiments and Analyses

We implement the proposed tracker on the PC with AMD Sempron X2 198 CPU (2.5 GHz) and 3 GB memory, using MATLAB 2011b, and compare with other 4 trackers (MIL [8], CT [9], L1APG [10] and MTT [12]) in 3 different scenes to validate the performance. It should be mentioned that all the compared 4 trackers provide source codes. The related parameters are set to: the number M of target templates is 10, and the number of candidates N is 600. Each candidate or target template is scaled to 32*32 pixels and is divided into 4*4 regions to extract the locally normalized feature. The regularization parameter λ is set to 0.0002. The number of iterations in the optimization of the objective function is set to 5. And threshold τ of template updating is set to 0.4.

Scene 1: We select 3 videos (Singer1, Sylvester2008b and Car4) to validate the performance in the scene of illumination variation as shown in Fig. 2. It can be seen that the proposed tracker can track robustly under various illumination scenes. The reasons are two-folds. For one thing, the proposed method takes into account that the target template similar with the candidate has a greater possibility to represent the candidate, thus has higher reconstruction accuracy. For the other thing, the locally normalized feature, as an illumination invariant feature, can well adapt to the illumination variation.

Scene 2: In the videos Dudek and Sylvester2008b, the pose change and rotation of the tracked target take place as shown in Fig. 3. From the tracking result, the proposed tracker is more accurate than MTT, which is attributed to the weighting step before joint sparse representation of candidates.

Scene 3: In the videos Caviar1, Caviar2, Occlusion1 and Deduk, the tracked target may be occluded partially or severely as shown in Fig. 4. It can be concluded that the proposed tracker can reliably track in various occlusion scenes. The reason is that the proposed tracker makes the target template similar to the candidate has the greater possibility to represent the candidate, and obtains higher representation accuracy.

In the process of tracking, the tracked target suffers from influence of all kinds of comprehensive conditions actually. From the position error curve of Fig. 5, the pro-

(a) Singer1

(b) Sylvester2008b

(c) Car4

Fig. 2. The tracking result in illumination variation

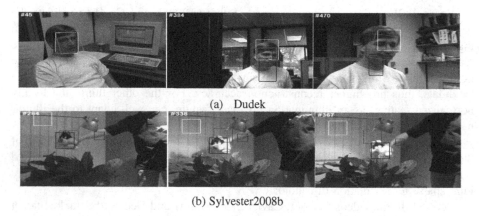

(a) Dudek

(b) Sylvester2008b

Fig. 3. The tracking result in the scene of the pose change and rotation

posed tracker can reliably track the target under various tracking conditions. Compared with other 4 trackers, the average tracking error of the proposed tracker is smallest, as shown in Table 1.

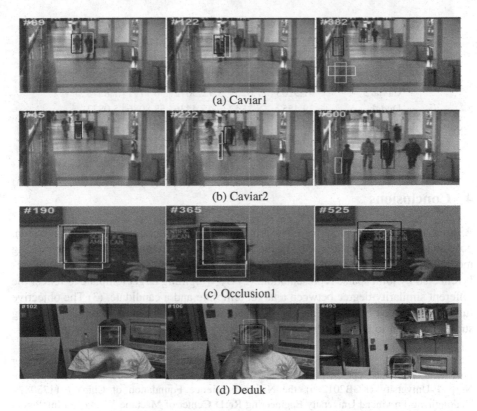

Fig. 4. The tracking result in the scene of occlusion

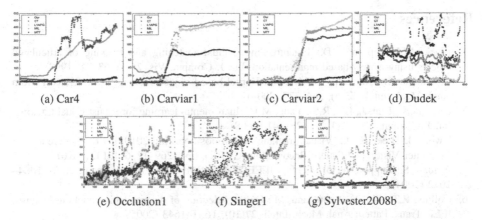

Fig. 5. Position error curve

Table 1. Average tracking error (center point error, unit: pixel)

Sequences	CT	L1APG	MIL	MTT	Our
Car4	217.6553	2.7661	191.7701	12.0800	2.0373
Caviar1	17.2037	89.9965	103.6075	53.1033	1.4344
Caviar2	61.9205	77.0259	72.5072	6.1791	1.8378
Dudek	39.4552	8.2842	45.4436	58.8086	5.0877
Occlusions1	20.1084	8.8239	39.3074	20.0080	5.7169
Singer1	14.4684	5.7062	18.1904	21.2293	3.9776
Sylvester2008b	18.2245	89.8053	128.9541	9.5268	4.0926

4 Conclusions

Aiming at the complex tracking environment such as illumination changes, occlusion, pose changes, rotation, and so on, a weighted joint sparse representation based tracking method is proposed with 4 characteristics: (1) The similarity relation among candidates is reflected by joint sparse representation. (2) By weighting target template to reflect the different similarity degree between a target template and a candidate. (3) The objective function is solved by modifying the APG algorithm. (4) A simple template updating strategy is used to adapt to the target appearance change.

Acknowledgment. This work is supported by Scientific Research Fund of Heilongjiang Provincial Education Department (NO: 12541238), Dr. Scientific Research Foundation of Harbin Normal University (KGB201216), the National Science Foundation of China (61173087), Heilongjiang Provincial University Engineering R&D Center of Machine Vision and Intelligent Detection, and Heilongjiang Provincial Education Department Key Laboratory of Intelligent Education and Information Engineering.

References

1. Black, M.J., Jepson, A.D.: Eigentracking: robust matching and tracking of articulated objects using a view-based representation. Int. J. Comput. Vis. **26**(1), 63–84 (1998)
2. Comaniciu, D., Ramesh, V., Meer, P.: Kernel-based object tracking. IEEE Trans. Pattern Anal. Mach. Intell. **25**(5), 564–577 (2003)
3. Ross, D.A., Lim, J., Lin, R.S., Yang, M.H.: Incremental learning for robust visual tracking. Int. J. Comput. Vis. **77**(1–3), 125–141 (2008)
4. Kwon, J., Lee, K.M.: Visual tracking decomposition. In: 2010 IEEE Conference on Computer Vision and Pattern Recognition (CVPR), pp. 1269–1276. IEEE (2010)
5. Avidan, S.: Support vector tracking. IEEE Trans. Pattern Anal. Mach. Intell. **26**(8), 1064–1072 (2004)
6. Collins, R.T., Liu, Y., Leordeanu, M.: Online selection of discriminative tracking features. IEEE Trans. Pattern Anal. Mach. Intell. **27**(10), 1631–1643 (2005)
7. Grabner, H., Leistner, C., Bischof, H.: Semi-supervised on-line boosting for robust tracking. In: Forsyth, D., Torr, P., Zisserman, A. (eds.) ECCV 2008, Part I. LNCS, vol. 5302, pp. 234–247. Springer, Heidelberg (2008)

8. Babenko, B., Yang, M.H., Belongie, S.: Robust object tracking with online multiple instance learning. IEEE Trans. Pattern Anal. Mach. Intell. **33**(8), 1619–1632 (2011)
9. Zhang, K., Zhang, L., Yang, M.H.: Real-time compressive tracking. In: European Conference on Computer Vision, pp. 864–877 (2012)
10. Mei, X., Ling, H.: Robust visual tracking using ℓ 1 minimization. In: 2009 IEEE 12th International Conference on Computer Vision, pp. 1436–1443. IEEE (2009)
11. Mei, X., Ling, H., Wu, Y., et al.: Minimum error bounded efficient ℓ 1 tracker with occlusion detection. In: 2011 IEEE Conference on Computer Vision and Pattern Recognition (CVPR), pp. 1257–1264. IEEE (2011)
12. Zhang, S., Yao, H., Zhou, H., et al.: Robust visual tracking based on online learning sparse representation. Neurocomputing **100**, 31–40 (2013)
13. Zhang, T., Ghanem, B., Liu, S., et al.: Robust visual tracking via structured multi-task sparse learning. Int. J. Comput. Vis. **101**(2), 367–383 (2013)
14. Tang, X., Feng, G., Cai, J.: Weighted group sparse representation for undersampled face recognition. Neurocomputing **145**, 402–415 (2014)
15. Yuan, X.T., Liu, X., Yan, S.: Visual classification with multitask joint sparse representation. IEEE Trans. Image Process. **21**(10), 4349–4360 (2012)

Single-Frame Super-Resolution via Compressive Sampling on Hybrid Reconstructions

Ji-Ping Zhang[✉], Tao Dai, and Shu-Tao Xia

Graduate School at Shenzhen, Tsinghua University,
Shenzhen 518055, Guangdong, China
{zjp13,dait14}@mails.tsinghua.edu.cn, xiast@sz.tsinghua.edu.cn

Abstract. It is well known that super-resolution (SR) is a difficult problem, especially the single-frame super-resolution (SFSR). In this paper, we propose a novel SFSR method, called compressive sampling on hybrid reconstructions (CSHR), with high reconstruction quality and relatively low computation cost. It mainly depends on the combination of the results of other SR methods, which are characteristic of high speed and low quality SR results alone. As a result, CSHR inherits the merit of low computation cost. We resample those low quality SR results in DCT domain instead of in pixel domain and regard the similar expansion coefficients as consensus which would be compressively sampled later. In CSHR, obtaining a high resolution image is only to solve a convex optimization program. We use compressed sensing theory to ensure the efficiency of our method. Also, we give some theoretic results. Experimental results show the effectiveness of the proposed method when compared to some state-of-the-art methods.

Keywords: Image processing · Single-frame super-resolution · Compressed sensing · Signal processing

1 Introduction

Image super-resolution (SR) is to recover a high resolution (HR) image from a series of low resolution (LR) images. SR is one of the most spotlighted research as it can overcome the limitation of hardware, such as the chip size, shot noise and diffraction in digital imaging system. Depending on the number of LR images, the recovery methods could be divided into two categories: single-frame super-resolution (SFSR) and multi-frame super-resolution (MFSR). Our research mainly focus on the SFSR problem.

The basic method for approximating a solution to SR recovery is through conventional linear interpolators, of which the bicubic interpolator is highly preferable. These SR methods and their variants have been elaborated by Park et al. [8] and Van [11]. Besides classical interpolation methods, another representative SR method is based on sparse representation of low and high resolution (LHR) patch-pairs over a dictionary pair. Yang et al. [13,14] propose a method working

© Springer International Publishing Switzerland 2015
S. Arik et al. (Eds.): ICONIP 2015, Part III, LNCS 9491, pp. 610–618, 2015.
DOI: 10.1007/978-3-319-26555-1_69

directly with the LR training patches and their features, which does not require any learning on the HR patches. The LR image is viewed as a downsampling version of the HR image, whose patches are assumed to have a sparse representation with respect to an over-complete dictionary of prototype signalatoms. Zeyde et al. [15] embark from Yang et al., and assume a local Sparse-Land model on image patches served as regularization.

The most important advantage of the interpolation based methods is that it contains low computation cost. However, the reconstruction quality is much poorer than sparse representation based methods, because the degradation models are limited: they are only applicable when the blur and the noise characteristics are the same for all LR images. Conversely, sparse representation of LHR patch-pairs over a dictionary has a better reconstruction quality with higher computational complexity. In practice, SFSR is an ill-posed problem due to the insufficient number of observations and the unknown registration parameters. No one knows what the original HR image exactly is. Fortunately, the only thing we can be sure of is that all of the SR reconstructions must be similar to each other, though different SR reconstructions are not exactly same to each other. Since the problem is that we have several SR reconstructions for the same HR image, yet how to use them to get a better reconstruction?

In this paper, we propose a novel compressive sampling on hybrid reconstruction (CSHR) method with a high reconstruction quality and low computation cost. It is mainly based on some low computation cost and low reconstruction quality methods' results. In order to focus on our method, we mostly deal with SFSR, although our method can be readily extended to handle multi-frame super-resolution. In CSHR, we are focusing on the common view (the same part) of different SR results. Even though there may be less consensus, compressed sensing (CS) [2,4] asserts that we can recover certain signals from many fewer samples or measurements than the traditional methods. Thus we could use CS to recover the HR by the limited consensus. CS is good at signal processing, however, it does not work well in image processing directly. Therefore, we transform other different SR results into an appropriate basis, which can be efficiently solved by CS. And we regard the similar expansion coefficients as consensus in that appropriate basis. Then we do compressive sampling on these consensus. Finally, we recover the HR image by solving a convex optimization program. Besides, we have proved that with a high probability, the coherence between sensing basis Φ and representation basis Ψ in CSHR is less than $\sqrt{2 \log N}$. Therefore, CS theory ensures the probability of CSHR's success. Unlike the aforementioned sparse representation of LHR, our method does not rely on any learning patches or over-complete dictionary. Consequently, CSHR works more efficiently.

The rest of this paper is organized as follows. Section 2 presents our CSHR method. We evaluate the performance of our CSHR method in Sect. 3 both in visual appearance and numerical criteria, and compare it with state-of-the-art methods for SFSR. Section 4 concludes the whole paper.

2   Compressive Sampling on Hybrid Reconstructions

2.1   Observation Model

Our observation model is based on Park's [8] multi-frame observation model. Supposing the resolution of the original HR image x is $L_1 N_1 \times L_2 N_2$, we rewrite it in the form of a vector $x = [x_1, x_2, \dots, x_N]^T$, where $N = L_1 N_1 \times L_2 N_2$. After downsampling the HR image x with horizontal scale factor L_1 and vertical scale factor L_2, we get the LR image f of which the resolution is $N_1 \times N_2$. Similarly, we rewrite it in the form of a vector $f = [f_1, f_2, \dots, f_M]^T$, where $M = N_1 \times N_2$. Now, the image acquisition process could be expressed as follows:

$$f = Wx + n \qquad (1)$$

where W is a $M \times N$ sampling matrix which denotes the warping, blurring and downsampling during the image acquisition and n is a $M \times 1$ vector which denotes the additive noise. There are two main differences between our model and Park's:

1. We convert the Park's MFSR model to our SFSR model.
2. We do not consider the sub-pixel movement in matrix W as Park did, for we focus on the SFSR problem. Hence, image registration is useless in our method.

2.2   Compressive Sampling on Consensus

Supposing we have known K different SR methods whose results are corresponding to: $\tilde{x}_1, \tilde{x}_2, \dots, \tilde{x}_K$. We regard $\tilde{x}_k$ as one of K different estimations for the HR x. Before we go further, let us take a look at the estimation $\tilde{x}_k$. Equation (1) tells us that the SFSR is an ill-posed problem: what we only have is an LR observation f. By estimating the sampling matrix W and modeling the noise n, we are able to find the HR image $\tilde{x}_k$ which satisfies (1). No one knows what the HR image x exactly is, however, the only thing we can be sure of is that all the $\tilde{x}_k$ must be similar to x, otherwise the kth SR method is not a good recovery method. Though different estimations are not exactly same to each other, they must be similar to each other because of the transitive relation. In this paper, we focus on the common part of different $\tilde{x}_k$. We will transform the $\tilde{x}_k$ into an appropriate basis and then use the similar expansion coefficients.

At the beginning, we transform the estimation $\tilde{x}_k$ into a different domain Ψ instead of in pixel domain:

$$\tilde{y}_k = \Psi \tilde{x}_k \qquad (2)$$

The basis Ψ has the size of $m \times N$ in (2). Ideally, $\tilde{x}_k$ has a sparse representation on the basis Ψ. The main role of the basis Ψ is to let the l_0-norm of the binary support vector $\|s_y\|_{l_0}$ in (3) as large as possible, that is non-zero elements in s_y as much as possible. It gives us representation of continuous image projected onto discrete domain. Different choices of Ψ allow us to choose the domain in which we decide to sample the image.

Then we look for a $m \times 1$ binary vector s_y as a support for all $\tilde{y}_k$:

$$\tilde{y}' = s_y \cdot \tilde{y}_k \qquad (k = 1, 2, \ldots, K) \tag{3}$$

In (3) denotes element-by-element multiplication and $\tilde{y}'$ is the consensus of all $\tilde{y}_k$. Equation (3) means that we could abstract the consensus of all $\tilde{y}_k$ to $\tilde{y}'$ by the support s_y. In other words, the support s_y is like a mask which retains all the similar expansion coefficients in K different $\tilde{y}_k$.

Supposing Φ_y is a $p \times m$ binary random matrix. In compressive sensing, we could regard Φ_y as a sensing basis, however, it does not work efficiently since it senses too much noise which is not the consensus we need. In fact, we hope to sample the consensus $\tilde{y}'$. Here, we could use a trick by the support vector s_y: let $S_y = [s_1, s_2, \ldots, s_p]^T$, where $s_1 = s_2 = \cdots = s_p = s_y$. Now we have got a $p \times m$ matrix Ψ which satisfies:

$$\Phi = \Phi_y \cdot S_y \tag{4}$$

If we use the Φ as a sensing basis:

$$\hat{y} = \Phi \tilde{y}_k \qquad (k = 1, 2, \ldots, K) \tag{5}$$

we could find that:

$$\hat{y} = \Phi_y \tilde{y}' \tag{6}$$

$\hat{y}$ is the sampling result on the consensus which is what we need. Equation (5) means that we could sample the consensus of all $\tilde{y}_k$ to $\hat{y}$.

For example, let $K = 4$, $m = 4$, we have $\tilde{y}_1 = \begin{bmatrix} 1 \\ 7 \\ 8 \\ 7 \end{bmatrix}$, $\tilde{y}_2 = \begin{bmatrix} 1 \\ 8 \\ 7 \\ 7 \end{bmatrix}$, $\tilde{y}_3 = \begin{bmatrix} 1 \\ 7 \\ 7 \\ 7 \end{bmatrix}$,

$\tilde{y}_4 = \begin{bmatrix} 1 \\ 8 \\ 8 \\ 7 \end{bmatrix}$, obviously, $s_y = \begin{bmatrix} 1 \\ 0 \\ 0 \\ 1 \end{bmatrix}$, and the consensus $\tilde{y}' = \begin{bmatrix} 1 \\ 0 \\ 0 \\ 7 \end{bmatrix}$. Supposing $p = 2$

and $\Phi_y = \begin{bmatrix} 1 & 0 & 1 & 0 \\ 1 & 1 & 0 & 1 \end{bmatrix}$, $S_y = \begin{bmatrix} 1 & 0 & 0 & 1 \\ 1 & 0 & 0 & 1 \end{bmatrix}$, then $\Phi = \begin{bmatrix} 1 & 0 & 0 & 0 \\ 1 & 0 & 0 & 1 \end{bmatrix}$, and we will get $\hat{y} = \begin{bmatrix} 1 \\ 8 \end{bmatrix}$.

Meanwhile, we will look for another $N \times 1$ support s_x which satisfies:

$$\tilde{x}' = s_x \cdot \tilde{x}_k \qquad (k = 1, 2, \ldots, K) \tag{7}$$

a binary random matrix Θ_x, the size of which is $r \times N$, let $S_x = [s_1, s_2, \ldots, s_r]^T$, where $s_1 = s_2 = \cdots = s_r = s_x$, and then we will get

$$\Theta = \Theta_x \cdot S_x \tag{8}$$

Θ in (8) satisfies:

$$\hat{x} = \Theta \tilde{x}_k \qquad (k = 1, 2, \ldots, K) \tag{9}$$

Equation (9) is similar to (5): $\hat{x}$ is the similar coefficients hybrid mixture abstracted from the K different $\tilde{x}_k$ by the $r \times N$ matrix Θ.

Finally, our problem is converted to recover a vector x from equation:

$$\hat{y} = Ax \tag{10}$$

where $A = \Phi\Psi$ is a $p \times N$ new sampling matrix telling us information about x on basis Ψ. Moreover, the solution x in (10) should obey (9), which means our result must be agree with the consensus of all $\tilde{x}_k$.

2.3   Recovery

CS recovery uses nonlinear approximation in the transform domain, although the measurements in (10) is linear. To solve (10) the recovery can be simplified to solving the following convex program:

$$\min_{\hat{x}} \|\Omega\hat{x}\|_{l_1} \qquad \text{subject to} \qquad A\hat{x} = \hat{y} \tag{11}$$

Equation (11) means that we find a N-dimention vector $\hat{x}$ which is the sparsest in the transform domain Ω that satisfies the measurements we observed. The constraint $A\hat{x} = \hat{y}$ means that we only consider the results which could produce the same measurements $\hat{y}$ which we have observed. The l_1 norm is the sum of magnitude:

$$\|z\|_{l_1} = \sum_{i=1}^{n} |z_i| \tag{12}$$

where $z = \Omega\hat{x}$. The reasons of using l_1 norm are: (i) sparse signal have small l_1 norm relative to its energy, (ii) it is convex that makes the optimization solvable [10].

2.4   Stability

The main result of CS theory [3,5] is that the number of measurements we need to recover the image does depend on the complexity of image representation in the domain we choose rather than the number of pixels we wish to recover.

CS is mainly concerned with low coherence pairs. The coherence between Φ and Ψ is defined as follows:

$$\mu(\Phi,\Psi) = \sqrt{N} \cdot \max_{1 \le i,j \le N} |\langle \phi_i, \psi_j \rangle| \tag{13}$$

which means the largest correlation between any two elements of Φ and Ψ [6]. If Φ and Ψ contain correlated elements, the coherence $\mu(\Phi,\Psi)$ is large, otherwise, it is small.

From (4) we know that Φ is the result of $\Phi_y \cdot S_y$,

$$\mu(\Phi,\Psi) \le \min(\mu(\Phi_y,\Psi), \mu(S_y,\Psi)) \tag{14}$$

Since Φ_y is a binary random matrix, with a high probability, the coherence between Φ_y and Ψ is about $\sqrt{2\log N}$. Hence with a high probability, the coherence between Φ and Ψ is less than $\sqrt{2\log N}$, which indicates that Φ and Ψ are incoherent. A most important theorem in CS [1] is that if x in our transform domain Ψ is S-sparse, select p measurements in the Ψ domain uniformly at random. Then if

$$p \geq C \cdot \mu^2(\Phi, \Psi) \cdot S \cdot \log N \tag{15}$$

for some positive constant C, the solution to (11) is exact with overwhelming probability. It is shown that the probability of success exceeds $1 - \delta$ if $p \geq C \cdot \mu^2(\Phi, \Psi) \cdot S \cdot \log(N/\delta)$.

In a word, CS theory preserves that x can be exactly recovered from our CSHR by minimizing a convex function. Solving the convex program does not need to assume any knowledge about the number of nonzero coordinates of $\hat{y}$, their locations, or their amplitudes which we assume all completely unknown priori.

3   Experimental Results

In this section, we simulated experiments to demonstrate the performance of our CSHR method by using several standard benchmark test images[1]. During the simulation scenario, we first downsample the HR test image x with both horizontal and vertical scale factor of 2 as the LR image f. Then, we recover the image $\tilde{x}_k$ by using Nearest neighbor, Bilinear, Bicubic, ScSR [14] and TIP14 [9]. Based on these $K = 5$ fundamental methods, we use DCT basis as Ψ and Ω. Figures 1 and 2 gives us a visual comparison on the partial of original images.

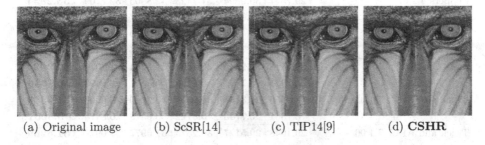

(a) Original image   (b) ScSR[14]   (c) TIP14[9]   (d) **CSHR**

Fig. 1. Visual comparison on Babbon.

In addition to subjective visual comparison, we also provide the peak-signal-to-noise ratio (PSNR) [7] as well as the structural similarity index measure (SSIM) [12] in Table 1 which are used quantitatively to measure the results of different SR methods[2]. PSNR is a traditional criterion that is widely used in signal fidelity, while SSIM is known to be more consistent with human visual system (HVS). Results with larger PSNR and SSIM are considered to have better results.

---

[1] http://sipi.usc.edu/database/database.php?volume=misc.
[2] More results are shown in https://gist.github.com/Brilliant/7472969d4020599a13d0.

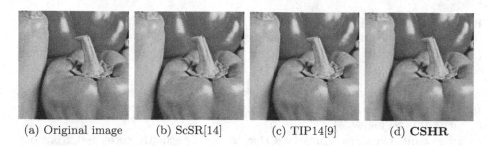

(a) Original image　　(b) ScSR[14]　　(c) TIP14[9]　　(d) **CSHR**

Fig. 2. Visual comparison on Peppers.

Table 1. Comparison based on PSNR (dB) and SSIM

Images	NN	Bilinear	Bicubic	ScSR	TIP14	CSHR
Airplane (U-2) 7.2.01	34.1285	34.0964	34.4150	34.7347	34.6681	**35.0588**
	0.8160	0.8031	0.8188	0.8383	0.8312	**0.8459**
Car and APCs 7.1.10	33.3599	33.9880	34.9014	35.8014	35.7503	**36.2948**
	0.8574	0.8605	0.8867	0.9106	0.9072	**0.9177**
Girl (Lena, or Lenna) 4.2.04	31.4187	32.6780	34.1092	36.1288	36.2255	**36.4926**
	0.8960	0.9028	0.9198	0.9367	0.9358	**0.9419**
Man 5.3.01	29.9675	30.8534	32.0520	33.8399	**33.8780**	33.8498
	0.8571	0.8599	0.8865	0.9133	0.9111	**0.9145**
Mandrill (a.k.a. Baboon) 4.2.03	23.1195	23.0403	23.6253	24.3864	24.3946	**24.8038**
	0.6979	0.6553	0.7116	0.7810	0.7775	**0.7882**
Peppers 4.2.07	29.7547	30.9894	31.7423	32.8905	32.7709	**33.2834**
	0.8624	0.8715	0.8843	0.8997	0.8973	**0.9019**
Tank 7.1.07	30.2140	30.4347	31.1732	31.9449	31.9254	**32.3743**
	0.7914	0.7793	0.8166	0.8544	0.8490	**0.8625**
Tank 7.1.09	30.0818	30.2545	30.9968	31.8264	31.7971	**32.3024**
	0.7921	0.7771	0.8150	0.8551	0.8493	**0.8636**
Truck and APCs 7.1.05	28.9856	29.3136	30.0591	30.9090	30.9007	**31.3353**
	0.7851	0.7745	0.8122	0.8518	0.8475	**0.8601**
Truck and APCs 7.1.06	29.1648	29.5127	30.2673	31.0977	31.0977	**31.5269**
	0.7880	0.7776	0.8157	0.8550	0.8508	**0.8629**

In Table 1, each cell 2 results shows: Top - image PSNR (dB), Bottom - SSIM index, which shows the improvement in PSNR and SSIM index by applying our method versus the other methods. The best result for each image are highlighted. It can be seen that the proposed CSHR outperforms all the other methods, including ScSR and TIP14 which stand for state-of-the-art method.

Also, we compare the running time of different state-of-the-art SR methods which is shown in Fig. 3:

In Fig. 3, x-axis means the size of image in pixel we use, y-axis is the average running time. It could be easily find that the running time of the baseline method

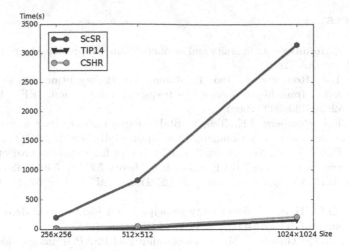

Fig. 3. Running time among the methods

ScSR is the longest. Obviously, TIP14 runs fastest, because it makes use of a statistical prediction model. Our proposed CSHR method works much faster than ScSR, while a little slower than TIP14. It means that the running time of CSHR is closing to the state-of-the-art method.

4 Conclusions

In this paper, in order to make use of the high speed and low quality SR results, we proposed a novel method CSHR for solving the SFSR problem. CSHR does not rely on any learning patches or over-complete dictionary, it is based on the consensus of other SR reconstructions. It makes a compressive sampling on these consensus hybrid in the pixel domain. By solving a convex optimization program, it recovers the HR image. We have also proved that with a high probability, the coherence between Φ and Ψ in CSHR is less than $\sqrt{2\log N}$. Thus CS theory will ensure the efficiency of CSHR. Experimental results demonstrate the effectiveness of the CSHR. CSHR is able to achieve the state-of-the-art SR results. Our future work will focus on the choice of Ψ, perhaps wavelet basis is a good choice. We know the fact that wavelets automatically adapt to singularities in the image; important wavelet coefficients tend to cluster around edge contours, while large smooth regions can be built up with relatively few terms.

Acknowledgements. The authors wish to thank the authors of [9,14] for generously sharing their code and data with them.

This research is supported in part by the Major State Basic Research Development Program of China (973 Program, 2012CB315803), the National Natural Science Foundation of China (61371078), and the Research Fund for the Doctoral Program of Higher Education of China (20130002110051).

References

1. Candes, E., Romberg, J.: Sparsity and incoherence in compressive sampling. Inverse Prob. **23**(3), 969 (2007)
2. Candès, E.J., Romberg, J., Tao, T.: Robust uncertainty principles: exact signal reconstruction from highly incomplete frequency information. IEEE Trans. Inf. Theory **52**(2), 489–509 (2006)
3. Candes, E.J., Romberg, J.K., Tao, T.: Stable signal recovery from incomplete and inaccurate measurements. Commun. Pure Appl. Math. **59**(8), 1207–1223 (2006)
4. Candes, E.J., Tao, T.: Near-optimal signal recovery from random projections: universal encoding strategies? IEEE Trans. Inf. Theory **52**(12), 5406–5425 (2006)
5. Donoho, D.L.: Compressed sensing. IEEE Trans. Inf. Theory **52**(4), 1289–1306 (2006)
6. Donoho, D.L., Huo, X.: Uncertainty principles and ideal atomic decomposition. IEEE Trans. Inf. Theory **47**(7), 2845–2862 (2001)
7. Huynh-Thu, Q., Ghanbari, M.: Scope of validity of PSNR in image/video quality assessment. Electron. Lett. **44**(13), 800–801 (2008)
8. Park, S.C., Park, M.K., Kang, M.G.: Super-resolution image reconstruction: a technical overview. IEEE Signal Process. Mag. **20**(3), 21–36 (2003)
9. Peleg, T., Elad, M.: A statistical prediction model based on sparse representations for single image super-resolution. IEEE Trans. Image Process. publ. IEEE Signal Process. Soc. **23**(6), 2569–2582 (2014)
10. Romberg, J.: Imaging via compressive sampling [introduction to compressive sampling and recovery via convex programming]. IEEE Signal Process. Mag. **25**(2), 14–20 (2008)
11. Van Ouwerkerk, J.: Image super-resolution survey. Image Vis. Comput. **24**(10), 1039–1052 (2006)
12. Wang, Z., Bovik, A.C., Sheikh, H.R., Simoncelli, E.P.: Image quality assessment: from error visibility to structural similarity. IEEE Trans. Image Process. **13**(4), 600–612 (2004)
13. Yang, J., Wright, J., Huang, T., Ma, Y.: Image super-resolution as sparse representation of raw image patches. In: IEEE Conference on Computer Vision and Pattern Recognition, CVPR 2008, pp. 1–8. IEEE (2008)
14. Yang, J., Wright, J., Huang, T.S., Ma, Y.: Image super-resolution via sparse representation. IEEE Trans. Image Process. **19**(11), 2861–2873 (2010)
15. Zeyde, R., Elad, M., Protter, M.: On single image scale-up using sparse-representations. In: Boissonnat, J.-D., Chenin, P., Cohen, A., Gout, C., Lyche, T., Mazure, M.-L., Schumaker, L. (eds.) Curves and Surfaces 2011. LNCS, vol. 6920, pp. 711–730. Springer, Heidelberg (2012)

Neuro-Glial Interaction: SONG-Net

Kirmene Marzouki[1,2(✉)]

[1] Informatics for Industrial Systems Laboratory,
University of Carthage, Carthage, Tunisia
kirmene@marzouki.tn
[2] Higher Institute of Applied Sciences and Technology of Sousse,
University of Sousse, Sousse, Tunisia

Abstract. More convincing evidence has proven the existence of a bidirectional relationship between neurons and astrocytes. Astrocytes, a new type of glial cells previously considered as passive support cells, constitute a system of non-synaptic transmission playing a major role in modulating the activity of neurons. In this context, this paper proposes to model the effect of these cells to develop a new type of artificial neural network operating on new mechanisms to improve the information processing and reduce learning time, very expensive in traditional networks. The obtained results indicate that the implementation of bio-inspired functions such as of astrocytes, improve very considerably learning speed.

The developed model achieves learning up to twelve times faster than traditional artificial neural networks.

Keywords: Brain · Neurons · Glial cells · Astrocytes · Artificial neural networks · SOM · MLP · Calcium waves

1 Introduction

At the beginning of their discovery in the second half of the 19th century, it was believed that glia cells play the role of "babysitters" of nerve cells. Recent findings suppose that these cells, despite lacking excitability, usually associated with most neurons, can be more actively involved in brain functions [1]. Glia are more and more appreciated as active participants in central neural processing via calcium waves, electrical coupling, and even synaptic-like release of "neuro"-transmitters [2, 3].

These findings indicate that the activation of calcium signaling in astrocytes, a type of glia cells, can regulate both excitatory and inhibitory synaptic transmission and mediate essential physiological functions. However, the exact mechanisms for how astrocytes exert influence over neuronal networks remain a matter of intense controversy [4].

One popular model known as the tripartite synapse suggests that astrocytes detect the release of neurotransmitters and actively modulate pre- and post-synaptic neurotransmission by the calcium dependent release of gliotransmitters [5].

By releasing gliotransmitters in synaptic clefts, astrocytes modulate the electrical activity of several neurons and plasticity of these connections.

Based on this assumption, this study proposes to model the effect of these cells and to develop a new type of artificial neural network operating on new mechanisms to

© Springer International Publishing Switzerland 2015
S. Arik et al. (Eds.): ICONIP 2015, Part III, LNCS 9491, pp. 619–626, 2015.
DOI: 10.1007/978-3-319-26555-1_70

improve the processing of information and reduce learning time, far very expensive in traditional networks.

The proposed model was tested with different problems: XOR, Iris Flower, Ionosphere and lung Cancer. Results show better learning performance than conventional MLP.

2 Related Works

Nervous cells compose higher brain's functions. They are divided into two kinds of cells: Neurons and Glia.

Recently, researchers discovered new glia functions [8, 9]. Glia cells transmit signals by changing concentrations of several ions [4]. That is why the study of the role of glia cells is becoming a new field of research. Some researchers consider that the features of glia cells can be applied to artificial neural networks. There are few papers dealing with this subject.

The first work was presented in [10], by Ikuta et al. who proposed the glial networks to improve the performance of Multi-Layer Perceptron (MLP). They proposed the MLP with the glial network generating chaotic oscillations.

In a second work [11], they investigated the MLP with the impulse glial network. Glia generate only impulse output, however they make the complex output by correlation with each other. They have investigated the proposed networks' parameter dependency.

The same team proposed in [12] a Multi-Layer Perceptron (MLP) with pulse glial chain from the features of biological glia. In this model, glia cells are connected to neurons, in a one-to-one scheme, in a hidden layer of the MLP.

Glia are excited by huge amount of output of connected neurons. The connected glia generates the pulse output. Moreover, the pulse excites the neighboring glia and affects the threshold of the connected neuron. In [13], they published a new study in which they applied the MLP with pulse glial chain to Two-Spirals Problem (TSP).

Finally, in [14], the same team proposed the MLP with positive and negative pulse glial chain, which is inspired from features of the biological glia. They added the MLP to the positive and negative pulse glial chain. In the positive and negative pulse glial chain, the glia cells are connected to neurons one-to-one. The glia generates pulse when it is excited by the connected neuron's output. If the connected neuron has large amount of output, the glia generates positive pulse. If the connected neuron has small amount of output, the glia generates the negative pulse. The positive and negative pulse are propagated to the connected neuron and neighboring glia. They considered that the positive and negative pulse glial chain give the relationships of position of neurons in a same layer.

In all models proposed by Ikuta et al. the glial network is connected to neurons in the last hidden layer only. On the other side, in the biological model astrocytes exist in all synaptic clefts. The mathematical proposed model does not match with the biological model.

Furthermore, we see that the error presents oscillations suggesting that the proposed algorithm is not stable.

In [16], Alvarellos et al. have investigated the consequences of including artificial astrocytes, which present the biologically defined properties involved in astrocyte-neuron communication, on artificial neural network performance. Using connectionist systems and evolutionary algorithms, they have compared the performance of artificial neural networks and artificial neuron-glia networks to solve classification problems.

In [16], they have investigated, using computational models, different astrocyte-neuron interactions for information processing; different neuron-glia algorithms have been implemented for training and validation of multilayer Artificial Neuron-Glia Networks oriented toward classification problem resolution.

It is noted that during their second published work, Alvarellos et al. are trying, in the training phase, to vary the values assigned to k and μ, but the question that arises here is the choice of these parameters.

We notice, also, that the proposed algorithm is based on the pruning idea: the least active unit is ignored comparing to other units, which is well-known as "Pruning Neural Networks" [17].

In addition, they used the genetic algorithm witch have no known relationship with the Neural Networks paradigm.

One other limitation of the proposed algorithm is that the biological inspiration does not match with the mathematical algorithm. In fact, each astrocyte is in interaction with a neuron, but there is no interaction between astrocytes.

3 Proposed Algorithm

Starting from the fact that neurons in the brain are interconnected in a layered manner, we decided to use MLP [6] given that its architecture coincides with the biological organization of neurons. Moreover, combined with Back Propagation, we insure that the error will converge to a minimum.

On the other side, it is now admitted that the modulation of neuronal activity in the biological system is much more profound than it is traditionally thought.

We can think that depending on the coming input signal, astrocytes insure a collaboration role between neurons. The neurons activity modulation can be thought as a selection job that astrocytes perform in order to have the "best" output(s) in terms of quality and quantity.

Besides, in all versions of SOM [7], learning is performed identically from the start of the training process till its end. The same adaptation equation is used all along the process without really considering the changes of the network and the information being "learned" so far, but for finding the Best Matching Unit (BMU).

This leads us to think to control the evolution of the network during the training process, by making it consistent with learning.

Motivated by these reasons we believe that the combination of MLP with SOM could be more effective given that MLP presents the neurons and SOM plays a selective job.

It should be mentioned that the resulted algorithm of combination of MLP and SOM is different from the Counter Propagation Network, CPN, as presented in [18] in many aspects.

In fact, in CPN, unsupervised learning is performed first on the hidden layer. Then, a supervised learning is done on the output layer. The two modes of learning are performed separately.

While in the proposed SONG-NET network, Fig. 1, the two learning modes are occurring at the same time.

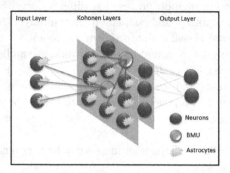

Fig. 1. Overall scheme of the SONG-NET network

3.1   SONG-Net Algorithm

The SONG-Net represents a combination between MLP and SOM. The main idea is to excite some neurons (called BMUs) comparing to some others when switching the information from one layer to another.

For reasons of simplicity and test, we chose to work with linear hidden layers.

The main steps in the proposed algorithm are:

Look for the BMUs:

- For each training vector we look for the neuron in the *(i + 1)* layer that is nearest to the *i*-layer: the nearest neuron is called the winner.
- Then, we perform statistics for all winners: a neuron is considered BMU if it is elected the maximum times as winner during the training of all vectors of the learning base. This property will be used the next epoch

To find the nearest neuron, we calculate the Euclidean distance to determine the distance of each node from the previous layer. Once the distances are calculated, we can now conclude the winner: this is actually the neuron that is closest to the desired layer.

Computing the Euclidean distance is done according to the following formula:

$$d = \sqrt{\sum_i (x_i - w_{ij})^2} \tag{1}$$

Updating Weights:

- Now, we apply the Back Propagation using two updating formulas:

If it is a BMU neuron that is determined in the previous epoch, we update its weights using the formula:

$$W^{jk} = W^{jk} + \eta \delta_k x_j + \eta^{**} \delta_k (x_j - W^{jk}) \tag{2}$$

Else: We keep the back propagation updating formula:

$$W^{jk} = W^{jk} + \eta \delta_k x_j \tag{3}$$

We note that the update formula is not the same for all neurons. Here the BMUs neurons (determined in the previous epoch) are favored by the addition of a second term when updating weights. In fact, the added term represents the SOM update formula multiplied by the contribution of the neuron in the final error.

Besides, the learning rate used is a decreasing function over time of the form

$$\eta_0 * (1 - (\text{current epoch} / \text{total epochs number})) \tag{4}$$

Using a learning rate in the previous form will guarantee a fast convergence in the beginning of learning (since its value is high), but also a speed at the end of learning (since its value becomes smaller and smaller).

4   Experiments and Results

The proposed model introduces the statistics notion to determine for each hidden layer the neuron that has been elected as a winner when passing all training vectors. Once the BMUs are determined, we applied the conventional BP for non-winning neurons and we changed the update formula for the winning neurons.

The algorithm was tested with XOR problem, classification of the iris flower problem, ionosphere problem and Lung cancer problem.

The various problems considered were tested with networks containing one, two, three and four hidden layers and with different numbers of neurons per hidden layers.

For all cases, the learning rate η chosen for updating non-winning neurons and in MLP algorithm with back propagation is of the order of 0.1.

The learning rate η_0 used to update winning neurons is of the order of 0.8 with respect to Eq. 4.

The following figures show examples of simulation of the proposed algorithm, and a comparison with the results given by the conventional MLP.

Figure 2, XOR problem, presents the Mean Square Error (RMSE) of a network with one hidden layer that contains 5 neurons. We stopped learning when the error reached 0.001. Our algorithm reached this error in 2 s. In return, the MLP has reached it after 20 s. SONG-Net is 10 times faster

In Fig. 3, Iris flower Classification, we used a network with three hidden layers. The first layer of the neural network contains 7 neurons, the second 5 and the third 7 neurons.

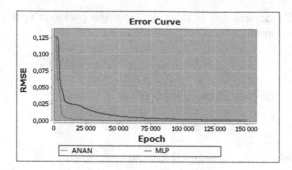

Fig. 2. RMSE of a network with one hidden layer (XOR)

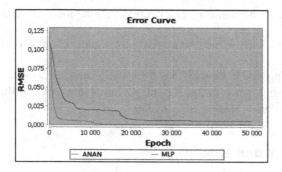

Fig. 3. RMSE of a network with three hidden layers (Iris Flower)

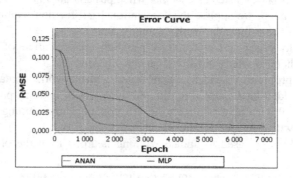

Fig. 4. RMSE of a network with two hidden layers (Lung Cancer)

Our algorithm reached the desired error 20 s, while the MLP stopped after 34 s.

Finally, in Fig. 4, Lung Cancer problem, the network contains two hidden layers. The first hidden layer contains 4 neurons and the second contains 5.

Our algorithm reached the desired error after 12 s while the MLP reached it at 45 s.

SONG-Net has been tested, firstly with the famous XOR problem. Very good results have been found. Our algorithm is 10 to 12 times faster than conventional MLP.

After that, our algorithm has been tested with the Iris flower problem. Again our algorithm is faster than the conventional MLP.

To go further, we tested SONG-Net with the problem of the Ionosphere and the lung cancer. As expected, our algorithm is faster. We even managed to have a gain of 6 times faster than conventional MLP.

Our algorithm was tested with diffrent number of hidden layers and neurons per layer, while Ikuta et al. [10–14] did perform in only one topology.

The results presented show that modeling the effect of astrocytes on neurons in neural networks suggests a rethinking in the way the existing algorithms are considered.

5 Conclusion

In this study, we have proposed the SONG-NET algorithm, a combination of SOM and MLP, with neuron-glial interaction. The astrocytes are involved in modulating synaptic activity by an increase of internal calcium levels. Glia cells involved in an indirect way, by release of neurotransmitters, to perform a certain selection based on various receptors located on the destination neuron. This selection is translated into either the excitation or inhibition of the recipient neuron.

Inspired by this idea, in our model some neurons, called Best Matching Units (BMUs), were excited in a selected manner.

Our algorithm were tested with four problems. We confirmed that the proposed model performs a better learning than conventional networks.

In the future works we will increase the number of winning neurons to observe the influence of that on the quality of the training.

Some perspectives shall be explored, in particular, we will try to test our algorithm in other, more powerful areas such as image processing field.

References

1. Barres, B.A.: New roles for glia. J. Neurosci. **11**(12), 3685–3694 (1991)
2. Pfrieger, F.W., Barres, B.A.: New views on synapse–glia interactions. Curr. Opin. Neurobiol. **6**, 615–621 (1996)
3. Nedergaard, M., Ransom, B., Goldman, S.: New roles for astrocytes: redefining the functional architecture of the brain. Trends Neurosci. **26**(10), 523–530 (2003)
4. Mattoson, M.P., Chan, S.L.: Neuronal and glial calcium signaling in Alzheimer's disease. Cell Calcium **34**, 385–397 (2003)
5. Pasti, L., Volterra, A., Pozzan, T., Carmignoto, G.: Intracellular calcium oscillations in astrocytes: a highly plastic, bidirectional form of communication between neurons and astrocytes in situ. J. Neurosci. **17**(20), 7817–7830 (1997)
6. Rumelhart, D.E., Hinton, G.E., Williams, R.J.: Learning representations by back-propagating errors. Nature **323–9**, 533–536 (1986)
7. Goyal, A., Lakhanpal, A., Goyal, S.: Learning of alphabets using Kohonen's self organized featured map. Int. J. Appl. Innov. Eng. Manage. **2**(12), 283–287 (2013)

8. Haydon, P.G.: Glia: listening and talking to the synapse. Nat. Rev. Neurosci. **2**, 844–847 (2001)
9. Ozawa, S.: Role of glutamate transporters in excitatory synapses in cerebellar Purkinje cells. Brain Nerve **59**, 669–676 (2007)
10. Ikuta, C., Uwate, Y., Nishio, Y.: Chaos glial network connected to multi-layer perceptron for solving two-spiral problem. In: Proceedings of the ISCAS 2010, pp. 1360–1363, May 2010
11. Ikuta, C., Uwate, Y., Nishio, Y.: Performance and features of multi-layer perceptron with impulse glial network. In: Proceedings of the IJCNN 2011, pp. 2536–2541, June 2011
12. Ikuta, C., Uwate, Y., Nishio, Y.: Multi-layer perceptron with impulse glial network. In: Proceedings of the NCN 2010, pp. 9–11, December 2010
13. Ikuta, C., Uwate, Y., Nishio, Y.: Multi-Layer perceptron with pulse glial chain for solving two-spiral problem. In: IEEE Workshop on Nonlinear Circuit Networks, December 2011
14. Ikuta, C., Uwate, Y., Nishio, Y.: Multi-layer perceptron with positive and negative pulse glial chain for solving two-spiral problem. In: WCCI 2012 IEEE World Congress on Computational Intelligence, June 2012
15. Porto-Pazos, A.B., Veiguela, N., Mesejo, P., et al.: Artificial astrocytes improve neural network performance. PLoS ONE **6**(4) (2011) (Article ID e19109)
16. Alvarellos-Gonzalez, A., Pazos, A., Porto-Pazos, A.B.: Computational models of neuron-astrocyte interactions lead to improved efficacy in the performance of neural networks, February 2012
17. Abdulla Saeed, R., Edwar George, L., et al.: Apply pruning algorithm for optimizing feed forward neural networks for crack identifications in francis turbine runner. Int. J. Soft Comput. Eng. **2**, 175–199 (2012)
18. Hecht-Nielsen, R.: Conterpropagation networks. In: Proceedings of the IEEE First International Conference on Neural Networks (1987)

Changes in Occupational Skills - A Case Study Using Non-negative Matrix Factorization

Wei Lee Woon[✉], Zeyar Aung, Wala AlKhader, Davor Svetinovic, and Mohammad Atif Omar

Institute Center for Smart and Sustainable Systems,
Masdar Institute of Science and Technology, P.O. Box 54224, Abu Dhabi, UAE
{wwoon,zaung,wabedalkhader,dsvetinovic,momar}@masdar.ac.ae

Abstract. Changes in the skill requirements of occupations can alter the balance in the numbers of high, middle and low-skilled jobs on the market. This can result in structural unemployment, stagnating income and other unforeseen social and economic side effects. In this paper, we demonstrate the use of a recent matrix factorization technique for extracting the underlying skill categories from O*NET, a publicly available database on occupational skill requirements. This study builds upon earlier work which also focused on this database, and which indicated that changes in skill requirements were in response to increased automation which unevenly affected different segments of the job market. In this paper we refine the methodological underpinnings of the earlier work and report some preliminary results which already show great promise.

Keywords: Non-negative matrix factorization · Data mining · Source separation · Empirical research · Job characteristics

1 Introduction

1.1 Background and Related Work

Technological advances have long had a disruptive effect on the job market. While the long term effects have generally been increased productivity and efficiency, this has often been at the expense of significant, if transient, social imbalance.

However, the pace of recent advances in digital technologies and automation have been unprecedented. IBM's "Watson", which competed with and defeated the best human competitors of the *Jeopardy* quiz show, is just one example of a range of technologies that are precipitating a broad shift towards increased automation in both a greater number and variety of jobs. Furthermore, it apparent that these changes will continue or even accelerate in the foreseeable future. The socio-economic impacts of these trends are far reaching - middle and low skill jobs are disappearing while labor force participation and median incomes have fallen [1]. It is unclear how these changes will develop in the long run but there is a pressing need to study the underlying factors so that informed mitigation strategies can be formulated.

© Springer International Publishing Switzerland 2015
S. Arik et al. (Eds.): ICONIP 2015, Part III, LNCS 9491, pp. 627–634, 2015.
DOI: 10.1007/978-3-319-26555-1_71

An earlier paper [2] (henceforth referred to as MC) examined how the skill content of jobs had evolved in the period between 2006 and 2014 using O*NET, a publicly available and comprehensive occupational skill requirements database[1] compiled by the US government. The findings of that study support the notion that substitution effects will remove some skills from occupations, complementarity effects will amplify other skills, and skills that are orthogonal will be amplified due to Baumols Cost Disease [3].

1.2 Motivations and Objectives

Key to successfully understanding these issues is the ability to detect and study the underlying skill "dimensions", i.e. groups of skill or ability elements which co-occur repeatedly across multiple occupations.

In [4], these were manually constructed based on domain knowledge, while in MC factor analysis (FA) was used to affect data driven skills aggregation. While the results of these studies were already very insightful, manual extraction is a highly subjective process while FA is based on a number of assumptions, in particular that the underlying factors are Gaussian distributed and zero mean. These are difficult to substantiate at best; in fact, the ratings provided in O*NET range from 1 to 5, while a summary inspection of the importance levels reveal a mix of different distributions, many of which are clearly not Gaussian (some examples are shown in Fig. 1). An additional issue is that the factor loadings include negative coefficients which are very difficult to interpret.

This paper addresses these concerns by, on the one hand, retaining the empirical approach to factor elicitation while on the other hand identifying and testing suitable alternatives to factor analysis. To evaluate the usefulness of the proposed method, we repeat elements of the analysis performed in MC and report on the findings and observations.

2   Methods and Data

2.1   O*NET

The Occupational Information Network (O*NET) is a publicly available database of occupations developed for the US Department of Labor and is the successor to the better-known Dictionary of Occupational Titles. O*NET contains detailed information on more than 950 US occupations, including the tasks associated with the various occupations, required knowledge, skills and abilities, typical work activities and the contexts in which the occupations are commonly performed. To facilitate comparison with earlier work, we utilize data matrices constructed using the combined importance levels for three main job characteristics: *Abilities*, *Skills* and *Work Activities*.

Also, note that O*NET only documents occupation types and not the demand for or frequency of these occupations. While this is a limitation, it also imposes

---

[1] http://www.onetonline.org.

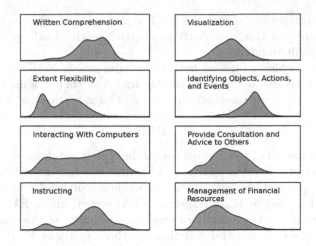

Fig. 1. Probability densities for a selection of skills/abilities/activities (2006 levels)

an emphasis on the study of *intensive* changes in occupations, *viz* how the actual composition of jobs is changing. This is interesting because most existing studies have focused on *extensive* changes even though it is known that both types of changes occur and are strongly affected by technological progress [5].

However, the data mentioned above is of extremely high dimensionality (even using only a subset of the job characteristics has resulted in a 128-dimensional data set). This is why it is particularly important to identify the most informative combinations of these dimensions if this data is to be effectively analyzed.

2.2 Factor Analysis

FA was used as a refinement to the approach taken in [4], where skills dimensions were empirically identified based on the statistical properties of the importance levels. FA models data as a linear combination of Gaussian distributed, uncorrelated *factors*. It is similar to PCA but with the addition of independent, Gaussian distributed error terms to each input variable. For a set of observed variables $x_i, \ldots, x_p$ with expected values $\mu_i, \ldots, \mu_p$:

$$x_i - \mu_i = l_{i1}F_1 + l_{i2}F_2 + \ldots + l_{ik}F_k + \epsilon_i, \tag{1}$$

where l_{ij} is the loading of the jth factor on the ith variable and F_j is factor j. To improve the readability of the resulting factor loadings, varimax rotation was subsequently applied to the loadings matrices. For conciseness, we will subsequently refer to this combination of factor analysis with varimax as FA.

2.3 Non-negative Matrix Factorization

As mentioned previously, the main objective of this work was to extend the methodological basis of MC by identifying and testing suitable alternatives to FA. The motivations for this are (i) The variables of interest (skill importance levels) are non Gaussian distributed (ii) Negative loadings produced by FA can be difficult to interpret.

One method which addresses both these concerns is Non-Negative Matrix Factorization or NMF [6]. NMF is similar to FA it that it aims to express a matrix of data as a linear combination of a set of basis vectors and a transformed data set:

$$V \approx WH, \tag{2}$$

where V is the matrix of row vectors containing the input data, W is the transformed data set and H is the basis set which defines a linear combination of the columns of W (to permit easier discussion we henceforth refer to W and H as the *factor* and *loading* matrices respectively). However, unlike FA, NMF makes no assumptions about Gaussianity or even orthogonality of the underlying factors. Instead, the only constraint is that all three matrices V, W and H are non-negative.

Since (2) does not have a unique or closed form solution, it is typically solved iteratively using the following multiplicative update terms [6]:

$$H_{bj}^{k+1} = H_{bj}^{k} \cdot \frac{((W^k)^T V)_{bj}}{((W^k)^T W^k H^k)_{bj}} \tag{3}$$

$$W_{ia}^{k+1} = W_{ia}^{k} \cdot \frac{(V(H^{k+1})^T)_{ia}}{(W^k H^{k+1}(H^{k+1}))_{ia}}. \tag{4}$$

In practice, the non-negative constraint encourages *additive* combinations of parts; for e.g., in face recognition, additive features include the nose, eyes and mouth. This characteristic often leads to overcomplete bases but also favors components that are relatively localized and easily interpreted over more compact representations. As such, it has been very useful in a variety of applications, examples of which include document clustering [7] and image representation [8].

We believe that this property will in turn be particularly useful in the present context as the factor loadings are equally important for the elucidation of occupational skill dimensions as for the subsequent statistical analysis.

2.4   Computational Considerations

All the analysis in this paper was conducted in the Python and R programming environments. The NMF implementation from the well known *scikit-learn*[2] was used while Factor analysis was performed in R and coordinated from Python using RPY2 (a Python variant provided by scikit-learn was evaluated but was rejected as it lacked a proper varimax implementation).

To retrieve the most representative skills for each factor, a procedure similar to that used in MC (but not identical) was adopted to facilitate comparison. Briefly, this worked as follows:

---

[2] http://www.scikit-learn.org.

1. All items with loadings below a threshold which was empirically set between the 90th to 95th percentile of factor loadings were retained.
2. All factors with at least three remaining items were retained.
3. Items not loading on any particular factor were discarded.
4. The procedure is iterated until all skills loaded on at least one factor.

While some "manual optimization" proved helpful in improving the clarity of results obtained, our experience was that these results were not overly sensitive to these parameters and were broadly consistent over a range of settings.

3   Results

The methods and approach discussed in the previous sections were applied to the data and the results will now be discussed. Due to space constraints most of the analysis focuses on the 2006 data, while for the regression analysis the 2014 loadings will be used as the dependent variable.

Figure 2 depicts factor loadings extracted from the 2006 data. From these figures, it is apparent that both methods were able to extract loading matrices that concentrated on a small subset of the skills. However, the factors extracted using FA tended to be weighted towards the first few factors, where the loadings were heavily concentrated on a larger number of coefficients, as may be expected based on the properties of the factor analysis algorithm. In contrast, NMF tended to extract factors with more evenly balanced loadings both across and within individual factors, which appears to stem from the lack of an imposed ordering on the factors.

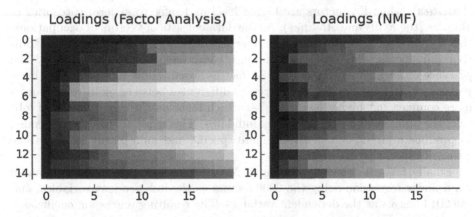

Fig. 2. Loading matrices for first 15 factors extracted using FA and NMF

To better understand the implications of this, a number of factors were extracted from the O*NET occupations data using the procedure described Sect. 2.4. When using FA, the extracted factors were:

1. **Leadership:** Instructing; Learning Strategies; Social Perceptiveness
2. **Manual:** Stamina; Gross Body Coordination; Trunk Strength
3. **Equipment:** Troubleshooting; Repairing; Equipment Maintenance
4. **Vehicle Operation:** Night Vision; Peripheral Vision; Glare Sensitivity
5. **Perception:** Flexibility of Closure; Perceptual Speed; Speed of Closure
6. **Mathematical:** Number Facility; Mathematical Reasoning; Mathematics

Using NMF, the factors extracted were as follows:

1. **Research:** Science; Analyzing Data or Information; Documenting/Recording Information
2. **Computing:** Interacting With Computers; Documenting/Recording Information; Processing Information
3. **Maintenance:** Repairing; Installation; Repairing and Maintaining Mechanical Equipment
4. **Design:** Drafting, Laying Out, and Specifying Technical Devices, Parts, and Equipment; Mathematics; Technology Design
5. **Perception:** Reaction Time; Hearing Sensitivity; Response Orientation
6. **Leadership:** Training and Teaching Others; Instructing; Coaching and Developing Others
7. **Manual:** Performing General Physical Activities; Stamina; Static Strength

Firstly, it is reassuring to note that the two methods produced factors that are broadly similar (it would be very disconcerting if completely different factors were produced), though there were also some very interesting differences. More comprehensive analysis is required to fully understand these, but it does appear that the NMF factors are more specific and emphasize the occupations (the *Work Activities*), while FA factors are higher level and refer to general categories of abilities (the *Skills* and *Abilities*). As the data consists of vectors of occupations, this is consistent with the main characteristic of NMF which is to produce additive components, where each occupation is composed of multiple activities. The presence of negative coefficients in FA factors would allow more compact basis vectors which, for example, could leverage differences between skills to produce a more compact but higher level basis set which focus on the *requirements* of jobs - i.e. the abilities and skills of ideal candidates. This explains why NMF tended to produce a larger, but generally more specific and more easily interpretable basis set.

Finally, following the methodology in MC, we now study the intensive changes via a linear regression using the 2006 factors as the independent variables, and the 2014 factors as the dependent variables. The resulting regression coefficients are shown in Table 1.

The main observations from Table 1 were as follows:

1. We note that the diagonal elements dominate Table 1, as would be expected. However, there are also substantial off-diagonal elements, which point to skill substitutions within occupations.

Table 1. Changes in Job Characteristics, analyzed using 2006 Factors

	Research	Computing	Maint'nce	Design	Perception	Leadership	Manual
Research	**0.6768**	0.0359	−0.0887	−0.0384	−0.0578	0.2462	−0.0376
Computing	0.0352	**0.7562**	−0.0711	−0.0068	−0.2041	−0.0767	−0.1303
Maintenance	−0.0115	0.0048	**0.6810**	−0.0186	0.1634	−0.0374	0.0167
Design	−0.0492	−0.0116	−0.0714	**0.4865**	−0.0246	0.0455	−0.0387
Machine	0.0074	−0.0001	0.0251	−0.0029	**0.5014**	−0.1054	−0.0782
Leadership	0.0276	−0.0809	−0.0418	0.0076	−0.0787	**0.5900**	−0.0203
Manual	−0.0137	−0.1075	−0.0140	−0.0344	−0.0478	−0.0614	**0.7914**
(Intercept)	−0.0107	0.0750	0.0135	−0.0104	0.1124	0.1029	0.0724
Rsq	0.8131	0.8159	0.7777	0.5605	0.6132	0.6110	0.8244

2. However it is difficult to analyze specific substitutions due to possible correlations between the components. In this aspect NMF is less useful than FA where, by definition, factors start out uncorrelated and any subsequent cross terms can be directly interpreted as the degree of skill substitutions.

3. As was explained in MC, negative intercepts actually imply an *increase* in the importance of particular skills. Here we see that two skills: *Research* and *Design*, both of which are intellectually demanding and resistant to automation, experienced increases in importance.

4. Similarly, the category with the largest decrease in importance was *Perception*. This also corroborates the results in MC, where the Perception component was the most negatively impacted over the same time period.

4   Conclusions and Future Plans

Broadly, the results presented here matched those in MC, which supports the validity and potential usefulness of NMF as a viable alternative to FA for studying the underlying skill dimensions within the O*NET database. However, both methods have their respective strengths and weaknesses. The key findings of this study are that:

1. While not a "drop-in" replacement, NMF provides a valuable addition to the analytical toolkit for studying changes in occupational skill compositions.

2. NMF seems to extract more specific factors which better reflect work activities, while FA tended to extract higher level factors consisting mainly of abilities and skills.

3. However, when performing regression analysis, for e.g. to study intensive changes in skills requirements, NMF factors produced results that were less clear. This could be because of residual correlations within the factors while, on the other hand, FA by definition produces factors which are orthogonal.

Acknowledgement. The authors would like to express their gratitude to the Masdar Institute of Science and Technology for supporting this research.

References

1. Autor, D.H., Dorn, D.: How technology wrecks the middle class. The New York Times, 24 August 2013
2. MacCrory, F., Westerman, G., AlHammadi, Y., Brynjolfsson, E.: Racing with and against the machine: changes in occupational skill composition in an era of rapid technological advance. In: Proceedings of the International Conference on Information Systems - Building a Better World through Information Systems, ICIS 2014, Auckland, New Zealand, 14–17 December 2014
3. Baumol, W.J., Bowen, W.G.: Performing Arts-the Economic Dilemma: A Study of Problems Common to Theatre, Opera, Music and Dance. MIT Press, Cambridge (1966)
4. Acemoglu, D., Autor, D.: Skills, tasks and technologies: implications for employment and earnings. Handb. Labor Econ. **4**, 1043–1171 (2011)
5. Autor, D., Levy, F., Murnane, R.: The skill content of recent technological change: an empirical exploration. Q. J. Econ. **118**(4), 1279–1333 (2003)
6. Lee, D.D., Seung, H.S.: Algorithms for non-negative matrix factorization. In: Advances in Neural Information Processing Systems, pp. 556–562 (2001)
7. Xu, W., Liu, X., Gong, Y.: Document clustering based on non-negative matrix factorization. In: Proceedings of the 26th Annual International ACM SIGIR Conference on Research and Development in Informaion Retrieval, pp. 267–273. ACM (2003)
8. Liu, H., Wu, Z., Li, X., Cai, D., Huang, T.S.: Constrained nonnegative matrix factorization for image representation. IEEE Trans. Pattern Anal. Mach. Intell. **34**(7), 1299–1311 (2012)

Constrained Non-negative Matrix Factorization with Graph Laplacian

Pan Chen, Yangcheng He, Hongtao Lu$^{(\boxtimes)}$, and Li Wu

Key Laboratory of Shanghai Education Commission for Intelligent Interaction
and Cognitive Engineering, Department of Computer Science and Engineering,
Shanghai Jiao Tong University, Shanghai 200240, China
{bzchenpan,clustering,htlu,beautifulgirl}@sjtu.edu.cn

Abstract. Non-negative Matrix Factorization (NMF) is proven to be a
very effective decomposition method for dimensionality reduction in data
analysis, and has been widely applied in computer vision, pattern recognition and information retrieval. However, NMF is virtually an unsupervised method since it is unable to utilize prior knowledge about data. In
this paper, we present *Constrained Non-negative Matrix Factorization
with Graph Laplacian* (CNMF-GL), which not only employs the geometrical information, but also properly uses the label information to enhance
NMF. Specifically, we expect that a graph regularized term could preserve the local structure of original data, meanwhile data points both
having the same label and possessing different labels will have corresponding constraint conditions. As a result, the learned representations
will have more discriminating power. The experimental results on image
clustering manifest the effectiveness of our algorithm.

Keywords: Non-negative matrix factorization · Dimensionality reduction · Semi-supervised learning · Graph Laplacian · Clustering

1 Introduction

Matrix factorization techniques are more and more popular as tools for dimensionality reduction. Using different criteria, researchers have proposed a number
of different matrix factorization methods including the most popular techniques:
singular value decomposition (SVD) [8], vector quantization (VQ) [1], and principal component analysis (PCA) [7].

Different from other matrix factorization methods, Non-negative matrix factorization (NMF) [5] requires that all elements of the decomposed matrices are
non-negative. NMF has shown its superiority to other dimensionality reduction
methods in document clustering [11], image processing, and face recognition
[5,10]. However, NMF is virtually an unsupervised learning method which means
that NMF cannot make use of any prior knowledge about data, while limited
label information is usually available and can produce considerable improvement in learning performance [4,12]. Therefore, NMF has been extended to
semi-supervised NMF to get a better result [3,6,13].

© Springer International Publishing Switzerland 2015
S. Arik et al. (Eds.): ICONIP 2015, Part III, LNCS 9491, pp. 635–644, 2015.
DOI: 10.1007/978-3-319-26555-1_72

Liu et al. [6] utilized the label information to present a constrained NMF (CNMF) method. CNMF requires that data points having the same label should be mapped to the same representation in the new space, thus it can possess more discriminating power. However, the main drawback of CNMF is that it just considers the data points having the same label without utilizing the information about data points possessing different labels. Another disadvantage is that it does not consider the local geometric structure in the data.

In this paper, we design a new semi-supervised matrix factorization algorithm which is called *Constrained Non-negative Matrix Factorization with Graph Laplacian* (CNMF-GL). In CNMF-GL, we solve the main weakness of CNMF [6] by incorporating the local geometric structure and cannot-link pairwise constraints between data points possessing different labels. The label information used in CNMF can be regarded as *hard* constraints, while cannot-link pairwise constraints can be regarded as *soft* constraints. We utilize both *hard* and *soft* constraints to deal with different data points. The experimental results on image clustering show the effectiveness of our algorithm.

We arrange this paper as follows: Sect. 2 reviews the NMF method and relative semi-supervised NMFs. In Sect. 3, we introduce our method. The experiment results on five datasets will be shown and discussed in Sect. 4. Finally, we summarize this paper in the last section.

2   Related Works

2.1   A Brief Review of NMF

Given a data matrix $\mathbf{X} = [\mathbf{x}_1, \mathbf{x}_2, \ldots, \mathbf{x}_n] \in \mathbb{R}^{m \times n}$, $\mathbf{x}_i$ denotes a data point. NMF [5] factorizes $\mathbf{X}$ into the product of two non-negative matrices $\mathbf{U} = [u_{ij}] \in \mathbb{R}^{m \times k}$ and $\mathbf{V} = [v_{ij}] \in \mathbb{R}^{n \times k}$, and the product of these two matrices can well approximate the matrix $\mathbf{X}$:

$$\mathbf{X} \approx \mathbf{U}\mathbf{V}^T \tag{1}$$

In order to quantify the quality of the approximation, we can minimize the most commonly used cost function (the square of the Euclidean distance), i.e.,

$$O = \|\mathbf{X} - \mathbf{U}\mathbf{V}^T\|^2 \tag{2}$$

where $u_{ij} \geq 0$ and $v_{ij} \geq 0$. In practice, k usually be set much smaller than m and n. If treating each column vector $\mathbf{u}_i$ of matrix $\mathbf{U}$ as a basis, we can use a linear combination of these k bases to approximate each data point $\mathbf{x}_i$, and elements of row vector $\mathbf{v}_i$ of matrix $\mathbf{V}$ are the coefficients of this combination. In this way, each row vector $\mathbf{v}_i$ can be regard as the k-dimensional representation of the original m-dimensional data point $\mathbf{x}_i$. That is how NMF reduces the dimension of original data from m to k. After NMF decomposition, we can apply K-means to the rows of matrix $\mathbf{V}$ for clustering.

2.2   Related Works to Enhance NMF

Cai et al. [3] applied the graph Laplacian into NMF and proposed the GNMF algorithm, whose objective function is:

$$O = \|\mathbf{X} - \mathbf{U}\mathbf{V}^T\|^2 + \lambda \mathrm{Tr}(\mathbf{V}^T\mathbf{L}\mathbf{V}) \tag{3}$$

where the Laplacian matrix is defined as $\mathbf{L} = \mathbf{D} - \mathbf{W}$. $\mathbf{W}$ is the similarity matrix, whose entry w_{ij} denotes the similarity between point $\mathbf{x}_i$ and $\mathbf{x}_j$. If $\mathbf{x}_j$ (or $\mathbf{x}_i$) is one of the p nearest neighbors of $\mathbf{x}_i$ (or $\mathbf{x}_j$) $(i \neq j)$, $w_{ij} = \exp(-\|\mathbf{x}_i - \mathbf{x}_j\|^2/\sigma^2)$, otherwise, $w_{ij} = 0$. $\mathbf{D}$ is a diagonal matrix with the diagonal entries $d_{jj} = \sum_{i=1}^{n} w_{ij}$.

Yang et al. [13] proposed the PCNMF method, where the pairwise constraints are used to enhance NMF. To be more specific, in the low-dimensional space, points having the same label will be very close, and points possessing different labels will be far apart.

Liu et al. [6] designed the CNMF method, where they build an auxiliary matrix $\mathbf{A}$ (shown in the next section) as additional *hard* constraints by utilizing the known label information. The objective function is:

$$O = \|\mathbf{X} - \mathbf{U}\mathbf{Z}^T\mathbf{A}^T\|^2 \tag{4}$$

We will describe the auxiliary matrix and the CNMF in detail in Sect. 3.1. CNMF method just considers the data points having the same label without making use of the information about data points possessing different labels, meanwhile it does not consider the local geometric structure in the data. Next, we will propose our new CNMF-GL method to handle those problems.

3   Constrained Non-negative Matrix Factorization with Graph Laplacian

3.1   The Objective Function

Given a data set matrix $\mathbf{X} = [\mathbf{x}_1, \mathbf{x}_2, \ldots, \mathbf{x}_n] \in \mathbb{R}^{m \times n}$, suppose the first l data points $\mathbf{x}_i$ $(i \leq l)$ are labeled and the rest points $\mathbf{x}_r (l < r \leq n)$ are unlabeled, and the data set $\mathbf{X}$ will be classified into c clusters. Each of the first l labeled data points is labeled with one of the clusters. Based on this, we can construct an indicator matrix $\mathbf{Y}_{l \times c}$, if $\mathbf{x}_i$ is labeled with the j-th cluster, $y_{ij} = 1$, otherwise, $y_{ij} = 0$. Next, we build the auxiliary matrix $\mathbf{A}$ as follows [6]:

$$\mathbf{A} = \begin{pmatrix} \mathbf{Y}_{l \times c} & \mathbf{0} \\ \mathbf{0} & \mathbf{I}_{n-l} \end{pmatrix} \tag{5}$$

where $\mathbf{I}_{n-l}$ is an identity matrix. To illustrate matrix $\mathbf{A}$, suppose that there are n data points and only the first 5 data points are labeled. Specifically, $\mathbf{x}_1$, $\mathbf{x}_2$

are labeled with cluster I, $\mathbf{x}_3$, $\mathbf{x}_4$ are labeled with cluster II, $\mathbf{x}_5$ is labeled with cluster III. Based on this, we can define the matrix $\mathbf{A}$ below:

$$\mathbf{A} = \begin{pmatrix} 1 & 0 & 0 & \mathbf{0} \\ 1 & 0 & 0 & \mathbf{0} \\ 0 & 1 & 0 & \mathbf{0} \\ 0 & 1 & 0 & \mathbf{0} \\ 0 & 0 & 1 & \mathbf{0} \\ \mathbf{0} & \mathbf{0} & \mathbf{0} & \mathbf{I}_{n-5} \end{pmatrix} \tag{6}$$

As we know in NMF, each m-dimensional data point $\mathbf{x}_i$ is mapped to k-dimensional representation $\mathbf{v}_i$. In order to employ the label information, we can substitute matrix $\mathbf{V}$ with the product of matrix $\mathbf{A}$ and matrix $\mathbf{Z}$, where $\mathbf{Z}$ is a non-negative matrix:

$$\mathbf{V} = \mathbf{A}\mathbf{Z} \tag{7}$$

It is not difficult to check that if $\mathbf{x}_i$ have the same label with $\mathbf{x}_j$, then $\mathbf{v}_i$ must be equal to $\mathbf{v}_j$. Thus the matrix $\mathbf{A}$ imposes *hard* constraints on the representation matrix $\mathbf{V}$. Replacing matrix $\mathbf{V}$ in NMF by $\mathbf{A}\mathbf{Z}$, the goal of the algorithm is to find two non-negative matrix $\mathbf{U}$ and $\mathbf{Z}$ such that the original matrix $\mathbf{X}$ can be approximated as

$$\mathbf{X} \approx \mathbf{U}(\mathbf{A}\mathbf{Z})^T \tag{8}$$

The above is the central idea of CNMF designed by Liu *et al.* [6].

In order to further use the label information of the first l labeled data points, we can also build a cannot-link pairwise constraint [9] matrix $\mathbf{C} = [c_{ij}] \in \mathbb{R}^{n \times n}$ as follows:

$$c_{ij} = \begin{cases} 1 & \text{if } \mathbf{x}_i, \mathbf{x}_j (i \neq j) \text{ possess different cluster labels} \\ 0 & \text{otherwise} \end{cases}$$

By incorporating the Graph Laplacian term [2,3] and cannot-link pairwise constraints into NMF, the objective function of our approach reduces to the following:

$$O = \|\mathbf{X} - \mathbf{U}\mathbf{V}^T\|^2 + \alpha \text{Tr}(\mathbf{V}^T \mathbf{L} \mathbf{V}) + \beta \text{Tr}(\mathbf{V}^T \mathbf{C} \mathbf{V}) \tag{9}$$

where $\mathbf{L} = \mathbf{D} - \mathbf{W}$, $\text{Tr}(\mathbf{V}^T \mathbf{C} \mathbf{V}) = \sum_{i=1}^{n} \sum_{j:c_{ij}=1} \mathbf{v}_i \cdot \mathbf{v}_j$, α and β are two positive parameters, controlling the trade-off of these three terms.

The first two terms in Eq. (9) are easy to understand. Considering the limited space, we only explain how the third term works: If two points $\mathbf{x}_i$ and $\mathbf{x}_j$ have a cannot-link constraint ($c_{ij} = 1$), we expect that $\mathbf{v}_i$ and $\mathbf{v}_j$, which are the new representation of $\mathbf{x}_i$ and $\mathbf{x}_j$ respectively, are as dissimilar as possible. That is why we minimize the inner product of $\mathbf{v}_i$ and $\mathbf{v}_j$.

To utilize the *hard* constraints of CNMF, we substitute $\mathbf{V} = \mathbf{A}\mathbf{Z}$ into the Eq. (9), the final objective function of our proposed approach (CNMF-GL) can be written as:

$$O = \|\mathbf{X} - \mathbf{U}\mathbf{Z}^T \mathbf{A}^T\|^2 + \alpha \text{Tr}(\mathbf{Z}^T \mathbf{A}^T \mathbf{L} \mathbf{A} \mathbf{Z}) + \beta \text{Tr}(\mathbf{Z}^T \mathbf{A}^T \mathbf{C} \mathbf{A} \mathbf{Z}) \tag{10}$$

$$\text{s.t. } u_{ij} \geq 0 \text{ and } z_{ij} \geq 0$$

3.2   The Updating Algorithm

The objective function O in Eq. (10) is not convex in both $\mathbf{U}$ and $\mathbf{Z}$. As a result, it is impractical for us to find the global minimum of O. In the following, we recommend an iterative updating algorithm, which can acquire the local minimum of O.

We can rewritten the objective function O in Eq. (10) as follows when applying the matrix property $\mathrm{Tr}(\mathbf{AB}) = \mathrm{Tr}(\mathbf{BA})$ and $\mathrm{Tr}(\mathbf{A}) = \mathrm{Tr}(\mathbf{A}^T)$:

$$
\begin{aligned}
O &= \mathrm{Tr}((\mathbf{X} - \mathbf{UZ}^T\mathbf{A}^T)^T(\mathbf{X} - \mathbf{UZ}^T\mathbf{A}^T)) + \alpha\mathrm{Tr}(\mathbf{Z}^T\mathbf{A}^T\mathbf{LAZ}) \\
&\quad + \beta\mathrm{Tr}(\mathbf{Z}^T\mathbf{A}^T\mathbf{CAZ}) \\
&= \mathrm{Tr}(\mathbf{XX}^T) - 2\mathrm{Tr}(\mathbf{XAZU}^T) + \mathrm{Tr}(\mathbf{UZ}^T\mathbf{A}^T\mathbf{AZU}^T) \\
&\quad + \alpha\mathrm{Tr}(\mathbf{Z}^T\mathbf{A}^T\mathbf{LAZ}) + \beta\mathrm{Tr}(\mathbf{Z}^T\mathbf{A}^T\mathbf{CAZ})
\end{aligned}
\tag{11}
$$

Let ϕ_{ij} and ψ_{ij} be the Lagrange multipliers for constraints $u_{ij} \geq 0$ and $z_{ij} \geq 0$, respectively, and $\boldsymbol{\Phi} = [\phi_{ij}]$, $\boldsymbol{\Psi} = [\psi_{ij}]$. The Lagrange function $\mathcal{L}$ is

$$
\mathcal{L} = O + \mathrm{Tr}(\boldsymbol{\Phi}\mathbf{U}^T) + \mathrm{Tr}(\boldsymbol{\Psi}\mathbf{Z}^T)
\tag{12}
$$

Working out the partial derivatives of $\mathcal{L}$ with respect to $\mathbf{U}$ and $\mathbf{Z}$, and letting them vanish, we have

$$
\frac{\partial\mathcal{L}}{\partial\mathbf{U}} = -2\mathbf{XAZ} + 2\mathbf{UZ}^T\mathbf{A}^T\mathbf{AZ} + \boldsymbol{\Phi} = 0
\tag{13}
$$

$$
\frac{\partial\mathcal{L}}{\partial\mathbf{Z}} = -2\mathbf{A}^T\mathbf{X}^T\mathbf{U} + 2\mathbf{A}^T\mathbf{AZU}^T\mathbf{U} + 2\alpha\mathbf{A}^T\mathbf{LAZ} + 2\beta\mathbf{A}^T\mathbf{CAZ} + \boldsymbol{\Psi} = 0
\tag{14}
$$

Using the KKT conditions $\phi_{ij}u_{ij} = 0$ and $\psi_{ij}z_{ij} = 0$, as well as $\mathbf{L} = \mathbf{D} - \mathbf{W}$, we can obtain the following multiplicative updating rules for u_{ij} and z_{ij}:

$$
u_{ij} \leftarrow u_{ij}\frac{(\mathbf{XAZ})_{ij}}{(\mathbf{UZ}^T\mathbf{A}^T\mathbf{AZ})_{ij}}
\tag{15}
$$

$$
z_{ij} \leftarrow z_{ij}\frac{(\mathbf{A}^T\mathbf{X}^T\mathbf{U})_{ij} + \alpha(\mathbf{A}^T\mathbf{WAZ})_{ij}}{(\mathbf{A}^T\mathbf{AZU}^T\mathbf{U})_{ij} + \alpha(\mathbf{A}^T\mathbf{DAZ})_{ij} + \beta(\mathbf{A}^T\mathbf{CAZ})_{ij}}
\tag{16}
$$

4   Experimental Results

Our experiments are based on image clustering tasks in this section, and we will organize the experiments from four parts: evaluation metrics, data sets, performance evaluations and comparisons, and parameters selection.

4.1   Evaluation Metrics

We compare the learned label of each data point with the label provided by the dataset so as to evaluate the clustering result. We use two metrics to measure the performance. The first one is accuracy (AC), which can measure the percentage of correct labels learned. The other one is the normalized mutual information (NMI), where we use mutual information to measure the similarity of two clusters. For the detailed definitions of these two metrics, please see [6].

4.2 Data Sets

Our experiments are conducted on five datasets, including ORL face database, Yale face database, AR face database, Caltech101 database and Event8 database.

1. The ORL[1] database contains 400 images, which are equally divided into 40 distinct subjects, namely each subject has 10 different images.
2. The Yale[2] database has 15 individuals, and each of them has 11 images, hence there are 165 grayscale images in total.
   Each image of the above two databases is 32×32 pixels.
3. The AR[3] database consists of over 4,000 facial images of 126 individuals. We select a subset which contains 50 male subjects and 50 female subjects, and the total number of images is 1,399.
4. The Caltech101[4] database consists of a total 9,146 images with 101 different object categories. We choose the 10 largest categories except the BACK-GROUNDGOOGLE category in our experiment with totally 3,044 images.
5. The fifth database is Event8[5] which includes 8 sports event categories with 1,579 images in total.

We have pre-processed every dataset and summarize their statistics in Table 1.

Table 1. Summary of the five datasets

Dataset	Size	Dimensionality	Clusters number
ORL	400	1024	40
Yale	165	1024	15
AR	1399	2580	100
Caltech101	3044	500	10
Event8	1579	1500	8

4.3 Performance Evaluations and Comparisons

We compare our proposed CNMF-GL method with the following four algorithms:

1. NMF based clustering [11].
2. Graph regularized NMF (GNMF) which integrates the graph Laplacian into NMF [3].
3. PCNMF which employs pairwise constraints information into NMF [13].
4. CNMF which uses label information as hard constraints [6].

---

[1] http://www.cad.zju.edu.cn/home/dengcai/Data/FaceData.html.
[2] http://www.cad.zju.edu.cn/home/dengcai/Data/FaceData.html.
[3] http://www-prima.inrialpes.fr/FGnet/data/05-ARFace/tarfd_markup.html.
[4] http://en.wikipedia.org/wiki/Caltech_101.
[5] http://vision.stanford.edu/lijiali/event_dataset/.

In each independent experiment, we set the number of categories used for clustering on ORL, Yale, AR, Caltech101 and Event8 to 20, 10, 20, 10, and 8 respectively, and set the number of labeled images of each category to 3, 3, 5, 10, and 10 respectively. Ten independent experiments are conducted on each dataset, and the selected categories or labeled images are different each time. Generally, the value of k has many choices, and we simply set k to the number of clusters in our experiments. We apply K-means method to the rows of matrix $\mathbf{V}$ for clustering. Our final results are the average AC and NMI as well as corresponding standard deviation over the ten experiments of each algorithm on each dataset. We appropriately choose the parameters for each algorithm so as to obtain its best performance. For GNMF and CNMF-GL, the number of the nearest neighbors p is fixed to 3 and the variance σ^2 is set to 1 on each dataset. The parameters used in CNMF-GL on each dataset are list in Table 2.

Table 2. Parameters used in CNMF-GL on each dataset.

	ORL	Yale	AR	Caltech101	Event8
α	0.2	0.2	0.4	1.2	35
β	7	6	7	9	7

Table 3. Clustering AC of each method on each dataset

Methods	Accuracy (AC) (%)				
	ORL	Yale	AR	Caltech101	Event8
NMF	56.4 ± 5.5	46.5 ± 3.4	49.6 ± 6.4	41.0 ± 0.1	38.6 ± 0.0
GNMF	68.5 ± 4.1	49.8 ± 3.3	48.1 ± 6.7	42.1 ± 0.7	41.7 ± 2.3
PCNMF	66.4 ± 3.8	50.6 ± 4.2	58.3 ± 7.1	39.8 ± 4.4	38.0 ± 3.8
CNMF	70.7 ± 4.6	53.4 ± 4.3	61.2 ± 4.9	42.8 ± 1.3	39.7 ± 2.9
CNMF-GL	$\mathbf{82.2 \pm 5.3}$	$\mathbf{64.5 \pm 4.8}$	$\mathbf{74.0 \pm 5.6}$	$\mathbf{46.6 \pm 2.7}$	$\mathbf{43.6 \pm 2.3}$

Table 4. Clustering NMI of each method on each dataset

Methods	Normalized Mutual Information (NMI) (%)				
	ORL	Yale	AR	Caltech101	Event8
NMF	69.4 ± 4.2	45.4 ± 2.8	63.1 ± 6.6	32.6 ± 0.0	23.7 ± 0.0
GNMF	78.1 ± 3.0	47.0 ± 3.7	60.3 ± 7.0	35.6 ± 0.3	25.0 ± 1.5
PCNMF	77.6 ± 3.1	47.8 ± 3.5	70.6 ± 5.5	31.1 ± 3.6	22.0 ± 2.9
CNMF	80.3 ± 2.8	53.3 ± 4.1	71.6 ± 3.8	36.5 ± 0.9	24.4 ± 1.7
CNMF-GL	$\mathbf{84.8 \pm 3.8}$	$\mathbf{58.1 \pm 4.3}$	$\mathbf{78.8 \pm 4.3}$	$\mathbf{37.4 \pm 0.9}$	$\mathbf{26.3 \pm 1.8}$

Tables 3 and 4 show the detailed clustering results of each algorithm on the five datasets. From the results, we can find that the clustering performance of each algorithm is related to specific dataset. For example, except our CNMF-GL, GNMF method gets the best result on the Event8 dataset, while it is the worst on the AR dataset. Besides, we can see that CNMF method performs outstandingly over other three methods, which manifests its effectiveness in clustering. However, in all datasets, our CNMF-GL is significantly better than the others including CNMF. Specifically, on ORL, CNMF-GL achieves 11.5 % improvement in AC and 4.5 % improvement in NMI over the second best method. On Yale, CNMF-GL gets 11.1 % improvement in AC and 4.8 % improvement in NMI over CNMF. On AR, CNMF-GL improves 12.8 % in AC and 7.2 % in NMI over CNMF. On the last two datasets, CNMF-GL also performs best and obtains a certain amount of improvement both in accuracy and NMI over other algorithms. Therefore, experiment results demonstrate that it is effective in clustering to utilize label information and the local structure of data.

4.4 Parameters Selection

There are two primary parameters in our CNMF-GL method: positive α and β. Next, we will illustrate how the two parameters affect the performance.

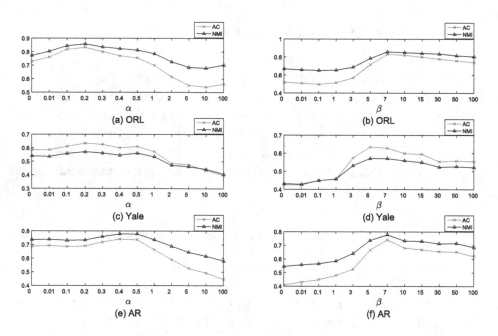

Fig. 1. The performance of CNMF-GL varies with α and β respectively

In this part, we conduct another set of experiments on each dataset with one parameter varying while another fixed. Due to the limited space, experiments on only three datasets (ORL, Yale, AR) are shown in Fig. 1. From Fig. 1(a), (c) and (e), we can see that when α is near to 0.2 or 0.5 with β fixed, the performance gets its maximum. Similarly, as described in Fig. 1(b), (d) and (f), the performance will be maximized when β is close to 6 or 7, meanwhile α is fixed. From the above analysis, we know that both the local structure and the cannot-link pair constraints are helpful to achieve a better performance.

5 Conclusions

A number of semi-supervised NMF methods have been proposed to improve the performance of NMF recently. However, some well-known algorithms can not take full advantage of prior label information. In this paper, we put forward CNMF-GL algorithm which not only utilizes the advantage of CNMF [6], but also appropriately employs cannot-link pairwise constraints and the local structure to strengthen the discriminating power of data. We compare our method with other four NMFs on five image datasets. The experimental results manifest the practicability of our approach.

Acknowledgments. This work is supported by NSFC (No. 61272247,61472075, 61533012), the Science and Technology Commission of Shanghai Municipality (No. 13511500200, 15JC1400103), 863 (No. SS2015AA020501, No. 2008AA02Z310) in China and Arts and Science Cross Special Fund of SJTU under Grant 13JCY14.

References

1. Gersho, A., Gray, R.M.: Vector Quantization and Signal Compression. Kluwer Academic Press, Boston (1992)
2. Belkin, M., Niyogi, P., Sinndhwani, V.: Manifold regularization: a geometric framework for learning from labeled and unlabeled examples. J. Mach. Learn. Res. **7**(11), 2399–2434 (2006)
3. Cai, D., He, X., Han, J., et al.: Graph regularized nonnegative matrix factorization for data representation. IEEE Trans. Pattern Anal. Mach. Intell. **33**(8), 1548–1560 (2011)
4. Chapelle, O., Scholkopf, B., Zien, A., et al.: Semi-supervised Learning, vol. 2. MIT Press, Cambridge (2006)
5. Lee, D., Seung, H., et al.: Learning the parts of objects by non-negative matrix factorization. Nature **401**(6755), 788–791 (1999)
6. Liu, H., Wu, Z., Li, X., Cai, D.: Constrained nonnegative matrix factorization for image representation. IEEE Trans. Pattern Anal. Mach. Intell. **34**(7), 1299–1311 (2012)
7. Jolliffe, I.: Principal Component Analysis. Springer, New York (1986)
8. Duda, R.O., Hart, P.E., Stork, D.G.: Pattern Classification. WileyInterscience, New York (2000)
9. Gao, S., Chen, Z., Zhang, D.: Learning mid-perpendicular hyperplane similarity from cannot-link constraints. Neurocomputing **113**(3), 195–203 (2013)

10. Li, S., Hou, X., Zhang, H., Cheng, Q.: Learning spatially localized, parts-based representation. In: Proceedings of the IEEE International Conference on Computer Vision and Pattern Recognition, pp. 207–212 (2001)
11. Xu, W., Liu, X., Gong, Y.: Document clustering based on non-negative matrix factorization. In: Proceedings of the Annual ACM SIGIR Conference on Research and Development in Information Retrieval (2003)
12. Zhu, X., Ghahramani, Z., Lafferty, J.: Semi-supervised learning using gaussian fields and harmonic functions. In: Proceedings of the 20th International Conference on Machine Learning (2003)
13. Yang, Y., Hu, B.: Pairwise constraints-guided non-negative matrix factorization for document clustering. In: IEEE/WIC/ACM International Conference on Web Intelligence, pp. 250–256. IEEE (2007)

Winner Determination in Multi-attribute Combinatorial Reverse Auctions

Shubhashis Kumar Shil, Malek Mouhoub[✉], and Samira Sadaoui

Department of Computer Science, University of Regina, Regina, SK, Canada
{shil200s,mouhoubm,sadaouis}@uregina.ca

Abstract. Winner(s) determination in online reverse auctions is a very appealing e-commerce application. This is a combinatorial optimization problem where the goal is to find an optimal solution meeting a set of requirements and minimizing a given procurement cost. This problem is hard to tackle especially when multiple attributes of instances of items are considered together with additional constraints, such as seller's stocks and discount rate. The challenge here is to determine the optimal solution in a reasonable computation time. Solving this problem with a systematic method will guarantee the optimality of the returned solution but comes with an exponential time cost. On the other hand, approximation techniques such as evolutionary algorithms are faster but trade the quality of the solution returned for the running time. In this paper, we conduct a comparative study of several exact and evolutionary techniques that have been proposed to solve various instances of the combinatorial reverse auction problem. In particular, we show that a recent method based on genetic algorithms outperforms some other methods in terms of time efficiency while returning a near to optimal solution in most of the cases.

Keywords: Combinatorial reverse auctions · Genetic algorithms · Winner determination · E-commerce

1 Introduction

The ultimate purpose of combinatorial auctions considering multiple items is to increase the efficiency of bid allocation, the latter corresponds to minimizing the procurement cost in a reasonable computation time when selecting the winner [3–5, 8, 14, 23]. Winner determination is an NP-complete problem [14, 23] and can be even more challenging if we consider additional constraints, such as seller's stocks and discount rate. Solving this problem with a systematic method will guarantee the optimal solution but comes with an exponential time cost. On the other hand, approximation techniques such as evolutionary algorithms are faster but trade the quality of the solution returned for the running time. In this regard, several exact and approximation methods have been proposed in the past to tackle this hard to solve problem. In [11, 16] combinatorial auctions have been applied to procurement scenarios such as travel packages and transportation. Hsieh and Tsai developed a Langrangian heuristic method to tackle combinatorial auctions [14]. Ant Colony algorithms have been defined by Sitarz for the same purpose [14].

© Springer International Publishing Switzerland 2015
S. Arik et al. (Eds.): ICONIP 2015, Part III, LNCS 9491, pp. 645–652, 2015.
DOI: 10.1007/978-3-319-26555-1_73

Combinatorial Reverse Auctions (CRAs) are a particular case of combinatorial auctions that can be of three types based on the number of attributes, items and instances: (1) single attribute multiple items with single instance per item, (2) single attribute multiple items with multiple instances per item and (3) multi-attribute multiple items with multiple instances per item. Unlike combinatorial auctions, fewer research works were dedicated to CRAs. These contributions can be summarized as follows.

In [20], we tackle the first type of CRAs by proposing a GA based method called Genetic Algorithms for Combinatorial Reverse Auctions (GACRA) to solve the winner determination problem. GACRA is based on Genetic Algorithms (GAs) and includes two repairing methods respectively called RemoveRedundancy and RemoveEmptiness as well as a modified two-point crossover operator. We conduct several experiments and based on the results we demonstrate that GACRA is better in terms of procurement cost and processing time comparing to another GA-based winner determination method [14]. Following the work reported in [20], we conduct several statistical experiments demonstrating that GACRA is a consistent method [18].

We address the second type of CRAs in [17] with a new GA-based method called Genetic Algorithms for Multiple Instances of Items in Combinatorial Reverse Auctions (GAMICRA). GAMICRA extends GACRA by improving the RemoveRedundancy and RemoveEmptiness methods. We update the chromosome representation as well as the definition of the fitness function to deal with multiple instances of items. Based on the results of several experiments on many CRAs instances, we show that GAMICRA is a consistent method which is capable of determining the winner in a very efficient computation time.

In [19], we address the third type of CRAs by considering two attributes, namely price and delivery rate, along with multiple instances of items. We also consider all-units discount strategy on both price and delivery rate which, undoubtedly, makes the problem even more challenging to solve as shown in [22]. In this latter paper, we treat the problem as multi-sourcing where supplier selection is an important task in a multi-criteria decision problem [6, 22]. Here, the buyer can order different items from different sellers. He can also purchase instances of the same item from different sellers. Moreover, we consider various situations which are related to the sellers' stock, such as the number of available instances of items provided by a given seller is (a) greater than or (b) less than the buyer's requirement or (c) the seller is out of that item. Also, the maximum price constraints of the buyer as well as the minimum price constraint of the sellers for each item instance were taken into consideration. To tackle these additional features, we define the chromosome representation based on the number of items and item instances. We also define the fitness function and keep it simple enough in order to maintain a reasonable processing time. Here, we propose a GA-based method to solve the winner determination problem. In order to evaluate the time performance of this method to return the best procurement cost, we conduct several experiments on randomly generated instances after tuning the parameters to their best. The results of these experiments clearly show that the proposed method is efficient and consistent.

The third type of CRAs is computationally complex. For instance, if there are j items, I instances (where $I = \sum_{J=1}^{j} i_J$ and i_J is the number of instances of item J) and k sellers, then the search space is k^I (if j = 10, I = 50 and k = 100, the number of potential solution space is 100^{50}). Moreover, to determine the winner, the solving procedure needs to satisfy other bidding, buyer and sellers' constraints.

While we address this latter problem with the performance results reported in [19], comparison with the existing methods has not been conducted. Moreover there is no evidence that the solutions returned in the experiments are the optimal ones.

In this paper, our goal is to address the above two issues. At first, we perform comparative experiments with a recent evolutionary technique proposed in [15] for solving a similar problem. This new method is an improved ant colony algorithm called Improved Ant Colony (IAC). IAC considers the Max-Min pheromone and dynamic transition strategy. The problem tackled by IAC is, however, different from the one addressed by GAMICRA as discount strategy, multiple instances of items and sellers' stock are not considered. In [15], IAC has been compared to two other methods: Enumeration Algorithm with Backtracking (EAB) and the traditional Ant Colony (AC) algorithm. It has been shown that IAC outperforms those two methods. We show that GAMICRA is superior in running time than AC, EAB and IAC. Besides evolutionary algorithms, we compare GAMICRA with a branch and bound technique proposed in [7] for solving multi-unit combinatorial auctions. Here again, based on the results returned we show that GAMICRA outperforms the branch and bound technique. To tackle the second issue, we implement an exact algorithm and use it to evaluate the optimality of the returned solutions. From the result of the comparative experiments (for both the exact method and GAMICRA), we demonstrate that GAMICRA is able to produce very close to optimal solutions.

The rest of this paper is organized as follows. In Sect. 2, we state the problem in details. We also present the problem formulation for GAMICRA and describe our algorithm. Section 3 reports the experimental study we conducted to address the above two issues. Finally in Sect. 4, concluding remarks and future research directions are depicted.

2 GAMICRA for Winner Determination

In a combinatorial reverse auction, there is one buyer and several sellers. The buyer specifies his requirements and the sellers compete to win. In [19], GAMICRA considers multiple attributes (price and delivery rate), multiple instances, multiple items and all-units discounts. It also considers the buyer's constraints such as maximum price of each instance of item and terminating condition. Bidders' constraints (minimum price for each instance of item, available stock and discount rates) are also taken into account. Given all these data, the goal is to determine the winner(s) in a reasonable computation time. Based on [19], this optimization problem can be formulated as shown in Table 1(a).

In order to check the quality of the solution returned by GAMICRA we implemented an Exact Algorithm (EA) that we present in Table 1(b) together with the pseudo code of GAMICRA.

Table 1(a). Problem formulation for GAMICRA

Variables	
nb_sellers	Number of sellers
nb_items	Number of items
nb_instances$_j$	Number of instances requested by the buyer for item *j*
capacity_instances$_{jk}$	Number of instances of item *j*, seller *k* has
minPrice$_{jk}$:	The lowest price, the *kth* seller can offer for the *jth* item
maxPrice$_j$	The highest price, the buyer can pay for the *jth* item
Bid(X$_{ijk}$)	Bid price, the *kth* seller bids for the *ith* instance of *jth* item
max_rounds	The maximum number of rounds used as a terminating condition

Constraints

- X_{ijk} : 1 if the *ith* instance of the *jth* item of the *kth* seller is selected and 0 otherwise.

$$1 \leq i \leq capacity\_instances_{jk}$$
$$1 \leq j \leq nb\_items$$
$$1 \leq k \leq nb\_sellers$$

- $\sum_{k=1}^{nb_{sellers}} \sum_{i=1}^{capacity_{instances\,jk}} X_{ijk} = nb_{instances\,j}$

$$1 \leq j \leq nb\_items$$

- $minPrice_{jk} \leq Bid(X_{ijk}) \leq maxPrice_j$

Objective Function

$$min \sum_{k=1}^{nb\_sellers} \sum_{j=1}^{nb\_items} \sum_{i=1}^{capacity\_instances_j} Bid(X_{ijk})$$

3 Experimentation

EA has been implemented in Java and executed on an Intel (R) Core (TM) i3-2330 M CPU with 4 GB of RAM and 2.20 GHz of processor speed. The experiments are conducted on the same data used in [19]. For GAMICRA, all results returned are the average values of 20 runs.

3.1 Experiment 1: Comparison with IAC, AC and EAB

In this experiment, we compare the computation time of GAMICRA with IAC, AC and EAB algorithms [15]. Here, ζ and δ denote number of chromosomes and number of generations respectively.

Table 2 presents the comparative computation time of these methods. In this table, "-" indicates a non-existing value. The test results for IAC, AC and EAB are taken

Table 1(b). Pseudocode of EA and GAMICRA

EA

Begin:
    generate solution space;
    initialize feasible solutions by considering bidding, buyer and sellers'
    constraints
    evaluate feasible solutions;
    return the winner(s);
End;

GAMICRA

Begin:
    generation ← 0;
    generate bids;
    initialize chromosomes X(*generation*);
    evaluate X(*generation*);
    while (not maximum generation) do
    Begin:
        generation ← *generation* + 1;
        select X(*generation*) from X(*generation* -1) by Gambling Wheel
        Disk method [10];
        recombine X(*generation*) by modified two-point crossover and
        mutation;
        evaluate X(*generation*);
    End;
    return the winner(s);
End;

Table 2. The Comparison of GAMICRA, IAC, AC and EAB

Algorithm	ζ	δ	nb_sellers	nb_items	Computation time (second)
EAB	–	–	100	30	3
AC	50	50	100	30	4
IAC	50	50	100	30	9
GAMICRA	50	50	100	30	0.83

from [15]. As we can easily see from the table, GAMICRA outperforms all these evolutionary techniques.

3.2   Experiment 2: Comparison with Branch and Bound

In this experiment, we compare the computation time of GAMICRA with a method based on Branch and Bound [7]. The tests have been conducted on the same instances but with a different computer (450 MHz of processor speed for Branch and Bound vs

Table 3. The Comparison of GAMICRA and branch and bound

Algorithm	nb_sellers	nb_items	$\sum nb\_instances_j$	Computation time (second)
Branch and Bound based Method	100	10	25	>100
GAMICRA	100	10	25	1.7

2.20 GHz for GAMICRA). Despite the difference in processor speed, we see again here that GAMICRA is superior to the branch and bound technique (Table 3).

3.3   Experiment 3: Comparison with the Exact Algorithm

In order to assess the quality of the solution returned by GAMICRA, we compare it with the EA reported in Algorithm 2 above. Two types of tests are conducted and the results are respectively reported in Tables 4 and 5.

Table 4. Comparison of GAMICRA and EA when varying the number of sellers

Algorithm	Bid price				Computation time (millisecond)			
	Test 1	Test 2	Test 3	Test 4	Test 1	Test 2	Test 3	Test 4
	Number of sellers				Number of sellers			
	10	20	30	40	10	20	30	40
EA	1472	1538	1412	915	60	392	4906	42137
GAMICRA	1474 (99.86 %)	1555 (98.91 %)	1462 (96.58 %)	988 (92.61 %)	39	46	55	62

Table 5. Comparison of GAMICRA and EA when varying the total number items instances

Algorithm	Bid price			Computation time (millisecond)		
	Test 1	Test 2	Test 3	Test 1	Test 2	Test 3
	Number of total instances			Number of total instances		
	2	3	4	2	3	4
EA	621	861	915	90	270	42137
GAMICRA	621 (100 %)	865 (99.54 %)	988 (92.61 %)	59	61.5	62

In the first type of tests, we measure bid price, computation time and the quality of the solution returned by GAMICRA and EA when varying the number of sellers. In these tests we use the following parameters: number of items = 3, total number of instances = 4 (2 instances of item1, 1 instance of item2, and 1 instance of item3), and number of sellers = 10, 20, 30, and 40. For example, in Test 4 (Table 4), there are 40 sellers and 4 instances of items, hence there are 40^4 (2560000) potential solutions. To determine the winner, EA takes 42.137 s while GAMICRA takes only 0.062 s. On the other hand, EA returns the optimal bid price of 915 and GAMICRA returns 988 which is close to the optimal solution. Hence, the accuracy ((EA result/GAMICRA result) × 100 %) of GAMICRA is 92.61 %. In the second type of tests, we vary the total number of instances of items from 2 to 4, and fix the number of sellers to 40. From the results of this second experiment reported in Table 5, we can easily see that, in most of the time, GAMICRA returns near to optimal solution. It also proves that GAMICRA is not trapped in local optima.

4 Conclusion and Future Work

In this paper, we compare a very recent GA-based solution for CRAs with exact and evolutionary methods. Based on the results of the experimental comparative study, we conclude that GAMICRA outperforms all these methods. Moreover, to assess the quality of the solution returned by GAMICRA we compare this latter with an exact method that we have implemented. The results of this comparative experiment clearly show that GAMICRA always produce near to optimal solutions and in some cases the optimum is reached.

Since parallel GAs [10, 12] are capable of providing the solutions in a better computation time [1, 2]; we will design, in the near future, a parallel version of GAMICRA. Another promising direction is to consider more attributes such as 'delivery time', 'seller reputation', and 'warranty' as reported in [21]. We will also investigate the applicability of other meta-heuristics such as Simulated Annealing, Hill Climbing, and Late Acceptance together with GAs [13]. Finally we will study and apply several diversity models such as Crowding, Restricted Mating, and Ranked Space with GAs [9].

References

1. Abbasian, R., Mouhoub, M.: An efficient hierarchical parallel genetic algorithm for graph coloring problem. In: 13th Annual GECCO, pp. 521–528 (2011)
2. Abbasian, R., Mouhoub, M.: A hierarchical parallel genetic approach for the graph coloring problem. Appl. Intell. **39**(3), 510–528 (2013). Springer
3. Avasarala, V., Mullen, T., Hall, D.L., Garga, A.: MASM: market architecture or sensor management in distributed sensor networks. In: SPIE Defense and Security Symposium, pp. 5813–5830 (2005)
4. Avasarala, V., Polavarapu, H., Mullen, T.: An approximate algorithm for resource allocation using combinatorial auctions. In: International Conference on Intelligent Agent Technology, pp. 571–578 (2006)
5. Das, A., Grosu, D.: A combinatorial auction-based protocols for resource allocation in grids. In: 19th IEEE International Parallel and Distributed Processing Symposium (2005)
6. Ebrahim, R.M., Razmi, J., Haleh, H.: Scatter search algorithm for supplier selection and order lot sizing under multiple price discount environment. Adv. Eng. Softw. **40**(9), 766–776 (2009)

7. Gonen, R., Lehmann, D.: Optimal solutions for multi-unit combinatorial auctions: branch and bound heuristics. In: 2nd ACM Conference on Electronic Commerce, pp. 13–20 (2000)

8. Gong, J., Qi, J., Xiong, G., Chen, H., Huang, W.: A GA based combinatorial auction algorithm for multi-robot cooperative hunting. In: International Conference on Computational Intelligence and Security, pp. 137–141 (2007)

9. Gupta, D., Ghafir, S.: An overview of methods maintaining diversity in genetic algorithms. Int. J. Emerg. Technol. Adv. Eng. **2**(5), 56–60 (2012)

10. Muhlenbein, H.: Evolution in time and space-the parallel genetic algorithm. In: Foundations of Genetic Algorithms, pp. 316–337 (1991)

11. Narahari, Y., Dayama, P.: Combinatorial auctions for electronic business. Sadhana **30**(Pt. 2 & 3), 179–211 (2005)

12. Nowostawski, M., Poli, R.: Parallel genetic algorithm taxonomy. In: 3rd International Conference on Knowledge-Based Intelligent Information Engineering Systems, pp. 88–92 (1999)

13. Ostler, J., Wilke, P.: Improvement by combination how to increase the performance of optimisation algorithms by combining them. In: 10th International Conference of the Practice and Theory of Automated Timetabling, pp. 359–365 (2014)

14. Patodi, P., Ray, A.K., Jenamani, M.: GA based winner determination in combinatorial reverse auction. In: 2nd International Conference on Emerging Applications of Information Technology (EAIT), pp. 361–364 (2011)

15. Qian, X., Huang, M., Gao, T., Wang, X.: An improved ant colony algorithm for winner determination in multi-attribute combinatorial reverse auction. In: IEEE Congress on Evolutionary Computation (CEC), pp. 1917–1921 (2014)

16. Rassenti, S.J., Smith, V.L., Bulfin, R.L.: A combinatorial auction mechanism for airport time slot allocation. Bell J. Econ. **13**, 402–417 (1982)

17. Shil, S.K., Mouhoub, M.: Considering multiple instances of items in combinatorial reverse auctions. In: Ali, M., Pan, J.-S., Chen, S.-M., Horng, M.-F. (eds.) IEA/AIE 2014, Part II. LNCS, vol. 8482, pp. 487–496. Springer, Heidelberg (2014)

18. Shil, S.K., Mouhoub, M., Sadaoui, S.: An approach to solve winner determination in combinatorial reverse auctions using genetic algorithms. In: 15th Annual GECCO, pp. 75–76 (2013)

19. Shil, S.K., Mouhoub, M., Sadaoui, S.: Evolutionary technique for combinatorial reverse auctions. In: 28th FLAIRS, pp. 79–84 (2015)

20. Shil, S.K., Mouhoub, M., Sadaoui, S.: Winner determination in combinatorial reverse auctions. In: Ali, M., Bosse, T., Hindriks, K.V., Hoogendoorn, M., Jonker, C.M., Treur, J. (eds.) Contemporary Challenges and Solutions in Applied AI. SCI, vol. 489, pp. 35–40. Springer, Heidelberg (2013)

21. Sadaoui, S., Shil, S.K.: Constraint and qualitative preference specification in multi-attribute reverse auctions. In: Ali, M., Pan, J.-S., Chen, S.-M., Horng, M.-F. (eds.) IEA/AIE 2014, Part II, LNCS(LNAI), vol. 8482, pp. 497–506. Springer, Heidelberg (2014)

22. Xia, W., Wu, Z.: Supplier selection with multiple criteria in volume discount environments. Omega **35**(5), 494–504 (2007)

23. Zhang, L.: The winner determination approach of combinatorial auctions based on double layer orthogonal multi-agent genetic algorithm. In: 2nd IEEE Conference on Industrial Electronics and Applications, pp. 2382–2386 (2007)

Real-Time Simulation of Aero-optical Distortions Due to Air Density Fluctuations at Supersonic Speed

Najini Harischandra[✉], Nihal Kodikara, K.D. Sandaruwan,
G.K.A. Dias, and Maheshya Weerasinghe

University of Colombo School of Computing, Colombo 07, Sri Lanka
{nsh,ndk,dsr,gkad}@ucsc.cmb.ac.lk, au.maheshya@gmail.com

Abstract. Implementations of visual simulations of shock phenomenon have been given significantly less-attention in last decades. We present a novel approach to simulate aero-optical distortions due to shock waves generated by a supersonic jet by considering the physics background of the shock phenomenon. The optical distortion is simulated by calculating the index of refraction for oblique shock waves. The refractive index for the shock wave was calculated, by considering the mean characteristics of supersonic flows. Even though the flow characteristics are not uniform across the shock wave the results shows that this approach is a better way to simulate aero-optical distortions in real time.

Keywords: Simulation · Shock waves · Supersonic · Distortion · Refraction

1 Introduction

In fluid dynamics the study of interaction between light and air is called as aero-optics [1]. In high-speed flow, optical aberrations induced by fluctuations of refractive index in turbulent flows are a serious concern in airborne communication and imaging systems. Such distortions can cause severe problems such as error in target location, beam jitter, image blur and loss of intensity in the far field [2].

The air in supersonic shock waves is optically denser than normal air. It produces a deviation of light rays that can optically displace objects from their true position. This phenomenon varies with speeds between Mach 1 and Mach 5; it is entirely absent below Mach 1 [3]. Figure 1 shows shock waves generated by a space shuttle vehicle.

This paper presents a method for efficient real time simulations of refraction occurs due to the shock waves generated by a supersonic jet considering uniform density, pressure and temperature distributions throughout the downstream of the shock waves. As this study concerns a jet which is moving at supersonic speed in air, to get the density variations in shock waves, equations for compressible flows have been used. In this approach we use several equations which describe the change in flow variables for flow across an oblique shock waves by National Aeronautics and Space Administration (NASA). Then we propose an approach to calculate refractive index across the shock wave according to the altitude and the speed of the jet.

© Springer International Publishing Switzerland 2015
S. Arik et al. (Eds.): ICONIP 2015, Part III, LNCS 9491, pp. 653–662, 2015.
DOI: 10.1007/978-3-319-26555-1_74

This proposed method is able to considerably reduce the computational complexity of these highly complex effects to achieve a real time simulation. Results shows that the accuracy of the simulation is high in the middle portion of the shock wave.

2   Related Work

Shock waves and aero-optical distortions have been a popular research topic for a half a decade. Most of these researches had been done focusing subsonic and transonic flows. Some have tried to measure the level of optical distortions using popular computation techniques [1, 4].

Direct observation of shock waves produced on the space shuttle vehicle had been done during Space Transportation System ascent imagery provided by NASA tracking cameras that are used to track the space shuttle vehicle [5]. The amount of refraction had been calculated over the position of one edge of the space shuttle main engine (SSME) nozzle. No. 1 nozzle has shifted by nearly 20 % of the actual diameter or about 20 in [5].

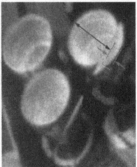

Fig. 1. Shock structure at 59.72 s mission elapsed time (left) [5], No. 1 nozzle refracted through shock boundary layer: bending due to refraction (red arrow) [5] (Color figure online)

Aero-optical distortions in turbulent flows has to be considered in velocimetry techniques such as PIV and PTV [4, 7]. Elsinga, Oudheusden, and Scarano had investigated aero-optical distortions effects on velocimetry techniques [4]. There are some presented efficient methods for visual simulations of shock phenomena in compressible, in viscid fluids [8, 9]. But these simulations had not addressed optical distortions in shock waves. In 2008, Sewall and et al. has presented a way to simulate shock waves. They have implemented a visual simulation of shock waves using ray tracing techniques which shows good performances. They have mainly focused on simulating explosions and thus optical distortions happening due to shock waves generated by a supersonic object has not been considered [8].

3   Methodology

The air in shock waves is compressed and these waves have different air densities. For air and many other fluids, the index of refraction is linearly related to the density of the fluid by the Gladstone-Dale relation [10]. Therefore the main reason for the optical distortions in supersonic speed is density variations in the flow field. Shock waves are very small regions in the gas where the gas properties change by a large amount. The static gas density, pressure and temperature increases almost instantaneously, across a shock wave [11]. There are two basic types of shock waves produced by a supersonic object [12]. Our main concern is about attached oblique shock waves as that is the type that typically occurs around supersonic airplanes and rockets. Therefore this research is focused only on simulating refraction occurred due to attached oblique shock waves.

For the mathematical convenience, the proposed solution assumes that the cockpit of the supersonic jet is rigid and impossible to deform and the shape of the nose cone is identical to a cone. Popular nose cone design for supersonic jet is called ¾ parabolic [13].

3.1   Calculate Properties of Shock Waves

Here, the oblique shock occurred around a finite-sized wedge whose half-angle (θ) is considered and the wedge is moving in M1 speed (Fig. 2). The wave angle of the shock is β. The starting point of the mathematical formulation of oblique shock wave is calculating shock wave angle. For a given Mach number M1 and wedge angle θ, the oblique shock angle β can be calculated using the θ-β-M equation [14].

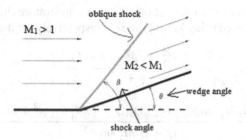

Fig. 2. Attached oblique shock wave

Given M_1 there is a maximum deflection angle (β_{max}) that can occur and after that only detached shocks are possible. When β is lower than β_{max} both strong and weak shocks are possible. The higher β value for the strong shock and lower β value for the weak shock. The downstream Mach number M2 can be calculated using (1). Though M2 is always less than M1, M2 can still be supersonic (weak shock wave) or subsonic (strong shock wave) [14].

$$M_2^2 \sin(\beta - \theta)^2 = \frac{(\gamma - 1) M_1^2 \sin \beta^2 + 2}{2\gamma M_1^2 \sin \beta^2 - (\gamma - 1)} \tag{1}$$

The shock angle can be found by solving θ-β-M equation using numerical methods. Solving an equation using numerical methods is of course very unhandy for automated lookup. Therefore instead of the θ-β-M equation, we used the equations set proposed by J. Anderson [15]. The positive roots correspond to the weak and strong shock solutions is found by substituting 1 and 0 as α respectively. This equation gives the same shock wave solutions which gives by the θ-β-M equation with better efficiency.

$$\tan \beta = \frac{M_1^2 - 1 + 2\lambda \cos\left(\frac{4\pi a + \cos^{-1} x}{3}\right)}{3\left(1 + \frac{\gamma-1}{2}M_1^2\right)\tan\theta}$$

$$\lambda = \sqrt{(M_1^2 - 1)^2 - 3\left(1 + \frac{\gamma-1}{2}M_1^2\right)\left(1 + \frac{\gamma+1}{2}M_1^2\right)\tan\theta^2} \tag{2}$$

$$X = \frac{1}{\lambda^3}\left[(M_1^2 - 1)^3 - 9\left(1 + \frac{\gamma-1}{2}M_1^2\right)\left(1 + \frac{\gamma-1}{2}M_1^2 + \frac{\gamma+1}{4}M_1^4\right)\tan\theta^2\right]$$

For a given θ and M_1, both strong and weak shock angles were calculated. Then for the strong shock angle, upstream and downstream density difference ratio is calculated using Eq. (3). The solution of this equation is the mean density difference ratio as air density varies in the downstream of the shock wave.

$$\frac{\rho_2}{\rho_1} = \frac{(\gamma + 1)M_1^2 \sin\beta^2}{(\gamma - 1)M_1^2 \sin\beta^2 + 2} \tag{3}$$

Also mean pressure difference ratio and mean temperature difference ratio across shock wave are calculated using Eqs. (4) and (5) respectively for the strong shock wave.

$$\frac{p_2}{p_1} = \frac{2\gamma M_1^2 \sin\beta^2 - (\gamma - 1)}{\gamma + 1} \tag{4}$$

$$\frac{T_2}{T_1} = \frac{[2\gamma M_1^2 \sin\beta^2 - (\gamma - 1)][(\gamma - 1)M_1^2 \sin\beta^2 + 2]}{(\gamma + 1)^2 M_1^2 \sin\beta^2} \tag{5}$$

The right hand side of all these equations depend only on the free stream Mach number and the shock angle. Hence knowing the Mach number and the wedge angle, all the conditions associated with the oblique shock can be determined [14].

Generally supersonic jets fly in Troposphere. Hence density, pressure and temperature values at 20,000 feet altitude are considered as upstream parameters. Using these values and already calculated ratio values, downstream parameters (density, pressure and temperature) are obtained for both strong and weak shock waves.

Now, since all the required properties of the downstream shock wave have been obtained we can calculate the refractive index for the downstream shock wave.

3.2 Calculate Refractive Index

Once the density and composition are specified everywhere in the fluid, the refractivity $(n-1)$ can be calculated for a given wavelength of light. In a typical gas the index of refraction is related to density by Gladstone-Dale relation.

When the Mach number of the supersonic object increases, the temperature of the air in downstream shock wave increases in a high rate. Gladstone-Dale constant for air in that much temperatures have not been calculated. Therefore to calculate the refractive index of the shock wave Gladstone-Dale relation cannot be used. Edlen and Metrologia formula is being used to calculate refractive index of the downstream of the shock wave [16]. For air at a temperature t °C and a pressure p Pa, the refractivity is given by (7). Ns is the refractive index of air in for standard air.

$$n_{tp} - 1 = (n_s - 1) \times \frac{p[1 + p(60.1 - 0.972t) \times 10^{-10}]}{96095.43(1 + 0.003661t)} \tag{6}$$

4 Implementation

4.1 Calculate Shock Wave Properties

The refractive index calculation and the creation of the grid of the shock wave is done using MATLAB by solving the above equations. Since air properties such as density, pressure and temperature vary according to the altitude, for the validation process we consider a constant altitude. Also we consider the half angle of the jet's cockpit as a constant. At high altitudes, air properties are different from air properties at sea level. Using the standard atmosphere table (ISA), atmosphere parameters (density, pressure and temperature) can be calculated for any altitude in the troposphere. We select 20,000 ft as the considering altitude and all the atmosphere parameters are mentioned below. Then using the Eq. (6), refractive index at this altitude is calculated (Tables 1 and 2).

Table 1. Variance of air parameters

	Sea level	Altitude 20,000 ft
Density (kg/m^3)	1.225	0.65268
Pressure (hPa)	1013.25	465.588375
Temperature (K)	288.15	248.55
Index of refraction (n)	1.00027712	1.0001476130

We consider the half angle of the cockpit as 20° and calculated shock wave solutions using Eq. (2) for several Mach numbers are mentioned below. Then the refractive indexes were calculated using Eq. (6). Those tables show how refractive index of air in downstream of the shock wave changes according to the speed of the jet at 20,000 ft altitude.

Table 2. Strong and weak shock wave parameters

Mach number	Strong shock wave		Weak shock wave	
	Angle (β) (degrees)	Index of refraction (n)	Angle (β) (degrees)	Index of refraction (n)
1.5	Detached		Detached	
2.0	74.270137042	1.000376378	53.422940527	1.000301265
2.5	80.069820045	1.000482401	42.890173846	1.000324513
3.0	82.146671023	1.000556931	37.763634148	1.000356477
3.5	83.216279333	1.000607177	34.602152399	1.000390111
4.0	83.854189326	1.000635829	32.463896850	1.000423544
4.5	84.269379813	1.000644041	30.935723966	1.000455786

4.2 Visual Simulation

For a supersonic jet pilot the visible part of the shock wave is limited by the window of the cockpit. Hence, although shock waves occur around the whole cockpit of a supersonic jet (Fig. 1), what matters is only the part above the cockpit window. Therefore in this research effort refraction is modeled only for a part of the shock wave.

The visualization module is implemented using OpenGL (Open Graphics Library) version 4.0 with GLFW library and GLSL high level shading language. Cube mapping is used to render the refracted texture. In the fragment shader program, first refracted texture for strong shock wave is generated. Then the refracted texture for weak shock wave is generated. Finally the output is produced by mixing bot textures together. Figure 3 shows refraction occur at the boundary layers of the shock waves for two different Mach numbers. Only the shock wave portion occurred around the window of

Fig. 3. Distortion happens at the boundary layer of shock wave Mach 3 (left), Mach 4 (right)

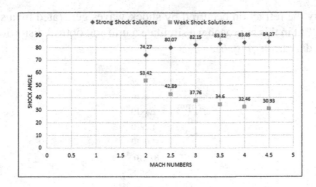

Fig. 4. Shock wave angles when deflection angle is 20°

the cockpit is simulated. We can clearly see that the distortion increases with the Mach number.

5   Evaluation

Strong shock wave angle increases with the Mach number while weak shock angle decreases with the Mach number. Figure 4 shows how the both strong and weak shock angles change according to the Mach number in this simulation. Figure 10.7 in the book called "Fundamentals of compressible Fluid Dynamics" by P. Balachandran shows the typical behavior of the shock angles when Mach number and deflection angle changes [17]. We can see that the plot in Fig. 4 coincides with the plot whose deflection angle equals to 20° in Fig. 10.7 in this book.

In this approach we assume the density distribution of the downstream of the shock wave is uniform. But in reality density distribution across the wave is not uniform [6]. Pressure distribution is also not uniform in the downstream shock wave [17]. This experiments show that there are several pressure regions occurred inside the downstream of the shock wave. Temperature distribution over the shock wave is also not uniform. But in this proposed approach we get the average pressure and temperature values to calculate the refractive index of air in the shockwave assuming the density, pressure and temperature distributions are uniform. As calculating exact values for these parameters for each pixel is very costly, this assumption is good for real time simulations.

5.1   Optical Distortion

Banakh et al., has presented studies of optical distortions in supersonic flows both experimentally and theoretically [6]. They have calculated the distortion of the laser beam by measuring the intensity of the laser beam in wind tunnels. The graphs in Fig. 5 shows the variance of intensity fluctuations along the jet axis and fluctuations of the refractive index.

According to the (a) chart variance of the intensity fluctuations increase with the jet axis. Hence the fluctuations of the refractive index also increases with the jet axis. Which means in reality the refraction occurred by shock waves generated by a supersonic jet is not uniform. But in this approach we simulate a uniform optical distortion throughout the shock boundary layer (Fig. 6).

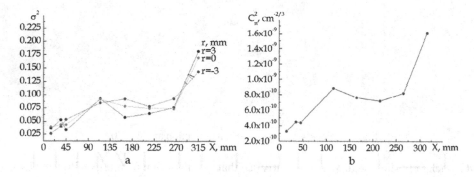

Fig. 5. Variance of intensity fluctuations along the jet axis (a), calculated values of the structure characteristic of fluctuations of the refractive index (b) [6]

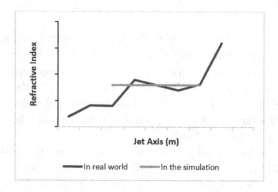

Fig. 6. Comparison of refractive index variance along the jet axis in real world and the simulation

According to the chart b in Fig. 5, the fluctuation of the refractive index is high at the nose front and end part of the nose cone. Refractive index is somewhat uniform in the middle part. There for this simulation we have modeled the refraction occurs in the middle part of the shock wave (Fig. 6). We cannot directly compare our simulation results with this wind tunnel experiment as they had used a very small model a jet in this wind tunnel.

6   Conclusions

In this research effort we proposed a simplified way to model refraction occurred due to the shock waves generated by a supersonic jet. As we focused on a higher level solution we only considered the major features.

Normally density, pressure and temperature distributions are not uniform throughout the shock waves. But we calculated an average refractive index for shock wave by assuming the previously mentioned parameters does not change throughout the shock wave. According to the experiments we can see that the refractive index fluctuations gets a uniform distribution over the middle part of the shock wave. Therefore the accuracy of this proposed approach is high in that region. We implemented a simple prototype to model static refraction occurred at the shock boundary layer.

7   Future Works

The implementation will be more beneficial, if it is integrated to achieve dynamic refraction in real time. This can be done using dynamic cube map technique. But dynamic cube refraction with low computational cost is still another research area. Further evaluation of the proposed model and rendering methodology is needed and there are additional future works such as evaluate computational efficiency/accuracy, design and conduct user tests to identify the degree of visual fidelity.

References

1. Mani, A., Wang, M., Moin, P.: Computational study of aero-optical distortion by turbulent wake. In: AIAA Paper, pp. 79–91 (2005)
2. Wang, K., Wang, M.: Aero-optics of subsonic turbulent boundary layers. J. Fluid Mech. **696**, 122–151 (2012)
3. Tredici, T.: Aerospace ophthalmology. USAF Flight Surg. Guid. (2005)
4. Elsinga, G., van Oudheusden, B., Scarano, F.: Evaluation of aero-optical distortion effects in PIV. Exp. Fluids **39**(1), 1–13 (2005)
5. O'Farrell, J.M., Rieckhoff, T.J.: Direct visualization of shock waves in supersonic space shuttle flight. Nat. Aeronautics and Space Admin., Marshall Space Flight Center, AL, Tech. memo. NASA/TM-2011–216455, M-1304, January 2011
6. Banakh, V., Marakasov, D., Tsvyk, R., Zapryagaev, V.: Study of turbulent supersonic flow based on the optical and acoustic measurements. In: Wind Tunnels and Experimental Fluid Dynamics Research, pp. 607–628. InTech (2011)
7. Scarano, F.: Overview of PIV in supersonic flows. In: Schroeder, A., Willert, C.E. (eds.) Particle Image Velocimetry, vol. 112, pp. 445–463. Springer, Heidelberg (2008)
8. Sewall, J., Lin, M., Galoppo, N., Tsankov, G.: Visual simulation of shockwaves. Graph. Models **71**(4), 126–138 (2009)
9. Van Rosendale, J., Ma, K., Vermeer, W.: 3D shock wave visualization on unstructured grids. In: Proceedings of the 1996 Symposium on Volume Visualization, pp. 87–95 (1996)
10. Zissis, G.J., Wolfe, W.L.: The Infrared Handbook (1978)
11. Benson, T.: Oblique shock wave (2014). http://www.grc.nasa.gov/WWW/k-12/airplane/oblique.html. [Accessed 12 June 2014]
12. Wikipedia, Shock wave (2014). http://en.wikipedia.org/w/index.php?title=Shockwave&oldid=610729841. [Accessed 15 April 2014]
13. Wikipedia, Nose cone design (2014). http://en.wikipedia.org/w/index.php?title=Nosecone design&oldid=611320846. [Accessed 10 August 2014]

14. Ames Research Staff, Equations tables and charts for compressible flow, Nat. Advisory Committee for Aeronautics, Calif (1953)
15. Anderson, J.: Modern Compressible Flow: With Historical Perspective: Aeronautical and Aerospace Engineering Series. McGraw-Hill Education (2003). http://books.google.lk/books?id=woeqa4-a5EgC
16. Edlen, B.: The refractive index of air. Metrologia **2**(2), 71 (1966)
17. Balachandran, P.: Fundamentals of Compressible Fluid Dynamics. Phi Learning (2006). https://books.google.lk/books?id=KEzdXmXgaHkC

Fine-Grained Risk Level Quantication Schemes Based on APK Metadata

Takeshi Takahashi[1](✉), Tao Ban[1], Takao Mimura[2], and Koji Nakao[1]

[1] National Institute of Information and Communications Technology, Tokyo, Japan
takeshi_takahashi@nict.go.jp
[2] Secure Brain Corporation, Tokyo, Japan

Abstract. The number of security incidents faced by Android users is growing, along with a surge in malware targeting Android terminals. Such malware arrives at the Android terminals in the form of Android Packages (APKs). Various techniques for protecting Android users from such malware have been reported, but most of them have focused on the APK files themselves. Unlike these approaches, we use Web information obtained from online APK markets to improve the accuracy of malware detection. In this paper, we propose category/cluster-based APK analysis schemes that quantify the risk of an APK. The category-based scheme uses category information available on the Web, whereas the cluster-based method uses APK descriptions to generate clusters of APK files. In this paper, the performance of the proposed schemes is verified by comparing their area under the curve values with that of a conventional scheme; moreover, the usability of Web information for the purpose of better quantifying the risks of APK files is confirmed.

Keywords: Android Package · APK · Malware · Static analysis · Security

1 Introduction

The number of incidents concerning malware is increasing with the number of Android terminal users. Malware arrives at an Android terminal in the form of an Android Package (APK). There are assorted channels through which terminals become infected. One common infection channel is the online Android market, where users download APK files that may contain malware. Vendors and carriers are working to sanitize the marketplace, but there are download sites that are not managed by vendors or carriers. Moreover, there are many other infection channels, such as email attachments and Web browsing, as is the case with personal computers. It is insufficient to sanitize the market—Android terminals should be equipped with a mechanism that prevents users from accessing malicious APK files. To this end, APK analysis techniques must be advanced.

Assorted related works have been reported. One such work is DroidRisk [8], which quantifies the risk of an APK file based on its permission request patterns.

© Springer International Publishing Switzerland 2015
S. Arik et al. (Eds.): ICONIP 2015, Part III, LNCS 9491, pp. 663–673, 2015.
DOI: 10.1007/978-3-319-26555-1_75

It calculates the risk of an APK by measuring the likelihood of misuse and the impact of the misuse of each permission. Sarma et al. [6] also proposed a scheme that determines whether an APK file is malware by introducing Category-based Rare Critical Permission (CRCP). A CRCP is a permission that should not usually be required for the applications of a category. When any of the CRCPs are called, the APK is regarded as risky. Various other APK analysis studies have been reported, but these need to improve their malware detection rate.

This paper introduces risk quantification schemes for APKs by extending DroidRisk. Unlike conventional schemes, we not only use APK files, but also certain metadata obtained from the Web, such as APK markets, to quantify the risk. We first introduce a category-based risk quantification scheme for APK files. This uses category information obtained from the Web to analyze the risk for each category, because the characteristics of APKs differ considerably from one category to another. Our performance evaluation shows that category-based risk quantification provides a better performance in terms of the area under curve (AUC) [3] values than a conventional scheme, but the advantage is insignificant. To bolster the advantage, we then propose another scheme that uses cluster information instead of category information. We generate clusters based on the application descriptions available on the Web, and quantify the risk of APK files for each cluster. Our performance evaluation shows that the cluster-based risk quantification scheme enjoys the best AUC values of the three schemes, outperforming the conventional scheme by 4.7 % on average.

The proposed schemes enable us to detect unknown malware, i.e., APKs that have not yet been reported as malware, and alert Android users to their purpose. Although a conventional scheme can also detect that, the proposed schemes improve the reliability of the alerts. We conclude that the metadata available online are useful for quantifying and analyzing the risks of APK files.

2   Conventional Scheme

This section analyzes DroidRisk, a conventional scheme, using our dataset.

2.1   Mechanism

DroidRisk [8] quantifies the risk of an APK file based on its permission request patterns. It first quantifies the security risk of each permission according to:

$$R(p) = L(p) \times I(p), \tag{1}$$

where $L(p)$ denotes the likelihood of permission p being used by malware, and $I(p)$ denotes the impact of permission p being misused by malware. DroidRisk then quantifies the security risk of an APK file by summing the quantified security risk of each permission used by the APK file:

$$R_A = \sum_i R(p_i). \tag{2}$$

R_A is the quantified risk value, and can be used to determine whether an APK file is malware by setting an appropriate threshold value.

2.2 Dataset

We collected 87,182 APK files from the Opera Mobile Store[1] over the period of January to September of 2014. Files from which we were unable to extract permission requests were excluded because permission requests are a necessary input for our risk quantification schemes. The files were then checked using VirusTotal[2] to determine whether they were malware. VirusTotal analyzes the risk of an APK file by using multiple evaluation engines from different vendors. If one or more of the results indicates that the file is malicious, we considered the APK file to be malware. Note that adware was not counted as malware, and the APK files that VirusTotal was unable to handle were excluded from the dataset in advance. As a result, we obtained a dataset of 78,649 APK files, consisting of 52,251 and 26,398 benign and malicious files, respectively.

Table 1. Overview of our dataset

Category	Benign	Malicious	Total
Business & Finance	4,495	938 (17.3%)	5,433
Communication	2,478	601 (19.5%)	3,079
eBooks	3,922	1,463 (27.2%)	5,385
Entertainment	15,580	6,696 (30.1%)	22,276
Games	11,762	5,500 (31.9%)	17,262
Health	1,687	657 (28.0%)	2,344
Languages & Translators	787	159 (16.8%)	946
Multimedia	2,607	1,165 (30.9%)	3,772
Organizers	1,402	252 (15.2%)	1,654
Ringtones	290	249 (46.2%)	539
Themes & Skins	4,221	7,969 (65.4%)	12,190
Travel & Maps	3,020	749 (19.9%)	3,769
Total	52,251	26,398 (33.6%)	78,649

We also collected metadata of APK files over the same period. The metadata include their application category, description, and number of downloads, and are available on the Opera Mobile Store along with the APK files. We stored the metadata in XML files, following the data structure defined in [7], although we used only the application category and description of the metadata in this paper. The breakdown of the APK files for each category is shown in Table 1.

---

[1] http://apps.opera.com/.
[2] http://www.virustotal.com/ja.

2.3   Evaluation Using Our Dataset

The performance of DroidRisk may vary according to the dataset. Thus, we first verified DroidRisk using our dataset. We extracted the permission patterns needed for running DroidRisk from the AndroidManifest.xml file of APK files[3].

We measured the quantification performance from the standpoint of AUC, as in [8]. The highest AUC value obtained with our dataset was 0.793, for which the value of $I(dp)$ was set to 1.0. Note that dp represents permissions whose protectionLevel[4] is set to "dangerous" by Android 4.4 r1, and np represents the other permissions. Instead of calculating $I(p)$ for each permission, we calculate $I(np)$ and $I(dp)$, as in [8]. We determine these parameter values using an empirical method with the Receiver Operating Characteristic (ROC) curve, as in [8], because it is difficult to determine the actual level of harm caused by normal or dangerous permissions[5]. Note that the evaluations in this paper takes care of the permissions defined by Android 4.4 r1. Other permissions are ignored for simplicity.

2.4   Category-Based Analysis

We have seen that the effectiveness of DroidRisk differs depending on the dataset. This indicates that DroidRisk may exhibit better AUC values for some parts of a dataset. We thus evaluated its effectiveness for each application category. The Opera Mobile Store organizes its APK files by classifying them into 12 categories. Using the category information, we measured the AUC values of DroidRisk in each category. We randomly chose 20 % of the dataset and used it for the evaluation, and the rest were used to study the optimal values for the DroidRisk parameters[6]. Note that we took the same data selection approach for all experiments reported in this paper.

Figure 1 shows the results. As can be seen, the effectiveness of DroidRisk differs with the APK category. APK files within some categories have higher AUC values, whereas the rest exhibit lower AUC values. This is because the permission patterns of benign APKs and those of malicious APKs overlap significantly for certain categories, whereas this is not the case for the other categories. If the overlaps are large, the effectiveness of DroidRisk will decrease.

---

[3] We believe that better quantification results are achieved if we consider the difference between the requested permissions and those actually used, i.e., permission gaps [1], because this removes noises added to the characteristics of the APK files. Nevertheless, this is beyond the scope of this paper.

[4] http://developer.android.com/reference/android/content/pm/PermissionInfo.html.

[5] We measure the AUC values by setting $I(np)$ to 1.0 and increasing the value of $I(dp)$ from 1.0 to 3.0 in increments of 0.1, and then choose the value that provides the highest AUC as the optimal value. Note that dangerous permissions are certainly more harmful than normal permissions.

[6] The evaluation following this procedure should be iterated to gain the average values of the studied values. Moreover, cross validation should be applied to the learning process. Our future work will cope with this.

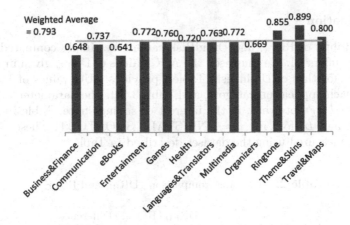

Fig. 1. AUC values of DroidRisk and $\mathrm{DR_{BD}}$

We use the above AUC values as the basis for the performance evaluation. For reference, we address the scheme that runs DroidRisk for each application category with the same parameter values used by DroidRisk as $\mathrm{DR_{BD}}$.

3   Category-Based Risk Quantification

This section introduces a risk quantification scheme that extends DroidRisk using application category information. We refer to the scheme as $\mathrm{RQ_{CG}}$.

3.1   Mechanism

$\mathrm{RQ_{CG}}$ extends DroidRisk to quantify security risks based on the context of APK files. DroidRisk measures the $L(p)$ and $I(p)$ values by studying the whole dataset. Nevertheless, the optimal values should differ depending on the type of application. For instance, many applications within the category of "Travel & Maps" request permission to use GPS, whereas few within the "Multimedia" category request such permission. Thus, $\mathrm{RQ_{CG}}$ sets different values for $L(p)$ and $I(p)$ for each application category by studying the data in each category.

As with DroidRisk, $\mathrm{RQ_{CG}}$ calculates the quantified value of an APK's risk, i.e., R_A, according to the following equations:

$$R_c(p) = L_c(p) \times I_c(p), \tag{3}$$

$$R_A = \sum_i R_c(p_i), \tag{4}$$

where $L_c(p)$ denotes the likelihood of permission p being used by malware when p is used within category c, and $I_c(p)$ denotes the impact of p being misused by malware when p is used within that category. Here, R_A can then be used to determine whether an APK file is malware by setting an appropriate threshold.

3.2 Evaluation

The AUC values of RQ_{CG} and DroidRisk cannot be simply compared because the context differs [3]. We thus used the AUC values of DR_{BD} given in Fig. 1, as a substitute for that of DroidRisk. Table 2 provides AUC values of DR_{BD} and RQ_{CG} for each application category, and Table 3 lists the parameter values, i.e., I, I_c, L and L_c[7]. Note that, in the interest of saving space, Table 3 shows the values of L and L_c for only $p = $ UNINSTALL_SHORTCUT. These values are empirically measured using the dataset described in Table 1.

Table 2. AUC value comparison (DR_{BD} and RQ_{CG})

	DR_{BD}	RQ_{CG}	Difference
Business & Finance	0.648	0.668	+0.021 (+3.2 %)
Communication	0.737	0.747	+0.010 (+1.3 %)
eBooks	0.641	0.680	+0.038 (+5.9 %)
Entertainment	0.772	0.775	+0.003 (+0.3 %)
Games	0.760	0.762	+0.002 (+0.3 %)
Health	0.720	0.721	+0.001 (+0.1 %)
Languages & Translators	0.763	0.778	+0.015 (+2.0 %)
Multimedia	0.772	0.780	+0.008 (+1.1 %)
Organizers	0.669	0.680	+0.012 (+1.8 %)
Ringtones	0.855	0.872	+0.017 (+2.0 %)
Themes & Skins	0.899	0.904	+0.006 (+0.6 %)
Travel & Maps	0.800	0.828	+0.028 (+3.5 %)
Average	0.753	0.766	+0.013 (+1.8 %)
Weighted average	0.768	0.777	+0.009 (+1.1 %)
Median	0.762	0.769	+0.007 (+0.9 %)
Minimum	0.641	0.668	+0.027 (+4.2 %)
Maximum	0.899	0.904	+0.006 (+0.6 %)
Standard deviation	0.074	0.071	−0.003 (−4.1 %)

As can be seen, RQ_{CG} optimizes the parameter values of DR_{BD} for each category and thus improves the AUC values of all categories. DR_{BD} uses a single pair of values for $I_c(p)$ and $L_c(p)$ for all APK files, but RQ_{CG} uses different parameter values for each application category. Note that the original DroidRisk and DR_{BD} are different. Indeed, we believe that DR_{BD} provides a better performance than DroidRisk, but AUC is not a suitable measure for evaluating this assertion. This paper retains AUC as the evaluation criteria, and considers the comparison of these two schemes to be beyond the scope of the current research. Thus, we

---

[7] The values for I and I_c were determined empirically, as with Sect. 2.3.

ignore the advantage of DR_{BD} over DroidRisk, and for simplicity, assume that they produce an identical performance.

Table 3. Value of parameters, where p = UNINSTALL_SHORTCUT

Category	$I(np)$	$I(dp)$	$I_c(np)$	$I_c(dp)$	$L(p)$	$L_c(p)$
Business & Finance	1.0	1.0	1.0	1.1	0.92	0.88
Communication	1.0	1.0	1.0	1.0	0.92	0.87
eBooks	1.0	1.0	1.0	1.4	0.92	0.97
Entertainment	1.0	1.0	1.0	2.9	0.92	0.93
Games	1.0	1.0	1.0	1.3	0.92	0.82
Health	1.0	1.0	1.0	1.2	0.92	0.91
Languages & Translators	1.0	1.0	1.0	1.3	0.92	1.00
Multimedia	1.0	1.0	1.0	1.0	0.92	0.86
Organizers	1.0	1.0	1.0	1.0	0.92	0.82
Ringtones	1.0	1.0	1.0	1.7	0.92	1.00
Themes & Skins	1.0	1.0	1.0	2.9	0.92	0.97
Travel & Maps	1.0	1.0	1.0	1.9	0.92	0.98

3.3   Toward Further Performance

Here, RQ_{CG} realizes a finer-grained risk level quantification of APK files by using the application category information. Nevertheless, the advantage of RQ_{CG} over DR_{BD}, shown in Table 2, is insignificant. Moreover, the AUC values of RQ_{CG} differ significantly for each category; the reliability of malware judgments with RQ_{CG} is low for certain categories. We hope to gain better AUC values without creating categories that show significantly low AUC values when compared to the other categories. We believe that the categorization can be modified further to achieve this.

The categories available on the Web are not intended for our scheme, and can occasionally be ambiguous. When registering a new APK, the registrar (who is often the developer) manually chooses the APK category. This could lead to a suboptimal classification because different people may classify the same application differently; the understanding of the category name may differ depending on the person. There could be a case in which an APK falls within more than one category; for instance, a healthcare-based quiz game may be classified as either "Games" or "Health" depending on the registrar's judgment. Moreover, manual operations could induce human errors when choosing categories. As a result, classification of the APK files may not be optimized. Furthermore, the categories were set by the Opera Mobile Store, and we assume that they were formed based on conceptual application categories set by humans. Hence, it is natural that the classification is unable to fully capture the characteristics of the permission patterns of the APK files. We thus aim to generate categories that capture the functional characteristics of the APK files.

4   Cluster-Based Risk Quantification

This section provides another risk quantification scheme. It generates application clusters that capture the characteristics the permission patterns of the APKs. Note that we use the term "cluster" hereafter to denote the categories automatically generated by an algorithm. We refer to this scheme as RQ_{CL}.

4.1   Mechanism

This scheme uses computer-generated clusters instead of human-made categories to avoid human judgment. The cluster is generated from the application descriptions available on the market website. We conduct a clustering operation using the steps described below: preprocess the application description information, build the topic model using a Latent Dirichlet Allocation (LDA) [2], and apply k-means clustering [5]. Note that these steps are based on the classification technique described in [4], which are adjusted for our scheme.

In the **preprocessing step**, we preprocess the application description information to produce words that are usable for LDA. First, we discard both non-English descriptions and non-text descriptions, i.e., numbers, HTML tags, Web links, and email addresses. Second, we extract stems from the description and truncate the stop words. Third, we count the number of words in the resultant description. If there are fewer than ten words, we discard the description. In the **topic model generation step**, we process the resultant words. We import the words in the description, and train a number of topics. We consider a total of 300 topics, with 0.05 as the threshold value of the topic proportion and a maximum of four topics per entry. Note that we used 300 topics because the MALLET documentation states the following: "The number of topics should depend to some degree on the size of the collection, but 200 to 400 will produce reasonably fine-grained results." As a result, this process outputs several {topic number, proportion value} pairs (up to four pairs). In the **clustering step**, we cluster the APKs according to the {topic number, proportion value} pairs for each description. The number of categories is set to 12, which is the same number of categories used in the Opera Mobile Store.

Finally, RQ_{CL} calculates the quantified value of risk of an APK file according to the following equations:

$$R_{c'}(p) = L_{c'}(p) \times I_{c'}(p), \tag{5}$$

$$R_A = \sum_i R_{c'}(p_i), \tag{6}$$

where $L_{c'}(p)$ denotes the likelihood of permission p being used by malware when p is used within cluster c', and $I_{c'}(p)$ denotes the impact of p being misused by malware when p is used within c'.

Table 4. Dataset breakdown by cluster

Cluster	Benign	Malicious	Total
cluster00	2,865	1,643 (36.4 %)	4,508
cluster01	3,159	1,613 (33.8 %)	4,772
cluster02	3,281	1,640 (33.3 %)	4,921
cluster03	4,232	2,221 (34.4 %)	6,453
cluster04	3,586	1,905 (34.7 %)	5,491
cluster05	3,151	1,746 (35.7 %)	4,897
cluster06	2,887	1,506 (34.3 %)	4,393
cluster07	4,422	3,006 (40.5 %)	7,428
cluster08	3,487	1,428 (29.1 %)	4,915
cluster09	3,679	1,499 (28.9 %)	5,178
cluster10	2,832	1,482 (34.4 %)	4,314
cluster11	2,897	1,563 (35.0 %)	4,460
All	40,478	21,252 (34.4 %)	61,730

4.2 Evaluation and Analysis

This section demonstrates the effectiveness of RQ_{CL} by comparing the AUC values of DroidRisk, RQ_{CG}, and RQ_{CL}. We used the dataset introduced in Sect. 2.2 for our experiment, but the number of valid applications was reduced during the preprocessing stage (see Sect. 4.1). Table 4 lists the number of APK files and malware for each cluster.

First, we empirically selected the optimal values for the DroidRisk and RQ_{CL} parameters (i.e., L, I, $L_{c'}(p)$, and $I_{c'}(p)$). Note that L and I remain constant regardless of the category because DroidRisk does not recognize the concept of application categories, whereas $L_c(p)$ and $I_c(p)$ are category-specific.

We then measured the AUC values of RQ_{CL} using the optimal parameter values. Note that the dataset used to evaluate RQ_{CL} differed from that described in Sect. 2.2 because some applications were unsuitable for generating clusters and were removed during the preprocessing stage (see Sect. 4.1). For a fair comparison, we measured the AUC values for DR_{BD} and RQ_{CG} using the dataset remaining after the preprocessing, i.e., using 61,730 APK files. For the purpose of a further analysis, we also measured the AUC values of a scheme running DroidRisk for each cluster using the optimal values of L and I. We refer to this scheme as RQ_{CL}-. The results are presented in Tables 5 and 6.

Table 6 indicates that RQ_{CL} produced the best performance among all of the schemes, with higher average and median scores; RQ_{CL} increased the AUC value by 0.036 points (+4.7 %) on average and exhibited a more stable performance with the smallest standard deviation. The table also indicates that proper classification, using categories or clusters, is more effective than parameter optimization in producing better AUC values. The advantage of RQ_{CG} over DR_{BD},

Table 5. AUC value comparison (RQ_{CL-} and RQ_{CL})

Category	RQ_{CL-}	RQ_{CL}	Difference
cluster00	0.780	0.784	+0.003(+0.4%)
cluster01	0.782	0.780	−0.002(−0.2%)
cluster02	0.831	0.834	+0.003(+0.4%)
cluster03	0.825	0.824	−0.002(−0.2%)
cluster04	0.799	0.800	+0.001(+0.2%)
cluster05	0.801	0.804	+0.003(+0.4%)
cluster06	0.794	0.795	+0.001(+0.2%)
cluster07	0.854	0.860	+0.005(+0.6%)
cluster08	0.770	0.776	+0.007(+0.8%)
cluster09	0.746	0.750	+0.004(+0.5%)
cluster10	0.808	0.815	+0.008(+0.9%)
cluster11	0.806	0.810	+0.003(+0.4%)
Average	0.800	0.803	+0.003(+0.4%)
Weighted average	0.797	0.800	+0.003(+0.4%)
Median	0.800	0.802	+0.002(+0.3%)
Minimum	0.746	0.750	+0.004(+0.5%)
Maximum	0.854	0.860	+0.005(+0.6%)
Standard deviation	0.028	0.028	+0.000(+0.4%)

Table 6. Comparison of AUC values for each scheme

	DR_{BD}	RQ_{CG}	RQ_{CL-}	RQ_{CL}
Average	0.767	0.782	0.800	0.803
Weighted average	0.781	0.789	0.797	0.800
Median	0.766	0.767	0.800	0.802
Minimum	0.666	0.692	0.746	0.750
Maximum	0.913	0.918	0.854	0.860
Standard deviation	0.075	0.073	0.028	0.028

as well as the advantage of RQ_{CL} over RQ_{CL-}, is relatively small compared to the advantage of RQ_{CL} over RQ_{CG}. Parameter optimization is more effective in categories or clusters that have lower AUC values. Tables 2 and 5 support this assertion, i.e., that the advantage of RQ_{CG} over DR_{BD} is greater than that of RQ_{CL} over RQ_{CL-}.

5   Conclusion and Future Work

This paper demonstrated the usefulness of APK metadata for risk analysis. The proposed schemes quantify the APK risk using category or cluster information obtained or generated from metadata available online. Both schemes show a better performance in terms of the AUC values than a conventional scheme. Indeed, the cluster-based scheme outperforms the category-based one because it captures functional characteristics more accurately than the other.

As future work, there are several issues we wish to approach toward the construction of better quantification schemes. For instance, another clustering method should be explored. As the k-means process uses cluster centers to model the data, it tends to find clusters of a comparable spatial extent, which matches our heuristic idea of an application category well. Nevertheless, when nonlinear structures and/or different cluster shapes are present, other advanced clustering methods, e.g., hierarchical clustering and density-based clustering, may be engaged. In addition, the parameters used for LDA and k-means clustering can be optimized. We expect that the AUC value will be further improved by coping with permission gaps, because this will decrease the overlap among the permission patterns of benign APKs and malicious APKs. We plan to take these issues into account and develop our scheme further.

References

1. Bartel, A., Klein, J., Le Traon, Y., Monperrus, M.: Automatically securing permission-based software by reducing the attack surface: an application to android. In: Proceedings of the 27th IEEE/ACM International Conference on Automated Software Engineering (2012)
2. Blei, D.M., Ng, A.Y., Jordan, M.I.: Latent dirichlet allocation. J. Mach. Learn. Res. **3**, 993–1022 (2003)
3. Brown, C.D., Davis, H.T.: Receiver operating characteristics curves and related decision measures: a tutorial. Chemometr. Intell. Lab. Syst. **80**(1), 24–38 (2006)
4. Gorla, A., Tavecchia, I., Gross, F., Zeller, A.: Checking app behavior against app descriptions. In: Proceedings of the 36th International Conference on Software Engineering (2014)
5. MacQueen, J.: Some methods for classification and analysis of multivariate observations. In: Proceedings of the Fifth Berkeley Symposium on Mathematical Statistics and Probability. Statistics, vol. 1, pp. 281–297 (1967)
6. Sarma, B.P., Li, N., Gates, C., Potharaju, R., Nita-Rotaru, C., Molloy, I.: Android permissions: a perspective combining risks and benefits. In: Proceedings of the 17th ACM Symposium on Access Control Models and Technologies (2012)
7. Takahashi, T., Nakao, K., Kanaoka, A.: Data model for android package information and its application to risk analysis system. In: Proceedings of the First ACM Workshop on Information Sharing and Collaborative Security (2014)
8. Wang, Y., Zheng, J., Sun, C., Mukkamala, S.: Quantitative security risk assessment of android permissions and applications. In: Wang, L., Shafiq, B. (eds.) DBSec 2013. LNCS, vol. 7964, pp. 226–241. Springer, Heidelberg (2013)

Opinion Formation Dynamics Under the Combined Influences of Majority and Experts

Rajkumar Das[1]([✉]), Joarder Kamruzzaman[2], and Gour Karmakar[2]

[1] Faculty of Information Technology, Monash University, Melbourne, VIC, Australia
rajkumar.das@monash.edu
[2] School of Engineering and Information Technology, Federation University
Australia, Gippsland Campus, Churchill, Australia
{joarder.kamruzzaman,gour.karmakar}@federation.edu.au

Abstract. Opinion formation modelling is still poorly understood due to the hardness and complexity of the abstraction of human behaviours under the presence of various types of social influences. Two such influences that shape the opinion formation process are: (i) the expert effect originated from the presence of experts in a social group and (ii) the majority effect caused by the presence of a large group of people sharing similar opinions. In real life when these two effects contradict each other, they force public opinions towards their respective directions. Existing models employed the concept of confidence levels associated with the opinions to model the expert effect. However, they ignored the majority effect explicitly, and thereby failed to capture the combined impact of these two influences on opinion evolution. Our model explicitly introduces the majority effect through the use of a concept called opinion consistency, and captures the opinion dynamics under the combined influence of majority supported opinions as well as experts' opinions. Simulation results show that our model properly captures the consensus, polarization and fragmentation properties of public opinion and reveals the impact of the aforementioned effects.

Keywords: Opinion · Majority effect · Expert effect · Consistency

1 Introduction

Opinion formation dynamics captures the evolution of individual's opinion in a social group to a collective opinion under social influences. Repeated interactions among the group members due to social influences enforce individuals to revise their thoughts, adopt new ideas, or refine beliefs. Modelling such dynamics ensures the better realization of individual behaviours and peer influences that encourage the group members to change their initial opinions and thereby helps better understanding of the nature and composition of the final opinion.

Opinion formation modelling has been established as an active research area, thanks to the seminal works of DeGroot [3], Clifford and Sudbury [4], Hegselmann

© Springer International Publishing Switzerland 2015
S. Arik et al. (Eds.): ICONIP 2015, Part III, LNCS 9491, pp. 674–682, 2015.
DOI: 10.1007/978-3-319-26555-1_76

and Krause [5], Deffuant and Weisbuch [6], and Sznajd [2] that have attracted researchers from multiple disciplines afterwards. The models either represent opinions using continuous values in a range [3,5,6] or consider discrete opinions [1,4]. Though a group of discrete opinion models have exploited majority effects [2] and some continuous models have considered expert effect to an extent, but none of the modelling approaches have considered majority and expert effects jointly. As both effects are concurrently present in the society with possible contradiction between them, we need to model their joint influence on an agent's opinion to better capture the dynamics.

Continuous opinion dynamics encode the expert effect by incorporating the concept of confidence of an agent on its expressed opinion [8,9]. Moussaid et al. [8] used confidence level in their modelling to distinguish experts as agents with high confidence levels. However, the efficacy of the approach is limited due to the use of: (i) a few intuitively defined rules considering a particular context and (ii) interactions are limited to pair-wise between agents. Consequently, this approach fails to capture the impact of the overall confidence levels of a group of agents having similar opinions. On the other hand, Cho et al. [9] uses uncertainty in beliefs to model the confidence of agents. Likewise [8], this model only considers pairwise interactions between the agents. As a result, both models fail to consider the conformity of an agent to a group with majority of them sharing similar opinions (majority effect) while in Asch experiment [10] conformity to majority group is exhibited as a strong social influencing factor. Consequently, their models ignore the presence of the joint influence of the majority and expert effects. Motivated by these research gaps, we assimilate the conformation to majority opinions in a group and the joint influence of the majority effect and the expert effect in continuous opinion formation modelling.

2 Proposed Opinion Formation Model

Consider a social network $G = (V, E)$ with $|V| = n$ agents forming a collective opinion. An agent $i \in V$ interacts to all its neighbours, $N_i = \{j|j \in V \wedge (i,j) \in E\}$ in each iteration. At time t, the opinion is denoted by $O_i(t) \in [0,1]$ for agent i. The range is called the Opinion Space (OS). Every agent also expresses a confidence level $C_i(t) \in [0,1]$ on its own opinion $O_i(t)$. A highly confident agent can be considered an expert whom other people can rely on. In contrast, individuals with very low confidence level represents lay people having poor knowledge and thus, very hard to be relied on. To jointly integrate the majority and expert effects, in our model an agent is usually influenced by four influencing factors: (i) its own opinion $O_i(t)$, (ii) its own confidence $C_i(t)$, (iii) neighbours' opinions $O_{N_i}(t) = \{O_j(t)|j \in N_i\}$, and (iv) neighbours' confidences $C_{N_i}(t) = \{C_j(t)|j \in N_i\}$. According to the literature [1,8], in real life opinion update process considers one of the three possible heuristics: (i) keep own opinion, (ii) make a compromise by averaging with neighbours' opinions and (iii) adopt neighbour's opinion. To make opinion update as realistic as possible, our model considers all of them. Our model captures the combined effects of majority and expert as described below.

2.1   Formulation of Majority and Expert Effects

To measure the combined impact of the majority and expert effect, an agent computes the credibility of its neighbours' opinions using the aforementioned four types of information. Here, the credibility score determines how much an agent can rely on its neighbours. To compute credibility, we use a particular measure called consistency for both opinions and confidence levels. Consistency ($\xi(A)$) of a set (A) of values indicates the overall similarity present in the set. The Shannon's entropy from information theory properly represents the consistency as a set with similar values results in a small entropy whereas diverge valued set has a large entropy. Consequently, a set's entropy normalized with the maximum possible entropy (e_m) [11] is a reasonable realization of its consistency.

Opinion Consistency (Majority Effect): Opinion consistency of a neighbourhood is calculated from the entropy value as per Eq. (1).

$$\xi(O_{N_i}(t)) = 1 - \frac{e(O_{N_i}(t))}{e_m} \text{ where,} \tag{1}$$

$$e(O_{N_i}(t)) = \text{entropy}(O_{N_i}(t)) = -\sum\nolimits_{O(t) \in O_{N_i}(t)} p(O(t)) \times \log(p(O(t)))$$

Here, $p(O(t))$ is the probability of opinion $O(t)$ in neighbours' opinion set $O_{N_i}(t)$. A high $\xi(O_{N_i}(t))$ indicates the presence of a large group of agents holding similar opinions, thus captures the majority effect.

Confidence Consistency (Expert Effect): We take into account the possibility of forming different opinion clusters and measure individual groups' expert effects because experts are likely to share similar thoughts. For simplicity, we assume the OS to be composed of a set $B = \{b_1, b_2, \cdots, b_z\}$ of equal sized blocks constructed from i's neighbourhood, and opinions inside a block form a cluster. We compute the confidence consistency of a block $b \in B$ as per Eq. (2).

$$\xi(C_{ib}(t)) = \log(\mathcal{E}(C_{ib}(t)) \times E(C_{ib}(t)) + \eta)/\log(1 + \eta) \text{ where,} \tag{2}$$

$$\mathcal{E}(C_{ib}(t)) = 1 - \frac{e(C_{ib}(t))}{e_m}, e(C_{ib}(t)) = -\sum\nolimits_{C(t) \in C_{ib}(t)} p(C(t)) \times \log(p(C(t))),$$

$$E(C_{ib}(t)) = \sum\nolimits_{C(t) \in C_{ib}(t)} p(C(t)) \times (C(t))$$

Here, $C_{ib}(t)$ is the set of confidence values within the block b and $E(C_{ib}(t))$ is its expected value. Unlike Eq. (1), normalized entropy $\mathcal{E}(C_{ib}(t))$ of the confidence of a block is further multiplied by its expected confidence $E(C_{ib}(t))$ to neutralize any undue impact of expert effect such as a group having similar but very low confidence levels. η is used to keep the values in a specific range. The expected confidence consistency as evaluated in Eq. (3) further validates the presence of such expert groups.

$$E(\xi(C_{ib}(t))) = \sum\nolimits_{b \in B} \xi(C_{ib}(t)) \times p_{ib}((t)) \text{ where, } p_{ib}((t)) = \frac{|O_{ib}(t)|}{|O_{N_i}(t)|} \tag{3}$$

Here, $O_{ib}(t)$ is the opinion set within b and $|.|$ computes the cardinality of a set.

So far, we have defined the theoretical underpinnings to capture majority and expert effects separately. However to realize their effects in opinion dynamics, Eqs. (2) and (3) can be used to represent the three major mutually exclusive real-world scenarios of opinion update process that are described below:

Scenario 1 (Joint Consideration of Majority and Expert Effects): This scenario captures the influence of a large group consisting of the majority neighbouring agents who are also experts. To represent the majority effect, the majority of the neighbouring agents have to possess similar opinions which yields high opinion consistency, while for the concurrent expert effect, their confidence levels are to be high and consistence. Therefore, their joint effects termed as the credibility score of neighbours can be calculated as the product of opinion and confidence consistencies, which is defined as:

$$\chi_{N_i}(t) = \xi(O_{N_i}(t)) \times E(\xi(C_{ib}(t))) \tag{4}$$

Scenario 2 (Influence of Small Groups Consisting of Experts): In the absence of such joint effects, an emerging group of expert agents in the neighbourhood having similar opinions influences the agent to update its opinion. The group exerts more influence as it grows larger, which is a characteristic reflecting majority phenomenon. The larger confidence consistency within group ensures greater influence according to the expert effect. Finally, based on the homophily principle [5], closer distant group have more impacts. Therefore, the influence of a group on an agent opinion is formulated as in Eq. (5).

$$\chi_{ib}(t) = \underset{b \in B}{\operatorname{argmax}}(\frac{\xi(C_{ib}(t)) \times p_{ib}(t)}{1 + abs(O_i(t) - \overline{O_{ib}(t)})}) \tag{5}$$

Here, $abs()$ denotes absolute value and $\overline{O_{ib}(t)}$ is the average opinion of block, b. $p_{ib}(t)$, $\xi(C_{ib}(t))$ and $abs(O_i(t) - \overline{O_{ib}(t)})$ represent the relative group size, its expert influence and the homophily principle, respectively.

Scenario 3 (Individual Expert's Influence): When there is no such group in the neighbour opinions, it is only the individual experts that influence an agent. In this situation, the amount of influence depends on their opinion distances [5] and relative confidence level [8] of an expert to an agent as captured by Eq. (6).

$$\zeta_{ij}(t) = \frac{C_j(t) - C_i(t)}{1 + abs(O_j(t) - O_i(t))}, \text{where } j \in N_i \text{ and } C_j(t) - C_i(t) \geq \zeta_{th}(t) \tag{6}$$

Here, $\zeta_{th}(t)$ determines the minimum level of confidence that an expert should be higher than an agent to impact on its decision. How an opinion be updated in our proposed model are presented in the following sections.

2.2 Model Description

While revising an agent's opinion at time t, it looks for the presence of majority and expert effects in the neighbourhood as measured by the neighbours' credibility score $\chi_{N_i}(t)$, small expert group influence $\chi_{ib}(t)$, and individual expert impact $\zeta_{ij}(t)$ and applies the opinion update process for each scenario. Here, the scenarios are confined using three bounding functions that are computed by applying bounding conditions on $\chi_{N_i}(t)$ and $\chi_{ib}(t)$ of Eqs. (4) and (5) respectively. The bounding conditions are: (i) $\chi_{N_i}(t) \geq \mathcal{F}_1(B, \eta)$ for Scenario 1 with $\overline{C_{N_i}(t)} \geq C_i(t)$, (ii) $\chi_{N_i}(t) < \mathcal{F}_1(B, \eta)$ and $\chi_{ib}(t) \geq \mathcal{F}_2(B, \eta)$ for Scenario 2 with $\overline{C_{ib}(t)} \geq C_i(t)$, (iii) $\mathcal{F}_3(B, \eta) \leq \chi_{N_i}(t)$ for Scenario 3 with not satisfying the conditions for Scenario 1 and 2 and (iv) $\chi_{N_i}(t) < \mathcal{F}_3(B, \eta)$ for others. The derivations of the bounding functions $\mathcal{F}_1(B, \eta)$, $\mathcal{F}_2(B, \eta)$ and $\mathcal{F}_3(B, \eta)$ are well formulated, however due to the page limitation, it have been presented in https://rajkumardas.files.wordpress.com/2015/08/iconip2015.pdf.

Here, we adopt weighted averaging as the compromise method. An agent choose from one of the following three compromise options:

– **Compromising with whole neighbourhood (Scenario 1):** As with the condition (i) for Scenario 1 defined above, the collective influence of neighbours is very strong and its average confidence level $\overline{C_{N_i}(t)}$ is higher than agent's own confidence. Thus, the neighbours are as good as one information source and agent i compromises with whole neighbourhood as per Eq. (7).

$$O_i(t+1) = (C_i(t) \times O_i(t) + \chi_{N_i}(t) \times \overline{O_{N_i}(t)})/(C_i(t) + \chi_{N_i}(t)) \quad (7)$$

Here, the neighbourhood opinion is represented as the average of their opinions, $\overline{O_{N_i}(t)}$ and the credibility score is as weight.
– **Compromising with a block (Scenario 2):** With $\chi_{N_i}(t) < \mathcal{F}_1(B, \eta)$, neighbours don't form one overall strong group. However, in absence of such larger block, there may exist some blocks consisting of experts that can influence an agent. $\chi_{ib}(t) \geq \mathcal{F}_2(B, \eta)$ and $\overline{C_{ib}(t)} \geq C_i(t)$ indicate the presence of such a group with higher average confidence than i, and enforce to compromise with that block as in Eq. (8).

$$O_i(t+1) = (C_i(t) \times O_i(t) + \chi_{ib}(t) \times \overline{O_{ib}(t)})/(C_i(t) + \chi_{ib}(t)) \quad (8)$$

where, $\overline{O_{ib}(t)}$ is the average opinion of block b.
– **Compromising with an expert (Scenario 3):** Condition (iii) implies that there is no majority block consisting of experts in the neighbourhood. In such scenario, an agent may search for individual experts to compromise its opinion with one of them chosen with probability $\zeta_{ij}(t)$ using Eq. (9).

$$O_i(t+1) = (C_i(t) \times O_i(t) + \zeta_{ij}(t) \times O_j(t))/(C_i(t) + \zeta_{ij}(t)) \quad (9)$$

As alluded at the beginning of Sect. 2, in addition to compromise, opinion update process also allows an agent either to keep own opinion or to adopt one of the neighbours' opinions which is discussed in the following section.

- **Other Scenarios:** $\chi_{N_i}(t) < \mathcal{F}_3(B, \eta)$ means neither a majority nor an expert effect is present. Therefore, the agent ignores them by keeping its opinion. However, we incorporate the free-will concept [7] in our model by making a probabilistic choice between the keep and adoption heuristic.

$$O_i(t+1) = \text{Choose randomly with probability p, } \{O_i(t), O_k(t)\} \qquad (10)$$

Here, $O_k(t) \in O_{N_i}(t)$ is chosen randomly from the neighbourhood.

3   Simulation Results and Analysis

We adopted the Erdos-Renyi random graph with 1000 nodes to represent a social group in our simulation. Both the agents' initial opinions and their declared confidence levels were randomly distributed in the range $[0, 1]$. We experimented with two random distributions for the initial values: (i) uniform random and (ii) normal random with particular mean (μ) and standard deviation (σ). We considered 10 equal length opinion blocks as per the options in a survey in a scale from 1 to 10. Initial confidence levels were assigned in one of the two ways: (i) uniformly random across the agents and (ii) separately for each block to differentiate their expertise level. To balance the randomness present in the free-will of human decisions, the probability p of selecting between keep and adoption heuristics was set with 0.5. Finally, $\{d_{th}, \lambda_{th}\}$ was assigned with $\{0.1, 0.85\}$ to make the confidence change condition more stringent.

3.1   Majority vs. Expert Effect

We examined the majority vs. expert effect through a controlled experiment. Agents started with a bimodal initial opinion distribution having two modes at $\mu_1 = 0.25$ (A) and $\mu_2 = 0.75$ (B) with a small standard deviation ($\sigma_1 = \sigma_2 = 0.05$) for both. The majority effect was created by increasing the number of opinions in mode A from 50 % to 95 %. However, their confidence was confined within $[0$–$0.2]$ to make them lay people. However, the agents in mode B were assigned confidence levels from 0.1 (very low) to 1 (very high) in experiment with all agents having the same confidence level for each assignment to reflect the expert effect. Figure 1 illustrates the findings by showing the predominant effects in the converged opinions. From Fig. 1 it is clear that consistent agents with high confidence level (0.8 or higher) can be considered experts and drive the majority towards them. As the confidence level decreases, majority effect becomes more prominent. However, there is another region of parameters where both the effects are exerted together which is identified as 'C' (combined). Figure 1(b) shows the results of Moussaid et al. [8] model for the same parameters. Their method fails to separate the majority and expert effects as indicated by the large number of grids with 'C'. It also couldn't accurately capture the expert effect as for very high confidence levels of 0.8–1 only 6 out of 18 cases i.e., (33.33 % cases) show the expert influences. In contrast, for confidence level 0.8–1, our model produces 94.44 % expert effect. Moreover, our model more accurately captures

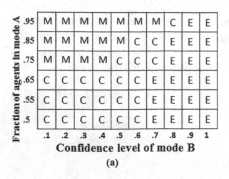
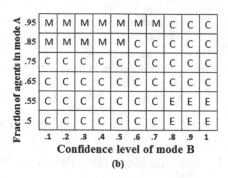

Fig. 1. Majority vs. expert effects observed at the end of opinion evolution. (a): Our model, (b): Moussaid et al. [8] Model. Here, E: Expert, M: Majority, and C: Combined.

the majority impacts as 75 % agents having the same initial opinion has produced majority effect in the absence of high confident agents. However, in such cases none of them has shown majority impact in their model.

3.2   Effect of Stubborn Agents

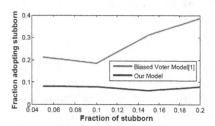

Fig. 2. Fraction adopting stubborn opinion

Stubborn agents don't change their opinions under any social influences. In Biased Voter Model [1], the proportion of agents adopting stubborn opinion is increased with the increase of the initial fraction of stubborn agents as shown in Fig. 2. In contrast, in our model this proportion remains steady. In society, stubbornness is not always viewed as an acceptable behaviour to others and is not conducive to influence people. Since our model captures expert effects, people are influenced by informative and reliable sources. Therefore, the effect of stubbornness is minimal, which is expected in a knowledge based society.

3.3   Consensus, Polarization and Fragmentation

Our model captures consensus, polarization, and fragmentation of opinion dynamics as shown in Fig. 3. From the figures we observe that a single confident group leads others to consensus around it (Fig. 3(a)) whereas the presence of more than one confident groups cause polarization (Fig. 3(b)) or fragmentation (Fig. 3(c)). Here, the reason behind the fragmentation of opinions is that the two confident groups reside at the two opposite end of the OS. Therefore, due to their distances and opposite forces, agents in the middle are not convinced by them, thus form their own groups. Without any experts, all agents are grouping in the middle which

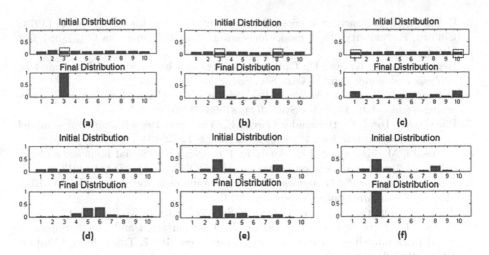

Fig. 3. The initial and final opinion distributions of all agents. Expert blocks are highlighted within rectangle. (a) Consensus (b) Polarization (c) Fragmentation (d) Big cluster in the middle of OS, (e)–(f) Majority effect in absence of experts

is an expected outcome of the dynamics (Fig. 3(d)). However, a majority group attract agents towards them in the absence of experts. The polarization around the group depends on the majority proportion. A group of 60 % majority attracts many agents to it (Fig. 3(e)), whereas a group of 70 % majority leads to consensus on its opinion (Fig. 3(f)).

4   Conclusion

In this paper we introduce an opinion formation model by considering the combined impact of majority and expert effects. We for the first time introduce the concept of consistency in opinion and associated confidence level of neighbouring agents and embed metrics derived from those to update own opinion as well as confidence level. The performance of our model is analysed through simulation and compared with recent existing opinion formation models. Results show that our model captures the effect of majority, expert and stubbornness more accurately compared to other models. Our future works include the mining of confidence level of agents from online social networks data and incorporation of thereof in our model.

References

1. Das, A., Gollapudi, S., Munagala, K.: Modeling opinion dynamics in social networks. In: Proceedings of WSDM (2014)
2. Xia, H., et al.: Opinion dynamics: a multidisciplinary review and perspective on future research. Intl. J. Knowl. Syst. Sci. **2**(4), 72–91 (2011)

3. DeGroot, M.H.: Reaching a consensus. J. Am. Stat. Assoc. **69**(345), 118–121 (1974)
4. Clifford, P., Sudbury, A.: A model for spatial conflict. Biometrika **60**(3), 581–588 (1973)
5. Hegselmann, R., Krause, U.: Opinion dynamics and bounded confidence: models, analysis and simulation. J. Artif. Soc. Soc. Sim. **5**(3), 1–24 (2002)
6. Deffuant, G., Neau, N., Amblard, F., Weisbuch, G.: Mixing beliefs among inter-acting agents. Adv. Complex Syst. **3**(1), 87–98 (2000)
7. Pineda, M., Toral, R., Hernandez-Garcia, E.: The noisy Hegselmann-Krause model for opinion dynamics. Eur. Phys. J. B **12**(86), 1–12 (2013)
8. Moussaid, M., Kaemmer, J.E., Analytis, P.P., Neth, H.: Social influence and the collective dynamics of opinion formation. PLoS ONE **8**(11), e78433 (2013)
9. Cho, J.-H., Swami, A.: Dynamics of uncertain opinions in social networks. In: IEEE Military Communications Conference (2014)
10. Asch, S.E.: Opinions and social pressure. Sci. Am. **193**(5), 31–35 (1955)
11. Hassan, R., Karmakar, G., Kamruzzaman, J.: Reputation and user requirement based price modelling for dynamic spectrum access. IEEE Trans. Mob. Comput. **13**(9), 2128–2140 (2014)

Application of Simulated Annealing to Data Distribution for All-to-All Comparison Problems in Homogeneous Systems

Yi-Fan Zhang[1], Yu-Chu Tian[1(✉)], Wayne Kelly[1], Colin Fidge[1], and Jing Gao[2]

[1] School of Electrical Engineering and Computer Science,
Queensland University of Technology, GPO Box 2434, Brisbane, QLD 4001, Australia
y.tian@qut.edu.au
[2] College of Computer and Information Engineering,
Inner Mongolia Agricultural University, 306 Zhaowuda Road,
Hohhot 010018, Inner Mongolia, China
gaojing@imau.edu.cn

Abstract. Distributed systems are widely used for solving large-scale and data-intensive computing problems, including all-to-all comparison (ATAC) problems. However, when used for ATAC problems, existing computational frameworks such as Hadoop focus on load balancing for allocating comparison tasks, without careful consideration of data distribution and storage usage. While Hadoop-based solutions provide users with simplicity of implementation, their inherent MapReduce computing pattern does not match the ATAC pattern. This leads to load imbalances and poor data locality when Hadoop's data distribution strategy is used for ATAC problems. Here we present a data distribution strategy which considers data locality, load balancing and storage savings for ATAC computing problems in homogeneous distributed systems. A simulated annealing algorithm is developed for data distribution and task scheduling. Experimental results show a significant performance improvement for our approach over Hadoop-based solutions.

1 Introduction

All-to-all comparison (ATAC) represents an important computing pattern in broad areas such as bioinformatics, biometrics and data mining. In the ATAC computing pattern, each file within a data set is pairwise compared with all other files. For example, Arora et al. evaluated the audio similarity of 3090 music pieces through ATACs [1].

Distributed systems with commodity resources have been widely used in processing data intensive ATAC problems [2–4]. However, the performance of an ATAC computation can be greatly affected by a number of factors, e.g., data transmission and system load balancing.

Strategies for load balancing have been reported in a number of references [2,3]. These have aimed to dispatch a similar number of comparison tasks to the computing nodes in a distributed system. However, because co-location of data needed

© Springer International Publishing Switzerland 2015
S. Arik et al. (Eds.): ICONIP 2015, Part III, LNCS 9491, pp. 683–691, 2015.
DOI: 10.1007/978-3-319-26555-1_77

for the comparisons is not considered during task allocation, poor 'data locality' in the initial distribution and massive consequent movement of data at runtime become major bottlenecks for the overall computation.

Recently, systematic frameworks have been developed to deal with data-intensive ATAC problems. Typical examples are Hadoop-based systems [4,5]. Hadoop simplifies the implementation of ATAC applications from the programmer's perspective and efficiently distributes the data across all nodes. However, its inherent MapReduce pattern does not match the ATAC computing pattern because MapReduce assumes that each data item can be processed independently, whereas ATAC comparisons each require two items. This leads to load imbalances and runtime inefficiencies due to poor data locality when Hadoop's data distribution algorithm is used for ATAC problems [6].

To solve these challenging issues, we present here a data distribution strategy for distributed computing of large-scale ATAC problems in homogeneous distributed computing systems. Our approach is scalable and efficient with full consideration of data locality, load balancing and storage usage. A simulated annealing (SA) algorithm is used for data distribution and static task scheduling.

The paper is organized as follows. Section 2 discusses related work and motivations. Section 3 describes the ATAC problem and its challenges. Our data distribution strategy using an SA algorithm is developed in Sect. 4. Experiments which validate the approach are described in Sect. 5. Finally, Sect. 6 concludes the paper.

2   Related Work and Motivations

Distributed computing systems have shown their advantages in processing ATAC problems in various domain areas. An efficient grid scheduler has been designed for parallel implementation of MSA algorithm on a computational grid [2]. It splits a single alignment task into optimal-size subtasks and then distributes the subtasks among multiple processors. However, it assumes that the task running time is much higher than the communication overhead in sending data of the sub-matrix, which makes the method unsuitable for large-scale ATAC problems.

Gunturu et al. [3] presented a load scheduling strategy with near optimal processing time for parallel DNA sequence alignment. In their study, the load distribution depends on the length of the sequence and the number of processors. They also assume that all the comparison tasks have local data, so how to achieve the data locality needed for ATAC problems in general is not answered.

Recently, computing frameworks have been developed for ATAC problems. CloudBlast [4] is a distributed implementation of NCBI BLAST. It integrates a number of technologies, e.g., Hadoop, virtual workspaces and ViNe, together to parallelize, deploy and manage bioinformatics applications. Compared to MPI-based solutions such as mpiBLAST, CloudBLAST shows improvement. It also simplifies the development and management of the computing applications.

Addressing Hadoop's weakness, improved methods have been proposed. Bi-Hadoop [7] is an extension of Hadoop to better support binary-input applications. By providing a binary-input aware task scheduler and a caching subsystem, it outperforms Hadoop by up to 3.3 times and a 48% reduction in remote data reads for binary-input applications. The caching mechanism improves the computing performance, but all data still needs to be distributed through Hadoop's data strategy. Thus, these improved methods still have limitations.

Among existing approaches for distributed processing of ATAC problems, many methods for load balancing do not simultaneously consider the need for data co-location. Also, Hadoop-based solutions are inefficient because of the unmatched data and task strategies. While attention has been paid to improving Hadoop for ATAC problems, the performance improvement is still limited due to Hadoop's fundamental basis in the MapReduce computing model.

To address these issues, we present here a substantial extension of our own previous work in the area [6]. A large-scale, high-performance data distribution strategy is developed for ATAC problems, based on simulated annealing.

3   Problem Statement and Challenges

An ATAC problem is a specific Cartesian product of a data set. Let A, A_i, $C(A_i, A_j)$, $M[i,j]$ represent the set of input data items, a single data item in set A, the comparison operation between data items A_i and A_j, and an output similarity matrix element, respectively. The ATAC problem is to calculate

$$M[i,j] = C(A_i, A_j) \quad \text{for} \quad i, j = 1, 2, \ldots, |A| \ . \tag{1}$$

For distributed processing of ATAC problems, both data set A and all comparison tasks $C(A_i, A_j)$ need to be distributed to different worker nodes. While different strategies have been developed previously, performance issues still exist.

Task Balancing Causes Data Storage Issues. Comparison tasks are usually allocated by rows or columns [8,9]. Though load balancing is considered, unoptimized data distribution causes severe data imbalances and high storage usage. Consider an example of 6 data items and 3 nodes. The workload is divided by the rows. The result in Fig. 1 shows that although each of the three nodes has 5 different comparison tasks, the data files are stored in inefficiently. Node 1 has to store copies of all the data items, but this should be avoided when for data-intensive computing. Moreover, three worker nodes have 6, 5 and 4 data items, respectively, implying a system data imbalance.

Storage Saving Causes Task Issues. When Hadoop-based solutions have been used to solve ATAC problems, each data item is randomly distributed to the worker nodes with a fixed number of replications. Although this achieves high data reliability due to replication, poor performance is inevitable due to the lack of consideration of comparison task allocation. Take 6 data items and 4 worker nodes for example. In Fig. 2, each data item has three copies in three different

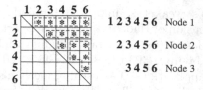

Fig. 1. A data imbalance

nodes for reliability. Each node stores 50 % of the data. However, 9 comparison tasks do not have local data, requiring massive data movement at runtime to complete the comparisons and consequently poor overall performance.

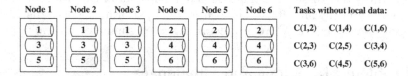

Fig. 2. Poor data locality for comparison tasks

4   Data Distribution Strategy with Simulated Annealing

This section presents our metaheuristic data distribution strategy for distributed computing of ATAC problems in homogeneous systems. For efficient derivation of data distribution and task scheduling, a simulated annealing (SA) algorithm is developed with specific methods for generating and selecting solutions.

Considering the challenges involved in solving ATAC problems, a data distribution strategy is presented below to meet the following requirements:

1. The system has good static load balancing (Load balancing);
2. All comparison tasks have the data they need locally (Data locality); and
3. The maximum number of data among all nodes is minimized (Storage saving).

Simulated Annealing [10] is a probabilistic optimization technique derived from the physical process of crystallization. It is widely used to solve global optimization problems. We adopt it here to solve the data distribution and task scheduling problems for ATAC computations. To use an SA approach, we must: (1) determine the Annealing module and Acceptance Probability module; and (2) determine an initial solution, the neighbourhood selection method and the fitness equation.

Annealing Module and Acceptance Probability Module. The setting of the SA module has significant effects on the final result [11]. As one of the fastest

Table 1. SA module parameter settings (k represents the iteration step)

Item	Setting
Temperature decreasing function	$t_k = T_0/k$
Starting temperature	1.0
Ending temperature	10^{-5}
Inner loop iteration threshold	100
Acceptance probability function	$P(\Delta E) = \exp(-\Delta E/t)$

decreasing temperature methods, we use Cauchy scheduling [12]. Parameters used for the example in Sect. 5 are shown in Table 1.

Initial Solution. For ATAC problems with M data items, $M(M-1)/2$ comparison tasks must be allocated. Hence, for a homogeneous system with N nodes, an initial solution can be generated by randomly and evenly allocating all comparison tasks and related data items. Let M, N, D_i, T_i and U represent the number of data files to be processed, the number of nodes in the system, the set of data files stored on node i, the comparison task set allocated to node i and the set of tasks that have not yet been scheduled, respectively. An initial solution is generated as follows:

1. Keep picking up comparison tasks from set U and allocating them to each node $i \in \{1, 2, \ldots, N\}$ until all have been allocated, i.e., $|T_i| = \left\lceil \frac{M(M-1)}{2N} \right\rceil$ or $U = \emptyset$; and
2. Based on each task set T_i, distribute all related data files to data set D_i.

The solution $S = \{(T_1, D_1), (T_2, D_2), ..., (T_N, D_N)\}$ is then a feasible solution, which meets both our initial requirements.

Neighbourhood Selection Method. Following the design of the initial solution, a new neighbourhood solution S' can be generated from a solution S from the following steps:

1. Randomly choose two nodes i and j;
2. Randomly pick up two comparison tasks $t_k \in T_i$ and $t_l \in T_j$ and swap them; and
3. Update related data files in data set D_i and D_j.

Considering that all the nodes in a homogeneous system can be treated as indistinguishable, this method promises that each new solution has the capability to solve the ATAC problem and all possible solutions can be generated theoretically.

Fitness Equation. Considering the requirements mentioned at the beginning of this section, the fitness equation $F(S)$ for a solution S is defined as the set of the number of data files allocated to each of the worker nodes:

$$F(S) = \{|D_1|, |D_2|, \ldots, |D_N|\}. \tag{2}$$

The difference ΔF between two different solutions S and S' is calculated as follows. Firstly, the elements in both $F(S)$ and $F(S')$ are sorted in descending order. Then, ΔF is obtained as:

$$\Delta F = F(S) - F(S') = \{(|D_1| - |D_1'|), \ldots, (|D_N| - |D_N'|)\}. \tag{3}$$

Finally, the value of the cost change Δf is defined as:

$$\Delta f = \begin{cases} \text{the 1st non-zero element value in Eq. (3),} & \text{if one exists} \\ 0, & \text{otherwise.} \end{cases} \tag{4}$$

This method promises that solution S with smaller maximum values in $F(S)$ always be accepted as required by SA. Moreover, unlike only comparing the maximum values in $F(S)$ and $F(S')$, this method utilizes much more information from other elements. Hence, the SA algorithm has a higher efficiency in searching for better solutions.

Data Distribution Algorithm. By integrating all the above designs, our data distribution algorithm using SA is presented as Algorithm 1.

Algorithm 1. Data Distribution Algorithm

Initialisation:
1: Randomly generate initial solution S using the method in **Initial Solution**;
2: Set parameters based on Table 1;
3: Set the current temperature t to be the starting temperature.
Distribution:
4: **while** The current temperature $t >$ the ending temperature **do**
5:     **while** The iteration step is below the inner loop iteration threshold **do**
6:         Generate a new solution S' from S (using the **Neighbourhood Selection Method**);
7:         Calculate the change of fitness, Δf, from Equation (4)
8:         (The fitness method for F, ΔF and Δf are developed in **Fitness Equation**);
9:         **if** $\exp(-\Delta f/t) > random[0,1)$ **then**
10:            Accept the new Solution: $S \leftarrow S'$
11:        Increment the iteration step by 1;
12:    Lower the current temperature t based on the function in Table 1;
13: Return final solution S.

---

5   Experiments

We conducted experiments to evaluate the following aspects of our algorithm: storage savings, task allocations, data scalability and computing performance.

Storage Savings, Task Allocations and Data scalability. An example with 4 nodes and 8 data items is used to show the effectiveness of our data distribution strategy. The results are summarized below:

It can be seen from these results that data balancing and static load balancing are achieved. Each worker node only stores 5 data items. Moreover, each node is allocated 7 comparison tasks all with good data locality.

Node	Distributed data files	Allocated comparison tasks
A	0, 2, 3, 6, 7	(0,2) (0,3) (0,6) (0,7) (2,3) (2,6) (2,7)
B	1, 3, 5, 6, 7	(1,3) (1,5) (1,6) (1,7) (3,7) (5,7) (6,7)
C	0, 1, 2, 4, 5	(0,1) (0,4) (0,5) (1,2) (1,4) (2,4) (2,5)
D	3, 4, 5, 6, 7	(3,4) (3,5) (3,6) (4,5) (4,6) (4,7) (5.6)

As the numbers of data items and nodes increases, Table 2 shows our strategy still achieves good results in storage saving, load balancing and data locality, compared with Hadoop's strategy (using 3 copies of each data item). Each node has an equal number of comparison tasks with good data locality in our solution. Although Hadoop's strategy uses less storage space overall, runtime performance issues are inevitable due to the poor data locality for comparison tasks. Figure 3 shows that our approach still has good data scalability, and is far better than the brute-force ATAC solution of copying all data items onto every node.

Table 2. Storage usage and storage savings of our work versus Hadoop for 256 files

No. of nodes	Max. # of files on a node		Storage space saving		# of tasks on each node
	This work	Hadoop	This work	Hadoop	This work
8	150	96	41 %	63 %	4080
16	116	48	55 %	81 %	2040
32	83	24	68 %	91 %	1020
64	56	12	78 %	95 %	510

Computing Performance. We also conducted experiments with a bioinformatics ATAC application. The experimental environment was set as:

- A homogeneous cluster with 5 machines, all running Redhat Linux. One acts as the master node, and all the nodes have one core and 64 GB of RAM.
- Sequential and distributed versions of the CVTree application, which is a typical ATAC problem in bioinformatics [13].
- DsDNA data files from the National Center for Biotech Information (NCBI).

Figure 4 shows the different computation times between our data distribution strategy and Hadoop's strategy. By considering the three requirements summarized in Sect. 4, our data distribution strategy achieves much higher computing performance than Hadoop's strategy. As we discussed in Sect. 3, this is because Hadoop's strategy needs to move numerous data items between nodes during the computation, due to poor data locality.

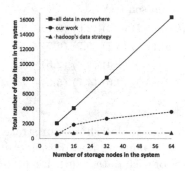
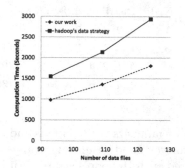

Fig. 3. Data Scalability   **Fig. 4.** Computation time performance

6   Conclusion

A scalable and efficient data distribution strategy using simulated annealing has been presented for distributed computing of all-to-all comparison problems in homogeneous distributed systems. It is designed to use as little storage space as possible while still achieving system load balancing and good data locality. Experiments have shown that although our approach uses more overall storage than Hadoop's, we achieve greatly reduced computation times.

Acknowledgments. Author J. Gao would like to acknowledge the support from the National Natural Science Foundation of China under the Grant Number 61462070, and the Inner Mongolia Government under the Science and Technology Plan Grant Number 20130364.

References

1. Arora, R., Gupta, M.R., Kapila, A., Fazel, M.: Similarity-based clustering by left-stochastic matrix factorization. J. Mach. Learn. Res. **14**, 1715–1746 (2013)
2. Somasundaram, K., Karthikeyan, S., Nayagam, M.G., RadhaKrishnan, S.: Efficient resource scheduler for parallel implementation of MSA algorithm on computational grid. In: International Conference on Recent Trends in Information, Telecommunication and Computing, Kochi, Kerala, pp. 365–368, IEEE, 12–13 March 2010
3. Gunturu, S., Li, X., Yang, L.: Load scheduling strategies for parallel DNA sequencing applications. In: Proceedings of the 11th IEEE International Conference on High Performance Computing & Communication Seoul, pp. 124–131. IEEE, 25–27 June 2009
4. Matsunaga, A., Tsugawa, M., Fortes, J.: Cloudblast: combining mapreduce and virtualization on distributed resources for bioinformatics applications. In: IEEE 4th International Conference on eScience, Indianapolis, IN, pp. 222–229, 7–12 December 2008
5. Pireddu, L., Leo, S., Zanetti, G.: Seal: a distributed short read mapping and duplicate removal tool. Bioinformatics **27**(15), 2159–2160 (2011)

6. Zhang, Y.F., Tian, Y.C., Fidge, C., Kelly, W.: A distributed computing framework for all-to-all comparison problems. In: The 40th Annual Conference of the IEEE Industrial Electronics Society (IECON 2014), Dallas, TX, USA. IEEE, 29 October–1 November 2014

7. Yu, X., Hong, B.: Bi-hadoop: Extending hadoop to improve support for binary-input applications. In: 13th IEEE/ACM International Symposium on Cluster, Cloud, and Grid Computing, Delft, IE, pp. 245–252, 13–16 May 2013

8. Moretti, C., Bui, H., Hollingsworth, K., Rich, B., Flynn, P., Thain, D.: All-pairs: an abstraction for data-intensive computing on campus grids. IEEE Trans. Parallel Distrib. Syst. **21**, 33–46 (2010)

9. Pedersen, E., Raknes, I.A., Ernstsen, M., Bongo, L.A.: Integrating data-intensive computing systems with biological data analysis frameworks. In: 23rd Euromicro International Conference on Parallel, Distributed, and Network-Based Processing, Turku, pp. 733–740. IEEE, 4–6 March 2015

10. Kirkpatrick, S., Gelatt, C.D., Vecchi, M.P.: Optimization by simulated annealing. Science **220**(4598), 671–680 (1983)

11. Wu, W., Li, L., Yao, X.: Improved simulated annealing algorithm for task allocation inreal-time distributed systems. In: IEEE International Conference on Signal Processing Communications & Computing (ICSPCC), Guilin, pp. 50–54. IEEE, 5–8 August 2014

12. Keikha, M.: Improved simulated annealing using momentum terms. In: 2011 Second International Conference on Intelligent Systems, Modelling and Simulation (ISMS), Kuala Lumpur, pp. 44–48. IEEE, 25–27 January 2011

13. Hao, B., Qi, J., Wang, B.: Prokaryotic phylogeny based on complete genomes without sequence alignment. Modern Phy. Lett. **2**(4), 14–15 (2003)

Cognitive Workload Discrimination in Flight Simulation Task Using a Generalized Measure of Association

Zhongxiang Dai[1], José C. Príncipe[2], Anastasios Bezerianos[1],
and Nitish V. Thakor[1(✉)]

[1] Singapore Institute for Neurotechnology, Centre for Life Sciences,
National University of Singapore, Singapore, Singapore
{daiz9109,sinapsedirector}@gmail.com,
tassos.bezerianos@nus.edu.sg
[2] Computational NeuroEngineering Laboratory, Department of Electrical
and Computer Engineering, University of Florida, Gainesville, USA
principe@cnel.ufl.edu

Abstract. Cognitive workload discrimination of pilots during flights can contribute to flight safety by preventing mental overloading of the aircraft crew. Research has been conducted to study the cognitive workload of pilots in flight simulation task. The estimation of the cortical connectivity is a critical step in mental workload assessment. Therefore, we adopted a novel, parameter-free method of evaluating the cortical connectivity, named Generalized Measure of Association (GMA), to assess and discriminate the mental workload of pilots using the Multi-Attribute Task Battery (MATB) flight simulation platform. A modified version of GMA (Time Series GMA) is applied on the pre-processed EEG time series recorded from eight subjects during the MATB experiment. Frobenius Norm is used to calculate the Euclidean distances between different TGMA series in order to assess the discriminability of the cognitive workloads. The results have shown clear distinction between the mean values of the inter-task and intra-task Euclidean distance series of TGMA matrices, but statistical significance is still lacking due to the relatively large standard deviations.

Keywords: Cognitive workload discrimination · Multi-attribute task battery · EEG · Generalized measure of association

1 Introduction

The assessment and discrimination of cognitive workloads has been recognized as a significant and promising research topic [1, 2]. Due to the limited capacity of human brain, excessive cognitive workload can arise under circumstances where the subjects are engaged in demanding tasks, causing negative impacts on both the human brain and the task performance [3]. Therefore, being able to evaluate and discriminate cognitive workload can be highly beneficial in various situations, especially in safety-critical tasks such as aircraft operation.

© Springer International Publishing Switzerland 2015
S. Arik et al. (Eds.): ICONIP 2015, Part III, LNCS 9491, pp. 692–699, 2015.
DOI: 10.1007/978-3-319-26555-1_78

Pilots are regularly involved in complicated and intensive tasks due to the complexity of the aircraft operating system [4]. Therefore, monitoring their mental workload and taking proper measures to prevent cognitive overloading during flight can reduce the possibility of operating errors and improve flight safety. Multi-Attribute Task Battery (MATB), which is a computer-based flight simulation system designed by NASA, contains a benchmark set of piloting tasks that emulate the activities performed by the aircraft crew during flight [5]. The MATB experimental protocol has been employed by researchers to study the mental workload of pilots from different perspectives such as the evaluation method [6], the interaction between the crew members [7], etc. The estimation of the cortical connectivity plays a critical role in cognitive workload classification of aircraft pilots [4]. Partial Directed Coherence (PDC), among others, has been adopted by researchers in estimating the cortical connectivity patterns of pilots during MATB experiments [7]. In this study, a novel, parameter-free method of evaluating the dependences between different cortical signals, Generalized Measure of Association (GMA) [8–11], is employed in cortical connectivity estimation using EEG signals and cognitive workload discrimination.

The remaining sections of the paper are organized as follows. GMA, including its advantages, calculation and modification, is briefly introduced in Sect. 2. In Sect. 3, the experimental protocol using the MATB simulation system and the method of data analysis are described. The results of cortical connectivity estimation using GMA and cognitive workload discrimination are produced in Sect. 4. Section 5 provides our discussion and conclusion, including our plan for future work.

2   Generalized Measure of Association

Generalized Measure of Association (GMA) measures the degree of dependences between two random variables. In recent years, GMA has been applied to estimate cortical connectivity by evaluating the dependences between EEG recordings [8–11]. When compared with conventional measures of dependences such as correlation, Mutual Information, Granger Causality and Partial Directed Coherence, GMA offers some major benefits including being free of parameters, the ability to detect nonlinear relationships, and the capability of reliably estimating the dependences with limited samples of time series.

GMA is calculated by assessing the degree to which the close distance between two realizations of one random variable is associated with the close distance between the two corresponding realizations of the other random variable. The GMA value between two random variables ranges between 0.5 and 1, with the values close to 0.5 indicating strong independence and the values close to 1 indicating strong dependence.

GMA has been modified to obtain (Time Series GMA) TGMA, in order to enhance its reliability in estimating the association between two time series [11]. TGMA minimizes the correlation over time by selecting a proper value for the time lag in a time delay embedded space, and discarding data vectors that are redundant when looking for the closest data point in terms of Euclidean distance. The time lag is usually selected to be the first minimum of the autocorrelation function [11]. TGMA therefore has two free parameters that need to be estimated from data (embedding dimension and delay).

3 Methodology

3.1 Experimental Protocol

Subjects. Eight healthy male subjects (mean age $\pm$ SD = 23.1 $\pm$ 2.7) participated in the MATB experiment. The study was approved by the Institutional Review Board of the National University of Singapore, and all the subjects gave written informed consent.

Experimental Procedure. During the experiments, the subjects are required to perform four subtasks shown on the computer screen that emulate the activities carried out by the pilots inside the aircraft cockpit: tracking a target (TRCK), communication task (COMM), a fuel resource management (RMAN) and system monitoring (SYSM) (Fig. 1).

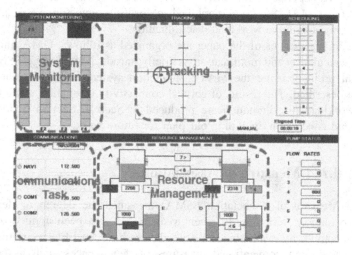

Fig. 1. Screenshot of Multi-Attribute Task Battery on computer screen. On the top in the center: tracking a target (TRCK); on the top left corner: system monitoring (SYSM); at the bottom left corner: communication tasks (COMM); at the bottom in the center: resource management (RMAN)

Two task conditions with different difficulty levels are defined: easy task and hard task. The two tasks differ in terms of the frequency of the stimuli and the required reaction time to the stimuli. Before the experiment, all subjects completed five consecutive days of training in the MATB tasks, with each training session lasting for half an hour. In the experiment, the duration of the entire recording session for every subject in each task condition was kept at 145 s.

EEG Data Recording and Pre-processing. Scalp EEG signals were recorded from 32 electrodes referenced to the average of both earlobes and grounded to the AFz electrode. The impedances of the EEG electrodes were kept below 10 kΩ. The sampling rate was 256 Hz.

Pre-processing is applied to the recorded EEG signals. The EEG signals are band-pass filtered (1-30 Hz). Artefacts associated with eye-blinks and eye movements are removed using techniques related to Independent Component Analysis (ICA).

3.2 EEG Data Analysis

TGMA Calculation. For further processing, the EEG signals need to be filtered in certain frequency bands. It has been proven in previous studies [10] that optimal performance of GMA can be achieved when the quality factor of the filter (Q) approaches unity. Therefore, several frequency bands are selected by keeping Q = 1: [6 Hz, 18 Hz], [8 Hz, 24 Hz], and [10 Hz, 30 Hz]. A time window is the duration of time in the EEG signals where a TGMA value is calculated. Two different sizes of the time window have been selected: two-second window with increment of one second (50 % overlap), and four-second window with increment of two seconds.

In order to minimize the temporal correlation of EEG time series, the TGMA algorithm is applied and an additional step needs to be implemented before the calculation of GMA. This is achieved by mapping the pre-processed EEG signals into a higher dimension through time delay embedding based on Takens' embedding theorem [12]. Two parameters need to be selected for the time delay embedding: the embedding dimension and the time lag. The time lag is chosen to be the first minimum of the autocorrelation of the EEG signals [11], which produces value L = 12. The optimal embedding dimension is obtained using the False Nearest Neighbor algorithm [13] and is found to be M = 6. The number of data samples used for TGMA calculation in a time window is 512 for 2-s window and 1024 for 4-s window.

After the time delay embedding, TGMA values are calculated using the EEG time series such that one TGMA value is obtained for each pair of EEG channels in every time window. As a result, for every subject at each task condition (easy task or hard task), a series of TGMA matrices are calculated with each time window and frequency band.

Cognitive Workload Discrimination. After obtaining the series of TGMA matrices, we attempt to discriminate the subjects' cognitive workload in the two task conditions by classifying the two series of TGMA matrices (corresponding to easy task condition and hard task condition) for each subject. Therefore, we calculated the inter-task, intra-easy-task and intra-hard-task series of Euclidean distances of the TGMA matrices using the Frobenius Norm [14].

The inter-task Euclidean distance series (each element being the Euclidean distance between the two corresponding TGMA matrices for the two tasks conditions), the intra-easy-task Euclidean distance series (each element being the Euclidean distance between two consecutive matrices in the TGMA matrix series obtained in the easy task condition), and the intra-hard-task Euclidean distance series (each element being the Euclidean distance between two consecutive matrices in the TGMA matrix series obtained in the hard task condition) are calculated for all the subjects when different frequency bands and time windows are applied. For each subject, the three Euclidean distance series are compared in an attempt to discriminate the subject's cognitive workloads when performing the MATB tasks at the two different difficulty levels.

4 Results and Analysis

4.1 Brain Networks in Different Task Conditions

By using TGMA to estimate the pairwise cortical connection, the TGMA matrices provide representation of the subjects' brain network. In fact, large values of TGMA mean that two channels are more dependent and the corresponding cortices are working together. By averaging over the TGMA matrix series and across all subjects, we observed clear differences between the brain networks of the subjects in the two different task conditions. Figure 2 shows the largest 10 % of the cortical connections in terms of TGMA values when the subject is performing the easy task and the hard task, together with their differences (with time window being 2 s, and the frequency band being between 8 Hz and 24 Hz).

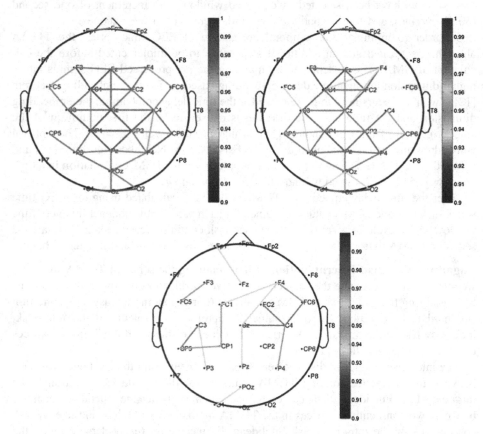

Fig. 2. Brain connectivity network showing significant cortical connections (largest 10 %) with 2\,s time window and the frequency band between 8 Hz and 24 Hz. Top left figure: easy task; top right figure: hard task; bottom figure: differences between the top 2 figures (insignificant edges omitted). The result is obtained by averaging over all TGMA series across all subjects.

4.2 Cognitive Workload Discrimination

The inter-task, intra-easy-task and intra-hard-task Euclidean distance series are calculated using the corresponding TGMA matrix series for all subjects. The means and standard deviations of series are used to evaluate the discriminability of the subjects' cognitive workload in the two task conditions.

The means and standard deviations of the resulted Euclidean distance series after averaging across all subjects are shown in Table 1.

Table 1. Means and standard deviations of inter-task, intra-easy-task and intra-hard-task Euclidean distance series (Averaged over all subjects)

Task	Time Window	2-Second Time Window			4-Second Time Window		
	Frequency Band (Hz)	[6, 18]	[8, 24]	[10, 20]	[6, 18]	[8, 24]	[10, 20]
inter-task	Mean	0.8219	0.9555	1.1860	0.8304	0.9740	1.1837
	STD	0.1691	0.1758	0.2291	0.1728	0.1786	0.2156
intra-easy-task	Mean	0.6085	0.6545	0.7658	0.5346	0.5474	0.6150
	STD	0.1195	0.1047	0.1328	0.1068	0.0895	0.1057
intra-hard-task	Mean	0.6231	0.6797	0.8002	0.5493	0.5751	0.6587
	STD	0.1243	0.1157	0.1455	0.1193	0.1092	0.1299

It can be observed from Table 1 that for all time windows and frequency bands, there are distinct differences between the mean values of the inter-task distance series and that of both the intra-easy-task and intra-hard task distance series. However, the results lack statistical significance because of the relatively large standard deviations. The statistical insignificance is corroborated by KS test, which produces very small p value (the order of -15 and -12 for easy and hard task conditions respectively) indicating non-discriminability between the inter-task and intra-task conditions.

Therefore, the method is sensitive to the cognitive workload of the subjects performing MATB tasks at different difficulty levels. However, the mean TGMA difference between conditions is not statistically significant, when all the channel pairs are used.

In addition to using TGMA, we also used correlation coefficient to represent cortical connection and calculated the corresponding Euclidean distance series. The results also indicate non-discriminability between the task conditions, which is also corroborated by KS test (the order of p value is -14 and -13 for easy and hard task conditions respectively).

5 Conclusion

In this study, a novel method of estimating cortical connectivity based on the Generalized Measure of Association, is applied to EEG signals, for the purpose of discriminating the cognitive workloads of subjects performing MATB tasks at different difficulty levels. The comparison between the inter-task and intra-task Euclidean distances of the TGMA matrices has shown that there are clear differences in the mean values over the scalp, but the results using all channels lack statistical significance due to the relatively large standard deviations. We suspect that the use of all channel pairs is increasing the variance substantially, therefore obscuring the networks that are discriminant between the tasks.

In our future work, we will explore methods to improve the results of the cognitive workload discrimination by identifying the active EEG channels that play significant roles in mental workload classification and discarding those channels that are not engaged in the mental workload.

References

1. Zarjam, P., Epsp, J., Chen, F., Lovell, N.H.: Estimating cognitive workload using wavelet entropy-based features during an arithmetic task. J. COMPUT. BIOL. MED. **43**, 2186–2195 (2013)
2. Hwang, T., Kim, M., Hwangbo, M., Oh, E.: Comparative analysis of cognitive tasks for modeling mental workload with electroencephalogram. In: 36th Annual International Conference of the Engineering in Medicine and Biology Society (EMBC), pp. 2661–2665. IEEE (2014)
3. Rebsamen, B., Kwok, K., Penney, T.B.: EEG-based measure of cognitive workload during a mental arithmetic task. In: Stephanidis, C. (ed.) Posters, Part II, HCII 2011. CCIS, vol. 174, pp. 304–307. Springer, Heidelberg (2011)
4. Borghini, G., Astolfi, L., Vecchiato, G., Mattia, D., Babiloni, F.: Measuring neurophysiological signals in aircraft pilots and car drivers for the assessment of mental workload, fatigue and drowsiness. J. Neurosci. Biobehav. Rev. **44**, 58–75 (2014)
5. Comstock, J.R., Arnegard, R.J.: The multi-attribute task battery for human operator workload and strategic behavior research. Technical report, National Aeronautics and Space Administration, Langley Research Center Hampton, VA, United States (1992)
6. Arico, P., Borghini, G., Graziani, I., Taya, F., Sun, Y., Bezerianos, A., Thakor, N.V., Cincotti, F., Babiloni, F.: Towards a multimodal bioelectrical framework for the online mental workload evaluation. In: 36th Annual International Conference of the Engineering in Medicine and Biology Society (EMBC, 2014), pp. 3001–3004. IEEE (2014)
7. Astolfi, L., Toppi, J., Borghini, G., Vecchiato, G., He, E.J., Roy, A., Cincotti, F., Salinari, S., Mattia, D., He, B., Babiloni, F.: Cortical activity and functional hyperconnectivity by simultaneous EEG recordings from interacting couples of professional pilots. In: Annual International Conference of the Engineering in Medicine and Biology Society (EMBC, 2012), pp. 4752–4755. IEEE (2012)
8. Fadlallah, B.H., Seth, S., Keil, A., Principe, J.C.: Robust EEG preprocessing for dependence-based condition discrimination. In: Annual International Conference of the Engineering in Medicine and Biology Society (EMBC, 2011), pp. 1407–1410. IEEE (2011)
9. Fadlallah, B.H., Seth, S., Keil, A., Principe, J.C.: Analyzing dependence structure of the human brain in response to visual stimuli. In: IEEE International Conference on Acoustics, Speech and Signal Processing (ICASSP, 2012), pp. 745–748 (2012)
10. Fadlallah, B., Seth, S., Keil, A., Principe, J.: Quantifying cognitive state from EEG using dependence measures. J. IEEE. Trans. Biomed. Eng. **59**(10), 2773–2781 (2012)

11. Fadlallah, B.H., Brockmeier, A.J., Seth, S., Li, L., Keil, A., Principe, J.C: An Association Framework to Analyze Dependence Structure in Time Series. In: Annual International Conference of the Engineering in Medicine and Biology Society (EMBC, 2012), pp. 6176–6179. IEEE (2012)
12. Takens, F.: Detecting strange attractors in turbulence. In: Rand, D.A., Young, L.-S. (eds.) Dynamical Systems and Turbulence. Lecture Notes in Mathematics, vol. 898, pp. 366–381. Springer, Heidelberg (1981)
13. Kennel, M.B., et al.: Determining embedding dimension for phase-space reconstruction using a geometrical construction. J. Phys. Rev. **45**(6), 3403–3411 (1992)
14. Wolfram MathWorld, http://mathworld.wolfram.com/FrobeniusNorm.html

Author Index